Elasticity and Geometry

Elasticity and Geometry

From hair curls to the non-linear response of shells

B. Audoly

Centre national de la recherche scientifique (CNRS)
and École polytechnique

Y. Pomeau

Centre national de la recherche scientifique (CNRS)
and École normale supérieure, and University of Arizona

OXFORD
UNIVERSITY PRESS

OXFORD

UNIVERSITY PRESS

Great Clarendon Street, Oxford, OX2 6DP,
United Kingdom

Oxford University Press is a department of the University of Oxford.
It furthers the University's objective of excellence in research, scholarship,
and education by publishing worldwide. Oxford is a registered trade mark of
Oxford University Press in the UK and in certain other countries

© Oxford University Press 2010

The moral rights of the authors have been asserted

First published 2010
First published in paperback 2018

Impression: 1

Published in the United States of America by Oxford University Press
198 Madison Avenue, New York, NY 10016, United States of America

British Library Cataloguing in Publication Data
Data available

Library of Congress Cataloging in Publication Data
Data available

ISBN 978–0–19–850625–6 (Hbk.)
ISBN 978–0–19–882626–2 (Pbk.)

Printed and bound by
CPI Group (UK) Ltd, Croydon, CR0 4YY

Links to third party websites are provided by Oxford in good faith and
for information only. Oxford disclaims any responsibility for the materials
contained in any third party website referenced in this work.

Foreword

Almost 400 years after the birth of Robert Hooke, the author of Hooke's Law of elasticity, the appearance of *Elasticity and Geometry* by Basile Audoly and Yves Pomeau testifies to the ongoing interest in elasticity theory and its extraordinary richness. Material elasticity throughout the book is linear and restricted to small strains. The richness of the topics covered by the book stems from the fact that geometric aspects are non-linear. The book is aptly titled. Its historical roots originate with Euler's Elastica and Gauss's theorem on developable surfaces. It deals with the deformation of rods, plates and shells, developing all the basic equations from first principles. The book includes many recent advances on these topics that focus on non-linear geometric aspects, including some of the authors' own seminal contributions. Throughout, the emphasis is on shapes and patterns—twists, curls, folds, wrinkles, buckles and more. The approach is largely analytical although numerical methods are introduced where needed with good effect. Contact with experimental observations is made for most of the topics. The authors have been true to their stated intent to introduce 'models as simple as possible without leaving aside anything fundamental'.

For almost two centuries, the primary motivation underlying developments in rods, plates and shells derived from the need to have quantitative methods to analyse man-made structures—buildings, bridges, dams, aircraft, etc. As a consequence, there was relatively little motivation to understand problems involving dramatic geometry changes such as a rod curling about itself or a hemispherical shell turned inside out. Such behaviour falls well outside the envelope of permissible deformations for conventional structures. Thus, while non-linear geometry does play an important role in the buckling of conventional structures, the research focus was almost always directed to questions such as the maximum load-carrying capacity of a structure and not, for example, its collapsed shape. And, usually, the shape at maximum load differs only slightly from the initial shape. In recent decades, new motivations began surfacing in fields such as bio-physics and bio-mechanics where extreme deformation shapes and patterns constitute normal behaviour. Nature is replete with examples, at the molecular scale and above. These modern motivations provide the background for *Elasticity and Geometry*. This new book provides an excellent complement to another recently published book, W. T. Koiter's *Elastic Stability of Solids and Structures* (A. M. A. van der Heijden (Ed.), Cambridge University Press, 2009), which also deals with the stability of rods, plates and shells but mainly from the structural vantage point just mentioned. For example, while *Elasticity and Geometry* investigates plates loaded far beyond the onset of the first buckles, *Elastic Stability* focuses on plate behaviour in the vicinity of the buckling load, determining whether buckling is benign or catastrophic. The former has application to patterns that develop in thin film delamination or in growing leaves, while the latter is used to predict load-carrying capacity and to understand structural performance.

The subject matter covered by Audoly and Pomeau's book has been drawn from a wide swath of literature in the engineering, mathematics and physics communities, and the book itself will be of interest to students and researchers from across these fields. For the most part, the mathematical level is at the level of graduate students, or possibly advanced undergraduates in France, with a firm grounding in methods of applied mathematics.

The developments are each self-contained. The book begins with an introduction to three-dimensional elasticity theory and has several stand-alone Appendices on topics such as the calculus of variations and boundary layers. The book emphasizes general results. The reader will be rewarded with a deep and informed insight into basic aspects of the large deformation behaviour underlying rods, plates and shells. In addition to coverage of these general aspects, the book gives detailed results on many specific problems such as the bending, twisting and writhing of rods; primary and secondary buckling under general biaxial stress conditions of plates; the peculiar deformation characteristics of toroidal shells, and the highly non-linear behaviour a spherical shell flattened against a rigid wall. Interesting applications include the mechanics of hair curls and the intriguing morphology of thin film and coating delaminations. *Elasticity and Geometry* will provide a fundamental foundation for students and researchers into fertile problems in elasticity for many years to come.

John W. Hutchinson

School of Engineering and Applied Sciences
Harvard University
November 19, 2009

Contents

1
Introduction

This book covers fundamental aspects of the mechanics of solids and, more specifically, of thin elastic solids. Its style is voluntarily informal: the methods of analysis are quite diverse and could have been put into a rigorous formalism only at the expense of the clarity of exposition. Keeping in mind that this branch of science aims at describing real world phenomena, we have put emphasis on the derivation of explicit solutions in specific geometries. Our hope is that a detailed analysis of these special cases will help the reader to develop a real intuition of the mechanical phenomena without being hindered by the formalism. Therefore, this book does not claim to be a comprehensive treatise on elasticity of rods, plates and shells. Instead, it focuses on some recent developments to which the authors have contributed. The problems that are investigated have been selected in part because they allow various important mathematical techniques to be introduced. These techniques, which include for instance boundary layer and bifurcation theories, have a range of applications that extends well beyond the theory of elasticity.

This book was designed to be as self-sufficient as possible and no prior knowledge of elasticity theory is required. It is intended for third year undergraduate and graduate students, for students in engineering schools, as well as for researchers from mechanical engineering and from other fields such as applied mathematics and numerical analysis, aerospace engineering and biophysics. A basic understanding of Euclidean geometry and of calculus with real variables is assumed. The reader does not need to be familiar with other, less classical tools: the basics of calculus of variation are given in a short Appendix A; a simple but illustrative example of a boundary layer is treated in Appendix B; the geometry of planar and space curves is summarized in Section 1.3 below, together with other geometrical notions; the geometry of surfaces is introduced when needed, in Part III of this book. Complex analysis is not required either.

When writing this book, both authors were employees of the French national institute for scientific research (CNRS), whose support is gratefully acknowledged. Many of the ideas and results presented here are the result of discussions and collaborations with colleagues who are credited in the relevant chapters. We shall express here our gratitude to Étienne Guyon, John Hutchinson, Roberto Toro, Sébastien Neukirch and Florence Bertails-Descoubes for their remarks and corrections on the manuscript.

1.1 Outline

Chapter 2 derives the basic equations of elasticity, and is a necessary starting point for the reader not yet familiar with continuum mechanics and with linear elasticity. What follows is organized into three mostly independent parts, each one being concerned with a family of thin elastic bodies, namely elastic rods, elastic plates and elastic shells. In each of these parts, the first chapter derives the governing equations (Kirchhoff equations for elastic rods, membrane and plate equations, etc.) and should be read first, while the remaining chapters focus on specific examples and can be skipped or read in any particular order.

We shall now outline in more details the structure of this book. Chapter 2 recalls the principles of the so-called Hookean elasticity in three dimensional solids, as formulated in a fully explicit form by Cauchy and Poisson. It serves as a starting point to build theories for one or two-dimensional elastic bodies in the following parts.

Chapter 3 is the first chapter of Part I, and contains a derivation of the equations for static elastic rods. The kinematical notions of bending and twist are first introduced. The equations of equilibrium are then derived, leading to the so-called Kirchhoff equations, which describe three-dimensional deformations of a rod. In Chapter 4 we look at an application of this theory for representing human hair. The mechanics of a single hair strand can be reduced, under reasonable assumptions, to a system with two dimensionless parameters, describing the effects of gravity and of permanent curling. In Chapter 5 we present another application of the theory of elastic rods. This is related to the spontaneous formation of ripples along the edge of a thin sheet of material when it is stretched. In a first attempt to understand these ripples, we analyse a simple rod model that describes what happens at the edge of such a sheet. This question will be reconsidered later on in Chapter 10 in a more geometrical spirit, using the full equations for plates.

Part II is devoted to the mechanics of elastic plates. We call a plate any piece of elastic material that has the shape of sheet (it has a small thickness in only one direction of space) and is planar in the absence of stress. In Chapter 6, we derive the equations of equilibrium for plates, the so-called Föppl–von Kármán equations (F.–von K. equations as a shorthand). These equations have an intricate mathematical structure—the source of in-plane stress is the Gauss curvature, a quadratic function of the out-of-plane deviation. A classical application of the F.–von K. equations is the analysis of the buckling of plates under transverse loading. We shall present various situations where this buckling phenomenon can be analysed, in connection with experiments. In Chapter 7, we consider buckling near threshold at length. Buckling at a finite distance from the threshold in loading space is studied in Chapter 8. Unlike mathematically related problems of instabilities in fluid mechanics, it makes sense to look at the buckling instability far above threshold. Folds and d-cones are a first type of solutions of the F.–von K. equations that can emerge at large loads, and are the building blocks for the polyhedral shape of crumpled paper. They are studied in Chapter 9. In Chapter 10, we derive a different kind of solution, relevant to the buckling of an elastic plate under large applied *edge* loads, which is self-similar.

Part III is devoted to the elasticity of thin shells. By definition a shell is a piece of thin elastic material that is curved at rest. The main contribution to the elastic energy of a shell comes from stretching effects. As a result, the mechanical behaviour of shells is related to a non-trivial question of geometry, namely whether a given surface can or cannot deform without stretching, that is without changing the lengths of curves drawn along it. This geometrical problem is considered in some detail in Chapter 11. Geometry is only one piece of the puzzle. A sphere, for instance, is found to be geometrically rigid but this does not tell how a spherical shell deforms when subjected to stress. Thanks to the existence of a small parameter, the ratio of thickness to radius of curvature, a consistent theory describes the deformations of shells at the dominant order. The so-called membrane theory is derived in

Chapter 12 for shells of revolution. We also show how it can be extended to take bending effects into account—the membrane equations in the general case, that is without symmetry, are given in Appendix 12 for the sake of completeness.

The symmetry of revolution applies for the examples we consider next, the elastic torus and the elastic ball. The elastic torus is the simplest shell geometry that is partly elliptic (on its outer part) and partly hyperbolic (on its inner part): its analysis, carried out in Chapter 13, provides insights into the mechanics of shells of mixed kind (i.e. shells that are neither fully hyperbolic nor fully elliptic), a class of shells for which very few general results are available. As we shall see, the mechanical stiffness of a torus is very different from that of more classical shells, an effect that we shall ultimately explain with the help of differential geometry. The calculation of the elastic response of a torus requires an elaborate boundary layer analysis. In the last chapter, Chapter 14, we focus on another simple shell geometry, namely a spherical elastic ball, and consider a question with an almost everyday relevance: how does a ball, such as a tennis or ping pong ball, flatten when hit, and how long does it take for it to bounce? This turns out to be a rather subtle question, because there are two regimes: at small forces, the sphere flattens at the contact with the pushing plane, while at larger forces it forms an inverted cap and contacts the pushing plane along a circular ridge.

1.2   Notations and conventions

We have tried to keep mathematical notations as simple and consistent as possible. Probably most of them are familiar to the reader.

1.2.1   Vectors

Real numbers, vectors and vector spaces

The set of real numbers is denoted $\mathbb{R}$. A real $x \in \mathbb{R}$ is a positive, null or negative number, $-\infty < x < +\infty$ ($\in$ means 'belongs to' but we shall rarely use this and other Bourbakist notations).

A *vector* $\mathbf{a}$ (which may be seen equivalently as a point here) in the three-dimensional Euclidean space is given by its coordinates (a_x, a_y, a_z) in an orthonormal basis $(\mathbf{e}_x, \mathbf{e}_y, \mathbf{e}_z)$. Using a particular basis, vectors can be identified with their coordinates, that is a collection of three real numbers (a_x, a_y, a_z). Therefore, the Euclidean three-dimensional space, which is the set of all vectors, is denoted $\mathbb{R}^3$. Vectors are denoted by *boldface* symbols such $\mathbf{a} \in \mathbb{R}^3$.

We shall also use the notation (a_1, a_2, a_3), or simply (a_i) $(i = 1, 2, 3)$ for vector coordinates:

$$a_x = a_1, \qquad a_y = a_2, \qquad a_z = a_3.$$

The notation first introduced, (a_x, a_y, a_z), is more readable and is preferred for explicit calculations. The second notation (a_i) allows implicit summation (see below), which yields compact notations.

Similarly, the orthonormal basis whose vectors define the x, y and z axes will be denoted $(\mathbf{e}_x, \mathbf{e}_y, \mathbf{e}_z)$ or $(\mathbf{e}_1, \mathbf{e}_2, \mathbf{e}_3)$.

The current point in physical space is denoted $\mathbf{r}$ and its coordinates are written (x, y, z), that is $x = r_x = r_1, y = r_y = r_2, z = \cdots$. We introduce yet another notation for the current point $\mathbf{r} = (x, y, z)$ in the form (x_1, x_2, x_3), or simply (x_i). This will be useful for implicit summation:

$$x_1 = x, \qquad x_2 = y, \qquad x_3 = z.$$

The presence of an index after the symbol x changes its meaning: x_i is a generic coordinate while x stands for the first coordinate. This does not introduce any ambiguity.

Operations on vectors

The scalar product of two vectors $\mathbf{a}$ and $\mathbf{a}'$ is denoted $\mathbf{a} \cdot \mathbf{a}'$:

$$\mathbf{a} \cdot \mathbf{b} = a_x\, b_x + a_y\, b_y + a_z\, b_z. \tag{1.1}$$

The condition that the basis $(\mathbf{e}_i)_{i=1,2,3}$ is orthonormal takes the form:

$$\mathbf{e}_i \cdot \mathbf{e}_j = \delta_{ij} \qquad \text{for } i, j = 1, 2, 3, \tag{1.2}$$

where we have introduced Kronecker's symbol δ_{ij} with integer indices, whose value is 1 if indices are equal and 0 otherwise.

When the vectors onto which the scalar product operates are themselves given by long expressions, we use an alternative notation for the scalar product which is easier to read, namely $\langle \mathbf{a} | \mathbf{b} \rangle$. The two notations are equivalent when applied to vectors:

$$\langle \mathbf{a} | \mathbf{b} \rangle = \mathbf{a} \cdot \mathbf{b}. \tag{1.3}$$

The notation with angular brackets is used in a more general context, for instance when we define the scalar product $\langle f | g \rangle$ of two functions f and g.

The norm of a vector is denoted like the absolute value of real numbers:

$$|\mathbf{a}| = \sqrt{\mathbf{a} \cdot \mathbf{a}}. \tag{1.4}$$

The cross product of two vectors $\mathbf{a}$ and $\mathbf{b}$ is a new vector $\mathbf{c}$ written as:

$$\mathbf{c} = \mathbf{a} \times \mathbf{b} \tag{1.5}$$

Let a_i, b_i and c_i be the coordinates of these three vectors ($i = 1, 2, 3$). The cross product operation is defined by:

$$c_1 = a_2\, b_3 - a_3\, b_2, \qquad c_2 = a_3\, b_1 - a_1\, b_3, \qquad c_3 = a_1\, b_2 - a_2\, b_1. \tag{1.6}$$

A double cross product can be expanded using the following identity, valid for any vectors $\mathbf{a}$, $\mathbf{b}$ and $\mathbf{c}$:

$$\mathbf{a} \times (\mathbf{b} \times \mathbf{c}) = (\mathbf{a} \cdot \mathbf{c})\, \mathbf{b} - (\mathbf{a} \cdot \mathbf{b})\, \mathbf{c}. \tag{1.7}$$

Note that on the right-hand side, every factor in parenthesis is given by a scalar product operation, yielding a real number, which is multiplied by a third vector. This identity can be checked by an explicit calculation, using the definition above for the cross product. It can also be established without calculation: this identity coincides with the definition of the cross product when $\mathbf{b} = \mathbf{e}_y$ and $\mathbf{c} = \mathbf{e}_z$. The equality for all $\mathbf{b}$ and $\mathbf{c}$ follows by bilinearity and antisymmetry of both sides of the equation with respect to $\mathbf{b}$ and $\mathbf{c}$.

Another useful identity for the cross product is the Jacobi identity:[1]

$$\mathbf{a} \times (\mathbf{b} \times \mathbf{c}) + \mathbf{b} \times (\mathbf{c} \times \mathbf{a}) + \mathbf{c} \times (\mathbf{a} \times \mathbf{b}) = \mathbf{0}, \tag{1.8}$$

valid for any vectors $\mathbf{a}$, $\mathbf{b}$ and $\mathbf{c}$. Again, this identity can be established by an explicit but tedious calculation, or simply by noting that it is true for $\mathbf{a} = \mathbf{e}_x$, $\mathbf{b} = \mathbf{e}_y$ and $\mathbf{c} = \mathbf{e}_z$, and by generalizing for all $\mathbf{a}$, $\mathbf{b}$ and $\mathbf{c}$ by bilinearity and antisymmetry of both sides with respect to all pairs of vectors.

Finally, the mixed product of three vectors $\mathbf{a}$, $\mathbf{b}$, $\mathbf{c}$ is defined as the scalar $(\mathbf{a} \times \mathbf{b}) \cdot \mathbf{c}$. By an explicit calculation in coordinates, this scalar can be shown to be equal to the determinant of the matrix with the coordinates of $\mathbf{a}$, $\mathbf{b}$ and $\mathbf{c}$ presented in rows in some orthonormal basis $(\mathbf{e}_x, \mathbf{e}_y, \mathbf{e}_z)$:

$$(\mathbf{a} \times \mathbf{b}) \cdot \mathbf{c} = \det \begin{pmatrix} a_x & b_x & c_x \\ a_y & b_y & c_y \\ a_z & b_z & c_z \end{pmatrix}. \tag{1.9}$$

As a result, the mixed product is invariant by direct permutation of the vectors:

$$(\mathbf{a} \times \mathbf{b}) \cdot \mathbf{c} = (\mathbf{b} \times \mathbf{c}) \cdot \mathbf{a} = (\mathbf{c} \times \mathbf{a}) \cdot \mathbf{b}. \tag{1.10}$$

These identities can also be shown geometrically, as the mixed product can be interpreted as the *signed* volume of the parallelepipedic 3D domain with edges $\mathbf{a}$, $\mathbf{b}$ and $\mathbf{c}$, positive if the vectors are in direct orientation and negative otherwise.

1.2.2 Tensors

Up to Part III (which is devoted to the elasticity of shells, i.e. to surfaces with an arbitrary shape at rest), we shall use tensors in a very naive manner: we make use only of orthonormal frames (making it unnecessary to distinguish covariant and contravariant indices) and introduce tensors merely as a convenient notation (Einstein summation) for linear algebra (matrix multiplication, etc.). A more intrinsic approach is followed in Part III and further details are given in due course.

The symbol i has already been used above in the expression of a generic vector coordinate x_i. More generally, we shall use Latin letters for indices of coordinates in the three-dimensional space. For instance, (x_i) means (x_1, x_2, x_3) unless a specific value of i was proposed in the context.

In contrast, Greek indices only run over the first two dimensions: (a_β) stands for the two-dimensional vector (a_1, a_2). This convention will be useful to distinguish between in-plane vectors and arbitrary vectors in plates and shells elasticity.

The range of indices can therefore be determined unambiguously from their reading. This allows us to introduce the Einstein summation convention. According to this convention:

- A sum sign is implicit for any index repeated twice in the same side of an equal sign.
- If an index appears once on either side of an equal sign (dangling index), the equality is true for all values of the index—unless a particular value of this index has been explicitly chosen.

[1] This equality defines what is called a Lie algebra when $\mathbf{a}$, $\mathbf{b}$ and $\mathbf{c}$ denote infinitesimal rotations.

- Similarly when there is no equal sign, indices that appear only once (and whose value is unspecified) are to be understood as generic indices.

For instance, a_i (with no repeated indices) collectively denotes the coordinates $(a_1, a_2, a_3) = (a_x, a_y, a_z)$ of a vector $\mathbf{a}$. In contrast, the notation $a_i\, b_i$ has a repeated index: a sum is implicit, making this expression a shorthand for

$$\sum_{i=1}^{3} a_i\, b_i,$$

which is the scalar product $\mathbf{a} \cdot \mathbf{b}$.

Another example is $a_i = b_{ij}\, c_j$, which stands for:

$$a_1 = \sum_{j=1}^{3} b_{1j}\, c_j, \qquad a_2 = \sum_{j=1}^{3} b_{2j}\, c_j, \qquad a_3 = \sum_{j=1}^{3} b_{3j}\, c_j.$$

Note that the index i appears once on either side of the equal sign, and so the equality must hold for all values of i: it is in fact an equality between vectors with index i. In contrast, j is repeated (in the right-hand side) and so must be summed. The above notation $a_i = b_{ij}\, c_j$ means that the vector $\mathbf{a}$ is obtained by multiplying the matrix B with entries (b_{ij}) by the vector $\mathbf{c}$.

The harmonic operator Δf acting on a scalar function $f(x, y, z)$ is defined as

$$\Delta f = \frac{\partial^2 f}{\partial x^2} + \frac{\partial^2 f}{\partial y^2} + \frac{\partial^2 f}{\partial z^2}. \tag{1.11}$$

Using Einstein summation, this can be put into the condensed form:

$$\Delta f = \frac{\partial^2 f}{\partial x_i{}^2}.$$

Here, the index i appears to be repeated when the denominator $\partial x_i{}^2 = \partial x_i\, \partial x_i$ is expanded, and so a sum is implicit. Because Greek indices scan only the first two directions, the operator $\partial^2 f / \partial x_\alpha{}^2$ stands for the *two-dimensional* harmonic operator, namely $\partial^2 f / \partial x^2 + \partial^2 f / \partial y^2$.

Both repeated and non-repeated indices are dumb: the actual symbol used is unimportant and can be changed as long as consistency is preserved. For instance, $a_i = b_{ij}\, c_j$ can be rewritten $a_p = b_{pq}\, c_q$. This is because the original equality is valid for all i and involves an implicit sum over j in the right-hand side.

The following operation, called the double contraction,[2] generalizes the dot product with two tensors with two indices:

$$a_{ij}\, b_{ij}.$$

Starting from two tensors with two indices, the result of the double contraction is a scalar. The density of elastic energy, for instance, involves the double contraction of the strain tensor by the stress tensor, as we shall see. The following property will be useful later on:

---

[2] More accurately, this is the double contraction of a tensor with the transpose of the other, as the double contraction is usually defined as $a : b = a_{ij}\, b_{ji}$.

the double contraction of a symmetric tensor by a skew-symmetric tensor is zero:

$$\text{if } a_{ij} = a_{ji} \text{ and } b_{ij} = -b_{ji}, \text{ then } a_{kl}\, b_{kl} = 0. \tag{1.12}$$

The proof comes by noticing that terms in the double contraction with $k = l$ are zero since any diagonal component b_{kl} is zero for an skew-symmetric tensor, while for $k \neq l$, the terms in the implicit sum with indices (k, l) and with the same indices permuted cancel each other since $a_{lk}\, b_{lk} = a_{kl}\, (-b_{kl}) = -(a_{kl}\, b_{kl})$ (no implicit summation here). It follows from this result that if a_{ij} is a symmetric tensor and b_{ij} and c_{ij} are two tensors with the same symmetric part, that is $(b_{ij} + b_{ji})/2 = (c_{ij} + c_{ji}/2)$, then b_{ij} and c_{ij} yield the same result by double contraction with a_{ij}: $a_{ij}\, b_{ij} = a_{ij}\, c_{ij}$. Indeed, when b_{ij} and c_{ij} have the same symmetric part, their difference is[3] a skew-symmetric tensor, and equation (1.12) applies.

1.2.3 Mechanical quantities

The current point in the reference configuration is written $\mathbf{r} = (x, y, z)$. Upon deformation, the associated material point moves to $\mathbf{r}' = (x', y', z')$ (the point in the actual configuration).

The displacement field will be denoted as $\mathbf{u}$. Its components (u_x, u_y, u_z) are also denoted (u, v, w) and, by definition, are equal to $(x' - x)$, $(y' - y)$ and $(z' - z)$. In our analysis of plates, we take (x, y) as the direction parallel to the centre plane in the reference configuration and assume small displacements (this is the framework of Föppl–von Kármán equations). As a result, the tangent plane to the plate remains everywhere close to the (x, y) plane upon deformation, and (u, v) are the in-plane components of the displacement. The transverse component, also called the deflection, is w. The coordinates along the plate, (x, y), are also written (x_α), with $\alpha = 1, 2$.

The strain tensor is written ϵ_{ij} and the stress tensor σ_{ij}. Volumic mass is denoted ρ and external forces per unit mass are written $\mathbf{g}$. Linear elastic materials are considered throughout this book: Young's modulus is denoted by E and Poisson's ratio by ν; Lamé coefficients are written according to the standard notation λ and μ.

In our analysis of elastic rods, the centre line has curvilinear coordinate s. The material frame is denoted $(\mathbf{d}_1, \mathbf{d}_2, \mathbf{d}_3)$ where $\mathbf{d}_1$ and $\mathbf{d}_2$ are along the principal direction of inertia within the cross-section and $\mathbf{d}_3$ is tangent to the centre line. Curvatures and twist are denoted κ_α and τ. Flexural and torsional stiffnesses are denoted EI_α and μJ respectively. The internal force and moment are denoted $\mathbf{N}$ and $\mathbf{M}$ respectively. Other notations are introduced in Chapter 3.

The geometry of surfaces is presented in Part III of this book. The principal curvatures are denoted κ_1 and κ_2 (there could be some ambiguity with the two curvatures of a Cosserat rod but we deal with rods and shells in different parts), the principal radii of curvatures are R_1 and R_2, the mean curvature is H, and the Gaussian curvature is K.

The thickness of a plate or a shell is written as h, a length that is by assumption much smaller than any other typical length scale in the problem. The bending stiffness is denoted D.

Elastic energies are denoted as $\mathcal{E}_{...}$ with various indices, such as $\mathcal{E}_{\mathrm{el}}$ (total elastic energy), $\mathcal{E}_{\mathrm{b}}$ (bending energy), $\mathcal{E}_s$ (stretching energy), etc.

---

[3] This can be shown by noticing that any tensor a_{ij} with two indices can be explicitly decomposed as the sum of a symmetric and a skew-symmetric tensor, which read $(a_{ij} + a_{ji})/2$ and $(a_{ij} - a_{ji})/2$ respectively.

1.2.4 Dimensional analysis, scaled variables

Let us consider a point-like mass m that is attached to a linear spring with an additional applied force $F(t)$. Its equation of motion is:

$$k\,x(t) + m\,\ddot{x}(t) = F(t),$$

where k is the spring stiffness, t the time, and the dots denote time derivation. We denote typical values with an asterisk: F^* for the force, t^* for the time, etc. These quantities can be chosen arbitrarily (also some choices are more helpful than others, see below). These typical quantities are used to introduce *rescaled*, dimensionless quantities, marked with a bar, such as

$$\bar{t} = \frac{t}{t^*}, \qquad \bar{F} = \frac{F}{F^*}.$$

A similar definition is used for other quantities.

By convention, *rescaled quantities* are implicitly considered to be *functions of rescaled parameters* and so $\bar{x}$ is a function of $\bar{t} = t/t^*$, unlike x, which is a function of t. As a result, derivatives acting on rescaled quantities are taken with respect to *rescaled variable*. Since

$$\bar{x}(\bar{t}) = \frac{x(t^*\,\bar{t})}{x^*}$$

we get that the derivative of a rescaled quantity rescales differently from the non-derived quantity:

$$\dot{\bar{x}}(\bar{t}) = \frac{\mathrm{d}\bar{x}}{\mathrm{d}\bar{t}} = \frac{t^*}{x^*}\frac{\mathrm{d}x}{\mathrm{d}t} = \frac{t^*}{x^*}\,\dot{x}(t) = \frac{\dot{x}(t)}{x^*/t^*}.$$

This shows that the natural scale for $\dot{x}(t)$ is x^*/t^*. A rule of thumb is that if a scales like a^* and b scales like b^*, the derivative $\mathrm{d}a/\mathrm{d}b$ scales like a^*/b^*. The only trick is not to forget scaling factors, such as $1/t^*$ in the previous example, associated with derivatives marked with a prime or dot shorthand.

As written above, scaling factors such as t^* and F^* can be chosen arbitrarily and independently of each other, but the trick is to choose them in such a way that the rescaled problem has as few dimensionless coefficients as possible in the end. In our example of the forced linear spring, the typical intensity of the force F^* provides a natural typical displacement in the form

$$x^* = \frac{F^*}{k}.$$

A typical time t^* can be proposed by balancing the scales associated with the first two terms in the equation of motion: $k\,x^* = m\,x^*/(t^*)^2$. By this equation, we are in fact trying to impose the condition that the coefficients in front of the first two terms become identical in the rescaled problem. This yields a natural choice for the time scale

$$t^* = \sqrt{\frac{m}{k}}.$$

Up to a factor 2π this is the period of the natural oscillations of the spring–mass system. Using these rescalings, the dimensionless problem becomes:

$$\bar{x}(\bar{t}) + \ddot{\bar{x}}(\bar{t}) = \bar{F}(\bar{t}).$$

The obvious advantage of the rescaled formulation is that it depends on fewer parameters. Here, the mass and the spring stiffness have been effectively set to one. The idea is to get rid of as many parameters as possible before actually solving equations. In several places in this book, the dimensional analysis absorbs all the parameters of the problem, making it possible to derive a universal solution by solving once and for all an equation with fixed numerical constants. Another important asset of introducing dimensionless equations is that it allows one to directly infer the dominant physical effect out of many possible effects:[4] the subdominant effects have a small coefficient in their mathematical representation once the rescaling has been made, although the coefficients of the leading terms remain of order one.

1.2.5 Miscellaneous notations

To allow compact notation, commas in indices denote partial derivatives. For instance,

$$f_{,x} \quad \text{stands for} \quad \frac{\partial f}{\partial x}.$$

The gradient of a scalar field $f(x, y, z)$ can then be written as the vector with coordinates $(f_{,x_i})$, or simply $(f_{,i})$. Using implicit summation over the index i, the harmonic operator introduced earlier can be written as $\Delta f = f_{,x^2} + f_{,y^2} + f_{,z^2} = f_{,x_i^2} = f_{,ii}$. Implicit summation over index i applies here.

In the Föppl–von Kármán theory of elastic plates, we use the differential operator $[f, g]$ acting on two smooth scalar fields $f(x, y)$ and $g(x, y)$ defined along the plane (x, y). This operator is a combination of products of second derivatives, and is known as the Monge–Ampère operator:

$$[f, g] = \frac{\partial^2 f}{\partial x^2} \frac{\partial^2 g}{\partial y^2} + \frac{\partial^2 g}{\partial x^2} \frac{\partial^2 f}{\partial y^2} - 2 \frac{\partial^2 f}{\partial x \partial y} \frac{\partial^2 g}{\partial x \partial y}. \tag{1.13}$$

In other places, we shall have to perform an expansion of various functions in powers of a small parameter. Let $f(\eta)$ be such a function of η, small. The Taylor expansion of $f(.)$ is written as

$$f(\eta) = f^{[0]} + \eta \, f^{[1]} + \dots.$$

Finally we shall use the shorthand ODE for Ordinary Differential Equations, PDE for Partial Differential Equations, l.h.s. and r.h.s. for the left-hand and right-hand side (of equations), F.–von K. as a shorthand for Föppl–von Kármán, b.c. for boundary conditions, as well as the classical 2D and 3D for 'two-dimensional' and 'three-dimensional'.

1.3 Mathematical background

1.3.1 Rigid-body rotations, infinitesimal rotations

Let $\mathbf{a}$ be a three-dimensional vector, $a = |\mathbf{a}|$ its norm, and $\mathbf{b}$ be an another arbitrary vector. When $\mathbf{a} \neq \mathbf{0}$, let $\mathbf{u} = \mathbf{a}/a$ be the unit vector parallel to $\mathbf{a}$, and $\mathbf{b}^{\|}$ and $\mathbf{b}^{\perp}$ be the

---

[4] See for instance our analyses of the hair in Chapter 4, and the regularization of the torus solution by non-linear versus bending effects based on the parameter η in Chapter 13.

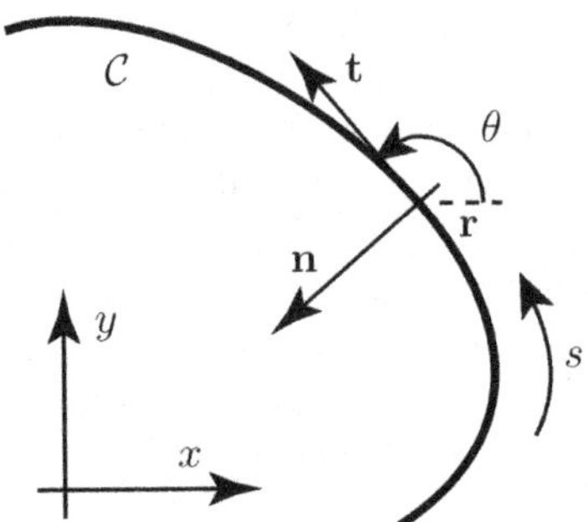

Fig. 1.1 Geometry of a planar curve. Curvature is associated with rotation $\theta'(s)$ of the frame $(\mathbf{t}, \mathbf{n})$ when one follows the curve.

longitudinal and perpendicular components of $\mathbf{b}$ with respect to the axis spanned by $\mathbf{u}$, respectively:

$$\mathbf{b}^{\parallel} = (\mathbf{u} \cdot \mathbf{b})\,\mathbf{u}, \qquad \mathbf{b}^{\perp} = \mathbf{b} - \mathbf{b}^{\parallel}.$$

When $\mathbf{a} = \mathbf{0}$, we set by convention $\mathbf{b}^{\parallel} = \mathbf{b}$ and $\mathbf{b}^{\perp} = \mathbf{0}$.

Let us call *rotation with vector* $\mathbf{a}$, denoted $R_{\mathbf{a}}$, the rotation with angle $a = |\mathbf{a}|$ in radians, whose axis is spanned by $\mathbf{u}$ and passes through the origin of the coordinates (the rotation with vector $\mathbf{a} = \mathbf{0}$ is by definition the identity mapping). The effect of a rotation $R_{\mathbf{a}}$ on a vector $\mathbf{b}$ is expressed by the following linear mapping:

$$R_{\mathbf{a}}(\mathbf{b}) = \mathbf{b}^{\parallel} + \mathbf{b}^{\perp} \cos a + (\mathbf{u} \times \mathbf{b}^{\perp}) \sin a.$$

Upon this rotation, the component $\mathbf{b}^{\parallel}$ of $\mathbf{b}$ along the axis of rotation remains unchanged while its normal component rotates by an angle a in the plane perpendicular to $\mathbf{u}$, spanned by $\mathbf{b}^{\perp}$ and $(\mathbf{u} \times \mathbf{b}^{\perp})$.

In the case of an infinitesimal rotation represented by a vector $\delta\mathbf{a}$ whose norm is small, $\delta a \ll 1$, the above formula becomes, at linear order with respect to δa:

$$R_{\delta\mathbf{a}}(\mathbf{b}) \approx \mathbf{b} + (\mathbf{u} \times \mathbf{b}^{\perp})\,\delta a + \dots$$

since $\mathbf{b}^{\parallel} + \mathbf{b}^{\perp} = \mathbf{b}$. Therefore, the change of $\mathbf{b}$ due to an infinitesimal rotation with vector $\delta\mathbf{a}$ reads

$$\delta\mathbf{b} = R_{\delta\mathbf{a}}(\mathbf{b}) - \mathbf{b} \approx \delta\mathbf{a} \times \mathbf{b}^{\perp},$$

using $\delta\mathbf{a} = (\delta a)\,\mathbf{u}$. Now, $\delta\mathbf{a} \times \mathbf{b}^{\parallel} = \mathbf{0}$ by construction and so an infinitesimal rotation is characterized by

$$\delta\mathbf{b} \approx \delta\mathbf{a} \times \mathbf{b}. \tag{1.14}$$

Consider the case where $\mathbf{b}$ is the position vector of a material point on a solid undergoing a rigid-body rotation with rotation vector $\boldsymbol{\omega}$. Then, $\delta\mathbf{b} = \mathbf{v}\,\delta t$ is the change of its position during a time interval δt, and the angle of rotation reads $\delta\mathbf{a} = \boldsymbol{\omega}\,\delta t$. The equation above yields the well-known form of the velocity field $\mathbf{v}$ in a rigid body in rotation with angular velocity $\boldsymbol{\omega}$, namely $\mathbf{v} = \boldsymbol{\omega} \times \mathbf{b}$.

1.3.2   Geometry of planar curves, curvature

We shall introduce here the geometry of curves and in particular the notion of curvature, which lies at the heart of the theory of rods. Let $\mathcal{C}$ be a smooth planar curve in the (x, y) plane, see Fig. 1.1. Let s be the curvilinear coordinate along this curve. We denote by $\mathbf{r}(s)$ its arc length parameterisation: $\mathbf{r}(s)$ is the current point along the curve in its actual configuration.[5] Since s is an arc length parameterisation, the tangent vector to the curve

$$\mathbf{t}(s) = \frac{\mathrm{d}\mathbf{r}}{\mathrm{d}s} \tag{1.15}$$

is unitary (its norm is one). Being a vector contained in the (x, y) plane, it can be written in the form:

$$\mathbf{t}(s) = \cos\theta(s)\,\mathbf{e}_x + \sin\theta(s)\,\mathbf{e}_y. \tag{1.16a}$$

Let us introduce the unitary normal vector to the curve (there is a unique normal vector because the curve is drawn on a plane):

$$\mathbf{n}(s) = -\sin\theta(s)\,\mathbf{e}_x + \cos\theta(s)\,\mathbf{e}_y. \tag{1.16b}$$

This vector is normal to the tangent, hence to the curve, and is such that $(\mathbf{t}(s), \mathbf{n}(s))$ is oriented in the trigonometric direction.

Assuming that the curve $\mathcal{C}$ is $\mathcal{C}^2$-smooth (the parameterisation $\mathbf{r}(s)$ twice differentiable with continuous second derivatives), it is possible to choose a continuously differentiable determination of the angle $\theta(s)$. By derivation of equations (1.16a) and (1.16b), we get:

$$\mathbf{t}'(s) = \theta'(s)\,\mathbf{n}(s) \quad \text{and} \quad \mathbf{n}'(s) = -\theta'(s)\,\mathbf{t}(s), \tag{1.17}$$

where primes denote derivation with respect to s. Introducing the vector

$$\mathbf{\Omega}(s) = \theta'(s)\,\mathbf{e}_z, \tag{1.18}$$

where z is the direction perpendicular to the plane, equations (1.17) can be put in the form:

$$\mathbf{t}'(s) = \mathbf{\Omega}(s) \times \mathbf{t}(s) \quad \text{and} \quad \mathbf{n}'(s) = \mathbf{\Omega}(s) \times \mathbf{n}(s). \tag{1.19}$$

Now consider a solid body in rotation with instantaneous rotation vector $\boldsymbol{\omega}(t)$. Let $\mathbf{r}$ be the actual position in space of an arbitrary point of this solid, with the origin $\mathbf{r} = \mathbf{0}$ lying on the axis of rotation. The velocity field due to this rotation is $\dot{\mathbf{r}} = \boldsymbol{\omega} \times \mathbf{r}$. Comparison with equation (1.19) yields a simple interpretation of the vector $\mathbf{\Omega}$: along the curve (that is when s varies) the frame $(\mathbf{t}(s), \mathbf{n}(s))$ rotates according to the rotation vector $\mathbf{\Omega}(s)$. As the frame is always contained in the (x, y) plane, the rotation vector has indeed to be perpendicular to this plane, that is along $\mathbf{e}_3$. The rotation 'velocity' around the direction $\mathbf{e}_3$ is $\theta'(s)$ from equation (1.18)—note that this is actually not a rotation velocity but instead a *rate of rotation* per unit curvilinear length. This is exactly the common sense notion of curvature: a curve is highly curved when the orientation of the tangent changes rapidly when one follows the curve. Therefore, we shall call curvature the quantity $\kappa(s) = \theta'(s)$.

---

[5] According to our general conventions, the current point along the curve in its *actual* configuration should be denoted by $\mathbf{r}'(s)$ and not by $\mathbf{r}(s)$. However, in the geometry of curves (and in the elasticity of rods), the current point in the reference configuration is directly parameterised by s and the reference configuration is of little use. Moreover, the prime notation in $\mathbf{r}'(s)$ could be misinterpreted as a derivation with respect to s. Therefore, in the particular case of rods, $\mathbf{r}$ shall refer to the actual configuration.

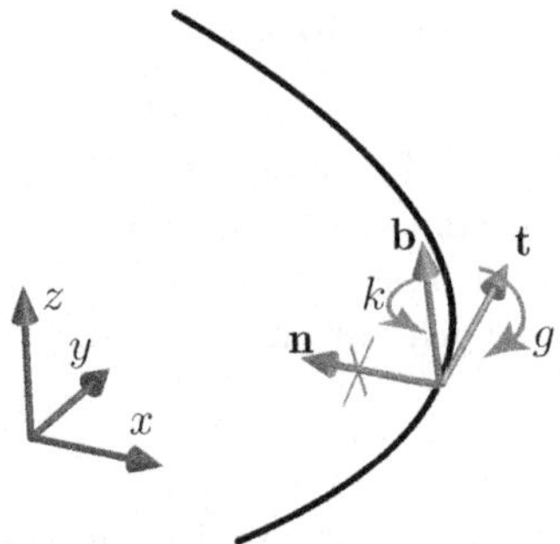

Fig. 1.2 A three-dimensional curve and the associated Serret–Frénet frame. Rotation of the Serret–Frénet frame, as described by the Darboux vector, involves a combination of a rotation about the tangent (torsion) and the binormal (curvature).

In the case of a circle of radius R, the parametric equation ($x = R\cos\phi, y = R\sin\phi$) yields $s = R\phi$ (curvilinear coordinate) and $\theta(s) = \phi + \pi/2 = s/R + \pi/2$, hence a constant curvature $\kappa(s) = 1/R$. Note that by following the circle in the opposite direction, we would have obtained $\kappa(s) = -1/R$.

This suggests the following *definitions*, valid for an arbitrary curve. The rate of rotation of the frame $(\mathbf{t}, \mathbf{n})$ is called the signed *curvature*:

$$\kappa(s) = \theta'(s). \tag{1.20}$$

Its inverse is called the signed *radius of curvature*:

$$R(s) = \frac{1}{\kappa(s)} = \frac{1}{\theta'(s)}. \tag{1.21}$$

In general, this quantity is not constant along the curve. It can be shown that this $R(s)$ is also the radius of the circle that best approximates the curve near a particular point $\mathbf{r}(s)$. The signs of both the curvature and the radius of curvature depend on the chosen orientation in the plane, as well as on the orientation of the curve set by the parameterisation s. Unsigned curvature and radius of curvature are defined as their absolute values, and are geometric quantities (they do not depend on choices of orientation).

1.3.3 Geometry of curves in 3D space, Serret–Frénet frame

The notion of curvature for a planar curve can be extended to the case of a space curve by means of the Serret–Frénet frame. Let again $\mathcal{C}$ be a space curve given by its arc length parameterisation $\mathbf{r}(s)$, and $\mathbf{t}(s) = \mathbf{r}'(s)$ its tangent. We shall first remark that:

$$\mathbf{t}'(s) \cdot \mathbf{t}(s) = \frac{1}{2}\frac{\mathrm{d}\,(\mathbf{t}(s) \cdot \mathbf{t}(s))}{\mathrm{d}s} = \frac{1}{2}\frac{\mathrm{d}\,(|\mathbf{t}(s)|^2)}{\mathrm{d}s} = 0, \tag{1.22}$$

since $\mathbf{t}(s)$ is a unitary vector: $|\mathbf{t}(s)|^2 = 1$ for all s. Therefore, the derivative of any unitary vector (in fact, of any vector of constant norm) with respect to some parameter yields a vector that is perpendicular to it. Whenever $\mathbf{t}'(s) \neq \mathbf{0}$, let us introduce the positive quantity $k(s) = |\mathbf{t}'(s)|$ and $\mathbf{n}(s)$ be the unitary vector parallel to it: $\mathbf{n}(s) = \mathbf{t}'(s)/|\mathbf{t}'(s)| = \mathbf{t}'(s)/k(s)$. This vector $\mathbf{n}$ is the *normal* vector. By construction, $\mathbf{n}(s)$ is perpendicular to $\mathbf{t}(s)$, and so the new vector $\mathbf{b}(s) = \mathbf{t}(s) \times \mathbf{n}(s)$ defines an orthonormal frame $(\mathbf{t}(s), \mathbf{n}(s), \mathbf{b}(s))$, called the

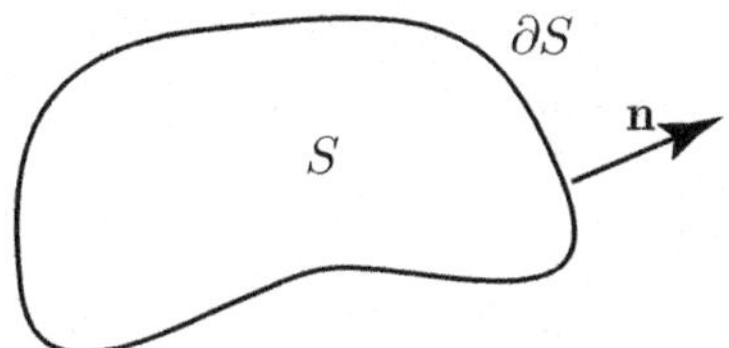

Fig. 1.3 Notations for the divergence theorem in two dimensions.

Serret–Frénet frame. The vector $\mathbf{b}$ is the *binormal vector*. Let us assume further that $\mathbf{n}(s)$ is differentiable at s and let us introduce $g(s) = \mathbf{n}'(s) \cdot \mathbf{b}(s)$. The functions $k(s)$ and $g(s)$ are called the curvature and the torsion respectively. Note that $k(s)$ and $g(s)$ are different, although related, to the curvatures and twist introduced in the context of a Cosserat curve, introduced in Chapter 3.

The derivatives of the vectors of the Serret–Frénet frame can be expressed in terms of the curvature and torsion. Indeed, by definition of $k(s)$, $\mathbf{t}'(s) = k(s)\,\mathbf{n}(s)$. By decomposition in the orthonormal Serret–Frénet frame, $\mathbf{n}' = (\mathbf{n}' \cdot \mathbf{t})\,\mathbf{t} + (\mathbf{n}' \cdot \mathbf{b})\,\mathbf{b}$, using $\mathbf{n}' \cdot \mathbf{n} = 0$ by the same argument as in equation (1.22). Now, by a similar argument again, we find:

$$\mathbf{n}' \cdot \mathbf{t} = \frac{\mathrm{d}(\mathbf{n} \cdot \mathbf{t})}{\mathrm{d}s} - \mathbf{n} \cdot \mathbf{t}' = -\mathbf{n} \cdot \mathbf{t}' = -k, \tag{1.23}$$

since $\mathbf{n}$ and $\mathbf{t}$ are perpendicular for all s. This yields $\mathbf{n}'(s) = -k(s)\,\mathbf{t}(s) + g(s)\,\mathbf{b}$. Similarly, one can show that $\mathbf{b}'(s) = -g(s)\,\mathbf{n}(s)$. Introducing the so-called *Darboux vector*:

$$\boldsymbol{\Omega}(s) = g(s)\,\mathbf{t}(s) + k(s)\,\mathbf{b}(s), \tag{1.24}$$

one can put all these equations into the compact form:

$$\mathbf{t}'(s) = \boldsymbol{\Omega}(s) \times \mathbf{t}(s), \quad \mathbf{n}'(s) = \boldsymbol{\Omega}(s) \times \mathbf{n}(s), \quad \mathbf{b}'(s) = \boldsymbol{\Omega}(s) \times \mathbf{b}(s). \tag{1.25}$$

By these relations, $\boldsymbol{\Omega}(s)$ can be interpreted as the local rate of rotation of the Serret–Frénet frame.[6] Note that, by construction, $\boldsymbol{\Omega}(s) \cdot \mathbf{n}(s) = 0$: the Serret–Frénet does not rotate about $\mathbf{n}(s)$, as indicated by the cross in Fig. 1.2.

The Serret–Frénet frame arises naturally when one deals with mathematical curves, but it is not well suited to the mechanics of rods. One of the problems associated with it is that it is not always well defined, even when the curve is smooth: this happens for a straight line, for instance, for which $\mathbf{t}'(s)$ vanishes everywhere and so $\mathbf{n}(s)$ is ill-defined. Even worse, the Serret–Frénet might not be continuous even though the curve is $\mathcal{C}^{\infty}$ (case of a planar curve with an inflexion point), and $k(s)$ and/or $g(s)$ may not go to zero even though the underlying curve converges uniformly to a straight line (the case of a helix with constant pitch and infinitesimal radius).

In the case of a planar curve, the Serret–Frénet frame is either $(\mathbf{t}_{2D}, \mathbf{n}_{2D}, \mathbf{e}_z)$ if $\kappa > 0$ or $(\mathbf{t}_{2D}, -\mathbf{n}_{2D}, -\mathbf{e}_z)$ if $\kappa < 0$, where $\mathbf{t}_{2D}$, $\mathbf{n}_{2D}$ refer to the quantities introduced in our analysis of planar curves above. Note that if $\kappa = 0$, the Serret–Frénet frame is undefined. By identification, one shows that $k(s) = |\kappa(s)|$ and $g(s) = 0$ for a planar curve. In fact, the

---

[6] The existence of the Darboux vector defined by equation (1.25) follows from the fact that the Serret–Frénet frame, being orthonormal, defines a rigid-body rotation in the Euclidean space with parameter s. By identifying the parameter s with time, the Darboux vector is just the instantaneous rotation velocity of the rigid body attached to the frame.

converse is true, and $g(s) = 0$ for all s implies that the curve is planar (provided the curve is smooth enough for the Serret–Frénet frame to exist).

1.3.4   Divergence theorem

We recall here the divergence theorem, which will be used later to derive the equations of mechanical equilibrium (2.51). Let V be a bounded region in three-dimensional space[7] with boundary $S = \partial V$. Let $\mathbf{a}(\mathbf{r})$ be a continuous, differentiable vector field defined over V. The divergence theorem relates the *volume* integral of the divergence of $\mathbf{a}$ over V with the *surface* integral of $\mathbf{a}$ over S:

$$\iiint_V (\operatorname{div} \mathbf{a})(\mathbf{r}) \, \mathrm{d}^3\mathbf{r} = \iint_S \mathbf{a}(\mathbf{r}) \cdot \mathbf{n}(\mathbf{r}) \, \mathrm{d}A, \tag{1.26}$$

where $\mathrm{d}^3\mathbf{r} = \mathrm{d}x \, \mathrm{d}y \, \mathrm{d}z$ is the element of volume, $\mathbf{n}$ is the outward pointing normal to S defined along S, $\mathrm{d}A$ is the element of area, and $\operatorname{div} \mathbf{a}$ denotes the differential operator acting on a vector field $\mathbf{a}$:

$$\operatorname{div} \mathbf{a} = a_{i,i} = \frac{\partial a_i}{\partial x_i} = \frac{\partial a_x}{\partial x} + \frac{\partial a_y}{\partial y} + \frac{\partial a_z}{\partial z}. \tag{1.27}$$

The right-hand side of equation (1.26) is called the flux[8] of vector $\mathbf{a}$ across surface S.

Note that the divergence theorem is a multi-dimensional generalization of the fundamental formula of calculus:

$$\int_p^q f'(x) \, \mathrm{d}x = f(q) - f(p) \tag{1.28}$$

in which $p < q$ defines a real interval $V = [p, q]$ with boundaries p and q. The outwards pointing normals at p and q are just $+\mathbf{e}_x$ and $-\mathbf{e}_x$ respectively, making the divergence theorem (1.26) an extension of equation (1.28). Equation (1.28) expresses the fact that the increase of f from x to $x + \mathrm{d}x$ is given to first order by $f'(x) \, \mathrm{d}x$. A very similar reasoning shows that the flux of $\mathbf{a}$ across the boundary of the infinitesimal cube defined by $(x, y, z) \in [x, x + \mathrm{d}x] \times [y, y + \mathrm{d}y] \times [z, z + \mathrm{d}z]$ is equal to first order to $(\operatorname{div} \mathbf{a}) \, \mathrm{d}x \, \mathrm{d}y \, \mathrm{d}z$, hence the form of the divergence theorem (1.26).

The divergence theorem can be extended to the case of a tensorial field $b_{ij}(\mathbf{r})$. By fixing the first index i in this tensor, we define the vector field $\mathbf{a}^{(i)}(\mathbf{r})$ given by the corresponding row in b_{ij}, such that $(a^{(i)})_j = b_{ij}$. Applying the divergence theorem to $\mathbf{a}^{(i)}$ yields:

$$\iiint_V \frac{\partial b_{ij}(\mathbf{r})}{\partial x_j} \, \mathrm{d}^3\mathbf{r} = \iint_S b_{ij} \, n_j \, \mathrm{d}A. \tag{1.29}$$

In this equation, the integrand in the left-hand side is the divergence of a tensor of rank two, this divergence being itself a vector (index i is dangling). In the right-hand side, $(b_{ij} \, n_j)$ is the vector obtained by contracting the tensor b_{ij} with the unit outwards normal n_j. Note that consistency requires that the *same index* of b_{ij} (here, the second one, labelled j) is

---

[7] The divergence theorem can be readily adapted to an arbitrary number of dimensions.

[8] As might be familiar to the reader, the divergence theorem can be used to write conservation laws in continuum mechanics: for instance, when $\mathbf{a} = \mathbf{v}$ is the Eulerian velocity of an incompressible fluid, the right-hand side of (1.26), which is the rate of variation of the mass contained in V, has to vanish. This holds for any volume V, and so the integrand in the left-hand side has to vanish too, yielding the classical condition of incompressibility $\operatorname{div} \mathbf{v} = 0$.

used for differentiation in the divergence of the left-hand side and for contraction with the normal in the right-hand side.

1.3.5   Constrained minimization

In elasticity, equilibrium configurations of a body minimize the elastic energy. Consider the simple case where the configuration of a body depends on only two discrete parameters, say X and Y. Let $\mathcal{E}(X, Y)$ be the total energy to be minimized. It is well-known that a necessary condition for (X, Y) to be a minimum is the stationarity condition $\nabla\mathcal{E} = 0$, where $\nabla\mathcal{E}$ is the gradient of $\mathcal{E}$ with respect to X and Y. In practice, this condition is enforced by computing the first-order variation $\delta\mathcal{E}$ for arbitrary changes of X and Y (denoted δX and δY) and imposing that $\delta\mathcal{E}$ vanishes for all δX and δY. For instance, $\mathcal{E}(X, Y) = X^2 + Y^2 + XY$ yields the condition $\delta\mathcal{E} = (2X + Y)\,\delta X + (2Y + X)\,\delta Y = 0$, hence $2X + Y = 0$ and $2Y + X = 0$. This linear system has the unique solution $(X, Y) = (0, 0)$, which is indeed the minimum of $\mathcal{E}$.

Let us now consider the more interesting case where the parameters X and Y are connected by a constraint of the form $g(X, Y) = 0$, and so cannot be varied independently. For instance, imagine (X, Y) is a parameterisation of a surface with elevation $z = \mathcal{E}(X, Y)$ along which a point mass is constrained to move. We consider the case where the mass is further constrained to move, say on the cylinder with equation $g(X, Y) = (X - 1)^2 + (Y + 1)^2 - 1 = 0$. The problem is now to find the minimum of $\mathcal{E}(X, Y)$ among allowed configurations, such that $g(X, Y) = 0$. For the particular function g chosen here, there is an obvious parameterisation of the allowed configurations (X, Y), namely $X(\theta) = 1 + \cos\theta$ and $Y(\theta) = -1 + \sin(\theta)$. The stationarity condition takes the form:

$$0 = \frac{\mathrm{d}\mathcal{E}(X(\theta), Y(\theta))}{\mathrm{d}\theta} = X'(\theta)\,\frac{\partial\mathcal{E}}{\partial X} + Y'(\theta)\,\frac{\partial\mathcal{E}}{\partial Y} = \mathbf{t}\cdot\nabla\mathcal{E}. \tag{1.30}$$

This equation has a very simple geometric interpretation: the gradient of $\mathcal{E}$ has to be orthogonal to the tangent $\mathbf{t}$ to the curve of allowed configurations. By definition, g is zero along this curve, and so equation (1.30) must also be satisfied when $\mathcal{E}$ is replaced by g. This shows that both $\nabla\mathcal{E}$ and ∇g are perpendicular to the tangent $\mathbf{t}$, and so are aligned: there should exist some scalar[9] λ such that

$$\nabla(\mathcal{E} + \lambda\, g) = 0.$$

This form is more convenient than (1.30), as it can be used directly from the implicit equation $g(X, Y) = 0$ and does not require an explicit parameterisation.

In practice, the stationarity condition of a constrained minimization problem is obtained by adding[10] to the variation of $\mathcal{E}$ a new term consisting of a yet undetermined scalar coefficient λ, multiplied by the function g expressing the constraint in an implicit form.

---

[9] The form of the new function to be minimized, $\mathcal{E} + \lambda g$, is reminiscent of the definition of the thermodynamic potentials. Indeed, a system that has a fixed volume and cannot exchange heat has an internal energy $U(S, V)$ that is minimum at equilibrium. When the condition of fixed volume is replaced by a condition of fixed pressure, it is the enthalpy, $H(S, P) = U + PV$, that is minimum at equilibrium. To deal with the constraint of a fixed pressure, a new term, PV, has been incorporated in the potential to be minimized. This new term is minus the product of the constraint, P, times a Lagrange multiplier, $(-V)$.

[10] By moving along the curve with equation $g = 0$, the function g does not change and so $\delta g = 0$. The new term added in the variation is therefore formally zero. Its only purpose is to introduce an additional parameter λ that can be tuned, making it possible for the constraint to be satisfied.

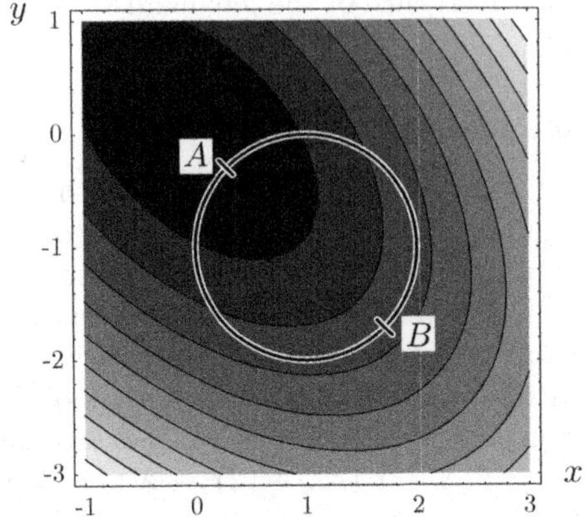

Fig. 1.4 Example of constrained minimization: finding the extrema of $\mathcal{E}(X,Y) = X^2 + Y^2 + XY$ (contours shown in background) along the circle with equation $g(X,Y) = 0$ with $g(X,Y) = (X-1)^2 + (Y+1)^2 - 1$.

The coefficient λ, called the Lagrange multiplier, is determined by requiring that the point obtained in this way does indeed satisfy the constraint $g = 0$. With our example functions,

$$\delta(\mathcal{E} + \lambda\, g) = (2X + Y + 2\lambda\,(X-1))\,\delta X + (2Y + x + 2\lambda\,(Y+1))\,\delta Y.$$

This leads to the equations $2X + Y + 2\lambda\,(X-1) = 0$ and $2Y + X + 2\lambda\,(Y+1)$. The solutions of these equations are

$$X(\lambda) = \frac{2\lambda}{1+2\lambda}, \qquad Y(\lambda) = \frac{-2\lambda}{1+2\lambda}.$$

They all lie on the line $X = -Y$, which is a bisector line of the coordinate axes. As explained above, λ is found by imposing that $(X(\lambda), Y(\lambda))$ satisfies the equation $g(X(\lambda), Y(\lambda)) = 0$. This yields $\lambda = (-1 \pm \sqrt{2})/2$. The two corresponding solutions A and B are shown in Fig. 1.4: A is a minimum of $\mathcal{E}$ along the circle while B is a maximum.

The present method can be extended to an arbitrary number c of scalar constraints $g_1(\mathbf{r}) = 0, \ldots, g_c(\mathbf{r}) = 0$ and when $\mathbf{r}$ is a vector of dimension n. Then, one has to introduce c Lagrange multipliers $(\lambda_1, \ldots, \lambda_c)$ and cancel the first-order variation of:

$$\delta_{\mathbf{r}}(\mathcal{E}(\mathbf{r}) + \lambda_1\, g_1(\mathbf{r}) + \cdots + \lambda_c\, g_c(\mathbf{r})) = 0, \tag{1.31}$$

leading to solutions in the form $\mathbf{r}(\lambda_1, \ldots, \lambda_c)$. As earlier, the values of the Lagrange multipliers are found by solving c equations:

$$g_i(\mathbf{r}(\lambda_1, \ldots, \lambda_c)) = 0 \qquad \text{for } 1 \leq i \leq c. \tag{1.32}$$

2
Three-dimensional elasticity

2.1 Introduction

In this chapter, we consider solid materials with volumic extension and derive the fundamental equations of the theory of elasticity in three dimensions. These equations will be specialized in the following chapters to the case of thin elastic bodies such as rods and plates.

We shall be concerned with two very different but not unrelated approximations throughout this book, which we introduce straight away: the *small strain* and the *small displacement* approximation. The small strain approximation, on the one hand, allows one to consider that the elastic response of a given material is linear (this contrasts with the case of finite, or large strain, where solids deform non-linearly and may even break). The small displacement approximation, on the other hand, is concerned with rotations of the material: volume elements may undergo rigid-body rotations, which are irrelevant for computing the elastic deformations. Keeping track of these rotations requires one to introduce nonlinearity into the equations. When the material is only slightly perturbed from rest, all such rotations remain infinitesimal. This brings many simplifications into the equations and defines what is called the approximation of small displacement.

Therefore, it appears that there are two sources of nonlinearity in the theory of elasticity: one comes from the elastic constitutive law and the other from geometry. The first one is associated with the intrinsic response of a material, that is how a volume element deforms under a given stress. It is well illustrated by stretching a piece of rubber: it becomes stiffer and stiffer and eventually breaks beyond a critical (and large—something that is rather specific to rubber) stretching. The other source of nonlinearity will be discussed in details in this book, and is tightly connected with the word *Geometry* in its title. These two kinds of non-linear contributions are directly related to the two families of approximations introduced above. When the strain remains small, elastic constitutive laws can be linearized; when displacements are small, geometrical nonlinearity is negligible. If only the strain is small, geometrical nonlinearity may still play a role.

Our approach in this book is to focus on geometrical nonlinearity and on important associated physical effects such as buckling. It turns out that geometrical nonlinearity is essential for predicting the elastic behaviour of thin bodies such as rods, plates and shells. Therefore, we do not limit our analysis to small displacements (we shall write that we consider *finite* displacement). However, to avoid adding too many difficulties to an already non-trivial topic, we always assume small strain and consider linear constitutive laws; this is the framework of Hookean elasticity. The equations of elasticity are given at the end of this chapter for small strain and finite displacement—to avoid a too steep presentation, we make an intermediate step and start by assuming small strain and small displacement, and generalize to finite displacement.

Classically the equations of elasticity in dimension three are introduced in two ways. The first is to derive the equations of equilibrium by balancing the forces and torques acting on a small element of the elastic material. This approach follows closely our physical intuition of forces and deformations. However it is sometimes more convenient, especially when considering thin elastic structures, to use a more abstract 'energy' approach: instead of balancing forces and torques, one derives the equilibrium equations for an elastic solid by imposing the condition that its configuration minimizes the elastic energy, a function of the displacements that is computed explicitly. In practice this 'energy' approach is often simpler than the force balance approach because the requirements coming from the symmetries or from constraints such as inextensibility (see the theory of rods in Chapter 3) can be taken into account more easily. Another advantage is that it displays more clearly the various physical effects at play: it is easier to identify the bending and stretching energies than the corresponding contributions to forces and torques. This ultimately makes clearer the analysis of the order of magnitude of these different physical effects. Such an analysis is used extensively throughout the following chapters. To illustrate that the minimization of energy gives the same result as the balance of forces we use both methods successively, in Sections 2.3.4 and 2.5.8.

In the following sections, 2.2 and 2.3, we introduce the fundamental notions of the theory of elasticity: the strain, which provides a measure of the deformation, and the stress, which accounts for the interior forces within a solid. The next step, addressed in Section 2.4, relates these quantities, by means of a so-called constitutive relation.

2.2 Strain

The theory of elasticity deals with solids that, in the presence of mechanical stress, depart from their 'natural' configuration. Unlike fluids, which can flow, the equilibrium configuration of an elastic solid is imprinted in matter, and recovered as soon as the stress is released. The strain provides a geometrical characterization of deformation: it measures by how much the solid departs from its natural configuration.

2.2.1 Transformation

The deformation of a solid is mathematically expressed[1] by a vector field $\mathbf{r}'(\mathbf{r})$ giving the final position in the physical space of any material point in the solid initially located at $\mathbf{r} = (x, y, z)$, see Fig. 2.1. By 'final', we mean the state of the solid *after* deformation, and by 'initial' *before* deformation. Making reference to time can be misleading and so we shall follow a classical terminology and call 'actual configuration' the deformed configuration (the 'final' one, which we are observing) and 'reference configuration' the initial one.

As implied by its name, defining the reference configuration is purely a matter of convention. It is often chosen to be a natural (that is, stress-free) configuration of the

---

[1] We recall that, in the context of three-dimensional elasticity, primes denote quantities in the deformed configuration (actual configuration), as opposed to the reference configuration.

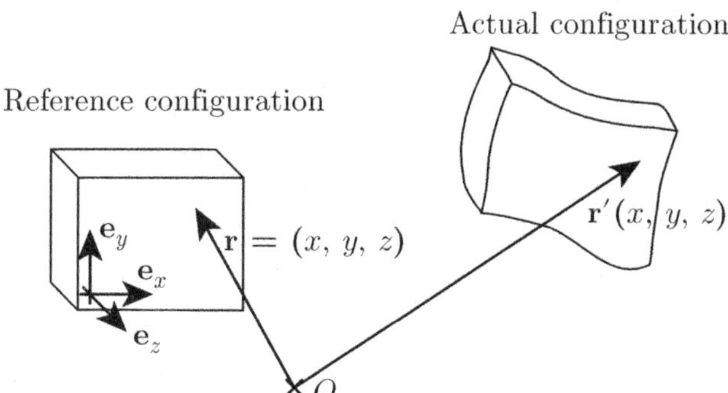

Fig. 2.1 Reference and deformed configurations of an elastic body. Material points are tagged using their position $\mathbf{r}$ in the reference configuration. Their final position $\mathbf{r}'$ yields a parameterisation $\mathbf{r}'(\mathbf{r})$ of the actual configuration.

body, but this is not mandatory.[2] The only purpose of the reference configuration is to tag material points[3] (atoms, molecules, etc.) with the aim of following them as the solid deforms: to this end, one can simply use their position $\mathbf{r} = (x, y, z)$ in the reference configuration. In mathematical terms, the only purpose of the reference configuration is to provide a parameterisation of the solid. It is not required to be a stress-free configuration, not even an equilibrium one. In fact, a more abstract parameterisation of the solid is possible.[4]

The vector field $\mathbf{r}'(\mathbf{r})$ introduced above is called the *transformation*. It defines the actual configuration by giving the position $\mathbf{r}'$ in actual configuration of any material point whose position in reference configuration was $\mathbf{r}$. Applying the mapping $\mathbf{r}'(\mathbf{r})$ upon the reference configuration has the effect of bringing the solid into the actual configuration, hence the name 'transformation'.

Here, we use a Cartesian (orthonormal) frame $\{\mathbf{e}_x, \mathbf{e}_y, \mathbf{e}_z\}$, call (x, y, z) and (x', y', z') the coordinates of the vectors $\mathbf{r}$ and $\mathbf{r}'$ respectively, and identify these vectors with their coordinates:

---

[2] There are two main situations where the reference configuration is not a natural (that is, not stress-free). Sometimes, there simply exists no stress-free configuration. This is the case for instance when a cube of ice is suddenly immersed in water. The temperature profile in ice soon becomes inhomogeneous, and the stress induced by thermal dilation cannot be fully removed by a change of shape of the ice. This is related to the question of strain compatibility (see Section 2.2.9). Another classical situation where it is convenient to choose a reference configuration that is not the natural one is in the analysis of stability of prestressed structures, as is done in many places in this book (buckling analysis). Then, one is interested in small deviations from a usually simple reference configuration, that is an equilibrium configuration but not a stress-free one. We use here the word 'equilibrium' in the sense of 'mechanical equilibrium', meaning that in such a state the forces are balanced everywhere in the steady situation. This does not necessarily mean that a system at equilibrium in this sense is in thermal equilibrium, as the example of the ice cube heated on the surface shows well.

[3] By material points, we mean points that have a concrete existence, and can be tagged and followed upon deformation.

[4] It is possible to derive the equations of elasticity while considering that the vectors $\mathbf{r}$ and $\mathbf{r}'$ live in different subspaces: the reference configuration (containing points $\mathbf{r}$) lives in an abstract space that parameterises material points in the solid while the actual configuration (containing points $\mathbf{r}'$) lives in the physical space.

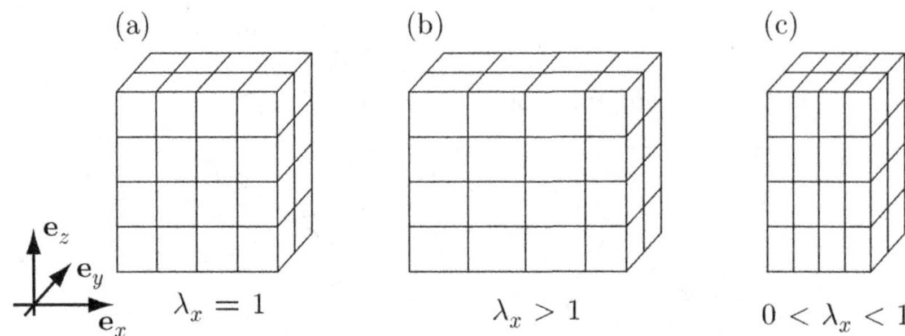

Fig. 2.2 A simple example of a transformation: uniaxial stretching/contraction as defined in equation (2.1). (a) Reference configuration, $\lambda_x = 1$. (b) Stretching, $\lambda_x > 1$. (c) Contraction, $0 < \lambda_x < 1$.

$$\mathbf{r} = \begin{pmatrix} x \\ y \\ z \end{pmatrix}, \qquad \mathbf{r}'(x,y,z) = \begin{pmatrix} x'(x,y,z) \\ y'(x,y,z) \\ z'(x,y,z) \end{pmatrix}.$$

The deformed configuration can be specified by the three scalar functions, $x'(x,y,z)$, $y'(x,y,z)$ and $z'(x,y,z)$. As explained above, an arbitrary point in the solid is parameterised using its position (x,y,z) in the reference configuration. This is the so-called Lagrangian description (also called the *referential* or *material* description), as opposed to the *Eulerian* (or *spatial*) description that is common in fluid mechanics.[5]

We shall introduce straight away a very simple example of transformation, the uniaxial stretching or contraction along the x axis:

$$x'(x,y,z) = \lambda_x\, x, \quad y'(x,y,z) = y, \quad z'(x,y,z) = z. \tag{2.1}$$

It is represented in Fig. 2.2. In this particular family of transformations, lines parallel to the x axis are stretched when $\lambda_x > 1$, or contracted when $0 < \lambda_x < 1$. The case $\lambda_x = 1$ corresponds to no deformation (the actual configuration coincides with the reference configuration, the transformation being the identity mapping). Negative values of λ_x are not possible.[6] This very simple family of transformations will be useful for illustrating the notions we shall introduce later on.

2.2.2   Displacement

The displacement field is defined as:

$$\mathbf{u}(\mathbf{r}) = \mathbf{r}'(\mathbf{r}) - \mathbf{r}. \tag{2.2}$$

---

[5] In the Eulerian description, material points are parameterised by their position in the actual configuration: there is no reference configuration.

[6] The determinant of the Jacobian of the transformation $\mathbf{r} \mapsto \mathbf{r}'$ cannot vanish, as otherwise elementary volume would be shrunk to a single point, requiring infinite work. Being one in the reference configuration and being a continuous function of the transformation, this determinant has to remain strictly positive. This means that the space orientation cannot be changed by the transformation. For uniaxial stretching or contraction, this determinant is λ_x, which must therefore be strictly positive.

It expresses how a given material point tagged by $\mathbf{r}$ moves upon transformation (from reference to actual configuration). In a Cartesian frame, the components of $\mathbf{u}$ are written (u_x, u_y, u_z):[7]

$$\mathbf{u}(x,y,z) = \begin{pmatrix} u_x(x,y,z) \\ u_y(x,y,z) \\ u_z(x,y,z) \end{pmatrix} = \begin{pmatrix} x'(x,y,z) - x \\ y'(x,y,z) - y \\ z'(x,y,z) - z \end{pmatrix}. \tag{2.3}$$

The deformed configuration can be either specified by the mapping $\mathbf{r}'(\mathbf{r})$ (Lagrangian parameterisation of the actual configuration) or by the displacement field $\mathbf{u}(\mathbf{r})$: the actual position can be expressed in terms of the displacement

$$\mathbf{r}'(x_j) = [x_i + u_i(x_j)]\,\mathbf{e}_i. \tag{2.4}$$

In this equation, Einstein summation, introduced in Section 1.2.2, is used for the first time in this book: j is a non-repeated dumb index and so $\mathbf{r}'(x_j)$ is a shorthand for $\mathbf{r}'(x_1, x_2, x_3) = \mathbf{r}'(x,y,z)$, with the convention that x_j denotes the j-th coordinate of $\mathbf{r}$. The index i is repeated twice, meaning that a summation over i is implicit: the right-hand side should be read as $\sum_i [x_i + u_i]\,\mathbf{e}_i$, where the sum runs over all possible values, $1 \leq i \leq 3$.

In the simple example of uniaxial stretching or contraction, the displacement field is given by:

$$u_x(x,y,z) = (\lambda_x - 1)\,x, \quad u_y(x,y,z) = 0, \quad u_z(x,y,z) = 0.$$

In the absence of deformation ($\lambda_x = 1$), the displacement is identically zero: $\mathbf{u}(\mathbf{r}) = \mathbf{0}$ for all $\mathbf{r}$.

2.2.3 Kinematics

In the theory of elasticity, one would like to associate a scalar, namely the elastic energy, to any configuration of a solid specified for instance by a displacement field $\mathbf{r}'$: our goal is to derive the equations of equilibrium by minimizing this energy. Upon deformation, any change of the relative position of atoms or molecules within a solid has to be made *against* the atomic or molecular interactions, which tend to bring them back to rest. We identify these interactions as the origin of the elastic response of the material. The elastic energy that we are seeking should therefore express how the potential energy of interaction of these atoms and molecules is affected by the deformation. Therefore, we need to characterize further the deformation and express how the relative positions of atoms or molecules within a solid change upon transformation. To start with, we leave aside the question of how the materials' angles are affected and focus on the changes in the *distances* between any pair of neighbouring material points: let $\mathbf{r}$ and $\mathbf{r} + \mathrm{d}\mathbf{r}$ be the positions of two such material points in the reference configuration, and $\mathbf{r}'$ and $\mathbf{r}' + \mathrm{d}\mathbf{r}'$ their positions in the actual configuration, see Fig. 2.3. We consider the variation of the *squared* distance, $(\mathrm{d}\mathbf{r}')^2 - (\mathrm{d}\mathbf{r})^2 = \mathrm{d}\mathbf{r}' \cdot \mathrm{d}\mathbf{r}' - \mathrm{d}\mathbf{r} \cdot \mathrm{d}\mathbf{r}$:

$$(\mathrm{d}\mathbf{r}')^2 - (\mathrm{d}\mathbf{r})^2 = (\mathrm{d}[x_i + u_i(x_j)]\,\mathbf{e}_i)^2 - (\mathrm{d}x_i\,\mathbf{e}_i)^2$$

$$= \left(\left[\mathrm{d}x_i + \frac{\partial u_i}{\partial x_j}\mathrm{d}x_j\right]\mathbf{e}_i\right)^2 - (\mathrm{d}x_i\,\mathbf{e}_i)^2 \tag{2.5}$$

---

[7] We shall sometimes write $(u, v, w) = (u_x, u_y, u_z)$ as the components of the displacement.

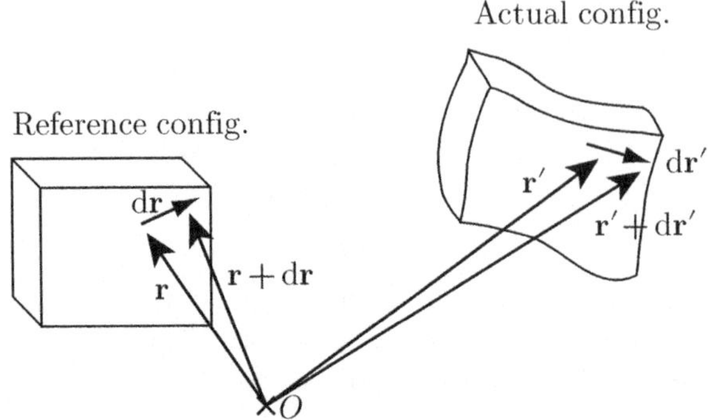

Fig. 2.3 The strain tensor expresses the variation of the squared distance $(\mathrm{d}\mathbf{r}')^2 - (\mathrm{d}\mathbf{r})^2$ between two neighbouring material points upon transformation.

In the right-hand side of this equation, the first square can be expanded as $\mathbf{a}^2 = \mathbf{a}\cdot\mathbf{a} = a_i\,\mathbf{e}_i \cdot a_k\,\mathbf{e}_k$ (one should take care to use different indices for each factor of the scalar product so as to remain consistent with Einstein notation):

$$A = \left[\left(\mathrm{d}x_i + \frac{\partial u_i}{\partial x_j}\mathrm{d}x_j\right)\mathbf{e}_i\right]^2 = \left(\mathrm{d}x_i + \frac{\partial u_i}{\partial x_j}\mathrm{d}x_j\right)\mathbf{e}_i \cdot \left(\mathrm{d}x_k + \frac{\partial u_k}{\partial x_l}\mathrm{d}x_l\right)\mathbf{e}_k \tag{2.6a}$$

$$= \left(\mathrm{d}x_i\,\mathrm{d}x_k + \frac{\partial u_i}{\partial x_j}\,\mathrm{d}x_j\,\mathrm{d}x_k + \ldots\right)\delta_{ik} = \left(\mathrm{d}x_i\,\mathrm{d}x_i + \frac{\partial u_i}{\partial x_j}\,\mathrm{d}x_j\,\mathrm{d}x_i + \ldots\right) \tag{2.6b}$$

$$= \left(\delta_{ij} + \frac{\partial u_i}{\partial x_j} + \frac{\partial u_j}{\partial x_i} + \frac{\partial u_k}{\partial x_i}\frac{\partial u_k}{\partial x_j}\right)\mathrm{d}x_i\,\mathrm{d}x_j. \tag{2.6c}$$

A reader not familiar with Einstein summation is encouraged to carry out this calculation explicitly. To derive equation (2.6b), we have used the fact that the frame is orthonormal: $\mathbf{e}_i.\mathbf{e}_k = \delta_{ik}$, where $\delta_{..}$ denotes Kronecker's symbol, equal to one when the two indices are equal and to zero otherwise. We have then suppressed the sum over the index k, using the fact that only the value $k = i$ yields non-zero contributions. In equation (2.6c), we have renamed the indices so as to collect coefficients of $\mathrm{d}x^2$, $\mathrm{d}x\,\mathrm{d}y$, etc.

This allows one to complete the calculation of the change of the squared distance, started in equation (2.5):

$$(\mathrm{d}\mathbf{r}')^2 - (\mathrm{d}\mathbf{r})^2 = \left(\left[\delta_{ij} + \frac{\partial u_i}{\partial x_j} + \frac{\partial u_j}{\partial x_i} + \frac{\partial u_k}{\partial x_i}\frac{\partial u_k}{\partial x_j}\right] - \delta_{ij}\right)\mathrm{d}x_i\mathrm{d}x_j$$

$$= \left(\frac{\partial u_i}{\partial x_j} + \frac{\partial u_j}{\partial x_i} + \frac{\partial u_k}{\partial x_i}\frac{\partial u_k}{\partial x_j}\right)\mathrm{d}x_i\,\mathrm{d}x_j. \tag{2.7}$$

This quantity $(\mathrm{d}\mathbf{r}')^2 - (\mathrm{d}\mathbf{r})^2$ appears to be a quadratic form of

$$\mathbf{dr} = (\mathrm{d}x, \mathrm{d}y, \mathrm{d}z) = (\mathrm{d}x_1, \mathrm{d}x_2, \mathrm{d}x_3)$$

at any point $\mathbf{r}$ in the solid. It reads

$$(\mathrm{d}\mathbf{r}')^2 - \mathrm{d}\mathbf{r}^2 = 2\,\epsilon_{ij}(\mathbf{r})\,\mathrm{d}x_i\,\mathrm{d}x_j. \tag{2.8}$$

From equation (2.7) the components of the tensor ϵ_{ij} are found to be:

$$\epsilon_{ij}(\mathbf{r}) = \frac{1}{2}\left(\frac{\partial u_i(\mathbf{r})}{\partial x_j} + \frac{\partial u_j(\mathbf{r})}{\partial x_i}\right) + \frac{1}{2}\frac{\partial u_k(\mathbf{r})}{\partial x_i}\frac{\partial u_k(\mathbf{r})}{\partial x_j}. \tag{2.9}$$

By construction and from equation (2.9) the indices i and j play a symmetric role:

$$\epsilon_{ij}(\mathbf{r}) = \epsilon_{ji}(\mathbf{r}) \tag{2.10}$$

for all i and j.

The quadratic form (2.8) defines a symmetric tensor, called the *Green–St. Venant strain tensor* or sometimes the *Green–Lagrange strain tensor*. We shall avoid introducing any other variant of the strain[8] and simply refer to it as the *strain tensor* in the following. The strain tensor characterizes the local deformation near a point tagged $\mathbf{r}$ in the solid, and varies with $\mathbf{r}$: the right-hand side of equation (2.9) may explicitly depend on (x, y, z) through the displacement $\mathbf{u}(\mathbf{r})$. By construction, the strain tensor vanishes for rigid-body displacements (rigid-body rotation and translation), which do not modify distances between material points.

The quantities introduced in equation (2.9) are the components of the strain tensor in the Cartesian frame $(\mathbf{e}_x, \mathbf{e}_y, \mathbf{e}_z)$. The component ϵ_{xx}, for instance, writen explicitly is

$$\epsilon_{xx} = \frac{\partial u_x}{\partial x} + \frac{1}{2}\left(\left(\frac{\partial u_x}{\partial x}\right)^2 + \left(\frac{\partial u_y}{\partial x}\right)^2 + \left(\frac{\partial u_z}{\partial x}\right)^2\right), \tag{2.11}$$

while a non-diagonal component such as ϵ_{xy} is:

$$\epsilon_{xy} = \frac{1}{2}\left(\frac{\partial u_x}{\partial y} + \frac{\partial u_y}{\partial x}\right) + \frac{1}{2}\left(\frac{\partial u_x}{\partial x}\frac{\partial u_x}{\partial y} + \frac{\partial u_y}{\partial x}\frac{\partial u_y}{\partial y} + \frac{\partial u_z}{\partial x}\frac{\partial u_z}{\partial y}\right).$$

We emphasize that the expressions (2.9) of the strain in terms of the displacement are *non-linear*. Geometry, through the squares in the definition of the Euclidean distance, brings in nonlinearity at the heart of the theory of elasticity.

The components of the strain are dimensionless numbers (also called *pure* numbers) as they are obtained by taking spatial derivatives of the displacement field, which is itself a length. Therefore, the magnitude of the strain has some absolute meaning, and the following approximation makes sense:

$$|\epsilon_{ij}| \ll 1 \text{ for all } i \text{ and } j \text{ (small strain approximation)}. \tag{2.12}$$

The point of this approximation is that it warrants that the elastic response of the material is linear, see Section 2.4. Moreover, because the strain is without physical dimension it can be small or large absolutely, without comparison with another physical quantity.

2.2.4 Length of a material curve

To show that the strain tensor fully characterizes the deformed geometry of a solid, we derive the final length of an arbitrary material curve drawn in the solid. In the reference configuration, the curve is given by a parameterisation: $\mathbf{r}(s) = (x_i(s))$, $0 \leq s \leq 1$.[9] The

---

[8] The Cauchy–Green tensor $C_{ij} = 2\,\epsilon_{ij} + \delta_{ij}$ is sometimes used in the analysis of finite displacement.

[9] As earlier, the notation $(x_i(s))$ with a dumb index i is a shorthand for the full list of coordinates, $(x_1(s), x_2(s), x_3(s)) = (x(s), y(s), z(s))$.

length element in the deformed configuration reads, by definition (2.8) of the strain tensor:

$$|\mathbf{dr'}| = \sqrt{(\mathbf{dr'})^2} = \sqrt{\mathbf{dr}^2 + 2\,\epsilon_{ij}\,\mathrm{d}x_i\,\mathrm{d}x_j}. \tag{2.13}$$

The squared length element in the reference configuration, $\mathbf{dr}^2$ can be rewritten in a tensorial form:

$$\mathbf{dr}^2 = \mathrm{d}x_i\,\mathrm{d}x_i = \delta_{ij}\,\mathrm{d}x_i\,\mathrm{d}x_j.$$

Therefore, the length of the curve in the deformed configuration becomes:

$$\ell' = \int_0^1 \mathrm{d}s\,\sqrt{(\delta_{ij} + 2\,\epsilon_{ij})\,\frac{\mathrm{d}x_i}{\mathrm{d}s}\,\frac{\mathrm{d}x_j}{\mathrm{d}s}}. \tag{2.14}$$

When extending this equation to non-Cartesian coordinate systems, one would need to replace δ_{ij} by the so-called *metric tensor*, usually noted g_{ij}. This tensor, which appears for the first time here, defines $\mathrm{d}s = \sqrt{g_{ij}\,\mathrm{d}x_i\,\mathrm{d}x_j}$ as the infinitesimal distance between two points in a curvilinear system. In Cartesian coordinates, the components of the metric tensor are simply given by Kronecker's symbol, $g_{ij} = \delta_{ij}$.

Equation (2.14) shows that the strain tensor can be used to compute the actual length of an arbitrary material curve drawn into the solid. In this sense, the strain tensor fully characterizes the change of lengths in a deformed solid.

2.2.5   Change of material angles

The above definition of strain was based on the change of distances between neighbouring material points upon transformation. We now show that the strain tensor also contains information about the change of angles between material points, although this is not yet obvious. Let $\mathbf{r}$, $\mathbf{r} + \mathbf{dr}_1$ and $\mathbf{r} + \mathbf{dr}_2$ be three neighbouring points in the reference configuration. These three material points are at positions $\mathbf{r'}$, $\mathbf{r'} + \mathbf{dr'}_1$ and $\mathbf{r'} + \mathbf{dr'}_2$ in the actual configuration. As they connect material points, and so flow along with them during deformation, both the vectors $\mathbf{dr}_1$ and $\mathbf{dr}_1$ are called material vectors. To compute the variation of the material angle θ between $\mathbf{dr}_1$ and $\mathbf{dr}_2$ (in the reference configuration) and θ' between $\mathbf{dr'}_1$ and $\mathbf{dr'}_2$ (in the actual configuration), it is sufficient to compute the variation of the scalar product:

$$\mathbf{dr'}_1 \cdot \mathbf{dr'}_2 - \mathbf{dr}_1 \cdot \mathbf{dr}_2. \tag{2.15}$$

Indeed, the angle θ' in the actual configuration can be recovered by the geometric formula:

$$\cos\theta' = \frac{\mathbf{dr'}_1 \cdot \mathbf{dr'}_2}{|\mathbf{dr'}_1|\,|\mathbf{dr'}_2|} = \frac{(\mathbf{dr}_1 \cdot \mathbf{dr}_2) + (\mathbf{dr'}_1 \cdot \mathbf{dr'}_2 - \mathbf{dr}_1 \cdot \mathbf{dr}_2)}{|\mathbf{dr'}_1|\,|\mathbf{dr'}_2|}. \tag{2.16}$$

The first term in the numerator concerns the reference configuration and so is known, while the second term in the numerator is just the one in equation (2.15), which we are going to compute. Either factor in the denominator is given by equation (2.13) in terms of the strain tensor.

Now, the quantity (2.15) is given by the following geometric identity:

$$\mathrm{d}\mathbf{r}_1' \cdot \mathrm{d}\mathbf{r}_2' - \mathrm{d}\mathbf{r}_1 \cdot \mathrm{d}\mathbf{r}_2 = \frac{1}{4}\left([\mathrm{d}\mathbf{r}_1' + \mathrm{d}\mathbf{r}_2']^2 - [\mathrm{d}\mathbf{r}_1' - \mathrm{d}\mathbf{r}_2']^2\right) - \frac{1}{4}\left([\mathrm{d}\mathbf{r}_1 + \mathrm{d}\mathbf{r}_2]^2 - [\mathrm{d}\mathbf{r}_1 - \mathrm{d}\mathbf{r}_2]^2\right)$$

$$= \frac{1}{4}\left([\mathrm{d}\mathbf{r}_1' + \mathrm{d}\mathbf{r}_2']^2 - [\mathrm{d}\mathbf{r}_1 + \mathrm{d}\mathbf{r}_2]^2\right) - \frac{1}{4}\left([\mathrm{d}\mathbf{r}_1' - \mathrm{d}\mathbf{r}_2']^2 - [\mathrm{d}\mathbf{r}_1 - \mathrm{d}\mathbf{r}_2]^2\right).$$

To first order in $\mathrm{d}\mathbf{r}_1$ and $\mathrm{d}\mathbf{r}_2$, the material point located at $\mathbf{r} + \mathrm{d}\mathbf{r}_1 \pm \mathrm{d}\mathbf{r}_2$ is located at $\mathbf{r}' + \mathrm{d}\mathbf{r}_1' \pm \mathrm{d}\mathbf{r}_2'$ in the actual configuration, as can be shown by linearizing the displacement $u_i(\mathbf{r})$ near $\mathbf{r}$. This means that both $\mathrm{d}\mathbf{r}_1 + \mathrm{d}\mathbf{r}_2$ and $\mathrm{d}\mathbf{r}_1 - \mathrm{d}\mathbf{r}_2$ are material vectors. By using the definition (2.8) of strain to evaluate either term in the last line of the above equation, one gets:

$$\mathrm{d}\mathbf{r}_1' \cdot \mathrm{d}\mathbf{r}_2' - \mathrm{d}\mathbf{r}_1 \cdot \mathrm{d}\mathbf{r}_2 = \frac{2\,\epsilon_{ij}\,\mathrm{d}(\mathbf{r}_1 + \mathbf{r}_2)_i\,\mathrm{d}(\mathbf{r}_1 + \mathbf{r}_2)_j}{4} - \frac{2\,\epsilon_{ij}\,\mathrm{d}(\mathbf{r}_1 - \mathbf{r}_2)_i\,\mathrm{d}(\mathbf{r}_1 - \mathbf{r}_2)_j}{4}.$$

Let us write the components of $\mathrm{d}\mathbf{r}_1$ and $\mathrm{d}\mathbf{r}_2$ as $\mathrm{d}x_i^1$ and $\mathrm{d}x_i^2$ respectively. By expanding the above expression and using the symmetry of the strain tensor, we obtain:

$$\mathrm{d}\mathbf{r}_1' \cdot \mathrm{d}\mathbf{r}_2' - \mathrm{d}\mathbf{r}_1 \cdot \mathrm{d}\mathbf{r}_2 = 2\,\epsilon_{ij}\,\mathrm{d}x_i^1\,\mathrm{d}x_j^2. \tag{2.17}$$

This equation generalizes equation (2.8), and shows that the strain not only characterizes the change of lengths within the solid but also the change of scalar products between different elementary material vectors. Combined with equation (2.16), it can be used to compute the actual angle θ' between two arbitrary material directions. In this sense, the strain tensor fully characterizes the relative positions of neighbouring material points in terms of both distances and angles.

2.2.6 Strain associated with uniaxial stretching

A simple class of transformations, namely the uniaxial stretching or contraction, has been introduced in equation (2.1). We shall consider a slightly more general transformation, which combines the former with a rigid-body rotation of angle ϕ around the z axis:

$$x'(x, y, z) = \lambda_x\, x\, \cos\phi - y\, \sin\phi$$

$$y'(x, y, z) = \lambda_x\, x\, \sin\phi + y\, \cos\phi \tag{2.18}$$

$$z'(x, y, z) = z.$$

The displacement then reads:

$$u_x(x, y, z) = x\,(\lambda_x\, \cos\phi - 1) - y\, \sin\phi$$

$$u_y(x, y, z) = \lambda_x\, x\, \sin\phi + y\,(\cos\phi - 1) \tag{2.19}$$

$$u_z(x, y, z) = 0.$$

The strain can be found either by using definition (2.8) or by using the explicit formula (2.9). The calculation, left to the reader, yields:

$$\epsilon_{ij} = 0 \text{ for all } i \text{ and } j, \text{ except for } \epsilon_{xx} = \frac{\lambda_x^2 - 1}{2}. \tag{2.20}$$

Note that the angle ϕ does not appear in this expression for the strain: by construction, the strain removes the rigid-body rotation applied after the uniaxial stretching or contraction.

The interpretation of the strain component ϵ_{xx} is obvious when one considers an elementary material vector aligned with the x axis in the reference configuration:

$$\mathrm{d}\mathbf{r}^{(x)} = \mathrm{d}x\,\mathbf{e}_x. \tag{2.21}$$

By equation (2.8), the length of this vector is scaled upon transformation by a factor:

$$\lambda^{(x)} = \frac{|(\mathrm{d}\mathbf{r}^{(x)})'|}{|\mathrm{d}\mathbf{r}^{(x)}|} = \sqrt{\frac{(\mathrm{d}\mathbf{r}^{(x)})'^2}{(\mathrm{d}\mathbf{r}^{(x)})^2}} = \sqrt{\frac{\mathrm{d}x^2 + 2\,\epsilon_{xx}\,\mathrm{d}x^2}{\mathrm{d}x^2}} = \sqrt{1 + 2\,\epsilon_{xx}}.$$

This number is $\lambda^{(x)} = 1$ when there is no stretching, which is indeed a scaling by 100%. We define the stretching rate of $\mathrm{d}\mathbf{r}^{(x)}$ as

$$\lambda^{(x)} - 1 = \sqrt{1 + 2\,\epsilon_{xx}} - 1, \tag{2.22}$$

which is positive for actual stretching, negative for contraction and zero when the length is not modified. Equation (2.22) shows that the strain component ϵ_{xx} is directly connected to the stretching rate of the material vectors aligned with the x axis in the reference configuration.[10] Combining equations (2.20) and (2.22), we find: $\lambda^{(x)} = \lambda_x$, that is the rate of stretching imposed when building the transformation is recovered when looking at elementary vectors.

For this particular transformation, the strain turns out to be uniform, that is independent of (x, y, z). This is by no means general: this property defines what is called a homogeneous transformation. It can be proved that homogeneous transformations are associated with a parameterisation $\mathbf{r}'(\mathbf{r})$ that is a *linear* mapping.[11]

Under the approximation (2.12) of small strain, $|\hat{\epsilon}_{xx} - 1| \ll 1$, the rate of stretching becomes

$$\lambda^{(x)} - 1 \approx \epsilon_{xx}. \tag{2.23}$$

The strain component ϵ_{xx} can then be directly interpreted as the rate of stretching ($\epsilon_{xx} > 0$) or contraction ($\epsilon_{xx} < 0$) of infinitesimal material vectors aligned with x in the reference configuration.[12]

2.2.7 Principal strains, principal strain directions

By its definition (2.9), the strain is a symmetric tensor of rank two (i.e. with two indices): it can be represented by a real symmetric 3×3 matrix. In general, the deformation is not uniform and this matrix depends on the position $\mathbf{r}$ in the body. With the purpose of interpreting the strain coefficients, we shall temporarily assume that the strain matrix is constant in space (we are in fact focusing on the neighbourhood of a particular point $\mathbf{r}$ and neglect the variations of the strain there). Such a symmetric matrix has six degrees of

---

[10] This is the x direction *in the reference configuration*: in the actual configuration, the direction of stretching has been rotated by an angle ϕ and is no longer along the x axis in general.

[11] This is obviously a sufficient condition, given that the strain depends on the gradients of the displacement $\mathbf{u}(\mathbf{r}) = \mathbf{r}'(\mathbf{r}) - \mathbf{r}$.

[12] This simple interpretation of ϵ_{xx} requires (and explains) the inclusion of the factor two in the right-hand side of definition (2.8) of the strain.

freedom: three diagonal entries, plus only three independent off-diagonal ones due to the symmetry. These degrees of freedom are interpreted as follows.

Any real symmetric matrix can be diagonalized by a suitable rotation of the coordinate system. Let $(\hat{\mathbf{e}}_x, \hat{\mathbf{e}}_y, \hat{\mathbf{e}}_z)$ be the orthonormal basis that diagonalizes the strain tensor ϵ_{ij} at a particular point $\mathbf{r}$, and $\hat{\epsilon}_{xx}$, $\hat{\epsilon}_{yy}$, $\hat{\epsilon}_{zz}$ the associated eigenvalues. In the diagonal basis, the strain matrix takes the form:

$$(\hat{\epsilon}_{ij}) = \begin{pmatrix} \hat{\epsilon}_{xx} & 0 & 0 \\ 0 & \hat{\epsilon}_{yy} & 0 \\ 0 & 0 & \hat{\epsilon}_{zz} \end{pmatrix}. \tag{2.24}$$

Now, this arbitrary diagonal strain matrix can be obtained by the following displacement field:

$$\begin{cases} \hat{u}_x = \left((1 + 2\,\hat{\epsilon}_{xx})^{1/2} - 1\right)\hat{x} \\ \hat{u}_y = \left((1 + 2\,\hat{\epsilon}_{yy})^{1/2} - 1\right)\hat{y}\,, \\ \hat{u}_z = \left((1 + 2\,\hat{\epsilon}_{zz})^{1/2} - 1\right)\hat{z} \end{cases} \tag{2.25}$$

as can be checked by plugging it into (2.9). This transformation is just an extension of our uniaxial stretching example:[13] it is a combination of stretching or contractions in the orthogonal directions $(\hat{\mathbf{e}}_x, \hat{\mathbf{e}}_y, \hat{\mathbf{e}}_z)$ that diagonalize the strain matrix. Note that $\lambda^{(x)}$, defined as a ratio of positive numbers, is positive and so, from equation (2.22):

$$\hat{\epsilon}_{xx} > -\frac{1}{2}.$$

The square roots in (2.25) are therefore well defined in these formulae.

Any deformation can be represented locally by a combination of stretching and contractions along three orthogonal directions,[14] called the principal directions of strain. The coefficients $\hat{\epsilon}_{xx}$, $\hat{\epsilon}_{yy}$ and $\hat{\epsilon}_{zz}$ in equation (2.25) above are called the principal strains. As revealed by our analysis of uniaxial contraction, the principal direction, say $\hat{x}$, is dilated if $\hat{\epsilon}_{xx} > 0$ and contracted if $-1/2 < \hat{\epsilon}_{xx} < 0$. The same holds for the other principal directions. Again, the basis $(\hat{\mathbf{e}}_x, \hat{\mathbf{e}}_y, \hat{\mathbf{e}}_z)$ is relevant to the reference configuration: the principal directions are parallel to these vectors only in the reference configuration.

At any point in the solid, the strain is therefore characterized by the three values of the principal strains, plus the three other parameters involved in the choice of the principal directions: the latter can be chosen as the three Euler angles of the 3D rotation bringing the strain matrix to its diagonal form (2.24). This is consistent with the fact that there are six independent parameters in the (symmetric) strain tensor. Note that both the principal directions of strain and the principal values of strain vary in general within the solid: the diagonalization outlined above has to be carried out at every point in the solid.

2.2.8 Shear deformations

We have just proposed an interpretation of the diagonal strain components: they yield the local stretching rate of material vectors aligned with the axes in the reference configuration.

---

[13] The coefficients in equation (2.25) can easily be guessed from equation (2.22): for $\phi = 0$, $u_x = x' - x = (\lambda_x - 1)\,x$.

[14] These principal directions may not be uniquely defined if two of the principal strains are equal at least—think of the case of an isotropic dilation or contraction $\hat{\epsilon}_{xx} = \hat{\epsilon}_{yy} = \hat{\epsilon}_{zz}$, for instance.

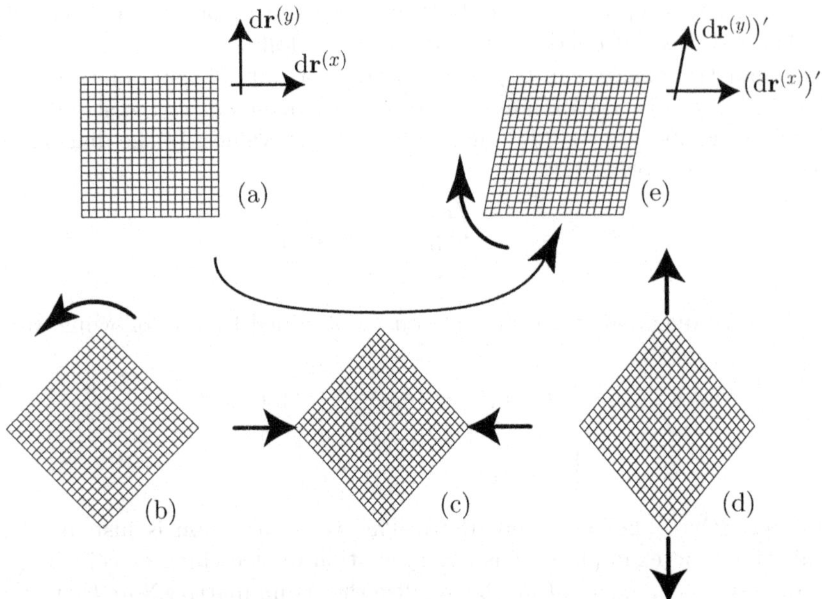

Fig. 2.4 A shear deformation [jump from (a) to (e)] amounts to combined compression (c) and dilation (d) along two oblique orthogonal axes. Intermediate steps (b) and (e) are rigid-body rotations transforming the axes into the principal strain directions.

Let us now consider off-diagonal elements. Recalling that the strain tensor is symmetric, we consider a matrix of the form:

$$(\epsilon_{ij}) = \begin{pmatrix} 0 & \epsilon_{xy} & 0 \\ \epsilon_{xy} & 0 & 0 \\ 0 & 0 & 0 \end{pmatrix} \tag{2.26}$$

in the Cartesian basis $(\mathbf{e}_x, \mathbf{e}_y, \mathbf{e}_z)$, with $|\epsilon_{xy}| < 1/2$.[15] This strain tensor can be derived from the following displacement field:

$$\begin{cases} u_x = 2\,\epsilon_{xy}\,y \\ u_y = -\left(1 - \sqrt{1 - 4\,\epsilon_{xy}{}^2}\right)y \;, \\ u_z = 0 \end{cases} \tag{2.27}$$

as can be checked by plugging it into equation (2.9). This displacement field yields a transformation sketched in Fig. 2.4(e), the reference configuration being given in (a). Visually, the main effect[16] comes from the fact that the displacement along x is proportional to y: this makes the lines parallel to x slide on to each other. This is a shear transformation.

---

[15] This condition comes from the inequality $|(\mathbf{dr}^{(x)})' \cdot (\mathbf{dr}^{(y)})'| < |(\mathbf{dr}^{(x)})'|\,|(\mathbf{dr}^{(y)})'|$ in equation (2.28), given that the norm of both $\mathbf{dr}^{(x)}$ and $\mathbf{dr}^{(y)}$ is unchanged by the transformation ($\epsilon_{xx} = \epsilon_{yy} = 0$).

[16] The component u_y of the displacement is second order in the strain, and so negligible compared with u_x in the small strain approximation $|\epsilon_{xy}| \ll 1$.

Using equation (2.17), we find that the scalar product of two elementary material vectors $\mathrm{d}\mathbf{r}^{(x)} = \mathrm{d}x\,\mathbf{e}_x$ and $\mathrm{d}\mathbf{r}^{(y)} = \mathrm{d}y\,\mathbf{e}_y$, which are perpendicular in the reference configuration, becomes in the actual configuration:

$$(\mathrm{d}\mathbf{r}^{(x)})' \cdot (\mathrm{d}\mathbf{r}^{(y)})' = 2\,\epsilon_{xy}\,\mathrm{d}x\,\mathrm{d}y. \tag{2.28}$$

Therefore the material vectors are no longer perpendicular to each other when $\epsilon_{xy} \neq 0$. This equation shows that the effect of a off-diagonal strain component, say ϵ_{xy}, is to change the angles between material directions that were aligned along x and y in the reference configuration. This is consistent with our explicit transformation in Fig. 2.4(e): there, $\epsilon_{xy} > 0$ and the vectors $(\mathrm{d}\mathbf{r}^{(x)})'$ and $(\mathrm{d}\mathbf{r}^{(y)})'$ indeed make an acute angle.

What are the principal strains and the principal strain directions in the case of a pure shear transformation? Diagonalization of the pure shear strain given in (2.26) is straightforward. The principal axes are the bisectors of the axes x and y, given by:

$$\hat{\mathbf{e}}_x = \frac{\mathbf{e}_x + \mathbf{e}_y}{\sqrt{2}}, \quad \hat{\mathbf{e}}_y = \frac{\mathbf{e}_x - \mathbf{e}_y}{\sqrt{2}}, \quad \hat{\mathbf{e}}_z = \mathbf{e}_z.$$

The associated strain principal values read:

$$\{\hat{\epsilon}_{xx}, \hat{\epsilon}_{yy}, \hat{\epsilon}_{zz}\} = \{\epsilon_{xy}, -\epsilon_{xy}, 0\}.$$

From our analysis of principal strains, this means that the pure shear transformation (2.26) with $\epsilon_{xy} > 0$ amounts to a stretching along $\hat{\mathbf{e}}_x$ and a contraction along $\hat{\mathbf{e}}_y$ (or the converse if $\epsilon_{xy} < 0$). This is illustrated in Fig. 2.4.

2.2.9 Strain compatibility

Sometimes one would like to specify the transformation by directly imposing the strain instead of the displacements.[17] The question then arises as to whether the strain field can be chosen arbitrarily, i.e. whether any strain field $\epsilon_{ij}(\mathbf{r})$ derives from at least one particular displacement field $\mathbf{u}(r)$. The answer is no: the strain field has six independent components, while the displacement field has only three. As a result, components of the strain are subject to compatibility conditions.[18] We shall not work out the compatibility conditions in the general case but limit ourselves to their small displacement form, derived in Section 2.2.12, which will be sufficient for our purpose.

2.2.10 Geometrical transformation of the strain tensor

As for any tensor, the components of the strain matrix have no absolute meaning: the value of any component, say ϵ_{xx}, will vary according to the orientation of the coordinate system. The fundamental property of the strain (2.8), or (2.17), can be used to derive the transformation rules for the strain tensor. The left-side of this equation is indeed scalar, i.e. has a value that is independent of the chosen coordinate system. As a result, between two different systems of coordinates $(x_i)_i$ and $(\tilde{x}_j)_j$, the following equality holds:

---

[17] Think of the former example of an ice cube dropped in water. Assuming that the temperature distribution is known within the ice, the stress-free configuration of the ice is given by prescribing the strain arising from thermal dilation given the local temperature value. As strain cannot be freely imposed, such a stress-free configuration does not exist in general.

[18] The compatibility conditions for the strain translate into compatibility conditions for the stress when the latter are chosen as unknowns. This remark yields one of the two Föppl–von Kármán equations in the theory of elastic plates (see Chapter 6).

$$\tilde{\epsilon}_{ij}\,\mathrm{d}\tilde{x}_i\,\mathrm{d}\tilde{x}_j = \epsilon_{ij}\,\mathrm{d}x_i\,\mathrm{d}x_j$$

$$= \epsilon_{kl}\left(\frac{\partial x_k}{\partial \tilde{x}_i}\,\mathrm{d}\tilde{x}_i\right)\left(\frac{\partial x_l}{\partial \tilde{x}_j}\,\mathrm{d}\tilde{x}_j\right). \tag{2.29}$$

Here, we have introduced the inverse $(\partial x_i/\partial \tilde{x}_j)_{ij}$ of the Jacobian matrix associated with the change of coordinates $(x_i)_i \mapsto (\tilde{x}_j)_j$. The two quadratic forms on both sides of equation (2.29) coincide for any choice of neighbouring material points, that is for any choice of the infinitesimal vector $\mathrm{d}\tilde{x}_i$. As a result, their coefficients must coincide in any particular coordinate system. This yields the transformation rules for the components of the strain tensor:

$$\tilde{\epsilon}_{ij} = \epsilon_{kl}\,\frac{\partial x_k}{\partial \tilde{x}_i}\,\frac{\partial x_l}{\partial \tilde{x}_j}. \tag{2.30}$$

These transformation rules characterize a *covariant* tensor[19] of rank 2.

2.2.11   The approximation of small displacement

The components of the strain tensor depend non-linearly of the displacement by equation (2.9). Assuming that the displacement field $\mathbf{u}(\mathbf{r})$ is small, the non-linear terms in this expression are negligible in front of the linear ones and the strain expression can be linearized:

$$\epsilon_{ij}(\mathbf{r}) \approx \frac{1}{2}\left(\frac{\partial u_i(\mathbf{r})}{\partial x_j} + \frac{\partial u_j(\mathbf{r})}{\partial x_i}\right). \tag{2.31}$$

The tensor on the right-hand side is called the *linearized* strain. It is an accurate approximation of the real strain when the following condition is satisfied:

$$\left|\frac{\partial u_i}{\partial x_j}\right| \ll 1, \text{ for all } i \text{ and } j \text{ (approximation of small displacement).} \tag{2.32a}$$

By integration of these conditions $\partial u_i/\partial x \approx 0$, $\partial u_i/\partial y \approx 0$, $\partial u_i/\partial z \approx 0$ for each component i of the displacement, one arrives at $u_i(\mathbf{r}) \approx a_i$, where a_i is a constant independent of $\mathbf{r}$. This can be rewritten $\mathbf{u}(\mathbf{r}) \approx \mathbf{a}$: the transformation is close to a translation by a vector $\mathbf{a}$. The approximation of small displacement specifically requires that this translation vector vanishes, $\mathbf{a} = \mathbf{0}$. Then, $\mathbf{u}(\mathbf{r}) \approx \mathbf{0}$, which means that the transformation is close to identity:

$$\mathbf{r}'(\mathbf{r}) \approx \mathbf{r}. \tag{2.32b}$$

Equations (2.32) define the approximation of small displacement. Whenever applicable, they allow important simplifications in the theory of elasticity: equation (2.31) removes non-linear effects from the calculation of the strain while equation (2.32b) warrants that the actual configuration is close to the reference configuration.

The distinction between the approximations of small strain (2.12) and of small displacement (2.32) is subtle but it can be grasped with the help of our family (2.19) of uniaxial stretching transformations. Let us focus on the calculation of the strain component ϵ_{xx} for instance, and split its expression (2.11) into a linear part and a non-linear one:

---

[19] The strain tensor is associated with a symmetric, bilinear application taking vectors of the reference configuration as arguments.

$$\epsilon_{xx} = \epsilon_{xx}^{\text{lin}} + \epsilon_{xx}^{\text{nl}},$$

where the linear part reads

$$\epsilon_{xx}^{\text{lin}} = \frac{\partial u_x}{\partial x} = \lambda_x \cos\phi - 1 \tag{2.33a}$$

and the non-linear one reads

$$\epsilon_{xx}^{\text{nl}} = \frac{1}{2}\left(\left(\frac{\partial u_x}{\partial x}\right)^2 + \left(\frac{\partial u_y}{\partial x}\right)^2 \right) = \frac{1}{2}\left((\lambda_x \cos\phi - 1)^2 + \lambda_x^2 \sin^2\phi \right) \tag{2.33b}$$

$$= -\lambda_x \cos\phi + \frac{\lambda_x^2 + 1}{2}. \tag{2.33c}$$

By combining these two terms, we indeed recover the physical strain given by equation (2.20):

$$\epsilon_{xx} = \frac{\lambda_x^2 - 1}{2}. \tag{2.34}$$

This is the only non-vanishing strain component, and so the approximation of small strain requires:

$$\lambda_x \approx 1.$$

On the other hand, the approximation of small displacement requires in particular $|\partial u_x/\partial x| = |\epsilon_{xx}^{\text{lin}}| \ll 1$ and $|\partial u_y/\partial y| = |\epsilon_{yy}^{\text{lin}}| \ll 1$, that is $\lambda_x \cos\phi \approx 1$ and $\cos\phi \approx 1$, which implies (the determination of the angle ϕ with the smallest absolute value is chosen):

$$\lambda_x \approx 1 \quad \text{and} \quad \phi \approx 0.$$

Therefore, the approximation of small displacement appears to be stronger than that of small strain, as it does not allow finite rotations.

We can gain further insight into these approximations by analysing what happens when the strain is small but the approximation of small displacement does not hold: we assume $\lambda_x \approx 1$ but let ϕ take an arbitrary value. Then, we find:

$$\epsilon_{xx}^{\text{lin}} \approx \cos\phi - 1, \qquad \epsilon_{xx}^{\text{nl}} \approx 1 - \cos\phi.$$

Then, although the strain given by the left-hand side of equation (2.9) is very small, in the right-hand side the linear terms and the non-linear terms are not small when evaluated separately; a cancellation takes place between them. Therefore, in the presence of finite rotations (here, when $\cos\phi$ is not close to unity), it makes no sense to use the linearized definition of the strain tensor *even though the physical strain may be very small*: by doing so, one would discard non-linear terms that are of the same order of magnitude as the linear ones. Stated differently, the equations of elasticity are no longer invariant by rotations when the linearized strain tensor is used.

To sum up, the approximation of small displacement is stronger than that of small strain: when equation (2.32a) holds, the non-linear terms in equation (2.9) are indeed negligible in front of the linear ones. The converse is not true as rigid-body rotations with a finite angle can only be captured with the help of the non-linear terms in the strain. This remark is particularly significant for elastic bodies with a small aspect ratio, like thin rods or thin plates. In our analysis of the bending of rods, for instance, we shall see that rotations by a finite angle are possible while the strain remains infinitesimally small everywhere: great care must be exercised when using the approximation of small displacement. This is not coincidental, as important physical effects are caused by the geometrical nonlinearity present in the definition of the strain.

2.2.12   Compatibility of strain (small displacement)

The compatibility conditions for the strain have been mentioned in Section 2.2.9. They can be worked out easily within the approximation of small displacement. Recall that 'compatibility' here means that a displacement field cannot be associated with an arbitrary strain field in general; this is possible only when the strain tensor satisfies some conditions, derived here.

To derive these conditions, we start from definition (2.31) of the linearized strain ϵ_{ij}:

$$2\,\epsilon_{ij} = \frac{\partial u_i}{\partial x_j} + \frac{\partial u_j}{\partial x_j},$$

which appears to be defined as the symmetric part of the gradient of the displacement. It is useful here to introduce an intermediate quantity, which we shall eliminate later, namely the antisymmetric part ϕ_{ij} of the gradient of displacement:

$$2\,\phi_{ij} = \frac{\partial u_i}{\partial x_j} - \frac{\partial u_j}{\partial x_j}.$$

For any pair of 3D indices (i,j) such that $i \neq j$, the components of this skew-symmetric tensor ϕ_{ij} can be interpreted as the average rotation in the plane spanned by directions i and j, when $i = j$, $\phi_{ij} = 0$ by skew-symmetry.

The gradient of the displacement can be found by combining the symmetric tensor ϵ_{ij} and the antisymmetric tensor ϕ_{ij}:

$$\frac{\partial u_i}{\partial x_j} = \epsilon_{ij} + \phi_{ij}. \tag{2.35}$$

Our goal being to eliminate the displacement in favour of the linearized strain, we can apply to the above equality the Leibniz identity:

$$\frac{\partial}{\partial x_k}\left(\frac{\partial u_i}{\partial x_j}\right) = \frac{\partial}{\partial x_j}\left(\frac{\partial u_i}{\partial x_k}\right), \tag{2.36}$$

which yields

$$\frac{\partial \phi_{ij}}{\partial x_k} - \frac{\partial \phi_{ik}}{\partial x_j} = -\frac{\partial \epsilon_{ij}}{\partial x_k} + \frac{\partial \epsilon_{ik}}{\partial x_j}. \tag{2.37a}$$

This equality implies a set of equations for all possible values of the triples of indices (i,j,k). In order to extract from this set a specific component of the gradient of ϕ_{ij}, it is helpful to rewrite equation (2.37a) with indices i and k permuted,

$$-\frac{\partial \phi_{jk}}{\partial x_i} + \frac{\partial \phi_{ik}}{\partial x_j} = -\frac{\partial \epsilon_{jk}}{\partial x_i} + \frac{\partial \epsilon_{ik}}{\partial x_j}, \tag{2.37b}$$

and then with indices i and j permuted:

$$-\frac{\partial \phi_{ij}}{\partial x_k} - \frac{\partial \phi_{jk}}{\partial x_i} = -\frac{\partial \epsilon_{ij}}{\partial x_k} + \frac{\partial \epsilon_{jk}}{\partial x_i}. \tag{2.37c}$$

In equations (2.37b) and (2.37c), indices have been reordered whenever convenient, using the skew-symmetry of ϕ_{ij} and the symmetry of ϵ_{ij}.

Summing now equations (2.37a) and (2.37b) and subtracting (2.37c) we can extract the gradient of the rotation:

$$\frac{\partial \phi_{ij}}{\partial x_k} = \frac{\partial \epsilon_{ik}}{\partial x_j} - \frac{\partial \epsilon_{jk}}{\partial x_i}.$$

The same trick can now be played again: elimination of the rotation ϕ_{ij} can be achieved by combining the Leibniz identity with this equation. This yields

$$\frac{\partial \epsilon_{ik}}{\partial x_j \, \partial x_l} - \frac{\partial \epsilon_{jk}}{\partial x_i \, \partial x_l} - \frac{\partial \epsilon_{il}}{\partial x_j \, \partial x_k} + \frac{\partial \epsilon_{jl}}{\partial x_i \, \partial x_k} = 0. \tag{2.38}$$

In three-dimensional elasticity, the indices can take on the values x, y or z, and the equation above yields six independent conditions of compatibility, as can be checked.

In two-dimensional elasticity, the indices can only take on the values x and y, and equation (2.38) yields a single equation of compatibility:

$$\frac{\partial^2 \epsilon_{xx}}{\partial y^2} - 2 \frac{\partial^2 \epsilon_{xy}}{\partial x \, \partial y} + \frac{\partial^2 \epsilon_{yy}}{\partial x^2} = 0, \tag{2.39}$$

obtained by setting $i = k = x$ and $j = l = y$, for instance. Any other choice of indices yields either $0 = 0$, or the same equation again.

The conditions of compatibility have been derived here under the approximation of small displacement. For finite displacement, a similar reasoning holds but the resulting explicit formula are much more cumbersome. In this book, we shall mainly need equation (2.39), which is the linearized condition of compatibility for the strain in two dimensions. By plugging the definition (2.31) of the linearized strain, it can easily be checked that the left-hand side of this equation (2.39) allows elimination of the displacement components u_x and u_y; this provides an elementary proof of this equation.

2.3 Stress

Unlike fluids in hydrostatic equilibrium,[20] small volume elements in an elastic solid are not only capable of transmitting forces normal to their surface, but also tangential forces. The magnitude of these normal forces may also explicitly depend on the direction of space, as happens for instance when a body is stretched along one particular direction (uniaxial stress). The notion of hydrostatic pressure from ordinary fluid mechanics has therefore to be extended to account both for the so-called shear stress and for the possible anisotropy of stress. This will lead us to the notion of the stress tensor, which we first introduce in the particular context of small displacements in the present section, and thereafter extend to

---

[20] *Flowing* liquids may generate tangential viscous stress, as does a classical Couette shear flow—this viscous stress, as it enters into the Navier–Stokes equations or into the linear Stokes equation valid for a very viscous fluid, can be conveniently introduced using the present formalism for the stress tensor. Anisotropic fluids, like liquid crystals, may also generate tangential stress without flow.

finite displacements in Section 2.5.2. The approximation of small displacements allows the stress tensor to be introduced in the first place without having to worry about the technical complications associated with finite displacements.

Because the present section is concerned with the particular case of small displacements, we shall make no distinction between actual and reference configurations: by equation (2.32b), $\mathbf{r}' \approx \mathbf{r}$.

2.3.1   Definition of the stress tensor

Because a solid is described as a *continuum*, it makes no sense to talk about forces acting between point-like elements in it: having no mass, such points can only stand vanishing forces.[21] The continuum description is an approximation that is valid at scales that are large compared with any characteristic microscopic length (microscopic lengths are inter-atomic spacing, diameter of molecules or grain size in polycrystalline materials, for instance, that are a length scale such that the material can no longer be considered as homogeneous) and it makes no sense to go to smaller length scales in the framework of continuum mechanics. To circumvent this difficulty, we shall allow ourselves to analyse and balance forces only over small but finite volume elements.

Let us therefore analyse the force applied on an arbitrary volume V in a solid. For the time being V does not need to be small. Two types of forces acting on this volume V need to be distinguished: remote forces on the one hand, which can act on every atom or molecule inside V, and interior forces on the other, which are transmitted across the boundary S of this arbitrary volume. Since we are making a global balance of force on this volume V, forces exerted between pairs of points located inside V are not counted. The distinction between remote and interior forces is based on their range. Long-range forces such as gravity, electric or magnetic forces concern any point within V, including those located in its interior, and are considered to be *remote forces*—we shall also refer to them as *volumic forces*. In contrast, forces that represent the effect of short-range interactions[22] between the elementary components of the solid (inter-atomic, inter-molecular or inter-grain forces) belong to *interior forces*. The distinctive feature of interior forces is that they concern only points located very close to the boundary of V. The simplest example of an interior force is hydrostatic pressure.

By assumption, the total volumic force $\mathbf{G}$ on the volume V is the sum of all contributions coming from elementary volumes:

$$\mathbf{G} = \iiint_V \rho\,\mathbf{g}(\mathbf{r})\,\mathrm{d}x\,\mathrm{d}y\,\mathrm{d}z, \tag{2.40}$$

where $\mathbf{g}(\mathbf{r})$ is the mass density of volumic forces, such as the acceleration due to gravity, and ρ is the volumic mass of the body: their product $\rho\,\mathbf{g}$ is the volumic density of remote forces. On the other hand, the net interior force, $\mathbf{F}$, comes from a contribution of all surface elements along the boundary $S = \partial V$ of V:

---

[21] We shall nevertheless meet in Chapter 9 and 14 non-zero forces localized at the contact between two solids, as an elastic rod is pushed on to a bulk solid for instance. Localized forces are represented by diverging stress. We shall then discuss how this infinite stress is actually smoothed out at micro-scales. Another classical instance of diverging stress in solid mechanics is at the tip of a crack, another well studied topic (L. B. Freund, 1998).

[22] A physical effect arising from molecular (short-range) forces is the capillary pressure. This pressure also exists in solids, although it is has fewer observable consequences in solids than in liquids.

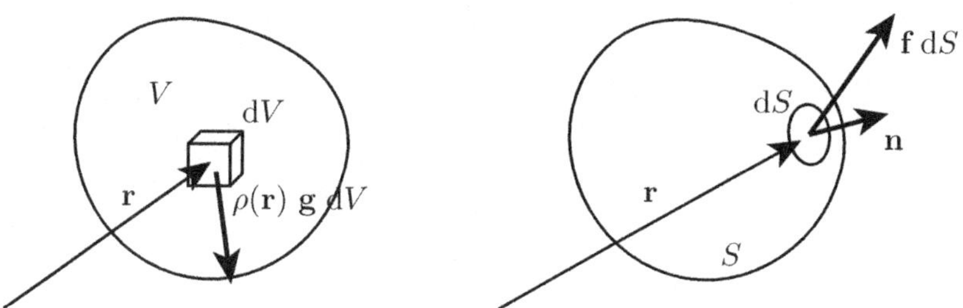

Fig. 2.5 Two types of forces can act on a volume element V, as expressed by equation (2.41): remote forces (left) have a volumic density $\rho\,\mathbf{g}$ while interior forces (right), which only concern points along the boundary $\mathrm{d}S$, have a surface density $\mathbf{f}$.

$$\mathbf{F} = \iint_S \mathbf{f}(\mathbf{r},\mathbf{n})\,\mathrm{d}S. \tag{2.41}$$

In this equation, we have introduced $\mathbf{f}(\mathbf{r},\mathbf{n})$, a vector, which is the density of the interior forces transmitted across a surface element, per unit area $\mathrm{d}S$. In this integral, $\mathbf{r}$ is the current point along the boundary S of V, and $\mathbf{n}(\mathbf{r})$ the local normal to S oriented outwards by convention. The interior force $\mathbf{f}$ is counted as 'view from V': it is the force applied from the exterior of V on to points located inside V. By Newton's third law (action–reaction) points inside V apply a density of force $-\mathbf{f}$ on to the exterior of V.

In the case of hydrostatic force, for instance:

$$\mathbf{f}(\mathbf{r},\mathbf{n}) = -p(\mathbf{r})\,\mathbf{n}. \tag{2.42}$$

The surface density of the interior force applied by the outer world is a normal vector pointing inwards (when $p > 0$) of magnitude p by definition. This example reveals that the density of force is a vector that depends explicitly on the orientation of the normal to the surface.

By equation (2.41), the interior force transmitted across a surface element $\mathrm{d}S$ reads $\mathbf{f}(\mathbf{r},\mathbf{n})\,\mathrm{d}S$. This form of the interior force is often referred to as *Cauchy's postulate* (see Fig. 2.5). In the following, we consider the net interior force over an elementary volume, and specifically require that the volumic density of this force is finite, which will lead us to a more explicit form of $\mathbf{f}$, and to the notion of the stress tensor.

Let us therefore make the volume V become small and apply Newton's fundamental law[23] of dynamics $\mathbf{G} + \mathbf{F} = m\,\mathbf{a}$, where m denotes the mass of this volume element, $\mathbf{a}$ its acceleration and $d \sim V^{1/3}$ its typical size. When d goes to zero, the terms $\mathbf{G}$ and $m\,\mathbf{a}$ both formally scale like the volume, $V \sim d^3$, while the interior forces $\mathbf{F}$ scale like the area of the boundary, that is like d^2, and so formally dominate. To order d^2, the fundamental law of dynamics reads:

$$\frac{\mathbf{F}}{d^2} \to \mathbf{0} \quad (\text{for } d \to 0)$$

and the net interior force, unbalanced at the dominant order, must therefore cancel—otherwise, the acceleration $\mathbf{a}$ would be infinite. Let us apply this reasoning to the infinitesimal tetrahedric volume shown in Fig. 2.6, which is such that one of its faces has a unit

---

[23] In the *Principia*, Newton attributes this law to Galileo, only claiming for himself the principle of action–reaction, today known as Newton's third law.

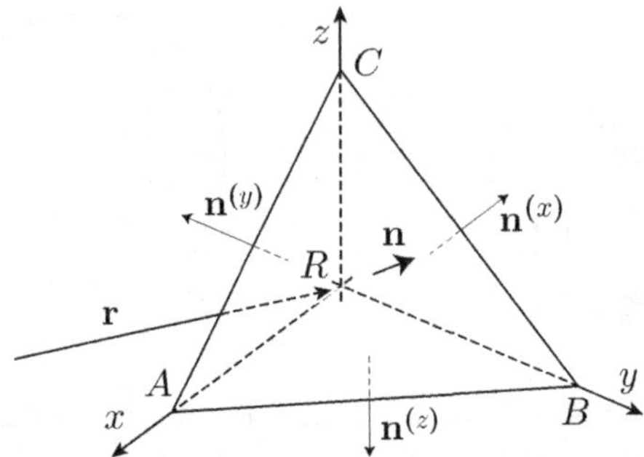

Fig. 2.6 Tetrahedron used to show that the force ($\mathbf{f}(\mathbf{r}, \mathbf{n})\,\mathrm{d}S$) transmitted across a surface element $(ABC) = \mathrm{d}S$ is linear with respect to the surface element vector, $\mathbf{n}\,\mathrm{d}S$.

normal $\mathbf{n}$ with an arbitrary orientation, while the other three faces are normal to the axes $\mathbf{e}_x$, $\mathbf{e}_y$ and $\mathbf{e}_z$ of the Cartesian frame. Adding the contribution to interior force from each face, we have:

$$\frac{\mathbf{F}}{d^2} = \mathbf{f}(\mathbf{r}, \mathbf{n})\,\frac{\mathrm{d}S}{d^2} + \mathbf{f}(\mathbf{r}, -\mathbf{e}_x)\,\frac{\mathrm{d}S^{(x)}}{d^2} + \mathbf{f}(\mathbf{r}, -\mathbf{e}_y)\,\frac{\mathrm{d}S^{(y)}}{d^2} + \mathbf{f}(\mathbf{r}, -\mathbf{e}_z)\,\frac{\mathrm{d}S^{(z)}}{d^2} \to \mathbf{0}. \tag{2.43}$$

Here, $\mathbf{r}$ denotes the centre of mass of the tetrahedron, $\mathrm{d}S$ denotes the area of the face (ABC) whose outer normal is $\mathbf{n}$, $\mathrm{d}S^{(x)}$ the area of the face (RBC) whose outer normal is $\mathbf{n}^{(x)} = (-\mathbf{e}_x)$ and so on. We label the faces of the tetrahedron that are normal to the Cartesian axes using the corresponding space directions, hence the notations $\mathrm{d}S^{(x)}$, $\mathrm{d}S^{(y)}$, etc. The present calculation need only be exact at leading order in d: the tetrahedron is small, and we can neglect the variations of $\mathbf{f}$ with respect to its first argument, $\mathbf{r}$, in equation (2.43).

The areas $\mathrm{d}S^{(x)}$, $\mathrm{d}S^{(y)}$ and $\mathrm{d}S^{(z)}$ are obtained by projection of the area $\mathrm{d}S$ along either axis of the Cartesian frame, and so they are given by:[24]

$$\mathrm{d}S^{(x)} = \mathrm{d}S\,\mathbf{n} \cdot \mathbf{e}_x, \qquad \mathrm{d}S^{(y)} = \mathrm{d}S\,\mathbf{n} \cdot \mathbf{e}_y, \qquad \mathrm{d}S^{(z)} = \mathrm{d}S\,\mathbf{n} \cdot \mathbf{e}_z. \tag{2.44}$$

As usual, let us write n_i the components of the vector $\mathbf{n}$ and define, for a reason that will soon become clear, the numbers σ_{ij}, for $1 \le i \le 3$ and $1 \le j \le 3$ by:

$$\sigma_{ij}(\mathbf{r}) = \mathbf{e}_i \cdot \mathbf{f}(\mathbf{r}, -\mathbf{e}_j), \tag{2.45}$$

that is the i-th Cartesian component of the force transmitted across the face RBC (if $j = 1$), RAC (if $j = 2$), or RAB (if $j = 3$) near the point $\mathbf{r}$. One can then rewrite equation (2.43) using equations (2.44) and (2.45) as:

$$\mathbf{f}(\mathbf{r}, \mathbf{n}) = \sigma_{ij}(\mathbf{r})\,n_j\,\mathbf{e}_i,$$

[24] An alternative (and more physical) derivation of equation (2.44) is to remark that the net interior force should be zero when the tetrahedron is embedded in a uniform hydrostatic pressure field p. From equation (2.42), it takes the form $\mathbf{F} = -p\,\mathrm{d}S\,\mathbf{n} - p\,\mathrm{d}S^{(x)}\,(-\mathbf{e}_x) - p\,\mathrm{d}S^{(y)}\,(-\mathbf{e}_y) - p\,\mathrm{d}S^{(z)}\,(-\mathbf{e}_z)$. Factoring out p and projecting over $\mathbf{e}_1$, $\mathbf{e}_2$ and $\mathbf{e}_3$ yields equation (2.44).

with implicit sums over i and j, according to Einstein summation. In the end, the elementary interior force becomes:

$$\mathrm{d}\mathbf{F} = \mathbf{f}(\mathbf{r}, \mathbf{n})\,\mathrm{d}S = \sigma_{ij}(\mathbf{r})\,(n_j\,\mathrm{d}S)\,\mathbf{e}_i, \tag{2.46}$$

that is, in coordinates:

$$\mathrm{d}F_i = \sigma_{ij}(\mathbf{r})\,n_j\,\mathrm{d}S. \tag{2.47}$$

This equation characterizes the interior force across the face (ABC), which has an arbitrary orientation in space. As a result, this equation (2.47) is not restricted to the tetrahedron and can be used to compute the interior force dF across an arbitrary surface element. This analysis shows that the the resultant of the interior forces over an element V with boundary S is given by:

$$F_i = \iint_S \sigma_{ij}(\mathbf{r})\,n_j(\mathbf{r})\,\mathrm{d}S. \tag{2.48}$$

This equation is a particular form of equation (2.41) which includes the requirement of finite accelerations.

As we showed, the density of interior force, $\mathbf{f}(\mathbf{r}, \mathbf{n})$, depends linearly on the unit normal vector $\mathbf{n}$, and the elementary interior force $\mathrm{d}\mathbf{F} = \mathbf{f}(\mathbf{r}, \mathbf{n})\,\mathrm{d}S$ depends linearly on the surface normal, $\mathbf{n}\,\mathrm{d}S$, as expressed by equation (2.47). Consider an infinitesimal surface $\mathrm{d}S$ in the neighbourhood of a fixed point $\mathbf{r}$, whose orientation, given by its outer normal $\mathbf{n}$, varies in space and probes the interior stress in the material. Then, there is a linear mapping that associates the interior force dF with any normal vector $\mathbf{n}\,\mathrm{d}S$. This linear mapping defines a tensor, called the stress tensor, at any point $\mathbf{r}$. By definition, the component (ij) of this tensor, written $\sigma_{ij}(\mathbf{r})$, is the coefficient by which the component j of the vector $\mathbf{n}\,\mathrm{d}S$ gets multiplied when one computes the component i of the interior force dF. By equation (2.47), this yields the quantities just introduced: $\sigma_{ij}(\mathbf{r})$ are the *components* of the stress tensor. They characterize the local state of stress in the material as they may depend on $\mathbf{r}$ but not on the orientation $\mathbf{n}$ of the surface element. The stress tensor extends the notion of pressure to solids, and allows for more general interior forces than the classical hydrostatic pressure in equation (2.42).

2.3.2   Symmetry of the stress tensor

We shall again use a similar argument and compute the net torque due to the interior forces acting on an infinitely small volume element, instead of the resultant of the interior force. This leads to a new constraint on the stress tensor, namely that it is represented by a symmetric matrix.

Let us consider a small parallelepipedic volume element of typical size d, as shown in Fig. 2.7, and compute the infinitesimal torque $\boldsymbol{\Gamma}_{\mathbf{F}}$ due to the interior forces, with respect to the centre of mass of the volume. For a given stress tensor, the interior forces on the faces are of order d^2 for small d, hence a torque $\boldsymbol{\Gamma}_{\mathbf{F}}$ of order d^3. In the dynamical balance of angular momentum for this volume, this leading order term cannot be balanced by the torque due to volumic forces nor by the acceleration of rotation, which are of order d^4 and d^5 respectively.[25] For the acceleration of rotation to remain finite, we require that $\boldsymbol{\Gamma}_{\mathbf{F}}/d^3$

---

[25] The torque due to volumic forces is the momentum of this force over the volume, proportional to a volumic integral of this force times the arm, a length. Therefore it is of order volume (d^3) times the size

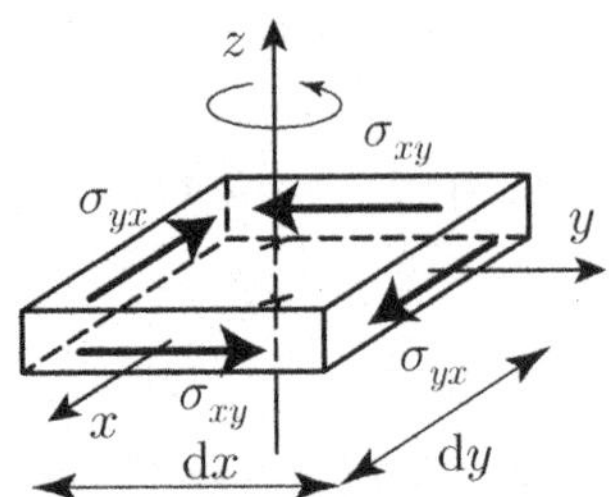

Fig. 2.7 Torque balance of the interior forces on a small rectangular element.

goes to zero for small d. As in the previous argument for the resultant of the interior force, this leading order calculation of the torque can be made by neglecting the variations of the stress tensor in the small volume: we assume that the density of interior forces is uniform over each face.

Let us focus on the z component of this torque, $(\Gamma_{\mathbf{F}})_z$: the other components can be recovered at the end by permutation of the indices. Since the density of interior force is assumed to be uniform over each face and the torque is computed with respect to the centre of mass of the parallelepiped, the contribution of faces perpendicular to $\mathbf{e}_z$ (upper and lower faces) to the torque is zero. The remaining contributions to the torque are due to the normal components of the interior forces over the four faces perpendicular to the x and y axes, as shown by the bold arrows in Fig. 2.7. Consider for instance the back face of the parallelepiped in the figure: its outer normal is $\mathbf{n} = -\mathbf{e}_x$, its area $\mathrm{d}y\,\mathrm{d}z$. By definition of the stress tensor, the y component of the interior force over this face is $\sigma_{yx}(-1)(\mathrm{d}y\mathrm{d}z)$. Its arm with respect to the axis is $\mathrm{d}x/2$ and a positive y component yields a negative contribution to the torque, hence a contribution $-(\mathrm{d}x/2)\,(-\sigma_{yx})\,\mathrm{d}y\,\mathrm{d}z$. Summing up the contributions over the four faces, we obtain:

$$\mathrm{d}\Gamma_z \approx \frac{\mathrm{d}x}{2}\,\sigma_{yx}\,\mathrm{d}y\,\mathrm{d}z + \frac{\mathrm{d}y}{2}\,(-\sigma_{xy})\,\mathrm{d}x\,\mathrm{d}z - \frac{\mathrm{d}x}{2}\,(-\sigma_{yx})\,\mathrm{d}y\,\mathrm{d}z - \frac{\mathrm{d}y}{2}\,\sigma_{xy}\,\mathrm{d}x\,\mathrm{d}z.$$

By cancelling this moment at order d^3 and restoring the implicit dependence on $\mathbf{r}$, we obtain:

$$\sigma_{xy}(\mathbf{r}) - \sigma_{yx}(\mathbf{r}) = 0.$$

After permutation of the indices, we conclude that the stress tensor is symmetric at equilibrium:

$$\sigma_{ij}(\mathbf{r}) = \sigma_{ji}(\mathbf{r}). \tag{2.49}$$

Stress tensors are therefore represented by a symmetric matrix—non-symmetric ones would yield infinite accelerations of rotation.

---

of this volume (d), hence of order d^4, as claimed. The inertial contribution to the torque is of order of an angular acceleration times the momentum of inertia of the volume. The angular acceleration is *a priori* independent on d, although the momentum of inertia is the volume integral of the tensor $r_i r_j$, where r_i is a Cartesian component of the position inside the volume element. Therefore this momentum of inertia is of order d^5, as stated. This simple estimate explains why, in continuum mechanics, there is no separate equation for the balance of angular momentum. At leading order this balance is a consequence of the balance of linear momentum.

2.3.3   Physical interpretation of the stress tensor

The interpretation of the components of the stress tensor follows closely that of the strain tensor. Both tensors are symmetric. By equation (2.47), the diagonal components of the stress tensor correspond to normal stress: the corresponding contribution to the interior force is aligned with the normal $\mathbf{n}$. Positive diagonal components[26] correspond to stretching (the interior force applied from the outer world on to the volume element is pointing outwards, like the normal $\mathbf{n}$), while negative diagonal elements correspond to compression (the interior force is pointing inwards). The case of hydrostatic pressure, for instance, is:

$$\sigma_{ij}(\mathbf{r}) = -p(\mathbf{r})\,\delta_{ij}, \tag{2.50}$$

where δ_{ij} is Kronecker's symbol. The stress tensor is then isotropic, being represented by a matrix that is proportional to the identity.

By equation (2.47), off-diagonal stress components represent shear stress for which the internal forces are parallel to the volume boundary.

Being represented by a symmetric matrix, the stress tensor can be diagonalized locally. This defines the principal stress directions and the principal values of stress, its eigenvalues (the principal values of stress are also simply called the principal stresses). Note that the principal stress directions do not coincide in general with the principal strain directions introduced previously. By a similar analysis as for shear strain, one can diagonalize a pure shear stress field ($\sigma_{xy} = \sigma_{yx}$, all other components of the stress tensor being zero). The result is similar: such a stress field is equivalent to a stretching and a compression with equal magnitudes applied in orthogonal directions, that is along the bisectors of the x and y axes.

2.3.4   Equation of mechanical equilibrium

Let us now study the condition of mechanical equilibrium, which imposes that the total force $\mathbf{F} + \mathbf{G}$ on any volume of solid should be zero. The resultant of the remote forces $\mathbf{G}$ was given in equation (2.40), and that for the net interior force $\mathbf{F}$ has been worked out in equation (2.48). The i-th component of the total force is:

$$F_i + G_i = \iint_S \sigma_{ij}\,n_j\,\mathrm{d}S + \iiint_V \rho\,g_i\,\mathrm{d}V.$$

Using the divergence theorem (1.29), we can put this into a single volume integral:

$$F_i + G_i = \iiint_V \left(\frac{\partial \sigma_{ij}}{\partial x_j} + \rho\,g_i \right)\mathrm{d}V,$$

where the explicit dependence of the integrand on $\mathbf{r}$ is implicit; ρ, $\mathbf{g}$ and σ_{ij} are functions of $\mathbf{r}$.

When a solid is in equilibrium, the balance of forces[27] has to hold for an arbitrary volume V. Then, the integrand in the equation above has to be identically zero:

$$\frac{\partial \sigma_{ij}(\mathbf{r})}{\partial x_j} + \rho(\mathbf{r})\,g_i(\mathbf{r}) = 0. \tag{2.51}$$

---

[26] The sign of non-diagonal stress components has no physical meaning and depends on the choice of the Cartesian frame.

[27] The balance of torques does not yield any new condition—we have already established the symmetry of the stress tensor.

This set of equations is called the Cauchy–Poisson condition of equilibrium for continuous media. The first term in this equation can be interpreted as the volumic density of net interior forces. It yields in particular the well-known $(-\operatorname{grad} p)$ term in the equation of motion for fluids when the stress tensor is a hydrostatic one, as given in equation (2.50). The second term, $\rho\, g_i$, is obviously the volumic density of remote forces. Recall that we consider here the case of small displacements. The extension to finite displacements will be considered later in Section 2.5.8.

The above equation of mechanical equilibrium applies to static problems. For dynamical problems, which we do not study in this book, this equation would have to be replaced by the balance of linear momentum. Then, the right-hand side would be an acceleration term: $\rho(\mathbf{r})\, a_i(\mathbf{r})$, where $\mathbf{a}(\mathbf{r})$ is the particle acceleration.

2.3.5   Elastic energy

In order to derive the density of elastic energy[28] of a solid inside a volume V, we study the work done by the interior forces when it undergoes an infinitesimal, quasi-static deformation: we assume that it undergoes a small change of the displacement field from $u_i(\mathbf{r})$ to $(u_i(\mathbf{r}) + \delta u_i(\mathbf{r}))$. This perturbation is virtual as δu_i does not need to describe an actual movement of the solid.[29] We continue to assume that displacements are small, which allows us to use the stress tensor defined above.

The work of the interior forces is:

$$\delta W_F = \iint_S \sigma_{ij}\, n_j\, \delta u_i\, \mathrm{d}S.$$

We can again use the divergence theorem to turn this into a volume integral:

$$\delta W_F = \iiint_V \frac{\partial(\sigma_{ij}\, \delta u_i)}{\partial x_j}\, \mathrm{d}V = \iiint_V \frac{\partial \sigma_{ij}}{\partial x_j}\, \delta u_i\, \mathrm{d}V + \iiint_V \sigma_{ij}\, \frac{\partial(\delta u_i)}{\partial x_j}\, \mathrm{d}V.$$

The first term is given by the equation of equilibrium (2.51). The second one can be rewritten using the symmetry of the stress tensor:

$$\sigma_{ij}\, \frac{\partial(\delta u_i)}{\partial x_j} = \frac{\sigma_{ij} + \sigma_{ji}}{2}\, \frac{\partial(\delta u_i)}{\partial x_j} = \sigma_{ij}\, \frac{1}{2}\left(\frac{\partial(\delta u_i)}{\partial x_j} + \frac{\partial(\delta u_j)}{\partial x_i}\right).$$

This yields, for the work of the interior forces:

$$\delta W_F = \iiint_V (-\rho\, g_i)\, \delta u_i\, \mathrm{d}V + \iiint_V \sigma_{ij}\left[\frac{1}{2}\left(\frac{\partial(\delta u_i)}{\partial x_j} + \frac{\partial(\delta u_j)}{\partial x_i}\right)\right]\, \mathrm{d}V. \qquad (2.52)$$

---

[28] The notion of elastic energy is relevant to static problems, but it can also serve as a basis for the derivation of the dynamic equations of elastic bodies. This is done using Lagrangian mechanics whereby one minimizes the action, which is a functional composed of a 'potential' energy part that includes the elastic energy, and a 'kinetic' energy part. This is also known as the principle of virtual work.

[29] The purpose of the virtual displacement δu_i is to compute the change of elastic energy and to derive the condition of mechanical equilibrium by demanding that the total energy is a minimum for *all* possible (virtual) perturbations, as done in Section 2.5.8. This perturbation can actually be given a physical meaning: in systems at finite temperature, there are always thermal fluctuations bringing perturbations to the equilibrium state (this equilibrium state is actually a minimum of the free energy but this does not make any difference). Therefore the condition of stationarity against small variations of the displacement field may be seen as a condition of stability against thermal fluctuations of the real system. This does not make any difference in the final results because the amplitude of the fluctuations is irrelevant in the stability condition.

In the first term we recognize the variation of the potential energy,[30] $\mathcal{E}_G$, associated with the remote forces $\mathbf{g}$, which is the opposite of the work done by these remote forces:

$$\delta\mathcal{E}_G = -\iiint_V \rho\, g_i\, \delta u_i \,\mathrm{d}V. \tag{2.53}$$

In the second term of equation (2.52), the factor in brackets is the variation of the linearized strain component ϵ_{ij}. We assumed that the displacements are small and so the non-linear terms in the definition (2.9) of strain are indeed negligible. This factor is then the variation of strain. One can therefore rewrite equation (2.52) as a balance of energy:

$$\delta W_F = \delta\mathcal{E}_G + \delta\mathcal{E}_{\mathrm{el}}, \tag{2.54}$$

the left-hand side being the work done by the interior forces, and the right-hand side is the sum of the potential energy associated with remote forces (such as gravity) and of the elastic energy, $\mathcal{E}_{\mathrm{el}}$. The latter is defined indirectly by its variation:

$$\delta\mathcal{E}_{\mathrm{el}} = \iiint_V \sigma_{ij}\, \delta\epsilon_{ij} \,\mathrm{d}V. \tag{2.55}$$

In order to show how to use equation (2.55), we consider a one-dimensional elastic medium, for example, a cylindrical spring that is stretched along the x axis with a force F. Let A be the area of its cross-section, L its natural length. We consider the case of small displacements, that is when its extension x is much smaller than L. We consider a uniform perturbation upon which the extension is perturbed by δx, while the transverse directions remain unperturbed. Then, the strain $\epsilon_{xx} = x/L$ varies by $\delta\epsilon_{xx} = \delta x/L$ and all other strain components are unperturbed. The associated stress component, σ_{xx}, has the average value F/A at equilibrium.[31] The integral in equation (2.55) is done over a volume $S\,L$, yielding:

$$\delta\mathcal{E}_{\mathrm{el}} = F\,\delta x. \tag{2.56}$$

Now, the elastic energy of the spring is a function of its extension x, and the above equality can be rewritten as:

$$F(x) = \frac{\partial\mathcal{E}_{\mathrm{el}}(x)}{\partial x}.$$

For a linear spring (such that $F(x) = k\,x$) or even a non-linear one ($F(x)$ is a non-linear function), this equation establishes a connection between the elastic response, namely its force–displacement curve $F(x)$, and the elastic energy: the elastic energy can be determined by integration of the force with respect to displacement, while the force is conversely given as the gradient of the elastic energy.

Equation (2.55) extends the notion of energy to the case of a continuous elastic medium. In an elastic medium, the elastic energy $\mathcal{E}_{\mathrm{el}}$ is assumed to be the integral of the density e_{el} of the elastic energy:

$$\mathcal{E}_{\mathrm{el}} = \iiint_V e_{\mathrm{el}}(\mathbf{r})\,\mathrm{d}V, \tag{2.57}$$

[30] We are implicitly assuming that all external forces are conservative, something that is indeed required for the equilibrium to be achieved at a minimum of some energy.

[31] In the approximation of small displacements, the cross-sectional areas A in the reference configuration and in the actual one are considered to be the same.

where this density of elastic energy is a function of the local strain $\epsilon_{ij}(\mathbf{r})$:

$$e_{\text{el}} = e_{\text{el}}(\mathbf{r}, \epsilon_{ij}(\mathbf{r})). \tag{2.58}$$

The additional explicit dependence of e_{el} on $\mathbf{r}$ accounts for possible material inhomogeneities. From equation (2.55), the variation of the density of elastic energy is:

$$\delta e_{\text{el}} = \sigma_{ij}\,\delta\epsilon_{ij}. \tag{2.59}$$

This equation plays exactly the same role as equation (2.56) for a spring: for a medium whose elastic response is known, the stress σ_{ij} is a known function of the local strain ϵ_{kl} and the equation (2.59) can be integrated to yield the elastic energy. Conversely, if the elastic energy is known, the calculation of its total variation yields by identification the stress σ_{ij} as a function of the strain ϵ_{kl}. All this will be put in practice in our analysis of linear isotropic elastic materials in Section 2.4, but the formalism is in fact general.

Note that we have used the condition of equilibrium (2.51) to derive equations (2.55) and (2.59) that define the elastic energy. Conversely, it is possible to derive the equations of equilibrium by starting from the elastic energy. This is what is done in Section 2.5.8: the equations for mechanical equilibrium are derived from the condition of stationarity of the total energy.

2.3.6   Validity

The above analysis of stress assumes small displacements as we did not mention whether geometrical quantities like the vectors $\mathbf{e}_x$ or the surface elements $dS^{(x)}$ were to be taken in the reference configuration or if they had 'flowed' upon deformation along with the material. When the displacements are finite, one needs to keep track of the quantities pertaining to the reference or actual configurations. This will be clarified in Section 2.5.

2.4   Hookean elasticity

Let us give a brief review of what we have learned so far, and how it applies to the particular case of an elastic spring: we have characterized the strain of the spring in terms of its extension x, the interior stress in the spring in terms of the applied force F, and we have explained how the elastic energy of the spring is connected to the force–displacement curve $F(x)$. However, we have not yet specified what this curve $F(x)$ is—as a matter of fact, different springs are characterized by different curves $F(x)$, some of them being linear while others exhibit various sorts of non-linear behaviours.

In order to close the equations of elasticity, one should specify the interior stress that take place in the material in reaction to mechanical strain or, conversely, by how much the material deforms under a given stress. This relation between the strain tensor and the stress tensor is called a *constitutive* law of the material.[32] For isotropic materials undergoing small strain, this constitutive law takes a very particular form, investigated below. For the

---

[32] Such constitutive laws depend on the details of the structure of the material, of the way that the atoms and/or molecules it is made of interact in the solid state to maintain its stability. It is a difficult task to try to relate the microscopic structure of the material to the macroscopic constitutive laws, a famous example being the case of glasses that look structurally very much like their melt although they have completely different constitutive laws: the melt is a regular liquid, although the glass at a slightly lower temperature is a solid for all practical purposes.

moment, we shall only consider the case of a *stress-free* reference configuration. The form of the constitutive equations is modified in the case of a prestressed reference configuration; this case arises in particular in the analysis of buckling, see for instance Section 7.3.3, and will be treated by introducing an auxiliary, stress-free reference configuration.

2.4.1 Constitutive relations for linear, isotropic materials

When subjected to small strain, most materials respond linearly and instantaneously to deformation. This is the so-called Hookean elasticity, discovered by Robert Hooke in 1660 and expressed by him in an anagram,[33] *ceiiinossssttuu* for *ut tensio sic vis*, that is 'the force (*vis* in Latin) is proportional to the extension' in modern language. In our formalism, this elastic response is expressed by a relation of proportionality between the stress tensor and the strain tensor. While this relation of proportionality can have as many as thirty-six[34] coefficients in the absence of symmetry, it is strongly constrained in isotropic materials as it should remain invariant by a change of coordinate frame. Any linear relation between two symmetric tensors of rank two that remains invariant by change of coordinates[35] takes the following form:

$$\sigma_{ij} = \lambda\, \epsilon_{kk}\, \delta_{ij} + 2\mu\, \epsilon_{ij}. \tag{2.60}$$

The coefficients λ and μ, which characterize the elasticity of the material, are called the Lamé coefficients. They have been introduced somewhat formally but shall soon be explained. The notation ϵ_{kk} implies a sum over the index k, and so $\epsilon_{kk} = \epsilon_{xx} + \epsilon_{yy} + \epsilon_{zz}$ is for the sum of all diagonal elements of the strain tensor, that is its so-called *trace*. On the other hand, the indices i and j, which occur once on each side of the equality sign, are dummy indices: equation (2.60) is valid for any value of i and j, and is in fact an equality between two matrices. The first term on the right-hand side of the constitutive law involves the trace of the strain, ϵ_{kk}, times the identity matrix δ_{ij}. The second term is just a multiplication of the stress tensor by a scalar.

It is obvious that the right-hand side of equation (2.60) is linear with respect to the strain tensor, and it can be checked straightforwardly that the stress–strain relation defined by this equation is invariant by a change of Cartesian frame. The converse, namely the fact that any stress–strain relation satisfying these conditions is of the form (2.60) for some λ and μ, is more difficult to prove.[36]

---

[33] An anagram collects all the letters of a sentence together. It is hard to decipher for short sentences and impossible if it is too long, even with the help of a big computer. In the present anagram there is one 'c', one 'e', three 'i's, etc. in the original Latin, whence the anagram 'ceiii...' Note that 'u' and 'v' are written the same way in Latin, whence the two u's at the end of the anagram.

[34] An arbitrary linear relation between two symmetric tensors with six parameters (recall the symmetry of both the strain and the stress tensors) depends itself on 6^2 coefficients.

[35] In fluid mechanics, viscous effects are taken into account via a stress tensor that is linear with respect to the rate of strain. One is then led to a very similar problem, finding the possible linear laws between two symmetric tensors that are furthermore invariant by a change of Cartesian frame. The resulting constitutive law for viscous stress is similar to (2.60) and indeed depends on *two* viscosities, the shear and the bulk viscosities (the latter is relevant for compressible fluids only).

[36] For the interested reader, a proof can be outlined as follows. First, one should notice that, the the only isotropic tensors being those proportional to the identity, the stress tensor associated with an identity strain field must itself be proportional to the identity, the proportionality constant being characteristic of the material. In a second step, one should study strain tensors of the form $\epsilon_{xx} = \epsilon$, $\epsilon_{yy} = -\epsilon$, all other components being zero; by studying the space invariances of this tensor (symmetry by reflection on the

It is often convenient to rearrange the right-hand side of equation (2.60) as follows:

$$\sigma_{ik} = \frac{E}{1+\nu}\left(\epsilon_{ik} + \frac{\nu}{1-2\nu}\,\epsilon_{jj}\,\delta_{ik}\right),\tag{2.61}$$

where the material coefficients E (this coefficient is traditionally denoted as E in honour of Euler) and ν are related to the Lamé coefficients by:

$$\lambda = \frac{E\,\nu}{(1+\nu)(1-2\nu)},\qquad \mu = \frac{E}{2(1+\nu)}.\tag{2.62}$$

The coefficient E appearing in this relation is Young's modulus of the material, and ν is its Poisson's ratio, whose interpretations are given below in Section 2.4.3. E and ν together characterize the elasticity of the material in the linear regime. Because ν is a dimensionless number (which cannot be very large or very small), Young's modulus is the only number with a physical dimension. Like stress, it is homogeneous to a pressure, and is measured in pascals (Pa) in the international unit system: a typical strain ϵ induces a typical stress of order $E\,\epsilon$.[37]

It is sometimes useful to express the strain as a function of stress by inverting the constitutive relations. To do so, one should multiply both sides of equation (2.61) by δ_{ik} and carry out the implicit summation over i (notice that $\delta_{kk} = 3$ is the dimension of space). This yields a relation between the trace of the stress tensor, σ_{jj}, and the trace of the strain tensor, σ_{jj}:

$$\sigma_{jj} = \frac{E}{1-2\nu}\,\epsilon_{jj}.\tag{2.63}$$

This relation can be used to express the last term of (2.61) in terms of the stress, yielding the inverse constitutive relation:

$$\epsilon_{ik} = \frac{1+\nu}{E}\,\sigma_{ik} - \frac{\nu}{E}\,\sigma_{jj}\,\delta_{ik}.\tag{2.64}$$

planes containing x and z, or y and z; antisymmetry by a rotation of angle $\pi/2$ with axis z, etc.) one can show that the associated stress tensor must be proportional to it component by component, hence a second scalar characteristic of the material. The third step is to notice that the latter strain tensor can be transformed into a strain tensor of the form (2.26) by a change of Cartesian frame—this is in fact the main result of our analysis of shear deformation in Section 2.2.8. Finally, an arbitrary strain tensor can be linearly decomposed into strain tensors of the above particular forms, yielding the associated stress tensor by linearity. We recover incidentally that any linear, isotropic constitutive relation depends on two scalars.

[37] This is true only as long as the material is well described by a linear constitutive relation, which requires a strain not much larger than $\epsilon \approx 10^{-2}$. For a larger strain, the materials leave the linear regime, either by breaking (the so-called brittle materials being either in the linear regime or broken—typically the glasses, ceramics, etc.) or by reaching a non-linear and non-reversible domain of deformations (plasticity for metals). Only the so-called rubber-like materials may undergo strain of order one, with the additional constraint that their volume does not change. Another interesting class of material is the non-cohesive ones, like dry sand and other granular materials. These materials, in a first approximation can stand compressive stress with a linear response, but cannot stand tensile stress. This is expressed mathematically by saying that the stress must have everywhere positive eigenvalues in the material. Elasticity problems in such materials involve typically a free boundary where one of the eigenvalue of the stress becomes zero.

2.4.2　Elastic energy of a Hookean material

In order to derive the elastic energy of a Hookean material, defined by the constitutive relation (2.60) or equivalently by (2.61), we shall need the following intermediate result, derived from the constitutive relation (2.61):[38]

$$\sigma_{ij}\,\epsilon'_{ij} = \sigma'_{ij}\,\epsilon_{ij}, \tag{2.65}$$

where ϵ_{ij} and ϵ'_{ij} are arbitrary symmetric strain tensors and σ_{ij} is the stress tensor associated with ϵ_{ij} using the constitutive relation, and σ'_{ij} that associated with ϵ'_{ij} using the *same* constitutive relation. Equation (2.65) can be checked straightforwardly by expressing σ_{ij} and σ'_{ij} in terms of ϵ_{ij} and ϵ'_{ij} using the constitutive relation.

Before deriving the elastic energy of a Hookean elastic medium, let us focus on the simple case of a linear spring. Its elastic energy is defined by equation (2.56), where the force F is defined by a linear constitutive relation $F(x) = k\,x$, k being the spring stiffness constant. Integration of equation (2.56) with respect to δx yields the well-known result:

$$\mathcal{E}_{\text{el}} = \frac{1}{2}k\,x^2 = \frac{1}{2}F(x)\,x.$$

This suggests the following form for the elastic energy of a Hookean material:[39]

$$\mathcal{E}_{\text{el}} = \frac{1}{2}\iiint_V \sigma_{ij}[\epsilon_{kl}(\mathbf{r})]\,\epsilon_{ij}(\mathbf{r})\,\mathrm{d}V, \tag{2.66}$$

where the factor $\sigma_{ij}[\epsilon_{kl}(\mathbf{r})]$ stands for the stress, as computed from local strain $\epsilon_{kl}(\mathbf{r})$ using the constitutive relation (2.64). The elastic energy appears to be a functional of strain and ultimately of the displacement. The factor $1/2$ in front of the integral in (2.66) is explained in the same way as the one in front of the energy of a linear spring: in order to bring a spring from its natural to its current state by quasi-static loading, the operator has to work against the spring tension, which increases linearly with stretching, and so the average spring tension is half its final value at full load.

The variation of the elastic energy (2.66) for a small perturbation of the configuration of the material is given by:

$$\delta\mathcal{E}_{\text{el}} = \frac{1}{2}\iiint_V \left((\delta\sigma_{ij})\,\epsilon_{ij} + \sigma_{ij}\,(\delta\epsilon_{ij})\right)\,\mathrm{d}V.$$

By linearity of the material, the stress perturbation $\delta\sigma_{ij}$ is that associated with the strain perturbation $\delta\epsilon_{ij}$ through the constitutive law. Therefore, using lemma (2.65), the first term in the integrand can be rewritten like the second one, that is $(\delta\sigma_{ij})\,\epsilon_{ij} = \sigma_{ij}\,(\delta\epsilon_{ij})$. As a result, the proposed elastic energy has the correct variation (2.55).

2.4.3　Interpretation of elasticity constants

In order to interpret Poisson's ratio and Young's modulus, we consider the so-called simple traction of a bar parallel to the axis z, with arbitrary but constant section. The bar is

---

[38] This property is known as the symmetry of the elasticity tensor.

[39] As obvious from the example of a spring, the form (2.66) of the energy is valid only for a linear material. For a non-linear one, one has to resort to the definition (2.55) of the elastic energy in variational form, and integrate it using whichever constitutive law is applicable.

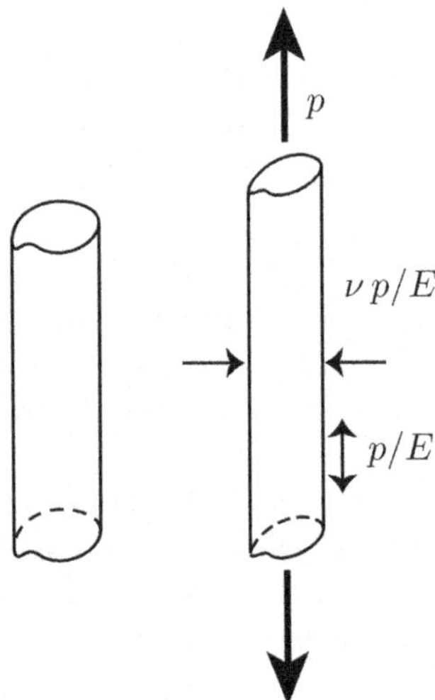

Fig. 2.8 Interpretation of Young's modulus and Poisson's ratio in a simple traction of a bar. When the sides of the bar are free of stress, the rate of longitudinal extension is simply the ratio of the applied traction p to Young's modulus, E, of the material, while the rate of contraction of the cross-section involves Poisson's ratio of the material, ν.

axially loaded with a traction p, defined as the ratio of the traction force to the area of the cross-section of the bar. This simple geometry yields homogeneous[40] stress and strain within the bar, making the solution of the elasticity equations particularly simple and illustrative. Indeed, the axial component of the stress tensor is $\sigma_{zz} = p$, while all the other components of the stress vanish due to the boundary conditions on the edge of the bar (see Chapter 3). By solving for the strain ϵ_{ij} using the inverse of the constitutive relation (2.64), one finds:[41]

$$\epsilon_{zz} = \frac{p}{E}, \quad \epsilon_{xx} = \epsilon_{yy} = -\nu\,\epsilon_{zz}, \tag{2.67}$$

the off-diagonal components being zero by symmetry. Therefore, a bar in simple traction is longitudinally stretched by a factor p/E, and its section is contracted. The ratio of the rate of contraction in the cross-section to the rate of elongation along the axis is precisely Poisson's ratio ν (see Fig. 2.8).

Mechanical stability requires that the density of mechanical energy in equation (2.58) be a definite positive quadratic form of the strain ϵ_{ij}. Otherwise, the energy could be lowered indefinitely by increasing deforming the material. In the framework of the Hookean linear elasticity, this yields constraints on the coefficients λ and μ or, equivalently, E and ν. As shown by the simple traction experiment, see equation (2.67), Young's modulus should be positive:

$$E > 0. \tag{2.68}$$

---

[40] Strictly speaking, the stress and strain are only homogeneous far from the ends of the bar, as they may be affected by the boundary conditions near these ends.

[41] Note that in equation (2.67), the indices x, y and z are repeated but, as they are not variables but instead specific space directions, there is no implicit summation.

Table 2.1 Poisson's ratio and Young's modulus of some materials, E is given in GPa, or gigapascals, that is 10^{12} N m^{-2}

Material	E (GPa)	ν
Aluminium	70	0.34
Concrete	48	0.20
Glass	65	0.23
Polyethylene (HDPE)	0.7	0.42
Steel	210	0.29

The admissible values of Poisson's ratio are in the range:

$$-1 \leq \nu \leq \frac{1}{2}. \tag{2.69}$$

Indeed, let us consider the effect of a uniform hydrostatic pressure p first, for which $\sigma_{ij} = -p\,\delta_{ij}$. By equation (2.63), the strain is isotropic and $\epsilon_{xx} = \epsilon_{yy} = \epsilon_{zz} = \epsilon_{jj}/3 = -(1 - 2\nu)\,p/E$. Mechanical stability requires that, if the pressure p is positive (compression), the material should not expand ($\epsilon_{xx} \leq 0$). Therefore, $\nu \leq 1/2$. The limit $\nu = 1/2$ corresponds to an incompressible material. Then, in equation (2.63), finite stress requires $\epsilon_{jj} = 0$; the trace of ϵ_{ij} being the relative change of volume caused by the transformation,[42] this means that the volume does not change under finite stress. The lower bound on Poisson's coefficient ν, $\nu \geq -1$, can be obtained by a similar argument, considering shear deformations instead.

Values of Poisson's ratio and of Young's modulus are given in Table 2.1 for typical materials. Poisson's ratio of rubber is very close to 0.5 and is therefore almost incompressible, something that can be explained by its microscopic structure. Materials with a negative Poisson's ratio can be found in nature, but they are all anisotropic—their Poisson's ratio is not negative along all space directions. Isotropic materials (recall that throughout this book we consider isotropic solids only) with a negative Poisson's ratio have been synthesized, the first example of which was a special foam (R. Lakes, 1987). Such materials have the striking property that they expand transversely when stretched. It is possible to design articulated systems displaying a negative Poisson's ratio at large scale.

2.5 Stress and strain for finite displacements

The stress tensor has been introduced in the previous sections under the somewhat restrictive approximation of small displacement (2.32). The aim of the present section is to extend the notion of stress when the material undergoes finite rotations. As we shall show,

---

[42] Consider the infinitesimal cube, with volume $(\mathrm{d}\hat{x}\,\mathrm{d}\hat{y}\,\mathrm{d}\hat{z})$, whose faces are perpendicular to the local principal directions of strain, $(\hat{\mathbf{e}}_x, \hat{\mathbf{e}}_y, \hat{\mathbf{e}}_z)$. Upon transformation, its volume is changed to

$$(1 + \hat{\epsilon}_x)\,\mathrm{d}\hat{x}\,(1 + \hat{\epsilon}_y)\,\mathrm{d}\hat{y}\,(1 + \hat{\epsilon}_z)\,\mathrm{d}\hat{z}$$

to first order in the strain, hence a relative change of volume $(\hat{\epsilon}_x + \hat{\epsilon}_y + \hat{\epsilon}_z)$. This number is the trace ϵ_{ii} of the strain tensor, a scalar that is invariant by change of Cartesian frame.

finite displacements introduce geometric nonlinearity at the heart of the theory, which are the origin of the complex phenomena investigated in this book. For instance, the nonlinear terms present in the Föppl–von Kármán equations for elastic plates, derived later in Chapter 6, stem from those derived in the present section.

We shall still consider that the strain remains small and that the material response is Hookean and isotropic, but we will no longer assume small displacements. Note that the assumption of small strain could be removed too, but we shall retain it as it is compatible with the main theme of this book, which is on geometrical nonlinearity. A typical situation where strain is small but displacement is not is in the buckling of slender structures, which may involve large rotations.

In the presence of finite displacements, the definition (2.47) for the stress tensor becomes ambiguous and one has to specify carefully whether the quantities entering into the equation pertain to the reference or to the actual configuration. The vector normal to the surface element appearing in the right-hand side could be either[43] that from the reference configuration $\mathbf{n}\,\mathrm{d}S$ or that from the actual configuration, $\mathbf{n}'\,\mathrm{d}S'$. Both conventions are possible, leading to different definitions of the stress tensor. Another ambiguity arises in the definition of the components $\mathrm{d}F_i$ of the interior force: they can be computed by projection in a fixed Cartesian frame, or in a material frame attached to the deformed body.

We shall only use a single set of conventions for the stress, that is introduce a single variant[44] of the stress tensor, namely the *second Piola–Kirchhoff stress tensor*, also called the *material stress* tensor. We shall simply refer to it as *the* stress tensor. The second Piola–Kirchhoff stress tensor is consistent with the stress tensor introduced previously in the case of small displacement, and so there is no ambiguity.

2.5.1   Local material frame

To define the Piola–Kirchhoff stress tensor, we first need to introduce the following local frame:

$$\mathbf{e}'_x(\mathbf{r}) = \frac{\partial \mathbf{r}'(x,y,z)}{\partial x}, \quad \mathbf{e}'_y(\mathbf{r}) = \frac{\partial \mathbf{r}'(x,y,z)}{\partial y}, \quad \mathbf{e}'_z(\mathbf{r}) = \frac{\partial \mathbf{r}'(x,y,z)}{\partial z}, \qquad (2.70)$$

which has a simple interpretation. The reference configuration is obviously parameterised by $\mathbf{r} = x_i\,\mathbf{e}_i$ and so by derivation:

$$\mathbf{e}_i = \frac{\partial \mathbf{r}(x,y,z)}{\partial x_i}.$$

This expression shows that there is only *one* coordinate system, (x,y,z), the orthogonal coordinates in the unperturbed configuration and a curvilinear system in general for the perturbed state. This means that $\mathbf{e}_x$, for instance, is given by the vector joining two

---

[43] The vector $\mathbf{n}$ is a unit vector normal to the surface element in reference configuration, while $\mathbf{n}'$ is a unit vector normal to the surface element in the deformed configuration. These vectors may differ significantly when the transformation involves finite rotations. Similarly, the areas $\mathrm{d}S$ and $\mathrm{d}S'$ measure the surface element in reference and actual configuration respectively, and may differ significantly if this area is stretched by the transformation for instance. The vectors $(\mathbf{n}\,\mathrm{d}S)$ and $(\mathbf{n}'\,\mathrm{d}S')$, although they refer to the 'same' surface element, are different vectors.

[44] Other conventions for defining the stress lead to the so-called Cauchy stress tensor, which is well suited for the Eulerian description of motion, and the first Piola–Kirchhoff stress tensor (also called nominal stress, or Boussinesq stress tensor in the French literature on the subject).

Fig. 2.9 Deformation of small volume element under a displacement field with finite amplitude. The material frame, $\mathbf{e}_i$ in the reference configuration, 'flows' along with the elastic body, yielding a frame $\mathbf{e}'_i$ in the deformed configuration given by equation (2.70).

neighbouring points in the reference configuration:

$$\mathbf{e}_x \approx \frac{\mathbf{r}(x+\delta x, y, z) - \mathbf{r}(x,y,z)}{\delta x}.$$

When the transformation is applied, these points $\mathbf{r}(x,y,z)$ and $\mathbf{r}(x+\delta x, y, z)$ move to $\mathbf{r}'(x,y,z)$ and $\mathbf{r}'(x+\delta x, y, z)$ respectively. One can therefore *define* a material vector[45] $\mathbf{e}'_x$ in the actual configuration by:

$$\mathbf{e}'_x \approx \frac{\mathbf{r}'(x+\delta x, y, z) - \mathbf{r}'(x,y,z)}{\delta x}.$$

Repeating this for all vectors of the initial Cartesian frame, and near every point $\mathbf{r}$ in the solid leads to the set of local material frames introduced in equation (2.70), and shown in Figure 2.9. In term of the displacement, the material frame is:

$$\mathbf{e}'_i(\mathbf{r}) = \frac{\partial \mathbf{r}'(\mathbf{r})}{\partial x_i} = \frac{\partial(x_j + u_j(\mathbf{r}))\,\mathbf{e}_j}{\partial x_i} = \left(\delta_{ij} + \frac{\partial u_j(\mathbf{r})}{\partial x_i}\right)\mathbf{e}_j. \tag{2.71}$$

Because the same coordinate system is used in the reference system and in the deformed state, the infinitesimal vector $\mathbf{e}'_x \mathrm{d}x$ is a vector transformed of the vector $\mathbf{e}_x \mathrm{d}x$ in the undisturbed frame. In general the frame $(\mathbf{e}'_x(\mathbf{r}), \mathbf{e}'_y(\mathbf{r}), \mathbf{e}'_z(\mathbf{r}))$ is not orthonormal: by setting $\mathrm{d}\mathbf{r}_1 = \mathbf{e}_i\,\mathrm{d}x_1$ and $\mathrm{d}\mathbf{r}_2 = \mathbf{e}_j\,\mathrm{d}x_2$ in equation (2.17), one obtains:

$$\mathbf{e}'_i \cdot \mathbf{e}'_j = \delta_{ij} + 2\,\epsilon_{ij}. \tag{2.72}$$

2.5.2 Definition of the second Piola–Kirchhoff stress tensor

For a reason explained below, the second Piola–Kirchhoff stress tensor $\sigma^{\mathrm{PK}}_{ij}$ is defined in such a way that the element of interior force can be written as:

$$\mathrm{d}\mathbf{F} = \mathbf{e}'_i\,[\sigma^{\mathrm{PK}}_{ij}\,(\mathbf{n} \cdot \mathbf{e}_j\,\mathrm{d}S)]. \tag{2.73}$$

As before, $\mathrm{d}\mathbf{F}$ is the interior force transmitted across a surface element, $(\mathbf{n}\,\mathrm{d}S)$ is the normal vector to the surface element *in the reference configuration*, $\mathbf{e}_j$ is a vector of the fixed Cartesian frame and $\mathbf{e}'_j$ is a vector of the local material frame[46] The term in brackets

---

[45] By *material* vector, we mean that this vector flows along with material points upon transformation.

[46] This is for the reader familiar with fluid mechanics. One could also introduce such a Piola–Kirchhoff tensor there, because no reference to the stress–strain or stress–velocity properties of the material is made when defining $\sigma^{\mathrm{PK}}_{ij}$. In fluid mechanics the reference coordinates would be the so-called Lagrangian coordinates, and the actual coordinates the Eulerian coordinates. However, because there is little use of the

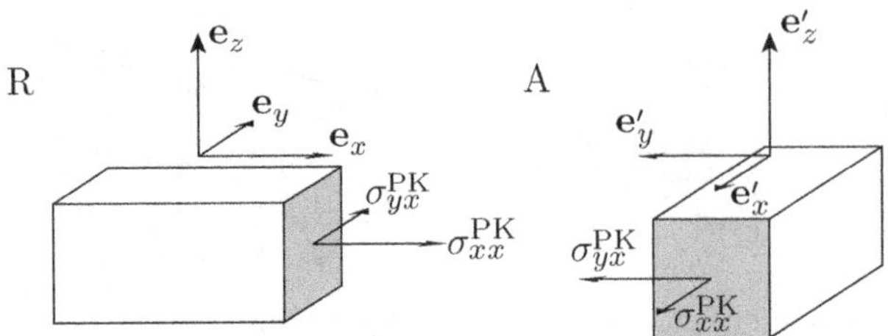

Fig. 2.10 Interpretation of the components of the stress tensor for finite displacement. We consider a volume element undergoing a rotation by $-\pi/2$ along $\mathbf{e}_z$ between the reference configuration (R) and the actual one (A), and focus on the face whose outer normal is $\mathbf{e}_x$ in the reference configuration. By definition (2.73), the component $\sigma_{xx}^{\mathrm{PK}}$ yields an interior force along $\mathbf{e}_x$ in the reference configuration (R) and along $\mathbf{e}_x'$ in the actual configuration (A): it follows the body upon rotation. The interior force corresponding to a component $\sigma_{yx}^{\mathrm{PK}}$ is also shown.

is therefore the component of the interior force in the local material frame, while that in parentheses is the component of the normal $\mathbf{n}\,\mathrm{d}S$ in the reference configuration, computed in the fixed Cartesian frame. The above definition of the stress tensor is based, as before, on the remark that there is a linear mapping from the components of the normal, in parentheses, to the components of the interior force, in brackets (this is still a linear mapping when these vectors are computed in two different coordinate systems).

Therefore, the normal $(\mathbf{n}\,\mathrm{d}S)$ is used in its reference configuration, while the interior force is measured in the local material frame. This has an important advantage. Start from a configuration of a body for which the approximation of small displacement is valid—then, the definition (2.73) of the stress becomes equivalent to that introduced earlier in the case of small displacement—and rotate both the body and the applied stress by a finite angle (this rotation amounts to a rotation of the observer in the opposite direction). Upon this rotation, the normal to an arbitrary surface element $(\mathbf{n}\,\mathrm{d}S)$ is unchanged as it is defined in the reference configuration; both the physical interior force $\mathrm{d}\mathbf{F}$ and the local frame $(\mathbf{e}_k')_k$ are rotated and so the components of $\mathrm{d}\mathbf{F}$ in this basis are *unchanged*. As a result, none of the components of the stress tensor $\sigma_{ij}^{\mathrm{PK}}$ is changed when the solid undergoes a rigid-body rotation (assuming of course that the forces follow this rotation): this tensor provides a measure of stress in a 'frame' attached to the body and follows closely the principles of Lagrangian mechanics, see Fig. 2.10. In this sense, it is a natural extension of the previously-defined stress tensor to the case of finite displacement.

2.5.3 Symmetry of the stress tensor

The stress tensor introduced just above is symmetric. This can be shown by a similar argument to that in Section 2.3.2, considering a cubic volume element whose faces are normal to the vectors $\mathbf{e}_x$, $\mathbf{e}_y$ and $\mathbf{e}_z$ in the reference configuration. The two faces that are normal to $\mathbf{e}_x$ yield a contribution to the torque that is, at dominant order:

$$\mathrm{d}x\,\mathbf{e}_x' \times (\mathbf{e}_i'\,\sigma_{ix}^{\mathrm{PK}}\,\mathrm{d}y\,\mathrm{d}z).$$

small strain limit in the fluid case, the Piola–Kirchhoff tensor is of very little used there, and one usually prefers to deal exclusively with the Eulerian coordinates.

The first factor is the arm of the torque, in the actual configuration, while the second one is the interior force as computed from the definition of the stress tensor. Note that areas of the faces of the cube are taken in the reference configuration, according to our conventions for the stress tensor. By permutation of indices, one gets a similar formula for the contribution of the two faces that were normal to $\mathbf{e}_y$ in the reference configuration:

$$\mathrm{d}y\,\mathbf{e}'_y \times (\mathbf{e}'_j\,\sigma_{jy}^{\mathrm{PK}}\,\mathrm{d}x\,\mathrm{d}z),$$

and a similar formula for the last faces perpendicular to $\mathbf{e}_z$. The projection of the total torque along the vector $\mathbf{e}'_z$ yields:

$$\mathrm{d}\Gamma_{z'} = \mathbf{e}'_z \cdot \left[\mathrm{d}x\,\mathbf{e}'_x \times (\mathbf{e}'_i\,\sigma_{ix}^{\mathrm{PK}}\,\mathrm{d}y\,\mathrm{d}z) + \mathrm{d}y\,\mathbf{e}'_y \times (\mathbf{e}'_j\,\sigma_{jy}^{\mathrm{PK}}\,\mathrm{d}x\,\mathrm{d}z) + \mathbf{e}'_z \times \cdots \right] \tag{2.74}$$

$$= \left[((\mathrm{d}x\,\mathbf{e}'_x) \times (\mathrm{d}y\,\mathbf{e}'_y)) \cdot (\mathrm{d}z)\,\mathbf{e}'_z\right] (\sigma_{yx}^{\mathrm{PK}} - \sigma_{xy}^{\mathrm{PK}}).$$

The first term in the right-hand size of the last equality is the mixed product of the vectors defining the edges of the cube in the actual configuration: it is the volume of the deformed cube, and is therefore non-zero. Therefore, the finiteness of acceleration requires, as in the case of small displacement, that the second factor, $(\sigma_{yx}^{\mathrm{PK}} - \sigma_{xy}^{\mathrm{PK}})$, vanishes. After permutation of the indices, we are led to the following symmetry condition for the stress tensor:

$$\sigma_{ij}^{\mathrm{PK}}(\mathbf{r}) = \sigma_{ji}^{\mathrm{PK}}(\mathbf{r}),$$

which extends equation (2.49). Note that this symmetry of the tensor depends crucially on the conventions used to define the stress tensor.[47]

2.5.4   Strain tensor for finite displacement

The definition of the strain tensor given in Section 2.2 did not rely on any particular approximation and can be used directly for transformations involving finite displacement. In fact, the strain arising from a finite rotation, introduced in equation (2.18), has already been studied.

From its definition (2.9), the strain tensor is a quadratic form that yields the change of squared length of a material vector (left-hand side) when applied on this vector taken in the *reference* configuration (right-hand side). As a result, the indices of the strain tensor refer to the reference configuration, and the components of the strain tensor are unaffected by a rigid-body rotation of the solid. Like the stress tensor, the strain tensor is expressed in a system of coordinates attached to the solid.

2.5.5   Description of volumic forces

For finite displacements, we continue to use equation (2.40) defining the mass density $\mathbf{g}$ of the non-interior force: this equation uses the volumic density $\rho(\mathbf{r})$ in the reference configuration, which is different in general from that in the actual configuration, $\rho'(\mathbf{r})$—this is consistent since $(\rho\,\mathrm{d}x\,\mathrm{d}y\,\mathrm{d}z)$ is the mass of an elementary volume, which is conserved upon transformation. We could decompose this density of force in the local material frame, or in the fixed Cartesian frame. Unlike with the stress tensor, this convention has limited consequences. For the sake of definiteness, we choose the latter:

---

[47] Some variants of the stress tensor used for finite displacements are indeed not symmetric, the symmetry being replaced by a more complicated condition.

$$G = \left(\iiint_V \rho(\mathbf{r})\, g_i(\mathbf{r})\, \mathrm{d}x\,\mathrm{d}y\,\mathrm{d}z \right) \mathbf{e}_i, \qquad (2.75)$$

where $\mathbf{G}$ is still the force on an arbitrary volume V due to non-interior forces.

2.5.6 Equation of equilibrium

The equation of equilibrium was derived in Section 2.3.4, under the approximation of small displacement. The argument can be extended straightforwardly to finite displacement. Let us consider again the total force acting on a given volume:

$$\mathbf{F} + \mathbf{G} = \iint_S \mathbf{e}_i'\, \sigma_{ij}^{\mathrm{PK}}\, n_j\, \mathrm{d}S + \iiint_V \rho\, \mathbf{g}\, \mathrm{d}V,$$

where, as earlier, the first contribution comes from the interior forces and is given by a surface integral on the boundary S of the volume. Some care must be taken, as the interior forces are now expressed in a local frame, the vectors $\mathbf{e}_i'$ depend on the point of integration $\mathbf{r}$. Using the same trick as before, we apply the divergence theorem to turn the surface integral giving the interior forces into a volume integral. This divergence theorem is written in the reference configuration: the element of integration $(n_j\,\mathrm{d}S)$ becomes a derivative with respect to x_j in the new volumic integral:

$$\mathbf{F} + \mathbf{G} = \iiint_V \left[\frac{\partial\left(\mathbf{e}_i'\, \sigma_{ij}^{\mathrm{PK}} \right)}{\partial x_j} + \rho\, \mathbf{g} \right] \mathrm{d}V.$$

The first term in the bracket is again interpreted as the volumic force due to unbalanced interior forces. The second is obviously the volumic density of remote force. The equilibrium condition requires that an arbitrary sub-volume V is at equilibrium, and so the integrand is identically zero:

$$\frac{\partial\left(\mathbf{e}_i'(\mathbf{r})\, \sigma_{ij}^{\mathrm{PK}}(\mathbf{r}) \right)}{\partial x_j} + \rho(\mathbf{r})\, \mathbf{g}(\mathbf{r}) = \mathbf{0}. \qquad (2.76)$$

This equation can be recovered directly by studying a balance of forces on the elementary volume whose edges are parallel to the Cartesian frame $(\mathbf{e}_i)_i$ in the reference configuration— in fact, this is exactly what the divergence theorem does. Note that the case of small displacements corresponds to $\mathbf{e}_i' \approx \mathbf{e}_i$: then, the equilibrium equation takes the simpler form (2.51).

With the aim of arriving at a more explicit expression, one can project this equation of equilibrium on to the directions of the fixed Cartesian frame. Along $\mathbf{e}_k$, this yields:

$$\frac{\partial\left((e_i')_k\, \sigma_{ij}^{\mathrm{PK}} \right)}{\partial x_j} + \rho\, g_k = 0,$$

where $(e_i')_k$ is the component of $\mathbf{e}_i'$ along $\mathbf{e}_k$, which can be expressed using equation (2.71) as:

$$(e_i')_k = \mathbf{e}_i' \cdot \mathbf{e}_k = \frac{\partial(x_k + u_k(\mathbf{r}))}{\partial x_i}.$$

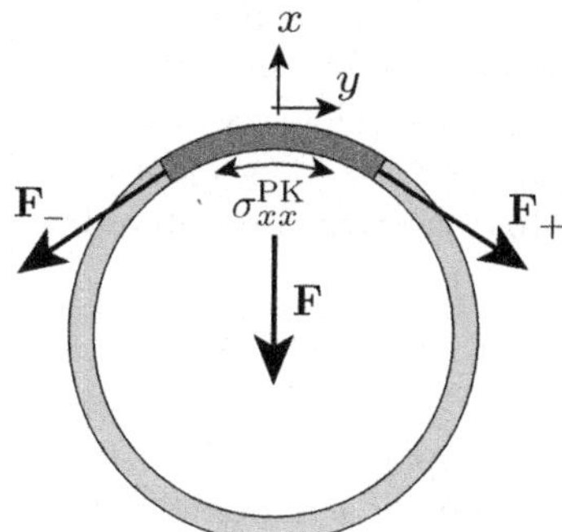

Fig. 2.11 Inflating a membrane: tangential stress $\sigma^{\mathrm{PK}}_{xx}$, when combined with curvature, can result in a net force **F** oriented *in the normal direction*. In the case of an inflating balloon, it is this net normal force that balances the (normal) pressure force of the gas inside the membrane. This coupled effect of stress and curvature is expressed by the second term in the equation (2.77) of equilibrium.

By expanding the partial derivative in the equation of equilibrium, it can be rewritten as:

$$\frac{\partial(x_k + u_k)}{\partial x_i}\,\frac{\partial \sigma^{\mathrm{PK}}_{ij}}{\partial x_j} + \sigma^{\mathrm{PK}}_{ij}\,\frac{\partial^2(x_k + u_k)}{\partial x_i\,\partial x_j} + \rho\, g_k = 0. \tag{2.77}$$

The first term is easy to interpret: its second factor, $\partial\sigma^{\mathrm{PK}}_{ij}/\partial x_j$, is the volumic resultant of interior forces, already present in the equation of equilibrium (2.51) for small displacements, while the first factor $\partial(x_k + u_k)/\partial x_i = (e'_i)_k$ defines the matrix associated with the change of frame between reference configuration and the actual one. This rotation is required to pass from the local frame, where we have decomposed the interior force, to the fixed Cartesian frame, where the balance of forces is written. The last term in the equation of equilibrium has an obvious interpretation in terms of the volumic density of external forces.

The second term in equation (2.77) is specific to finite displacements,[48] and couples[49] the stress with the curvature (the second derivative of the displacement). This effect is interpreted graphically in Fig. 2.11 from the analysis of an inflating membrane. Since both

[48] This term is the product of stress times curvature: in general, it is a second-order quantity that can be neglected. However, there are two important circumstances where it must be retained. The first is when infinitesimal perturbations near a prestressed configuration are studied, as happens in the analysis of vibrations or in the analysis of the linear stability of structures under imposed stress—then, the stress is no longer a first-order quantity but a 'zeroth-order' one, and our term becomes first-order like the other ones in the equation. The second circumstance is when the tangential and normal displacement are of different orders of magnitude. Then, a more careful evaluation of the orders of magnitudes of the different terms of the equation is required. As we shall see, the scalings associated with the Föppl–von Kármán theory of elastic plates imply that this term is comparable to the other ones—in fact, this term is the one that brings in nonlinearity in the theory and is at the source of the interesting, non-linear behaviour of plates such as buckling.

[49] The same effect is known as Laplace's law for capillary fluids: capillarity gives rise to a surface tension at an interface between different fluids and/or gas; this tension is similar to our stress σ^{PK}_{xx}. When the interface is curved, this surface tension γ results in a net normal force per unit area, proportional to γ times the mean curvature of the surface. For the interface to be in equilibrium, this net normal force has to be balanced by a jump in pressures across the interface.

the stress and the displacements are unknowns in a problem of elasticity, this term introduces a nonlinearity into the equation of equilibrium (this nonlinearity is absent in the case of small displacement). In Section 2.5.6 we shall carry out the variation of the elastic energy for finite displacements, and show that this non-linear term in the equations of equilibrium is tightly connected to the non-linear terms encountered earlier in the geometric definition of strain.

2.5.7   Elastic energy

In this section, we show that the equation (2.55) defining the elastic energy extends to finite displacement, when the second Piola–Kirchhoff stress tensor is used. We proceed as in Section 2.3.5, and consider the work of the interior forces upon an arbitrary perturbation of the body:

$$\delta W_F = \iint_S \left[(\mathbf{e}'_i\, \sigma_{ij}^{\mathrm{PK}}\, \mathrm{n}_j\, \mathrm{d}S) \cdot \delta \mathbf{r}' \right].$$

In this equation, the term in parenthesis is the interior force, as computed from the definition of the stress tensor. Applying the divergence theorem, we obtain:

$$\delta W_F = \iiint_V \frac{\partial}{\partial x_j} \left(\sigma_{ij}^{\mathrm{PK}}\, \mathbf{e}'_i \cdot \delta \mathbf{r}' \right)\, \mathrm{d}V,$$

which is split into two terms, as before:

$$\delta W_F = \iint_V \left\{ \sigma_{ij}^{\mathrm{PK}} \left[\mathbf{e}'_i \cdot \frac{\partial(\delta \mathbf{r}')}{\partial x_j} \right] + \delta \mathbf{r}' \cdot \frac{\partial \left(\sigma_{ij}^{\mathrm{PK}}\, \mathbf{e}'_i \right)}{\partial x_j} \right\}\, \mathrm{d}V.$$

The factor in brackets can be rewritten as:

$$\left[\mathbf{e}'_i \cdot \frac{\partial(\delta \mathbf{r}')}{\partial x_j} \right] = \mathbf{e}'_i \cdot \delta \left(\frac{\partial \mathbf{r}'}{\partial x_j} \right) = \mathbf{e}'_i \cdot \delta \mathbf{e}'_j,$$

whose symmetric part is:

$$\frac{\mathbf{e}'_i \cdot \delta \mathbf{e}'_j + \mathbf{e}'_j \cdot \delta \mathbf{e}'_i}{2} = \frac{\delta(\mathbf{e}'_i \cdot \mathbf{e}'_j)}{2} = \frac{\delta(\mathbf{e}_i \cdot \mathbf{e}_j + 2\,\epsilon_{ij})}{2} = \delta \epsilon_{ij}.$$

As in the case of finite displacement, only this symmetric part is relevant since this tensor gets multiplied by a symmetric tensor, $\sigma_{ij}^{\mathrm{PK}}$.

Using the equation of equilibrium (2.76), the second term in δW_F yields the perturbation to the potential energy associated with the remote force $\mathbf{g}$, $\delta \mathcal{E}_G$. This potential energy was introduced in equation (2.53).

Using the same balance of energy as in equation (2.54), one is finally led to the following definition of the elastic energy:

$$\delta \mathcal{E}_{\mathrm{el}} = \iiint_V \sigma_{ij}^{\mathrm{PK}}\, \delta \epsilon_{ij}\, \mathrm{d}V, \tag{2.78}$$

which is valid for finite displacement. This definition extends the one obtained previously for small displacement, in equation (2.55). In fact, both equations are formally identical. As a result, the elastic energy is a well-defined functional $\mathcal{E}_{\mathrm{el}}(\epsilon_{ij})$ of the strain, which characterizes the material properties—very much like specifying the energy of a spring in the form $\mathcal{E}(x) = kx^2/2$ implies that it is linear, and sets its stiffness—and encompasses both the cases of small and finite displacement. For practical purposes, this means that the constitutive equation (2.60) introduced in Section 2.4 for an isotropic, linear material undergoing small displacement can also be used for finite displacement. Indeed the relation (2.78) does define the small increment of energy, although the expression of the energy itself should follow from an integration of this small increment with respect to the amplitude of the strain (for instance) going from zero to finite values. This is valid if an energy can be attributed unambiguously to a given state of deformation, which assumes reversible transformations.

2.5.8 Mechanical equilibrium: a variational view

In the previous section and in Section 2.3.5, we have introduced the elastic energy by starting from the constitutive relations and from the equations of mechanical equilibrium. The concept of elastic energy is often used in the reverse direction: instead of deriving the value of the energy for a given equilibrium state from the constitutive relations and from the balance of forces, one directly uses the energy itself as a Lagrange functional to derive the differential equations for stress and strain, by writing that the energy is stationary at equilibrium (see Appendix A for an outline of the general method of variational calculus). This is particularly useful in the context of thin elastic bodies, characterized by a small thickness h, in front of the other typical dimensions, $h \ll L$. To study the elasticity of such bodies the easiest way is to introduce an approximate elastic energy, which is more and more accurate in the limit of a small aspect ratio, $h/L \to 0$. By variation, one can derive the (approximate) equations of equilibrium for any h, and study their (exact) limit for $h/L \to 0$. It would be much more cumbersome to derive the equations of equilibrium without using this variational formulation: for small but non-zero h, the strain and stress fields can only be guessed with some approximation and so cannot be expected to satisfy the equations of equilibrium.

Let us derive the equations of equilibrium by variation of the elastic energy. We again use the definition of the elastic energy (2.78), and expand the term $\delta\epsilon_{ij}$ using the definition of strain given in (2.9):

$$
\begin{aligned}
\delta\mathcal{E}_{\mathrm{el}} &= \iiint \sigma_{ij}^{\mathrm{PK}} \left(\delta\left(\frac{\partial u_i}{\partial x_j}\right) + \frac{\partial u_k}{\partial x_i} \, \delta\left(\frac{\partial u_k}{\partial x_j}\right) \right) \, \mathrm{d}x\,\mathrm{d}y\,\mathrm{d}z \\
&= \iiint \sigma_{ij}^{\mathrm{PK}} \left(\delta_{ik} + \frac{\partial u_k}{\partial x_i} \right) \delta\left(\frac{\partial u_k}{\partial x_j}\right) \, \mathrm{d}x\,\mathrm{d}y\,\mathrm{d}z \\
&= \iiint \sigma_{ij}^{\mathrm{PK}} \frac{\partial(x_k + u_k)}{\partial x_i} \, \delta\left(\frac{\partial u_k}{\partial x_j}\right) \, \mathrm{d}x\,\mathrm{d}y\,\mathrm{d}z. \\
&= \iiint (e_i')_k \, \sigma_{ij}^{\mathrm{PK}} \frac{\partial(\delta u_k)}{\partial x_j} \mathrm{d}V.
\end{aligned}
\tag{2.79}
$$

The next step is to apply the divergence theorem. After adding the variation of the potential energy due to interior forces and to volumic forces and cancelling the variation of the total energy, one obtains the equations of equilibrium already given in equation (2.76):

$$
\frac{\partial\left(\mathbf{e}_i'(\mathbf{r})\,\sigma_{ij}^{\mathrm{PK}}(\mathbf{r})\right)}{\partial x_j} + \rho(\mathbf{r})\,\mathbf{g}(\mathbf{r}) = \mathbf{0}.
$$

The calculation above is similar to the one in the previous section, but goes in the reverse direction. It provides an opportunity to stress a subtle but important point: comparison of the two equations above reveals that the nonlinearity in the equations of equilibrium ($\mathbf{e}'_i \neq \mathbf{e}_i$ in the first term) *derives* from those in the definition of strain. Indeed, for a small displacement, the non-linear terms in the definition (2.9) of strain disappear, as do the non-linear contributions in the equation of equilibrium, see equation (2.51). This remark will be particularly useful in our analysis of plates: the Föppl–von Kármán equations involve a small displacement approximation in the tangent directions, but assume finite displacement in the transverse one—as shown by the above analysis, the resulting equations of equilibrium are non-linear, but in the transverse direction only.

2.6   Conclusion

We have derived the equations governing three-dimensional elastic bodies both in the limit of a small displacement, a convenient approximation, and in the more general case of finite displacement. We have always assumed that the strain remains small everywhere, which constitutes the so-called Hookean elasticity. By taking the limit of small thickness in these general equations, we shall derive in the following chapters the equations for thin elastic bodies, such as rods, plates and shells. As we will show, the geometric nonlinearity present in the definition (2.9) of strain plays an important role in this limit.

References

L. B. Freund. *Dynamic Fracture Mechanics*. Cambridge Monographs on Mechanics. Cambridge University Press, 1998.

R. Lakes. Foam structures with a negative Poisson's ratio. *Science*, 235(4792):1038–1040, 1987.

Part I

Rods

Long, slender elastic bodies, rods in short, can be found in many places in the natural world, the most obvious example being perhaps the human hair. Trees or weeds at large scale can be seen as rods that are strong enough to stay straight under the opposite effects of gravity that tends to pull them down and their stiffness that keeps them upright, as fixed by the root system. Rods are also found in all sorts of technological applications. A recent major improvement in a classical technology is the use of very long rails made in one piece for fast trains, made possible by preventing deformation (buckling) by thermal dilation. These two examples illustrate the fundamental role played by the so-called Elastica problem, first solved by Euler, which addresses the stability of an elastic bar under compressive stress. This stability analysis is based on a simple premise: shortening a bar by longitudinal compression requires a lot of elastic energy; it is far less costly, energy-wise, for a bar to deviate from its straight shape by bending, which is another way for the rod to bring its two ends closer to each other. For the rail problem for instance, bending is avoided by well-designed straighteners at short intervals along the rail. For elastic bars in general, the fact that bending deformations are naturally selected by energy minimization puts geometry at the heart of the theory.

The next part of this book is devoted to various aspects of the elasticity of rods. As stated in the general introduction, this book does not intend to be a general treatise of elasticity theory. In this part, as elsewhere in the book, we focus on various applications of the classical theory of elastic rods with an emphasis on the geometrical aspects, and introduce various mathematical methods that should prove useful in general situations. Chapter 3 derives the rod equations from 'first principles', namely from the equation of three-dimensional elasticity solved in the relevant limit, for an elastic body that is much longer in one direction than in the two other directions. The resulting equations, due to Kirchhoff, mix in a rather tangled way the geometry of three-dimensional curves and the conditions of mechanical equilibrium. The case of rods with equal moments of inertia, which includes rods with an axisymmetric section, leads to some simplifications and is also discussed. A simple version of the rod equations valid for the so-called Elastica, which describes *planar* deformations, was used by Euler long before Kirchhoff derived the general equations. As an example of the use of the simple Elastica equations, in Chapter 4 we look at a simple model for the elasticity of a single human hair. Its shape is the result of three competing effects. One is its elasticity that resists bending. Some curling (possibly natural but sometimes resulting from human trickery) makes the equilibrium rest state naturally curved. Finally the hair 'lives' in a gravity field that tends to pull it downwards. These effects can be taken into account with two dimensionless parameters, which fit rather nicely the space of all possible shapes: straight up, bending, curled all along the length and finally bending with curls near the free end.

The last chapter of this part, Chapter 5, is devoted to the so-called 'godets' problem. It aims at explaining the undulating patterns observed near the edge of thin surfaces with stretched edges, in different mechanical systems. An example is a lettuce leaf with an edge growing quicker than needed for a simple, planar growth. This edge tends to buckle out-of-plane to accommodate the overgrowth. The resulting 'godets' patterns are investigated using a simple elastic model based on the Kirchhoff equations. This yields the so-called 'godet ribbon', a model that describes the neighborhood of the leaf edge. This model yields a parameterless problem that can be fully solved and compares well with the experimental patterns. It is based on a particular type of rod, which features both spontaneous curvature and also has a very large bending stiffness in one direction: the 'godet ribbon' is described by a simple variant of the Kirchhoff equations that is amenable to solutions, or at least leads to a simple set of ODEs that is easily solvable numerically and in physically relevant situations.

3
Equations for elastic rods

3.1 Introduction

This chapter is devoted to the derivation of the equilibrium equations for elastic rods. An elastic rod is a slender elastic body: its length in one space direction is much greater than its length in the two other perpendicular directions, which define the cross-section. The theory for the finite displacement of thin rods has been developed by Kirchhoff and Clebsch—for a detailed historical discussion of this derivation, see the paper by Dill 1992. The theory of elastic rods has many applications in engineering, for instance to predict and try to prevent the unwanted pop-out of sub-oceanic cables (E. E. Zajac, 1962), but also in biology, where many structures display rod-like elasticity, such as DNA (W. R. Bauer, R. A. Lund, and J. H. White, 1993; D. Bensimon *et al.*, 1995; P. Cluzel *et al.*, 1996; S. B. Smith, Y. Cui, and C. Bustamante, 1996; T. R. Strick *et al.*, 1996; J. F. Marko and E. D. Siggia, 1995; B. D. Coleman and D. Swigon, 2004; N. Clauvelin, B. Audoly, and S. Neukirch, 2008), other polymers (R. E. Goldstein and S. A. Langer, 1995; J. V. Selinger, F. C. MacKintosh, and J. M. Schnur, 1996), bacterial fibres (I. Klapper, 1996) and tendril bearers in climbing plants (T. McMillen and A. Goriely, 2002). In the following chapters, we shall explain in detail how the theory of elastic rods can be applied to two simple systems arising in biology: we shall study the mechanics of the human hair in Chapter 4, and the wrinkles observed along the edge of many plant leaves in Chapter 5. In the latter problem, we consider a rod with a very flat cross-section, the so-called elastic ribbon (see Fig. 3.1); this offers a smooth transition to Part II on elastic plates.

The derivation of the equilibrium equations of rods follows the general pattern of the analysis of equilibrium in elasticity theory: an expression for the energy of an arbitrary configuration of the rod is found and the equations for equilibrium are obtained by assuming that this energy is stationary under small deformations for given boundary conditions—and geometrical constraints, if any, as in the case of a ribbon. The resulting equations, due to Kirchhoff (G. R. Kirchhoff, 1859), are non-linear, not because of the material response but because of the geometry. The importance of geometric nonlinearity for rods is clearly pointed out in the short account of Euler's Elastica in Appendix A, a limit case of the more general Kirchhoff equations.[1]

At large scale the rod can simply be seen as a curve, its *centre line*, in the Euclidean 3D space. However a complete description of a rod requires one to introduce quantities that are not present in the geometrical Serret–Frénet theory of smooth curves in 3D space. This

[1] Before reading the present chapter, the reader may start with Appendix A. Euler's *Elastica* problem addresses the deformations of a thin rod that remains contained within a plane. Its importance can hardly be overemphasized since this was the first rational approach to a problem of elasticity as well as the first example of analysis of a bifurcation problem. It provides a gentle introduction to the more general Kirchhoff equations.

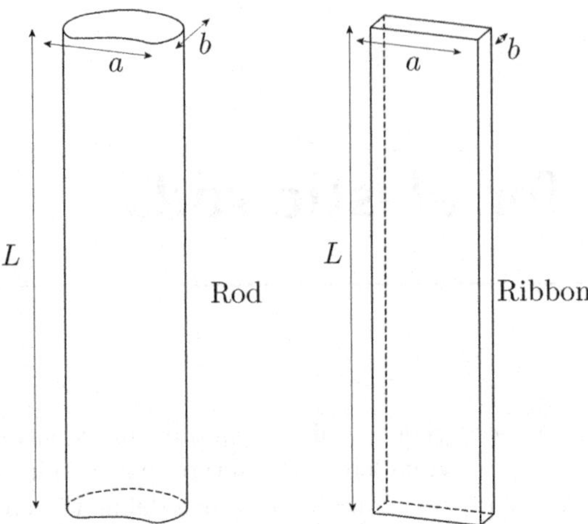

Fig. 3.1 An elastic rod (left) and the particular case of a ribbon (right). A rod is such that the length L is much larger than the transverse sizes, a and b. In the case of a ribbon, the cross-section is flat: it has a large aspect ratio, $a/b \gg 1$.

is because a configuration of the rod is not only characterized by the path of its centre line, but also by how much it twists around this centre line. Consider for instance a straight but twisted rod; its centre line is straight, as in the reference configuration. By twisting one end with respect to the other, one has to work against the reaction torque of the rod and to put some elastic energy into it. Obviously, this energy does not depend only on the configuration of the centre line, which is unchanged by twisting. As we shall show, it is proportional to the square of the twist rate of the material frames[2] attached to the cross-section, as expressed by the term $\mu J \tau^2$ in equation (3.40). This example illustrates that the twist should be taken into account whenever one seeks to describe the shape of an elastic rod.

The theory of thin rods is often presented starting from the assumption that every cross-section deforms in a rigid-body manner (this is the case of quite a few textbooks and research papers). Even though this kinematic postulate formally leads to the correct equations of equilibrium, it is extremely misleading, to say the least, as it does not yield the correct value of the twist modulus. We shall indeed show in Section 3.4 that the cross-sections do not remain planar in general[3] when the rod is twisted, a phenomenon called warping. For extensible rods, the assumption of rigid cross-sections predicts an erroneous stretching modulus too, since it neglects contraction of the cross-section in response to extension of the centre line, a phenomenon known as the Poisson effect, see Section 2.4.3 on a bar loaded in simple traction. To avoid these difficulties, we shall start our derivation with Saint-Venant's solutions describing a uniformly twisted or bent rod. These solutions reveal the actual three-dimensional form of the displacement field in a rod, which is more complex than a combination of rigid-body rotations in every

---

[2] By material frame, we mean a frame that follows the elastic body upon deformation, as introduced in Chapter 2.

[3] The only case when there is no warping is for a rod with circular cross-section, as the axial displacement happens to vanish.

cross-section. A 1D description of the rod is proposed next, based on these solutions of the equations for 3D elasticity. Historically, the derivation of the equations for rods by Kirchhoff, Clebsch and then Love has used Saint-Venant's solutions as a starting point too (E. H. Dill, 1992).

3.2   Geometry of a deformed rod, Darboux vector

We consider a reference configuration such that the rod is straight and has cylindrical symmetry, see Fig. 3.1: it is assumed that the rod has uniform properties, both geometrical and mechanical, along its length, and that all cross-sections are identical.[4] This excludes the case of a rod with a variable thickness. Note that the cylindrical configuration of reference does not need to be free of mechanical stress; in our later analysis of the human hair, for instance, we consider rods whose cross-sections have all the same geometries but can have some natural curvature or twist, and are not cylindrical in their natural configuration. In this reference configuration, let us call x and y the coordinates in the plane of the cross-sections, see Fig. 3.2. The coordinate along the rod length is denoted s or z (these two variables are equivalent in the context of rods).

We call the *centre line* a material curve parallel to the axis z of the rod in the reference configuration. Its equation is therefore $(x = x_0, y = y_0)$. At the moment, this centre line is not fully defined and we shall explain later how the coordinates x_0 and y_0 are fixed. We consider *inextensible* rods to be rods whose centre line remains inextensible upon deformation.[5] Then, the variable s is also the curvilinear coordinate along the centre line *in the actual configuration*.

We shall keep track of the twist by introducing the material frame $(\mathbf{d}_1(s), \mathbf{d}_2(s), \mathbf{d}_3(s))$ in the deformed configuration. It is attached to the centre line of the rod (s being, again, the curvilinear distance along the centre line). This centre line, together with the set of material frames, form what is sometimes called a *Cosserat curve*. We choose the orientation of these material frames in such a way that $\mathbf{d}_1$ and $\mathbf{d}_2$ lie in the plane of the cross-section, while $\mathbf{d}_3$ is a tangent to the center line (see Fig. 3.2); the orientations of $\mathbf{d}_1$ and $\mathbf{d}_2$ in the plane of the cross-section is arbitrary for the moment but will be specified later on. Upon deformation, the material vectors $(\mathbf{d}_1, \mathbf{d}_2, \mathbf{d}_3)$ are given by equation (2.70) as the gradients of the transformation. In fact, these vectors $\mathbf{d}_i$ are the same mathematical entity as the vectors of the local frame $\mathbf{e}_i$, except that they are defined along the centre line only:

$$\mathbf{d}_i(s) = \mathbf{e}_i(x_0, y_0, s),$$

where $(x = x_0, y = y_0)$ defines the centre line, (x, y) being the coordinates in the plane of the cross-section.

In a deformed state, the centre line has no particular reason to remain straight and, in general, $\mathbf{d}_1$ and $\mathbf{d}_2$ will twist along the centre line. However, in the case of small strain

---

[4] This approximation is for the sake of simplicity only, and could be removed without difficulty to the case of slowly varying mechanical and/or geometrical properties—slowing meaning at a length scale much longer than the thickness.

[5] As explained in detail in Section 3.7 and later on in our account of plate elasticity, this assumption is physically justified for a wide range of loading conditions, provided the aspect ratio of the rod h/L is a small number. Here h is the typical size of the cross-section and L the length of the rod.

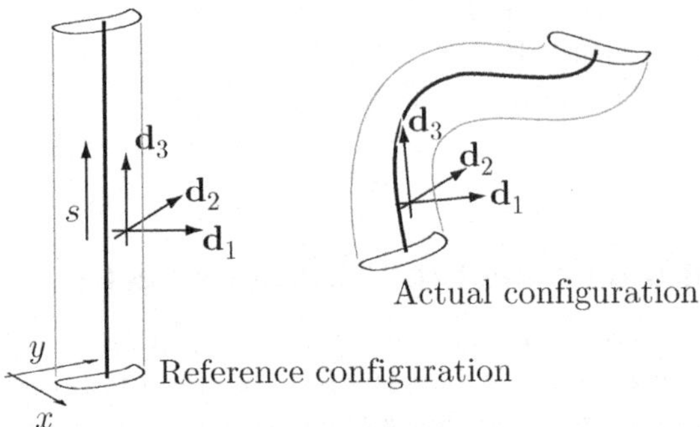

Fig. 3.2 Centre line, coordinate system in the reference configuration, and material frame attached to the centre line of the rod.

that we consider, the triad $(\mathbf{d}_1, \mathbf{d}_2, \mathbf{d}_3)$ remains approximately orthonormal,[6] provided it has been chosen orthonormal in the reference configuration. This is known as the Euler–Bernoulli or Navier–Bernoulli kinematical hypothesis,[7] or sometimes as the assumption of unshearable rods. Indeed, the strain tensor measures the changes in the scalar products of material vectors by equation (2.72), and a small strain implies that these scalar products are only slightly changed.

We consider two types of deformation, not mutually exclusive. The first are deformations involving *small strain*, for which the material frame $(\mathbf{d}_1(s), \mathbf{d}_2(s), \mathbf{d}_3(s))$ remains approximately orthonormal. The second case involves *uniform* deformations with arbitrary strain. By uniform, we mean that the strain does not depend explicitly on the longitudinal coordinate s. Then, equation (2.72) shows that the scalar products $\mathbf{d}_i(s) \cdot \mathbf{d}_j(s)$ are independent of s. In both cases, the frame $(\mathbf{d}_1(s), \mathbf{d}_2(s), \mathbf{d}_3(s))$ deforms in a rigid manner when s is changed. Let us compute the derivatives of these vectors with respect to s, noted $\mathrm{d}(\mathbf{d}_i(s))/\mathrm{d}s$, $\mathbf{d}_{i,s}(s)$ or simply $\mathbf{d}_i'(s)$ (these quantities are the rate of variation of these material vectors when one follows the centre line). By an argument similar to that used in Section 1.3.3, the condition of orthonormality yields the following relations:

$$\mathbf{d}_i'(s) \cdot \mathbf{d}_i(s) = 0 \quad \text{and} \quad \mathbf{d}_i'(s) \cdot \mathbf{d}_j(s) = -\mathbf{d}_j'(s) \cdot \mathbf{d}_i(s),$$

for all indices i and j (there is no implicit sum over i in the first equation). As a result, there exist three scalar functions $\kappa^{(1)}(s)$, $\kappa^{(2)}(s)$ and $\tau(s)$ such that the derivatives of the material frame can be put in the form:

$$\mathbf{d}_1'(s) = \tau(s)\,\mathbf{d}_2(s) - \kappa^{(2)}(s)\,\mathbf{d}_3(s) \tag{3.1a}$$

$$\mathbf{d}_2'(s) = -\tau(s)\,\mathbf{d}_1(s) + \kappa^{(1)}(s)\,\mathbf{d}_3(s) \tag{3.1b}$$

$$\mathbf{d}_3'(s) = \kappa^{(2)}(s)\,\mathbf{d}_1(s) - \kappa^{(1)}(s)\,\mathbf{d}_2(s). \tag{3.1c}$$

---

[6] In the previous chapter, we used primes to denote quantities pertaining to the present configuration as opposed to the reference configuration. From now on, primes are instead used to mark spatial derivatives along the centre line, that is with respect to the variable s. We shall specify whether a quantity refers to the present or reference configuration only in cases of a possible ambiguity.

[7] The word hypothesis should not be misleading: these assumptions can actually be proved by taking the limit of the 3D elasticity theory in the limit of a small aspect ratio, at least in simple cases. This is discussed in Appendix D for the case of plates.

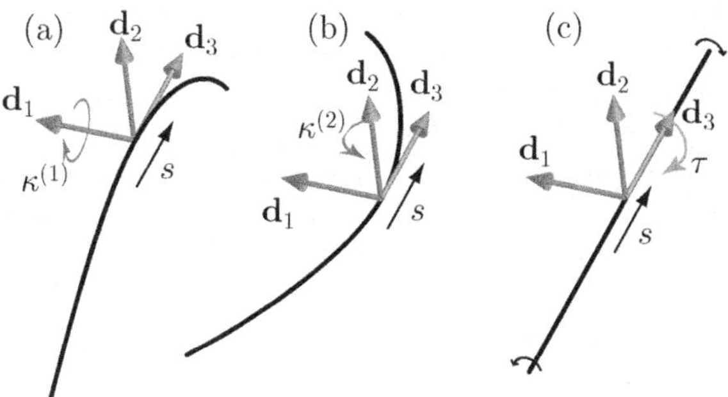

Fig. 3.3 The three local modes of deformation of an elastic rod, associated with the change of (a) material curvature $\kappa^{(1)}$ associated with the direction $\mathbf{d}_1$ of the cross-section, (b) material curvature $\kappa^{(2)}$ associated with $\mathbf{d}_2$, and (c) twist.

This is the most general set of first-order linear equations conserving the orthonormal character of the material frame $(\mathbf{d}_1, \mathbf{d}_2, \mathbf{d}_3)$. These equations describe the rigid-body rotation of the frame.

Using the notation $\mathbf{u} \times \mathbf{v}$ for the cross product of two vectors, the equations (3.1) can be rewritten

$$\mathbf{d}_1'(s) = \mathbf{\Omega}(s) \times \mathbf{d}_1(s) \qquad \mathbf{d}_2'(s) = \mathbf{\Omega}(s) \times \mathbf{d}_2(s) \qquad \mathbf{d}_3'(s) = \mathbf{\Omega}(s) \times \mathbf{d}_3(s), \qquad (3.2)$$

where we have introduced the Darboux vector $\mathbf{\Omega}(s)$:

$$\mathbf{\Omega}(s) = \kappa^{(1)}(s)\,\mathbf{d}_1(s) + \kappa^{(2)}(s)\,\mathbf{d}_2(s) + \tau(s)\,\mathbf{d}_3(s). \qquad (3.3)$$

The interpretation of equation (3.2) is that the material frame rotates with a rotation 'velocity' $\mathbf{\Omega}(s)$ when the centre line is followed at unit speed—the vector $\mathbf{\Omega}(s)$ is in fact a rate of rotation per unit length along the centre line. From equation (3.3), the numbers $\kappa^{(1)}$ and $\kappa^{(2)}$ express by how much the material frame rotates around the directions $\mathbf{d}_1$ and $\mathbf{d}_2$ of the cross-section. They are called the material curvatures. The number τ expresses how much the material frame rotates around the tangent $\mathbf{d}_3$, and is called the material twist of the rod, or the twist rate. Materials curvatures and twist are illustrated in Fig. 3.3. This Darboux vector is different from the one defined in Section 1.3.3 that characterizes the Serret–Frénet frame. Likewise, the material curvatures $\kappa^{(i)}$ and twist τ defined above are different from the geometrical curvature k and torsion g introduced in the analysis of the Serret–Frénet frame.[8]

3.3 Flexion

In this section and in the next one, we analyse the fundamental modes of deformations of a rod, namely pure and uniform flexion (parameters $\kappa^{(1)}$ and $\kappa^{(2)}$) and pure and uniform twist (parameter τ). To study these fundamental modes, we can assume that the deformation is invariant along the rod: the equations of 3D elasticity reduce to a set of

---

[8] Only two numbers, k and g characterize the derivatives of the Serret–Frénet frames, instead of three here. This is related to the fact that the Serret–Frénet frame does not contain any information on mechanical twist. Nevertheless, the two set of numbers are not independent: for instance, $k = ((\kappa^{(1)})^2 + (\kappa^{(2)})^2)^{1/2}$.

linear partial differential equations defined over the cross-section, which is a planar domain. These equations, together with their analytical solution due to Saint-Venant, are derived in this section (bending modes) and in the next one (twisting mode). We also derive the elastic energies of these fundamental modes deformations; as shown in Section 3.5, the energy of an arbitrary configuration (mixing all three modes, and non-uniform along the length of the rod) is simply a local superposition of the energy of three modes of deformation, two bending modes and one twist mode.

3.3.1   Kinematics of a bent rod

Let us first consider a rod undergoing uniform flexion[9] with a radius of curvature R. In the present section, we build a parameterisation of the bent rod based on symmetry considerations only, and propose explicit expressions for the transformation bringing the rod from reference to its bent configuration. The aim is to guess the form of the displacement field and to prepare the ground for the subsequent solution of the equations of elasticity. As we shall show later, the assumption of small strain requires that R be much larger than h, the typical radius of the cross-section. Although we could assume small strain ($R \gg h$) from the beginning, we delay this assumption until the next section.

The orientation of the material frame $(\mathbf{d}_1, \mathbf{d}_2)$ in the cross-section has not yet been specified: we can freely assume that $\mathbf{d}_1$ lies in the plane of curvature, that is that the material curvatures in equation (3.1) are of the form $\kappa^{(2)} = -1/\hat{R}$ and $\kappa^{(1)} = 0$ and the twist $\tau = 0$ (the exact relation between the radius of curvature R and the parameter $\hat{R}$ will be given later, but we anticipate that $R \approx \hat{R}$ for small strain). Then, by equation (3.1b), $\mathbf{d}_2(s)$ is a constant vector, around which $\mathbf{d}_1(s)$ and $\mathbf{d}_3(s)$ rotate. As a result, the centre line describes an arc of a circle in the plane perpendicular to $\mathbf{d}_2$.

There are two important symmetries for a uniformly bent rod. The first is an invariance by mirror symmetry with respect to any cross-section. The second is the invariance by translation along the centre line. Upon mirror symmetry with respect to the plane of a cross-section, the material vectors $\mathbf{d}_1$ and $\mathbf{d}_2$ are conserved, while the material tangent $\mathbf{d}_3$ is inverted. As a result, the invariance requires $\mathbf{d}_1(s) \cdot \mathbf{d}_3(s) = 0$ and $\mathbf{d}_2(s) \cdot \mathbf{d}_3(s) = 0$ for all s. These scalar products are zero, as in the reference configuration, and so the strain components $\epsilon_{13} = 0$ and $\epsilon_{23} = 0$ (recall that the off-diagonal stress components measure the change in the scalar product between material vectors). In other words, the material tangent $\mathbf{d}_3$ remains everywhere exactly[10] perpendicular to the plane of the cross-section.

Since the vectors $\mathbf{d}_2$ and $\mathbf{d}_3$ are unit vectors in the reference configuration, their norm is given in the actual configuration by equation (2.8):

$$|\mathbf{d}_2(0)| = \sqrt{1 + 2\,\epsilon_{22}^0}, \qquad |\mathbf{d}_3(0)| = \sqrt{1 + 2\,\epsilon_{33}^0},$$

where ϵ_{22}^0 and ϵ_{33}^0 are the strain components computed at the intersection ($x = x_0, y = y_0, s = 0$) of the centre line with the cross-section $s = 0$ (by invariance with respect to s,

---

[9] We shall use equivalently the words 'flexion' or 'bending': both refer to a mode of deformation associated with a change of the material curvatures $\kappa^{(1)}$ or $\kappa^{(2)}$ upon transformation.

[10] In the small strain approximation, the material tangent turns out to be always *approximately* perpendicular to the plane of the cross-section. Here, this result is exact, due to the assumed symmetries.

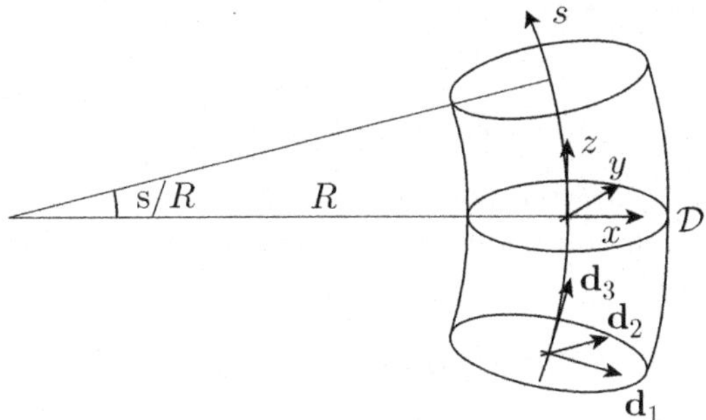

Fig. 3.4 In a rod that is bent uniformly, two arbitrary cross sections are connected by an isometry—a rotation whose axis $\mathbf{e}_y$ is perpendicular to the plane containing the center line. This property allows one to reconstruct the displacement field in the whole rod from its components $X(x,y)$, $Y(x,y)$ in any particular cross section.

these strains are in fact independent of s). For small strain, these numbers ϵ_{22}^0 and ϵ_{33}^0 would be small.

Let us define the following orthonormal Cartesian frame:

$$\mathbf{e}_y = \frac{\mathbf{d}_2(0)}{\sqrt{1 + 2\,\epsilon_{22}^0}}, \quad \mathbf{e}_z = \frac{\mathbf{d}_3(0)}{\sqrt{1 + 2\,\epsilon_{33}^0}}, \quad \mathbf{e}_x = \mathbf{e}_y \times \mathbf{e}_z$$

and an associated system of coordinates, (x, y, z), as sketched in Fig. 3.4. By construction, $\mathbf{e}_x$ and $\mathbf{e}_y$ span the cross-section at $s = 0$ and those vectors are close to $\mathbf{d}_1(0)$ and $\mathbf{d}_2(0)$ for small strain. The frame $(\mathbf{e}_x, \mathbf{e}_y, \mathbf{e}_z)$ is orthonormal by construction, while $(\mathbf{d}_1(0), \mathbf{d}_2(0), \mathbf{d}_3(0))$ is only close to orthonormal, when the strain is small.

Integration of the kinematic relation (3.1b) shows that $\mathbf{d}_2(s)$ is independent of s. The Darboux vector is then given by:

$$\boldsymbol{\Omega}(s) = -\frac{\sqrt{1 + 2\,\epsilon_{22}^0}}{\hat{R}}\,\mathbf{e}_y$$

and is itself a constant vector. Let us identify the prefactor in the equation above as the physical curvature $1/R$:[11]

$$\frac{1}{R} = \frac{\sqrt{1 + 2\,\epsilon_{22}^0}}{\hat{R}}. \tag{3.4}$$

Equations (3.2) can then be integrated. The result is that every vector $\mathbf{d}_i(s)$ relative to the cross-section s is obtained from the one relative to the initial cross-section $s = 0$, $\mathbf{d}_i(0)$, by a rotation with angle s/R around the axis supported by $\mathbf{e}_y$.

Now, let us introduce the unknown displacement field in the particular cross-section $s = 0$. By invariance under mirror symmetry, the displacement is perpendicular to $\mathbf{e}_z$ (it has no

[11] Except for small strain, $1/R$ is not strictly speaking the physical curvature of the centre line. From equation (3.9), the angle of tangent to the curve varies as s/R. Since the centre line is stretched, the curvilinear abscissa of the centre line is given by $\hat{s} = s/(1 + 2\,\epsilon_{33}^0)^{1/2}$ and not by s, and so the physical curvature is $R/(1 + 2\,\epsilon_{33}^0)^{1/2}$. This complication is overlooked here as it becomes irrelevant in the limit of small strain anyway.

component perpendicular to the cross-section). The deformed cross-section $s = 0$ is thus parameterised by:

$$\mathbf{r}(x, y, 0) = \mathbf{r}(x_0, y_0, 0) + (x - x_0 + X(x, y))\,\mathbf{e}_x + (y - y_0 + Y(x, y))\,\mathbf{e}_y, \qquad (3.5)$$

where $X(x, y)$ and $Y(x, y)$ are the two unknown components of the displacement in this cross-section: the above relation between the transformation and the displacement can be inverted as

$$X(x, y) = \mathbf{e}_x \cdot (\mathbf{r}(x, y, 0) - \mathbf{r}(x_0, y_0, 0)) - (x - x_0),$$

and a similar formula for $Y(x, y)$. These functions X and Y will be determined later on by solving the equations of elasticity; for the moment, our point is simply to show that knowing $X(x, y)$ and $Y(x, y)$ in a particular cross-section is enough to reconstruct the displacement field in the whole rod.

By translational invariance along the centre line, all cross-sections should deform 'identically'. More accurately, there must exist in the deformed configuration an isometric mapping between the cross-section $s = 0$ and an arbitrary cross-section at another s, see Fig. 3.4. We already know that the material vectors $\mathbf{d}_1(s)$ and $\mathbf{d}_2(s)$ that span this cross-section are obtained by a rotation from $\mathbf{d}_1(0)$ and $\mathbf{d}_2(0)$. Then, the mapping between the corresponding cross-sections must be the same rotation, that is a rotation of angle s/R and axis parallel to $\mathbf{e}_y$. Upon this rotation, the vector $\mathbf{r}(x, y, 0) - \mathbf{r}(x_0, y_0, 0)$, which describes an arbitrary vector contained in the cross-section, is transformed into $\mathbf{r}(x, y, s) - \mathbf{r}(x_0, y_0, s)$. The former is given by equation (3.5) in terms of the functions X and Y. This yields, for an arbitrary cross-section:

$$\mathbf{r}(x, y, s) = \mathbf{r}(x_0, y_0, s)$$
$$+ (x - x_0 + X(x, y)) \left(\cos \frac{s}{R}\,\mathbf{e}_x + \sin \frac{s}{R}\,\mathbf{e}_z \right) + (y - y_0 + Y(x, y))\,\mathbf{e}_y. \qquad (3.6)$$

The parameterisation of a uniformly bent rod is now almost complete. It remains to find the equation for the centre line $\mathbf{r}(x_0, y_0, s)$. In equation (2.71), we defined the vectors of the material frame as the gradient of the position in the actual configuration with respect to the coordinates of the parameterisation. The vector $\mathbf{d}_3$ is associated with the longitudinal coordinate s. This yields:

$$\frac{\partial \mathbf{r}(x_0, y_0, s)}{\partial s} = \mathbf{d}_3 = \sqrt{1 + 2\epsilon_{33}^0}\left(-\sin \frac{s}{R}\,\mathbf{e}_x + \cos \frac{s}{R}\,\mathbf{e}_z \right). \qquad (3.7)$$

Integration of this equation is straightforward and yields the parameterisation of the centre line:

$$\mathbf{r}(x_0, y_0, s) = \mathbf{r}(x_0, y_0, 0) + R\sqrt{1 + 2\epsilon_{33}^0}\left(\left(\cos \frac{s}{R} - 1\right)\mathbf{e}_x + \sin \frac{s}{R}\,\mathbf{e}_z \right). \qquad (3.8)$$

By combining equations (3.6) and (3.8), we derive the parameterisation of the deformed configuration:

$$\mathbf{r}(x, y, s) = \mathbf{r}(x_0, y_0, 0) + R\sqrt{1 + 2\epsilon_{33}^0}\left(\left(\cos \frac{s}{R} - 1\right)\mathbf{e}_x + \sin \frac{s}{R}\,\mathbf{e}_z \right)$$
$$+ (x - x_0 + X(x, y)) \left(\cos \frac{s}{R}\,\mathbf{e}_x + \sin \frac{s}{R}\,\mathbf{e}_z \right) + (y - y_0 + Y(x, y))\,\mathbf{e}_y. \qquad (3.9)$$

This equation expresses that the centre line follows an arc of circle (first two terms in the right-hand side) and might be stretched (square root factor), while all sections deform identically, up to a rigid-body rotation. This particular form of displacement results merely from symmetry considerations. It is also valid for finite strain—that is for radii of curvature comparable to the radius h of the cross-section. The aim of the following subsection is to determine the remaining parameters (the strain ϵ_{33}^0 and the functions X and Y) by solving the equations of elasticity.

3.3.2 Calculation of strain

So far, we have built a parameterisation of the rod. It remains to compute the associated strain and stress. Once again we use the translational invariance of the rod and notice that the strain and stress tensor cannot depend explicitly on s: we only need to compute them in the particular section $s = 0$. We could undertake the calculation of the strain without any further assumption, but the calculation is much easier in the particular case of small strain (which is the only case we consider in the end anyway): the strain near $s = 0$ can then calculated using the approximation of small displacement in the frame $(\mathbf{e}_x, \mathbf{e}_y, \mathbf{e}_z)$. Therefore, we assume small strain, which requires that the imposed radius of curvature R remains much larger than the typical radius h of the cross-section:

$$h \ll R. \tag{3.10}$$

The variables x and y parameterise the cross-section and are of magnitude of order h: we have $|x| \ll R$ and $|y| \ll R$. In this approximation:

$$\sqrt{1 + 2\,\epsilon_{33}^0} - 1 \approx \epsilon_{33}^0 = \eta,$$

where we have renamed $\eta = \epsilon_{33}^0$, the stretching rate of the centre line.

The components (u, v, w) of the displacement are found by identification of equation (3.9) with

$$\mathbf{r}(x, y, s) = (x + u)\,\mathbf{e}_x + (y + v)\,\mathbf{e}_y + (z + w)\,\mathbf{e}_z. \tag{3.11}$$

Upon expansion of equation (3.9) for small s, namely near the reference cross-section $s = 0$, this yields:

$$u(x, y, z) = (x_0^{\mathrm{a}} - x_0) + X(x, y) + \mathcal{O}(z^2/R), \tag{3.12a}$$

$$v(x, y, z) = (y_0^{\mathrm{a}} - y_0) + Y(x, y), \tag{3.12b}$$

$$w(x, y, z) = (z_0^{\mathrm{a}}) + z\left(\frac{x - x_0}{R} + \eta\right) + \mathcal{O}(z^3/R^2), \tag{3.12c}$$

where $(x_0^{\mathrm{a}}, y_0^{\mathrm{a}}, z_0^{\mathrm{a}})$ are the coordinates of the point $\mathbf{r}(x_0, y_0, 0)$ in the actual configuration. The constants $(x_0^{\mathrm{a}} - x_0)$, $(y_0^{\mathrm{a}} - y_0)$ and z_0^{a} in the functions u, v and w correspond to a rigid-body translation, and disappear in the subsequent calculation of the strain. Recall that the variables s or z are equivalent in the above equation, $s = z$. This redundancy is because the notation s is traditionally for the curvilinear abscissa in the context of rods, while z denotes the third direction of space, x and y being along the cross-section $s = 0$. Notations such as $\mathcal{O}(z^2/R)$ stand for small terms that are at most of order z^2/R for small z.

We shall assume that the displacement (X, Y), which is expected to be small in the limit of small strain, $R \gg h$, has small gradients with respect to x and y as well. This reasonable assumption can be checked *a posteriori*. Then, all the gradients with respect to x, y and z of u, v and w given in equations (3.12), are small, and the approximation of small displacement applies. The strain can then be computed using the simplified expressions (2.31):

$$\epsilon_{xx}(x, y) = \frac{\partial X}{\partial x}, \qquad \epsilon_{yy}(x, y) = \frac{\partial Y}{\partial y}, \qquad \epsilon_{xy}(x, y) = \frac{1}{2}\left(\frac{\partial X}{\partial y} + \frac{\partial Y}{\partial x}\right),$$

$$\epsilon_{zz}(x, y) = \eta + \frac{x - x_0}{R}, \qquad \epsilon_{xz}(x, y) = 0, \qquad \epsilon_{yz}(x, y) = 0. \tag{3.13}$$

Note that the small terms $\mathcal{O}(z^2/R)$ and $\mathcal{O}(z^3/R^2)$ in equations (3.12) yield contributions that vanish exactly for $s = 0$. These strain components have been computed at $s = 0$ for convenience but are actually independent of s.

Equation (3.13) reduces the problem of a uniformly bent rod to a problem of two dimensional elasticity (determining the unknown two dimensional displacement X and Y in a particular cross-section), in which the bending appears only through the imposed strain component ϵ_{zz}. This strain component is linear with respect to x, something than can easily be interpreted: the effect of bending is to stretch lines parallel to the centre line on the 'outer' side of the rod, $x > (x_0 - \eta R)$, and to compress those on its 'inner' side, $x < (x_0 - \eta R)$, as shown in Fig. 3.5. The coordinate $(x_0 - \eta R)$ will be interpreted as the position of the centre line, defined below.

3.3.3   Solution of the equations of elasticity

It remains to solve the equations of elasticity for the displacement $X(x, y)$ and $Y(x, y)$ in the cross-section and for the unknown stretch rate η of the centre line. The frame $(\mathbf{e}_x, \mathbf{e}_y, \mathbf{e}_z)$ has been set up in such a way that the approximation of small displacement holds near the cross-section $s = 0$. One has to solve the equations of elastic equilibrium $\partial \sigma_{ij}/\partial x_j = 0$ for small displacement in this cross-section, $s = 0$. This solution describes a rod that is in equilibrium in every cross-section, by invariance with respect to s. Let $\mathcal{D}$ be the planar domain such that $(x, y) \in \mathcal{D}$ spans the cross-section. The stress components σ_{ij} must be computed from the strain components given above using Hooke's law, in equation (2.61).

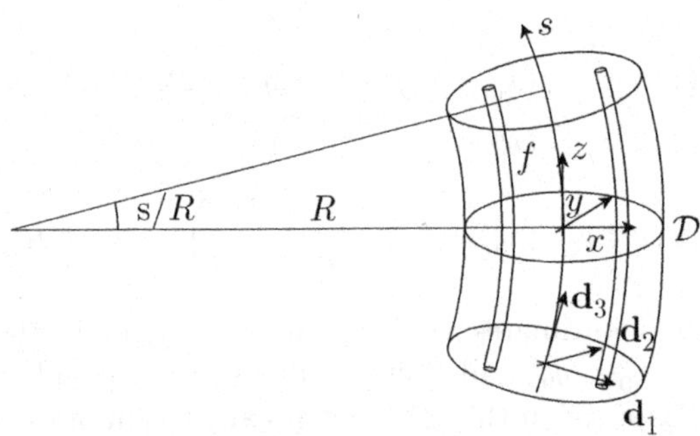

Fig. 3.5 In a rod, bending energy comes from the stretching of filaments in the 'outer' side of the centre line and compression of those in the 'inner' side. Note that all the filaments are drawn for the purpose of the demonstration only.

The boundary conditions to be imposed on the boundary of the cross-section are those for free edges:

$$\sigma_{ij}\, n_j = 0 \qquad \text{on the boundary of the cross section } \mathcal{D}, \tag{3.14}$$

where n_j is a vector perpendicular to the boundary of the cross-section, in the reference configuration. This equation expresses the fact that the contact force transmitted across a surface element lying on the boundary of the rod must vanish, this boundary being free of applied forces.

There is a very simple solution to this set of equations, inspired from the analysis of a bar under simple traction, outlined in Section 2.4. This solution is constructed by assuming that any filament f parallel to the centre line in the reference configuration (see Fig. 3.5) undergoes a simple traction along z. The filament is under tension, and by using our former analysis of the simple traction, equation (2.67):

$$\sigma_{zz} = E\,\epsilon_{zz} = E\left(\eta + \frac{x - x_0}{R}\right), \tag{3.15}$$

while all the other stress components vanish by assumption. In particular, all the components of the form σ_{ix} and σ_{iy} vanish, which make this solution satisfy the boundary condition (3.14), the normal vector being contained in the plane (x, y). The associated strain can be computed by inverting Hooke's relations, which can be seen as a simple linear system between the non-vanishing components of the strain and stress tensors: $\epsilon_{xx} = -\frac{\nu}{E}\,\sigma_{zz}$, $\epsilon_{yy} = -\frac{\nu}{E}\,\sigma_{zz}$ and $\epsilon_{zz} = \frac{1}{E}\,\sigma_{zz}$. This yields

$$\epsilon_{xx} = -\nu\left(\eta + \frac{x - x_0}{R}\right), \qquad \epsilon_{yy} = \epsilon_{xx}, \qquad \epsilon_{zz} = \left(\eta + \frac{x - x_0}{R}\right), \tag{3.16}$$

all the off-diagonal components being zero.

Usually, one runs into two difficulties when trying to guess the strain field directly, as we do here. Finding a stress field that satisfies the equations of equilibrium is not so easy. Here, the only equation of equilibrium that involves non-zero components of the stress is $\partial \sigma_{zz}/\partial z = 0$ and is indeed verified inside $\mathcal{D}$, as σ_{zz} does not depend on z. The second difficulty is that the strain field associated with an arbitrary stress field is not 'compatible' in general, as defined in Section 2.2.9. Here, we are lucky enough that the proposed strain field derives precisely from our particular displacement field.

By identification[12] of equations (3.13) and (3.16), one determines the displacement field in a cross-section:

$$X(x, y) = -\frac{\nu}{2R}\left((x - x_0 + R\,\eta)^2 - (R\,\eta)^2 - (y - y_0)^2\right), \tag{3.17a}$$

$$Y(x, y) = -\frac{\nu}{R}(x - x_0 + R\,\eta)\,(y - y_0). \tag{3.17b}$$

The gradients of the functions $X(x, y)$ and $Y(x, y)$ are small, of order $d/R \ll 1$ at most, in agreement with the approximation of small displacement.

---

[12] Solving for X and Y involves a 2D strain compatibility problem, see equation (3.13). Its solution is unique, up to a rigid body transformation made of a combined rotation and translation. The translation is fixed by noticing that, by equation (3.5), $X(x_0, y_0) = 0$ and $Y(x_0, y_0) = 0$. The rotation is fixed by requiring that $\mathbf{e}_y$ and $\mathbf{d}_2(0)$ are aligned by construction, the latter being given as the gradient of $\mathbf{r}(x, y, 0)$ with respect to y. This yields the constants of integration given in equation (3.17).

3.3.4   Bending energy

In the previous section, we have solved the equations of elasticity for a rod under flexion. In order to compute the elastic energy of this configuration, one can simply use its general definition in equation (2.66). In this equation, the strain is with respect to the natural configuration of the rod. For we shall assume that the reference, cylindrical configuration of the rod is free of stress—we shall explain in due course how to account for rods with a spontaneous curvature. In the formula for the elastic energy, all the terms implied by the summation with respect to i and j disappear but one, corresponding to the only non-vanishing component of the stress:

$$\mathcal{E}_{\text{el}} = \frac{1}{2} \iiint \sigma_{zz}\,\epsilon_{zz}\,\mathrm{d}x\,\mathrm{d}y\,\mathrm{d}z.$$

Using equation (3.13) for ϵ_{zz} and equation (3.15) for σ_{zz}, we derive the bending energy in the form:

$$\mathcal{E}_{\text{b}} = \frac{E\,L}{2} \iint_{\mathcal{D}} \left(\frac{x - x_0}{R} + \eta \right)^2 \mathrm{d}x\,\mathrm{d}y. \tag{3.18}$$

The coefficient η, namely the rate of stretching of the centre line, has not yet been determined. It is such that the elastic energy given above is minimum:

$$\frac{1}{E\,L} \frac{\partial \mathcal{E}_{\text{b}}}{\partial \eta} = 0.$$

Using the above equation for the elastic energy, this leads to:

$$\iint_{\mathcal{D}} \left(\frac{x - x_0}{R} + \eta \right) \mathrm{d}x\,\mathrm{d}y = 0.$$

Noticing that η and x_0 are independent of x and y, this can be rewritten

$$\frac{\langle x \rangle_{\mathcal{D}} - x_0}{R} + \eta = 0, \tag{3.19}$$

where $\langle f(x,y) \rangle_{\mathcal{D}}$ denotes the average of a quantity f over the cross-section. Here, $\langle x \rangle_{\mathcal{D}}$ is the x coordinate of the centre of mass of the cross-section. Recall that the centre line has not yet been completely defined. Let us therefore define it as the material line parallel to the z direction in the reference configuration, which moreover *passes through the centre of mass of each cross-section*. This amounts to set

$$x_0 = \langle x \rangle_{\mathcal{D}} \quad \text{and} \quad y_0 = \langle y \rangle_{\mathcal{D}} \tag{3.20}$$

in the above calculations. More explicitly, the centre line is fixed by

$$x_0 = \frac{1}{S} \iint_{\mathcal{D}} x\,\mathrm{d}x\,\mathrm{d}y \quad \text{and} \quad y_0 = \frac{1}{S} \iint_{\mathcal{D}} y\,\mathrm{d}x\,\mathrm{d}y, \tag{3.21}$$

where S is the area of the cross-section.

The interpretation of equation (3.19) then becomes obvious. The condition of stationarity of the energy with respect to the stretching rate η of the centre line is simply that the latter is zero, $\eta = 0$. *The centre line defined above is inextensible* at equilibrium, as long as the strain remains small. It is remarkable that there exists a single material line that deforms in an inextensible manner for *any* possible type of flexural deformations. This centre line is sometimes called the *neutral line* of the rod. Note that it does not need to be located inside the elastic material itself. For instance, in the case of a hollow cylinder (a pipe) this centre line is the axis of the cylinder, where there is no matter.

The result of the above analysis is that the bending energy per unit length of the rod takes the simple form:

$$\frac{\mathcal{E}_\mathrm{b}}{L} = \frac{E}{2} \iint_\mathcal{D} (\kappa^{(2)} (x - x_0))^2 \, \mathrm{d}x \, \mathrm{d}y, \tag{3.22}$$

where we have identified $1/R$ with $-\kappa^{(2)}$.

The above derivation assumed for the sake of simplicity that the second curvature $\kappa^{(2)}$ only is non-zero. If the first curvature is also non-zero but still constant along the rod, the above formula can be extended to

$$\frac{\mathcal{E}_\mathrm{b}}{L} = \frac{E}{2} \iint_\mathcal{D} (\kappa^{(2)} (x - x_0) - \kappa^{(1)} (y - y_0))^2 \, \mathrm{d}x \, \mathrm{d}y. \tag{3.23}$$

Now, let us consider the following quadratic form that takes two arbitrary numbers a and b as arguments:

$$(a, b) \mapsto \frac{1}{2} \iint_\mathcal{D} (a (x - x_0) - b (y - y_0))^2 \, \mathrm{d}x \, \mathrm{d}y. \tag{3.24}$$

As any quadratic form, it can be diagonalized by a suitable choice of the orthonormal frame in the cross-section, $(\mathbf{d}_1, \mathbf{d}_2)$, that is by a suitable choice of the directions x and y attached to the cross-section. These directions have not been specified so far. The frame that diagonalizes the quadratic form above is such that the cross term vanishes:

$$\iint_\mathcal{D} (x - x_0) (y - y_0) \, \mathrm{d}x \, \mathrm{d}y = 0.$$

The directions $\mathbf{d}_1$ and $\mathbf{d}_2$ defined in this way are called the *principal directions of curvature* in the plane of the cross section, and are orthogonal. In the frame $(\mathbf{d}_1, \mathbf{d}_2)$, the bending energy (3.23) takes the form:

$$\frac{\mathcal{E}_\mathrm{b}}{L} = \frac{E \, I^{(1)}}{2} (\kappa^{(1)})^2 + \frac{E \, I^{(2)}}{2} (\kappa^{(2)})^2, \tag{3.25}$$

where the so-called principal moments of inertia[13] are defined by

$$I^{(1)} = \iint_\mathcal{D} (y - y_0)^2 \, \mathrm{d}x \, \mathrm{d}y, \qquad I^{(2)} = \iint_\mathcal{D} (x - x_0)^2 \, \mathrm{d}x \, \mathrm{d}y. \tag{3.26}$$

In the case of a rectangular cross-section with sides of lengths h_1 along $\mathbf{d}_1$ and h_2 along $\mathbf{d}_2$, the moments of inertia can be computed as:

$$I^{(1)} = \frac{h_1{}^3 h_2}{12}, \qquad I^{(2)} = \frac{h_1 h_2{}^3}{12}, \tag{3.27}$$

while in the case of a circular cross-section of radius r, for example,

$$I^{(1)} = I^{(2)} = \frac{\pi \, r^4}{4}. \tag{3.28}$$

---

[13] These principal moment of inertia have no physical connection with their usual dynamical definition. They are just named this way as they happen to be given by a similar formula to the moment of inertia of a rotating body.

3.3.5   Bending moment

We shall now compute the bending moment $\mathbf{M}(s)$, defined as the moment of the contact forces transmitted across the cross-section with curvilinear abscissa s. More accurately, the so-called internal moment, $\mathbf{M}(s)$, is defined[14] as the moment of the contact forces applied from the part of the rod that is downstream of the cross-section s, on to the part that is upstream of the cross-section s. Here, 'downstream' and 'upstream' refer to the chosen orientation for the centre line, and correspond to curvilinear abscissas q such that $q > s$ and $q < s$ respectively. Consider the upstream half of the rod ($q < s$, located below the cross-section in Fig. 3.5). The outward normal for this domain is $(+\mathbf{d}_3(s))$ on the cross-section s in the actual configuration, and $\mathbf{e}_z$ in the reference configuration. By the definition (2.73) of the stress tensor, the elementary contact force applied by the downstream part is $\mathrm{d}\mathbf{F} = \sigma_{iz}(x,y)\,\mathbf{d}_i(s)\,\mathrm{d}x\,\mathrm{d}y$. The arm of this moment with respect to the intersection of the centre line and the cross-section s, with coordinates (x_0, y_0, s), is the vector joining this intersection to the current point of the cross-section, in actual configuration, that is $(x - x_0)\,\mathbf{d}_1(s) + (y - y_0)\,\mathbf{d}_2(s)$ (we neglect the small variations of the local frame with x and y). The elementary moment is therefore

$$\mathrm{d}\mathbf{M} = [(x - x_0)\,\mathbf{d}_1(s) + (y - y_0)\,\mathbf{d}_2(s)] \times \sigma_{iz}(x,y)\,\mathbf{d}_i(s)\,\mathrm{d}x\,\mathrm{d}y.$$

The stress field is given by equation (3.15). The components σ_{iz} are zero for $i = x$ and $i = y$ and the only non-zero contribution to $\mathrm{d}\mathbf{M}$ comes from $i = z$, for which $\sigma_{zz}(x,y) = E\,(x - x_0)/R$. This expression was derived in the particular case where the rod is bent in the plane perpendicular to $\mathbf{d}_2$, that is for $\boldsymbol{\Omega} = -\mathbf{d}_2/R$. In the general case, that is for uniform bending in an arbitrary direction, $\boldsymbol{\Omega} = \kappa^{(1)}\,\mathbf{d}_1 + \kappa^{(2)}\,\mathbf{d}_2$, the longitudinal stress is given by:

$$\sigma_{zz}(x,y) = -E\,\kappa^{(2)}\,(x - x_0) + E\,\kappa^{(1)}\,(y - y_0).$$

When plugged into the elementary moment, this yields for the full bending moment:

$$\mathbf{M} = E\,I^{(1)}\,\kappa^{(1)}\,\mathbf{d}_1 + E\,I^{(2)}\,\kappa^{(2)}\,\mathbf{d}_2, \tag{3.29}$$

where we have used the definition of the principal moments of inertia $I^{(i)}$. This equation is the constitutive relation for a bent rod: it relates the material curvatures, which characterize the geometry of the bent rod, to the moment of the internal stress that take place in it.

There exists a much quicker—although more abstract—derivation of the constitutive equation (3.29), which runs as follows. Consider a rod of length L, for which s varies in the range $0 \leq s \leq L$, which is clamped at its section $s = 0$, and bent by applying a moment at the section $s = L$. We assume that the loading is such that the rod deforms by uniform bending (the loading should involve only a moment, the resulting force being zero). Integration of the kinematical equations show that the end section $s = L$ is rotated by a rotation vector[15] $(\kappa^{(1)}\,\mathbf{d}_1(0) + \kappa^{(2)}\,\mathbf{d}_2(0))\,L$. Upon an infinitesimal change of the loading, the curvatures $\kappa^{(i)}$ are changed by $\delta\kappa^{(i)}$, which corresponds to an infinitesimal rotation of the end section by the vector $[(\delta\kappa^{(1)})\,\mathbf{d}_1(0) + (\delta\kappa^{(2)})\,\mathbf{d}_2(0)]\,L$. Upon this rotation, the operator, which applies a moment $\mathbf{M}(L)$ to maintain equilibrium, perform the mechanical work

---

[14] By the law of action and reaction, the moment of the contact force applied from the upstream part of the rod on to the downstream one is just the opposite.

[15] The vector defining a rigid body rotation has been introduced in Section 1.3.1. In the present case, this vector is independent of s, as can be shown by deriving with respect to s and using the relations (3.2).

$[\mathbf{M}(L) \cdot ((\delta\kappa^{(1)})\,\mathbf{d}_1(0) + (\delta\kappa^{(2)})\,\mathbf{d}_2(0))\,L]$. By equation (3.25), the elastic energy changes by $L\,(E\,I^{(1)}\,\kappa^{(1)}\,\delta\kappa^{(1)} + E\,I^{(2)}\,\kappa^{(2)}\,\delta\kappa^{(2)})$. By definition of the elastic energy, the work of the operator should equal the change of elastic energy. By identifying the two corresponding expressions, one derives equation (3.29) again.

Classically, the two components of the internal moment $\mathbf{M}$ along the cross-section are called the flexural couples, and are denoted $G^{(1)}$ and $G^{(2)}$:

$$\mathbf{M}(s) = G^{(1)}(s)\,\mathbf{d}_1(s) + G^{(2)}(s)\,\mathbf{d}_2(s).$$

The constitutive relations can be expressed in terms of these flexural couples as

$$G^{(1)} = E\,I^{(1)}\,\kappa^{(1)}, \qquad G^{(2)} = E\,I^{(2)}\,\kappa^{(2)}. \tag{3.30}$$

By solving the equations of three-dimensional elasticity, we have been able to derive the bending energy of a rod in equation (3.25). In the following section, we consider another fundamental type of deformation, involving twist. Later on, we shall extend this calculation to the case of non-uniform deformations, provided the curvatures and twist change over a length scale much larger than h, the typical radius of the cross-section.

3.4   Twist

3.4.1   Kinematics of a twisted rod

In this section, the case of a uniform twist is considered: the curvatures $\kappa^{(1)}$ and $\kappa^{(2)}$ defined in equation (3.1) vanish while the material twist, noted $\hat{\tau}$, is a constant, see Fig. 3.6. In this case, the Darboux vector $\boldsymbol{\Omega}$ is parallel to the tangent $\mathbf{d}_3$ and the material frame attached to the centre line rotates around the tangent. As a result, this tangent $\mathbf{d}_3(s)$ is a constant vector.

As in the case of flexion, the invariance along s imposes that the deformed cross-sections are all isometric. The difference is that the axis of rotation is now parallel to the centre line, as shown in Fig. 3.6, instead of being contained in the plane of the cross-section. We introduce a Cartesian frame $(\mathbf{e}_x, \mathbf{e}_y, \mathbf{e}_z)$ in the deformed configuration, such that $\mathbf{e}_z$ is aligned with the axis of rotation between different cross-sections. We proceed as in the analysis of bending and introduce the parameterisation of the particular cross-section $s = 0$:

$$\mathbf{r}(x, y, 0) = \mathbf{r}(x_0, y_0, 0) + (x - x_0 + X(x, y))\,\mathbf{e}_x + (y - y_0 + Y(x, y))\,\mathbf{e}_y + Z(x, y)\,\mathbf{e}_z,$$

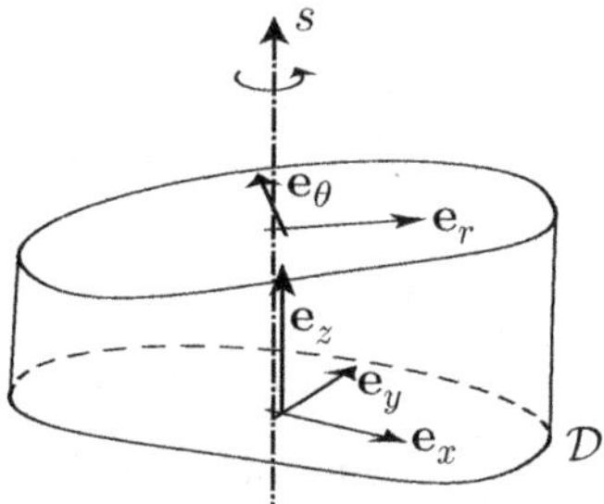

Fig. 3.6 Twist of a rod.

where $X(x, y)$, $Y(x, y)$ and $Z(x, y)$ are the unknown components of the displacement. The invariance by mirror symmetry with respect to a cross-section is no longer present and a transverse displacement is now possible, through the unknown function Z. As a result, the cross-sections do not necessarily remain planar upon twisting, an effect called *warping*.

We are building a solution that is invariant with respect to s. Therefore, the rate of twist $\hat{\tau}$ is uniform along the centre line, and the angle by which the section at $s = 0$ is rotated to give the section at s is a linear function of s, namely $\hat{\tau} s$. As in the case of bending, this known rigid-body rotation between cross-sections allows one to reconstruct any particular cross-section from the functions $X(x, y)$, $Y(x, y)$ and $Z(x, y)$:

$$\mathbf{r}(x, y, z) = \mathbf{r}(x_0, y_0, z) + (x - x_0 + X(x, y))\, \mathbf{e}_r(\hat{\tau}\, z)$$

$$+ (y - y_0 + Y(x, y))\, \mathbf{e}_\theta(\hat{\tau}\, z) + Z(x, y)\, \mathbf{e}_z, \tag{3.31}$$

where we have introduced the polar frame $(\mathbf{e}_r(\theta), \mathbf{e}_\theta(\theta))$ in the plane (x, y), shown in Fig. 3.6. It is obtained by rotating the frame $(\mathbf{e}_x, \mathbf{e}_y)$ by an angle θ:

$$\mathbf{e}_r(\theta) = \mathbf{e}_x \cos\theta + \mathbf{e}_y \sin\theta, \qquad \mathbf{e}_\theta(\theta) = -\mathbf{e}_x \sin\theta + \mathbf{e}_y \cos\theta.$$

Its rate of variation is given by

$$\mathrm{d}\mathbf{e}_r(\theta) = \mathbf{e}_\theta(\theta)\, \mathrm{d}\theta, \quad \mathrm{d}\mathbf{e}_\theta(\theta) = -\mathbf{e}_r(\theta)\, \mathrm{d}\theta.$$

The angle θ is given, for pure and uniform twist $\hat{\tau}$, by $\theta = \hat{\tau}\, z$.

There remains to find the parameterisation of the centre line $(x = x_0, y = y_0)$ in the actual configuration. Note that we use the centre line defined in the analysis of bending, namely the material curve passing through the centre of mass of every cross-section. The material vector $\mathbf{d}_3(s)$ attached to the centre line has been defined earlier by

$$\mathbf{d}_3(s) = \frac{\partial \mathbf{r}(x_0, y_0, s)}{\partial s}.$$

As explained at the beginning of this section, this vector is independent of s. By construction, it is moreover aligned with $\mathbf{e}_z$. Let ϵ_{ss}^0 be the unknown longitudinal strain measured along the centre line, which is independent of s by translational invariance. Knowing its norm and direction, one can write the vector $\mathbf{d}_3$ as

$$\mathbf{d}_3 = \sqrt{1 + 2\,\epsilon_{ss}^0}\; \mathbf{e}_z.$$

The above differential equation for $\mathbf{r}(x_0, y_0, s)$ can then be integrated as

$$\mathbf{r}(x_0, y_0, s) = \mathbf{r}(x_0, y_0, 0) + s\,\sqrt{1 + 2\,\epsilon_{ss}^0}\; \mathbf{e}_z.$$

Combining this equation with equation (3.31), one obtains a parameterisation of the twisted rod:

$$\mathbf{r}(x, y, z) = \mathbf{r}(x_0, y_0, 0) + s\,\sqrt{1 + 2\,\epsilon_{ss}^0}\; \mathbf{e}_z + (x - x_0 + X(x, y))\, \mathbf{e}_r(\hat{\tau}\, s)$$

$$+ (y - y_0 + Y(x, y))\, \mathbf{e}_\theta(\hat{\tau}\, s) + Z(x, y)\, \mathbf{e}_z. \tag{3.32}$$

This equation plays a similar role to equation (3.9) for bending: it provides a general parameterisation for a uniformly twisted rod depending on few parameters and unknown functions, based on general symmetry considerations.

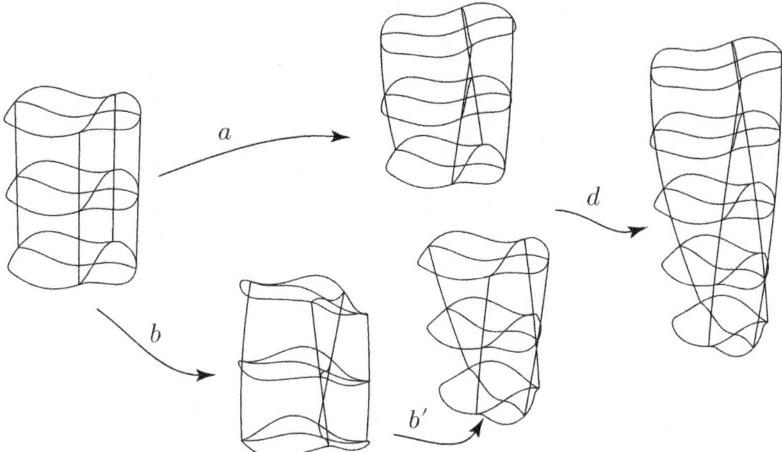

Fig. 3.7 Invariance of a twisted rod by mirror symmetry with respect to a section. It is equivalent to applying (a) a positive twist, or (b) the opposite twist followed by (b'), a mirror symmetry. The two resulting cross-sections are identical, and the resulting deformed rods can be patched (d).

In our analysis of bending, we noted that the rod was invariant by mirror symmetry with respect to any cross-section. This invariance is not present in twisted rods and would be incompatible with warping. It is replaced by a different invariance, which involves changing the sign of the twist $\hat{\tau}$, as explained in Fig. 3.7. Applying a twist, $\hat{\tau}$, or the opposite twist, $-\hat{\tau}$, yields two rod configurations that are not identical: the transverse displacement $Z(x, y)$ has a different sign and the extremal cross-sections cannot be applied one onto the other. However, these two configurations are mirror-symmetric, as illustrated in the figure. The cross-sections corresponding to $\hat{\tau}$ and $-\hat{\tau}$ being mirror symmetric, they are in particular isometric: the strains in the directions tangent to the cross-section, $\epsilon_{xx}(x, y)$, $\epsilon_{xy}(x, y)$ and $\epsilon_{yy}(x, y)$ remain the same functions of x and y when the sign of $\hat{\tau}$ is changed—they are even functions of $\hat{\tau}$. The same holds[16] for the strain component perpendicular to the cross-section, $\epsilon_{ss}(x, y)$.

3.4.2   Case of a weakly twisted rod

The parameterisation of a twisted rod derived above is valid for finite strain. We now consider the case of small strain, for which important simplifications can be made. First, anticipating that the strain along the centre line ϵ^0_{ss} is small in this limit, the parameter $\hat{\tau}$ can be identified with the physical twist—a similar identification was possible between $\hat{R}$ and R in the analysis of bending in the same limit.

Let h be the typical radius of the cross-section: the strain is of order $|\tau|\,h$, as we shall show. We are now considering the limit $|\tau|\,h \ll 1$. The displacements $X(x, y)$, $Y(x, y)$ and $Z(x, y)$ are vanishing in the reference configuration ($\tau = 0$), and so must be small when the applied twist rate is small: one is interested in the main contribution to these displacements, expected to be of order τ at least for small τ.

---

[16] The shear components $\epsilon_{xz}(x, y)$ and $\epsilon_{xz}(x, y)$ have their sign reversed upon a change of sign of the twist.

As explained above, some strain components are even functions of τ. Their expansion with respect to τ yields a vanishing linear term: by symmetry we have

$$\epsilon_{xx}(x,y) = 0, \quad \epsilon_{xy}(x,y) = 0, \quad \epsilon_{yy}(x,y) = 0, \quad \epsilon_{ss}^0 = 0.$$

Note that in the limit of small strain, the material vectors $\mathbf{d}_1(s) = \mathbf{r}_{,x}(x_0, y_0, s)$ and $\mathbf{d}_2(s) = \mathbf{r}_{,y}(x_0, y_0, s)$ define an almost orthonormal frame for any s, and are both almost orthonormal to $\mathbf{e}_z \approx \mathbf{d}_3(s)$: choosing the vectors $\mathbf{e}_x$ and $\mathbf{e}_y$ to be close to $\mathbf{d}_1(0)$ and $\mathbf{d}_2(0)$, respectively, allows one to make rotations small, and so to use the approximation of small displacement near the section $s = 0$. Inserting the definition of linearized in-plane strain into equation above, we have

$$X_{,x}(x,y) = 0, \quad \frac{X_{,y}(x,y) + Y_{,x}(x,y)}{2} = 0, \quad Y_{,y}(x,y) = 0.$$

The displacement field (X, Y) characterizes the transformation of the cross-section *projected in the plane* (x, y). By equation above, this displacement is associated with zero strain: it is a two-dimensional rigid-body transformation, i.e. a translation combined with a rotation. The translation mode can be discarded by noticing that $X(x_0, y_0) = 0$ and $Y(x_0, y_0) = 0$ by equation (3.31). The rotation can be discarded by our choice of $\mathbf{e}_x$ and $\mathbf{e}_y$, which coincide with $\mathbf{d}_1(0)$ and $\mathbf{d}_2(0)$ respectively. These constraints imply that the rigid-body transformation described by the cross-sectional displacement X and Y is null: we have

$$X(x,y) = 0 \quad \text{and} \quad Y(x,y) = 0 \tag{3.33}$$

at order in τ as a result of the symmetries of the problem.

The parameterisation given in equation (3.32) can then be simplified:

$$\mathbf{r}(x,y,z) = \mathbf{r}(x_0, y_0, 0) + z\,\mathbf{e}_z + (x - x_0)\,\mathbf{e}_r(\tau z) + (y - y_0)\,\mathbf{e}_\theta(\tau z) + Z(x,y)\,\mathbf{e}_z. \tag{3.34}$$

In the following section, it is used to compute the strain in a twisted rod, and to solve the equations of elasticity for the unknown function $Z(x,y)$.

3.4.3 Calculation of strain

By combining these equations as for the case of bending, one derives the strain in any cross-section:

$$\epsilon_{xz} = \frac{1}{2}\left(Z_{,x} - \tau\,(y - y_0)\right), \qquad \epsilon_{yz} = \frac{1}{2}\left(Z_{,y} + \tau\,(x - x_0)\right), \tag{3.35}$$

while all other components vanish.[17] In equation (3.35), the notations $Z_{,x}$ and $Z_{,y}$ are for partial derivatives, as usual. Using Hooke's relations (2.60) expressed in terms of Lamé coefficients,[18] one finds that two components of the stress tensor only, σ_{xz} and σ_{yz}, are non-zero:

$$\sigma_{xz} = 2\,\mu\,\epsilon_{xz}, \quad \sigma_{yz} = 2\,\mu\,\epsilon_{yz}, \tag{3.36}$$

---

[17] The twist has almost disappeared from the explicit writing of the displacement in equation (3.34). But it remains present in the dependence of the base vectors $\mathbf{e}_r(\tau z)$ and $\mathbf{e}_\theta(\tau z)$, whence the τ dependence in the expressions of the strain.

[18] We are here assuming implicitly that the reference, cylindrical configuration is free of stress. We shall explain later how to account for rods with a non-zero natural twist.

where $\mu = E/2(1+\nu)$, as defined in equation (2.62), is the shear modulus of the material. The unknown transverse displacement Z must be such that the equation of equilibrium $\sigma_{xz,x} + \sigma_{yz,y} = 0$ holds. Using the constitutive relations (3.36), this yields:

$$\epsilon_{xz,x} + \epsilon_{yz,y} = 0. \tag{3.37}$$

3.4.4   Solution of the equations of elasticity

In contrast with the analysis of bending, no simple trick saves one from solving the equations of elasticity. The rest of this section is devoted to the solution of the equations of equilibrium.

We note that equation (3.37) is a partial differential equation in the (x, y) plane, which implies that the vector field $(-\epsilon_{yz}(x, y)\, \mathbf{e}_x + \epsilon_{xz}(x, y)\, \mathbf{e}_y)$ has a zero curl. Therefore, this vector field derives[19] from a potential $\chi(x, y)$, such that:

$$\epsilon_{xz}(x, y) = \frac{\partial \chi(x, y)}{\partial y}, \qquad \epsilon_{yz}(x, y) = -\frac{\partial \chi(x, y)}{\partial x}. \tag{3.38}$$

This kind of potential functions will again be used later in the Föppl–von Kármán equations for plates. The equation (3.37) becomes the condition of equality of the cross derivatives of the potential χ: $\chi_{,xy}(x, y) = \chi_{,yx}(x, y)$. By defining a rescaled potential, $\bar{\chi} = \chi/\tau$, and by eliminating Z from equations (3.35) and (3.38), one derives the following equation for the potential $\bar{\chi}$:

$$\Delta\bar{\chi}(x, y) = -1. \tag{3.39a}$$

Like the equation of equilibrium, this equation has to be satisfied in the interior of a cross-section, a domain denoted $\mathcal{D}$. The harmonic operator is the two-dimensional one, $\Delta = \partial^2/\partial x^2 + \partial^2/\partial y^2$.

The boundary condition (3.14) holds along the boundary of the cross-section $\mathcal{D}$. Using the constitutive relation, it can be rewritten as $\epsilon_{xz}\, n_x + \epsilon_{yz}\, n_y = 0$, where (n_x, n_y) is the normal to the curve bounding the domain $\mathcal{D}$. In terms of the potential, this yields $\bar{\chi}_{,y}\, n_x - \bar{\chi}_{,x}\, n_y = 0$. Introducing the tangent $(t_x, t_y) = (-n_y, n_x)$ to the boundary, we rewrite this as $\bar{\chi}_{,x}\, t_x + \bar{\chi}_{,y}\, t_y = 0$. The latter expression is the derivative of $\bar{\chi}$ along the curve bounding the domain $\mathcal{D}$. It must be vanishing, which means that $\bar{\chi}(x, y)$ has to be a constant function along this boundary. The constant value of $\bar{\chi}$ along the boundary can be arbitrarily set to zero, given that only the gradient of $\bar{\chi}$ is physical:

$$\bar{\chi} = 0 \quad \text{along the boundary of the cross section } \mathcal{D}. \tag{3.39b}$$

This potential $\bar{\chi}$ is the solution[20] of the Poisson equation (3.39a) inside $\mathcal{D}$ with the boundary condition (3.39b). These equations depend only on the geometry of the cross-section. As a result, the potential $\bar{\chi}$ is itself a function of the geometry of the cross-section only. It allows one to compute the twist rigidity of the rod, as shown in the next section.

In the case of a circular section with radius h, for instance, the Poisson equation has the following solution:

$$\bar{\chi}(x, y) = \frac{h^2 - (x^2 + y^2)}{4}.$$

---

[19] We assume for sake of simplicity that the cross-section is simply connected.

[20] Solving the Poisson equation in the plane with a so-called Dirichlet boundary condition is a well-posed mathematical problem, with a unique solution. This means that our parameterisation of a twisted bar was general enough to include the actual solution of the full three-dimensional problem.

3.4.5   Energy of twist

The energy per unit length of a twisted bar can be expressed by once again using the general formula for the elastic energy of a Hookean material:

$$\frac{\mathcal{E}_t}{L} = \frac{1}{2} \iint_{\mathcal{D}} \sigma_{ij}\,\epsilon_{ij}\,\mathrm{d}x\,\mathrm{d}y = \frac{1}{2} \iint_{\mathcal{D}} (2\,\sigma_{xz}\,\epsilon_{xz} + 2\,\sigma_{yz}\,\epsilon_{yz})\,\mathrm{d}x\,\mathrm{d}y$$

$$= \frac{1}{2} \iint_{\mathcal{D}} 4\mu\,(\epsilon_{xz}{}^2 + \epsilon_{yz}{}^2)\,\mathrm{d}x\,\mathrm{d}y = 2\,\mu\,\tau^2 \iint_{\mathcal{D}} (\bar{\chi}_{,x}^2 + \bar{\chi}_{,y}^2)\,\mathrm{d}x\,\mathrm{d}y.$$

We put the above expression in a form similar to the bending energy (3.25):

$$\frac{\mathcal{E}_t}{L} = \frac{\mu\,J}{2}\,\tau^2, \tag{3.40}$$

where the moment of twist, J, is a geometrical coefficient defined by:

$$J = 4 \iint_{\mathcal{D}} (\bar{\chi}_{,x}^2 + \bar{\chi}_{,y}^2)\,\mathrm{d}x\,\mathrm{d}y, \tag{3.41}$$

which, like the moments of inertia $I^{(i)}$ defined in the analysis of bending, depends only on the geometry of the cross-section.

  With equation (3.40), we have almost completed the analysis of twist: this equation gives the elastic energy of any twisted configuration. This energy depends only on the value of the material twist, defined along the centre line. Any dependence on the cross-sectional variables x and y is captured in the coefficient J.

  For instance, a rod with a circular cross-section with radius h has the twist rigidity:

$$J = \frac{\pi}{2}\,h^4.$$

This formula can be obtained by carrying out the double integral in the definition of J using polar coordinates, with the expression for $\bar{\chi}(x, y)$ for a circular cross-section, given earlier.

  According to this analysis, the centre line, with equation $(x = x_0, y = y_0)$, remains straight in actual configuration, for any shape of the cross-section. All other material lines that were parallel to it in the reference configuration are deformed into helices that wind around the centre line: the centre line seems to deform in a very peculiar manner when the bar is twisted. This leads to an apparent contradiction when one recalls that the choice of the centre line is arbitrary: in the above analysis of twist, x_0 and y_0 can be any numbers. As a result, a different twist transformation, denoted $\mathbf{r}^{\dagger}(x, y, z)$, could be introduced based on a different choice of centre line, that is by repeating the above construction with different values of x_0 and y_0. Then, the material curves that are parallel to the axis in the reference configuration wind around different centre lines for each of these two solutions, $\mathbf{r}^{\dagger}(x, y, z)$ and $\mathbf{r}(x, y, z)$. Now, the two solutions involve the same strain, as the strain tensor, defined by equations (3.38) and by the solution of the geometrical Dirichlet problem (3.39), is independent of the choice of x_0 and y_0; the two solutions must therefore be identical, up to a rigid body translation and rotation. This paradox is resolved by noticing that the two solutions $\mathbf{r}$ and $\mathbf{r}^{\dagger}$ are isometric, but *only up to terms that are first order* in the small parameter $(\tau\,h)$. One cannot claim that the twisted rod winds around its centre line, with equation $x = x_0$ and $y = y_0$, based on the above analysis: this property is not robust upon the addition of next-order terms.

3.4.6 Moment of twist

The internal moment $\mathbf{M}(s)$ has been calculated in Section 3.3.5 for a uniformly bent rod. To address the case of uniform twist, we can use a similar argument. Consider a twisted rod of length L with uniform twist τ, whose lower end $s = 0$ is clamped. The other end $s = L$ is rotated by a vector $\mathbf{d}_3(0)\,\tau\,L$, that is by an angle $\tau\,L$ around the direction parallel to the axis of the rod $\mathbf{d}_3$. Upon a small perturbation $\delta\tau$, this section undergoes an infinitesimal rotation by an angle $(\delta\tau)\,L\,\mathbf{d}_3$. Meanwhile, the elastic energy varies by $(\mu\,J\,\tau\,\delta\tau)$, and the external operator performs the work $\mathbf{M}\cdot(\delta\tau)\,L\,\mathbf{d}_3$. By identification, we obtain the constitutive relation for a twisted rod:

$$\mathbf{M}(s) = \mu\,J\,\tau\,\mathbf{d}_3(s).$$

Traditionally, the component of the internal moment $\mathbf{M}$ along the axis $\mathbf{d}_3$ of the rod is called the torsional couple, and is denoted $H(s)$:

$$\mathbf{M}(s) = H(s)\,\mathbf{d}_3(s). \tag{3.42}$$

The constitutive relation for a twisted rod can then be written as:

$$H(s) = \mu\,J\,\tau(s), \tag{3.43}$$

which states the proportionality between the twist rate τ and the moment of twist H.

3.5 Energy

The aim of this section is to go beyond the simple configurations studied above: we study configurations of the rod for which both curvatures $\kappa^{(i)}$ and twist τ may be non-zero. Furthermore, we allow these quantities to vary along the rod. We shall nevertheless remain within the approximation of small strain.

3.5.1 Superposition of bending and twist

In a first step, the previous analyses of pure uniform bending *or* pure uniform deformations can be extended to the case of deformations involving both flexion and twist. The material curvatures $\kappa^{(i)}$ and twist τ are fixed to constant values[21] for the moment.

With the aim of representing such deformations, we consider the strain tensor obtained by linear superposition of the strain fields constructed in Sections 3.3 and 3.4, with the relevant values $\kappa^{(1)}$, $\kappa^{(2)}$ and τ, respectively. We shall show that this superposition derives from an actual displacement field (geometric compatibility problem) and describes an equilibrium configuration: it is the relevant solution to the equations of elasticity. We shall also compute the associated internal moment, and the elastic energy.

This strain is geometrically compatible if the approximation of small displacement holds. Indeed equation (2.39) derived in Section 2.2.12 is linear. The elementary strain fields associated with pure bending and with twist deformations are compatible since they derive from an explicit displacement field. As a result, the strain field obtained by linear

---

[21] This particular case of constant material curvatures and twist yields a centre line that is an arc of a helix, as shown later on in Section 5.3.1.

superposition is geometrically compatible as well,[22] and it is consistent to define the rod configuration by this strain field.

Near an arbitrary section, we use the 'adapted' system of coordinates introduced earlier in the analysis of bending and twist deformations, see equation (3.4). In this frame, the approximation of small displacement holds. As a result, the equations of equilibrium are linear with respect to the stress. For Hookean materials the stress tensor resulting from the superposition of the elementary strains is itself the linear superposition of the elementary stress tensors, which all satisfy the equations of equilibrium. This means that the stress field associated with the proposed configuration satisfies the equations of equilibrium.

The internal moment $\mathbf{M}(s)$ has been calculated in Section 3.3.5 as an integral of this stress field over a cross-section (a similar calculation can be made for the internal moment due to twist). Therefore, the internal moment is also a linear superposition of the internal moments due to elementary modes of deformation. From the constitutive equations for a bent rod (3.29) and for a twisted rod (3.42), we derive

$$\mathbf{M}(s) = G^{(1)}(s)\,\mathbf{d}_1(s) + G^{(2)}(s)\,\mathbf{d}_2(s) + H(s)\,\mathbf{d}_3(s), \tag{3.44}$$

where the flexural and torsional couples, defined as the components of $\mathbf{M}(s)$ in the local material frame, are given by

$$G^{(1)}(s) = EI^{(1)}\,\kappa^{(1)}(s) \qquad G^{(2)}(s) = EI^{(2)}\,\kappa^{(2)}(s) \qquad H(s) = \mu J\,\tau(s). \tag{3.45}$$

These relations between the flexural and twist couples on the one hand and the material curvatures and twist on the other are called the constitutive relations for the rod. They are the translation at large scale of the constitutive relation of the material, as revealed by the prefactors E (Young's modulus) and μ (shear modulus). That they are linear is the consequence of the fact that we considered a Hookean material.[23]

Let us now examine the elastic energy of the rod, given by equation (2.66) for a Hookean material with arbitrary geometry. The non-vanishing components of the strain and stress tensors split into σ_{zz} and ϵ_{zz}, on the one hand coming from the purely flexural solution, and ϵ_{xz}, ϵ_{yz} and ϵ_{xz}, σ_{yz} on the other, coming from the purely twisted solution. Dropping terms that are zero, this yields, for the elastic energy:

$$\frac{\mathcal{E}_{\mathrm{rod}}}{L} = \frac{1}{2} \iint_{\mathcal{D}} \left([2\,\sigma_{xz}\,\epsilon_{xz} + 2\,\sigma_{yz}\,\epsilon_{yz}] + [\sigma_{zz}\,\epsilon_{zz}] \right)\,\mathrm{d}x\,\mathrm{d}y,$$

where the first group of terms in square brackets refer to the displacements fields determined above in our analysis of bending (Section 3.3), and the last term to twist (Section 3.4). As a result, there is no cross-term that is proportional, e.g., to $(\tau\,\kappa^{(1)})$ in the energy, something that could have been derived directly by symmetry consideration.[24] In the absence of

---

[22] This argument shows that the compatibility holds under the approximation of small displacement. If displacements are not small but strain nevertheless remains small (for a slender rod twisted by a finite angle for instance), the compatibility can be established by using the 'adapted' coordinate system defined next, in which the approximation of small displacement holds.

[23] The present analysis can be extended to rods made of a non-Hookean or non-isotropic material by modifying the effective constitutive relations of the rod, or by using a different, possibly non-quadratic, energy in equation (3.46).

[24] In the elastic energy we shall derive, there is no direct coupling term between bending and twist— apart from the indirect coupling due to geometrical effects. This holds true in general for rods without intrinsic helicity, that is rods that are invariant in their natural state upon mirror symmetry: the latter changes the sign of the twist but not that of the curvatures. A well-known counter-example of such a coupling is the DNA double helix that, on large scale and in its extended form, can be modelled by an

coupling between flexion and twist, the elastic energy of the rod is expressed simply as a sum of the bending and twist contributions:

$$\frac{\mathcal{E}_{\text{rod}}}{L} = \frac{E\,I^{(1)}}{2}\,(\kappa^{(1)})^2 + \frac{E\,I^{(2)}}{2}\,(\kappa^{(2)})^2 + \frac{\mu\,J}{2}\,\tau^2, \tag{3.46}$$

when $\kappa^{(1)}$, $\kappa^{(2)}$ and τ are independent of s for the moment. The first two terms in the right-hand side are the bending terms given earlier in equation (3.25) and the last term comes from equation (3.40).

3.5.2 Non-uniform bending and twist

It remains to extend this calculation of the rod energy to arbitrary configurations, such that $\kappa^{(1)}$, $\kappa^{(2)}$ and τ depend on s. As shown by equation (3.13), the flexural strain is of order $\kappa^{(i)}\,h$, and the torsional strain of order $\tau\,h$—see equation (3.35)—where h is the typical radius of the cross-section. We shall consider only the case of small strain, that is when $\kappa^{(1)}$, $\kappa^{(2)}$ and τ remain much smaller than $1/h$: the theory of rods[25] is concerned with loadings such that the centre line changes over a 'large scale', of order $L \gg h$. For slowly varying material curvatures and twists, we consider the three-dimensional displacement field constructed by using *locally* the homogeneous solution derived above with the local values of the curvatures and twist. This yields a configuration that does not exactly satisfy the equations of equilibrium, but is a good approximation[26] of the actual solution when these curvatures and twist vary *slowly* with s. Under these assumptions, the energy of the rod can be written as:

$$\mathcal{E}_{\text{rod}} = \int \mathrm{d}s \left(\frac{E\,I^{(1)}}{2}\,(\kappa^{(1)}(s))^2 + \frac{E\,I^{(2)}}{2}\,(\kappa^{(2)}(s))^2 + \frac{\mu\,J}{2}\,(\tau(s))^2 \right). \tag{3.47}$$

Corrections to this energy account for the non-vanishing derivatives of $\kappa^{(i)}(s)$ and $\tau(s)$, and yield terms that are negligible since they come with an additional factor which is a power of $h/L \ll 1$, where h is the typical radius of the cross-section and L is the typical distance of variation of $\kappa^{(i)}(s)$ and $\tau(s)$ along the centre line.

Owing to the separation of scales in an elastic rod, the three-dimensional equations of elasticity can be solved in two separate steps: at the scale h of the cross-section first, as we did in our analysis of bending and twist, and at the scale L relevant for the longitudinal direction, as we do next. Equation (3.47) indeed yields the energy of an elastic rod: it is based on a 'partial' solution of the equations of elasticity, along the planes of the cross-sections only; in the next section, we shall solve for the equilibrium in the remaining axial direction, and derive the equations of equilibrium. This is done by variation of the energy derived above.

elastic rod (S. Neukirch, 2004) with some built-in helicity. Contrary to the equations for cylindrical rods, derived below, the equilibrium equations for a helical rod include in general a cross-term proportional to the product of a curvature by the twist (T. Lionnet *et al.*, 2006).

[25] If the material curvatures or twist vary over length scales comparable to the thickness, the present approach, based on the separation of scales, is not applicable and one has to revert to the full equations of 3D elasticity.

[26] In mathematical terms, the convergence of the trial solution constructed by using locally the homogeneous solution, towards the actual solution of the problem can be established in different frameworks of asymptotic analysis, see for instance (J.-J. Marigo and N. Meunier, 2006; A. Cimetière *et al.*, 1988; R. Jamal and É. Sanchez Palencia, 1996; J. Sanchez-Hubert and É. Sanchez Palencia, 1999; M.-G. Mora and S. Müller, 2002).

3.6   Equilibrium: Kirchhoff equations

The elastic energy derived in equation (3.47) allows one to consider an elastic rod as a one-dimensional elastic object. There are essentially two possible approaches to derive the equilibrium equations for a rod, the so-called Kirchhoff equations. The first one, followed here, is to write the condition of stationarity of the energy. Another possible approach is to write directly a balance of forces and torque over an element of the rod. While the second approach deals with quantities closer to the physics of the problem, its drawback is that it forces one to introduce some essential quantities (like the internal force $\mathbf{F}$ in the rod) in a somewhat arbitrary manner. In contrast, the derivation of the equations for mechanical equilibrium by variation of the energy naturally leads to such quantities, and also helps to clarify the question of boundary conditions. Moreover, this more formal derivation emphasizes that the complexity of the Kirchhoff equations is fundamentally of geometric origin. In any case, preferring one approach over the other one is ultimately a matter of personal taste, and the reader interested in the 'physical' derivation of the Kirchhoff equations can freely jump ahead to Section 3.6.7.

3.6.1   Variation of the energy

We proceed to derive the equations of equilibrium of a rod by variation of the elastic energy, thereby following a classical approach—see for instance references (J.-F. Bourgat, P. Le Tallec, and S. Mani, 1988; P. Le Tallec, S. Mani, and F. Alves Rochinha, 1992; D. J. Steigmann and M. G. Faulkner, 1993; J. Langer and D. A. Singer, 1996). Our presentation is inspired by that of Chouaieb (N. Chouaïeb, 2003), and emphasizes the geometric origin of the coupling between bending and twist, arising from the fundamental geometric identity (3.51) characterizing rotations in the 3D Euclidean space. An even clearer geometrical picture of the equations for rods can be obtained by using the equivalent notions of a Bishop frame or parallel transport (R. Bishop, 1975; J. Langer and D. A. Singer, 1996; R. E. Goldstein, T. R. Powers, and C. H. Wiggins, 1998; M. Bergou *et al.*, 2008) but this is beyond the scope of this book.

The variational approach is based on the elastic energy in equation (3.47). As usual, we consider a perturbation on top of an arbitrary configuration, the perturbed quantities being denoted by the symbol δ. The material curvatures and twist are affected by a perturbation of the rod, and the corresponding variation of the energy $\mathcal{E}_{\mathrm{rod}}$ is:

$$\delta\mathcal{E}_{\mathrm{rod}} = \int \mathrm{d}s \left[\left(EI^{(1)}\,\kappa^{(1)} \right) \delta\kappa^{(1)} + \left(EI^{(2)}\,\kappa^{(1)} \right) \delta\kappa^{(2)} + \left(\mu J\,\tau \right) \delta\tau \right].$$

The factors in parentheses are the flexural and torsional couples introduced in equation (3.44). This allows one to rewrite the variation of the elastic energy as

$$\delta\mathcal{E}_{\mathrm{rod}} = \int \mathbf{M}(s) \cdot \left(\mathbf{d}_1(s)\,\delta\kappa^{(1)} + \mathbf{d}_2(s)\,\delta\kappa^{(2)} + \mathbf{d}_3(s)\,\delta\tau \right) \mathrm{d}s. \tag{3.48}$$

This variation does not yield the equations of equilibrium directly. Indeed, a local perturbation of $\kappa^{(1)}$, $\kappa^{(2)}$ or τ yields a perturbation to the centre line that is not local, as illustrated Fig. 3.8. In contrast, the condition of stationarity for the energy in equation (3.48) only

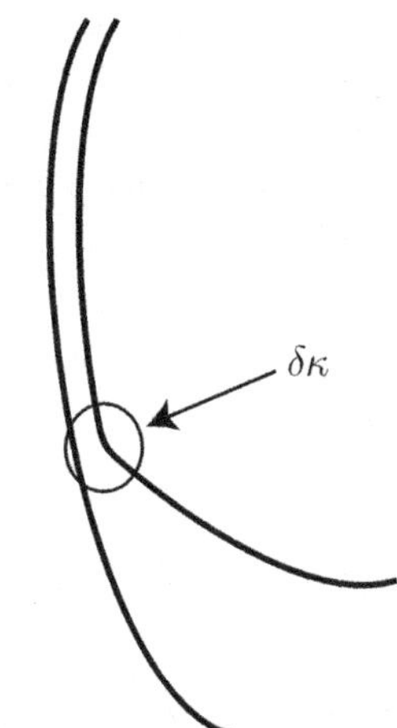

Fig. 3.8 A local perturbation of the curvatures $\kappa^{(i)}$ (or twist τ) yields a *non-local* perturbation to the centre line and to the attached material frame.

holds[27] with respect to *localized* perturbations of the displacement field. Some care must therefore be taken to derive the equations of equilibrium by variation, using the variables $\kappa^{(1)}$, $\kappa^{(2)}$ and τ. Equation (3.48) must first be rewritten in terms of quantities which remain locally perturbed when the rod undergoes a similarly localized perturbation in space.

To this end, we introduce perturbations $\delta\phi(s)$ to the orientation of the material frame, defined below. Recall that the material frame has to remain orthonormal: any change in its orientation is locally characterized by three degrees of freedom $\delta\phi_1$, $\delta\phi_2$, $\delta\phi_3$, each being a small angle associated with a small rotation about $\mathbf{d}_1$, $\mathbf{d}_2$ or $\mathbf{d}_3$ respectively:

$$\delta\mathbf{d}_1 = \delta\phi_3\,\mathbf{d}_2 - \delta\phi_2\,\mathbf{d}_3 \tag{3.49a}$$

$$\delta\mathbf{d}_2 = -\delta\phi_3\,\mathbf{d}_1 + \delta\phi_1\,\mathbf{d}_3 \tag{3.49b}$$

$$\delta\mathbf{d}_3 = \delta\phi_2\,\mathbf{d}_1 - \delta\phi_1\,\mathbf{d}_2. \tag{3.49c}$$

In vector form, this becomes:

$$\delta\mathbf{d}_1 = \delta\phi \times \mathbf{d}_1 \qquad \delta\mathbf{d}_2 = \delta\phi \times \mathbf{d}_2 \qquad \delta\mathbf{d}_3 = \delta\phi \times \mathbf{d}_3, \tag{3.50}$$

where the infinitesimal rotation vector $\delta\phi(s)$ was defined in Section 1.3.1 and reads:

$$\delta\phi(s) = \delta\phi_1(s)\,\mathbf{d}_1(s) + \delta\phi_2(s)\,\mathbf{d}_2(s) + \delta\phi_3(s)\,\mathbf{d}_3(s)$$

in the present case. Equation (3.49) yields the most general infinitesimal perturbation that conserves the scalar products between the vectors $\mathbf{d}_i$, to linear order in the perturbation. The small angles of rotation $\delta\phi_i$ are related to the changes of material curvatures and twist.

We shall now express the variation of the elastic energy in equation (3.48) in terms of the quantities $\delta\phi_i$'s. To do so, one needs to compute the perturbations to the curvatures $\kappa^{(1)}$, $\kappa^{(2)}$ and twist τ caused by the perturbations to the orientation of the material frame, $\delta\phi_i$. The

---

[27] A difficulty appears when one tries to compute the work of the external force using the variables $\kappa^{(1)}$, $\kappa^{(2)}$ and τ: the perturbation being non-local in the physical space, the work of the external forces cannot be written in a straightforward manner.

trick is to notice that the perturbation to the derivatives of the material frames, $\delta(\mathbf{d}_i')$, should be equal with the derivatives of its perturbations, $(\delta\mathbf{d}_i)'$: by definition, $\delta(\mathbf{d}_i') = (\mathbf{d}_i + \delta\mathbf{d}_i)' - (\mathbf{d}_i)' = (\delta\mathbf{d}_i)'$. This equation $\delta(\mathbf{d}_i') = (\delta\mathbf{d}_i)'$ leads to the following compatibility condition between the Darboux vector $\mathbf{\Omega}$ and the infinitesimal material rotation $\delta\boldsymbol{\phi}$:

$$\delta(\mathbf{\Omega} \times \mathbf{d}_i) = \frac{\mathrm{d}(\delta\boldsymbol{\phi} \times \mathbf{d}_i)}{\mathrm{d}s}.$$

After expansion of the δ in the left-hand side and of the derivative in the right-hand side, and after using equations (3.2) and (3.50) several times, this yields:

$$(\delta\mathbf{\Omega}) \times \mathbf{d}_i + \mathbf{\Omega} \times ((\delta\boldsymbol{\phi}) \times \mathbf{d}_i) = (\delta\boldsymbol{\phi})' \times \mathbf{d}_i + (\delta\boldsymbol{\phi}) \times (\mathbf{\Omega} \times \mathbf{d}_i).$$

Using the Jacobi identity to collect the two terms involving double cross products, we are led to:

$$((\delta\mathbf{\Omega}) - (\delta\boldsymbol{\phi})' + \mathbf{\Omega} \times \delta\boldsymbol{\phi}) \times \mathbf{d}_i = \mathbf{0}.$$

This equation must be valid for any vector $\mathbf{d}_i$ of the material frame, which implies the equality between vectors:

$$(\delta\mathbf{\Omega}) - (\delta\boldsymbol{\phi}) \times \mathbf{\Omega} = (\delta\boldsymbol{\phi})'. \tag{3.51}$$

This geometric identity expresses the compatibility of infinitesimal changes of a rotation, this change being obtained either by increasing the arc length ($\mathbf{\Omega}$) or the perturbation parameter ($\delta\boldsymbol{\phi}$); this identity reveals the geometrical structure of the group of rotations in the 3D Euclidean space.

The vector in the left-hand side can be expressed using the decomposition (3.3) of the Darboux vector $\mathbf{\Omega}$:

$$\begin{aligned}
(\delta\mathbf{\Omega}) - (\delta\boldsymbol{\phi}) \times \mathbf{\Omega} &= \delta(\kappa^{(1)}\,\mathbf{d_1} + \dots) - (\delta\boldsymbol{\phi}) \times (\kappa^{(1)}\,\mathbf{d_1} + \cdots) \\
&= ((\delta\kappa^{(1)})\,\mathbf{d_1} + \kappa^{(1)}\,(\delta\boldsymbol{\phi}) \times \mathbf{d_1} + \cdots) - (\delta\boldsymbol{\phi}) \times (\kappa^{(1)}\,\mathbf{d_1} + \cdots) \\
&= (\delta\kappa^{(1)})\,\mathbf{d_1} + \cdots = (\delta\kappa^{(1)})\,\mathbf{d_1} + (\delta\kappa^{(2)})\,\mathbf{d_2} + (\delta\tau)\,\mathbf{d_3}.
\end{aligned}$$

In this calculation, the ellipses are for contributions in the directions $\mathbf{d}_2$ and $\mathbf{d}_3$ that are similar to the contributions along $\mathbf{d}_1$ written explicitly. The right-hand side of equation (3.51) is identical to the first factor in the integrand of the variation of the elastic energy given in equation (3.48). Using the identity (3.51), the latter can be rewritten as:

$$\delta\mathcal{E}_{\mathrm{rod}} = \int \mathbf{M}(s) \cdot \frac{\mathrm{d}\,\delta\boldsymbol{\phi}}{\mathrm{d}s} \cdot \mathrm{d}s.$$

Upon integration by parts, we finally obtain the variation of elastic energy in a compact form:

$$\delta\mathcal{E}_{\mathrm{rod}} = [\mathbf{M}(s) \cdot \delta\boldsymbol{\phi}(s)] - \int \mathbf{M}'(s) \cdot \delta\boldsymbol{\phi}(s)\,\mathrm{d}s, \tag{3.52}$$

where the bracket stands for the variation of its argument over the entire interval and is the boundary term from the integration by parts:

$$[\mathbf{M}(s) \cdot \delta\boldsymbol{\phi}(s)] = \mathbf{M}(L) \cdot \delta\boldsymbol{\phi}(L) - \mathbf{M}(0) \cdot \delta\boldsymbol{\phi}(0),$$

assuming that the rod extends from $s = 0$ to $s = L$.

The interpretation of equation (3.52) is as follows. With our downstream/upstream conventions for defining $\mathbf{M}(s)$, the moment applied by the operator on the end $s = L$ is $\mathbf{M}(L)$ and $(-\mathbf{M}(0))$ at the end $s = 0$. The term $[\mathbf{M} \cdot \delta\phi]$ is therefore the work done by the operator upon a change of orientation of the ends. Similarly, for a small piece of the rod located between the cross-sections s and $s + \mathrm{d}s$, the applied moment is $\mathbf{M}(s + \mathrm{d}s)$ on the end located at $s + \mathrm{d}s$ and $(-\mathbf{M}(s))$ at s. The net moment is therefore $\mathbf{M}' \, \mathrm{d}s$, and the elementary contribution to the integral, $\mathbf{M}' \cdot \delta\phi \, \mathrm{d}s$, appears to be the elementary work of the internal moment (flexural and torsional couples) upon a rotation by a vector $\delta\phi$.

To get the Kirchhoff equations out of this variation, one more step is necessary. Indeed, we have not yet considered the distributed torques and forces applied to the rod, whose work must balance the variation of the elastic energy at equilibrium.

3.6.2 Work of applied forces and torques

We consider two types of external forces: point forces $(\mathbf{P}(0), \mathbf{P}(L))$ and torques $(\mathbf{Q}(0), \mathbf{Q}(L))$ are applied at the two ends $s = 0$ and $s = L$. We also take in account the possibility that a force and a torque are applied over the whole length of the rod, with linear densities $\mathbf{p}(s)$ and $\mathbf{q}(s)$ respectively. The density of force $\mathbf{p}(s)$ can represent for instance the weight of the rod, and the density of moment $\mathbf{q}(s)$, the result of viscous stress due to a swirling flow around the rod. The total work done by these external forces upon a small arbitrary perturbation of the rod is:

$$\delta W = \mathbf{P}(0) \cdot \delta\mathbf{r}(0) + \mathbf{Q}(0) \cdot \delta\phi(0) + \mathbf{P}(L) \cdot \delta\mathbf{r}(L) + \mathbf{Q}(L) \cdot \delta\phi(L) +$$

$$+ \int \mathrm{d}s \, \left(\mathbf{p}(s) \cdot \delta\mathbf{r}(s) + \mathbf{q}(s) \cdot \delta\phi(s) \right), \tag{3.53}$$

where $\mathbf{r}(s)$ denotes the position of the centre line.

It remains to express $\delta\mathbf{r}(s)$ in terms of the perturbations to the orientation of the material frame, as given by $\delta\phi(s)$. In fact, the differential equation allowing one to compute the position of the centre line has already been given in (3.7). It can be rewritten as

$$\mathbf{r}'(s) = \mathbf{d}_3(s) \tag{3.54}$$

and integrated into:

$$\mathbf{r}(s) = \mathbf{r}(0) + \int_0^s \mathrm{d}s' \, \mathbf{d}_3(s'). \tag{3.55}$$

By variation, using (3.49c) to compute $\delta\mathbf{d}_3$, one obtains $\delta\mathbf{r}(s)$ as a function of $\delta\phi(s)$:

$$\delta\mathbf{r}(s) = \delta\mathbf{r}(0) + \int_0^s \mathrm{d}s' \, \left(\delta\phi_2(s') \, \mathbf{d}_1(s') - \delta\phi_1(s') \, \mathbf{d}_2(s') \right)$$

$$= \delta\mathbf{r}(0) - \int_0^s \mathrm{d}s' \, \mathbf{d}_3(s') \times \delta\phi(s'). \tag{3.56}$$

After putting this expression into equation (3.53) and integrating by parts, one finds:

$$\delta W = (\mathbf{P}(0) + \mathbf{F}(0)) \cdot \delta\mathbf{r}(0) + \mathbf{Q}(0) \cdot \delta\phi(0) + \mathbf{Q}(L) \cdot \delta\phi(L)$$

$$+ \int \mathrm{d}s \, [\mathbf{q}(s) + \mathbf{d}_3(s) \times \mathbf{F}(s)] \cdot \delta\phi(s). \tag{3.57}$$

In this equation, we have introduced the auxiliary quantity:[28]

$$\mathbf{F}(s) = \left(\int_s^L \mathrm{d}s' \, \mathbf{p}(s') \right) + \mathbf{P}(L). \tag{3.58}$$

For any fixed s, $\mathbf{F}(s)$ is the total external force applied on that part of the rod defined by $s \le s' \le L$. By writing the balance of force on this part of the rod, which is assumed to be in equilibrium, it turns out that $\mathbf{F}(s)$ is also the force transmitted across the section located at s by the upstream side of the rod ($s' \le s$) on to the downstream one ($s' \le s$). This quantity is called the *internal force* in the rod. Note that its sign depends on the choice of orientation of the rod.

3.6.3   Equation of equilibrium

At equilibrium, the variation of the potential energy is given by the work done by the external forces for an arbitrary perturbation:

$$\delta \mathcal{E}_{\mathrm{rod}} - \delta W = 0.$$

Combining equations (3.52) and (3.57), we rewrite this variation as

$$-(\mathbf{P}(0) + \mathbf{F}(0)) \cdot \delta \mathbf{r}(0) \;\; -(\mathbf{M}(0) + \mathbf{Q}(0)) \cdot \delta \boldsymbol{\phi}(0) + (\mathbf{M}(L) - \mathbf{Q}(L)) \cdot \delta \boldsymbol{\phi}(L)$$

$$- \int [\mathbf{M}'(s) + \mathbf{d}_3(s) \times \mathbf{F}(s) + \mathbf{q}(s)] \cdot \delta \boldsymbol{\phi}(s) \, \mathrm{d}s = 0. \tag{3.59}$$

By requiring that the integral in this variation vanishes for an arbitrary perturbation $\delta \boldsymbol{\phi}(s)$, we find the Kirchhoff equation:

$$\frac{\mathrm{d}\mathbf{M}(s)}{\mathrm{d}s} + \mathbf{d}_3(s) \times \mathbf{F}(s) + \mathbf{q}(s) = \mathbf{0}, \tag{3.60}$$

an equation that we shall interpret in Section 3.6.7 as a balance of moments. This vector equation can be projected along the three directions of the local material frame, using the constitutive equation (3.44) and the kinematics of the material frame in equation (3.2). This yields a set of three coupled differential equations:

$$\frac{\mathrm{d}G^{(1)}}{\mathrm{d}s} - G^{(2)}\,\tau + H\,\kappa^{(2)} - F_2 + q_1 = 0 \tag{3.61a}$$

$$\frac{\mathrm{d}G^{(2)}}{\mathrm{d}s} - H\,\kappa^{(1)} + G^{(1)}\,\tau + F_1 + q_2 = 0 \tag{3.61b}$$

$$\frac{\mathrm{d}H}{\mathrm{d}s} - G^{(1)}\,\kappa^{(2)} + G^{(2)}\,\kappa^{(1)} + q_3 = 0. \tag{3.61c}$$

In these differential equations, the unknown are the material curvatures $\kappa^{(1)}(s)$, $\kappa^{(2)}(s)$ and twist $\tau(s)$, the flexural and torsional couples $G^{(1)}(s)$, $G^{(20)}(s)$ and $H(s)$ being given directly in terms of the latter by the constitutive equations (3.45). The quantities F_1 and F_2 are the components of $\mathbf{F}$ in the local frame: $F_1 = \mathbf{F} \cdot \mathbf{d}_1$ and $F_2 = \mathbf{F} \cdot \mathbf{d}_2$. These equations are

---

[28] The internal force can also be viewed as a Lagrange multiplier in the different but equivalent approach where the conditions of adaptivity of the tangent, $\mathbf{r}'(s) = \mathbf{d}_3(s)$ and of inextensibility are viewed as constraints (N. Chouaïeb, 2003).

non-linear, not as a result of the material response, which is described by linear constitutive relations, but as a result of geometry.

3.6.4   Boundary conditions

In equation (3.59), we have only considered the integral term so far. The boundary terms should vanish as well since this equation is to be satisfied for perturbations localized near its ends as well. The first boundary terms, associated with $\delta\mathbf{r}(0)$ yields

$$(\mathbf{F}(0) + \mathbf{P}(0)) \cdot \delta\mathbf{r}(0) = 0. \tag{3.62a}$$

One of the remaining boundary terms is associated with rotations $\delta\boldsymbol{\phi}(0)$ of the end of the rod at $s = 0$ and yields

$$(\mathbf{M}(0) + \mathbf{Q}(0)) \cdot \delta\boldsymbol{\phi}(0) = 0. \tag{3.62b}$$

The other term concerns the other end, $s = L$:

$$(-\mathbf{M}(L) + \mathbf{Q}(L)) \cdot \delta\boldsymbol{\phi}(L) = 0. \tag{3.62c}$$

The interpretation of these boundary conditions is fairly straightforward. Consider equation (3.62a) first. If the endpoint $s = 0$ is free to move, the vector $\delta\mathbf{r}(0)$ is arbitrary and one is led to the boundary condition $\mathbf{F}(0) + \mathbf{P}(0) = \mathbf{0}$. This is the total force applied on the section $s = 0$, which is the sum of the internal force $\mathbf{F}(0)$ transmitted by the downstream part of the rod, $s > 0$, and of the force $\mathbf{P}(0)$ applied by the operator. This total force should obviously cancel at equilibrium when the end is free to move. If the endpoint $s = 0$ is fixed, the perturbations that are consistent with the kinematics are such that $\delta\mathbf{r}(0) = \mathbf{0}$ and the equation is automatically satisfied[29] (the boundary condition $\mathbf{F}(0) + \mathbf{P}(0) = \mathbf{0}$ is then replaced by the one imposing the position of the fixed end, which leaves the total number of boundary conditions unchanged).

The two other conditions written in equations (3.62b) and (3.62c) can be handled similarly. Near an end whose orientation is unconstrained, the total torque should be zero. This total torque is $\mathbf{M}(0) + \mathbf{Q}(0)$ near the end $s = 0$, by a similar reasoning as above. However, the total torque is $-\mathbf{M}(L) + \mathbf{Q}(L)$ near the opposite end, $s = L$, as the moment due to internal forces is now applied by the downstream part of the rod, $s < L$. When the orientation of the end is imposed by boundary conditions, as happens for instance for a clamped end, these boundary conditions do not hold (they are replaced by boundary conditions imposing the orientation of the material frame).

3.6.5   Case of two clamped ends, Lagrange multipliers

In the previous section, we derived the equations for an elastic rod by variation. To do so, we have implicitly assumed that the perturbation $\delta\boldsymbol{\phi}(s)$ can be chosen freely. This is not the case when the rod is subject to kinematical constraints. The particular case that we consider here is when both endpoints of the rod are fixed. In the present section, we investigate how the equations of equilibrium are modified in the presence of such a constraint. The final result brings no big surprise: the present section is given for the sake of completeness, but can be skipped in a first reading.

---

[29] This is the case of a clamped end, for instance, for which a non-zero value of $\mathbf{F}(0)$ can simply be interpreted as the opposite of the reaction force of the support.

Minimization with constraints is introduced in Section A.4 in the case of a finite number of parameters. The case of an infinite dimensional space of parameters (minimization with respect to *functions* as we do here) is studied in Appendix A, Section A.4.

Consider a rod with no applied external force, whose ends have a prescribed position but a free orientation. Such ends are said to be *simply supported*, *pinned* or *hinged*. To derive the equations of equilibrium, we need to add into the variation of the elastic energy a Lagrange multiplier $\boldsymbol{\lambda}$ to impose the relative position $\mathbf{r}(L) - \mathbf{r}(0) = \mathbf{R}$ of the two ends. This yields

$$\delta\mathcal{E}_{\mathrm{rod}} - \boldsymbol{\lambda} \cdot \delta(\mathbf{r}(L) - \mathbf{r}(0)) = 0.$$

The sign in front of $\boldsymbol{\lambda}$, chosen to be negative here, is a mere matter of convention. Note that there is no work from external forces since the endpoints do not move and the simple support is assumed to be a perfect one.

Using equation (3.52) for the variation of the elastic energy and equation (3.56) to express the variation of the constraint, we obtain

$$[\mathbf{M} \cdot \delta\boldsymbol{\phi}] - \int \mathbf{M}' \cdot \delta\boldsymbol{\phi}\,\mathrm{d}s + \boldsymbol{\lambda} \cdot \int \mathbf{d}_3 \times \delta\boldsymbol{\phi}\,\mathrm{d}s = 0.$$

After permutation of the vectors in the mixed product by means of equation (1.10), this equation becomes

$$[\mathbf{M} \cdot \delta\boldsymbol{\phi}] - \int (\mathbf{M}' + \mathbf{d}_3 \times \boldsymbol{\lambda}) \cdot \delta\boldsymbol{\phi}\,\mathrm{d}s = 0.$$

This equation is similar to equation (3.59), with the internal force $\mathbf{F}(\mathbf{s})$ equal to a constant, the Lagrange multiplier $\boldsymbol{\lambda}$. This constant internal force can be interpreted as the force applied by the support at $s = L$ which is, at equilibrium, the opposite of the force applied by the support at $s = 0$. When both ends are clamped, this force is an unknown of the problem; apart from this minor difference, the equations are the same as in the unconstrained case.

3.6.6   Kirchhoff equations: summary

The Kirchhoff equations are traditionally written as the derivative of equation (3.58), together with equation (3.60):

$$\mathbf{F}'(s) + \mathbf{p}(s) = \mathbf{0}, \tag{3.63a}$$

$$\mathbf{M}'(s) + \mathbf{d}_3(s) \times \mathbf{F}(s) + \mathbf{q}(s) = \mathbf{0}. \tag{3.63b}$$

To these equations, one must add the kinematical relations for the centre line in equations (3.2) and (3.3):

$$\mathbf{d}_i'(s) = \boldsymbol{\Omega}(s) \times \mathbf{d}_i(s) \quad \text{where } \boldsymbol{\Omega}(s) = \kappa^{(1)}(s)\,\mathbf{d}_1 + \kappa^{(2)}(s)\,\mathbf{d}_2 + \tau(s)\,\mathbf{d}_3 \tag{3.64}$$

together with the constitutive relations (3.45):

$$G^{(1)}(s) = EI^{(1)}\,\kappa^{(1)}(s) \qquad G^{(2)}(s) = EI^{(2)}(s)\,\kappa^{(2)}(s) \qquad H(s) = \mu J\,\tau(s) \tag{3.65a}$$

$$\mathbf{M}(s) = G^{(1)}(s)\,\mathbf{d}_1(s) + G^{(2)}(s)\,\mathbf{d}_2(s) + H(s)\,\mathbf{d}_3(s). \tag{3.65b}$$

3.6.7 Interpretation of Kirchhoff equations

The boundary conditions at the ends of the rod have already been interpreted in Section 3.6.4, based on the fact that the vector $\mathbf{M}(s)$ measures the moment of the contact forces transmitted through a cross-section and $\mathbf{F}(s)$ the resultant of these contact forces.

Let us compute the forces acting on a small cylindrical element of the rod of length $\mathrm{d}s$, as in Fig. 3.9. The element is submitted to the contact forces $\mathbf{F}(s+\mathrm{d}s)$ and $-\mathbf{F}(s)$ from the neighbouring elements, and to the external force $(\mathbf{p}\,\mathrm{d}s)$. At equilibrium, the total force $(\mathbf{F}'\,\mathrm{d}s + \mathbf{p}\,\mathrm{d}s)$ is zero. This yields equation (3.63a).

Similarly, let us compute the net torque on this cylindrical element, with respect to the point A located on the centre line, halfway between the sections at s and $(s+\mathrm{d}s)$. Let $\mathrm{d}\mathbf{r} \approx (\mathbf{d}_3(s)\,\mathrm{d}s)$ be the vector joining the centres of mass of these two cross-sections, as shown in Fig. 3.9. The moment of the contact forces applied on the 'downstream' cross-section, at $s+\mathrm{d}s$, is $\mathbf{M}(s+\mathrm{d}s)$. By convention, this moment is calculated with respect to the centre of mass of the cross-section. The moment calculated with respect to the point A is different, as the contact forces over this face have a non-zero resultant $\mathbf{M}(s+\mathrm{d}s)$: the moment with respect to the point A reads $\mathbf{M}(s+\mathrm{d}s) + (\mathrm{d}\mathbf{r}/2) \times \mathbf{F}(s+\mathrm{d}s)$. The same reasoning yields for the contribution of the other cross-section: $-\mathbf{M}(s) + (-\mathrm{d}\mathbf{r}/2) \times (-\mathbf{F}(s))$, where the signs in front of $\mathbf{M}$ and $\mathbf{M}$ have changed since the rod element is now downstream with respect

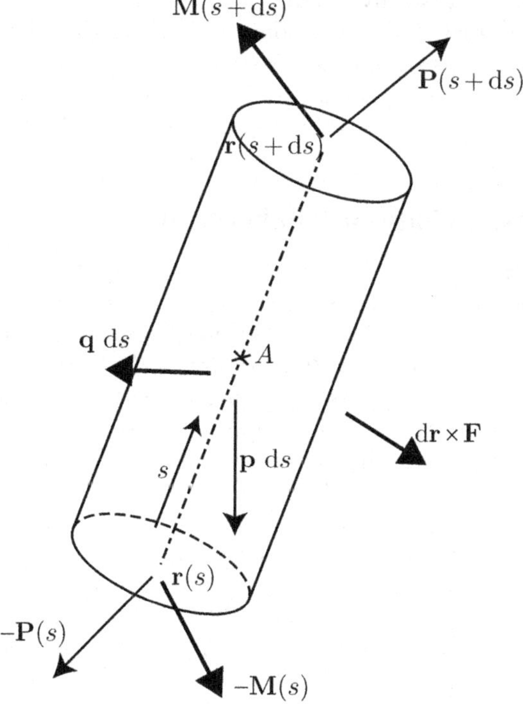

Fig. 3.9 Forces (shown by small arrows) and moments (bold arrows) acting on a small cylindrical rod element: $\mathbf{F}$ and $\mathbf{M}$ are the internal force and moments, transmitted by the neighbouring elements, $\mathbf{p}$ and $\mathbf{q}$ the lineic density of the external force and moment. Note the moment $\mathrm{d}\mathbf{r} \times \mathbf{M}$ due to the internal force applied on the endpoints, which yields the second term in equation (3.63b).

to the cross-section. Adding the torque $\mathbf{q}(s)\,\mathrm{d}s$ due to external forces, this again yields equation (3.63b), at first order in $\mathrm{d}s$.

We have rederived the Kirchhoff equations directly: they express the balance of force and of torque on a small element of the rod. This is consistent with the fact that the left-hand side of equation (3.63b) gets multiplied by $\delta\phi$ in the variation of the energy.

3.6.8   Solving the Kirchhoff equations

In general, the static Kirchhoff equations (3.63) are solved according to the following pattern. First, one computes the internal force $\mathbf{F}(s)$ in the full length of the rod by integration of equation (3.63a) starting from a free edge whose position is not fixed (the case of *two* fixed ends is discussed below). Then, one solves the second Kirchhoff equations (3.63b) with the boundary conditions discussed above, using the constitutive relations (3.65) to eliminate the bending moments $G^{(1)}$, $G^{(2)}$ and the torque H in favour of the unknowns $\kappa^{(1)}$, $\kappa^{(2)}$ and τ.

In equation (3.63b), the orientation $\mathbf{d}_3(s)$ of the tangent in not known in advance.[30] The tangent must be determined by integration of the kinematical equations (3.64), which depend themselves on the unknown material curvatures and twist. As a result, the Kirchhoff equations do not fall into the general category of differential equation with a well-posed Cauchy initial condition. Moreover, these equations being non-linear, the equilibrium solutions of a rod are not necessarily unique.

Explicit solutions of the Kirchhoff equations are given in the following chapters for some particular geometries. A class of interesting solutions that can be derived analytically is presented in Section 3.8; it takes advantage of the similarity of the equations for a rod in equilibrium with those for a spinning table top, as noticed by Kirchhoff.

3.7   Inextensibility, validity of Kirchhoff model

In this section, we briefly discuss the assumption of inextensibility of the centre line of the rod. Elastic rod models combining extension of the centre line *and* bending (or twisting) are very popular in the literature. We present a simple scaling argument showing that such models are in fact of very limited use. Indeed, for a slender rod, only two cases are possible generically,[31] as we show below: when the loading is moderate, the centre line remains inextensible and the classical inextensible rod model applies; when the loading is stronger, the centre line stretches, and bending and twisting forces become negligible.

Consider an elastic rod with thickness h and length L. Let F be the typical total load applied on the rod—for instance, $F \sim \rho g S L$ if the rod is subjected to its own weight, has volumic mass ρ and cross-sectional area S. We shall assume that this load F varies over the macroscopic length scale L. Then, the typical moment applied on the rod is $F L$. By the fundamental law for the bending of bars, see equation (3.30), we can estimate

$$F L \sim E I \kappa,$$

---

[30] The same problem manifests itself when the second Kirchhoff equation is written in the local frame by the fact that the projections F_1 and F_2 of the internal force on to the cross-section are not known in advance in equation (3.61).

[31] This reasoning is only for a generic cross-section and can be contradicted in the case of 'exotic' rods, such as rods made by combining materials with a very large elastic contrast.

where $E\,I$ is the bending modulus of the rod, $I \sim h^4$ is moment of inertia of the cross-section as defined in equation (3.26), E is Young's modulus of the material. Consider the ratio of the typical curvature κ induced by the loading to the typical curvature $1/L$ associated with the length of the rod:

$$\frac{\kappa}{1/L} \sim \frac{F\,L^2}{E\,I} \sim \frac{F}{E\,h^2}\left(\frac{L}{h}\right)^2. \tag{3.66a}$$

The same reasoning holds for twisting, which obeys exactly the same scaling laws as bending (unless the cross-section is very flat, as in the case of a ribbon, something we do not consider here).

Let us now estimate the amount of stretching induced by the loading. The fundamental law for the stretching of bars yields the typical stretching deformation ϵ of the centre line (axial strain) as

$$\epsilon \sim \frac{F}{E\,S},$$

where $S \sim h^2$ is the area of the cross-section—see equation (2.67) and Fig. 2.8. Therefore, we have

$$\epsilon \sim \frac{F}{E\,h^2}. \tag{3.66b}$$

Combining equations (3.66a) and (3.66b) to eliminate F, we find

$$\epsilon \sim \frac{\kappa}{1/L}\left(\frac{h}{L}\right)^2, \quad \text{where} \quad \frac{h}{L} \ll 1. \tag{3.67}$$

The condition $h \ll L$ is the fundamental assumption underlying the theory of rods.

The classical theory of the elastic rod assumes that the radius $1/\kappa$ is comparable to the macroscopic dimension L. By equation (3.67), this implies that the axial strain ϵ is negligible, $\epsilon \sim (h/L)^2$. This provides a simple justification of the assumption of inextensibility.

The opposite case is when the loading is such that the extension of the centre line becomes larger than a very small number, that is if $\epsilon \gg (h/L)^2$. Then, equation (3.67) implies $\kappa \sim 1/L\,\epsilon/(h/L)^2 \gg 1/L$. This means that the rod makes sharp angles at localized points, as happens for instance when an elastic string is stretched and applied around the edge of a table, making a sharp angles at the points of contact with the edge of the table.[32] In this case, the equilibrium of the rod is dominated almost everywhere by stretching forces; bending (and twisting) can be neglected. The sharp angles can be understood to be the consequence of the bending modulus being effectively zero.

To sum up, different models of thin rods must be used depending on the intensity of the loading. The two main models are:[33] the model of an inextensible rod discussed in this

---

[32] This also applies to the following geometry: when a rod is glued on a flat surface by a strong adhesion force, and peeled off this surface by a strong pulling force, a localized region with large curvature is created near the point of debonding.

[33] Some degenerate variants of these two models are possible too. For very small applied forces, the classical inextensible rod model still holds, but it can be simplified by linearizing the equations (small displacement approximation). In addition, a hybrid variant of the classical inextensible rod model and of the extensible string model is relevant for intermediate loading; this is the inextensible string model (it is inextensible like the classical rod model, but has zero bending modulus like the extensible string model).

present chapter (with finite bending and twisting moduli), and the model of an extensible string (with zero bending and twisting moduli). For a more rigorous derivation of this result, see for instance reference (J.-J. Marigo and N. Meunier, 2006). Combining extension of the centre line with a bending or twisting effect in a single model make it look more general... and therefore appealing. However, the above argument shows that the resulting 'do-it-all' model is of limited use. This should remind us that not all generalizations are useful.

Similarly, some models for elastic rods attempt to generalize the classical Euler–Bernoulli assumption by introducing shearing. This involves relaxing the kinematical hypothesis (3.54), and allowing $\mathbf{d}_3$ to point in a different direction from the tangent $\mathbf{r}'(s)$ to the centre line. Even though this generalization looks interesting, it is unclear how much it helps. A similar dimensional argument as above shows that the intensities of forces required to shear the cross-section are much larger than those consistent with the bending and twisting of rods. As a result, the bending and twist moduli can effectively be set to zero whenever the loading is actually such that shear is non-negligible.

An important consequence of the above argument is that description of a given mechanical system—an elastic rod—may require a different model depending on the intensity of the applied forces. This explains why some scaling assumptions have to be made as a starting point of the derivation of the equations for plate, see Appendix D.

3.8 Mathematical analogy with the spinning top

In this section, we establish the *Kirchhoff kinetic analogy*: the equations for the static equilibria of an elastic rod with symmetric cross-section are the same as those for the motion of a spinning top, with time replaced by the curvilinear coordinate s. Since the spinning top had a long history of mathematical investigations, this Kirchhoff analogy has been widely used to obtain several exact solutions for the Kirchhoff equations, as well as to provide methods for classifying the possible solutions when formal integration cannot be carried out (S. Antman, 2005). This section follows loosely the presentation of two recent papers on the topic (M. Nizette and A. Goriely, 1999; G. H. M. van der Heijden and J. M. T. Thompson, 2000).

Let us write again the equations for a rod. The internal moment $\mathbf{M}$ is related to the curvature and twist by:

$$\mathbf{M} = G^{(1)}\,\mathbf{d}_1 + G^{(2)}\,\mathbf{d}_2 + H\,\mathbf{d}_3, \tag{3.68}$$

where $G^{(1)} = EI^{(1)}\,\kappa^{(1)}$, $G^{(2)} = EI^{(2)}\,\kappa^{(2)}$ and $H = \mu J\,\tau$ (linear constitutive relations). The analogue of the equation (3.68) in the spinning top problem is the relation between the angular momentum (mathematically equivalent to $\mathbf{M}$) and the three Cartesian components $(G^{(1)}, G^{(2)}, H)$ of the instantaneous angular velocity along directions $(\mathbf{d}_1, \mathbf{d}_2, \mathbf{d}_3)$ fixed rigidly to the spinning top. These directions are chosen along the principal axes of the spinning top.[34] Therefore $EI^{(1)}$, $EI^{(2)}$ and μJ play the role of the principal moments of inertia of the top.

---

[34] These principal axes diagonalize the symmetric quadratic form (or tensor of inertia): $\boldsymbol{\Omega} \mapsto \langle \boldsymbol{\Omega} | \mathbf{J}(\boldsymbol{\Omega}) \rangle$, where $\boldsymbol{\Omega}$ is the instantaneous rotation vector and $\mathbf{J}$ the angular momentum of the rotating top.

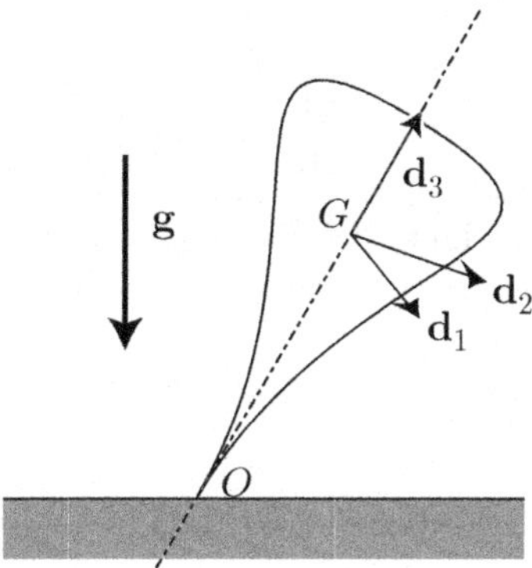

Fig. 3.10 Kirchhoff analogy of the equations for an elastic rod with a spinning top.

Consider equation (3.63a) with no external force $\mathbf{p}(s)$:

$$\mathbf{F}'(s) = 0, \tag{3.69}$$

This is solved as $\mathbf{F}(s) = \mathbf{F}$, a constant vector that denotes the force exerted at one end of the rod, which is also the opposite force being exerted at the other end to maintain global equilibrium. This constant vector $\mathbf{F}$ is identified, up to a negative multiplicative constant, to the constant pull of gravity of the spinning top, $m_{\text{top}}\,\mathbf{g}$:

$$\mathbf{g} = -\frac{\mathbf{F}}{m_{\text{top}}\,|\mathbf{OG}|}, \tag{3.70}$$

where $|\mathbf{OG}|$ denotes the distance from the fixed point of contact of the top with the plane, O, to the centre of gravity G of the top, and m_{top} its mass. All the coming developments assume that the vector $\mathbf{OG}$ is aligned with $\mathbf{d}_3$, namely that the point of contact O lies on the third principal axis, as in the geometry of Fig. 3.10 for instance:

$$\mathbf{OG} = |\mathbf{OG}|\,\mathbf{d}_3.$$

The balance of torque in equation (3.63b) becomes, in the absence of applied moment $\mathbf{q}$:

$$\mathbf{M}' + \mathbf{d}_3 \times \mathbf{F} = \mathbf{0}. \tag{3.71}$$

This equation, together with the definition (3.70), yields the fundamental law of dynamics for solid bodies in rotation, $\mathbf{M}' = \mathbf{OG} \times (m_{\text{top}}\,\mathbf{g})$: the time derivative of the angular momentum $\mathbf{M}$ is the torque due to gravity.

Below we shall give some properties of the equations for elastic rods with constant internal force (i.e. without applied force $\mathbf{p}(s)$) that naturally follow from the Kirchhoff analogy.

3.8.1 Conserved quantities

As usual in this class of problem many methods of solution, at least in the classical analytical sense, are based upon the existence of conserved quantities, this concerning equation (3.71). These conserved quantities are Noether invariant, given that the equations of motion of a spinning top can be derived by Lagrangian mechanics. We do not rely on the general expressions for the Noether invariants and propose a direct derivation. Two conservation laws are found at once: by taking the scalar product of the equation (3.71) with the constant vector $\mathbf{F}$, one gets:

$$\frac{\mathrm{d}\langle \mathbf{M}|\mathbf{F}\rangle}{\mathrm{d}s} = 0. \tag{3.72}$$

For the spinning top, the invariant $\langle \mathbf{M}|\mathbf{F}\rangle$ is the vertical component of the angular momentum. This is the Noether invariant associated with the invariance of the problem by rotation about the vertical axis.

Another invariant is found by taking the scalar product of equation (3.71) with $\mathbf{d}_3$. This yields:

$$0 = \langle \mathbf{M}'|\mathbf{d}_3\rangle = (\langle \mathbf{M}|\mathbf{d}_3\rangle)' - \langle \mathbf{M}|\mathbf{d}_3'\rangle. \tag{3.73}$$

The derivative $\mathbf{d}_3'$ is obtained from the following kinematical expression (3.1c):

$$\mathbf{d}_3' = \left(\kappa^{(1)}\mathbf{d}_1 + \kappa^{(2)}\mathbf{d}_2 + \tau\mathbf{d}_3\right) \times \mathbf{d}_3.$$

We consider the case of a rod with equal moments of inertia, as in the case of an axisymmetric cross-section: $I^{(1)} = I^{(2)} = I$. The constitutive relations $G^{(1)} = EI\kappa^{(1)}$ and $G^{(2)} = EI\kappa^{(2)}$ together with equation (3.68) show that one can write the vector $\left(\kappa^{(1)}\mathbf{d}_1 + \kappa^{(2)}\mathbf{d}_2 + \tau\mathbf{d}_3\right)$ as the sum of a vector proportional to $\mathbf{M}$ plus an irrelevant part proportional to $\mathbf{d}_3$:

$$\kappa^{(1)}\,\mathbf{d}_1 + \kappa^{(2)}\,\mathbf{d}_2 + \tau\,\mathbf{d}_3 = \frac{\mathbf{M}}{EI} + \alpha\,\mathbf{d}_3.$$

Therefore

$$\langle \mathbf{M}|\mathbf{d}_3'\rangle = \langle \mathbf{M}|\frac{\mathbf{M}}{EI} \times \mathbf{d}_3\rangle = 0.$$

From equation (3.73), this yields:

$$\frac{\mathrm{d}\langle \mathbf{M}|\mathbf{d}_3\rangle}{\mathrm{d}s} = 0 \tag{3.74}$$

Therefore, the component of $\mathbf{M}$ along $\mathbf{d}_3$ is constant, independent of s, although $\mathbf{d}_3$ depends on s in general. This Noether invariant reflects the invariance of the problem by a rotation of the top around the axis $\mathbf{d}_3$.

The third constant of the motion is found by taking the scalar product of equation (3.71) with the Darboux vector $\boldsymbol{\Omega} = \left(\kappa^{(1)}\mathbf{d}_1 + \kappa^{(2)}\mathbf{d}_2 + \tau\mathbf{d}_3\right)$. This gives:

$$0 = \langle \mathbf{M}'|\boldsymbol{\Omega}\rangle + \langle \boldsymbol{\Omega}|\mathbf{d}_3 \times \mathbf{F}\rangle. \tag{3.75}$$

From equation (3.68), $\mathbf{M} = EI^{(1)}\kappa^{(1)}\,\mathbf{d}_1 + EI^{(2)}\kappa^{(2)}\,\mathbf{d}_2 + \mu J\tau\,\mathbf{d}_3$. Therefore the derivative $\mathbf{M}'$ will include derivatives of the basis vectors, namely quantities like $EI^{(1)}\kappa^{(1)}\,\mathbf{d}_1' = EI^{(1)}\kappa^{(1)}\,\boldsymbol{\Omega} \times \mathbf{d}_1$ and terms proportional to the derivatives of the curvatures, like $EI^{(1)}(\kappa^{(1)})'\,\mathbf{d}_1$. Terms proportional to derivatives such as $\mathbf{d}_1'$ do not contribute to the scalar product with $\boldsymbol{\Omega}$, because of relations like $\langle \mathbf{d}_1'|\boldsymbol{\Omega}\rangle = \langle \boldsymbol{\Omega} \times \mathbf{d}_1|\boldsymbol{\Omega}\rangle = 0$. Only the contributions of the derivatives of the scalar curvatures remain:

$$\langle \mathbf{M}'|\boldsymbol{\Omega}\rangle = EI^{(1)}\,\kappa^{(1)}(\kappa^{(1)})' + EI^{(2)}\,\kappa^{(2)}(\kappa^{(2)})' + \mu J\,\tau\,\tau' = \frac{1}{2}\langle \mathbf{M}|\boldsymbol{\Omega}\rangle'.$$

Putting this into equation (3.75) one obtains the third invariant:

$$\frac{1}{2}\langle \mathbf{M}|\boldsymbol{\Omega}\rangle' + \langle \boldsymbol{\Omega}|\mathbf{d}_3 \times \mathbf{F}\rangle = \frac{1}{2}\langle \mathbf{M}|\boldsymbol{\Omega}\rangle' + \langle \boldsymbol{\Omega} \times \mathbf{d}_3|\mathbf{F}\rangle = \left(\frac{1}{2}\langle \mathbf{M}|\boldsymbol{\Omega}\rangle + \langle \mathbf{d}_3|\mathbf{F}\rangle\right)' = 0 \quad (3.76)$$

In the case of the spinning top, this invariant is the total energy, the sum of the kinetic energy and potential energy in the gravitational field.

Thanks to the existence of three invariants when $I^{(1)} = I^{(2)}$, the equations can be solved by quadrature as was shown long ago by Lagrange (another general solution by quadrature was found later on by Kowalevskaya in the rather special case $EI^{(1)} = EI^{(2)}/2 = \mu J$). Below we shall focus on the so-called *symmetric* (or 'Lagrange') case $I^{(1)} = I^{(2)}$, realized for axisymmetric rods, for instance, but not exclusively:[35] for symmetric rods, the quantity $\langle \mathbf{M}|\mathbf{d}_3\rangle$ is conserved.

3.8.2 Special cases

Simple solutions of the above equations can be derived for particular choices of the constant of the motion and/or of the internal force $\mathbf{F}$. Let us consider first the case $\mathbf{F} = \mathbf{0}$: the rod is loaded by a pure torque at its ends, and this corresponds to switching off gravity for the spinning top. Then the integration for $\mathbf{M}$ becomes trivial: $\mathbf{M}$ is a constant. Since the scalar product of $\mathbf{M}$ with $\mathbf{d}_3$ is also constant, the twist τ of the rod is constant.

Let us write explicitly the derivative $\mathbf{M}'$ in the local coordinate system $(\mathbf{d}_1, \mathbf{d}_2, \mathbf{d}_3)$, as given in equation (3.68):

$$\mathbf{M}' = EI\left(\mathbf{d}_1'\kappa^{(1)} + \mathbf{d}_1(\kappa^{(1)})' + \mathbf{d}_2'\kappa^{(2)} + \mathbf{d}_2(\kappa^{(2)})'\right) + \mu J \tau\, \mathbf{d}_3'. \tag{3.77}$$

The derivatives like $\mathbf{d}_1'$ are computed from

$$\mathbf{d}_1' = (\kappa^{(1)}\mathbf{d}_1 + \kappa^{(2)}\mathbf{d}_2 + \tau\mathbf{d}_3) \times \mathbf{d}_1.$$

This yields

$$\mathbf{M}' = \mathbf{d}_1(EI\,(\kappa^{(1)})' + EI\,\kappa^{(2)}\tau - \mu J\,\tau\,\kappa^{(2)})$$
$$+\mathbf{d}_2(EI\,(\kappa^{(2)})' - EI\,\kappa^{(1)}\tau + \mu J\,\tau\,\kappa^{(1)}). \tag{3.78}$$

Because $\mathbf{M}$ is constant, the components of $\mathbf{M}'$ on $\mathbf{d}_1$ and $\mathbf{d}_2$ should cancel. This yields two linear coupled ODEs with constant coefficients:

$$(\kappa^{(1)})' + \kappa^{(2)}\,\tau\left(1 - \frac{\mu J}{EI}\right) = 0$$

$$(\kappa^{(2)})' - \kappa^{(1)}\,\tau\left(1 - \frac{\mu J}{EI}\right) = 0.$$

Eliminating one unknown function by derivation and noticing that the coefficient $\tau(1 - \mu J/EI)$ is constant, one obtains the equation of the harmonic oscillator that can be solved immediately. The rod described by these equation is a twisted helix with curvilinear period $2\pi/|\tau(1 - \mu J/EI)|$. In the special case $\tau = 0$, $\mathbf{M} = EI\boldsymbol{\Omega}$ is constant and the rod takes the shape of an arc of circle, as studied in Section 5.3.1.

---

[35] Note that a rod with a square or triangular cross-section is symmetric in this sense.

Another situation where the Kirchhoff equations can be solved is Euler's planar Elastica. For planar configurations of the rod, $\mathbf{d}_1$ and $\mathbf{d}_3$ stay in the (fixed) plane of the rod, $\mathbf{d}_2$ is a constant vector perpendicular to this plane and $\mathbf{F}$, a force, lies in the same plane as the deformed Elastica. From the kinematic equation (3.1b) this requires $\tau = \kappa^{(1)} = 0$. A calculation very similar to the one yielding equation (3.78) gives:

$$\mathbf{M}' = EI\,\mathbf{d}_2\,\frac{\mathrm{d}\kappa^{(2)}}{\mathrm{d}s}. \tag{3.79}$$

The balance of momentum $\mathbf{M}' + \mathbf{d}_3 \times \mathbf{F} = \mathbf{0}$ becomes:

$$EI\,\mathbf{d}_2\,\frac{\mathrm{d}\kappa^{(2)}}{\mathrm{d}s} + (\mathbf{d}_1 \times \mathbf{d}_2) \times \mathbf{F} = \mathbf{0}.$$

This vector equation has components along $\mathbf{d}_2$ only. It yields the following scalar equation after projection:

$$EI\,\frac{\mathrm{d}\kappa^{(2)}}{\mathrm{d}s} + \langle \mathbf{d}_1 | \mathbf{F} \rangle = 0.$$

Denoting now as $\theta(s)$ the angle between the direction of $\mathbf{F}$ and $\mathbf{d}_3$, and recalling that $\kappa^{(2)} = \frac{\mathrm{d}\theta}{\mathrm{d}s}$, with $F\cos(\theta) = \langle \mathbf{d}_3 | \mathbf{F} \rangle$, one finds the equation of Euler's Elastica:

$$EI\,\frac{\mathrm{d}^2\theta(s)}{\mathrm{d}s^2} + F\sin(\theta(s)) = 0, \tag{3.80}$$

an equation that is studied in Appendix A and Chapter 4.

3.9   The localized helix: an explicit solution

We present here a classical solution of the Kirchhoff equations for symmetric rods, relevant to the loading geometry shown in Fig. 3.11: an infinitely long, naturally straight rod is subjected to combined axial tension T and twist M on its endpoints. We consider here a symmetric rod, that is we assume that the two principal bending moduli are equal. Remarkably, an exact solution[36] of the non-linear equations for rods can be derived in this particular geometry (J. Coyne, 1990), thanks to the existence of a sufficient number of invariants. This solution describes a helix with a modulated amplitude, see figure. Our presentation is adapted from reference (G. H. M. van der Heijden and J. M. T. Thompson, 2000) with some simplifications; additional references can be found therein.

The z axis is chosen such such that it coincides with the direction of the applied force, and with the asymptotically straight shape of the rod at infinity, see Fig. 3.11. Then, the unit tangent $\mathbf{t}(s) = \mathbf{d}_3(s)$ satisfies:

$$\mathbf{t}(s) \to \mathbf{e}_z \quad \text{for } s \to \pm\infty. \tag{3.81a}$$

Let us denote $T\,\mathbf{e}_z$ the force applied on the endpoint on the positive side ($s = +\infty$). The global force balances requires that the opposite force ($-T\,\mathbf{e}_z$) is applied on the negative

---

[36] Another example of a non-linear solution is given elsewhere in this book, for the Euler column that arises in the context of the equations for plates, see Section 8.5.

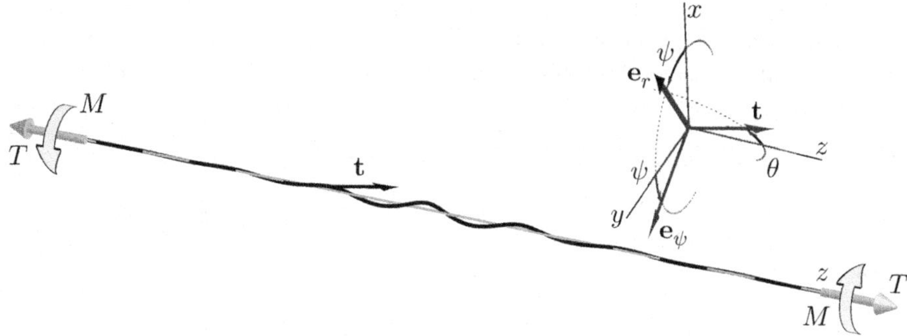

Fig. 3.11 Localized helical buckling of a naturally straight, symmetric rod under combined tension and twist: geometry, notations.

side ($s = -\infty$). Similarly, the applied twist is $M\,\mathbf{e}_z$ on the positive size, and $(-M\,\mathbf{e}_z)$ on the negative side. Equilibrium of the endpoints dictates the following conditions:

$$\mathbf{F}(s) \to T\,\mathbf{e}_z \quad \text{and} \quad \mathbf{M}(s) \to M\,\mathbf{e}_z \quad \text{for } s \to \pm\infty.$$

Since there is no applied force, $\mathbf{p}(s) = \mathbf{0}$ in equation (3.63a) and the internal force $\mathbf{F}(s)$ does not depend on s, its value is fixed by the above asymptotic condition:

$$\mathbf{F}(s) = T\,\mathbf{e}_z.$$

The values of the three invariants (3.72), (3.74) and (3.76) identified in Section 3.8.1 can be calculated using the boundary conditions (3.81):

$$\langle \mathbf{M}(s)|\mathbf{e}_z \rangle = M \tag{3.82a}$$

$$\langle \mathbf{M}(s)|\mathbf{t}(s) \rangle = M \tag{3.82b}$$

$$\frac{1}{2}\,\langle \mathbf{M}(s)|\mathbf{\Omega}(s) \rangle + \langle T\,\mathbf{e}_z|\mathbf{t}(s) \rangle = \frac{M^2}{2\,\mu J} + T, \tag{3.82c}$$

where $\mathbf{\Omega}(s)$ is the Darboux vector defined in equation (3.2). In this last equation, we have used the constitutive law for twist (3.65), with $\tau(s) = \mathbf{\Omega}(s) \cdot \mathbf{t}(s)$: when combined with the asymptotic conditions (3.81), it yields $\mathbf{\Omega} \cdot \mathbf{e_z} \to M/\mu J$ for $s \to \pm\infty$. These equalities (3.82) hold for any solution of the equations of equilibrium, and for any s.

For isotropic rods, it is possible to formulate the equations of equilibrium without any reference to the normal directors $\mathbf{d}_1(s)$ and $\mathbf{d}_2(s)$—the twist enters into the equations for the centre line in the form of a parameter, M. To show this, let us parameterise the unit tangent $\mathbf{t}(s) = \mathbf{d}_3(s)$ using Euler angles $\theta(s)$ and $\psi(s)$:

$$\mathbf{t}(s) = \mathbf{e}_r(\psi(s))\,\sin\theta(s) + \mathbf{e}_z\,\cos\theta(s), \tag{3.83}$$

where $\mathbf{e}_r(\psi)$ is part of the polar basis in the plane (x, y), defined by

$$\mathbf{e}_r(\psi) = \mathbf{e}_x\,\cos\psi + \sin\psi\,\mathbf{e}_y \tag{3.84a}$$

$$\mathbf{e}_\psi(\psi) = -\sin\psi\,\mathbf{e}_x + \cos\psi\,\mathbf{e}_y. \tag{3.84b}$$

This polar basis $(\mathbf{e}_r, \mathbf{e}_\psi)$ and the angles θ and ψ are represented in Fig. 3.11.

The normal curvature can be calculated by derivation of the tangent $\mathbf{t}(s)$ with respect to arc length s (this normal curvature has been noted $\mathbf{t}'(s) = k(s)\,\mathbf{n}(s)$ in Section 1.3.2). Using the classical relations for the polar basis, $\mathbf{e}'_r(\psi) = \mathbf{e}_\psi(\psi)$ and $\mathbf{e}'_\psi(\psi) = -\mathbf{e}_r(\psi)$, it reads:

$$\mathbf{t}'(s) = \mathbf{e}_\psi(\psi(s))\,\sin\theta(s)\,\psi'(s) + (\mathbf{e}_r(\psi(s))\,\cos\theta(s) - \mathbf{e}_z\,\sin\theta(s))\,\theta'(s). \tag{3.85}$$

Since $\mathbf{t}(s)$ is a unit vector for all s, its derivative $\mathbf{t}'(s)$ is everywhere orthogonal to $\mathbf{t}(s)$. As a result, we have the identity $|\mathbf{t}(s) \times \mathbf{t}'(s)| = |\mathbf{t}'(s)|$, and the square norm of this vector is therefore

$$|\mathbf{t}(s) \times \mathbf{t}'(s)|^2 = |\mathbf{t}'(s)|^2 = \theta'^2(s) + \sin^2\theta(s)\,\psi'^2(s), \tag{3.86}$$

a result that will be useful later.

The definition (3.2) of the Darboux vector $\boldsymbol{\Omega}(s)$ imposes $\mathbf{t}'(s) = \boldsymbol{\Omega}(s) \times \mathbf{t}(s)$, and the general solution of this equation for $\boldsymbol{\Omega}(s)$ is

$$\boldsymbol{\Omega}(s) = \mathbf{t}(s) \times \mathbf{t}' + \tau(s)\,\mathbf{t}(s),$$

where the tangential component of $\boldsymbol{\Omega}(s)$, denoted $\tau(s) = \langle \boldsymbol{\Omega}(s)|\mathbf{t}(s)\rangle$, is unknown for the moment.

Identifying $\kappa^{(1)}$, $\kappa^{(2)}$ and τ by comparing expression above for $\boldsymbol{\Omega}(s)$ with equation (3.64), we can rewrite the constitutive relation (3.65) for a symmetric rod with bending modulus EI and twist modulus μJ in the condensed form:

$$\mathbf{M}(s) = EI\,\mathbf{t}(s) \times \mathbf{t}'(s) + \mu J\,\tau(s)\,\mathbf{t}(s). \tag{3.87}$$

We shall now make use of the three invariants so as to eliminate all unknowns except the function $\theta(s)$. Using the invariant (3.82b), we find the value of the quantity

$$\tau(s) = \frac{M}{\mu J}.$$

Plugging this into the first invariant (3.82a), and using the explicit expression of $\mathbf{t}'(s)$ in equation (3.85), we have

$$EI\,\psi'(s)\,\sin^2\theta(s) + M\,\cos\theta(s) = M.$$

This equation can be used to eliminate the function $\psi(s)$:

$$\psi'(s) = \frac{M}{EI}\,\frac{1 - \cos\theta(s)}{\sin^2\theta(s)}. \tag{3.88}$$

We have not yet used the last invariant (3.82c), associated with the energy of the top. Evaluation of its left-hand side requires the calculation of $|\mathbf{t} \times \mathbf{t}'|^2$, which has already been done in equation (3.86). The invariant yields

$$\frac{EI\,(\theta'^2(s) + \sin^2\theta(s)\,\psi'^2(s))}{2} + \frac{M^2}{2\,\mu J} + T\,\cos\theta(s) = \frac{M^2}{2\,\mu J} + T.$$

Eliminating ψ' with the help of equation (3.88), we arrive at an equation for the function $\theta(s)$ only:

$$\frac{1}{2}\theta'^2(s) + \frac{T}{EI}\,V_\gamma(\theta(s)) = 0, \tag{3.89a}$$

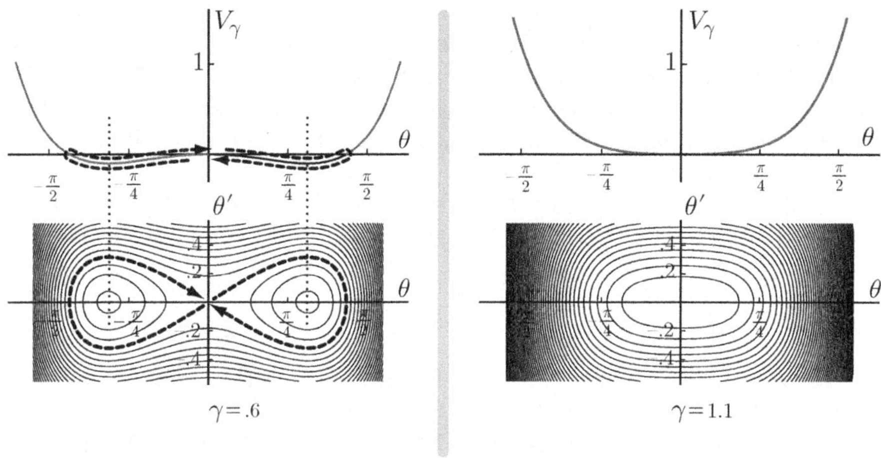

Fig. 3.12 Potential (top) and phase diagram (bottom) of the virtual pendulum associated with the envelope $\theta(s)$ of the localized helix solution, for $\gamma = 0.6 < \gamma_c$ (left) and for $\gamma = 1.1 > \gamma_c$ (right).

where

$$V_\gamma(\theta) = 2\gamma \left(\frac{1 - \cos\theta}{\sin\theta} \right)^2 - (1 - \cos\theta) \tag{3.89b}$$

and

$$\gamma = \frac{M^2}{4\,EI\,T}. \tag{3.89c}$$

Equation (3.89a) is the equation of motion for a virtual pendulum in a non-linear potential $V_\gamma(\theta)$—more accurately, this is the equation for the trajectory *with zero total energy*. The potential V_γ is plotted in Fig. 3.12 for two particular values of γ. As indicated by the Taylor expansion of $V_\gamma(\theta)$ near $\theta = 0$

$$V_\gamma(\theta) = \frac{\gamma - 1}{2}\,\theta^2 + \cdots,$$

the number of equilibria changes for the critical value γ_c of the parameter γ:

$$\gamma_c = 1, \tag{3.90a}$$

corresponding to a critical value M_c of the applied twist M (for a given tension T):

$$M_c = 2\sqrt{EI\,T}. \tag{3.90b}$$

Indeed, for $0 < \gamma < \gamma_c$, as in the left part of the figure, there are two symmetric stable equilibria, and an unstable equilibrium $\theta = 0$. For $\gamma > \gamma_c$, the only equilibrium is $\theta = 0$ and it is stable. The non-linear ordinary differential equation (3.89a) for $\theta(s)$ can be integrated graphically in phase space, by representing the level curves of the left-hand side of equation (3.89a) in the (θ, θ') plane. These curves are shown in the lower part of Fig. 3.12. In this plane, solutions $(\theta(s), \theta'(s))$ of equation (3.89a) draw a parametric curve that is contained in the level curve corresponding to the value 0; this is the curve touching the origin, since $V_\gamma(0) = 0$. For $\gamma > \gamma_c$, this level curve is reduced to a point, which shows that the only solution is the unbuckled solution, $\theta(s) = 0$ for all s. However, for $0 < \gamma < \gamma_c$, the level curve corresponding to the value 0 is not trivial: either half of the figure of eight drawn in Fig. 3.12 with dashes defines a so-called *homoclinic* curve (the two corresponding solutions correspond to the symmetry by a change of sign of $\theta(s)$). Note that it takes an infinite time to land on, and to depart from the origin for these homoclinic solutions: the origin $\theta = 0$ is reached asymptotically only, for $s \to \pm\infty$.

An explicit form of these homoclinic solutions can be found by quadrature, starting from equation (3.89a). We omit the details of the calculation, and the final result is

$$\theta(s) = \cos^{-1}\left(2\gamma - 1 + 2\left(1 - \gamma\right)\tanh^2\left(\sqrt{1 - \gamma}\,\frac{s - s_0}{\sqrt{EI/T}}\right)\right), \tag{3.91}$$

where the free parameter s_0 corresponds to the invariance of the equations by translation along the centre line. This classical non-linear solution is found in many places, see for instance equation (69) in reference (G. H. M. van der Heijden and J. M. T. Thompson, 2000).

Equation (3.91) characterizes the envelope of the modulated helix: by equation (3.83), the angle $\theta(s) = \cos^{-1}\langle\mathbf{t}(s)|\mathbf{e}_z\rangle$ yields the deviation of the tangent from the z axis. The other Euler angle $\psi(s)$ can be reconstructed from equation (3.88); the aspect of the solution is shown in Fig. 3.11.

The formation of a localized helix is confirmed in Fig. 3.13 by numerical simulations of the dynamics of rods with contact, based on the numerical model of 'discrete elastic rods' described in reference (M. Bergou *et al.*, 2008). A naturally straight rod is twisted, and then clamped at both endpoints. As the clamping frames are brought closer to each other, a helical pattern appears. For small displacements of the endpoints, the helix is weakly localized and has a small amplitude, as the parameter γ is slightly below 1; when the displacement of the endpoints is increased, the parameter γ decreases, and both the amount of localization and the amplitude of the helix increase. At some point, the helical solution becomes unstable and the rod jumps to a configuration with contact—see frames (e–f) of the simulation.

The stability of the helix and the formation of a loop have been investigated both experimentally and theoretically by Thompson and Champneys (J. M. T. Thompson and A. R. Champneys, 1996) in the case of an infinite rod, and by van der Heijden and collaborators (G. H. M van der Heijden *et al.*, 2003) for a rod of finite length. In the case of an infinitely long rod, the modulated helix becomes unstable when the maximum deflection of the tangent from the z axis reaches $\pi/2$ (S. Neukirch, G. H. M. van der Heijden, and J. M. T. Thompson, 2002); by equation (3.91), this happens for $s = s_0$, $\cos\theta_0 = 0$ and so $\gamma = 1/2$, that is $M = M_c/\sqrt{2}$.

Even though our derivation of the localized helix solution used the presence of three invariants to start with, the main features of this family of solutions is actually independent of these invariants. The localized helix indeed belongs to a broader family of patterns that appear generically near a threshold for linear stability. Such patterns are obtained by slowly modulating the amplitude of the linearly unstable mode, and are ruled by amplitude equations.[37] A weakly non-linear analysis of a rod under combined tension and twist is performed in reference (B. Audoly and S. Neukirch, 2009), and leads to a localized helix

---

[37] Amplitude equations will be discussed in Section 7.5.3 in the context of buckling analysis of a long rectangular plate.

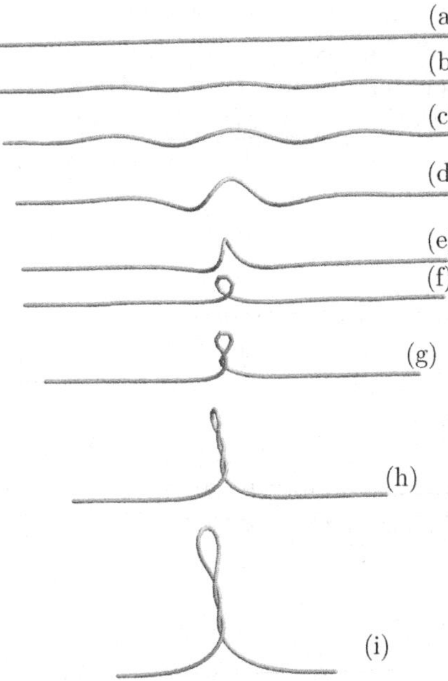

Fig. 3.13 From helical buckling to plectonemes. A relative rotation of the ends is imposed before the simulation starts; subsequently, in frames (a) to (i), the endpoints are brought closer to each other. The unbuckled configuration (a) evolves to a helix, which becomes more and more localized as the amount of shortening increases, see frames (b) to (e). At some point, the helix solution (e) becomes unstable, and the rod jumps to a configuration with a loop and one point of contact, (f). Upon further shortening, an entangled structure, known as *plectonemes* or *localized writhing* is formed and grows, see frames (g) to (i).

solution that is identical to that described by equation (3.91) in the limit $\gamma \to \gamma_c$, and does not make use of the invariants. In a more elementary manner, we should also note that the expression (3.90b) of the critical twist, as well as the wavelength of the helix at the threshold of instability can be obtained directly by a classical analysis of *linear* stability.

The buckling of a planar Elastica, studied in Appendix A, is governed by competition between bending energy and the work of the external force; twist plays no role. The helical buckling of a rod under mixed tension and twist is a more interesting example as it mixes bending and twist deformations. This helical buckling can be more easily interpreted by considering mixed—half-dead, so to say—boundary conditions, where the traction force is imposed and the orientation of the end cross-sections is fixed. The bending energy is zero in the straight, unbuckled configuration, and so increases upon buckling. When the rod buckles, it accepts expending some extra bending energy only as long as this allows it to lower its total energy: the helical pattern has to help decrease the twisting energy. Why is this so, and how can a change in the centre line affect the amount of twist stored in the rod? This is because the orientation of the cross-sections, the twist τ and the shape of the centre line are all dependent: one cannot modify one without changing the others—this is the meaning of the fundamental geometrical identity (3.51). As a result, the twist will be affected by any perturbation of the centre line, when the end cross-sections have a prescribed orientation as we assumed here. This is classically explained as the conversion between twist and writhe (G. H. M. van der Heijden and J. M. T. Thompson, 2000). An equivalent view based on the geometrical notion of parallel transport is proposed in reference (M. Bergou *et al.*, 2008).

3.10   Conclusion

The Kirchhoff equations have a rather complex structure due to the inter-twinning of geometry and elasticity. Being non-linear they have explicit solutions in a few special cases only. Nowadays the search for explicit solutions is perhaps not as crucial as it used to be because of the possibility of finding easily accurate numerical solutions. Nevertheless, it is often instructive to have some idea of the 'physics' behind the equations and classical analysis is helpful for this. In the coming chapters, we shall use the Kirchhoff equations to get insight into various problems, like the structure of the edge of leaves and the shape of human hair.

References

B. Audoly and S. Neukirch. Instability of an elastic knot. In preparation, 2009.

S. Antman. *Nonlinear Problems of Elasticity*. Springer, 2nd edition, 2005.

R. Bishop. There is more than one way to frame a curve. *The American Mathematical Monthly*, 83:246–251, 1975.

J.-F. Bourgat, P. Le Tallec, and S. Mani. Modélisation et calcul des grands déplacements de tuyaux élastiques en flexion torsion. *Journal de Mécanique Théorique et Appliquée*, 7(4):379–408, 1988.

W. R. Bauer, R. A. Lund, and J. H. White. Twist and writhe of a DNA loop containing intrinsic bends. *Proceedings of the National Academy of Sciences*, 90(3):833–837, 1993.

D. Bensimon, A. J. Simon, V. Croquette, and A. Bensimon. Stretching DNA with a receding meniscus: Experiments and models. *Physical Review Letters*, 74(23):4754–4757, June 1995.

M. Bergou, M. Wardetzky, S. Robinson, B. Audoly, and E. Grinspun. Discrete elastic rods. *ACM Transactions on Graphics*, 27(3):63, 2008.

N. Clauvelin, B. Audoly, and S. Neukirch. Mechanical response of plectonemic DNA: An analytical solution. *Macromolecules*, 41(12):4479–4483, 2008.

A. Cimetière, G. Geymonat, H. Le Dret, A. Raoult, and Z. Tutek. Asymptotic theory and analysis for displacements and stress distribution in nonlinear elastic straight slender rods. *Journal of Elasticity*, 19:111–161, 1988.

N. Chouaïeb. *Kirchhoff's problem of helical solutions of uniform rods and stability properties*. PhD thesis, École polytechnique fédérale de Lausanne, Lausanne, Switzerland, 2003.

P. Cluzel, A. Lebrun, C. Heller, R. Lavery, J.-L. Viovy, D. Chatenay, and F. Caron. DNA: An extensible molecule. *Science*, 271(5250):792–794, 1996.

J. Coyne. Analysis of the formation and elimination of loops in twisted cable. *IEEE Journal of Oceanic Engineering*, 15(2):72–83, 1990.

B. D. Coleman and D. Swigon. Theory of self-contact in Kirchhoff rods with applications to supercoiling of knotted and unknotted DNA plasmids. *Philosophical Transactions of the Royal Society A: Mathematical, Physical and Engineering Sciences*, 362(1820):1281–1299, 2004.

E. H. Dill. Kirchhoff's theory of rods. *Archive for History of Exact Sciences*, 44(1):1–23, 1992.

R. E. Goldstein and S. A. Langer. Nonlinear dynamics of stiff polymers. *Physical Review Letters*, 75(6):1094–1097, Aug. 1995.

R. E. Goldstein, T. R. Powers, and C. H. Wiggins. Viscous nonlinear dynamics of twist and writhe. *Physical Review Letters*, 80(23):5232–5235, June 1998.

R. Jamal and É. Sanchez Palencia. Théorie asymptotique des tiges courbes anisotropes. *Comptes Rendus de l'Académie des Sciences - Series I - Mathematics*, 322:1099–1106, 1996.

G. R. Kirchhoff. Über das Gleichgewicht und die Bewegung eines unendlich dünnen elastischen Stabes. *Journal für die reine und angewandte Mathematik. Journal de Crelle (Berlin)*, 56:285–313, 1859.

I. Klapper. Biological applications of the dynamics of twisted elastic rods. *Journal of Computational Physics*, 125(2):325–337, 1996.

T. Lionnet, S. Joubaud, R. Lavery, D. Bensimon, and V. Croquette. Wringing out DNA. *Physical Review Letters*, 96(17):178102, 2006.

P. Le Tallec, S. Mani, and F. Alves Rochinha. Finite element computation of hyperelastic rods in large displacements. *M2AN*, 26(5):595–625, 1992.

J. Langer and D. A. Singer. Lagrangian aspects of the Kirchhoff elastic rod. *SIAM Review*, 38(4):605–618, 1996.

T. McMillen and A. Goriely. Tendril perversion in intrinsically curved rods. *Journal of Nonlinear Science*, 12(3):241–281, 2002.

M.-G. Mora and S. Müller. Derivation of the nonlinear bending-torsion theory for inextensible rods by gamma-convergence. *Calculus of Variations and Partial Differential Equations*, 18:287–305, 2002.

J.-J. Marigo and N. Meunier. Hierarchy of one-dimensional models in nonlinear elasticity. *Journal of Elasticity*, 83:1–28, 2006.

J. F. Marko and E. D. Siggia. Statistical mechanics of supercoiled DNA. *Physical Review E*, 52(3):2912–2938, September 1995.

S. Neukirch. Extracting DNA twist rigidity from experimental supercoiling data. *Physical Review Letters*, 93(19):198107, 2004.

M. Nizette and A. Goriely. Towards a classification of Euler–Kirchhoff filaments. *Journal of Mathematical Physics*, 40(6):2830–2866, 1999.

S. Neukirch, G. H. M. van der Heijden, and J. M. T. Thompson. Writhing instabilities of twisted rods: from infinite to finite length. *Journal of the Mechanics and Physics of Solids*, 50(6):1175–1191, 2002.

T. R. Strick, J.-F. Allemand, D. Bensimon, A. Bensimon, and V. Croquette. The elasticity of a single supercoiled DNA molecule. *Science*, 271(5257):1835–1837, 1996.

S. B. Smith, Y. Cui, and C. Bustamante. Overstretching B-DNA: The elastic response of individual double-stranded and single-stranded DNA molecules. *Science*, 271(5250):795–799, 1996.

D. J. Steigmann and M. G. Faulkner. Variational theory for spatial rods. *Journal of Elasticity*, 33(1):1–26, 1993.

J. Sanchez-Hubert and É. Sanchez Palencia. Statics of curved rods on account of torsion and flexion. *European Journal of Mechanics. A. Solids*, 18:365–390, 1999.

J. V. Selinger, F. C. MacKintosh, and J. M. Schnur. Theory of cylindrical tubules and helical ribbons of chiral lipid membranes. *Physical Review E*, 53(4):3804–3818, April 1996.

J. M. T. Thompson and A. R. Champneys. From helix to localized writhing in the torsional post-buckling of elastic rods. *Proceedings of the Royal Society A: Mathematical, Physical and Engineering Sciences*, 452(1944):117–138, 1996.

G. H. M van der Heijden, S. Neukirch, V. G. A. Goss, and J. M. T. Thompson. Instability and self-contact phenomena in the writhing of clamped rods. *International Journal of Mechanical Sciences*, 45:161–196, 2003.

G. H. M. van der Heijden and J. M. T. Thompson. Helical and localised buckling in twisted rods: A unified analysis of the symmetric case. *Nonlinear Dynamics*, 21(1):71–99, 2000.

E. E. Zajac. Stability of two planar loop Elasticas. *Journal of Applied Mechanics*, 29:136–142, 1962.

4
Mechanics of the human hair

The human hair is a good example of a system with a simple geometry but a complex mechanical behaviour. Because its length (a few centimetres to a few tens of centimetres) is much larger than its radius (about a tenth of a millimetre), it can be described by the equations for elastic rods, already derived in Chapter 3. The human hair will be the first example of application of the Kirchhoff equations.[1] To a large extent the geometric nonlinearity in these equations is responsible for the complex mechanical behaviour of hairs. Besides its interest as a subject of purely scientific investigation, the hair is at the core of several industries. The most obvious example is the haircare industry (making and selling shampoo and conditioners but also hair colourants and styling agents) but it is also present in the entertainment industry, for video games and feature animation movies; until recently a visually realistic computer model for hair was lacking (N. Magnenat-Thalmann, S. Hadap, and P. Kalra, 2000; D. K. Pai, 2002; K. Ward *et al.*, 2007). Such a realistic model of hair requires a good understanding of its mechanics, which is the topic of the present chapter.

The following development will explain one important issue, namely that some people have almost straight hairs, although others have wavy, curled hairs. Of course, the variety we can observe is often the result of more or less complex processing, including rather sophisticated chemistry, with the purpose of changing the physical properties as well as the colour. But there is already a great deal of variety in natural hairstyles and we do not plan to explain here the ways of the cosmetic industry. Instead we shall try to understand with a rather simple model why hair can be straight, curled from the scalp to the end, or have curls only at their end (ringlets). By 'understand' we mean that we shall reduce the different looks (excluding colour) to quantitative changes of the mechanical parameters in a single mathematical model. This simple model has the virtue of putting together the main effects present in the mechanics of the single hair: permanent/natural waving or curling (if there is one), gravity and length. Another phenomenon at work in the hairdo is the interaction between hairs; this is out of the scope of the present development because it would bring us too far from the mainstream of this book, the relationship between elasticity and geometry. The interaction between hair strands is the result of a rather complex process and is quite sensitive to added chemicals, to humidity, static electricity and so on.

The net outcome of the theory presented here (which includes, to be sure, some major simplifications) is that hairs can be described physically by two dimensionless parameters, with consequences that are in fair agreement with casual observations.

---

[1] The more classical example of Euler's Elastica is presented in Appendix A.

4.1 Dimensional analysis

In Section 4.1.1 below, we introduce our model for a single hair strand. This model is based on the elastic response of the hair, as described by the Kirchhoff equations for elastic rods, with two additional ingredients: gravity (the hair falls under its own weight) and spontaneous curvature (natural tendency to curve). In Section 4.1.1, it is shown that all these ingredients can be collected in an energy functional, see equation (4.1), which is the starting point of our analysis. Then, in Section 4.1.2, we present a qualitative classification of the hair shapes based on a dimensional analysis of this energy. The equations of equilibrium for the hair are derived only later in Section 4.2 and their solutions in various limits are derived in Sections 4.3 and 4.4.

4.1.1   A 2D mechanical model for the single hair

The assumptions that we shall make are for the purpose of obtaining a model that is as simple as possible, without leaving aside anything fundamental. To start with, we shall assume that the hair, curled or not, is two dimensional: until the end of the present chapter, in Section 4.5.1, we shall not attempt to describe truly 'helical' curling. In the following 'two dimensional' theory, we shall allow self-intersection of the curve that describes a single hair: such an intersection is obviously unlikely in 3D and therefore irrelevant. Interaction with obstacles, such as the head and shoulders, are not considered in our simple model—they are considered in the extension of the model used to simulate a full head of hair, as discussed in the end of this chapter.

Geometrically the hair strand will be described by an inextensible planar curve, eventually self-intersecting, with s as the curvilinear abscissa, and $\mathrm{d}s$ as the element of arc length. The total length, L, is fixed. The hair has a natural curvature κ_0, although in its deformed state (that is once gravity sets in) it has curvature $\kappa(s)$. This natural curvature κ_0 is taken to be uniform over the hair length for the sake of simplicity, although the extension of the present results to a non-constant natural curvature, $\kappa_0(s)$, does not raise any difficulty. Finally, let ρ and S be its volumic mass and its section, which are also assumed constant. The energy of an arbitrary hair configuration is:

$$\mathcal{E}_{\mathrm{hair}} = \int_0^L \mathrm{d}s \left[\frac{E\,I}{2}(\kappa(s) - \kappa_0)^2 + \rho\,g\,S\,z(s) \right].\tag{4.1}$$

In this expression, $(E\,I)$ is the bending stiffness of the hair and $z(s)$ is the vertical coordinate along the hair, $\rho g S z(s)\,\mathrm{d}s$ being the potential energy of the length element $\mathrm{d}s$ in the vertical gravity field $\mathbf{g} = -g\mathbf{e}_z$ (see Fig. 4.1).

The explanation of the various terms in equation (4.1) is as follows. The first term accounts for the elasticity of the hair: it is proportional to $(\kappa(s) - \kappa_0)^2$, a quantity that is minimum (and in fact equal to zero) if the actual curvature $\kappa(s)$ is equal to the natural one, κ_0. This term is the elastic energy for a rod with natural curvature; with κ_0 and g set to zero, the energy $\mathcal{E}_{\mathrm{hair}}$ becomes the energy of the familiar Euler's *Elastica*, studied in Appendix A.

By elastic effects, the rod tends to recover a naturally circular shape (uniform κ_0). This is balanced by the downward pull of gravity represented by the second term, $\rho g S z(s)$. The gravitational contribution to energy is made the lowest, as expected, when $z(s)$ is as negative as possible, that is when the hair bends downwards. Assuming the hair to be a

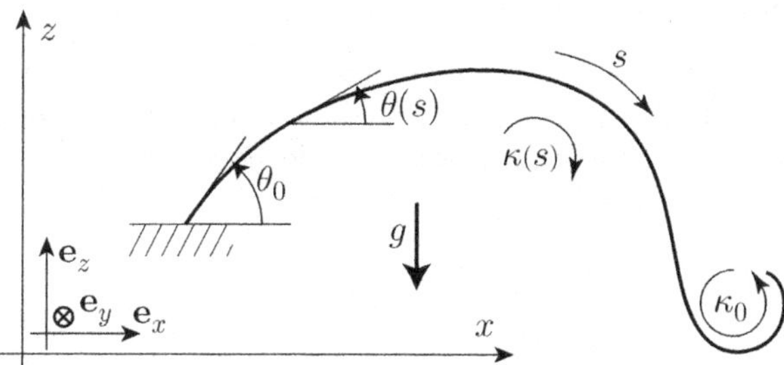

Fig. 4.1 Elastic parameters and geometric definitions for our two-dimensional model of a hair.

rod of circular section made of homogeneous Hookean material,[2] we derived in Chapter 3, in equation (3.28), the following expression for the bending modulus:

$$E\,I = \frac{E\,\pi\,r^4}{4}.$$

(4.2)

Note that for planar deformations, as investigated here, there is only one relevant elastic modulus for the rod, the bending moment (EI).

We shall be concerned with the minimization of the energy $\mathcal{E}_{\mathrm{hair}}$ under the constraint that one end of the hair is rooted in the scalp and starts with a fixed angle θ_0. The shape of the hair is parameterised using the function $\theta(s)$, which is the angle of the local tangent to the hair with respect to the horizontal direction, see Fig. 4.1. At equilibrium, the function $\theta(s)$ is such that the total energy $\mathcal{E}_{\mathrm{hair}}$ is minimum.[3] The parameterisation by the tangent direction $\theta(s)$ incorporates the inextensibility of the hair: any smooth function $\theta(s)$ satisfying the relevant boundary condition at the scalp yields a physically possible hair configuration. As a result, the extensibility constraint does not need to be considered during the minimization of the energy. Recall that the curvature $\kappa(s)$ has been defined in Section 1.3.2 as the rate of change of the direction $\theta(s)$ of the tangent along the hair: $\kappa(s) = \mathrm{d}\theta/\mathrm{d}s$. The boundary condition at the root is $\theta(0) = \theta_0$. The vertical position $z(s)$, needed to compute the potential energy, is a secondary quantity that is given in terms of the main unknown $\theta(s)$ by the integral:

$$z(s) = \int_0^s \mathrm{d}s' \sin\theta(s').$$

(4.3)

[2] A real human hair strand has a complex structure, made of bundles of intertwined fibres at many length scales. It is therefore far from being uniform in the radial direction. The estimation of the bending modulus in equation (4.2), based on the approximation of a uniform and isotropic material, gives an order of magnitude estimate at best. This does not prevent the present theory being applicable, however, using an *effective* bending modulus (EI). The latter cannot be easily computed from a micromechanical model of the hair, but it can be directly measured in simple bending experiments (G. V. Scott and C. R. Robbins, 1969; J. A. Swift, 1995) or by analysing the vibration modes of a hair fibre (G. C. Garson *et al.*, 1980) for instance.

[3] Given the non-linear character of the Kirchhoff equations, there can in fact be several local minima of energy for fixed values of the parameters. This will be briefly discussed in Section 4.5.1 and in more detail in Chapter 5.

4.1.2   Classification of hair shapes in two dimensions

With the aim to determine the relevant parameters for the hair shape, let us introduce dimensionless variables. The dimensionless curvilinear coordinate is defined as $\bar{s} = s/L$ and varies between 0 (root) and 1 (free end). Similarly let $\bar{\kappa}(\bar{s}) = L\,\kappa(s)$ and $\bar{z} = z/L$. In terms of the overlined quantities, the elastic energy becomes:

$$\mathcal{E}_{\text{hair}} = \frac{E\,I}{L} \int_0^1 \mathrm{d}\bar{s} \left[\frac{1}{2}(\bar{\kappa}(\bar{s}) - \alpha)^2 + \frac{\bar{z}(\bar{s})}{\beta} \right]. \tag{4.4}$$

Here, we have introduced two dimensionless parameters, α and β, which will play a crucial role in the following:

$$\alpha = L\,\kappa_0 \qquad \beta = \frac{E\,I}{\rho\,g\,S\,L^3} \tag{4.5}$$

The equilibrium shape of the hair, given by minimization of the energy (4.4), depends on θ_0 and on these two parameters, which turn out to vary quite widely between humans. Typically a human hair has radius $r \sim 50\,\mu\mathrm{m}$, length $L \sim 10\,\mathrm{cm}$, natural curvature $\kappa_0 \sim 0.05$ to $5\,\mathrm{cm}^{-1}$, average modulus $\mathrm{E} \sim 1\,\mathrm{G\,Pa}$ and density $\rho \sim 1.3\,10^3\,\mathrm{kg.m}^{-1}$. As a result, α varies typically in the range 0.1 to 1.0, and β in the range 0.01 to 10 (β varies greatly with hair length and humidity condition).

The parameter α is the ratio of the total length L to the natural radius of curvature, $1/\kappa_0$. This is 2π times the number of curls made by the hair in the absence of gravity. If α is large, spontaneous curling is important, while for small α the hair is almost straight. The other parameter, β, measures by how much the pull of gravity changes the natural shape of the hair: its shape is not too affected if β is large[4] but is significantly affected if β is small. The influence of gravity on hair shape is not so obvious: if gravity is important (β small), curling may be removed almost everywhere (the hair is straight under its own weight) but persist near the ends only. There is a third dimensionless parameter, the root angle θ_0, in this problem. However, its range of values is almost the same from one hairstyle to another and it has a weaker influence on the global aspect of the hair.[5] Therefore, we shall mainly concentrate on the influence of the two parameters α and β.

The existence of two dimensionless parameters suggests a classification of the hairs in a two-dimensional plane (α,β), as shown in Fig. 4.2. For β and α small, the hair is straight and flexible: it will tend to fall vertically in the gravity field. For β still small but α large, the hair will tend to fall vertically in the gravity field but its end will be curled because the spontaneous radius of curvature is much less than the total length. When both α and β are large, gravity is unimportant and the hair is curled along its length. For β large and α small, gravity is unimportant as well as curling: the hair tends to spontaneously pop-up all

---

[4] In the case of weak gravity, the small parameter $1/\beta$ has a simple physical interpretation. Indeed, consider a hair that is naturally straight ($\alpha = 0$) and is clamped horizontally in the scalp ($\theta_0 = 0$). Then, under weak gravity, this initially straight hair will slightly depart from its horizontal configuration. Its weight, $P \sim (\rho\,S\,L\,g)$, indeed induces the typical bending torque $M \sim P\,L \sim (\rho\,S\,L^2\,g)$, hence the typical, small curvature $\kappa \sim M/(E\,I) \sim (\rho\,S\,L^2\,g)/(E\,I)$. This means that the direction of the hair near the free end will change by an angle of order $\theta(L) \sim \kappa\,L \sim 1/\beta$ in radians due to gravity. The detailed calculation is given in Section 4.3.

[5] The implantation of hair on the scalp is non-uniform. This is revealed by vortex patterns in short hairs. For an unknown, probably genetic, reason those patterns happen to run is most cases in the clockwise direction.

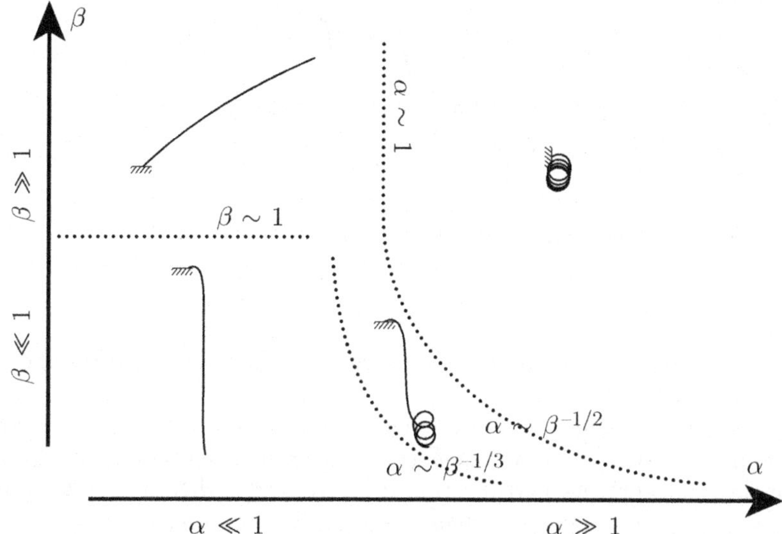

Fig. 4.2 Classification of hair shapes in the plane of dimensionless parameters (α, β): α measures the amount of spontaneous curling, and β the importance of gravity on the equilibrium shape. The upper part of the diagram (weak gravity) is studied in Section 4.3: there, the hair adopts a configuration that is close to its natural shape, which is almost straight for $\alpha \ll 1$, and curled for $\alpha \gg 1$. The lower part of the diagram (strong gravity) is studied in Section 4.4, where we show that for small spontaneous curvatures ($\alpha \ll \beta^{-1/3}$), the hair falls almost vertically, while for $\beta^{-1/3} \ll \alpha \ll \beta^{-1/2}$, the best compromise between elastic and gravitational energy is to keep curls, but only near the free end.

along the direction shown by its root angle. We shall make this discussion more quantitative by explicitly solving the minimization problem in these various limits.

4.2 Equilibrium equations

The aim of the present section is to derive the equilibrium equations for a hair strand. These equilibrium equations are first deduced in Section 4.2.1 by imposing that the total energy in equation (4.1) be minimum at equilibrium. Doing so, we are essentially repeating the derivation of the Kirchhoff equations presented in Chapter 3 but in a much simpler, two-dimensional geometry. In fact, the reader who is already familiar with the equations for rods presented in Chapter 3 can jump ahead to Section 4.2.2: there, the equations for a hair strand are deduced directly from the Kirchhoff equations. For a first insight into the equations of rods, the self-contained derivation in Section 4.2.1 might be helpful.

4.2.1 Direct minimization of the energy

Let $\overline{\mathcal{E}}_{\text{hair}} = E I / L \; \mathcal{E}_{\text{hair}}$ be the dimensionless elastic energy of the single hair given in equation (4.4). The form of equation (4.4) is not very convenient as it involves $\overline{z}(\overline{s})$, which is related to $\theta(\overline{s})$ by an integral (4.3). To make things more explicit, we rewrite the contribution of gravitational energy as follows.:

$$\frac{1}{\beta} \int_0^1 d\overline{s}\ \overline{z}(\overline{s}) = \frac{1}{\beta} \int_0^1 d\overline{s} \int_0^{\overline{s}} d\overline{s}'\ \sin\theta(\overline{s}').$$

A change of order of integration transforms this gravitational energy into:

$$\frac{1}{\beta} \int_0^1 d\overline{s}'\ (1 - \overline{s}')\ \sin\theta(\overline{s}').$$

Plugged back into the dimensionless energy, this yields:

$$\overline{\mathcal{E}}_{\text{hair}} = \int_0^1 d\overline{s} \left[\frac{1}{2}(\overline{\kappa}(\overline{s}) - \alpha)^2 + \frac{(1 - \overline{s})}{\beta} \sin\theta(\overline{s}) \right]. \tag{4.6}$$

For the sake of readability, we shall drop the overlines in the following, although we continue to deal with rescaled quantities.

Let us now derive the condition of stationarity of this energy, using the Euler–Lagrange method, presented in Appendix A. Let $\delta\theta(s)$ be a small perturbation to $\theta(s)$. From the definition of the curvature, its perturbation is $\delta\kappa(s) = \delta\theta'$, where $\delta\theta' = d/ds\ \delta\theta(s)$. After integration by parts, one obtains the change of $\mathcal{E}_{\text{hair}}$ to first order in $\delta\theta$:

$$\delta\mathcal{E}_{\text{hair}} = [(\theta' - \alpha)\delta\theta]_0^1 + \int_0^1 ds\ \delta\theta(s) \left[-\frac{d}{ds}(\theta' - \alpha) + \frac{(1 - s)}{\beta} \cos\theta(s) \right]. \tag{4.7}$$

In this equation, the boundary terms of the integration by parts, $[(\theta' - \alpha)\,\delta\theta]_0^1$, are treated as explained in Section 3.6.4: at the clamped end $s = 0$, the only boundary condition is $\theta(0) = \theta_0$. At the free end $s = 1$, the equilibrium requires the additional condition $\theta'(1) = \alpha$. This condition can be interpreted as the cancellation of the internal moment, $M_y(L) = EI\,(\kappa(L) - \alpha)$.

In-between the two ends the integrand in (4.7) must be zero for any value of $\delta\theta$, which imposes that $\theta(s)$ is the solution of the second-order non-autonomous differential equation:

$$\theta''(s) - \frac{(1 - s)}{\beta} \cos\theta(s) = 0, \tag{4.8}$$

subject to the boundary conditions:

$$\theta(0) = \theta_0, \qquad \theta'(1) = \alpha. \tag{4.9}$$

This non-linear equation is not integrable analytically for arbitrary values of α and β.

4.2.2 Derivation from the general equations for rods

A direct minimization of the elastic energy (4.1) has yielded the equation of equilibrium for an isolated hair strand. For the sake of completeness, we derive below the same equation from the general theory of rods presented in Chapter 3.

The general formalism of three-dimensional rods can be particularized to the planar case by choosing for the material frame $\mathbf{d}_1(s) = \mathbf{e}_y$, $\mathbf{d}_2(s) = \mathbf{n}(s)$ and $\mathbf{d}_3(s) = \mathbf{t}(s)$, where $\mathbf{t}(s) = \cos\theta(s)\,\mathbf{e}_x + \sin\theta(s)\,\mathbf{e}_z$ are the tangent and $\mathbf{n}(s) = -\sin\theta(s)\,\mathbf{e}_x + \cos\theta(s)\,\mathbf{e}_z$ the normal to the curve in the plane of the deformation. Then, from the general definition of the Darboux vector, we find $\kappa^{(1)}(s) = 0$ (the curve does not bent out of its plane), $\tau(s) = 0$ (no twist) and $\kappa^{(2)}(s) = -\theta'(s)$. The latter equation means that the second material curvature is *minus* the signed curvature of the planar centre line. The minus sign results from the geometrical conventions and is without special meaning.

In the present case, the applied force is the weight, whose lineic density reads:

$$\mathbf{p} = -\rho\,g\,S\,\mathbf{e}_z,\tag{4.10}$$

where $\mathbf{e}_z$ is a unit vector oriented upwards, as shown in Fig. 4.1. Note that we have temporarily restored the physical quantities in the present section. This lineic weight is uniform: equation (3.63a) for the tension can readily be integrated as

$$\mathbf{F}(s) = -\rho\,g\,S\,(L-s)\,\mathbf{e}_z.\tag{4.11}$$

Recalling that $\mathbf{n}_2(s) = \mathbf{e}_y$, the balance of torque in equation (3.63b), projected along the transverse direction $\mathbf{e_y}$ yields:

$$\frac{\mathrm{d}G^{(2)}}{\mathrm{d}s} + \langle \mathbf{d}_3 \times \mathbf{F} | \mathbf{e}_y \rangle = 0,\tag{4.12}$$

where the notation $\langle \mathbf{a}|\mathbf{b}\rangle$ is for the scalar product of two vectors.

The constitutive relations (3.45) with spontaneous curvature is

$$G^{(2)} = E\,I^{(2)}\left(\kappa^{(2)} - \kappa_0^{(2)}\right) = EI\left(-\theta'(s) + \alpha\right).\tag{4.13}$$

Note the presence of a minus sign in this formula, due to the fact that an *increase* of θ corresponds to a rotation of the material frame attached to the centre line with a rotation vector *opposed* to $\mathbf{e}_y$—see Fig. 4.1.

Combining the two equations above, and replacing $\mathbf{d}_3$ and $\mathbf{F}$ by their expressions, one obtains

$$L^2\,\theta'' - \frac{\rho\,g\,S\,L^3}{E\,I}\left(1 - \frac{s}{L}\right)\cos\theta = 0,\tag{4.14}$$

which is just equation (4.8) in physical, non-rescaled variables. The equation of equilibrium for a hair expresses the balance of torques over a small hair element of length $\mathrm{d}s$, whereby a change of curvature, proportional to θ'' is the response of the rod to the *transverse* component of the local tension: this local tension is just the weight of the half of the hair located on the free hanging side of the current hair element. This weight is proportional to $(L-s)$, the factor $\cos\theta$ coming from the transverse projection.

Likewise, the boundary conditions can be derived from the general equations derived in the last chapter. The first boundary condition in equation (4.9) is simply the clamping condition at $s = 0$. The second one expresses the cancellation of the internal moment at a free edge, as derived by the principle of energy minimization in Section 3.6.4.

4.3　Weak gravity

The equation of equilibrium (4.8) is both non-linear and non-autonomous. Therefore it is not solvable analytically for arbitrary values of the parameters α and β. However, we show in this section and in the next one that solutions can be obtained in various limits, by means of asymptotic techniques.

The limit $\beta \to \infty$ is the easiest one to handle analytically. From the definition (4.5) of β, this corresponds to a hair that is very stiff (large EI), very light (small ρgSL) or/and short (small L), in such a way that the influence of gravity is weak. In this limit, $(1-s)/\beta\cos\theta(s)$ becomes negligible in front of θ'' in equation (4.8), which reduces to $\theta'' = 0$ at dominant order. The solution satisfying the boundary conditions is simply $\theta(s) = \theta_0 + \alpha s$. The centre

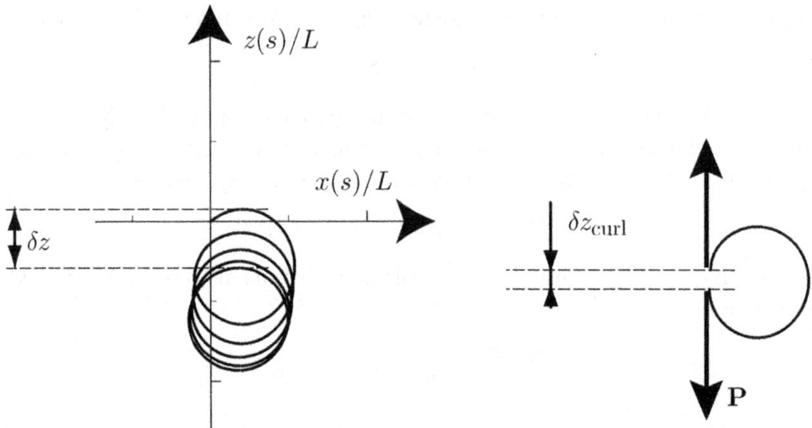

Fig. 4.3 The limit of large β (weak gravity) and large α (curly hair): every curl deforms as a circular spring loaded by the weight of the part of the hair that is located on the free hanging side as shown on the right. The vertical displacement of the free end, δz, is computed in equation (4.18).

line is then an arc of a circle, as imposed by our assumption that κ_0 is constant along the hair.

This solution can further be refined when β is large but finite. We shall restrict this analysis to the simple case where α is also large, although the results could be extended to arbitrary values of α. Then, in its natural configuration, the hair coils around a circle, making approximately $N \approx \alpha/2\pi$ turns, N being the number of times that θ changes by 2π as s goes from 0 to 1. A quantity of interest is the displacement of the free end of this curled hair due to gravity, see Fig. 4.3—this displacement is much smaller than L when β is large. Below, we calculate this displacement to first order in $1/\beta$.

To do so we expand the solution $\theta(s)$ of equation (4.8) in powers of $1/\beta$. The dominant order,[6] $\theta_{(0)}$, is the natural configuration without gravity:

$$\theta_{(0)}(s) = \theta_0 + \alpha\,s, \tag{4.15}$$

which satisfies the boundary conditions by construction.

The first correction is obtained by inserting the solution at the dominant order, $\theta_{(0)}$, into the small term in the equation, that is into $(1 - s)/\beta \cos\theta(s)$. This first-order correction, denoted as $\theta_{(1)}(s)$, is therefore the solution of

$$\frac{\mathrm{d}^2\theta_{(1)}(s)}{\mathrm{d}s^2} = \frac{1 - s}{\beta}\,\cos(\theta_0 + \alpha\,s).$$

Its general solution reads

$$\theta_{(1)}(s) = -\frac{(1 - s)}{\alpha^2\,\beta}\,\cos(\theta_0 + \alpha\,s) + \frac{u\,s + v}{\beta} + \mathcal{O}(1/\alpha^3).$$

For α large, the terms of order $1/\alpha^3$ are negligible compared with the one written explicitly, of order $1/\alpha^2$. The homogeneous solution $(us + v)$, where u and v are arbitrary constants,

---

[6] Note that $\theta_{(0)}$ stands for the dominant term in the expansion of $\theta(s)$, while θ_0 is the clamping angle in the scalp.

should be such that $\theta_{(0)} + \theta_{(1)}$ satisfies the boundary conditions, hence $\theta_{(1)}(0) = 0$ and $\theta'_{(1)}(1) = 0$. This yields:

$$\theta_{(1)}(s) = \frac{1}{\alpha^2 \beta} \left[-(1-s)\cos(\theta_0 + \alpha s) + \cos\theta_0 - s\cos(\theta_0 + \alpha) \right] + \mathcal{O}(1/\alpha^3). \qquad (4.16)$$

Using this expansion of $\theta(s)$, one can compute the vertical position of the free end:

$$z(1) = \int_0^1 \sin(\theta(s))\,\mathrm{d}s \approx \int_0^1 \sin(\theta_{(0)}(s))\,\mathrm{d}s + \int_0^1 \left[\cos\theta_{(0)}(s) \right] \theta_{(1)}(s)\,\mathrm{d}s. \qquad (4.17)$$

The last member of this equation was obtained by expanding the argument of the sine in the integral. Its first term, $\int \sin\theta_{(0)}\mathrm{d}s$, is the position of the free end in the absence of gravity. The displacement of this end is given by the last term that, using the expressions of $\theta_{(0)}$ and $\theta_{(1)}$ above, becomes:

$$\delta z = \frac{1}{\alpha^2 \beta} \int_0^1 \left[-(1-s)\cos(\theta_0 + \alpha s) + \cos\theta_0 - s\cos(\theta_0 + \alpha) \right] \cos(\theta_0 + \alpha s)\,\mathrm{d}s,$$

up to quantities of order $1/\alpha^3$.

There are two length scales involved in this integral: the first one, $1/\alpha$, for terms like $\cos(\theta_0 + \alpha s)$, corresponds to the curvilinear length of a single curl; the other, 1 (or L in physical variables), for terms like $(1-s)$, is the length of the hair. Now, by expanding the bracket, we get three terms: the first one, proportional to $-(1-s)\cos^2(\theta_0 + \alpha s)$, has the value $-(1-s)/2$ when averaged over the short length scale $2\pi/\alpha \ll 1$ while the two others have a dependence on the short length scale proportional to $\cos(\theta_0 + \alpha s)$ without the square, which is zero on average. Therefore, in the limit of large α, only the first term of the expansion contributes to δz at dominant order, and

$$\delta z = \frac{1}{\alpha^2 \beta} \int_0^1 -\left(\frac{1-s}{2}\right)\mathrm{d}s = -\frac{1}{4\,\alpha^2\,\beta}.$$

After restoring physical quantities, this becomes

$$\delta z = -\frac{\rho\,g\,S\,L^2}{4\,E\,I\,(\kappa_0)^2}. \qquad (4.18)$$

This is the vertical displacement of the free end of the hair due to gravity.

This expression, valid for both α and β large, can be extended without difficulty to large β but arbitrary α by retaining all the terms in the above expansions for large α.

There is a simple interpretation of equation (4.18): from Fig. 4.3, every curl behaves like a circular spring opened by a force that is the weight of the part of hair located on the free hanging side. Every spring has a length of order $1/\kappa_0$. The only stiffness that can be built using this length and the bending stiffness of the hair, $E\,I$, is, by dimensional analysis: $k_{\text{curl}} \sim E I(\kappa_0)^3$. Every such curl is loaded by a weight of order $\rho g S L$ (in fact, an individual curl can only support a fraction of this weight, but this is unimportant for the present dimensional analysis). Therefore a curl opens by a typical vertical gap $\delta z_{\text{curl}} \sim \rho g S L/k_{\text{curl}}$. The vertical displacement at the free end of the hair results from N such contributions, where $N \sim L\,\kappa_0$, since there are N curls, hence:

$$\delta z \sim N\,\delta z_{\text{curl}} \sim \frac{\rho\,g\,S\,L^2}{E\,I\,(\kappa_0)^2},$$

which is just equation (4.18) up to a numerical factor, 1/4. This factor can be recovered by computing the stiffness of a circular spring that is slightly opened by some tensile force at its ends.

4.4 Strong gravity

Let us now examine the opposite limit of small β. This describes a hair that is compliant, long, and/or heavy (recall that the parameter β combines the length, the elastic modulus and the mass density) as at the bottom of figure 4.2. The dominant term in equation (4.8) is now $(1 - s)/\beta \cos \theta(s)$ which, once set to zero, no longer yields a differential equation and so cannot satisfy the boundary conditions.[7] In order to satisfy the condition $(1 - s)/\beta \cos \theta(s) = 0$, the dominant order solution is such that $\cos(\theta) = 0$. This is a downward pointing hair with $\theta = -\pi/2$, a configuration that obviously minimizes the gravitational energy, which becomes dominant as $\beta \to 0$. This first-order solution $\theta(s) = -\pi/2$ is not consistent with the boundary conditions, except for the rather uninteresting case $\theta_0 = -\pi/2$ and $\kappa_0 = 0$, the solution then being a perfectly straight and vertical hair. Two boundary layers[8] are needed to accommodate this formal solution $\theta = -\pi/2$ with the boundary conditions: one near the root where the hair changes its orientation from θ_0 at $s = 0$ to $-\pi/2$ slightly further, and another one near the end $s = 1$, where the curvature goes from zero to α at the free end. This qualitative picture of a mostly vertical hair surrounded by two boundary layers near the clamped and free ends was first described by Prost (J. Prost, 1994) to the best of our knowledge.

4.4.1 Boundary layer near scalp

The boundary layer near $s = 0$ is fairly simple to analyse. For β small, the term proportional to s is negligible compared with 1 in the boundary layer, so that $(1 - s)$ can be replaced by 1 in (4.8). The thickness λ of this boundary layer can then be estimated as follows. In equation (4.8), the first term is of order $1/\lambda^2$ since θ changes its value by a finite amount, from θ_0 to $-\pi/2$, over a typical length scale λ; the second term is of order $1/\beta$. In the boundary layer, these two terms should balance[9] and must therefore be of the same order of magnitude, from which we deduce that λ should be of order $\beta^{1/2}$ in rescaled variables, that is $L\beta^{1/2}$ in physical variables. Therefore, one introduces[10] the *stretched* variable $\tilde{s} = s/\beta^{1/2}$ and derives from (4.8) the so-called *inner* equation:

$$\frac{\mathrm{d}^2 \theta_{\mathrm{sc}}}{\mathrm{d}\tilde{s}^2} - \cos \theta_{\mathrm{sc}}(\tilde{s}) = 0, \tag{4.19}$$

in which the small parameter β is no longer present. In this equation, θ_{sc} stands for the function $\theta(\tilde{s})$ near the scalp $\tilde{s} = 0$, solution of the inner problem. The inner stretched variable

---

[7] The order of the differential equation is decreased by formally taking the limit $\beta = 0$. This is the typical mathematical situation where boundary layers occur, see Appendix B.

[8] The analysis of boundary layers is introduced in Appendix B.

[9] At scales much larger than λ, the first term becomes negligible and one recovers the straight, vertical configuration.

[10] We have previously introduced a first rescaling to get rid of units. The s in the right-hand side of the equation defining the new rescaling, $\tilde{s} = s/\beta^{1/2}$ is in fact the rescaled length $\overline{s} = s/L$. These successive rescalings stack up, and so $\tilde{s} = s/(L\beta^{1/2})$ in physical units.

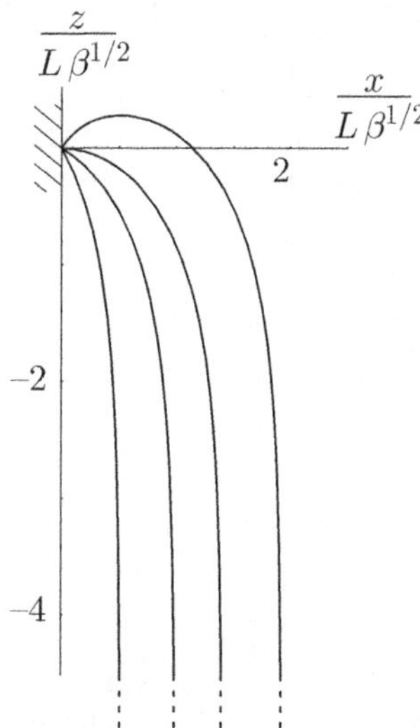

Fig. 4.4 Boundary layer solution near the scalp in stretched coordinates for a heavy/long/ compliant hair ($\beta \ll 1$), as given by equation (4.23). The plots are for different values of the clamping angle θ_0. These universal profiles must be magnified by a scaling factor $L\,\beta^{1/2}$ to obtain the real profiles.

$\tilde{s}$ varies from 0 to ∞. The solution of (4.19) has to satisfy the boundary conditions

$$\theta_{\mathrm{sc}}(\tilde{s} = 0) = \theta_0, \qquad \theta_{\mathrm{sc}}(\tilde{s} = \infty) = -\frac{\pi}{2}. \tag{4.20}$$

Indeed, when β is small, large values of the *stretched* coordinate $\tilde{s} = s/\beta^{1/2}$ correspond to points that are outside of the boundary layer ($s \gg \beta^{1/2}$) but can still be close to the clamped end ($s \ll 1$, that is $s \ll L$ in physical variables), as explained in Appendix B. This defines the so-called intermediate or matching region, $\beta^{1/2} \ll s \ll L$, where the boundary layer solution (the inner solution) should match the outer solution.

Equation (4.19) can be integrated explicitly:

$$\theta_{\mathrm{sc}}(\tilde{s}) = \frac{\pi}{2} - 4\tan^{-1}\left[\tanh\left(\frac{\tilde{s} + \tilde{s}_0}{2}\right)\right]. \tag{4.21}$$

This solution tends to $-\pi/2$ as $\tilde{s}$ tends to $+\infty$. The constant of integration $\tilde{s}_0$ must be adjusted to satisfy the boundary condition at $s = 0$. This yields:

$$\tilde{s}_0 = 2\tanh^{-1}\left[\tan\left(\frac{\pi/2 - \theta_0}{4}\right)\right]. \tag{4.22}$$

The shape of the centre line near the scalp can then be determined by integration of the kinematical equation $(\tilde{x}'(\tilde{s}), \tilde{z}'(\tilde{s})) = (\cos\theta_{\mathrm{sc}}(\tilde{s}), \sin\theta_{\mathrm{sc}}(\tilde{s}))$, where $\tilde{x} = x/\beta^{1/2}$ and $\tilde{z} = z/\beta^{1/2}$. The integration can be performed explicitly:

$$\tilde{x}(\tilde{s}) = \tilde{x}_0 - 2\left(1 + \frac{1}{\cosh(s + \tilde{s}_0)}\right), \quad \tilde{z}(\tilde{s}) = \tilde{z}_0 + 2\tanh(s + \tilde{s}_0) - (s + s_0), \tag{4.23}$$

where $(\tilde{x}_0, \tilde{z}_0)$ is the position of the clamping point. This yields the profiles shown in Fig. 4.4.

Note that the boundary layer equations we have just solved are in fact those for a classical rod that is clamped on one end and pulled in an oblique direction (taken here as the vertical direction) with a force (here, the weight $\rho S g L$) much larger than the typical force EI/L^2: the two ingredients that we have added to the general theory of rods, namely a distributed weight and the spontaneous curvature, play no role in this boundary layer analysis.

4.4.2 Boundary layer near the free end

The other end of the hair is more interesting, both from the point of view of the analysis and for the look of the hairdo. The estimate of the size λ' of this boundary layer near the free end $s = L$ is similar to that for the boundary layer near the scalp except that the factor $1 - s$, which is $L - s$ in physical coordinates, is now of order λ'. Balancing the two terms in equation (4.8) yields: $1/\lambda'^2 \sim \lambda'/\beta$, hence $\lambda' \sim \beta^{1/3}$. We therefore define the stretched variable:

$$\tilde{s} = \frac{1 - s}{\beta^{1/3}}. \tag{4.24}$$

The same notation is used as for the stretched variable of the previous boundary layer, but the ambiguity should be easy to resolve from the context. In terms of the stretched variable, equation (4.8) can be rewritten without any small parameter:

$$\frac{\mathrm{d}^2\theta_{\mathrm{fr}}}{\mathrm{d}\tilde{s}^2} - \tilde{s}\,\cos\theta_{\mathrm{fr}}(\tilde{s}) = 0, \tag{4.25}$$

where θ_{fr} is the function $\theta(\tilde{s})$ near the free end. For $\tilde{s}$ tending to plus infinity, $\theta_{\mathrm{fr}}(\tilde{s})$ should tend to $-\pi/2$: by the same argument as for the boundary layer near the scalp, this condition allows matching with the outer solution in the intermediate region.

The second boundary condition is $\theta'_{\mathrm{fr}}(0) = -\gamma$, where $\gamma = \alpha\beta^{1/3}$. Although β is small by assumption, $|\gamma|$ can be either large or small, or take finite values since it is the product of a small number $\beta^{1/3}$ times α that is arbitrary. If $|\gamma|$ is finite, nothing more can be said and one has to resort to a numerical solution of the boundary layer equation, leading to a profile similar to that shown in Fig. 4.5(b).

Almost vertical free end ($\gamma \ll 1$)

For γ small, the hair deviates only slightly from the vertical orientation near the free end, as in Fig. 4.5(a). The solution can then be expanded in powers of γ:

$$\theta_{\mathrm{fr}}(\tilde{s}) \approx -\frac{\pi}{2} + \gamma\,\theta_{(1)}(\tilde{s}) + \dots$$

where the dots are for quantities that are of order γ^2 or beyond, and where $\theta_{(1)}(\tilde{s})$ is the solution of the linear equation:

$$\frac{\mathrm{d}^2\theta_{(1)}}{\mathrm{d}\tilde{s}^2} - \tilde{s}\,\theta_1(\tilde{s}) = 0. \tag{4.26}$$

The boundary condition are $\theta_{(1)} \to 0$ as $\tilde{s} \to +\infty$ and $\theta'_{(1)}(\tilde{s}) = -1$ for $\tilde{s} = 0$. An explicit solution can be given in terms of the Airy function $\mathcal{A}_i$:

$$\theta_{(1)} = 3^{1/3}\,\Gamma(1/3)\,\mathcal{A}_i(\tilde{s}),$$

where Γ is Euler's gamma function. As near the scalp, the profile near the free end can then be determined by integration of the kinematical equation for the centre line. This yields the parametric equation:

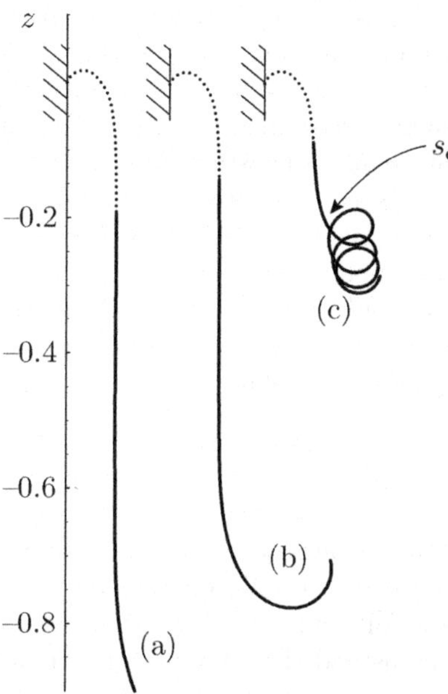

Fig. 4.5 Influence of the parameter γ when β is large: $\beta = 730$ for all plots while (a) $\gamma = 0.25$, (b) $\gamma = 2.0$, (c) $\gamma = 3.45$.

$$x(s) = x_{\mathrm{v}} + \gamma\, L\, \beta^{1/3}\, X\left(\frac{L-s}{L\,\beta^{1/3}}\right), \quad z(s) = z_L + (L-s), \tag{4.27}$$

where x_{v} is the x-coordinate of the central region of the hair ($\tilde{s} \to \infty$) that is vertical and y_L the coordinate of the free end of the hair. The universal profile X is given by:

$$X(\tilde{s}) = 3^{1/3}\,\Gamma(1/3) \int_{\tilde{s}}^{+\infty} \mathcal{A}_i(u)\,\mathrm{d}u,$$

an integral that could be expressed in terms of special functions. The displacement field given in equation (4.27) is along x at the dominant order, which is consistent with $|\theta - (-\pi/2)| \sim |\gamma| \ll 1$. This yields a profile similar to that obtained near the free end in Fig. 4.5(a).

Curled free end ($\gamma \gg 1$)

The other limit, $\gamma \to \infty$, is less easy to handle. A typical numerical solution in this limit is shown in Fig. 4.5(c). The equation we start from is the parameterless ODE (4.25) with the variable $\tilde{s}$ defined in equation (4.24):

$$\frac{\mathrm{d}^2 \theta_{\mathrm{fr}}}{\mathrm{d}\tilde{s}^2} - \tilde{s}\cos\theta_{\mathrm{fr}}(\tilde{s}) = 0. \tag{4.28}$$

The boundary conditions are: $\theta'_{\mathrm{fr}}(0) = -\gamma$ and $\theta_{\mathrm{fr}}(\tilde{s}) \to -\pi/2$ as $\tilde{s} \to +\infty$. In the limit γ large, the frequency of oscillation of the cosine function is much larger than 1 since $|\theta'_{\mathrm{fr}}(1)| = |\gamma| \gg 1$: the coefficient $\tilde{s}$ in front of the cosine in equation (4.28) changes much more slowly than the cosine itself. The existence of two very different length scales is the basis of an analytical solution of these equations in the limit $\gamma \to \infty$, called the adiabatic oscillator approximation. Such an oscillator is defined by the fact that its parameters, here the factor $\tilde{s}$ in the term $\tilde{s}\cos\theta$, change much more slowly than its oscillatory phase, θ.

The time[11] evolution of an adiabatic oscillator is monitored by the constraint of conservation of the adiabatic invariant, according to a general result (L. D. Landau and E. M. Lifshitz, 1982). For the present oscillating system, the adiabatic invariant is the integral over one period of oscillation of $\mathrm{d}\theta/\mathrm{d}\tilde{s}\,\mathrm{d}\theta$, as derived in Section 4.4.3 below. In the remaining part of this section, we shall drop the index 'fr' below θ: θ will always be a function of $\tilde{s}$ in the following analysis of the boundary layer.

Supposing first that $\tilde{s}$ that multiplies $\cos\theta(\tilde{s})$ in equation (4.28) is a constant, an integral of this equation can be obtained by multiplying both terms by $\mathrm{d}\theta/\mathrm{d}\tilde{s}$, yielding the conservation of the energy of the oscillator:

$$\frac{1}{2}\left(\frac{\mathrm{d}\theta}{\mathrm{d}\tilde{s}}\right)^2 - \tilde{s}\sin\theta(\tilde{s}) = C. \tag{4.29}$$

Since $\tilde{s}$ varies over the slow timescale, the constant of integration C is only constant on the fast time scale (the period of the oscillations) but varies on the slow timescale. It is therefore written as $C(\tilde{s})$ in the following.[12] As shown in Section 4.4.3, the quantity that is constant over this slow scale is instead the adiabatic invariant A, defined as:

$$A = \frac{1}{2\pi}\int_{\theta(\tilde{s})}^{\theta(\tilde{s})+2\pi}\mathrm{d}\theta\,\frac{\mathrm{d}\theta}{\mathrm{d}\tilde{s}} = \frac{1}{2\pi}\int_{\theta(\tilde{s})}^{\theta(\tilde{s})+2\pi}\mathrm{d}\theta\,\sqrt{2\left[C(\tilde{s}) + \tilde{s}\sin\theta\right]},$$

where the integral is taken over one arbitrary (fast) period, that is any interval during which the solution $\theta(\tilde{s})$ of (4.28) changes by 2π. The square root in the last integrand has been taken with a plus sign; this assumes $\theta' > 0$, that is $\gamma < 0$ or $\kappa_0 < 0$. The case $\kappa_0 > 0$ is treated similarly.

Since θ is only present in the argument of the sine function in the integral and since the $\tilde{s}$ in front of $\sin\theta(\tilde{s})$ can be seen as a constant over the integration range, which is one period of the short time scale, this definition of the adiabatic invariant can be rewritten using a dumb variable, ϕ:

$$A(\tilde{s}, C(\tilde{s})) = \frac{1}{2\pi}\int_0^{2\pi}\sqrt{2\left[C(\tilde{s}) + \tilde{s}\sin\phi\right]}\,\mathrm{d}\phi. \tag{4.30}$$

This integral can be expressed by means of elliptic integrals of the second kind, but we shall leave it in this integral form.

The value of the adiabatic invariant A can be computed near $\tilde{s} = 0$ because there the boundary condition makes the derivative term $\mathrm{d}\theta/\mathrm{d}\tilde{s}$ equal to $(-\gamma) = |\gamma|$, so that

$$A = \frac{1}{2\pi}\int_0^{2\pi}\mathrm{d}\theta\,\frac{\mathrm{d}\theta}{\mathrm{d}\tilde{s}} \approx \frac{-\gamma}{2\pi}\int_0^{2\pi}\mathrm{d}\theta \approx |\gamma|.$$

---

[11] In the development below, we use the standard wording for time-dependent oscillators referring to the variable s as 'time'. This is for the purpose of making more obvious the connection with the theory of adiabatic invariants of dynamical systems.

[12] As explained later, C also has a small dependence on the fast timescale, not written here. This subdominant term is important as it explains why the adiabatic invariant involves an integral over one period of this fast timescale.

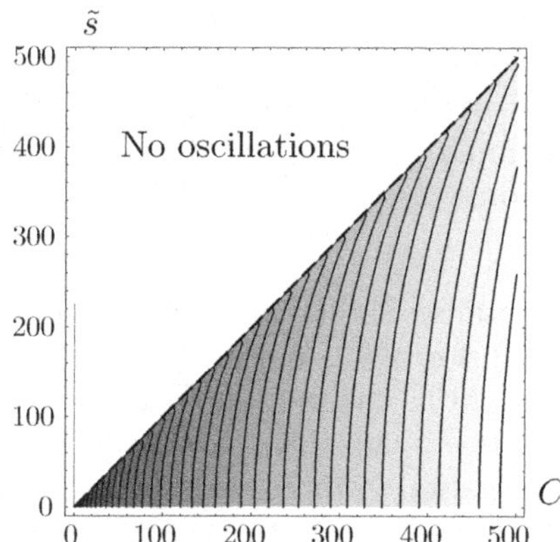

Fig. 4.6 Isolevel contours of the function $A(\tilde{s}, C)$, computed by numerical integration of equation (4.30). By conservation of the adiabatic invariant, the energy of the oscillator varies with $\tilde{s}$ in such a way that $(C(\tilde{s}), \tilde{s})$ moves along an isolevel of the function A in the plane $(C, \tilde{s})$ until it reaches the white region where the last curl opens up and matches with the straight, intermediate portion of the hair. Mathematically, $A(C, \tilde{s})$ is undefined in the white region, where the argument of the square root in equation (4.30) takes up negative values.

Now, the equation:

$$A(\tilde{s}, C(\tilde{s})) = |\gamma|, \tag{4.31}$$

where the left-hand side is given in equation (4.30), is an implicit relation between the variables evolving according to the slow time scale, namely $C(\tilde{s})$ and $\tilde{s}$. When the function $A(\tilde{s}, C)$ is tabulated numerically, the numbers $(\tilde{s}, C(\tilde{s}))$ must evolve along a curve of isolevels of A, see Fig. 4.6. Note that the orders of magnitude of the quantities in this 'adiabatic' domain with fast oscillations are $C \sim \tilde{s} \sim \gamma^2$, as shown by equation (4.29).

In Fig. 4.5(c), curls are present only in the vicinity the free end. The hair is made up of two regions, one with curls and one that is essentially straight. This can be understood as a change of regime of this adiabatic oscillator when $\tilde{s}$ is increased. Notice that the square root in the integrand of $A(\tilde{s}, C)$ in equation (4.30) is only defined for $\tilde{s} \leq C$; otherwise, there are no oscillations and our assumption that there are two different scales for $\tilde{s}$ breaks down. As $\tilde{s}$ increases starting from 0 (free end), it reaches a value $\tilde{s}_c$ such that $C(\tilde{s}_c) = s_c$, a point that is located along the edge of the white region in Fig. 4.6. Since $\tilde{s}$ and C are related through the adiabatic invariant $A(\tilde{s}_c, C(\tilde{s}_c)) = A(\tilde{s}_c, \tilde{s}_c) = \gamma$, this yields:

$$\frac{A}{\sqrt{\tilde{s}_c}} = \frac{1}{2\pi} \int_0^{2\pi} \mathrm{d}\phi \sqrt{2\,(1 + \sin\phi)} = \frac{4}{\pi},$$

hence

$$\tilde{s}_c = \frac{\pi^2 \gamma^2}{16} \quad i.e. \quad s_c \approx L\left(1 - \frac{\pi^2}{16}\alpha^2\beta + \cdots\right) \tag{4.32}$$

after restoring the original quantities. This s_c is the curvilinear abscissa where the curls merge with the straight hair, see Fig. 4.5(c).

In equation (4.32), take β small but fixed and increase α, as when going from left to right in Fig. 4.5. The present section is concerned with $\gamma \gg 1$, that is $\alpha \gg \beta^{-1/3}$. When

α increases from $\beta^{-1/3}$ up towards values of order $\beta^{-1/2}$, s_c decreases from L to 0 and the region with curls, initially concentrated near the free end, progressively invades the full length of the hair. When α actually becomes of order $\beta^{-1/2}$, the assumption that the curls are concentrated in a small boundary layer breaks down, and one recovers a geometry very similar to that studied in Section 4.3.

Discussion of the merging near s_c

This section discusses in more details what happens near the point s_c where curls merge with the straight region in Fig. 4.5(c), for $\beta^{-1/3} \ll |\alpha| \ll \beta^{-1/2}$. The results presented here are more advanced than in the rest of this chapter, and are not essential for the understanding of the following: the reader may freely skip ahead to the derivation of the adiabatic invariant in Section 4.4.3 or to the investigation of three-dimensional effects in Section 4.5.1.

To analyse the merging of curls with a vertical part of the hair, we study the solution of the differential equation (4.25) in the region when $\tilde{s}$ crosses the critical value $\tilde{s}_c$. To do so, we analyse the trajectory followed by the solution of this equation in the phase space (θ, θ'), shown in Fig. 4.8 by a bold curve.

Before deriving the trajectory of the actual solution of equation (4.25), we first consider the set of trajectories obtained by taking the coefficient $\tilde{s}$ of the cosine function to be constant in this equation, as we did earlier in the adiabatic approximation:

$$\frac{\mathrm{d}^2\theta}{\mathrm{d}\tilde{s}^2} - \tilde{s}_0 \cos\theta(\tilde{s}) = 0, \quad \tilde{s}_0 \text{ fixed.} \tag{4.33}$$

In this approximation, the equations of motion are simply those for a simple pendulum, see for instance (M. Nizette and A. Goriely, 1999); there are four sets of trajectories, shown by thin curves in the phase portrait in Fig. 4.7. Two sets of open trajectories run either from large negative θ to large positive θ or the converse, labelled (a) and (b). Trajectories that are closer to the $\theta' = 0$ axis follow nested, closed curves centred on points of equilibrium of the ODE, namely the points of coordinate $(\theta = \pi/2 + 2k\pi, \theta' = 0)$, k being being a positive or negative integer. The latter points describe a hair pointing upwards $(\theta = \pi/2)$, something that is obviously an unstable equilibrium.[13] The three sets of trajectories are separated by a fourth type of solution, along the so-called separatrices, running from one saddle point $(\theta = -\pi/2 + 2k\pi, \theta' = 0)$ to one of its two neighbours $(\theta = -\pi/2 + 2(k \pm 1)\pi, \theta' = 0)$. As shown in Fig 4.7, for a hair, the open trajectories correspond to curls with a vertical drift. The closed trajectories correspond to undulated hairs pointing upwards on average $(\theta = +\pi/2 + 2k\pi)$. Near the saddle points, the hair is vertical $(\theta = -\pi/2 + 2k\pi)$. The separatrix represents a localized buckling mode (G. H. M. van der Heijden and J. M. T. Thompson, 2000), whereby two asymptotically vertical segments are connected by one loop; this localized buckling mode is the solution derived for strong gravity near the scalp in Section 4.4.1. For a pendulum, the open trajectories correspond to the case where the mechanical energy is large enough for the pendulum to wind around its axis. Close trajectories correspond to oscillatory motion near the the equilibrium configuration. The separatrix is the so-called heteroclinic orbit that is run in an infinite

---

[13] Note that the stability of the equilibrium positions of the dynamical system in equation (4.33) on the one hand, and the corresponding hair configurations on the other are reversed. The equilibrium $\theta = \pi/2 + 2k\pi$ corresponds to a *stable* equilibrium for the dynamical system with respect to the variable s (see Fig. 4.7), but to an *unstable* equilibrium for the hair: for $\theta = \pi/2$, the hair is pointing upwards (when followed from scalp to end). Equation (4.33) is the equation of motion for a simple pendulum with the unusual convention that $\theta = \pi/2$ is in its stable equilibrium and $\theta = -\pi/2$ in its unstable equilibrium.

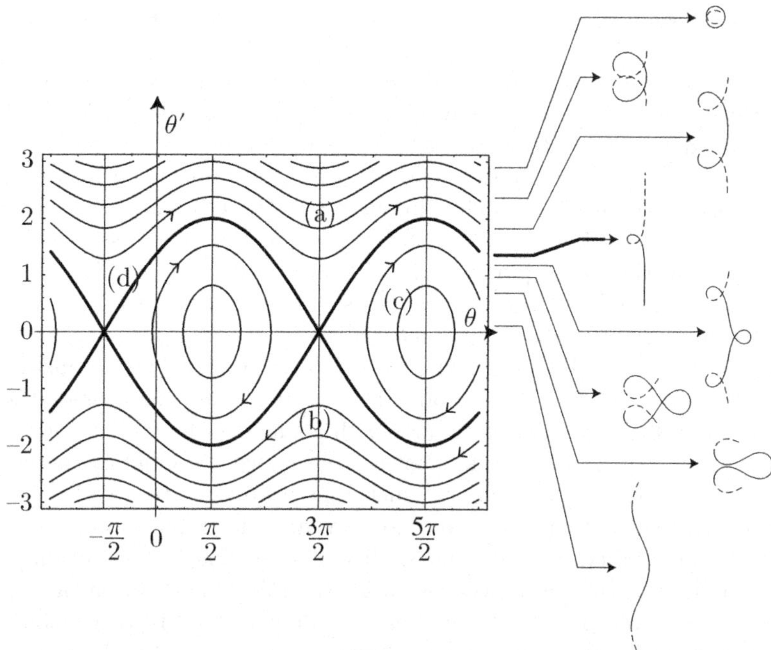

Fig. 4.7 Phase portrait of the pendulum equation (4.33) obtained by taking the prefactor $\tilde{s}$ in equation (4.25) constant. The pendulum has a conserved mechanical energy and its trajectories in this phase space fall into four categories, see for instance (M. Nizette and A. Goriely, 1999): open trajectories (a) and (b), oscillatory solutions (c) and the so-called separatrix (d). On the right, configurations of the hair corresponding to some representative trajectories are shown. Our aim in this section is to match the curl type behaviour (shown top right) on the free side with the vertical configuration on the clamped side, as obtained along the separatrix (shown right, middle). Note that the patterns of type (c) are irrelevant for the mechanics of hair.

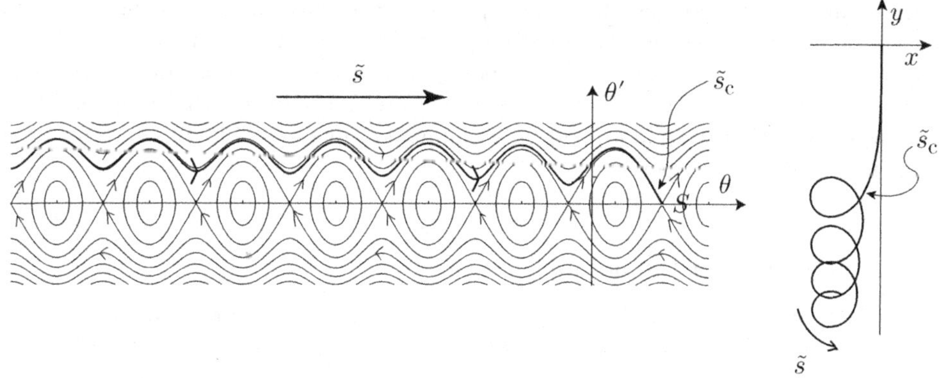

Fig. 4.8 Trajectory of a hair with curls on the free end merging with a vertical segment (bold curve) in the phase space (θ, θ'), left, and in physical space, right. As explained in the text, the solution of equation (4.25) starts along open trajectories of type (a) near the free end ($s \ll s_{\rm c}$). As $\tilde{s}$ increases, it drifts towards the separatrix. At about $\tilde{s} \sim \tilde{s}_{\rm c}$, it lands on the saddle point S where it spends a long time ($\tilde{s} \gg \tilde{s}_{\rm c}$). Pendulum solutions derived in Fig. 4.7 are shown by thin curves.

time when the pendulum is released from its unstable equilibrium position with a vanishing velocity.

In equation (4.25), the coefficient $\tilde{s}$ of the cosine is in fact not constant. As a result, the trajectory corresponding to an actual hair ($\tilde{s}$ not constant, bold curve) will slowly drift with respect to the pendulum trajectories ($\tilde{s}$ constant, thin lines) in Fig. 4.8. This slow drift is such that at any given $\tilde{s}$, the adiabatic invariant A has the value given in equation (4.31)— the current energy $C(\tilde{s})$ specifies which thin curve (for a pendulum) the bold curve should intersect. Near the free end, $\tilde{s} \ll s_c$, the hair shown in Fig. 4.8 is such that $\theta' > 0$: the relevant trajectory in the phase portrait of the simple pendulum in Fig. 4.7 are the open trajectories with $\theta' > 0$, labelled (a). As $\tilde{s}$ increases the bold trajectory in this figure drifts toward the separatrix (downward in the phase space (θ, θ')) until it reaches the separatrix. There two possibilities may happen: generically (and as was shown for instance in (J. L. Tennyson, J. R. Cary, and D. F. Escande, 1986)) the trajectory crosses the separatrix and then starts spiralling inwards as it further drifts across the closed trajectories of the autonomous system. This is not the kind of trajectory we are looking for: we are after a trajectory asymptotically vertical, namely such that $\theta(\tilde{s})$ tends to $-\pi/2 + 2k\pi$ as $\tilde{s}$ tends to plus infinity. Instead a trajectory spiralling inward along the closed set would reach $\theta(\tilde{s}) \approx \pi/2 + 2k\pi$ at infinity. The only possibility of satisfying the condition $\theta(\tilde{s})$ tends to $-\pi/2 + 2k\pi$ at infinity is that, as $\tilde{s}$ becomes large, the solution settles on the separatrix and stops finally at the saddle point $\theta = -\pi/2$, $\theta' = 0$ after an infinite travel time, as $\tilde{s} \to \infty$. This peculiar solution is not the one usually sought in studies of adiabatic invariants with separatrices (as in reference (J. L. Tennyson, J. R. Cary, and D. F. Escande, 1986)) because it is non-generic. As shown however in (J. L. Tennyson, J. R. Cary, and D. F. Escande, 1986), the dynamics near the separatrix is a tricky question as the 'fast' period of the oscillations becomes the infinite travel time from one saddle point to the other and this period cannot be assumed to be small compared with the typical time of variation of the slow parameter. A detailed analysis can be made of the corrections to the adiabatic theory relevant for the 'landing' of the trajectory on the separatrix, this is rather complicated and out of the scope of the present developments. However it is interesting to point out that the last curls, namely the ones just below $\tilde{s} = \tilde{s}_c$, correspond to trajectories close to the separatrix, and so have longer and longer parts with θ close to $-\pi/2$, corresponding to longer and longer travel close to the saddle point. Moreover this explains well the casual observation that curls at the end of a long hair merge with the straight uncurled part along an exponential tail, without oscillations. A naive linear theory on the other hand would predict oscillations that were more and more damped as one moves higher and higher along the hair, not at all what is observed.

4.4.3   A derivation of the adiabatic invariant

This subsection aims at deriving the expression of the adiabatic invariant pertinent to equation (4.28). A general (and far more sophisticated) derivation of the adiabatic invariant can be found in (L. D. Landau and E. M. Lifshitz, 1982). We start by rewriting equation (4.28) for $\theta(\tilde{s})$ in the form:

$$\frac{\mathrm{d}^2\theta}{\mathrm{d}\tilde{s}^2} + \frac{\partial V(\theta, \lambda)}{\partial \theta} = 0, \tag{4.34}$$

where we have introduced the potential $V(\theta, \lambda) = -\lambda \sin(\theta)$. In the usual theory of adiabatic oscillators, θ is the angle in a pair of angle-action variables, $\tilde{s}$ is the time, and λ a parameter depending slowly on time. In the hair curls problem, λ happens to be equal to the 'time' itself, $\lambda = \tilde{s}$, but we shall not rely too much on this particular dependence. Compared with the phase of $\sin(\theta)$, the quantity λ is a slowly varying function, since the phase θ changes by 2π when $\lambda = \tilde{s}$ changes by a quantity of order $2\pi/|\gamma|$, $|\gamma|$ being large by assumption.

Now we try to solve equation (4.34) by successive approximations, the lowest- order term being obtained with λ constant (i.e. independent on $\tilde{s}$), then slowly varying, etc. Assuming first that λ is constant, one has by simple integration:

$$\frac{1}{2}\left(\frac{\mathrm{d}\theta}{\mathrm{d}\tilde{s}}\right)^2 + V(\theta, \lambda) = C, \tag{4.35}$$

where the constant C of integration is the energy of the oscillator, sum of the kinetic and potential energies. This equation is similar to (4.29) and again C is practically constant over the fast time scale owing to the slow variation of $V(\theta, \lambda)$ with respect to λ.

Over the slow timescale, the energy is not conserved as the change of λ amounts to a forcing of the system by an external operator. Then, C, defined by (4.35), becomes a function of time. This dependence of C on the slow timescale is expressed by writing $C(\lambda)$. There is an additional subtlety, however: as explained below, one has to take into account that, on top of this dependence on the slow time scale through λ, C also depends on the fast time scale through θ *because of a small correcting term.* Therefore, we shall write C in the form:

$$C(\lambda, \theta) \approx C_{(0)}(\lambda) + C_{(1)}(\lambda, \theta),$$

where $C_{(1)}$ is small compared with $C_{(0)}$ but eventually includes a dependence on the fast timescale. Equation (4.35) becomes:

$$\frac{1}{2}\left(\frac{\mathrm{d}\theta}{\mathrm{d}\tilde{s}}\right)^2 + V(\theta, \lambda) = C_{(0)}(\lambda) + C_{(1)}(\lambda, \theta). \tag{4.36}$$

Take now the derivative of this equation with respect to $\tilde{s}$. Two terms cancel because $\theta(\tilde{s})$ is a solution of equation (4.34). The remaining terms read

$$\frac{\partial\left[C_{(0)} - V(\theta, \lambda)\right]}{\partial\lambda}\frac{\mathrm{d}\lambda}{\mathrm{d}\tilde{s}} + \frac{\partial C_{(1)}}{\partial\theta}\frac{\mathrm{d}\theta}{\mathrm{d}\tilde{s}} = 0, \tag{4.37}$$

where we have neglected the $\partial C_{(1)}/\partial\lambda$ in front of the term $\partial C_{(0)}/\partial\lambda$. We have nevertheless retained the term with $\partial C_{(1)}/\partial\theta$ which, although much smaller than $\partial C_{(0)}/\partial\lambda$, is multiplied by a much larger factor, $\mathrm{d}\theta/\mathrm{d}\tilde{s}$ versus $\mathrm{d}\lambda/\mathrm{d}\tilde{s}$.

This equation can now be read on the slow timescale. To do so, we average it over one period of oscillations, that is from $\tilde{s}_n$ to $\tilde{s}_{n+1}$. The integral of $\partial C_{(1)}/\partial\theta \; \mathrm{d}\theta/\mathrm{d}\tilde{s}$ yields a small quantity of second order: the integrand, $C_{(1)}$, is much smaller than $C_{(0)}$; moreover, after one period, $C_{(1)}$ almost comes back to the same value and so the integral is proportional to the small drift of λ over a period of the fast oscillations. This term can therefore be neglected provided the integral is taken *over a period of the fast oscillations.* This yields:

$$\int_{\tilde{s}_n}^{\tilde{s}_{n+1}} \frac{\mathrm{d}\lambda}{\mathrm{d}\tilde{s}} \frac{\partial\left[C_{(0)} - V(\theta, \lambda)\right]}{\partial\lambda}\,\mathrm{d}\tilde{s} = 0.$$

One can pull $\mathrm{d}\lambda/\mathrm{d}\tilde{s}$ out of the integral, since λ is almost constant over one period (it is in fact exactly constant and equal to one in the present case since $\lambda = \tilde{s}$). From this we deduce:

$$\frac{\mathrm{d}}{\mathrm{d}\lambda} \int_{\tilde{s}_n}^{\tilde{s}_{n+1}} \left(C_{(0)} - V(\theta,\lambda)\right) \, \mathrm{d}\tilde{s} = 0.$$

On the slow time scale this yields

$$\int_{\tilde{s}_n}^{\tilde{s}_{n+1}} \left(C_{(0)} - V(\theta,\lambda)\right) \, \mathrm{d}\tilde{s} = \pi A,$$

A being a constant, called the adiabatic invariant. The π factor added in front of A is a matter of convention. At the dominant order, we have

$$\mathrm{d}\tilde{s} = \frac{\mathrm{d}\tilde{s}}{\mathrm{d}\theta}\mathrm{d}\theta = \frac{\mathrm{d}\theta}{\sqrt{2\left(C_{(0)} - V(\theta,\lambda)\right)}},$$

where we assume, as earlier, $\mathrm{d}\theta/\mathrm{d}\tilde{s} > 0$. The case $\mathrm{d}\theta/\mathrm{d}\tilde{s} < 0$ would be treated similarly, except for a minus sign in front of the square root.

The final expression for A reads

$$A = \frac{1}{2\pi} \int_{\theta_n}^{\theta_{n+1}} \mathrm{d}\theta \, \sqrt{2\left(C - V(\theta,\lambda)\right)}. \tag{4.38}$$

In the expression above, we used C for $C_{(0)}$, which is justified because $C_{(0)}$ is the dominant contribution to C. We also put the traditional factor $1/2\pi$ in front of the definition of A. Finally the integration limits in (4.38) have been written as θ_n and θ_{n+1}, which are by definition separated by one period: $\theta_{n+1} = \theta_n + 2\pi$. The integrand being the number $\mathrm{d}\theta/\mathrm{d}\tilde{s}$, the integrand can be written in a shorthand form as

$$A = \frac{1}{2\pi} \int_{\theta_n}^{\theta_{n+1}} \mathrm{d}\theta \, \frac{\mathrm{d}\theta}{\mathrm{d}\tilde{s}}.$$

As explained above, the adiabatic approximation becomes invalid near the separatrix, because there the period of the fast oscillations diverge, and become of the same order as the rate of variation of the slow variable.

4.5   Extensions of the model

In this section, two extensions of the theory presented above are presented: three-dimensional effects including the possibility of helical curling often observed in real hair, and a short account of the description of hair motion.

4.5.1   Three-dimensional effects

The Kirchhoff theory for elastic rods has been presented in the most general three-dimensional case in Chapter 3. Let $\kappa^{(1)}$ and $\kappa^{(2)}$ be the material curvatures associated with the principal directions of the cross-section and τ the material twist. The hair energy is, in three dimensions:

$$\mathcal{E}_{\text{hair}} = \int_0^L \mathrm{d}s \Big[\frac{E\,I^{(1)}}{2} (\kappa^{(1)}(s) - \kappa_0^{(1)})^2$$

$$+ \frac{E\,I^{(2)}}{2} (\kappa^{(2)}(s) - \kappa_0^{(2)})^2 + \frac{\mu\,J}{2} (\tau(s) - \tau_0)^2 + \rho\,g\,S\,z(s) \Big]. \quad (4.39)$$

This formula extends the two-dimensional version given in equation (4.1). It is again the sum of the Kirchhoff elastic energy and a potential energy accounting for the weight of the hair. The possibility of non-zero natural curvatures $\kappa_0^{(1)}, \kappa_0^{(2)}$ and twist τ_0 of the hair is considered; the natural twist introduces a parameter with the physical dimension of a length, $1/\tau_0$, typically of the same order of magnitude as the natural radius of curvature.

For the sake of simplicity, we assume that these natural curvatures and twist are constant along the hair length.[14] Then, the natural shape of the hair is a helix, as shown in Appendix C.

The equilibrium equations resulting from the minimization of this energy functional by the Euler–Lagrange method (see Appendix A) have already been derived in Chapter 3. The equation (3.63a) for the tension $\mathbf{F}(s)$ can be readily integrated:

$$\mathbf{F}(s) = -\rho\,g\,S\,(L - s)\,\mathbf{e}_z. \quad (4.40)$$

At equilibrium, the balance of torques yields the Kirchhoff equation (3.63b):

$$\frac{\mathrm{d}\mathbf{M}(s)}{\mathrm{d}s} + \mathbf{d}_3(s) \times \mathbf{F}(s) = 0,$$

submitted to the boundary condition for a free end, $\mathbf{M}(L) = \mathbf{0}$ at $s = L$. In this equation, the tension $\mathbf{F}$ is known in advance, but not the unit tangent $\mathbf{d}_3$. As explained in Section 3.6.8, the internal moment $\mathbf{M}(s)$ depends on the orientation of the tangent $\mathbf{d}_3(s)$ through the equation above while the orientation of the tangent $\mathbf{d}_3(s)$ itself depends globally on $\mathbf{M}(s')$ through the Darboux vector, related to $\mathbf{M}$ by the constitutive relations:

$$G^{(1)}(s) = E\,I^{(1)}\,(\kappa^{(1)}(s) - \kappa_0^{(1)})$$

$$G^{(2)}(s) = E\,I^{(2)}\,(\kappa^{(1)}(s) - \kappa_0^{(2)})$$

$$H(s) = \mu\,J\,(\tau(s) - \tau_0).$$

Note that these constitutive relations have been modified to take into account the natural curvatures and twist.

To sum up, the internal moment $\mathbf{M}(s)$ and the orientation of the material frame $(\mathbf{d}_1(s), \mathbf{d}_2(s), \mathbf{d}_3(s))$ satisfy a system of coupled differential equations with boundary conditions on both ends $s = 0$ and $s = L$. The structure of the equations defines a boundary value problem, which is very similar to that for $\theta(s)$ in two dimensions—except that in three dimensions there are far more unknowns and the equations are more intricate. This is why no analytical solution can be obtained in general, except in some limits, similar to those that have been investigated in two dimensions in Sections 4.3 and 4.4.

---

[14] This holds if the natural curvatures and twist are produced during the growth of the strands, which takes place at the follicle, and this process is time independent. In fact, other factors influence the natural curvatures and twists (such as into what shape the hair was last dried) and are neglected here.

A numerical solution of the equations of equilibrium for the hair in three dimensions can follow three main approaches. Recall that, as in two dimensions, a direct integration over s of the equations of equilibrium is not possible, as one boundary condition is imposed on each end: there is no starting point where a Cauchy initial condition of the differential equations is known. To solve this boundary value problem, one can use a shooting method (D. K. Pai, 2002), or a relaxation method. Both methods are discussed in detail in Section 13.6.2 of Chapter 13.

However, when the applied forces are conservative, the most convenient numerical method is not to solve the equations of equilibrium but instead to minimize the elastic energy. The advantage of dealing directly with the energy functional is that it is much simpler to implement than the equations of equilibrium, although they have the same physical content. Moreover, the positiveness of the energy makes the search for minimum a stable method, which is not always the case with shooting or relaxation methods. For the present problem, the minimization of the energy is implemented using one-dimensional finite elements.[15] The hair is divided into n small segments, usually taken of equal length for the sake of simplicity. In each element, the curvatures and twist are prescribed to be constant. The configuration of the hair is therefore prescribed by these n curvatures $\kappa_i^{(1)}$, $\kappa_i^{(2)}$ and twist τ_i in each segment i, $0 \leq i \leq n - 1$. The fact that the curvatures and twist are piecewise constant functions constrains the set of allowed configurations of the hair—this is the principle of a Ritz method. However, it does so in a less and less severe manner as n is increased. The degrees of freedom of the present model are the curvatures and twist over each element. This choice is particularly well suited to Kirchhoff rods, whose energy depends directly on these quantities. In Ref. (Y. Yang, I. Tobias, and W. K. Olson, 1993), a different discretization has been introduced, which uses the positions of the nodes and the Euler's angles of the material frames as degrees of freedom, leading to a significantly more involved implementation.

Having chosen the discretization variables, the main task of the numerical implementation is to map an energy to an arbitrary configuration:

$$\mathcal{C}_{3n} = \{\kappa_0^{(1)}, \kappa_0^{(2)}, \tau_0, \ldots \kappa_{n-1}^{(1)}, \kappa_{n-1}^{(2)}, \tau_{n-1}\}. \tag{4.41}$$

Calculation of the elastic energy is straightforward as the first three terms in the integrand of equation (4.39) are piecewise constant functions. Calculation of the gravitational energy requires integration of the kinematic relations (3.1), in order to determine the function $z(s)$. A key remark is that the centre line is piecewise a helix as the curvatures and twist are piecewise constant, see Appendix C.

Equilibrium configurations are finally obtained by minimizing this total energy with respect to the discretized curvatures and twist, $\mathcal{C}_{3n}$. Typical configurations obtained in this way are shown in Fig. 4.9.

4.5.2 Full head of hair

The method described above is at the basis of a simulator for natural hairstyles, described in reference (F. Bertails *et al.*, 2005). To simulate a full head of hair, the hair is divided into wisps. Each of these wisps is build around a skeleton whose shape is determined by minimization of the three-dimensional energy of the previous section. Interactions between

---

[15] Finite elements are another instance of the Ritz method discussed in detail in Chapter 8.

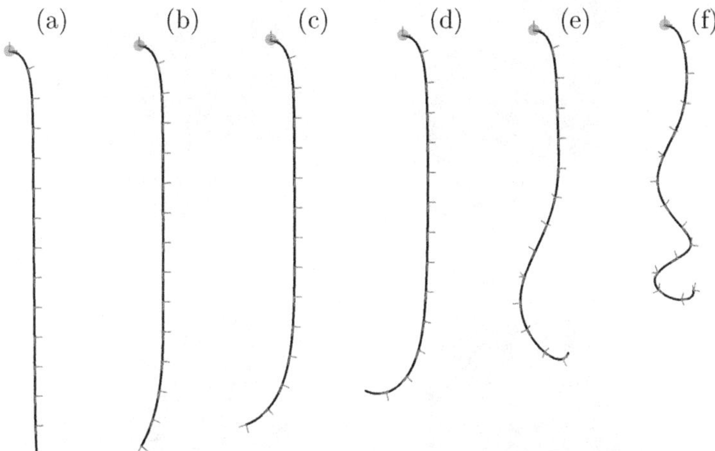

Fig. 4.9 Typical three-dimensional configurations obtained by numerical minimization of the hair energy (4.39). The material frames attached to the centre line of the hair are shown in grey to allow visualization of twist. The natural curvature $\kappa_0^{(2)}$ and twist τ_0 were set to zero, while the other spontaneous curvature, $\kappa^{(1)}$, is progressively increased from left to right. As in Fig. 4.5 for the 2D case, spontaneous curvature concentrates near the free end in order to lower the potential energy. At frame (c), a three-dimensional instability takes place: twist becomes non-zero and the shape is no longer planar. This instability results from a trade-off between twist and gravity.

different wisps and between wisps and the body (shoulders and head) are set up using penalty forces: when penetration is detected, a force is applied so as to move the elements apart. It is quite difficult to detect collisions in a reasonable computation time and a specific algorithm had to be designed. Some care must be taken to apply penalty forces that are smooth enough to allow the energy minimization to converge but be strong enough to enforce the constraint of non-penetrability.

In Fig. 4.10, this method was used to make synthetic images mimicking various natural hairstyles (long, short, straight or curly hairs). The method provides an intuitive control to the user that is based on very few relevant parameters (the natural curvatures and twist, the length and bending and torsional stiffnesses of the hair strands). As a result, realistic hairstyles can be produced quickly. This is an advantage commonly associated with a physics-based method, such as this one: the physics is captured in the algorithm and about any set of parameters produces a realistic result; this allows the operator to focus on a small set of physically relevant parameters.

4.5.3 Dynamics

All the methods presented in this chapter were concerned with the equilibrium configurations of an elastic rod. The dynamics of rods, namely the prediction of their motion under the combined effect of applied forces and inertia, is a considerably more difficult problem. The dynamic Kirchhoff equations have just a single additional term compared with the static case, the balance of force being replaced by the balance of momentum: $\partial \mathbf{F}(s,t)/\partial s + \mathbf{p}(s,t) = \rho S\, \ddot{\mathbf{r}}(s,t)$, where t is the time variable, ρS is the lineic mass of

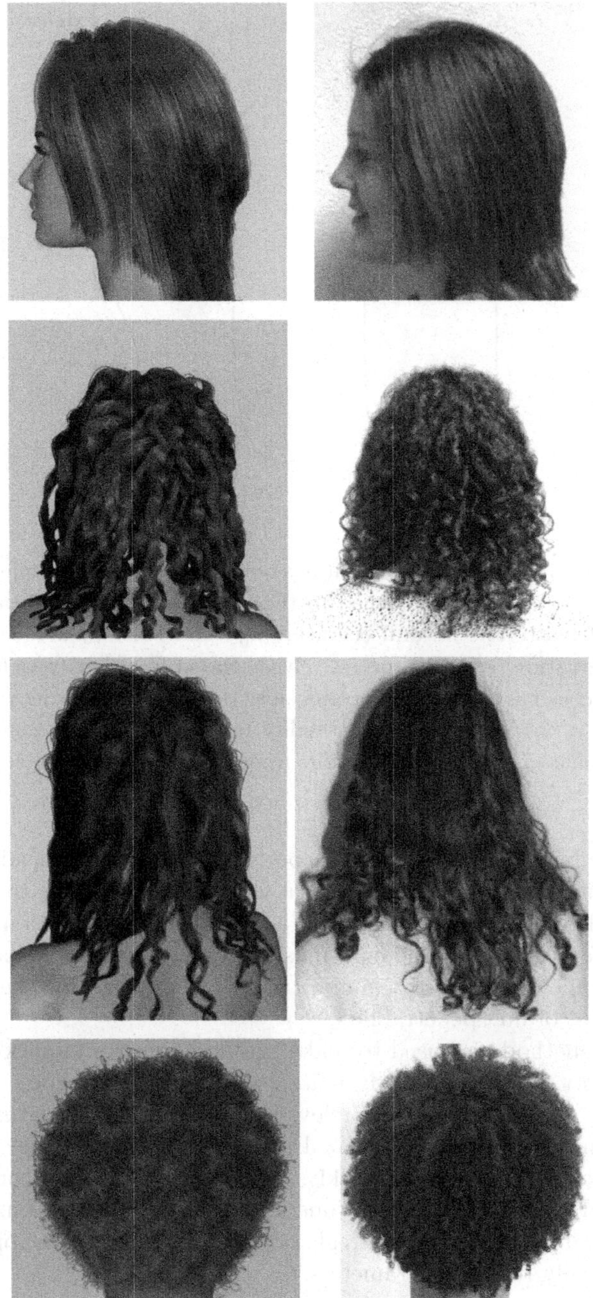

Fig. 4.10 Static simulations of natural hairstyles based on the three-dimensional model of Section 4.5.1. Simulated hairstyles and real ones are represented side-by-side. Typical patterns studied in the two-dimensional model, such as ringlets, are recovered for some values of the parameters. Models (from top to bottom): Alba Ferrer-Biosca, Antoine Bouthors, Florence Bertails-Descoubes and Brandon Michael Arrington. All computer images as well as the three upper photographs are courtesy of Florence Bertails-Descoubes; the bottom-most photograph is courtesy of Eitan Grinspun. Some of these images appeared in reference (F. Bertails *et al.*, 2005) © Eurographics Association 2005; reproduced by kind permission of Eurographics Association.

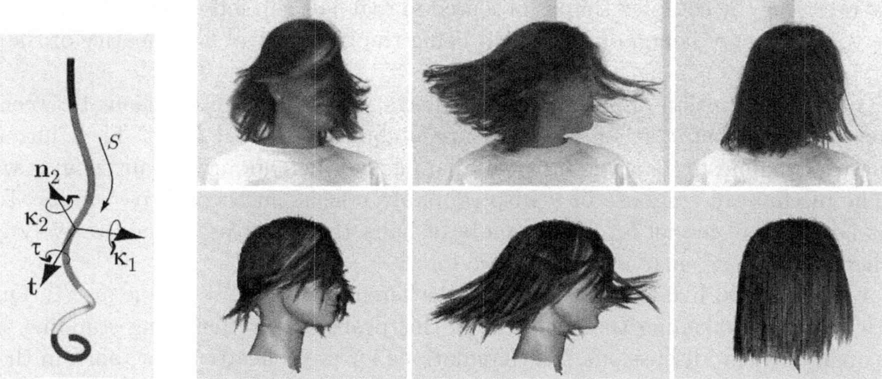

Fig. 4.11 Dynamic simulations of a full head of hair. *Left*: the Super-Helix is a numerical model for predicting the dynamics of elastic rods that is based on piecewise helical curves—here, with five helical elements. *Right*: frames from an animation of a full head of hair using this method (bottom) compared with real hair movement (top). Model: Alba Ferrer-Biosca. Computer images and photographs courtesy of F. Bertails (INRIA, Grenoble). These images appeared in reference (F. Bertails *et al.*, 2006) available from http://dx.doi.org/10.1145/1360612.1360662 © 2006 by the Association for Computing Machinery, Inc., and in reference (F. Bertails, B. Audoly, and M.-P. Cani, 2008) © Techniques de l'ingénieur 2008, http://www.techniques-ingenieur.fr.

the rod and $\ddot{\mathbf{r}}$ the acceleration of the centre line. This single term changes the whole mathematical structure of the equations. For instance, the internal force $\mathbf{F}(s,t)$ cannot be determined in advance as it becomes coupled with the acceleration. There are important numerical difficulties associated with the simulations of the dynamic Kirchhoff equations, which are numerically stiff, a point that is very well discussed in Reference (T. Y. Hou, I. Klapper, and H. Si, 1998). A fast and robust method for simulating hair based on the Kirchhoff equations was recently proposed (F. Bertails *et al.*, 2006). Being based on the same kind of approach as the static model of the previous section, this approach offers similar advantages in terms of user control and realism, and is applicable to a wide range of hairstyles. A few frames taken from movies generated with this method are shown in Fig. 4.11.

4.6 Conclusion

We have presented the mechanics of the human hair as an application of the theory of elastic rods. This problem provided us with an opportunity to cover a wide range of techniques of applied mathematics that are relevant for the analysis of various limits in the parameter space. By taking into account a few simple ingredients (gravity, spontaneous curvature and the elasticity of rods), we were able to recover many typical shapes observed in natural hairs, some of which are rather complex, and to outline a classification of natural hair shapes based on two-dimensionless parameters. Probably the practitioners in the field have noticed for some time that gravity tends to pull down the hairs into dull hairdos and that curls spontaneously appear near the free end rather than near the scalp. The present approach was an attempt to bring casual observation of this sort into a consistent framework and to demonstrate that the complex shapes observed in real hair can follow from a combination of relatively simple ingredients, one of which being the geometrical nonlinearity of the theory of thin elastic bodies.

The current description of other physical effects, such as the interactions between hairs, is rather limited. Contacts, for instance, are ubiquitous in real hairs. They introduce a coupling that manifests itself in the local alignment of neighbouring hair strands within a wisp. The mechanical response of a wisp probably reflects such collective effects. To date, there is no accurate model for an assembly of hairs that addresses quantitatively contact and other interactions such as electrostatic forces.

The authors would like to thank L'Oréal Recherche, and J.-L. Lévêque and B. Querleux in particular, for awakening their interest in this problem, for their long standing support and for many fruitful discussions. The simulations for a full head of hair, both in the static and dynamic cases, have been obtained in a joint work with F. Bertails-Descoubes (INRIA Grenoble, France).

References

F. Bertails, B. Audoly, M-P. Cani, B. Querleux, F. Leroy, and J.-L. Lévêque. Super-helices for predicting the dynamics of natural hair. In *ACM Transactions on Graphics*, pages 1180–1187, August 2006.

F. Bertails, B. Audoly, and M.-P. Cani. Chevelures numériques. *Techniques de l'ingénieur*, August 2008.

F. Bertails, B. Audoly, B. Querleux, F. Leroy, J.-L. Lévêque, and M.-P. Cani. Predicting natural hair shapes by solving the statics of flexible rods. In J. Dingliana and F. Ganovelli, editors, *Eurographics (short papers), Dublin (IR)*. Eurographics, August 2005.

G. C. Garson, M. Vidalis, P. Roussopoulos, and J.-L. Lévêque. Les propriétés vibratoires transversales des fibres de kératine. Influence de l'eau et d'autres agents. *International Journal of Cosmetic Science*, 2:231–241, 1980.

T. Y. Hou, I. Klapper, and H. Si. Removing the stiffness of curvature in computing 3-D filaments. *Journal of Computational Physics*, 143(2):628–664, 1998.

L. D. Landau and E. M. Lifshitz. *Mechanics (Course of Theoretical Physics)*. Butterworth-Heinemann, 3rd edition, 1982.

N. Magnenat-Thalmann, S. Hadap, and P. Kalra. State of the art in hair simulation. In *International Workshop on Human Modeling and Animation, Seoul, Korea*, pages 3–9, Korea Computer Graphics Society, 2000.

M. Nizette and A. Goriely. Towards a classification of Euler–Kirchhoff filaments. *Journal of Mathematical Physics*, 40(6):2830–2866, 1999.

D. K. Pai. STRANDS: Interactive simulation of thin solids using Cosserat models. *Computer Graphics Forum*, 21(3), 2002.

J. Prost. Vers la modélisation d'une chevelure. Technical Report 1654, Science & Tech., 1994.

G. V. Scott and C. R. Robbins. A convenient method for measuring fiber stiffness. *Textile Research Journal*, 39:975–976, Oct. 1969.

J. A. Swift. Some simple theoretical considerations on the bending stiffness of human hair. *International Journal of Cosmetic Science*, 17:245–253, 1995.

J. L. Tennyson, J. R. Cary, and D. F. Escande. Change of the adiabatic invariant due to separatrix crossing. *Physical Review Letters*, 56(20):2117–2120, May 1986.

G. H. M. van der Heijden and J. M. T. Thompson. Helical and localised buckling in twisted rods: A unified analysis of the symmetric case. *Nonlinear Dynamics*, 21(1):71–99, 2000.

K. Ward, F. Bertails, T.-Y. Kim, S. R. Marschner, M.-P. Cani, and M. Lin. A survey on hair modeling: Styling, simulation, and rendering. *IEEE Transactions on Visualization and Computer Graphics (TVCG)*, 13(2):213–234, 2007.

Y. Yang, I. Tobias, and W. K. Olson. Finite element analysis of DNA supercoiling. *The Journal of Chemical Physics*, 98(2):1673–1686, 1993.

5

Rippled leaves, uncoiled springs

We analyse rippled patterns that can be observed along the edge of thin sheets that have a stretched edge. Examples of systems displaying such ripples are given in Fig. 5.1 and include the edge of some plant leaves, flowers, clothes, and even torn plastic sheets. In this chapter we present an analogy between the analysis of such patterns and that for the unfolding of a spiral elastic spring; the problem of unfolding a spiral spring is solved using the theory of elastic rods. These ripples are further investigated in Chapter 10 where we address an extreme case of the same rippling phenomenon: when the edge of the sheet is strongly stretched, *self-similar* ripples can be observed; they will be studied using the theory of thin elastic plates.

5.1   Introduction

The similarity of the patterns shown in Fig. 5.1 in systems as diverse as skirts, flowers and plastic bags is striking. These systems are all similar from a mechanical perspective: they behave like elastic materials at least in a first approximation, have the geometry of a thin sheet, and their edge has been stretched.

In some of plants species, the growth of the leaves or petals of flowers is indeed enhanced near their edge where biological tissues tend to be produced in excess (M. Marder *et al.*, 2003). This induces compressive stress in a band near the edge, as sketched in Fig. 5.2. By a general instability mechanism known as buckling, such compressive stress can lead to undulated patterns.

In the garment industry, such ripples are sometimes used to decorate the lower edge of skirts, called *godet* skirts. A mathematical analysis of these ripples has been made by Nechaev and Voituriez (S. Nechaev and R. Voituriez, 2001) who studied mathematical surfaces featuring the same patterns. Extending the fashion terminology, they coined the word 'godet surfaces'.[1] Continuing on this track, we shall call our mechanical model a 'godet ribbon', a typical configuration of which is shown in Fig. 5.13.

In the present chapter, a simple mechanical model based on the theory of rods, which accounts for these undulations is presented. It uses three main ingredients: the thin geometry of the sheet, its elastic behaviour, and the presence of a stretched edge. This rod model provides a gentle introduction to plate buckling, studied in the second part of this book.

---

[1] *Oxford English* uses 'godet', a word of French origin, as in 'godet skirt', although the original French would be 'jupes à godets'.

Fig. 5.1 Ripples can be observed near the edge of many types of thin 'sheets': simple undulations (a) in leaves of Cally lily (*Zantedeschia aethiopica*), (b) in fern leaves, (c) in a flower of Paphiopedilum orchid; (d) multi-scale undulations in cyclamen flowers (*Cyclamen coum*). An extreme case of this phenomenon is self-similar rippling, studied later in Chapter 10, observed in picture (e) along the cut of a torn garbage bag. (f) A godet skirt. In all these systems the edge is effectively stretched, by differential biological growth, by plasticity, or by the insertion of extra fabric. Image (d) courtesy of Calliope Audoly.

5.1.1 Non-Euclidean surfaces with ripples

Godet skirts can be crafted according to the process depicted in Fig. 5.2(a), which is very illustrative: one starts from a regular, flat piece of fabric, makes evenly spaced cuts perpendicular to an edge and inserts small triangular patches[2] into these cuts. The excess of fabric near the edge effectively increases the length of the edge compared with

---

[2] Godets are the technical name for these inserts in the fashion industry.

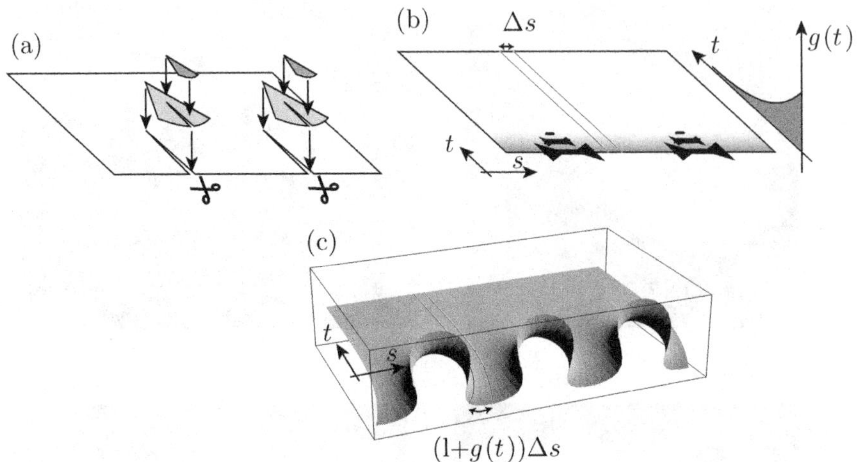

Fig. 5.2 A sheet with residual stress near its edge, after Nechaev and Voituriez (S. Nechaev and R. Voituriez, 2001). (a) Construction of a godet fabric by inserting triangular patches. (b) Associated abstract manifold: the non-Euclidean metric (5.2) on the manifold reflects the differential strain due to the insertion of patches. (c) An embedding of this metric is a three-dimensional surface having the same metric: because of the ripples, the curves near the edge are stretched, $g(y) > 0$.

the planar, initial configuration—this added material is represented by dark tones in Fig. 5.2(b). This effect is localized near the edge as the cuts are short. Effectively, the edge length has been increased, as if it had been stretched, although the rest of the fabric remains unchanged. Ripples localized near the edge allow the sheet to balance the excess of length available there. Near the edge of leaves, flowers and of plastic sheets, the excess of material arises from different physical processes but the rippling mechanism is otherwise similar. In torn plastic sheets, for instance, the stress focusing near the moving tip of the crack induces a plastic deformation of the material (ductile fracture). This flow creates permanent deformation and elongates the newly formed edge (E. Sharon *et al.*, 2002). In plant leaves and flowers, enhanced biological growth near the edges has a similar effect (U. Nath *et al.*, 2003).

The patch construction shown in Fig. 5.2(a) suggests a geometrical approach to the ripples problem, explored in reference (S. Nechaev and R. Voituriez, 2001) and outlined below. We shall consider the continuum limit whereby infinitely many microscopic patches are inserted. The point in considering this limit is that one can assume that the properties of the resulting surface become invariant by translation along its edge, which makes the analysis much simpler. The effect of the triangular inserts can be represented by a non-Euclidean metric on the surface, as sketched in part (b) of this figure. We call s the coordinate along the free edge, and t that perpendicular to the edge in the flat configuration of reference, see Fig. 5.2(b). Mathematically, the effect of the inserts is to redefine the natural distance $\mathrm{d}\ell$ between two points with coordinates (s, t) and $(s + \mathrm{d}s, t + \mathrm{d}t)$. In the original, planar configuration, this distance is given by the Euclidean metric $\mathrm{d}\ell^2 = \mathrm{d}s^2 + \mathrm{d}t^2$. This is no longer the case after inserting the patches, and the fundamental mathematical object we consider is the one expressing the infinitesimal *natural* distances between neighbouring points. This object is called an abstract metric, and is imposed by the problem at hand, i.e. by the microscopic details of tissue growth for plants or plasticity for plastic bags. By natural distance, we mean the length of the shortest curve joining two neighbouring points, when a small piece of sheet surrounding this two points has been cut out and the stress has been relaxed. The word *abstract* is used as there does not necessarily exist a global configuration of the sheet such that the physical distance between any pair of neighbouring points is equal to its natural distance; instead, the natural distances are reset blindly by a microscopic process in the first place, and it is well possible that no actual configuration of the sheet in space is capable of bringing any pair of points at a distance exactly given by their natural distance–this is the mathematical problem of embedding in Euclidean space an abstract surface, to which we shall come back later.

Consider two neighbouring lines that are perpendicular to the edge in the reference configuration, as shown in part (b) Fig. 5.2. They are the lines with coordinates s and $s + \Delta s$. Consider a pair of neighbouring points on each of these lines, with coordinates (s, t) and $(s + \Delta s, t)$. In the 3D configuration with undulations, the separation between the two curves increases near the edge as a result of the localized inclusions (representing the localized differential strain). Assuming invariance of the metric with respect to the direction along the edge, the natural distance $\Delta \ell(\Delta s, t)$ between the points with coordinates (s, t) and $(s + \Delta s, t)$ is a function of t for fixed Δs: it will increase closer to the edge (smaller t, t being the distance to the edge $t = 0$). This effect can represented by the equation:

$$\Delta \ell(\Delta s, t) = (1 + g(t)) \, \Delta s, \tag{5.1}$$

where by convention:

$$g(t) > 0 \text{ for all } t, \qquad g(t) \to 0 \text{ for } t \to +\infty.$$

Here, $g(t_0)$ represents the rate of elongation of the curve of equation $t = t_0$, which is parallel to the free edge (this rate of elongation is assumed to be uniform along the edge). The reference (no elongation, $g = 0$) is the separation Δs measured far from the edge ($t \to +\infty$), where no stretching has been applied.

From equation (5.1), the squared natural distance corresponding to a displacement parallel to the edge, such that $dt = 0$, reads: $d\ell^2 = (1 + g(t))^2 \, ds^2$. We shall assume that the inclusion of patches elongates curves parallel to the edge, but brings in no shear strain.[3] As a result, the set of coordinate s-curves and t-curves remain everywhere perpendicular and the squared length element has no cross-term of the form $ds \, dt$. We shall moreover assume that the length of curves perpendicular to the edge is not modified by the patches: along these curves, $d\ell^2 = dt^2$. Finally, the squared infinitesimal natural length between arbitrary points can be written in the form

$$d\ell^2 = (1 + g(t))^2 \, ds^2 + dt^2, \tag{5.2}$$

where we recall that s is the coordinate along the free edge and t is normal to the free edge. Equation (5.2) expresses the natural length between any pair of neighbouring points along the stretched edge, as redefined by the microscopic process; the unstretched case $g = 0$ corresponds for a classical Euclidean metric $d\ell^2 = ds^2 + dt^2$. Note that, for an arbitrary

---

[3] As explained in Chapter 10, the other components of the metric are irrelevant under the assumption of invariance along the t direction: they can be relaxed by an in-plane displacement.

(a)

(b)

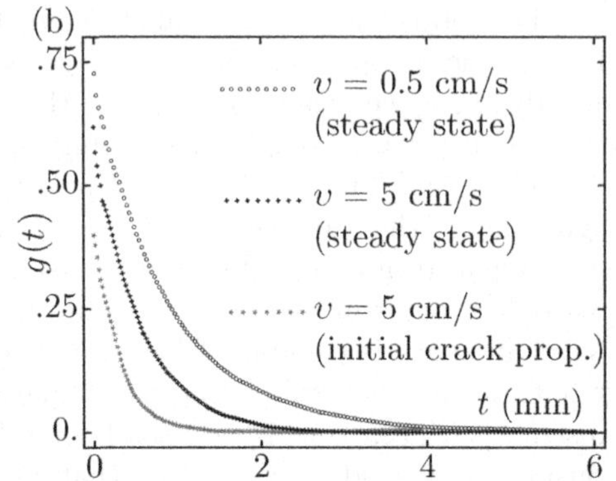

Fig. 5.3 Irreversible stretching induced by tearing a polyethylene sheet. (a) In this minimalistic experiment, stretching is visualized by the deformation of a grid drawn on the sheet (after E. Sharon and B. Roman), and by the presence of a lighter region surrounding the edge of the crack, where irreversible deformations make the material opaque. (b) Measurements of Sharon and collaborators of the metric function $g(t)$ as a function of the distance to the edge t for a 0.20 mm-thick sheet for various crack velocities v, reproduced from reference (E. Sharon, B. Roman, and H. L. Swinney, 2007).

function g, it is not possible to revert to the Euclidean metric by a simple change of coordinates in general.[4]

By printing a grid with a small mesh size on a sheet of plastic, by tearing the sheet, and by measuring the 3D distances between grid points once it has been traversed by the crack, it is possible to estimate the irreversible stretching of the edge. This allows one to measure experimentally the function $g(t)$ (E. Sharon *et al.*, 2002). Another possibility is to scan the deformed sheet using a profilometer (E. Sharon, B. Roman, and H. L. Swinney, 2007), as shown in Fig. 5.3(b). The function $g(t)$ could also be derived from first principles using a biological growth model for foliages or a model of plastic flow for torn plastic sheets. We shall not attempt to do this and consider the function $g(t)$ as given—in the simplified model of the present chapter, the only useful information about $g(t)$ will be its derivative at the edge, $g'(0)$, a quantity with the physical dimension of an inverse length that will disappear by rescalings.

In the classical geometry of surfaces, lengths of curves drawn along a surface are measured using the classical Euclidean metric in the 3D space. In equation (5.2), we have taken a very different approach as we specify an abstract metric, that is we use a somewhat arbitrary rule for measuring the lengths of curves drawn along the surface, without making reference to the surrounding three-dimensional space. This intrinsic approach goes back to Gauss and Riemann. It is at the heart of modern differential geometry and is famous for having found applications in the theory of general relativity. Specifying a geometrical object, here a half-plane, together with an arbitrary metric such as (5.2) defines what is called an *abstract metric manifold*. Some abstract manifolds, those with a non-positive metric for instance, like the Minkowski one, $d\ell^2 = ds^2 - dt^2$, cannot be realized by an actual surface in 3D Euclidean space.

---

[4] This can be shown by computing the perimeter of small circles with the new metric, which is not given by $2\pi r$; see Chapter 6.

The geometrical properties of a surface, including its metric tensor and curvature, can be calculated as a function of its configuration in the 3D Euclidean space. Here, we would like to solve the inverse problem: given an abstract manifold whose metric is imposed by, e.g., some biological growth process, we seek the 3D representation of the surface that will realize this metric—that is, we seek concrete surfaces in the Euclidean space $\mathbb{R}^3$, whose metric coincides with the given abstract metric. Such a surface is called an *embedding* of the abstract manifold. Finding embeddings is a famous problem of differential geometry, discussed in Chapter 11, which has no general solution, even for the simple family of metrics (5.2). An embedding is, by definition, a configuration of the sheet where in-plane stress has been removed altogether. If such embeddings exist, they will be observed at equilibrium: the analysis of embeddings provides a purely geometric approach of the buckling problem.[5]

This brings us to the model of a 'godet surface', that is a surface with a stretched edge described by a metric (5.2). This model was introduced and studied by Nechaev and Voituriez (S. Nechaev and R. Voituriez, 2001): with the aim of describing the shape of the edge of plant, they constructed an explicit embedding for a particular metric $g(t)$. The search for embeddings of this kind is a topic with a long history; a related question, the patching of flat pieces of fabric to build up a cloth with an arbitrary three-dimensional shape, was discussed by the Russian mathematician P. L. Tchebytchev (P. L. Tchebychef, 1878) in a talk delivered in Paris as early as on 28 August 1878.

5.1.2 From the godet surface to the godet ribbon model

We have introduced enough geometry for the needs of the present chapter, and we shall not discuss these matters any further until Chapter 10. For the purpose of explaining the *undulations* observed along the plant leaves, one can get rid of almost all the complexities of the elasticity of surfaces and focus on the narrow band near the edge of the sheet. This region can be seen as a very flat rod, which we call a godet *ribbon*.[6] The elastic model of a godet ribbon was first proposed by Marder *et al.* (M. Marder *et al.*, 2003) to describe the edge of plant leaves; our presentation follows the solution given in reference (B. Audoly and A. Boudaoud, 2002). This model is based on the equations for rods and is simple enough to be solved. Its connection with the buckling of elastic sheets is established below. The equations that govern a godet ribbon are derived next, in Section 5.2. Two types of solution are then derived, namely helical solutions in Section 5.3 and those describing godet patterns in Section 5.4.

The idea behind the godet ribbon model is that the stretch rate $g(t)$ goes rapidly to zero away from the edge, that is when t becomes larger than a typical length scale much smaller than the size of the leaf. Consider a narrow band of width a obtained by cutting the leaf along its edge, as sketched in Fig. 5.4. When cut, the band can be straighten out, as in part (b) of the figure, and appears to be longer than the width of the sheet. This is

---

[5] The analysis of embeddings is in fact more of a preliminary step taking place before the mechanical analysis of a plate or shell problem than a complete answer. It does not give the final answer as the embeddings are seldom unique—most often, they are either absent or non-unique, see Chapter 10.

[6] In Chapter 10, the analysis of these ripples will be pushed much further as we shall explain not only the presence of ripples but also the fact that some patterns are self-similar. This is beyond the grasp of the godet ribbon model and requires the introduction of the equations for plates. The experimental conditions in which a self-similar pattern is observed instead of simple undulations will also be elucidated–roughly speaking, the elastic object must be thin enough and the stretching of the edge must be concentrated in a region that is narrow enough.

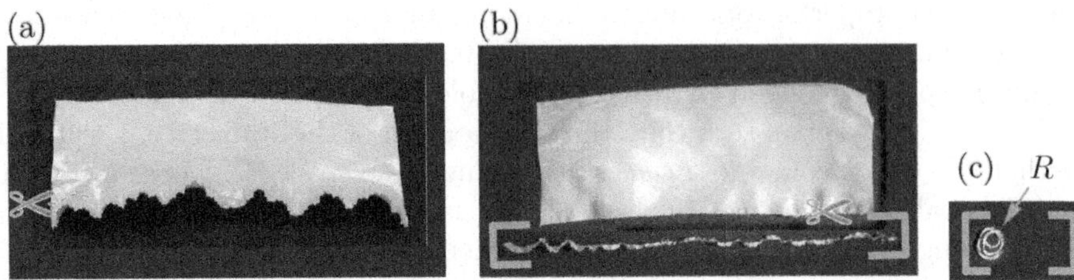

Fig. 5.4 (a) Cutting a narrow band along the stretched edge of a torn plastic sheet, as done by Sharon *et al.* (E. Sharon, M. Marder, and H. L. Swinney, 2004) for plant leaves. (b) The natural length of the band is larger than the width of the sheet, as revealed by the cut. (c) Because of the transverse gradient of stretching in the sheet, this band coils spontaneously into a ring when flattened. The idea of the godet ribbon model is to stretch the flat ring in (c) back into a straight-on-average configuration.

because the edge has been stretched longitudinally, and so $g(t) > 0$ near the edge. Moreover, when flattened, this band coils spontaneously into a ring shape shown in part (c)—this spontaneous coiling of ribbons cut along the edge has been beautifully demonstrated in plant leaves, see (E. Sharon, M. Marder, and H. L. Swinney, 2004). This effect reveals the presence of a transverse gradient of stretching, such that $g'(0) < 0$. Indeed, $g(0) > 0$ at the edge while g decays to zero far from the edge; as a result, one edge of the band is naturally longer than the other, and the band spontaneously coils into a ring.

The godet ribbon model makes use of the physical ingredients revealed by the experiment in Fig. 5.4(c) and focuses on the mechanics of the narrow band that has been cut out. Starting from the natural configuration of a flat ring (c), one applies stress so as to unfold it into a shape that is straight on average. It turns out that the patterns obtained by this process are qualitatively similar to those observed at the edge of the original sheet. The godet ribbon is built upon the following ingredients.

- It has a *flat cross section*. Its width a, although much smaller than its length, is much larger than its thickness h. This separation of scales allows one to retain at least qualitatively the two-dimensional character of the original leaf while using the Kirchhoff equations, which is in a much simpler framework than the full equations for plates.
- A godet ribbon displays some *natural curvature*: as noted in Fig. 5.4(c), the difference of natural lengths of the sides of the band make it coil into a flat ring. Let ρ be the natural radius of curvature of the ribbon (ρ is directly connected to $g'(0)$ as explained next).
- The third feature in the ribbon model has to do with *topology*. The two lateral edges of the ribbon do not wind around each other before it is cut out from the sheet—this is no longer so when the ribbon is released and coils. This can be formalized using the notion of linking number (G. H. M. van der Heijden and J. M. T. Thompson, 2000), a topological invariant for ribbon with fixed ends.

This last remark about the linking number dictates the loading to be imposed on the godet ribbon. The natural ring shape of the ribbon is a periodic configuration that has a non-zero linking number and so cannot represent the edge of a leaf. External forces or moments must be applied to the ribbon to impose the correct topology, as sketched in Fig. 5.5. In this figure,

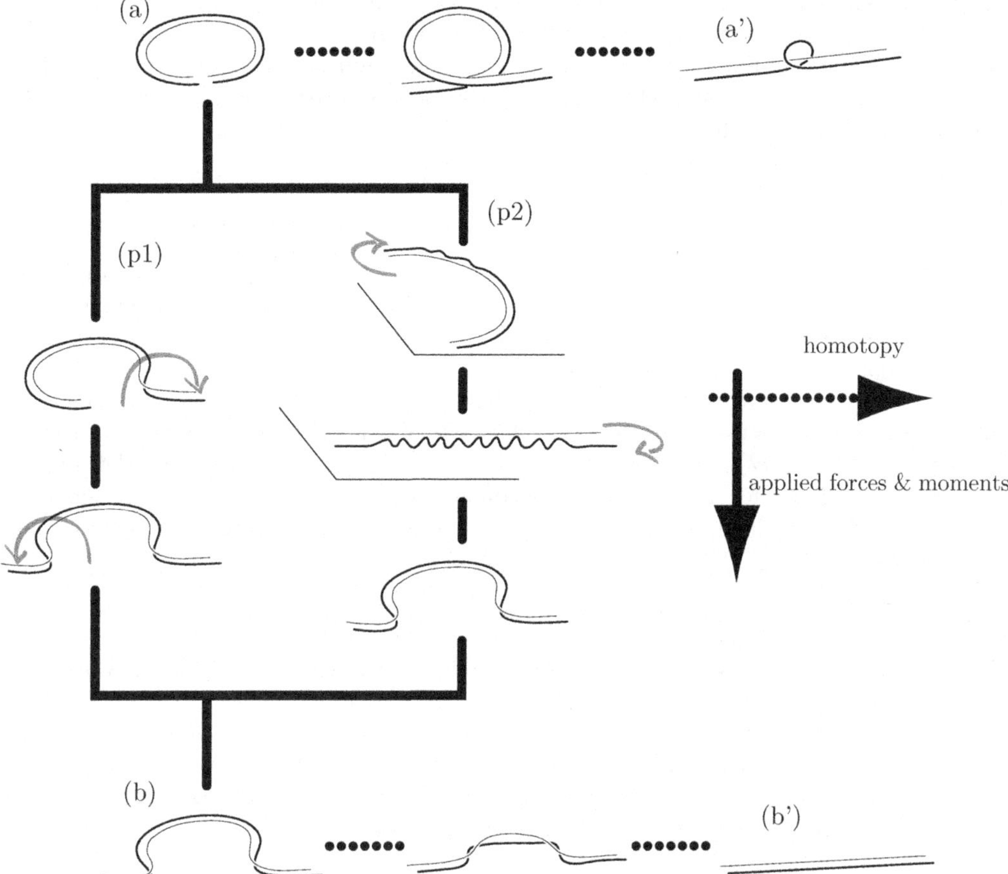

Fig. 5.5 Topology of a ribbon. The topology of the lateral edge curves is characterized by the linking number, which measures how much these curves wind around each other. Horizontal moves in the figure denote homotopies, which are transformations leaving unchanged the topology of these curves. (a) The natural, planar configuration of the ribbon has a non-zero linking number per period, as revealed by the transformation (a'). To remove it, one has to apply forces and or moments. The godet configuration (b)-(b') has the required topology (vanishing linking number). Two different transformations can be used to obtain the godet configuration, represented by the vertical thick paths in black.

horizontal dots represent an homotopy, which is by definition a deformation preserving the topology of the two edge curves of the ribbon: two configurations of these edge curves are said to be homotopic if one can pass continuously from one to the other without crossing. To transform the natural ring-like shape of the ribbon into a godet configuration, one can use two strategies, shown by the two vertical paths in the figure. Along the first path (p1), the ends of the ring are flipped and go out of the plane. Along the alternative path (p2), in-plane stress is applied, as if one tried to straighten the ring while keeping it planar; the ribbon cannot remain planar and the applied stress induces the wrinkles shown in the intermediate configuration (p2). This out-of-plane buckling yields a configuration (b) with godet that is similar[7] to that obtained when following (p1).

In the following, the flat ring shown in (a) is unfolded by following the path (p1)—this is the easier way, as there is no buckling and the solution changes smoothly with the applied loading. The applied loading is shown in Fig. 5.9, p. 149: the ribbon is loaded with an internal moment $\mathbf{M}$ lying in the initial plane of the ribbon.

5.1.3   Connection between the godet ribbon and the leaf edge

This section explains how the natural radius of curvature, ρ, of the godet ribbon can be related to the metric properties of the original leaf. As it is mostly independent from the rest of this chapter, the reader can freely jump ahead to the derivation of the equations for a godet ribbon in Section 5.2 in a first reading.

The spontaneous curvature of the godet ribbon, ρ, is derived by identifying the metric of a planar ring with curvature ρ and width a, with the metric of the stretched sheet proposed in equation (5.2). The coordinate along the edge of the sheet, s, becomes the orthoradial (or azimuthal) coordinate for the ring. The coordinate transverse to the edge, t, becomes the radial one for the ring. Therefore, the (s, t) parameterisation of the ring in polar coordinates (r, θ) reads

$$r(s,t) = \rho + \frac{a}{2} - t, \qquad \theta(s,t) = \beta\, s.$$

This parameterisation is such that the former edge of the sheet, $t = 0$, is the outer edge and has a radius $\rho + a/2$, while the cut line ($t = a$) is the inner edge of the ring and has a radius $\rho - a/2$. The centre line, $t = a/2$, is a circle of radius ρ, which is consistent with the definition of ρ, and the width of the ribbon is a. The coefficient β describes the stretching of the centre line, and will be determined below. The metric of the ring is found next by computing the distance between two neighbouring points:

$$\mathrm{d}\ell^2 = \mathrm{d}r^2 + r^2\, \mathrm{d}\theta^2 = \mathrm{d}t^2 + \beta^2 \left(\rho + \frac{a}{2} - t\right)^2 \mathrm{d}s^2.$$

By identification with the metric of equation (5.2), we find that the ring is isometric to the edge of the sheet, provided

$$\beta\left(\rho + a/2 - t\right) = 1 + g(t) \qquad \text{for } 0 \le t \le a. \tag{5.3}$$

Those two functions of t cannot match exactly, unless the growth or plasticity profile $g(t)$ happens to be linear near the edge.[8] Nevertheless, if the ribbon is narrow enough (small a), $g(t)$ can be reasonably approximated by its linear expansion near the edge $t = 0$.

---

[7] The fact that two different transformations can balance the non-zero linking number found in (a) and ultimately yield the same configuration (b) at the end has to do with the fact that the linking number can manifest itself as writhe (relating to the number and types of crossings of the center line when it is projected in a plane) or twist (in the usual sense). While the writhe and twist of a ribbon is not a topological invariant, their sum is the linking number which is an invariant. As a result, writhe and twist can be traded—see for example (G. H. M. van der Heijden and J. M. T. Thompson, 2000). The path (p1) can be viewed as an attempt to change twist and (p2) as an attempt to writhe.

[8] This argument shows that the edge of the sheet is a developable surface, i.e. it can mapped onto a plane, when the profile $g(t)$ is a linear function of t. In fact, the Gauss curvature associated with the abstract metric (5.2) can be computed and it is proportional to $g''(t)$: the linearity of the profile $g(t)$ with t is also a necessary condition for developability.

This writes $g(t) \approx g(0) + g'(0)\, t$. From equation (5.3), this yields the following isometry condition:

$$\beta\,(\rho + a/2) = 1 + g(0) \qquad \text{and} \qquad -\beta = g'(0).$$

By inverting these relations, one obtains the radius of curvature ρ of the ring and the rate of stretching β of its centre line as

$$\rho = \frac{1 + g(0)}{-g'(0)}, \qquad \beta = (-g'(0)), \tag{5.4}$$

where we have used $a \ll \rho$. A typical profile $g(t)$ is given in Fig. 5.3: $g'(0)$ is a negative number as $g(t)$ goes from $g(0) > 0$ at the edge to zero far away from the edge. The equation above shows that the radius of the ring, ρ, is of the same order of magnitude as the typical length scale over which the function $g(t)$ goes to zero. The length ρ is the only length that enters in the ribbon model. Therefore the wavelength of the undulations is of order ρ: it is fixed by the microscopic process (biological growth or plasticity) that stretches the free edge. Using a more elaborate plate model, we shall see in Chapter 10 that the typical length scale in the profile $g(t)$ sets the smallest wavelength in a cascade of ripples.

5.2   Governing equations

The equations of equilibrium for elastic rods have been derived in Chapter 3: established by Kirchhoff (G. R. Kirchhoff, 1859) in 1859 they have been the subject of numerous investigations since then. For circular cross-sections, the Kirchhoff equations are integrable (see Section 3.8). However, they are not integrable for arbitrary sections and spatial chaos may even occur (A. Mielke and P. Holmes, 1988). This is probably the reason why rods with non-circular cross-sections have received comparatively little attention until recently. Goriely *et al.* (A. Goriely and P. Shipman, 2000; A. Goriely, M. Nizette, and M. Tabor, 2001) investigated the dynamic instabilities of initially straight or helical ribbons.

The geometry of a godet ribbon is shown in Fig. 5.6, which shows the reference configuration, left, and a typical deformed configuration, right. As above, ρ is the natural radius of curvature of the ribbon, a its width, s the curvilinear coordinate along the centre line. A material frame $(\mathbf{d}_1(s), \mathbf{d}_2(s), \mathbf{d}_3(s))$ is attached to the centre line, and remains orthonormal

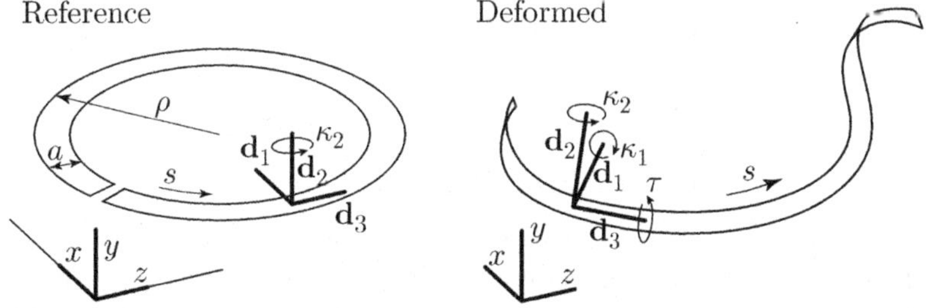

Fig. 5.6 Geometry of a godet ribbon: reference configuration (left) and deformed configuration (right). The curvature κ_2 associated with the rotation of the normal vector $\mathbf{d}_2$ is frozen when the aspect ratio of the cross-section is large, $a \gg h$.

for all s (small strain approximation). As in Chapter 3, we take $\mathbf{d}_3$ as tangent to the centre line. By convention $\mathbf{d}_1$ is parallel to the long dimension of the cross-section while $\mathbf{d}_2$ is the local normal to the ribbon. A fixed orthonormal frame $(\mathbf{e}_x, \mathbf{e}_y, \mathbf{e}_z)$ is also used.

The kinematics of the orthonormal frame is characterized by the Darboux vector $\mathbf{\Omega}(s)$, such that

$$\mathbf{d}_i'(s) = \mathbf{\Omega}(s) \times \mathbf{d}_i(s) \qquad i = 1, 2, 3 \tag{5.5}$$

for all vectors of the basis $(i = 1, 2, 3)$. As earlier, primes denote derivatives with respect to curvilinear length, s. The centre line can be reconstructed by integration of the tangent:

$$\mathbf{r}'(s) = \mathbf{d}_3(s). \tag{5.6}$$

The Kirchhoff equations are the equilibrium conditions written for the internal moment $\mathbf{M}(s)$ and force $\mathbf{F}(s)$:

$$\mathbf{F}'(s) = \mathbf{0}$$

$$\mathbf{M}'(s) = -\mathbf{d}_3(s) \times \mathbf{F}(s).$$

Note that there is no distributed force or torque: stress is applied at the ends of the rod at infinity and we neglect the weight of the ribbon. The two equations can be integrated. Let $\mathbf{P}$ be the force applied by the operator on the end where s reaches its maximum value, which may be at plus infinity. Then, the internal force is constant and equal to $\mathbf{P}$:

$$\mathbf{F}(s) = \mathbf{P}. \tag{5.7a}$$

Let $\mathbf{M}_0$ be the internal moment $\mathbf{M}(0)$ at $s = 0$, which is yet unknown. The second Kirchhoff equation can be integrated as

$$\mathbf{M}(s) = \mathbf{M}_0 - (\mathbf{r}(s) - \mathbf{r}_0) \times \mathbf{F}, \tag{5.7b}$$

where $\mathbf{r}_0 = \mathbf{r}(0)$ is the position of the cross-section with $s = 0$.

To make this system of equations complete and to compute the shape of the ribbon for a given loading, one has to consider the constitutive relations, which locally relate the flexural and torsional couples, $G_1(s)$, $G_2(s)$ and $H(s)$, to the material curvatures and twist, $\kappa_1(s)$, $\kappa_2(s)$ and $\tau(s)$. The flexural and torsional couples are the components of the internal moment $\mathbf{M}(s)$ in the local material frame $(\mathbf{d}_1(s), \mathbf{d}_2(s), \mathbf{d}_3(s))$, while the material curvature and twist are those of the Darboux vector, $\mathbf{\Omega}(s)$. This yields, in a condensed form:

$$\mathbf{M}(s) \cdot \mathbf{d}_1(s) = EI_1\, \mathbf{\Omega}(s) \cdot \mathbf{d}_1(s) \tag{5.8a}$$

$$\mathbf{M}(s) \cdot \mathbf{d}_2(s) = EI_2 \left(\mathbf{\Omega}(s) \cdot \mathbf{d}_2(s) - \frac{1}{\rho} \right) \tag{5.8b}$$

$$\mathbf{M}(s) \cdot \mathbf{d}_3(s) = \mu J\, \mathbf{\Omega}(s) \cdot \mathbf{d}_3(s). \tag{5.8c}$$

Equation (5.8a), for instance, is the constitutive relation $G_1 = EI_1\, \kappa_1$ with $G_1 = \mathbf{M} \cdot \mathbf{d}_1(s)$ and $\kappa_1 = \mathbf{\Omega} \cdot \mathbf{d}_1$. The spontaneous curvature $1/\rho$ of the ribbon has been taken into account in the constitutive relation (5.8b); it concerns the curvature $\kappa_2 = \mathbf{\Omega} \cdot \mathbf{d}_2$ associated with the director $\mathbf{d}_2$: in the absence of applied forces, the ribbon is a planar ring and the material frame then rotates around the normal $\mathbf{d}_2$ to the plane of the ribbon, hence $\mathbf{\Omega} = 1/\rho\, \mathbf{d}_2$.

It remains to take into account the large aspect ratio $a/h \gg 1$ of the ribbon. The bending stiffnesses of a rod with rectangular cross-section is given in equation (3.27). Its twist stiffness will be derived later in equation (6.108) in the limit of a large aspect ratio $a/h \gg 1$. Collecting these results, we have:

$$E I_1 = \frac{E\, a\, h^3}{12}, \qquad E I_2 = \frac{E\, a^3\, h}{12}, \qquad \mu\, J = \frac{E\, a\, h^3}{6\,(1+\nu)},$$

where E denotes Young's modulus and ν is Poisson's ratio of the ribbon. In the limit $a/h \gg 1$, one bending modulus, EI_1, and the twist modulus, μJ, are of the same order of magnitude while the other bending modulus, EI_2, is much larger, by a factor $(a/h)^2$.

The consequence is that bending in the direction associated with the large bending modulus EI_2 is strongly penalized for a ribbon. Because the order of magnitude of the left-hand side (internal moment $\mathbf{M}$) is fixed by the applied loading in the constitutive equations (5.8), taking the limit $EI_2 \gg EI_1 \sim \mu J$ in the right-hand side leads to $|\mathbf{\Omega} \cdot \mathbf{d}_2 - 1/\rho| \ll |\mathbf{\Omega} \cdot \mathbf{d}_1| \sim |\mathbf{\Omega} \cdot \mathbf{d}_3|$: the second material curvature κ_2 remains very close[9] to $1/\rho$. This mode of deformation is frozen for a ribbon, which has locally one mode of curvature only, in addition to the twist mode. One can therefore rewrite the constitutive equations in the following condensed form:

$$\mathbf{\Omega}(s) = \frac{\mathbf{M}(s) \cdot \mathbf{d}_1(s)}{EI_1}\, \mathbf{d}_1(s) + \frac{1}{\rho}\, \mathbf{d}_2(s) + \frac{\mathbf{M}(s) \cdot \mathbf{d}_3(s)}{\mu J}\, \mathbf{d}_3(s). \tag{5.9}$$

Note that the second constitutive relation (5.8b), associated with the frozen curvature mode, is no longer used.[10]

The Kirchhoff equations for a ribbon are the set of differential equations (5.5) and (5.6) for the functions $\mathbf{r}(s)$ and $\mathbf{d}_i(s)$, which involve the functions $\mathbf{\Omega}(s)$, $\mathbf{M}(s)$ and $\mathbf{F}(s)$ defined in equations (5.9) and (5.7). The parameters $\mathbf{P}$ and $\mathbf{M}_0$ in these equations are related to the force and moment applied on the ends. It must be emphasized that the validity of these equations, which are based on the theory of rods, is limited to moderately wide ribbons, that is to $a \ll \sqrt{\rho h}$, as discussed in detail in Section 6.7.

The above equations for elastic ribbons can be found for instance in references (L. Mahadevan and J. B. Keller, 1993; B. Audoly and A. Boudaoud, 2002; M. Marder *et al.*, 2003). The present derivation is more straightforward and yields the final equations and conditions in a compact form. Two classes of solutions of these equations are derived in the following sections: helical solutions in Section 5.3 and godet solutions in Section 5.4.

5.3   Helical solutions

Before we focus on undulating patterns, we present in this section a simple example of application of the equations for a ribbon, namely the analysis of its *helical* solutions. Helical

---

[9] In some geometries, the constraint $\hat{\kappa}(s) = \hat{\kappa}_0$ may happen to be inconsistent with the boundary conditions for the ribbon. Then, the ribbon would follow a different regime than that discussed here. As happens for a rod that is severely stretched longitudinally, or for a plate that is inhibited (i.e. such that its mean surface has no mode of isometric deformation, see Chapter 11), the ribbon would indeed simply minimize the elastic energy associated with the most penalizing mode, that is the bending mode $\hat{\kappa}(s)$ here.

[10] Because this second constitutive relation is no longer used, the flexural couple $\hat{G} = \mathbf{M} \cdot \mathbf{d}_2$ can be computed but this does not affect the ribbon shape. This quantity $\hat{G}(s)$ can in fact be interpreted as the Lagrange multiplier associated with the constraint $\hat{\kappa}(s) = \hat{\kappa}_0$.

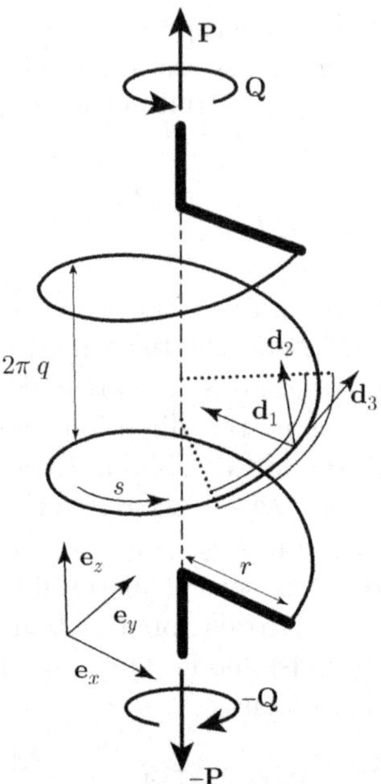

Fig. 5.7 Helical conformation of a godet ribbon. This simple geometry allows one to derive analytical solutions to the equations of equilibrium, and prepares the analysis of Section 5.4.

solutions for generic rods are studied in reference (A. Goriely and M. Tabor, 1997), for instance, and the present chapter focuses more specifically on the case of naturally curved ribbons. Such helical conformations are invariant with respect to the curvilinear coordinate s: the geometry is simple enough and it is possible to obtain analytical solutions.

5.3.1 Geometry of an helix

The geometry of the helical conformation is shown in Fig. 5.7. Owing to the spontaneous curvature $1/\rho$ of the ribbon, the material vector $\mathbf{d}_1$ parallel to the long edge of the cross-section has to point towards the axis. This configuration is called a normal helix.[11] The helix has the following parametric equation in Cartesian coordinates:

$$x(\theta) = r\,\cos\theta \tag{5.10a}$$

$$y(\theta) = r\,\sin\theta \tag{5.10b}$$

$$z(\theta) = q\,\theta, \tag{5.10c}$$

where the azimuthal angle θ is related to the curvilinear coordinate s by:

$$\mathrm{d}s = \sqrt{r^2 + q^2}\;\mathrm{d}\theta. \tag{5.11}$$

---

[11] In ribbons or rods without spontaneous curvature, there exists a second helical conformation of the rod, called the binormal helix, in which this long edge of the cross-section is parallel to the axis, but this binormal helix does not exist as an equilibrium solution when $1/\rho \neq 0$.

Here, r is the radius of the cylinder onto which the helix is drawn and $2\pi q$ is the pitch of the helix. In the planar configuration, the quantity r would be equal to the natural radius of curvature, ρ. When the helix is stretched along its axis, ρ and r differ. This helix defines the centre line of the ribbon. The attached orthonormal material frame shown in the figure is given by

$$\mathbf{d}_1 = \begin{pmatrix} -\cos\theta \\ -\sin\theta \\ 0 \end{pmatrix} \quad \mathbf{d}_2 = \frac{1}{\sqrt{r^2 + q^2}} \begin{pmatrix} q\sin\theta \\ -q\cos\theta \\ r \end{pmatrix} \quad \mathbf{d}_3 = \frac{1}{\sqrt{r^2 + q^2}} \begin{pmatrix} -r\sin\theta \\ r\cos\theta \\ q \end{pmatrix}. \quad (5.12)$$

By computing the derivative of this material frame with respect to s and by identification with the definition of the Darboux vector in equation (5.5), we obtain

$$\boldsymbol{\Omega} = \frac{1}{\sqrt{r^2 + q^2}} \, \mathbf{e}_z. \quad (5.13)$$

This rotation vector is aligned with the axis of the helix, and independent of s, a property reflecting the symmetries of the helix. The components of $\boldsymbol{\Omega}$ in the local frame yield the material curvatures and twist:

$$\kappa_1 = \boldsymbol{\Omega} \cdot \mathbf{d}_1 = 0, \quad \kappa_2 = \boldsymbol{\Omega} \cdot \mathbf{d}_1 = \frac{r}{r^2 + q^2}, \quad \tau = \boldsymbol{\Omega} \cdot \mathbf{d}_1 = \frac{q}{r^2 + q^2}. \quad (5.14)$$

The vanishing of the first curvature, κ_1, is also a consequence of the symmetries of the helix.

Let us now define the rate of stretching of the helix, ϵ, by:

$$\epsilon = \frac{dz}{ds} = q\,\frac{d\theta}{ds} = \frac{q}{\sqrt{r^2 + q^2}}. \quad (5.15)$$

This number ϵ is independent of s. Its value is $\epsilon = 0$ when the helix coils into a circle in the plane ($q = 0$), while it is $\epsilon = \pm 1$ when its centre line is stretched to a straight line ($r = 0$). The helical configurations form a one-parameter family of solutions of the ribbon, indexed by $\epsilon \in [-1, +1]$. Indeed, one mode of curvature is frozen and the condition $\kappa_2 = 1/\rho$ yields, with equation (5.14), a relation between r, q and ρ. Using equation (5.15), one can therefore eliminate the geometrical parameters of the helix q and r in favour of ϵ and ρ:

$$r = (1 - \epsilon^2)\,\rho, \qquad q = \epsilon\,\sqrt{1 - \epsilon^2}\,\rho. \quad (5.16)$$

This analysis of the geometry of a helical ribbon must be supplemented by the equations of mechanical equilibrium derived previously.

5.3.2 Torques and forces on a godet ribbon in a helical conformation

As in Figure 5.7, we call $\mathbf{P}$ and $\mathbf{Q}$ the force and moment applied along the axis of the helix at the 'downstream' end s^+ (the one with large s). The overall equilibrium of the helix requires that the force and moment applied along the axis at the 'upstream' end are $-\mathbf{P}$ and $-\mathbf{Q}$. By symmetry, we choose $\mathbf{P}$ and $\mathbf{Q}$ to be along the axis of the helix: $\mathbf{P} = P\,\mathbf{e}_z$ and $\mathbf{Q} = Q\,\mathbf{e}_z$. The equilibrium of the subsystem composed of the part of the ribbon $s' > s$ that is downstream of a given section s, including the small bar (in thick black in the figure) that transmits the applied force $\mathbf{P}$ and torque $\mathbf{Q}$ to the endpoint of ribbon, yields:

$$\mathbf{F}(s) = \mathbf{P}, \qquad \mathbf{M}(s) = \mathbf{Q} - \mathbf{r}(s) \times \mathbf{P}.$$

Indeed, the forces and torques applied on this subsystem are $\mathbf{P}$ and $\mathbf{Q}$ along the axis and, by the principle of action–reaction, $-\mathbf{F}(s)$ and $-\mathbf{M}(s)$ at the point $\mathbf{r}(s)$—recall that $\mathbf{F}(s)$ and $\mathbf{M}(s)$ are the internal force and moment in the ribbon, defined using specific upstream–downstream conventions. The above expressions for $\mathbf{F}(s)$ and $\mathbf{M}(s)$ are equivalent to the equations (5.7) derived earlier, but use $\mathbf{P}$ and $\mathbf{Q}$ as parameters instead of $\mathbf{P}$ and $\mathbf{M}_0$. Using the explicit expression for $\mathbf{r}(s)$, we obtain

$$\mathbf{M}(s) = r\,P\,(-\sin\theta\,\mathbf{e}_x + \cos\theta\,\mathbf{e}_y) + Q\,\mathbf{e}_z.$$

By projection on the local material frame, this yields the flexural and torsional couples:

$$G_1 = \mathbf{M}\cdot\mathbf{d}_1 = 0$$

$$G_2 = \mathbf{M}\cdot\mathbf{d}_2 = \sqrt{1-\epsilon^2}\,Q - \epsilon\,(1-\epsilon^2)\,\rho\,P$$

$$H = \mathbf{M}\cdot\mathbf{d}_2 = \epsilon\,Q + (1-\epsilon^2)^{\frac{3}{2}}\,\rho\,P.$$

The vanishing of G_1, like that of κ_1, is due to the symmetries of the helix: the constitutive relation $G_1 = E\,I_1\,\kappa_1$ is automatically satisfied. The second flexural couple, G_2, is given here for reference but has no effect on the associated curvature, which is frozen. The torsional couple H has to satisfy the remaining constitutive relation, $H = \mu\,J\,\tau$, where τ was given in terms of r and q in equation (5.14). This yields the equation

$$\epsilon\,Q + (1-\epsilon^2)^{\frac{3}{2}}\,\rho\,P = \frac{\mu\,J}{\rho}\,\frac{\epsilon}{\sqrt{1-\epsilon^2}}. \tag{5.17}$$

This implicit equation for the rate of elongation ϵ characterizes the response of the ribbon to an applied force P and torque Q. This implicit equation for ϵ is solved graphically in Fig. 5.8. There, the dimensionless applied moment, $Q\rho/(\mu\,J)$ is plotted as a function of the dimensionless applied tension, $P\rho^2/(\mu\,J)$, for various values of elongation rates ϵ in the range $-1 \le \epsilon \le 1$. For any value of the loading (P,Q), the line passing through the corresponding point in the diagram yields the value of the rate of elongation, ϵ,

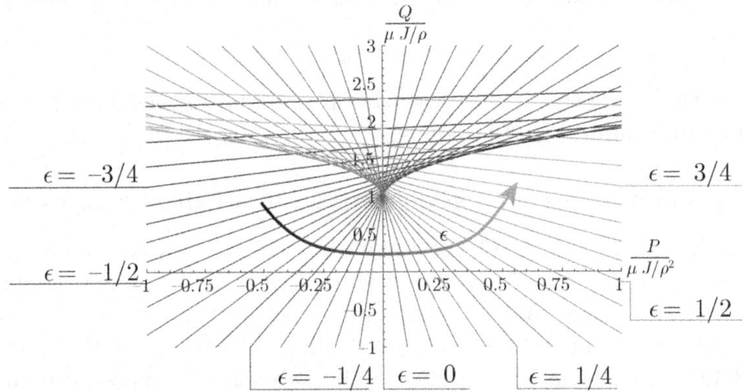

Fig. 5.8 Graphical solution of the rate of elongation ϵ of a helical godet ribbon, as a function of the rescaled applied loading P and Q, by equation (5.17). Each line corresponds to a fixed value of ϵ. Notice the presence of a caustic, indicating a sub-critical bifurcation for large positive Q, that is for an applied moment that tends to *increase* the natural curvature.

of a helical solution. This value of ϵ allows one to determine the helix parameters by equation (5.16).

In this plot, there are two regions, separated by a caustic: below the caustic, that is for Q lower than a critical value $M_c(P)$, there exists a unique solution ϵ to the implicit equation (5.17). Along the caustic, a bifurcation towards a different helical solution takes place and above the caustic (that is for large enough Q), three helical solutions are present. These helical solutions were simply derived to provide an example of applications of the equations for ribbons; the analysis of their stability is beyond the scope of this chapter. We shall simply note that this bifurcation is sub-critical, i.e. the solution jumps to a stable solution by a perturbation having a finite amplitude at the onset. Generally speaking, helical solutions can loose stability in a number of ways (A. Goriely and M. Tabor, 1997): by a static bifurcation towards a different helical shape, by a static bifurcation towards a non-helical shape, or even by a dynamical instability.

The particular case of pure traction, $Q = 0$ (that is when the end of the helix is free to wind around the axis while the latter is stretched), leads to a simple expression:

$$P = \frac{\mu J}{\rho^2} \frac{\epsilon}{(1 - \epsilon^2)^2}. \tag{5.18}$$

For weak normal tension forces ($|\epsilon| \ll 1$), the ribbon behaves like a linear spring with stiffness k:

$$P \approx k\, z(L) \qquad \text{where } k = \frac{\mu J}{\rho^2 L}, \tag{5.19}$$

where L is the curvilinear length of the ribbon and $z(L)$ its total elongation. For larger tension forces, the ribbon develops a non-linear response. The origin of the nonlinearity lies in the geometry of the helix, studied in Section 5.3.1—because the ribbon is inextensible, the tensile force P diverges as ϵ approaches ± 1, a behaviour that is clearly non-linear.

We emphasize that this geometrically non-linear response of the ribbon can be observed while the material remains well in the linear regime (Hookean elasticity). Indeed, from Section 3.5 the strain in the ribbon is of order $a\,\tau$. The order of magnitude of τ can be found by equations (5.14) and (5.16). This yields:

$$a\,|\tau| \sim \frac{a}{\rho} \frac{|\epsilon|}{\sqrt{1 - \epsilon^2}} \ll 1 \qquad \text{(validity of Hookean elasticity)}. \tag{5.20}$$

A very narrow band has been sliced off along the edge of the sheet and so $a \ll \rho$. Therefore, the material can be assumed to respond linearly unless the helix is almost fully stretched, $|\epsilon \mp 1| = \mathcal{O}((a/\rho)^2)$. In contrast, the geometrical nonlinearity enters into play as soon as $|\epsilon|$ is not a very small number. For loadings such that $|\epsilon|$ is not very close to one, the Hookean elasticity combined with the geometrically non-linear Kirchhoff theory provide an accurate description of the elasticity of the ribbon: the response of the material is locally linear, even though the ribbon displays a response that is globally non-linear.

5.4 Godet solutions

The helical geometry studied above is presumably the simplest one for which the equations for a godet ribbon can be solved. These helical solution do not display godets because an axial loading cannot get rid of the non-zero linking number found in the natural, planar configuration. In the present section, we return to the analysis of ripples. A different loading geometry is used, leading to non-helical shapes: instead of being stretched perpendicularly to its rest plane, the ribbon is loaded by forces and moments that unfold it, providing it with an overall shape consistent with the edge of plate.

5.4.1   Dimensionless equations

To ease the numerical solution of the equations for a ribbon, the latter are first made dimensionless. Following our conventions, we introduce starred quantities that define new units. All lengths are rescaled using the natural radius of curvature, ρ:

$$s^* = \rho \quad \kappa^* = 1/\rho \quad r^* = \rho, \tag{5.21a}$$

and the bending and twist stiffnesses using EI. The rescaling of the other quantities follows by balancing all terms in the equations of equilibrium:

$$(EI)^* = EI, \quad (\mu J)^* = EI, \quad F^* = P^* = \frac{EI}{\rho^2}, \quad M^* = \frac{EI}{\rho}. \tag{5.21b}$$

The rescaled quantities are temporarily marked with an overline, that will soon become implicit: $\bar{s} = s/s^*, \bar{\kappa} = \kappa/\kappa^*$, etc. Thanks to the choice of reference values in equation (5.21), the rescaled bending stiffness is one, $\overline{EI} = 1$, as is the rescaled natural curvature $1/\bar{\rho} = 1$. The rescaled twist stiffness can be computed from equation (5.9):

$$\overline{\mu J} = \frac{\mu J}{EI} = \frac{2}{1+\nu}. \tag{5.22}$$

The dimensionless vectors of the local material frame $\mathbf{d}_i$ are not rescaled, $\mathbf{d}_i(s) = \overline{\mathbf{d}}_i(\bar{s})$. Note that the rescaled quantities are implicitly a function of rescaled variables, such as $\bar{s}$, not of the original variables. Finally, $\overline{\mathbf{r}}(\bar{s}) = r^* \, \mathbf{r}(s) = \rho \, \mathbf{r}(\bar{s}s^*)$.

The equations for our ribbon can then be rewritten in a dimensionless form:

$$\overline{\mathbf{r}}'(\bar{s}) = \overline{\mathbf{d}}_3(\bar{s}) \tag{5.23a}$$

$$\overline{\mathbf{d}}_i'(\bar{s}) = \overline{\boldsymbol{\Omega}}(\bar{s}) \times \overline{\mathbf{d}}_i(\bar{s}) \quad \text{for } i = 1, 2, 3, \tag{5.23b}$$

where the auxiliary variables are given by:

$$\overline{\boldsymbol{\Omega}}(\bar{s}) = \left(\overline{\mathbf{M}}(\bar{s}) \cdot \overline{\mathbf{d}}_1(\bar{s})\right) \overline{\mathbf{d}}_1(\bar{s}) + \overline{\mathbf{d}}_2(\bar{s}) + \frac{1+\nu}{2} \left(\overline{\mathbf{M}}(\bar{s}) \cdot \overline{\mathbf{d}}_3(\bar{s})\right) \overline{\mathbf{d}}_3(\bar{s}) \tag{5.23c}$$

$$\overline{\mathbf{M}}(\bar{s}) = \overline{\mathbf{M}}_0 - (\overline{\mathbf{r}}(\bar{s}) - \overline{\mathbf{r}}_0) \times \overline{\mathbf{P}}. \tag{5.23d}$$

This set of differential equations are the rescaled equations for a godet ribbon. It remains to set the free parameters and the initial conditions of these equations, which is the subject of the next section.

5.4.2   Symmetries, parameters and initial conditions

In order to integrate the above differential system, one needs to specify the initial values $\overline{\mathbf{r}}(0)$ and $\overline{\mathbf{d}}_i(0)$, as well as the applied loading through $\overline{\mathbf{P}}$ and $\overline{\mathbf{M}}_0$. These parameters cannot be set arbitrarily if a periodic solution is to be obtained in the end. In the experiment shown in Fig. 5.13, one obtains a periodic pattern far from the edges when the slinky is long, independent of how the boundary forces are applied. A periodic pattern is observed as well

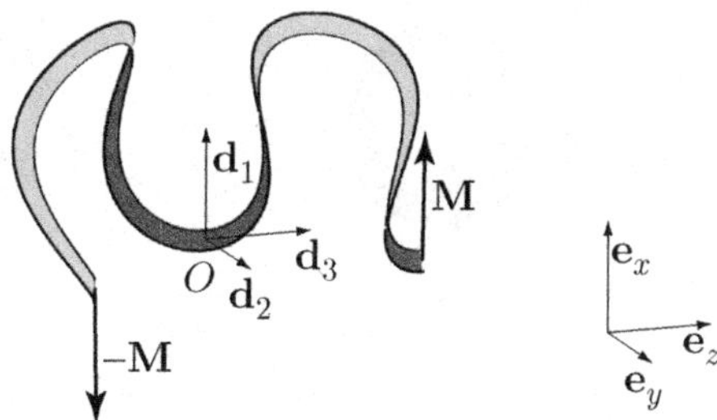

Fig. 5.9 Loading geometry: a godet ribbon is unfolded by applying a moment co-planar with the initial plane, (Oxz).

when one stretches the whole structure. This suggests that there exist periodic solutions to the equations above for an arbitrary values of $\overline{\mathbf{P}}$. However, we consider only the easier case $\overline{\mathbf{P}} = \mathbf{0}$ in the following. This corresponds to the path (p1) in Fig. 5.5, whereby the ribbon is unfolded by applying a moment only.

In what follows, we continue to deal with rescaled quantities implicitly, although we drop the overlines for sake of readability. Let us choose the system of coordinates such that $\mathbf{e}_z$ is aligned with the edge of the original sheet, which is also the direction in which the sought after godet pattern stretches out. We shall call this direction $\mathbf{e}_z$ the 'axis' of the pattern. In the absence of internal force $\mathbf{P} = \mathbf{0}$, the internal moment $\mathbf{M}$ is uniform in the ribbon. By symmetry, this vector $\mathbf{M}$ has to be aligned with or be perpendicular to the axis of the pattern. The path (p1) in Fig. 5.5 suggests that $\mathbf{M}$ is in fact perpendicular to $\mathbf{e}_z$ and we choose the coordinates such that $\mathbf{e}_x$ is aligned with $\mathbf{M}$. Calling M the unknown magnitude of $\mathbf{M}$, we have

$$\mathbf{M}(s) = M\,\mathbf{e}_x.$$

Let us moreover choose the origin $s = 0$ of the curvilinear coordinate to be the bottommost point of a buckle, as in Fig. 5.9. This point is a point of mirror symmetry of the final pattern. As a result, the tangent $\mathbf{d}_3(0)$ there is aligned with the axis of the pattern. By the same argument, the director $\mathbf{d}_1(0)$ lying in the tangent plane of the ribbon is aligned with $\mathbf{M}$. Choosing this point as the origin of the coordinates yields $\mathbf{r}(0) = \mathbf{0}$. Therefore, we complement the differential system (5.23) with the following initial conditions and parameters values:

$$\mathbf{r}(0) = \mathbf{0}, \quad \mathbf{d}_1(0) = \mathbf{e}_x, \quad \mathbf{d}_2(0) = \mathbf{e}_y, \quad \mathbf{d}_3(0) = \mathbf{e}_z, \quad \mathbf{M}(s) = M\,\mathbf{e}_x, \quad \mathbf{P} = \mathbf{0}, \quad (5.24)$$

For any value of the parameters ν and M, this forms a well-posed Cauchy problem and the differential system can be integrated. Poisson's ratio, ν, which enters in the rescaled value of the twist modulus, reflects the material properties of the ribbon—we shall consider only the typical case $\nu = 0.3$ as we expect a weak dependence of the final pattern on the value of ν. It remains to adjust the value of M so as to obtain a periodic pattern, as explained in the next section.

5.4.3   Unfolding into a periodic wavy pattern

When the applied moment M is increased from zero, the ribbon progressively unfolds, as shown in Fig. 5.10. When the applied moment M reaches a particular value M_g, the

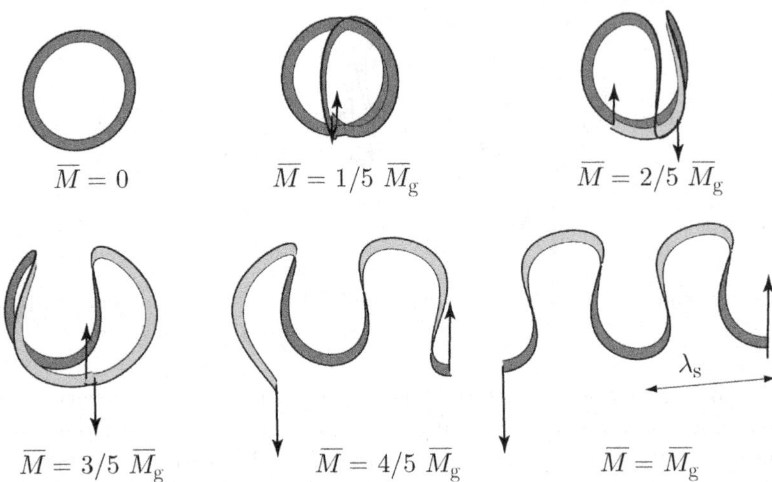

Fig. 5.10 Simulation of the unfolding of a ribbon with $\nu = 0.3$ by out-of-plane deformations—path (p1) in Fig. 5.5. This unfolding is achieved by integrating the differential system (5.23) with loading and initial conditions given by equation (5.24), for increasing values of M. For a value $M = M_{\mathrm{g}} \approx 1.115$ of the applied moment, a periodic godet pattern, shown bottom right, is obtained.

pattern shown in the figure becomes periodic by translation along the axis. The resulting undulations are qualitatively similar to the godet pattern observed in plant leaves. We call it the *godet* configuration of a ribbon. This solution was first reported in references (B. Audoly and A. Boudaoud, 2002) and (M. Marder *et al.*, 2003) but the present derivation is much more direct.

The value M_{g} of the applied moment corresponding to a periodic pattern satisfies the implicit equation:

$$\tau(\lambda_{\mathrm{c}}, M_{\mathrm{g}}) = 0 \tag{5.25a}$$

$$y(\lambda_{\mathrm{c}}, M_{\mathrm{g}}) = 0, \tag{5.25b}$$

where $\tau(s, M) = (M\,\mathbf{e}_x)\cdot\mathbf{d}_3(M, s)$ is the twist at s, the tangent $\mathbf{d}_3(M, s)$ being computed by integration of the differential system with the parameter M, and $y(s, M) = \mathbf{e}_y\cdot\mathbf{r}(M, s)$ is the y coordinate of the centre line. Recall that all these quantities are implicitly rescaled. In this equation, we have introduced λ_{c}, the *curvilinear* wavelength of the pattern, not be confused with the *spatial* wavelength λ_{s}—for instance, in the planar rest configuration, $\lambda_{\mathrm{c}} = 2\pi$ while $\lambda_{\mathrm{s}} = 0$. By symmetry, the twist vanishes at the topmost and bottommost points of the ribbon, as shown in the figures. Therefore, $\tau = 0$ when s is a multiple of $\lambda_{\mathrm{c}}/2$, hence equation (5.25a). The second equation (5.25b) imposes that the periodic displacement leaving the pattern invariant is parallel to $\mathbf{e}_z$.

Equations (5.25) are a set of implicit equations for the unfolding moment M_{g} and wavelength λ_{c}. The roots $(\lambda_{\mathrm{c}}, M_{\mathrm{g}})$ of this set of equations can be computed numerically

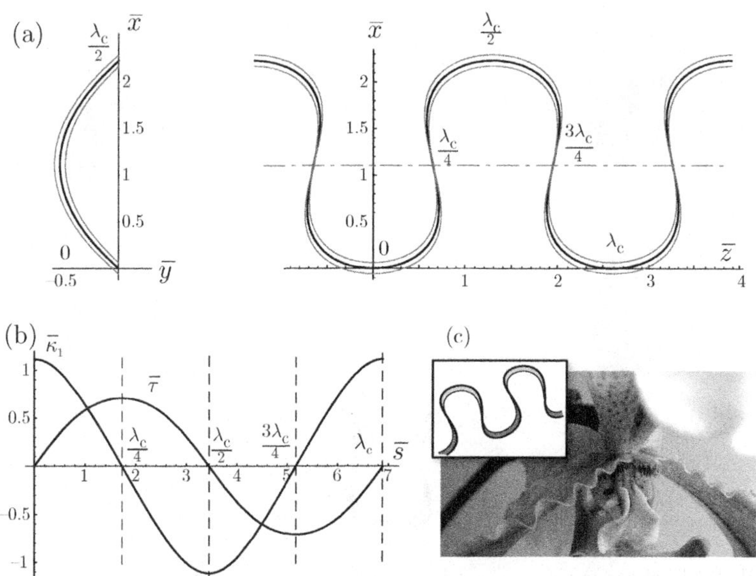

Fig. 5.11 Periodic pattern obtained by unfolding a ribbon with $M = M_{\mathrm{g}}$. (a) Projection of the centre line in the planes (x, y) perpendicular to the axis and in the plane (x, z). The edges of the ribbon are shown to give a sense of the 3D shape of the ribbon. (b) Curvature κ_1 and twist τ of the solution. (c) Visual comparison of the unfolded ribbon with the edge of a flower of *Paphiopedilum* orchid, the same plant as in Fig. 5.1(c).

by integration of the equations of equilibrium for a ribbon, for various values of M. This yields

$$\lambda_{\mathrm{c}} = 6.894\,\rho, \qquad M_{\mathrm{g}} = 1.115\,\frac{E\,I}{\rho} \qquad \text{for } \nu = 0.3, \tag{5.26}$$

after restoring the physical units. The corresponding periodic solution is shown in Fig. 5.11. The curvilinear wavelength λ_{c} is slightly larger than $2\,\pi\,\rho \approx 6.28\,\rho$: one wavelength in the final pattern stems from slightly more than one turn in the original, flat configuration. The spatial wavelength of the pattern is $\lambda_{\mathrm{s}} = 2.606\,\rho$.

The shape of the godet patterns can be understood qualitatively by calling up the notion of perversion. Perversions are typical patterns that can be observed in tendrils of climbing plants or in telephone cords, see Fig. 5.12. Made of a buckle connecting two helices with opposite chiralities (T. McMillen and A. Goriely, 2002; G. Domokos and T. J. Healey, 2005) they are found in rods that have some intrinsic curvature. The resemblance of the buckle in perversions with those found in godet ribbons is striking: one wavelength of the godet pattern analysed above can be viewed as composed of two mirror-symmetric perversions. The similarity between the two patterns can be explained from geometry. A perversion is invariant by mirror symmetry with respect to a plane passing through the centre of the buckle. As a result, it has a vanishing overall linking number: it satisfies the topological

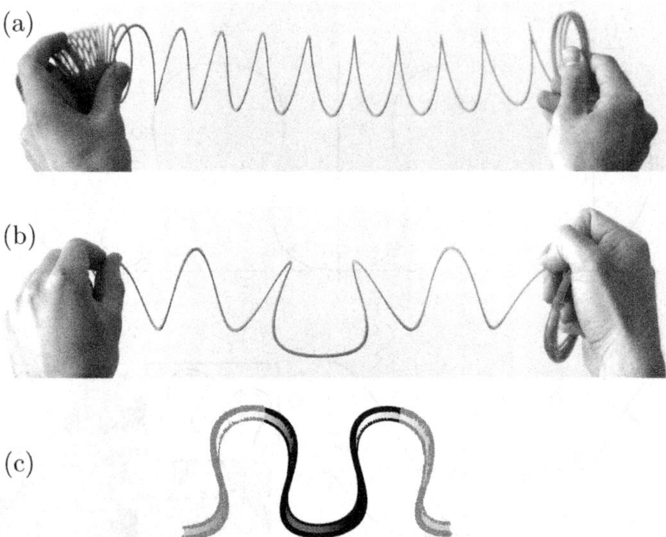

Fig. 5.12 A helical spring, shown in undeformed configuration (a), is deformed into a pattern called a perversion (b): this pattern connects a left and a right helix (T. McMillen and A. Goriely, 2002). (c) The godet pattern is made up of wrinkles that are similar to the central buckle in a perversion. (a) and (b) are reprinted from M. Bergou *et al.*, "Discrete Elastic Rods", ACM Transactions on graphics, Vol. 27:3, © 2008 ACM, Inc., by permission from ACM.

constraint studied in Section 5.1.2. However, the helices connected by the buckling both have non-zero twist densities, with opposite signs: the linking number is zero globally but not locally. By replicating and patching the small part of the perversion pattern that surrounds the buckle, the godet pattern avoids local accumulation of the linking number, making it possible for the rippled edge to merge smoothly with the flat region of the leaf lying far from the edge.

In the next section, we present a simple experiment based on a man-made ribbon, that allows one to observe these godet patterns solutions.

5.4.4   Experiments

It is possible to observe the godet ribbon experimentally using a US toy called the *Slinky*,[12] as shown in Fig. 5.13. This slinky is a flexible spring with a flat cross-section.[13] When properly launched it can hop down staircases. It features both the flat cross-section and the spontaneous curvature of our rod model. In Fig. 5.13, we show a typical godet configurations than can be obtained by opening a slinky, following the unfolding path sketched on the left-hand side of Fig. 5.10. The agreement between the numerically computed shape

---

[12] Slinky is a trademark of James Industries.

[13] As an alternative to using a Slinky, one can print a few spires of a spiral with a small step on a sheet of overhead transparencies and cut along the spiral. The resulting spring has an almost uniform spontaneous curvature ρ and is a good approximation to our godet ribbon—it is not as convenient to use as the Slinky, however, because the material used for transparencies is not very stiff, which makes the resulting spiral sensitive to gravity.

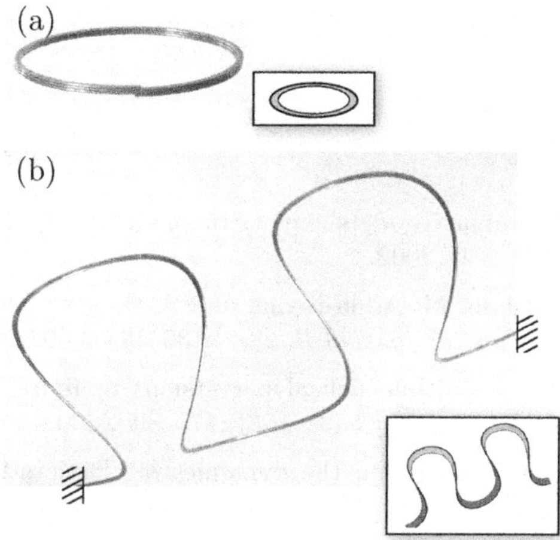

Fig. 5.13 Experimental observation of godets using a slinky: (a) natural shape, (b) periodic godet configuration. Comparison with the numerical solution is shown in the insets.

and experimentally observed one is visually good. A quantitative comparison is made in reference (M. Marder *et al.*, 2003).

5.5   Conclusion

The comparison presented in Fig. 5.11(c) shows that the proposed model reproduces and qualitatively explains the formation of ripples in plant leaves and other elastic sheets that have a stretched edge. This model captures the basic rippling mechanism while avoiding the difficulties associated with the full equations for elastic plates. The key features that have been retained are the differential growth near the edge and the small thickness of the initial elastic plate. With the help of geometrical nonlinearity, this is sufficient to account for the rippled patterns. In plants, the details of the ripples geometry are probably not coded genetically, only the differential growth near the edge being encoded (E. Sharon *et al.*, 2002). The coding at the microscopic level remains very simple, most of the final complexity coming from non-linear mechanics; this general point of view that has been explored for instance by Y. Couder and coworkers (Y. Couder *et al.*, 2002).

Obviously, some features of the patterns in plant leaves are beyond the reach of such a simple model. Most notably, the possibility of self-similar patterns with smaller and smaller length scales near the edge can only be fully understood with the help of the theory of elastic plates, as is done in Chapter 10. Based on the present rod model, a simple argument can nevertheless explain qualitatively the making of self-similar patterns by elastic effects. The wavelength of the godet patterns in a ribbon was found in the present chapter to be of order $1/|g'(t)|$, that is the inverse of the transverse derivative of the stretch rate of the edge, g. According to the experimental measurements shown in Fig. 5.3, g is a function that is peaked at the edge $t = 0$ and decays to zero away from it. As a result, the natural length scale of the pattern $1/|g'(t)|$ might be a rapidly increasing function of the distance t to the edge. Self-similar patterns can be understood as godets patterns of varying wavelength, which remains close to the local optimal one, $1/|g'(t)|$. A more quantitative version of this argument, including a discussion of the conditions in which the cascade appear, is the subject of Chapter 10.

References

B. Audoly and A. Boudaoud. 'Ruban à godets': an elastic model for ripples in plant leaves. *Comptes Rendus Mécanique*, 330(12):831–836, 2002.

Y. Couder, L. Pauchard, C. Allain, M. Adda-Bedia, and S. Douady. The leaf venation as formed in a tensorial field. *European Physical Journal B*, 28(2):135–138, 2002.

G. Domokos and T. J. Healey. Multiple helical perversions of finite, intrinsically curved rods. *International Journal of Bifurcation and Chaos*, 15(3):871–890, 2005.

A. Goriely, M. Nizette, and M. Tabor. On the dynamics of elastic strips. *Journal of Nonlinear Science*, 11(1):3–45, 2001.

A. Goriely and P. Shipman. Dynamics of helical strips. *Physical Review E*, 61(4):4508–4517, April 2000.

A. Goriely and M. Tabor. Nonlinear dynamics of filaments. III. Instabilities of helical rods. *Proceedings of the Royal Society A: Mathematical, Physical and Engineering Sciences*, 453(1967):2583–2601, 1997.

G. R. Kirchhoff. Über das Gleichgewicht und die Bewegung eines unendlich dünnen elastischen Stabes. *Journal für die reine und angewandte Mathematik. Journal de Crelle (Berlin)*, 56:285–313, 1859.

T. McMillen and A. Goriely. Tendril perversion in intrinsically curved rods. *Journal of Nonlinear Science*, 12(3):241–281, 2002.

A. Mielke and P. Holmes. Spatially complex equilibria of buckled rods. *Archive for Rational Mechanics and Analysis*, 101(4):319–348, 1988.

L. Mahadevan and J. B. Keller. The shape of a Möbius band. *Proceedings of the Royal Society A: Mathematical, Physical and Engineering Sciences*, 440, 1993.

M. Marder, E. Sharon, S. Smith, and B. Roman. Theory of edges of leaves. *Europhysics Letters*, 62(4):498–504, 2003.

U. Nath, B. C. W. Crawford, R. Carpenter, and E. Coen. Genetic control of surface curvature. *Science*, 299(5611):1404–1407, 2003.

S. Nechaev and R. Voituriez. On the plant leaf's boundary, 'jupe à godets' and conformal embeddings. *Journal of Physics A: Mathematical and General*, 34:11069–11082, 2001.

E. Sharon, M. Marder, and H. L. Swinney. Leaves, flowers and garbage bags: Making waves. *American Scientist*, 92(3):254, 2004.

E. Sharon, B. Roman, M. Marder, G.-S. Shin, and H. L. Swinney. Mechanics: Buckling cascades in free sheets. *Nature*, 419(6907):579–579, 2002.

E. Sharon, B. Roman, and H. L. Swinney. Geometrically driven wrinkling observed in free plastic sheets and leaves. *Physical Review E (Statistical, Nonlinear, and Soft Matter Physics)*, 75(4):046211, 2007.

P. L. Tchebychef. On the cut of clothes. Talk at l'Association Française pour l'avancement des sciences (seems to have remained unpublished after Tchebychef's will), 28 Aug. 1878.

G. H. M. van der Heijden and J. M. T. Thompson. Helical and localised buckling in twisted rods: A unified analysis of the symmetric case. *Nonlinear Dynamics*, 21(1):71–99, 2000.

Part II

Plates

A thin sheet of elastic material that is flat at rest is called an elastic plate. The mechanics of plates has many applications in everyday life. For instance, it can explain why blisters sometimes tend to detach a layer of paint from its substrate. A similar problem arises in the industrial applications of thin coated films. Plates have a striking mechanical property that can be easily demonstrated: one can crumple a sheet of paper, uncrumple it and observe that the remaining scars tend to be localized on it and to have a typical crescent shape. This simple observation is not so easy to understand and will be the matter of the coming Chapter 9.

In Chapter 6, we first derive the relevant equations, the Föppl–von Kármán equations (abbreviated as F.–von K.), a set of two non-linear equations coupling the out-of-plane deviation $w(x, y)$ and the in-plane stress. This stress derives from the Airy potential, denoted as $\chi(x, y)$. There is a small parameter in the equations, which is related to the small aspect ratio of the plate. It was the great discovery of Föppl that bending effects have to be retained in the equations, even though they are formally of a higher order in h, the small quantity, than the formally dominant stretching effects.

The validity of this claim is made clear when analysing the buckling of a long rectangular plate under in-plane stress. This classical example of bifurcation theory is studied in Chapters 7, 8 and 10. The buckling of thin plates is interesting because it provides a rather unique opportunity of exploring the solutions of a bifurcation problem in the weakly and strongly non-linear regime, and because this problem can also be approached by experiments.

Thin elastic plates are subject to buckling in mainly two types of situations. One possibility is that a freely hanging plate is loaded in some way or another and bends when the load exceeds a threshold. A familiar example is thermal buckling, as can be observed in modern buildings covered with thin plates of glass. Those plates are held on their edges and may buckle when they expand too much under heating. Sometimes they even drop off their cask and fall to the ground. This type of situation can be well represented in the laboratory by loading a thin metallic plate held by its sides and by looking at the way it deforms. In Chapter 8 we report on some recent experiments on this topic; a Ritz-like theory is presented next, which provides a way to draw the domain in the parameter space where various buckling patterns are observed, in fair agreement with the experiments. There is another important class of problems where buckling of thin plates occurs: for practical applications, many materials are coated with thin layers of a different materials, with the goal of improving one or another of their properties such as their resistance to wear, to surface melting, etc. This has been done for a long time: walls covered with plaster have been painted, and still are, to get rid of the dust and to harden their surface. When a substrate and its coating are made of different materials, it happens fairly often that so-called residual stress takes place in the film—for instance, when the film is deposited at high temperature onto the substrate and their thermal expansion coefficients differ, or when some chemical reaction takes place in the film or at the interface, which induces stress. Film under compressive stress has long been observed to buckle in various ways, depending on the specific system considered. Blisters are also sometimes observed, in which case the film debonds from the substrate. It is important to characterize the mechanical conditions in which the film–substrate assembly can fail, such blisters being unwanted in most situations. This is a complicated question as its answer depends on many parameters, including the film's and substrate's mechanical and geometrical properties as well as the bonding properties of the interface between them. We shall focus on a specific buckling pattern, the so-called telephone cord, and show that it can be understood based on the equations for elastic plates, as studied in Chapter 8.

The following chapters in Part II will be concerned with plate solutions for large loads, *i.e.* when the applied loading is much larger than the critical load for buckling. These solutions display singularities which can be of various types. In Chapter 9, we look at one class of solutions of the F.-von K. equations, the so-called d-cone and the fold: those are the basic blocks that build up the polyhedral geometry of crumpled paper. In Chapter 10, we investigate a different type of solutions, arising when a large deformation is imposed along a boundary, either externally or through residual stress localized along an edge (as happens along the edge of lettuce leaves). We show that self-similar solutions are selected in this situation: a cascade of smaller and smaller structures accumulates near the boundary, a phenomenon that has connections with other physical systems.

6
The equations for elastic plates

A plate is a thin and flat piece made of an elastic material: it has a very small thickness, its dimension in one direction, called the transverse direction, being much smaller than in the two other directions;[1] it is also flat in its stress-free configuration.[2] This definition is reminiscent of that of an elastic rod, although a rod is of course of dimension one and a plate is of dimension two. The equations for elastic plates that we shall derive indeed appear as an extension of the equations for elastic rods with two space variables instead of one. However, this extension is far from straightforward: in the following chapters, we point out several geometrical effects, such as the role of isometric embeddings, that are specific to elastic plates.

The aim of the present chapter is to derive the equations for elastic plates. Several variants of these equations are found in the literature, which are applicable to finite or small rotations, to small or large loads, etc. To address a plate problem that involves only small deflections from the planar state, for instance, it makes little sense to resort to the finite displacement version of the plate equations, which are quite intricate.[3]

In this book, we focus on the Föppl–von Kármán equations for elastic plates (later abbreviated as F.–von K. equations), which suit all our needs for the problems studied in the following chapters. These equations are derived under the assumption of small deflections (every tangent plane to the plate is assumed to undergo small rotations) although some non-linear terms are retained. They are suited for the analysis of buckling, not too far from the threshold. A second assumption is that the strain remains everywhere small: the F.–von K. equations assume a linear elastic response of the material (Hookean elasticity). The tricky question of the validity of these equations is discussed in detail in several places in this book. At this point, we shall simply stress that the F.–von K. equations are not applicable to all problem in plates elasticity. Once the assumptions behind them have been identified and understood, one can easily determine whether they are suited or not to a specific problem.

Going from the general theory of 3D elasticity derived in Chapter 2 to the two dimensional F.–von K. equations requires one to eliminate any explicit dependence of the fields with respect to the transverse direction of the plate. Although it might seem easy in the first

---

[1] The thickness of the plate is assumed uniform for the sake of simplicity. This assumption can be relaxed without major difficulty.

[2] We shall consider later on the possibility of residual stress in the flat, reference configuration.

[3] This situation is similar to that encountered with elastic rods: the Kirchhoff equations were derived under the assumption that the rod deforms in an inextensible manner. While this assumption holds true for many problems, like those discussed in Part I of this book, some loading geometries involve significant longitudinal stretching and call for different equations.

place, this reduction is quite difficult to achieve in practice: the equations of elasticity for thin, planar rods were derived by Euler in the eighteenth century but it took another century and a half to obtain the equation of elasticity of thin plates, something achieved in the early twentieth century by Föppl and von Kármán. The resulting equations are notorious for their intricate mathematical structure, as they are a set of coupled non-linear partial differential equations. The origin of this complexity lies in the geometry of surfaces; the structure of the equations for plates reflects the deep connection between elasticity and geometry.

A central problem in differential geometry is to find the conditions for a given surface to be able to deform without changing the lengths of all curves drawn onto it, that is without changing the distances measured *along* the surface. These deformations of an unstretchable surface are called isometric deformations. A related question is to describe these deformations when they exist. This geometric problem is deeply connected to the mechanics of plates. When a piece of paper is rolled in one way or another without tearing, the length of any curve drawn on it obviously changes by a very small amount. The measurement of lengths along a surface is formalized by the mathematical notion of a *metric*. This notion is relatively straightforward in the case of a flat surface but becomes far more subtle for non-planar surfaces such as a sphere. Riemannian geometry is precisely the geometry of surfaces and their generalization to higher dimensions, the so-called manifolds, such that these distances are defined in an abstract way.

In Section 6.1, we explain in more detail how isometric deformations are connected to the elasticity of plates: such deformations involve an unusually low amount of elastic energy and so are favoured when the goal is to minimize the elastic energy. In the following sections, we proceed to the geometrical problem itself. In Section 6.2, the existence of an invariant found by Gauss is established, which concerns isometric deformations. This invariant is called the Gauss curvature; in Section 6.3, this invariant is used to characterize the isometric deformations of a plane—this is the theory of developable surfaces. The next step is to incorporate the equations of elasticity into this geometrical analysis, with the aim of deriving the equations for elastic plates. This is achieved in two steps: in Section 6.4, the simple case where the flexural energy can be neglected is considered, and this leads to the so-called *membrane* equations; flexural[4] effects are finally taken into account and, in Section 6.5, we derive the full F.–von K. equations for elastic plates. Finally, in Section 6.6, we show that the plate equations follow from a variational principle, and compute the elastic energy of a plate.

6.1   Bending versus stretching energy

In this section, we shall estimate the elastic energy involved in the deformation of a plate. We point out a particular class of deformations namely the pure bending deformations[5] that, in the limit of a small thickness, involve a very low amount of elastic energy. This section is based on heuristic arguments and order of magnitude estimates and is intended to prepare the subsequent, more formal derivations.

---

[4] We shall use as equivalent words 'flexion' and 'bending'.

[5] These pure bending deformations, although isometric deformations, play a similar role to the inextensional deformations of a rod, see Section 3.7. Although it is quite obvious that a curve in the Euclidean space can always deform isometrically, the existence of isometric deformations for a plane or a surface is a much more tricky question; it is studied later.

Hooke's assumption of linear elastic response, which is valid for small strain in general, yields a local proportionality relation between strain and stress. As explained in Chapter 2, an equivalent statement is that the density of elastic energy is a *quadratic* function of the strain components. In the framework of Hookean elasticity, the elastic energy of the plate P reads, from equation (2.66):

$$\mathcal{E} = \frac{1}{2} \iint_P \mathrm{d}x \, \mathrm{d}y \int_{-h/2}^{h/2} \mathrm{d}z \; (\sigma_{ij}\, \epsilon_{ij}), \tag{6.1}$$

where x and y denote coordinates along the centre surface of the plate, while z is in the transverse direction. Since the top and bottom surfaces of the plate are assumed to be free boundaries, any out-of-plane component of the stress vanishes there. As a result[6] the implicit sum in the equation above runs only over the in-plane indices $i, j = \alpha$, $\beta \in \{x, y\}$.

When the thickness h is small, an obvious simplification in (6.1) is to neglect any dependence of the energy density on the transverse coordinate z running from $-h/2$ to $+h/2$, where $z = 0$ is by definition the centre (or neutral) surface. This gives the elastic energy in the 'membrane' approximation, $\mathcal{E}_{\mathrm{mb}}$ as:

$$\mathcal{E}_{\mathrm{mb}} = \frac{1}{2} \iint_P \mathrm{d}x \, \mathrm{d}y \; h \, (\sigma_{\alpha\beta}\, \epsilon_{\alpha\beta})_{\mathrm{cs}}$$

$$\sim E\, h \iint_P \mathrm{d}x \, \mathrm{d}y \; (\epsilon_{\alpha\beta})_{\mathrm{cs}}^2, \tag{6.2}$$

where the subscript 'cs' means that we can evaluate the density of the elastic energy along the centre surface, $z = 0$. The second line in this equation is an estimation that is only true up to a numerical factor of order one: in the present Hookean framework, the in-plane stress components $\sigma_{..}$ are linear functions of the strain $\epsilon_{..}$ with a coefficient that is Young's modulus E, up to numerical factors of order one that depend on Poisson's ratio. Equation (6.2) reveals the structure of the membrane energy: it is the surface integral of the squared, two-dimensional strain along the centre surface, with a prefactor $E\, h$ that is proportional to the thickness h and to Young's modulus E of the material. Any reference to the transverse direction has been eliminated by integration across the plate thickness. We have just outlined the so-called *membrane* theory of plates, derived later in Section 6.4.

This estimate of the elastic energy is not always relevant. When the deformation is such that the centre surface deforms isometrically, the two-dimensional strain $\epsilon_{\alpha\beta}$ in equation (6.2) vanish—by definition, the in-plane strain measures the change in length of curves drawn along the centre surface, see Section 2.2. Then, the membrane energy (6.2) above is zero. There are many ways to deform a plane isometrically, as we shall see, hence many configurations having zero membrane energy. To lift this degeneracy, one has to consider the bending energy of the plate, whose form is very similar to the bending energy of rods. This is demonstrated in Fig. 6.1(b). Physically, the bending contribution to the elastic energy comes from the fact that a curved plate is under compression on one side of the centre surface while the other half is under extension. This effect is missed by the

---

[6] This argument is made more precise later.

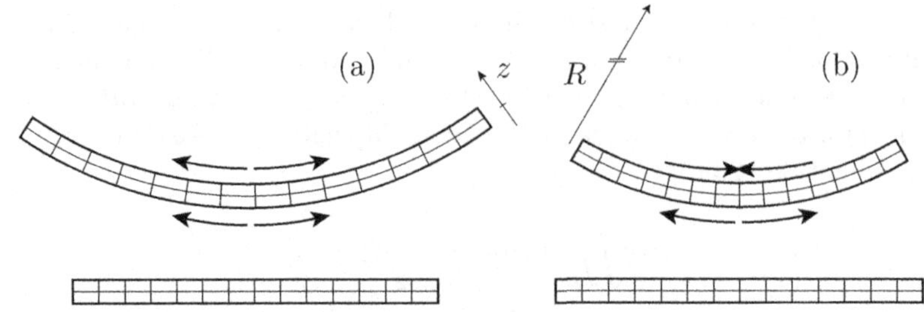

Fig. 6.1 For a deformation of the plate involving extension (or compression) of the centre surface, as in (a), the elastic energy can be considered as uniform through the thickness in a first approximation. However, for isometric (also called *pure bending*) deformations as in (b), the membrane energy of the plate vanishes and a flexural energy, connected to z-dependent strain, needs be considered.

membrane theory, whereby the stress is averaged over the whole thickness. As a result, the curvature energy shows up in the equations through the *linear* dependence of the strain as a function of the transverse coordinate z. As we shall see, this introduces a prefactor h^3 in front of the bending contribution to the elastic energy, where h is the small thickness of the plate, compared with a prefactor h for the stretching contribution.

Then let $R(x, y)$ be the typical radius of curvature of the plate. To estimate the bending energy, we shall first analyse the strain induced by bending. In a cut such as in Fig. 6.1(b), the inner edge, the centre surface and the outer edge are approximately given by concentric arcs of circle with radii $R - h/2$, R and $R + h/2$ respectively. Since they belong to roughly the same angular sector, their lengths are proportional to these radii, and so the outer and inner edges are compressed and stretched, respectively, by a factor of order h/R with respect to the centre surface. For isometric deformations,[7] the lengths along the centre surface are not changed upon deformation. As a result, the length of the middle arc in the actual configuration is also the reference length of all the three arcs. This means that the in-plane components of strain, $\epsilon_{\alpha\beta}$, which have been identified as those responsible for the bending energy, are of order $\pm h/R$ on the top and bottom sides of a curved plate. By conducting the same reasoning for a plane with a constant coordinate z, with $-h/2 \leq z \leq h/2$, we have, up to numerical factors of order one:

$$\epsilon_{\alpha\beta}(z) \sim \frac{z}{R}. \tag{6.3}$$

When plugging the associated energy density $E\sigma_{\alpha\beta}u_{\alpha\beta}$ into equation (6.1) we obtain, after integration over z, an estimate for the *flexural* (or bending) energy:

$$\mathcal{E}_{\mathrm{b}} \sim \frac{1}{2} \iint_P \mathrm{d}x\,\mathrm{d}y \,\frac{E\,h^3}{12}\left(\frac{1}{R_{\mathrm{cs}}(x, y)}\right)^2, \tag{6.4}$$

---

[7] It is important to estimate accurately the bending energy in the case of isometric deformations of the centre surface *only*: when it is significantly stretched, the bending energy is negligible anyway.

an expression that is only valid in order of magnitude:[8] it yields the correct structure of the bending energy, which is the surface integral of the squared curvature[9] of the centre surface. This energy comes with a prefactor, called the bending stiffness, that is of order $E\,h^3$.

Comparison of the stretching energy (6.2) with the bending energy (6.4) reveals that the small thickness h comes in the flexural energy (6.4) with a larger power than in the stretching energy, namely h^3 in place of h. For very thin plates, this makes the energy of isometric deformations much lower than those involving significant stretching of the center surface. By this remark, isometric deformations play a very special role. The low energy involved in pure bending deformations indeed explains many specific features of plates: we know from everyday experience that it is much easier to roll a piece of paper into a conical or cylindrical shape than to pull it on to a sphere for instance. Later in this book, the formation of singularities during crumpling of a piece of paper will be connected to the peculiar structure of the elastic energy of an elastic plate. The bending energy becomes important whenever the solution of the membrane equations becomes singular, with infinite curvature localized on lines or at well-defined points. There, bending is needed to regularize the shape of the surface near points with singular of curvature. The interplay of stretching and bending effects is important in many plate problems studied in the forthcoming chapters.

6.2   Gauss' Theorema egregium

We have just shown that a particular type of deformations, such that the centre surface remains unstretched, yields a very low elastic energy, and will therefore be favoured at mechanical equilibrium, where the energy is minimum. Such deformations are characterized by the fact that the in-plane[10] components of the strain tensor $\epsilon_{\alpha\beta}$ remains zero everywhere along this centre surface. In this section and in the next one, we characterize these isometric deformations and end up by listing all the surfaces that are isometric to a plane. Such surfaces are called *developable*. To do so, we essentially have to solve for the deformations of a piece of a plane such that

$$\epsilon_{xx}(x,y) = 0 \quad \epsilon_{xy}(x,y) = 0 \quad \epsilon_{yy}(x,y) = 0.$$

The first step is, in the present section, to establish Gauss' Theorema egregium, which puts a strong constraint on the candidate surfaces: they must everywhere have a principal direction of vanishing curvature. In a second step, in Section 6.3, we proceed to the actual characterization of developable surfaces. These two sections address the question of isometric deformations and we do not yet make use of the equations of elasticity. Interestingly, this purely geometrical problem contains all the non-linear contributions found later in the F.–von K. equations. Elasticity will be added on top of this geometrical approach next, starting from Section 6.4.

---

[8] This simple derivation does not yield the correct numerical factors in the bending modulus, which in fact depends on Poisson's ratio. Moreover it does not specify which combination of the two curvatures of the surface should be used for $1/R$. These questions are answered below in Section 6.5.4.

[9] Recall from Section 1.3 that the curvature is defined as the inverse of the radius of curvature R, in such a way that a straight curve, whose radius of curvature is infinite, has a vanishing curvature.

[10] Recall that Greek indices run over the two dimensions that are along the surface, $\alpha, \beta \in \{x, y\}$, while Latin ones run over the three dimensions of space.

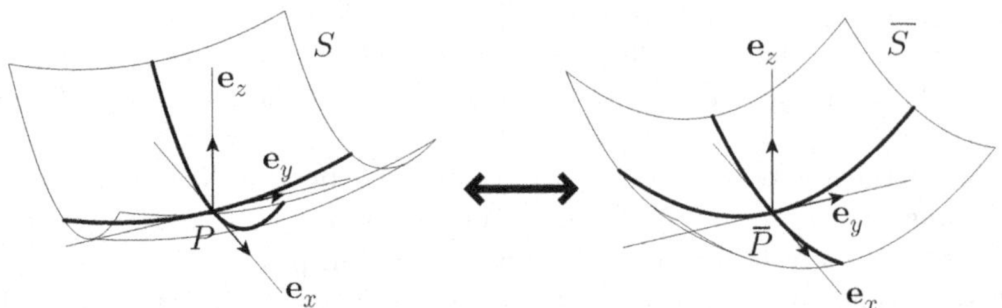

Fig. 6.2 Two isometric surfaces. The Gauss curvature, defined as the product of the two principal curvatures, is conserved: one of the principal curvature decreases although the other one increases. By applying a rigid-body rotation, one can assume that the two surfaces have the same tangent plane at the origin, as well as the same set of tangent vectors $\mathbf{e}_x$ and $\mathbf{e}_y$.

6.2.1 A direct derivation

The famous Gauss' Theorema egregium can be derived in many ways. The following derivation is based on a particular coordinate system and on expansions of various quantities near the origin. Although it might seem surprising to use expansions to establish the Theorema egregium, which is of course independent of the coordinates, this derivation is probably the easiest and is very instructive for applications to elastic plates and shells. A derivation that avoids reference to any particular coordinate system is possible, and can indeed be found in many textbooks, such as (M. Spivak, 1979). We also propose in Sections 6.2.2 and 6.2.3 a geometric interpretation of the conservation of Gauss curvature, which explains why Gauss curvature is an intrinsic quantity.

The present derivation of Gauss' Theorema egregium is done in a more general framework that the deformation of flat surfaces: we consider the isometric deformations of an arbitrary smooth (possibly curved) surface altogether. This will prove to be useful for the analysis of elastic shells in Part III of this book. Let us therefore consider two smooth surfaces, S and $\bar{S}$, and study under what conditions they can be connected by an isometry, that is a mapping from S to $\bar{S}$ that conserves the lengths of in-plane curves. We focus on an particular point P of S, which is mapped by the isometry to a point $\bar{P}$ of $\bar{S}$, and attempt to characterize the isometry in the neighbourhood of P, see Fig. 6.2. Doing so, we shall point out a local quantity, the Gauss curvature, which is conserved by the isometry.

We introduce a coordinate system and an orthonormal frame $(\mathbf{e}_x, \mathbf{e}_y, \mathbf{e}_z)$ such that $\mathbf{e}_z$ is perpendicular to the surface S at P. Then $(\mathbf{e}_x, \mathbf{e}_y)$ is an orthonormal frame for the tangent plane $T_P S$ at the point P to S. We consider a Cartesian parameterisation $z = f(x, y)$ of S in the vicinity of P in this frame. By construction,

$$f_{,x}(0,0) = 0 \quad \text{and} \quad f_{,y}(0,0) = 0, \tag{6.5}$$

where the commas in the indices denote partial derivatives.

Let us assume that S and $\bar{S}$ are connected by an isometry. This isometry is specified by the components $u(x, y)$, $v(x, y)$ and $w(x, y)$ of the associated displacement field in this frame.[11] Then, the surface $\bar{S}$ follows the parametric equation:

$$(x, y) \mapsto \bar{\mathbf{r}}(x, y) = (x + u(x, y), y + v(x, y), f(x, y) + w(x, y)), \tag{6.6a}$$

where the point $\bar{\mathbf{r}}(x, y)$ on $\bar{S}$ is the image by the isometry of the point $\mathbf{r}(x, y)$ on S, with coordinates:

$$(x, y) \mapsto \mathbf{r}(x, y) = (x, y, f(x, y)). \tag{6.6b}$$

By the isometric mapping, the tangent vectors at the origin, $\mathbf{r}_{,x}(0,0) = \mathbf{e}_x$ and $\mathbf{r}_{,y}(0,0) = \mathbf{e}_y$ are transported into two units, normal vectors $\bar{\mathbf{r}}_{,x}(0,0)$ and $\bar{\mathbf{r}}_{,y}(0,0)$, as explained in Section 2.5.1. By applying a rigid-body rotation on $\bar{S}$, it is possible to bring back these two vectors onto $\mathbf{e}_x$ and $\mathbf{e}_y$, as shown in Fig. 6.2. This allows one to assume without loss of generality that $\bar{\mathbf{r}}_{,x}(0,0) = \mathbf{e}_x$ and $\bar{\mathbf{r}}_{,y}(0,0) = \mathbf{e}_y$. This yields, in coordinates:

$$u_{,x} = u_{,y} = v_{,x} = v_{,y} = w_{,x} = w_{,y} = 0 \quad \text{at } (x, y) = (0, 0). \tag{6.7a}$$

Similarly, by applying a rigid-body translation on $\bar{S}$, the point $\bar{P}$, which is the image of P by the isometry, can be brought back onto P. This yields:

$$u = v = w = 0 \quad \text{at } (x, y) = (0, 0). \tag{6.7b}$$

The equations (6.5) and (6.7) have no physical content and simply reflect a choice of a convenient coordinate system.

We now proceed to write the condition of isometry between the surfaces S and $\bar{S}$. To do so, one has to compute the strain due to the displacement field (u, v, w), that measure the relative changes of lengths of small curves along the surface, and set them to zero. Calculation of this strain is essentially a repetition of the calculation of Section 2.2.3, with a slight modification to account for the curvature in the reference configuration. To avoid repeating this calculation with minor changes, we use a little trick and consider the strain $\epsilon_{\alpha\beta}$ and $\bar{\epsilon}_{\alpha\beta}$ measuring the change of lengths along S or $\bar{S}$ with respect to the *flat* configuration defined by $f(x, y) = 0$. The strain tensor $\epsilon_{\alpha\beta}$ is that associated with the transformation from the planar configuration to S given by equation (6.6b), while the strain tensor $\bar{\epsilon}_{\alpha\beta}$ corresponds to the transformation from the planar configuration to $\bar{S}$ given by equation (6.6a). We consider the general case of surfaces that are not necessarily developable: the strains tensors $\epsilon_{\alpha\beta}$ and $\bar{\epsilon}_{\alpha\beta}$ may not be zero—recall that they are defined with reference to a plane. The condition that S and $\bar{S}$ are connected by an isometric mapping can be written as:

$$\epsilon_{\alpha\beta}(x, y) = \bar{\epsilon}_{\alpha\beta}(x, y) \quad \text{for } \alpha, \beta = x, y. \tag{6.8}$$

These strains $\epsilon_{\alpha\beta}$ and $\bar{\epsilon}_{\alpha\beta}$ use the planar configuration as a reference and so can be computed from the general formula (2.9) given in Section 2.2.3, using the relevant displacement

---

[11] These components of the displacement field were denoted u_x, u_y and u_z in the previous chapters, the vector $\mathbf{u}$ having Cartesian components (u_x, u_y, u_z). Here, we use a different notation, namely (u, v, w) for (u_x, u_y, u_z), to avoid proliferation of indices (these two notations are used indifferently in this book). The boldface notation removes any possible ambiguity between the full displacement vector $\mathbf{u}$ and its x-component u.

fields, which is $(0, 0, f(x, y))$ and $(u(x, y), v(x, y), \overline{f}(x, y))$ respectively, where $\overline{f}(x, y)$ $= f(x, y) + w(x, y)$ denotes the profile[12] in the deformed configuration:

$$\epsilon_{xx} = \frac{f_{,x}^{\,2}}{2} \qquad\qquad \overline{\epsilon}_{xx} = u_{,x} + \frac{u_{,x}^{\,2} + v_{,x}^{\,2} + \overline{f}_{,x}^{\,2}}{2} \tag{6.9a}$$

$$\epsilon_{yy} = \frac{f_{,y}^{\,2}}{2} \qquad\qquad \overline{\epsilon}_{yy} = v_{,y} + \frac{u_{,y}^{\,2} + v_{,y}^{\,2} + \overline{f}_{,y}^{\,2}}{2} \tag{6.9b}$$

$$\epsilon_{xy} = \frac{f_{,x}\, f_{,y}}{2} \qquad\qquad \overline{\epsilon}_{xy} = \frac{u_{,y} + v_{,x}}{2} + \frac{u_{,x}\, u_{,y} + v_{,x}\, v_{,y} + \overline{f}_{,x}\, \overline{f}_{,y}}{2}. \tag{6.9c}$$

This is the general, non-linear formula for strain, for configuration S (left column) and $\bar{S}$ (right column).

Before attempting to solve the equations for an isometry $\epsilon_{\alpha\beta} = \overline{\epsilon}_{\alpha\beta}$, we shall first simplify these expressions for the strain by using the coordinate system introduced above, in which many non-linear terms can be neglected in the neighbourhood of P. This gives us the opportunity of introducing for the first time the simplified strain definitions that underly F.–von K.'s theory of plates.

In order to motivate this simplification, let us start by considering a very simple example of isometry, the rolling of a plane into a cylinder. The surface S is the plane, with equation $f(x, y) = 0$, while $\bar{S}$ is the cylinder, assumed of radius R. It is not difficult to find the explicit displacement field associated with this isometry:

$$\begin{cases} x + u = R \sin x/R \\ y + v = y \\ \quad w = R\,(1 - \cos x/R) \end{cases} \qquad \text{hence} \qquad \begin{cases} u = R \sin x/R - x \\ v = 0 \\ w = R\,(1 - \cos x/R) \end{cases}.$$

Taylor expansions of the functions given above reveal that the in-plane displacement $u(x, y)$ is cubic at dominant order in the variables x and y, while $w(x, y)$ is quadratic, for this particular example. As we now show, this behaviour is perfectly generic and derives in fact from the condition of isometry.

The proof that u and v have a cubic expansion near P for arbitrary S and $\bar{S}$ is as follows. From equation (6.9a), the condition $\epsilon_{xx} = \overline{\epsilon}_{xx}$ reads, after expanding $\overline{f} = f + w$ and cancelling $f_{,x}^{\,2}/2$, which appears on both sides of the equation:

$$u_{,x} + \frac{u_{,x}^{\,2} + v_{,x}^{\,2} + 2\,f_{,x}\, w_{,x} + w_{,x}^{\,2}}{2} = 0.$$

We shall now express this equation to first order in the variables x and y. In the numerator, $u_{,x}^{\,2}$ is a second-order quantity since it is the square of $u_{,x}$, which vanishes at the origin by equation (6.7a) and so it is at most linear near P. The same reasoning shows that $v_{,x}^{\,2}$ is a second-order quantity too. Both factors in the following term $(f_{,x}\, w_{,x})$ vanish at the origin by equations (6.5) and (6.7a); as a result, the term $(f_{,x}\, w_{,x})$ is a second-order quantity near

---

[12] Note that $\overline{f}(x, y)$ is not strictly speaking the graph of $\bar{S}$, as $\overline{f}(x, y)$ is the z coordinate of the point whose in-plane coordinates are $(x + u(x, y), y + v(x, y))$, and not (x, y). However, since u and v will appear to be small quantities of the third order near P, one can check that one can safely assimilate $\overline{f}(x, y)$ with the graph of $\bar{S}$.

the origin. Again, $w_{,x}$ vanishes at the origin, and so $w_{,x}$ is at least first order and the last term in the numerator is at most second order. We conclude that the whole fraction in the equation above is a second-order quantity, and so is $u_{,x}$ by the isometry condition:

$$u_{,x}(x,y) = \mathcal{O}(x^2, x\,y, y^2),$$

where the $\mathcal{O}$ notation stands for a quantity bounded by linear coefficients of the arguments x^2, $x\,y$ and y^2, i.e. a quantity that is at most second order. If we write the Taylor expansion of $u(x,y)$ near the origin to second order, and use the conditions $u(0,0) = 0$ by equation (6.7b), $u_{,y}(0,0) = 0$ by equation (6.7a) and the equation above for $u_{,x}$, we find that the expansion has to be of the form:

$$u(x,y) = \frac{u_{,yy}(0,0)}{2}\, y^2 + \mathcal{O}(x^3, x^2\,y, x\,y^2, y^3).$$

A similar reasoning with the equation $\epsilon_{yy} = \bar{\epsilon}_{yy}$ shows that $v(x,y)$ is of the form:

$$v(x,y) = \frac{v_{,xx}(0,0)}{2}\, x^2 + \mathcal{O}(x^3, x^2\,y, x\,y^2, y^3).$$

Finally, using the condition $\epsilon_{xy} = \bar{\epsilon}_{,xy}$ as earlier, we find that $u_{,y} + v_{,x} = 0$, up to first order included. Plugging the above expansions for $u(x,y)$ and $v(x,y)$ into this equation, we obtain $v_{,xx}(0,0)\,x + u_{,yy}(0,0)\,y = 0$, which implies

$$u_{,yy}(0,0) = 0 \quad \text{and} \quad v_{,xx}(0,0) = 0.$$

The final result is that both $u(x,y)$ and $v(x,y)$ are actually third-order quantities near the origin:

$$u(x,y) = \mathcal{O}(x^3, x^2\,y, x\,y^2, y^3) \quad \text{and} \quad v(x,y) = \mathcal{O}(x^3, x^2\,y, x\,y^2, y^3). \tag{6.10}$$

In contrast, the vertical component f is a second-order quantity, since the second derivatives of f may be non-zero. The expansion of the in-plane displacement for the case of a plane rolled onto a cylinder has been found to start with cubic terms; we just showed that this is always so for an isometric transformation.

Knowing the actual order of magnitude of the various terms in the non-linear strain given by equation (6.9) is useful as it allows one to simplify the expression near the origin. Indeed, all terms are second order near the origin, except the non-linear terms involving the in-plane displacement, such as $u_{,x}{}^2$, $v_{,x}{}^2$, $(u_{,x}\,u_{,y})$, which are fourth order. Neglecting the latter, we introduce an approximation for the strain, valid near the origin up to third order included:

$$\epsilon_{xx} = \frac{f_{,x}{}^2}{2} \qquad\qquad \bar{\epsilon}_{xx} = u_{,x} + \frac{\bar{f}_{,x}{}^2}{2} \tag{6.11a}$$

$$\epsilon_{yy} = \frac{f_{,y}{}^2}{2} \qquad\qquad \bar{\epsilon}_{yy} = v_{,y} + \frac{\bar{f}_{,y}{}^2}{2} \tag{6.11b}$$

$$\epsilon_{xy} = \frac{f_{,x}\,f_{,y}}{2} \qquad\qquad \bar{\epsilon}_{xy} = \frac{u_{,y} + v_{,x}}{2} + \frac{\bar{f}_{,x}\,\bar{f}_{,y}}{2}. \tag{6.11c}$$

This same approximate formulae for the strain $\bar{\epsilon}_{\alpha\beta}$ will be recovered under slightly different assumptions and will be at the heart of the F.–von K. theory of plates.

With these simplifications, it is now much easier to solve the condition of isometry, $\epsilon_{\alpha\beta} = \bar{\epsilon}_{\alpha\beta}$. Indeed, since the in-plane strain has been linearized with respect to the in-plane displacement, $\bar{\epsilon}_{\alpha\beta}$ depends on the functions (u, v) exactly as in the framework of small displacements, see Section 2.2.11. In Section 2.2.12, we addressed the compatibility of strain in this framework, and found that the functions u and v can be eliminated from the strain by forming the quantity

$$2\frac{\partial^2 \bar{\epsilon}_{xy}}{\partial x\,\partial y} - \frac{\partial^2 \bar{\epsilon}_{xx}}{\partial y^2} - \frac{\partial^2 \bar{\epsilon}_{yy}}{\partial x^2}.$$

We can use the same trick again to eliminate u and v from the conditions of isometry, and compute $2\,\bar{\epsilon}_{xy,xy} - \bar{\epsilon}_{xx,yy} - \bar{\epsilon}_{yy,xx}$ using our new definition of strain in equation (6.11). As expected, the in-plane displacements is eliminated and the result is a function of the transverse displacement only:

$$2\frac{\partial^2 \bar{\epsilon}_{xy}}{\partial x\,\partial y} - \frac{\partial^2 \bar{\epsilon}_{xx}}{\partial y^2} - \frac{\partial^2 \bar{\epsilon}_{yy}}{\partial x^2} = \bar{f}_{,xx}\,\bar{f}_{,yy} - \bar{f}_{,xy}{}^2.$$

We have a similar formula, without the bars, connecting $\epsilon_{\alpha\beta}$ and f. This is a fundamental equation, which shows that when the longitudinal strain is linearized with respect to in-plane displacement, as in equations (6.11), the resulting components are subjected to a compatibility condition. The right-hand side of this equation is called the Gauss curvature and is denoted $K(x, y)$:

$$2\frac{\partial^2 \bar{\epsilon}_{xy}}{\partial x\,\partial y} - \frac{\partial^2 \bar{\epsilon}_{xx}}{\partial y^2} - \frac{\partial^2 \bar{\epsilon}_{yy}}{\partial x^2} = \overline{K}(x, y), \quad \text{where } \overline{K} = \bar{f}_{,xx}\,\bar{f}_{,yy} - \bar{f}_{,xy}{}^2. \tag{6.12}$$

Here again, we have a similar equation for the surface S but without the bars. The equation above holds at P since the expressions we used for the strain were valid up to second order near P, and have been derived twice.

By the fundamental equation (6.12), the condition of isometry $\epsilon_{\alpha\beta}(x, y) = \bar{\epsilon}_{\alpha\beta}(x, y)$ implies $K(0, 0) = \overline{K}(0, 0)$:

$$f_{,xx}\, f_{,yy} - f_{,xy}{}^2 = \bar{f}_{,xx}\,\bar{f}_{,yy} - \bar{f}_{,xy}{}^2 \quad \text{for } (x, y) = (0, 0). \tag{6.13}$$

We have just arrived at the first formulation of Gauss' Theorema egregium: the Gauss curvature is conserved by isometries. In a coordinate system such that the coordinates (x, y) span the tangent plane at a point P of a surface S, the Gauss curvature K is given by the determinant of the Hessian matrix, which is by definition made up of the second derivatives of $f(x, y)$:

$$K(P) = \det H(P) \quad \text{where } H(P) = \begin{pmatrix} f_{,xx} & f_{,xy} \\ f_{,yx} & f_{,yy} \end{pmatrix}_{(0,0)}. \tag{6.14}$$

In this definition of the Hessian matrix, the right-hand side is evaluated at the point P, chosen as the origin of the coordinate system (x, y). Note that the Hessian matrix H is symmetric by Leibnitz identity, $(f_{,x})_{,y} = (f_{,y})_{,x}$. A more general expression for the Gauss curvature will be given in equation (6.42).

This first definition of the conserved quantity, $K(P)$, although complete, is rather impractical as it involves a different system of coordinates (x, y) attached to every point P on the surface S. In the following sections, we give a more geometrical interpretation of the Gauss curvature. By doing so, we shall be able to put Gauss' *Theorema egregium* in an intrinsic form and avoid any explicit reference to a coordinate system.

6.2.2   Geometrical interpretation of Gauss curvature

The geometric interpretation of the Gauss curvature $K(P)$ defined in (6.14) is based on the principal radii of curvature of the surface, which we define first. Let us pick an arbitrary point P on the surface. We use the same adapted coordinate system (x, y, z) as in the previous section, which is such that (x, y) spans the tangent plane to S at P—we shall soon be able to remove any reference to this *ad hoc* frame. With this choice of coordinates, there are no constant and linear terms in the expansion of $f(x, y)$ near P, that is near $(x, y) = (0, 0)$:

$$f(x, y) = \frac{1}{2}(f_{,xx}\, x^2 + 2\, f_{,xy}\, x\, y + f_{,yy}\, y^2) + \cdots$$

$$= \frac{1}{2}\,{}^{t}(x, y) \cdot H(P) \cdot (x, y) + \cdots . \tag{6.15}$$

Here, we denote ${}^{t}\mathbf{x}$ as the transpose of any vector (or matrix) $\mathbf{x}$. The first equality comes from the Taylor expansion of a function $f(x, y)$ with two variables while the second one is a rewriting of this expansion as a quadratic form of the variables x and y. The coefficients of this quadratic form are the entries of the Hessian matrix, H. This symmetric matrix can be diagonalized by a suitable rotation in the plane (x, y). Its eigenvectors define two orthogonal directions in the tangent plane, the so-called *principal directions of curvature*, denoted x' and y', see Fig. 6.3. The associated eigenvalues $\kappa_{x'}$ and $\kappa_{y'}$ are called the *principal curvatures*, for a reason that we explain below.

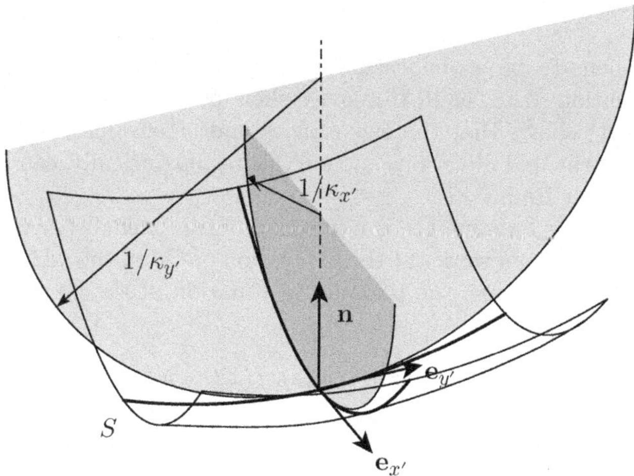

Fig. 6.3 Principal curvatures $\kappa_{x'}$ and $\kappa_{y'}$ and principal directions of curvature at a point P on a surface S. Note that both these principal directions and curvatures depend on P and therefore vary along the surface.

Let (x', y') be the coordinate system oriented using the principal directions at P. In this frame, the expansion (6.15) of f becomes:

$$f(x', y') = \frac{1}{2}\left(\kappa_{x'}(P)\, x'^2 + \kappa_{y'}(P)\, y'^2\right) + \cdots$$

$$= \frac{1}{2}\begin{pmatrix} x' \\ y' \end{pmatrix} \cdot \begin{pmatrix} \kappa_{x'}(P) & 0 \\ 0 & \kappa_{y'}(P) \end{pmatrix} \cdot (x', y') + \cdots, \tag{6.16}$$

since the Hessian matrix is diagonal by construction.

In equation (6.14), the Gauss curvature K was defined as the determinant of the Hessian. This determinant can be computed by using the diagonal form (6.16) of the Hessian matrix: at any point P, the Gauss curvature is just the *product* of the two principal curvatures:

$$K(P) = \kappa_{x'}(P)\, \kappa_{y'}(P). \tag{6.17}$$

It remains to give a geometrical interpretation of these principal directions and curvatures. To this end, we consider an arbitrary smooth curve $\mathbf{r}(s)$ drawn on the surface S passing through P for $s = 0$. Then, $\mathbf{r}_{,s}(s)$ is the tangent to the curve at s. In particular, $\mathbf{r}_{,s}(0)$ lies in the tangent plane $T_P S$ to S at P. Let $\mathbf{n}$ be the normal to the surface S at P. We consider the derivative of the tangent vector at P, namely $\mathbf{r}_{,ss}(0)$, and call $C\{\mathbf{r}(s)\}$ its component[13] along the normal to the surface, $\mathbf{n}$:

$$C_P\{\mathbf{r}(s)\} = \mathbf{r}_{,ss}(0) \cdot \mathbf{n}. \tag{6.18}$$

This number, $C_P\{\mathbf{r}(s)\}$ depends *a priori* on the full curve $\mathbf{r}(s)$, as indicated by the braces. In fact, we shall show that C_P depends on *the value of the tangent at the origin*, $\mathbf{r}_{,s}(0)$, *only*. As a result, C_P is a scalar valued[14] function defined over the tangent plane $T_P S$. The tangent to the curve, $\mathbf{t} = \mathbf{r}_{,s}(0) \in T_P S$, is mapped to

$$\mathbf{t} \mapsto C_P(\mathbf{t}). \tag{6.19}$$

This mapping happens to be a quadratic form whose matrix coincides with the Hessian matrix given in equation (6.15) or (6.16), as we show below. This intrinsic definition of the quadratic form $C_P(\mathbf{t})$ shows that the eigenspaces and eigenvalues of the Hessian matrix, which are called the principal directions and curvatures intrinsically, can be defined without reference to a particular frame.

To show that the scalar function C_P is a function of the tangent $\mathbf{t} = \mathbf{r}_{,s}(0)$ only, and that it is a quadratic form, one can work out the expression of C_P in the adapted frame (x', y', z) introduced above. In this frame, the parametric equation of the curve $\mathbf{r}(s)$ can be put in the form:

$$s \mapsto \mathbf{r}(s) = (x'(s), y'(s), f(x'(s), y'(s))), \tag{6.20}$$

---

[13] Among the two other components of $\mathbf{r}_{,ss}(0)$, the one along the tangent $\mathbf{t} = \mathbf{r}_{,s}(0)$ to the curve is unphysical and can be set to zero by using the arc length parameterisation. The last component, along the direction of the tangent plane perpendicular to the tangent $\mathbf{t}$ to the curve, defines what is called the geodesic curvature of the curve, a quantity that can be shown to be invariant by isometric deformations of the surface—it measures in an intrinsic manner how much the curve is curved within the surface.

[14] Note that the image of C_P is a scalar, being defined without reference to any particular coordinate system.

where $(x'(s), y'(s))$ is a parameterisation of the curve projected onto the tangent plane $T_P S$. The normal vector $\mathbf{n}$ has coordinates $(0, 0, 1)$ by construction. The number C_P is then computed as:

$$C_P = \mathbf{n} \cdot \frac{\mathrm{d}^2 \mathbf{r}}{\mathrm{d}s^2}\bigg|_{s=0} = \frac{\mathrm{d}^2 f(x'(s), y'(s))}{\mathrm{d}s^2}\bigg|_{s=0}$$

$$= x'^2_{,s}\, f_{,x'x'} + 2\, x'_{,s}\, y'_{,s}\, f_{,x'y'} + y'^2_{,s}\, f_{,y'y'} + x'_{,ss}\, f_{,x'} + y'_{,ss}\, f_{,y'}, \tag{6.21}$$

where all the quantities must be evaluated at $(x, y) = (0, 0)$. The right-hand side in this equation has been obtained by evaluating the derivative twice using the chain rule. In this particular frame, $f_{,x'}(0,0) = 0$ and $f_{,y'}(0,0) = 0$ and the last two terms are zero, while $f_{,x'x'}(0,0) = \kappa_{x'}$, $f_{,x'y'}(0,0) = 0$, $f_{,y'y'}(0,0) = \kappa_{y'}$. Introducing the components of the tangent vector $\mathbf{t} = \mathbf{r}_{,s}(0)$:

$$t_{x'} = \langle \mathbf{t} | \mathbf{e}_{x'} \rangle = \frac{\mathrm{d}x'}{\mathrm{d}s}(0,0) \qquad t_{y'} = \langle \mathbf{t} | \mathbf{e}_{y'} \rangle = \frac{\mathrm{d}y'}{\mathrm{d}s}(0,0),$$

we find

$$C_P = \kappa_{x'}(P)\, t_{x'}{}^2 + \kappa_{y'}(P)\, t_{y'}{}^2. \tag{6.22}$$

As announced, this shows that C_P is a function of the tangent at P, $\mathbf{t} = \mathbf{r}_{,s}(0)$, only, and this mapping is a quadratic form represented by the Hessian matrix given in equation (6.16). The intrinsic nature of the principal curvatures and principal directions has therefore been established. For more details on the definition of curvatures on abstract surfaces, see for instance reference (M. Spivak, 1979). In the rest of this section, we give a geometrical interpretation of these principal curvatures and justify their names.

Consider the curve $\mathbf{r}(s)$ drawn on the surface whose projection onto the tangent plane $T_P S$ yields the x' axis. Such a curve is shown in thick lines in Fig. 6.3. This curve belongs to the family of curves $\mathbf{r}(s)$ introduced above, with $x'(s) = s$ and $y'(s) = 0$. By construction, it has a tangent $\mathbf{t} = \mathbf{r}_{,s}(0) = \mathbf{e}_{x'}$ at P, which spans a principal direction. The second derivative at P of this curve can be calculated with the help of equation (6.22). This yields, with $t_{x'} = \partial x' / \partial s = 1$ and $t_{y'} = \partial y' / \partial s = 0$:

$$\mathbf{r}_{,ss}(0) = \kappa_{x'}(P)\, \mathbf{n}, \tag{6.23}$$

where $\mathbf{n}$ is the normal to S at P. Note that this vector has vanishing tangential components owing to the fact that $x'(s)$ and $y'(s)$ are linear functions of s. Equation (6.23) shows that the radius of curvature of this curve is $R_{x'} = 1/\kappa_{x'}$. This number $R_{x'}$ is called a principal *radius* of curvature. We arrive to the following interpretation of the principal curvatures: as sketched in Fig. 6.3, the intersection of the surface S with the plane containing the normal $\mathbf{n}$ and the principal direction $\mathbf{e}_{x'}$ at P yields a curve that has a radius of curvature $R_{x'} = 1/\kappa_{x'}$ at P. In other words, the surface has an osculating circle of radius $R_{x'}$ at P contained in the normal plane along the first principal direction. A similar reasoning holds for the second principal direction. We summarize this as:

$$R_{x'} = \frac{1}{\kappa_{x'}} \qquad R_{y'} = \frac{1}{\kappa_{y'}}, \tag{6.24}$$

where the R_i are the radii of the osculating circles along the principal directions, shown in Fig. 6.3.

To sum up, we have shown in this section that the conserved quantity K, defined formally in equation (6.14), could be interpreted geometrically in terms of the principal radii of curvature:

$$K(P) = \frac{1}{R_{x'}(P)} \, \frac{1}{R_{y'}(P)}. \tag{6.25}$$

This quantity is conserved upon isometric deformations, that is:

$$K(P) = K(\Phi(P)) \tag{6.26}$$

for any point P on a surface S, and for any isometric deformation Φ of S, $\Phi : P \mapsto \Phi(P)$. Note however that neither the principal directions nor the principal curvatures $\kappa_{x'}$ and $\kappa_{y'}$ are individually conserved (generally they are not mapped into each other by the isometric deformation). Only the product $K = \kappa_{x'}\kappa_{y'}$ is.

6.2.3   Gauss curvature and the perimeter of circles

In this section, we propose a different derivation of Gauss' Theorema egregium: we show that the Gauss curvature K enters into the value of the perimeters of small circles (defined precisely below) drawn on the surface around a point. These perimeters being conserved by isometries, this provides a direct proof of the conservation of K by isometries.

Let us define a circle of radius r drawn on a surface as the set of points lying at a fixed distance r from its centre, this distance r being measured *along* the surface.[15] When the surface is curved, the perimeter of a circle of radius r centred on a point P is not given by $2\pi r$, as in Euclidean geometry. Below, we compute this perimeter $\mathcal{P}(P,r)$ for small r to third order, and show that this function $\mathcal{P}(P,r)$ can be used to compute the Gauss curvature $K(P)$ at the centre of the circles. We again use the adapted triad and system of coordinates (x',y',z). The calculation of the perimeter is done in two steps. The first step is to find the equation for this circle, which is a curve drawn on the surface S. The second step is to compute the length of this curve by integration of the length element.

The equation for the circle of radius r centred at P is derived as follows. We first consider a radial curve drawn on the surface, which we define as a curve whose projection on the tangent plane is a straight line passing through P making a direction θ with the coordinate axes $(\mathbf{e}_{x'}, \mathbf{e}_{y'})$. This radial curve and its projection are shown in Fig. 6.4 using thick lines. It has the following parametric equation:

$$\mathbf{r}_{\mathrm{rad}}(\theta, s) = \begin{pmatrix} x'(s) = s\cos\theta \\ y'(s) = s\sin\theta \\ z(s) = f(x'(s), y'(s)) \end{pmatrix}. \tag{6.27}$$

Note that the parameter s is the curvilinear abscissa *of the projection on the tangent planes,* and not of the curve itself: the tangent vector $\mathbf{r}_{\mathrm{rad},s}$ is not a unit vector.

---

[15] Note that this definition is intrinsic: it makes reference to the metric of the surface only, not to its shape in the three-dimensional space. As a result, the image of the circle of radius r by an isometric mapping is the circle of radius r drawn around the image of the centre.

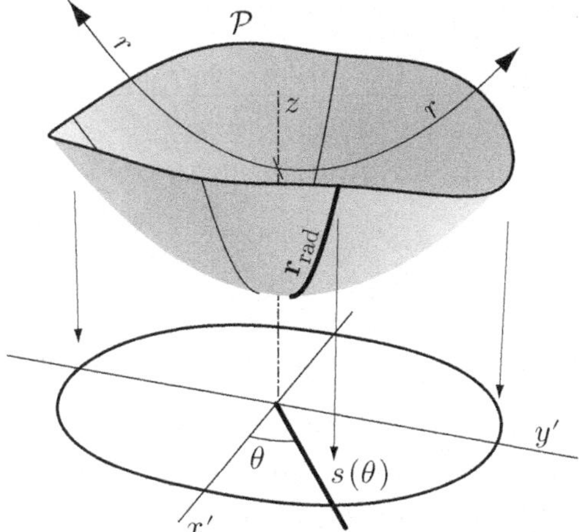

Fig. 6.4 In the presence of curvature, the perimeter $\mathcal{P}$ of a circle with radius r drawn on a surface is not given by the formula $\mathcal{P}_{\mathrm{Eucl}} = 2\pi\, r$. Here, circles are defined intrinsically as the set of points at a fixed distance r from the centre P, this distance being measured *along* the surface (see curved arrow on top of drawing). For small circles, the deviation from $\mathcal{P}$ to $2\pi\, r$ is related to the Gauss curvature at the centre.

The intersection of this curve $\mathbf{r}_{\mathrm{rad}}(\theta,.)$ with the circle of radius r can be computed by the condition that the curvilinear length along the radial curve between P and intersection should be equal[16] to the circle radius, r:

$$\int_0^{s(\theta)} \left| \frac{\partial \mathbf{r}_{\mathrm{rad}}(\theta, s')}{\partial s'} \right| \mathrm{d}s' = r, \tag{6.28}$$

where $s(\theta)$ is the coordinate of the intersection, and the vertical bar denotes the norm of a vector.

To compute the integral above, we first calculate the 'acceleration' along the radial curve at $s = 0$ by equation (6.22), with $t_{x'} = \cos\theta$ and $t_{y'} = \sin\theta$:

$$\left. \frac{\partial^2 \mathbf{r}_{\mathrm{rad}}(\theta, s)}{\partial s^2} \right|_{s=0} = \left(\kappa_{x'} \cos^2\theta + \kappa_{y'} \sin^2\theta \right) \mathbf{n}.$$

The tangent vector along the radial curve in equation (6.28) can then be computed by a Taylor expansion:

$$\frac{\partial \mathbf{r}_{\mathrm{rad}}(\theta, s)}{\partial s} = (\cos\theta\, \mathbf{e}_{x'} + \sin\theta\, \mathbf{e}_{y'}) + \left[\left(\kappa_{x'} \cos^2\theta + \kappa_{y'} \sin^2\theta \right) s + \mathcal{O}(s^2) \right] \mathbf{n}.$$

---

[16] By doing so, we are implicitly assuming that the radial curve $s \mapsto \mathbf{r}_{\mathrm{rad}}(\theta, s)$ is a geodesic on the surface, that is the shortest path from P to its intersection with the circle. This radial curve was defined by the fact that its projection yields a straight line, and the curve so defined is not a geodesic in general. However, for small values of r, the actual geodesic is very close to our radial curve. Using this curve instead of this geodesic in equation (6.28) yields an error in the final formula (6.32) that is beyond the third order in r.

This yields in turn the integrand in equation (6.28) as:

$$\left| \frac{\partial \mathbf{r}_{\text{rad}}(\theta, s)}{\partial s} \right| = 1 + \frac{\left(\kappa_{x'} \cos^2 \theta + \kappa_{y'} \sin^2 \theta \right)^2}{2} s^2 + \mathcal{O}(s^3).$$

Upon integration, as in equation (6.28), this yields the coordinate $s(\theta)$ of the intersection in an implicit form:

$$s(\theta) + \frac{\left(\kappa_{x'} \cos^2 \theta + \kappa_{y'} \sin^2 \theta \right)^2}{6} s^3(\theta) + \mathcal{O}(s^4) = r, \tag{6.29}$$

a relation that can be inverted to yield the parametric equation of the circle:

$$s(\theta) = r - \frac{\left(\kappa_{x'} \cos^2 \theta + \kappa_{y'} \sin^2 \theta \right)^2}{6} r^3 + \cdots . \tag{6.30}$$

This equation is valid to third order for small radius r.

The equation of a small circle of radius r centred at P is now given by equation (6.27) with s given by equation (6.30) above. The length of this curve can be computed by carrying out a simple integral:

$$\mathcal{P}(P, r) = 2\pi r \left(1 - \frac{\kappa_{x'} \kappa_{y'}}{6} r^2 + \cdots \right) = 2\pi r \left(1 - \frac{K(P)}{6} r^2 + \mathcal{O}(r^3) \right), \tag{6.31}$$

where $K(P)$ is the Gauss curvature at the centre of the small circle. The Gauss curvature can therefore be calculated at any point on a surface by:[17]

$$K(P) = -6 \lim_{r \to 0} \left[\frac{1}{r^2} \left(\frac{\mathcal{P}(P, r)}{2\pi r} - 1 \right) \right]. \tag{6.32}$$

This formula shows that, in order to compute the Gauss curvature $K(P)$ on a surface, it suffices to measure the perimeters (or the area) of small circles drawn on the surface.[18] This geometric interpretation yields a direct proof of the *Theorema egregium*, since these perimeters are by definition invariant by isometric deformations of a surface. It also shows that the perimeter of a circle of radius r is less than $2\pi r$ on elliptic surfaces such as

---

[17] A similar formula relates the Gauss curvature to the deviation in the area of a disc of radius r from its Euclidean value, πr^2. This area is

$$\mathcal{A}(P, r) = \pi r^2 \left(1 - \frac{K(P)}{12} r^2 + \mathcal{O}(r^3) \right).$$

[18] Imagine flat, two dimensional animals living on a surface without any sense of the third dimension. For them, quantities like $\kappa_{x'}$ or $\kappa_{y'}$ make no sense. However equation (6.32) provides a mean to tell whether the surface they live on has non-zero Gauss curvature or not, without them even looking out of their two dimensional world. All they have to do is to measure perimeters of circles. In this sense, the notion of curvature can be defined on a space that is not embedded in some higher dimensional space: this is called an abstract manifold. Concerning the geometry of abstract manifolds, see Chapter 10, Section 10.1.2. This is how one should understand the famous statement by Einstein that space-time is curved by the presence of mass according to his theory of general relativity. Note that Gauss himself was aware of the possibility that our three-dimensional world could be curved and tried to measure the sum of the angles of a large triangle: this is π only when the space is flat ($K = 0$). Even nowadays, the curvature of space by gravitational effects is too small to be measured by experiments on the surface of the Earth.

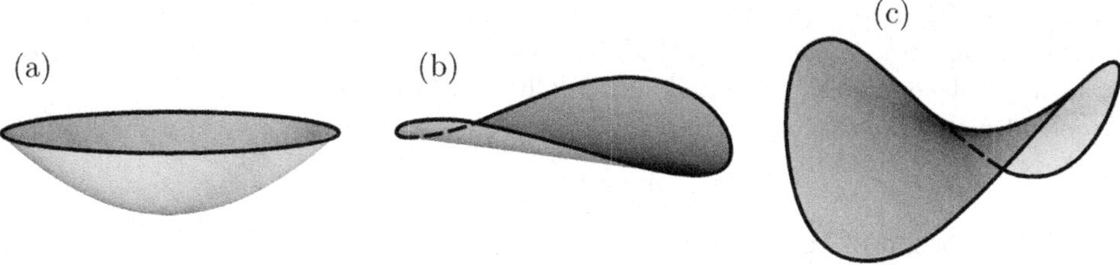

Fig. 6.5 Interpretation of equation (6.32): compared with the developable case (b) ($K = 0$, parabolic surface), the perimeter of a circle of fixed radius drawn on a surface is smaller for a locally convex surface (a) ($K > 0$, elliptic surface), and larger for a saddle-like surface (c) ($K < 0$, hyperbolic surface).

a paraboloid ($K > 0$), but more than that on hyperbolic surfaces such that ($K < 0$), as illustrated in Fig. 6.5.

6.2.4 A historical note

The conservation of the Gauss curvature upon isometric deformations of a surface, as established by equation (6.26) or (6.32), was first proved by C. F. Gauss in the Theorema egregium of his *Disquisitiones generales circa superficies curvas* (General investigations of curved surfaces), published in 1827. The Latin adjective *egregium* derives from 'ex grege', meaning 'out of the flock'. It means 'remarkable' or 'extraordinary'. Gauss indeed estimated that his formula 'may be counted among the most productive theorems in the theory of curved surfaces' (*'inter fertilissima theoremata in doctrina de superficiebus curvis referenda est'*).

To derive this result, Gauss proceeded technically by expressing the Gauss curvature, defined as the determinant of the curvature form, in terms of coefficients of the metric tensor:[19]

$$E = \left|\frac{\partial \mathbf{r}}{\partial u}\right|^2 \qquad F = \left|\frac{\partial \mathbf{r}}{\partial v}\right|^2 \qquad G = \frac{\partial \mathbf{r}}{\partial u} \cdot \frac{\partial \mathbf{r}}{\partial v}, \tag{6.33}$$

and their partial derivatives, where $\mathbf{r}(u, v)$ is any parameterisation of the surface. By definition, the components E, F and G of the metric tensor are unchanged upon isometric deformation of the surface.

6.2.5 Practical calculation of Gauss curvature

In the previous sections, we computed the Gauss curvature $K(P)$ at a point P of a surface S given as a graph $(x, y) \mapsto (x, y, f(x, y))$ in a Cartesian system of coordinate (x, y, z) such that the tangent plane at P is horizontal. In this section, we relax this assumption and compute the Gauss curvature when the orientation of the tangent plane is arbitrary. The surface is still given as the graph of a function $f(x, y)$. We start from the intrinsic definition of K as the determinant of the quadratic form C_P introduced in the previous section.

---

[19] The notations E, F and G for the components of the metric tensor were introduced by Gauss himself, and were still used up until recently.

A similar reasoning allows one to compute the Gauss curvature along surfaces defined by a parameterisation $(x(U, V), y(U, V), z(U, V))$, such as surfaces of revolution, see Section 11.5, or by an implicit equation (M. Spivak, 1979)[Chap. 3 Vol 3].

We consider an arbitrary smooth graph:

$$(x, y) \mapsto (x, y, f(x, y)) \tag{6.34}$$

and choose the origin of the coordinates (x, y) to be at the point P at which we want to compute K, without loss of generality. The quadratic form C_P introduced in the previous section involves the calculation of the acceleration normal to the surface of a path drawn on it that passes through P. Let us introduce a family of such curves:

$$s \mapsto \mathrm{r}^{(\lambda,\mu)}(s) = (\lambda\, s, \mu\, s, f(\lambda\, s, \mu\, s)), \tag{6.35}$$

where λ and μ parameterise the family of curves drawn on S while s is the parameterisation of a given curve with fixed λ and μ. The tangent vector at P reads:

$$\mathbf{r}^{(\lambda,\mu)}_{,s}(0) = (\lambda, \mu, \lambda\, f_{,x} + \mu\, f_{,y}), \tag{6.36}$$

where the right-hand side must be evaluated at the origin $(x, y) = (0, 0)$. Let $\mathbf{t}^{(1,0)} = (1, 0, f_{,x}(0, 0))$ be the tangent vector for $\lambda = 1$ and $\mu = 0$, and $\mathbf{t}^{(0,1)} = (0, 1, f_{,y}(0, 0))$ that for $\lambda = 0$ and $\mu = 1$. These two vectors span the tangent plane at P. We denote E, F and G the following scalar products of $\mathbf{t}^{(1,0)})$ and $\mathbf{t}^{(0,1)}$, which are known as the components of the metric tensor as defined earlier in equation (6.33):

$$E = \mathbf{t}^{(1,0)} \cdot \mathbf{t}^{(1,0)} = 1 + (f_{,x})^2$$

$$F = \mathbf{t}^{(1,0)} \cdot \mathbf{t}^{(0,1)} = f_{,x}\, f_{,y}$$

$$G = \mathbf{t}^{(0,1)} \cdot \mathbf{t}^{(0,1)} = 1 + (f_{,y})^2,$$

where all quantities are again evaluated at P. A unit vector that is normal to the surface at P is given by:

$$\mathbf{n}(P) = \frac{\mathbf{t}^{(1,0)} \times \mathbf{t}^{(0,1)}}{|\mathbf{t}^{(1,0)} \times \mathbf{t}^{(0,1)}|} = \frac{1}{\sqrt{1 + f_{,x}^2 + f_{,y}^2}} \begin{pmatrix} -f_{,x} \\ -f_{,y} \\ 1 \end{pmatrix}. \tag{6.37}$$

Using this normal vector we can evaluate C_P by computing the acceleration $\mathbf{r}^{(\lambda,\mu)}_{,ss}$ and projecting over the normal $\mathbf{n}$. The calculation is identical to that leading to equation (6.21), the only difference being in the presence of a denominator in the above expression for $\mathbf{n}$. This leads to

$$C_P(\lambda\, \mathbf{t}^{(1,0)} + \mu\, \mathbf{t}^{(0,1)}) = \frac{\lambda^2 f_{,xx} + 2\lambda\mu\, f_{,xy} + \mu^2 f_{,yy}}{\sqrt{1 + f_{,x}^2 + f_{,y}^2}}. \tag{6.38}$$

The argument of this quadratic form is a vector in the tangent plane, $\mathbf{r}_{,s}(\lambda, \mu, 0) = \lambda\, \mathbf{t}^{(1,0)} + \mu\, \mathbf{t}^{(0,1)}$. We rewrite equation (6.38) in a matrix form:

$$C_P(\lambda\, \mathbf{t}^{(1,0)} + \mu\, \mathbf{t}^{(0,1)}) = {}^t\mathbf{\Lambda} \cdot \frac{\begin{pmatrix} f_{,xx} & f_{,xy} \\ f_{,xy} & f_{,yy} \end{pmatrix}}{\sqrt{1 + f_{,x}^2 + f_{,y}^2}} \cdot \mathbf{\Lambda}, \tag{6.39}$$

where $\mathbf{\Lambda} = (\lambda, \mu)$ is a two-dimensional vector. The right-hand side of this equation expresses the quadratic form C_P using the frame $(\mathbf{t}^{(1,0)}, \mathbf{t}^{(0,1)})$ of the tangent plane. This frame is not orthonormal and so the determinant of the quadratic form is not equal to the determinant of the matrix in the right-hand side of equation (6.39). To compute the Gauss curvature, which is the determinant of the quadratic form,[20] we first need to express C_P using an orthonormal frame in the tangent plane.

Let T be the matrix that transforms the coordinates $\mathbf{\Lambda} = (\lambda, \mu)$ in the original frame into the coordinates $\mathbf{\Lambda}' = (\lambda', \mu')$ in some orthonormal frame: $\mathbf{\Lambda}' = T \cdot \mathbf{\Lambda}$. We are going to show how equation (6.39) is transformed into the new frame. To this end, we consider the scalar product of two arbitrary tangent vectors $\mathbf{\Lambda}^{(1)}$ and $\mathbf{\Lambda}^{(2)}$:

$$\langle \lambda^{(1)} \mathbf{t}^{(1,0)} + \mu^{(1)} \mathbf{t}^{(0,1)} | \lambda^{(2)} \mathbf{t}^{(1,0)} + \mu^{(2)} \mathbf{t}^{(0,1)} \rangle = {}^{t}\mathbf{\Lambda}^{(1)} \cdot \begin{pmatrix} E & F \\ F & G \end{pmatrix} \cdot \mathbf{\Lambda}^{(2)}$$

$$= {}^{t}(T^{-1} \cdot \mathbf{\Lambda}'^{(1)}) \cdot \begin{pmatrix} E & F \\ F & G \end{pmatrix} \cdot (T^{-1} \cdot \mathbf{\Lambda}'^{(2)}) = {}^{t}\mathbf{\Lambda}'^{(1)} \cdot \left[{}^{t}T^{-1} \cdot \begin{pmatrix} E & F \\ F & G \end{pmatrix} \cdot T^{-1} \right] \cdot \mathbf{\Lambda}'^{(2)}.$$

In the new system of coordinates, this scalar product reads $\lambda'^{(1)} \lambda'^{(2)} + \mu'^{(1)} \mu'^{(2)} = {}^{t}\mathbf{\Lambda}'^{(1)} \cdot \mathbf{\Lambda}'^{(2)}$, since the frame is orthonormal. By requiring that these two calculations yield the same result for two arbitrary vectors in the tangent plane, we find that the matrix in square brackets in the equation above is the identity. By taking its determinant, we obtain:

$$(\det T^{-1})^2 = \frac{1}{\det \begin{pmatrix} E & F \\ F & G \end{pmatrix}} = \frac{1}{1 + f_{,x}^2 + f_{,y}^2}. \tag{6.40}$$

This is all we need to know to compute the determinant of the quadratic form C_P: in the new orthonormal frame, the equation (6.39) for C_P can be re-expressed as

$$C_P(\mathbf{\Lambda}') = {}^{t}\mathbf{\Lambda}' \cdot \left[{}^{t}T^{-1} \cdot \frac{\begin{pmatrix} f_{,xx} & f_{,xy} \\ f_{,xy} & f_{,yy} \end{pmatrix}}{\sqrt{1 + f_{,x}^2 + f_{,y}^2}} \cdot T^{-1} \right] \cdot \mathbf{\Lambda}', \tag{6.41}$$

The quadratic form C_P being expressed by the above matrix in square brackets in an orthonormal basis, its determinant is the Gauss curvature:

$$K = (\det T^{-1})^2 \, \frac{\det \begin{pmatrix} f_{,xx} & f_{,xy} \\ f_{,xy} & f_{,yy} \end{pmatrix}}{1 + f_{,x}^2 + f_{,y}^2}.$$

---

[20] The determinant of the matrix expressing a quadratic form is not invariant by a change of frame. For instance, the quadratic form $x^2 + y^2$ is expressed by the identity matrix in the orthonormal frame $(\mathbf{e}_x, \mathbf{e}_y)$ but by the matrix $\begin{pmatrix} 4 & 0 \\ 0 & 1 \end{pmatrix}$ in the non-orthonormal frame $(2\,\mathbf{e}_x, \mathbf{e}_y)$ since $x^2 + y^2 = 4\,(x/2)^2 + y^2$. The determinants differ, as they are 1 and 4. However, this determinant is invariant by rotations: it is the same in any orthonormal frame. This number is what is called the determinant of the quadratic form itself.

Using equation (6.40) to get the Jacobian of the change of coordinates, we find the Gauss curvature in an explicit form:

$$K(x,y) = \frac{f_{,xx}\, f_{,yy} - f_{,xy}^{\,2}}{\left(1 + f_{,x}^{\,2} + f_{,y}^{\,2}\right)^2},\tag{6.42}$$

a formula that generalizes equation (6.14) when the tangent plane is not horizontal ($f_{,x}$ and $f_{,y}$ are arbitrary). Note that equation (6.42) was derived at the origin of the coordinates for the sake of simplicity but it actually holds for any point on a surface.

6.3   Developable surfaces

In this section, we use the *Theorema egregium* just established to study the surfaces that are isometric to a plane. Such surfaces are called developable. Isometric deformations of a curved surface is a much more difficult problem, which will be studied in Chapter 11, with a particular emphasis on surfaces of revolution.

The Gauss curvature of a planar surface is zero everywhere. By the *Theorema egregium*, a developable surface also has a vanishing Gauss curvature everywhere. It turns out that the local constraint $K(x,y) = 0$ satisfied by a developable surface can be integrated into a global constraint, namely the existence of generatrices.[21] In this section, we establish this result that will allow us to derive an explicit list of all developable surfaces.

6.3.1   Developable surfaces are ruled

Consider a surface given by its Cartesian parameterisation $z = f(x,y)$. By equation (6.42), the *Theorema egregium* implies that the determinant[22] of the Hessian form is zero:

$$f_{,xx}\, f_{,yy} - f_{,xy}^{\,2} = 0.\tag{6.43}$$

By the geometrical interpretation of the Gauss curvature given in Section 6.2.2, this means that the surface has a principal curvature that vanishes at every point: at every point, there exists a principal direction along which the surface has vanishing curvature.[23] In other words, a developable surface looks locally like a cylinder, the axis of the cylinder being the local direction of zero principal curvature.

It can easily be checked that any symmetric 2×2 matrix with a vanishing determinant and a trace equal to one depends on one parameter only[24] and can be put in the form:

---

[21] Generatrices are sometimes called rulings.

[22] A classical problem in analysis, called the Monge–Ampère equation, is to find the unknown function(s) $f(x,y)$ such that $f_{,xx}\, f_{,yy} - f_{,xy}^{\,2}$ is a prescribed function in some domain. The equation (6.43) for developable surfaces appears as a particular case of this general problem.

[23] On a developable surface, the direction with vanishing curvature is also a principal direction. This constraint is much more severe than the mere existence of a direction with vanishing curvature. In fact, for any surface that is hyperbolic ($K < 0$) such as the one shown in Fig. 6.5(c), the quadratic form C_P has a negative determinant at every point and is therefore non-definite: there are locally two directions in the tangent space that make C_P vanish. Along these two directions, the curvature of the surface vanishes; such directions are called asymptotic. By integration, this leads to the existence of two one-parameter families of asymptotic curves on any hyperbolic (saddle-like) surface.

[24] A 2×2 matrix has four parameters. A symmetric one has only three as the off-diagonal elements are equal. The condition that the determinant is zero and the trace one add two more constraints.

$$\begin{pmatrix} \sin^2 \phi & -\sin \phi \cos \phi \\ -\sin \phi \cos \phi & \cos^2 \phi \end{pmatrix},$$

where ϕ is a parameter. Applying this to $H(x,y)/(\operatorname{tr} H(x,y))$, where $H(x,y)$ is the Hessian matrix filled with the second derivatives of $f(x,y)$ defined in equation (6.14), we find that $H(x,y)$ can be put in the form

$$\begin{pmatrix} f_{,xx} & f_{,xy} \\ f_{,xy} & f_{,yy} \end{pmatrix}_{(x,y)} = L(x,y) \begin{pmatrix} \sin^2 \phi(x,y) & -\sin \phi(x,y) \cos \phi(x,y) \\ -\sin \phi(x,y) \cos \phi(x,y) & \cos^2 \phi(x,y) \end{pmatrix}, \quad (6.44)$$

where $\phi(x,y)$ and $L(x,y)$ are free functions, $L(x,y) = \operatorname{tr} H(x,y)$ being the trace of H.

The two functions $\phi(x,y)$ and $L(x,y)$ are not independent as they express the second derivatives of a single function $f(x,y)$. This is very similar to the remark made in Section 2.2.12 that led to the condition of the compatibility of strain. Indeed, according to Leibniz identity the partial derivatives can be permuted:

$$\frac{\partial f_{,xx}}{\partial y} = \frac{\partial f_{,xy}}{\partial x} \quad \text{and} \quad \frac{\partial f_{,xy}}{\partial y} = \frac{\partial f_{,yy}}{\partial x},$$

and we obtain, using equation (6.44):

$$\frac{\partial}{\partial x} \left([L \cos \phi].[\sin \phi]\right) + \frac{\partial}{\partial y} \left([L \sin \phi].[\sin \phi]\right) = 0$$

$$\frac{\partial}{\partial x} \left([L \cos \phi].[\cos \phi]\right) + \frac{\partial}{\partial y} \left([L \sin \phi].[\cos \phi]\right) = 0.$$

By expanding the partial derivatives of the bracketed terms,[25] multiplying the first equation by $\cos \phi$, the second one by $\sin \phi$, subtracting the results and factoring out $L(x,y)$, which we assumed to be non-zero,[26] one obtains a compatibility condition

$$\left[\cos \phi(x,y) \frac{\partial}{\partial x} + \sin \phi(x,y) \frac{\partial}{\partial y} \right] \phi(x,y) = 0, \quad (6.45)$$

which depends on ϕ only. This partial differential equation can easily be integrated if we introduce the so-called characteristic functions $(x_g(s), y_g(s))$ of the partial differential equation, satisfying:

$$\frac{\mathrm{d}x_g(s)}{\mathrm{d}s} = \cos[\phi(x_g(s), y_g(s))], \qquad \frac{\mathrm{d}y_g(s)}{\mathrm{d}s} = \sin[\phi(x_g(s), y_g(s))]. \quad (6.46)$$

We then define $\mathbf{g}_M^{2D}(s)$ to be the curve in the (x,y) plane defined by this parameterisation, $(x_g(s), y_g(s))$, when the differential system is started with the initial condition $M = (x_g(0), y_g(0))$. Here M is a point in the (x,y) plane, as shown in Fig. 6.6. By varying M, we obtain a family of curves $\mathbf{g}_M^{2D}$ covering the whole projection of S onto the (x,y)

---

[25] We shall consider only the case of a smooth surface, such that the functions $L(x,y)$ and $\phi(x,y)$ are continuously differentiable.

[26] We do not consider the degenerate case in which the Hessian form is zero, that is when both principal radii of curvature vanish. If this degeneracy occurs at a point or along a curve, we simply exclude this region from the domain first and in a second time extend the results derived above by continuity. If the degeneracy occurs in a domain whose interior is non-empty, the demonstration below can be adapted to show that the surface is locally a plane.

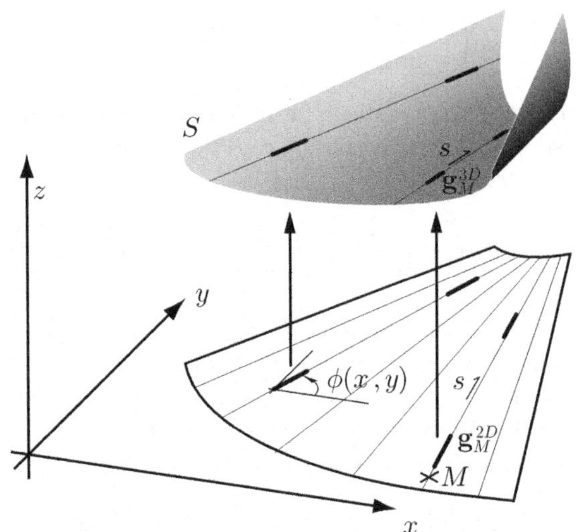

Fig. 6.6 In Section 6.3.1, we show than any developable surface is ruled. To do so, we introduce a direction $\phi(x,y)$ in the plane (x,y), define the flow-lines $\mathbf{g}_M^{2D}(s)$ of the vector field $(\cos\phi, \sin\phi)$, show that ϕ is constant along any flow-line, which implies that all these flow-lines, are straight lines; then, we define the curves $\mathbf{g}_M^{3D}$ obtained by lifting these flow-lines on to the surface, and show that the surface is flat along the tangents $(\mathbf{g}_M^{3D})'(s)$ and, finally, that the lifted curves are actually straight lines spanning the whole surface.

plane. By definition, these characteristic curves are also the flow-lines of the vector field $(\cos\phi, \sin\phi)$.

The compatibility condition (6.45) implies that:

$$\frac{\mathrm{d}\phi(x_g(s), y_g(s))}{\mathrm{d}s} = 0,$$

which means that ϕ takes a constant value along any flow-line $\mathbf{g}_M^{2D}$. By equation (6.46), the direction of the tangent to the flow-line is given by ϕ. We conclude that the flow-lines are *straight* lines.

We shall now look at the curves $\mathbf{g}_M^{3D}$ that are obtained by lifting these flow-lines from the plane (x,y) on to the surface S:

$$\mathbf{g}_M^{3D}(s) = (x_g(s), y_{g(s)}, f(x_g(s), y_g(s))).$$

By construction, the vector tangent to the curves $\mathbf{g}_M^{3D}$, namely $(\mathbf{g}_M^{3D})'(s)$, is also tangent to the surface, and its coordinates in the tangent plane are $\mathbf{\Lambda} = (\cos\phi, \sin\phi)$, using the same basis $((1, 0, f_{,x}), (0, 1, f_{,y}))$ of the tangent plane as earlier.

We shall now compute how the surface is curved along these flow-lines $\mathbf{g}_M^{3D}$. To do so, we first notice that:

$$\begin{pmatrix} \sin^2\phi & -\sin\phi\cos\phi \\ -\sin\phi\cos\phi & \cos^2\phi \end{pmatrix} \cdot \begin{pmatrix} \cos\phi \\ \sin\phi \end{pmatrix} = \begin{pmatrix} 0 \\ 0 \end{pmatrix}.$$

By equation (6.44), this yields $H \cdot \mathbf{\Lambda} = 0$, which implies that $C_P((\mathbf{g}_M^{3D})'(s)) = 0$ along any flow-line. In other words, the surface S is flat along any tangent $(\mathbf{g}_M^{3D})'(s)$ to a curve $\mathbf{g}_M^{3D}$: this tangent is therefore a direction of vanishing curvature (the quadratic form C_P being degenerate by assumption, this direction of zero curvature is also a principal direction).

At this point, it appears that the curves $\mathbf{g}_M^{3D}(s)$ are all straight lines. Indeed, the second derivatives reads

$$\left(\mathbf{g}_M^{3D}\right)''(s) = \frac{\mathrm{d}}{\mathrm{d}s}\left(\cos\phi, \sin\phi, f_{,x}\cos\phi + f_{,y}\sin\phi\right)$$

$$= (0, 0, {}^t(\cos\phi, \sin\phi) \cdot H \cdot (\cos\phi, \sin\phi)) = \mathbf{0},$$

since ϕ is constant along characteristic lines, $\mathrm{d}\phi/\mathrm{d}s = 0$.

Therefore, there exists a one-parameter family of straight lines $\mathbf{g}_M^{3D}(s)$ spanning any developable surface S, as sketched in Fig. 6.6. A surface that can be spanned by straight lines is called a ruled surface—we have just shown that any developable surface is ruled. The straight lines are called generatrices or sometimes rulings. The *Theorema egregium*, which yields a local constraint $K(x, y) = 0$ for a developable surface, has been transformed into a global constraint,[27] namely the existence of straight generatrices.

6.3.2 Developability of a ruled surface

In the previous section, we proved that any developable surface is ruled, which means that it can be swept out by a straight line moving in the three-dimensional space. The converse is not true[28]. By studying what ruled surfaces are actually developable, we are now going to obtain the complete list of developable surfaces.

Let us consider an arbitrary ruled surface, such as the catenoid shown in Fig. 6.7.

It is parameterised by two real variables, (u, v):

$$\mathbf{r}(u, v) = \mathbf{a}(u) + v\,\mathbf{g}(u), \tag{6.47}$$

where $\mathbf{a}(u)$ is a three-dimensional curve called the base curve, $\mathbf{g}(u)$ defines the direction of the generatrix and v is the coordinate along this generatrix. For the surface to be $\mathcal{C}^2$ smooth, we will impose that both $\mathbf{a}(u)$ and $\mathbf{g}(u)$ are $\mathcal{C}^2$-smooth functions, i.e. are twice continuously differentiable.

Many possible choices of $\mathbf{a}(u)$ and $\mathbf{g}(u)$ yield equivalent parameterisations of the same ruled surface. To reduce this indeterminacy, we shall impose some constraints on these functions. First, note that changing the length of the vector $\mathbf{g}(u)$ amounts to rescaling the parameter v, so we can freely impose that $\mathbf{g}(u)$ has unit length: $|\mathbf{g}(u)| = 1$. Similarly, replacing $\mathbf{a}(u)$ by any other curve $\mathbf{a}(u) + \mu(u)\,\mathbf{g}(u)$, where μ is an arbitrary function of u, amounts to changing the origin of v. By adjusting the function μ, it is always possible to satisfy the constraint $\mathbf{a}'(u) \cdot \mathbf{g}(u) = 0$. Finally, by choosing u to be the curvilinear length

---

[27] For surfaces with non-zero Gauss curvature, there is no equivalent of this global constraint.

[28] On an arbitrary ruled surface, a generatrix defines a direction along which the surface has zero curvature, that is $C_P(\mathbf{g}(M)) = 0$. As a result, the quadratic form C_P cannot be positive definite and its determinant satisfies $K \leq 0$. A ruled surface is therefore hyperbolic ($K < 0$) in general; it can also be developable ($K = 0$) but this is not the generic case. The paraboloid, for instance, is hyperbolic (it is doubly ruled; is has two generatrices passing through every point, so that C_P has two distinct asymptotic directions and therefore a non constant sign, which shows that $K < 0$).

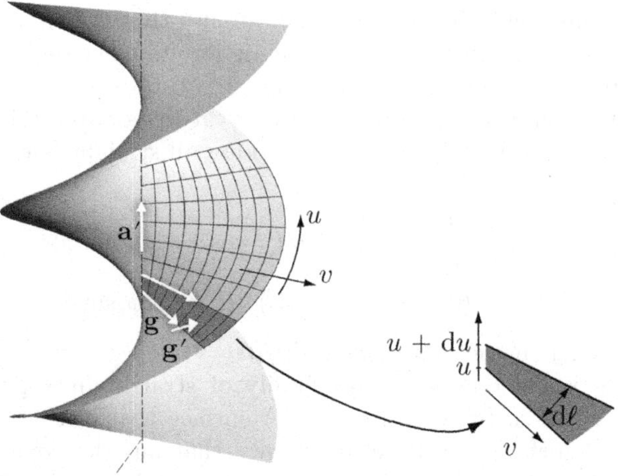

Fig. 6.7 The catenoid is a typical ruled surface that is *not* a developable surface. The tangent along the base curve, $\mathbf{a}'(u)$, the direction of the generatrix, $\mathbf{g}(u)$ and its derivative $\mathbf{g}'(u)$ are not coplanar. This property, reflected in the profile of the separation $\mathrm{d}\ell(u, v, \mathrm{d}u)$ between two neighbouring generatrices as a function of v (see main text), prevents the catenoid from being developable onto a plane

along the base curve $\mathbf{a}$ (normal parameterisation), one has $|\mathbf{a}'(u)| = 1$. To summarize, we can impose:

$$|\mathbf{g}(u)| = 1 \tag{6.48a}$$

$$|\mathbf{a}'(u)| = 1 \tag{6.48b}$$

$$\mathbf{a}'(u) \cdot \mathbf{g}(u) = 0 \tag{6.48c}$$

without loss of generality. Then, by a trick used several times before, $\mathbf{g}(u) \cdot \mathbf{g}'(u) = 1/2\, \mathrm{d}|\mathbf{g}(u)|^2/\mathrm{d}u = 0$ and similarly $\mathbf{a}'(u) \cdot \mathbf{a}''(u) = 0$.

To study the developability of a given ruled surface, we could calculate the Gauss curvature by computing the quadratic form C_P explicitly, as we did before in Section 6.2.5 for a surface given as the graph of a function $f(x, y)$. Imposing that its determinant, the Gauss curvature, is zero leads to a condition for the ruled surface to be developable. Here we propose an equivalent but more geometric reasoning that leads to the same condition. To this end, we shall attempt to characterize how fast two neighbouring generatrices are moving away from each other along a ruled surface, when one follows one of these generatrices. Therefore, we consider a fixed generatrix with coordinate u, and a neighbouring generatrix with coordinate $u + \mathrm{d}u$. For any point (u, v) on the first generatrix, we seek the point $(u + \mathrm{d}u, v'(u, v))$ on the second generatrix that is the closest to the first point, as shown in Fig. 6.7. By differentiation of the parametric equation of the surface, we find the vector joining these two points

$$\mathbf{q}(u, v, \mathrm{d}u, v') = (\mathbf{a}'(u) + v\,\mathbf{g}'(u))\,\mathrm{d}u + \mathbf{g}(u)\,(v' - v). \tag{6.49}$$

Now, the point on the second generatrix is indeed the closest one to the point (u, v) when v' minimizes the norm of $\mathbf{q}$. This yields $\mathbf{q} \cdot \partial \mathbf{q}/\partial v' = 0$. Using $\mathbf{a}' \cdot \mathbf{g} = 0$ and $\mathbf{g}' \cdot \mathbf{g} = 0$, as noted earlier, we find $v'(v) = v$. Then, the norm of $\mathbf{q}$ reads $|\mathbf{a}' + v\,\mathbf{g}'|\,|\mathrm{d}u|$. This is the shortest distance between a point (u, v) on the ruled surface and a given neighbouring generatrix. Let us note $\mathrm{d}\ell(u, v, \mathrm{d}u)$ this distance: we have just shown

$$\left(\frac{\mathrm{d}\ell(u, v, \mathrm{d}u)}{\mathrm{d}u}\right)^2 = (\mathbf{a}'(u) + v\,\mathbf{g}'(u))^2 = 1 + 2v\,\mathbf{a}'(u) \cdot \mathbf{g}'(u) + v^2\,[\mathbf{g}'(u)]^2. \tag{6.50}$$

We shall now assume that the ruled surface is developable, which means that there exists some mapping that takes any point $\mathbf{r}(u, v)$ of the ruled surface onto a point $\bar{\mathbf{r}}(u, v)$ within the (x, y) plane, and does not change the distances measured along the surface. The generatrices of the ruled surface are straight lines in the Euclidean space; as a result, any generatrix defines the shortest path along the surface between any pair of its points—such a curve is called a geodesic curve. Their definition being only in terms of metric properties of the surface, the image of a geodesic by an isometry remains a geodesic: as a result, the image of a generatrix by our assumed isometric mapping onto a plane must define the shortest path in the (x, y) plane between any pair of its points. We conclude that the image of generatrices are straight lines: the generatrix passing through the point (u, v) with direction $\mathbf{g}(u)$ is transformed by the isometry into a straight line with direction $\bar{\mathbf{g}}(u) = \partial\bar{\mathbf{r}}(u, v)/\partial u$, which does not depend on v as we have just shown. We also define $\bar{\mathbf{a}}(u) = \bar{\mathbf{r}}(u, 0)$ to be the image of the base curve $\mathbf{a}(u) = \mathbf{r}(u, 0)$.

Recall that equation (6.50) characterizes how fast two neighbouring generatrices move away from each other when one walks along them (consider that u is fixed while v varies). The left-hand side of the equation involves only quantities measured along the surface (it is the length of the shortest path between an arbitrary point and a generatrix passing nearby) and so will be invariant by the isometry. The right-hand side is then invariant too, which implies[29] that:

$$\mathbf{a}'(u) \cdot \mathbf{g}'(u) = \bar{\mathbf{a}}'(u) \cdot \bar{\mathbf{g}}'(u) \quad \text{and} \quad [\mathbf{g}'(u)]^2 = [\bar{\mathbf{g}}'(u)]^2. \tag{6.51}$$

Now, the equations (6.48) characterize the dot products and the norms of tangent vectors and so are conserved by the isometry: the vectors $\bar{\mathbf{g}}(u)$ and $\bar{\mathbf{a}}'(u)$ define an orthonormal basis in the (x, y) plane. The vector $\bar{\mathbf{g}}'(u)$, which has to be both perpendicular to $\bar{\mathbf{g}}(u)$ (so that the norm of $\bar{\mathbf{g}}(u)$ remains one for any u) and contained in the (x, y) plane, must be aligned with $\mathbf{a}'(u)$. Since $\mathbf{a}'(u)$ is a unit vector, this can be expressed by writing

$$(\bar{\mathbf{a}}'(u) \cdot \bar{\mathbf{g}}'(u))^2 = [\bar{\mathbf{g}}'(u)]^2,$$

the left-hand side of the equation being the square norm of the vector $\mathbf{g}'$ projected along $\mathbf{a}'$. Using equation (6.51), we find that this equation must also hold with the original ruled surface:

$$(\mathbf{a}'(u) \cdot \mathbf{g}'(u))^2 = [\mathbf{g}'(u)]^2.$$

---

[29] That the dot product between the vectors $\mathbf{a}'(u)$ and $\mathbf{g}'(u)$ is conserved upon isometry is not obvious. Indeed, $\mathbf{g}'(u)$ is not tangent to the ruled surface in general. As a result, the conservation of the metrics of the surface cannot be invoked directly to show that $\mathbf{g}' \cdot \mathbf{a}'$ is conserved.

By the same interpretation as above, this means that $\mathbf{g}'(u)$ has to be aligned with the unit vector $\mathbf{a}'(u)$ for the ruled surface to be developable. This is the condition for a ruled surface to be developable that we were seeking. If we choose a parameterisation of the ruled surface that is not of the particular form (6.48), $\mathbf{g}'(u)$ might have a component along $\mathbf{g}(u)$ too—recall that $\mathbf{g}' \cdot \mathbf{g} = 1/2\, \mathrm{d}[\mathbf{g}]^2/\mathrm{d}s$ is only zero when $|\mathbf{g}|$ does not depend on u. With an arbitrary parameterisation of a ruled surface, the developability condition does not require that $\mathbf{g}'$ is along $\mathbf{a}'$, but simply that $\mathbf{g}'$ has no component normal to the plane spanned by $\mathbf{g}$ and $\mathbf{a}'$:

$$(\mathbf{a}'(u) \times \mathbf{g}(u)) \cdot \mathbf{g}'(u) = 0 \quad \text{for all } u. \tag{6.52}$$

This is the mixed product of the three vectors $\mathbf{a}'$, $\mathbf{g}$ and $\mathbf{g}'$. The condition means that $\mathbf{g}'$ has to be contained in the local tangent plane, which is indeed spanned by $\mathbf{g}$ and $\mathbf{a}'$.

Equation (6.52) yields a necessary condition for a ruled surface to be developable. This is in fact a sufficient condition; we shall simply outline the demonstration here. Recall that, for all u, $((\overline{\mathbf{g}}(u), \overline{\mathbf{a}}'(u))$ has to be an orthonormal frame in the plane (x, y). Let us define $\Omega(u)$ as its rate of rotation: $\overline{\mathbf{g}}'(u) = \Omega(u)\,\mathbf{e}_z \times \overline{\mathbf{g}}(u)$ and $\overline{\mathbf{a}}''(u) = \Omega(u)\,\mathbf{e}_z \times \overline{\mathbf{a}}'(u)$. It appears that $\Omega(u)$ is given by $\Omega(u) = \overline{\mathbf{g}}'(u) \cdot \overline{\mathbf{a}}'(u)$, which is known by equation (6.51): the dot product $\mathbf{g}'(u) \cdot \mathbf{a}'(u)$ directly yields the rate of rotation of the frame $(\overline{\mathbf{g}}(u), \overline{\mathbf{a}}'(u))$ transported by the isometry. Knowing the rate of rotation, one can determine by integration with respect with u the orientation of $(\overline{\mathbf{g}}(u), \overline{\mathbf{a}}'(u))$ for all u, up to a global rotation. By another integration, one can determine the image $\overline{\mathbf{a}}(u)$ of the base curve. Since the image of generatrices are straight lines, we find then that the isometry is necessarily given by $\overline{\mathbf{r}}(u, v) = \overline{\mathbf{a}}(u) + v\,\overline{\mathbf{g}}(u)$. It is not difficult to check that this explicit construction indeed defines an isometry.

6.3.3   Listing all developable surfaces

At this point, we know that the smooth developable surfaces are the ruled surfaces that satisfy equation (6.52) above. Below, we show that only three types of surfaces, listed in figure (6.8), are solutions of this equation: the cones, the cylinders, and the so-called *tangent developables*.[30] The latter are defined as the ruled surfaces swept by the set of all tangents to a smooth curve in the Euclidean space $\mathbb{R}^3$.

The case of a cylinder corresponds to $\mathbf{g}'(u) = 0$ for all u in equation (6.52) above: then, $\mathbf{g}(u)$ defines a constant direction in space. When $\mathbf{g}'(u)$ is non-zero, it is orthogonal to $\mathbf{g}(u)$ when $\mathbf{g}(u)$ is chosen unitary. Then, equation (6.52) imposes that $\mathbf{a}'(u)$ is a linear combination of $\mathbf{g}(u)$ and $\mathbf{g}'(u)$: $\mathbf{a}'(u) = \lambda(u)\,\mathbf{g}(u) + \mu(u)\,\mathbf{g}'(u)$, which can be rewritten as:

$$\tilde{\mathbf{a}}'(u) = [\lambda(u) - \mu'(u)]\,\mathbf{g}(u) \quad \text{where } \tilde{\mathbf{a}}(u) = \mathbf{a}(u) - \mu(u)\,\mathbf{g}(u). \tag{6.53}$$

The new base curve $\tilde{\mathbf{a}}(u)$ provides an equivalent parameterisation of the ruled surface. When $\lambda(u) = \mu'(u)$ for all u, then $\tilde{\mathbf{a}}(u)$ does not depend on u, which corresponds to a cone—the base curve is shrunk to a fixed point. When $\lambda(u) \neq \mu'(u)$, the generatrix $\mathbf{g}(u)$ is aligned with the tangent to the base curve $\tilde{\mathbf{a}}(u)$. In this case, the surface is the tangent developable to the curve $\tilde{\mathbf{a}}(u)$, i.e. it is swept out by all its tangents. All these developable surfaces are not necessarily representable by a sheet of paper in the real world as they may cross themselves or exhibit singularities.

---

[30] Note that the cones can be considered as a degenerate case of tangent developables.

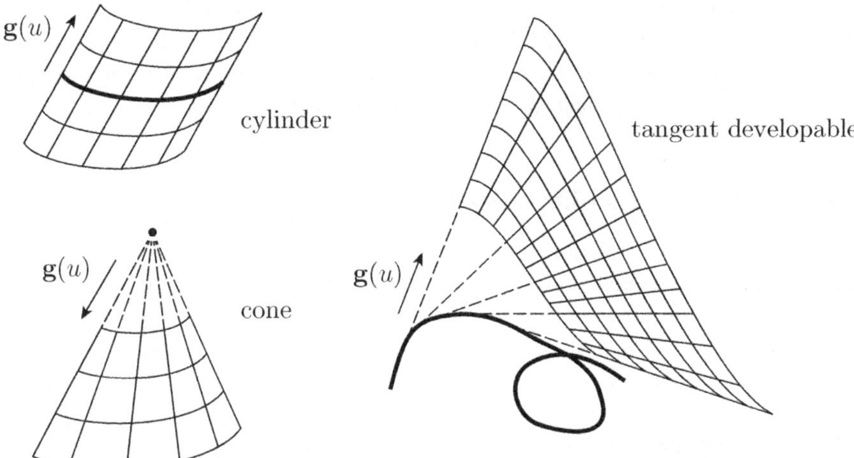

Fig. 6.8 Developable surfaces are of three types: cones, cylinders and tangent developables. The latter are the ruled surfaces, the generatrices of which consist of the tangents to a curve in $\mathbb{R}^3$. The curve $\mathbf{a}(u)$ is shown using a thick line.

6.3.4   From the geometry of surfaces to the elasticity of plates

The *Theorema egregium* gives the condition under which a surface is mapped to another surface while keeping the distances unchanged. This has an obvious connection with the theory of thin elastic plates and shells, which deform so as to avoid stretching whenever possible: if the lengths along their centre surface are unchanged, any energy of deformation arises from a bending effect, that is from slight inhomogeneities of the deformation field throughout the small plate thickness. As suggested by equation (6.4), this bending energy will probably be some sort of small correction in a systematic expansion with respect to the small parameter represented by the plate thickness—or more accurately by the dimensionless aspect ratio of the plate.

The theory of elasticity of plates, although built on the geometrical concepts above, is not a straightforward application of them, for two main reasons. First, the energy of deformation may involve at the same order in-plane stretching effects and bending effects, the latter being absent in a purely geometrical description. The Gauss-inspired differential geometry relies upon the crucial assumption that the surface is $\mathcal{C}^2$ smooth[31] to establish the condition of isometric mapping. Unfortunately, it happens quite often that a sheet or a shell under strong deformation will minimize its elastic energy by becoming singular in the sense of the 2D geometry of surfaces, the regularization being due to the three-dimensional structure. For instance, we will show in Chapter 9 that the radius of curvature of a crumpled plate can become comparable to the thickness h. In the thin plate limit, $h \to 0$, this would appear as a singular ridge or a singular conical point.

Secondly, there are often many ways for a plate or a shell to deform isometrically: depending on the boundary conditions and applied forces, the state with minimal stretching

---

[31] $\mathcal{C}^2$ smooth means that the surface has a tangent plane everywhere, and that this tangent plane has a continuously differentiable normal vector. This is equivalent to requiring that the two curvatures $\kappa_{x'}$ and $\kappa_{y'}$ are well-defined and continuous.

energy can be degenerate. In this case, prediction of the actual shape of a plate or a shell relies on the consideration of the bending energy, which requires more work than the geometric considerations above. A typical example of such a degeneracy, present in the geometrical limit but lifted by a proper elastic theory, is given in Chapter 10.

6.4   Membranes: stretching energy

In the rest of this chapter, we derive the equations of equilibrium for a thin elastic plate whose thickness, h, is assumed much smaller than any other length in the problem. In the present section, the stretching energy is considered but not the bending energy. This yields a *membrane*-like[32] theory of plates. In Section 6.5, we shall incorporate a bending energy and arrive at the Föppl–von Kármán equations.

6.4.1   Underlying assumptions

We shall start by defining the approximations made in the present section and in the following one. The membrane approximation means that we neglect the bending energy and consider stretching effects only. This approximation is relevant whenever the neglected bending energy is indeed small compared with the stretching energy. Using our estimates for the bending and stretching energies in equations (6.4) and (6.2), this yields

$$E\,h^3 \left(\frac{1}{R}\right)^2 \ll E\,h\,\left(\epsilon_{\alpha\beta}^{\mathrm{cs}}\right)^2, \quad \text{that is} \quad \left(\frac{h}{R}\right)^2 \ll \left(\epsilon_{\alpha\beta}^{\mathrm{cs}}\right)^2, \tag{6.54}$$

where R is the typical radius of curvature of the centre surface and $\epsilon_{\alpha\beta}^{\mathrm{cs}}$ its typical in-plane strain. For the approximation to be valid, these in-plane strains $\epsilon_{\alpha\beta}^{\mathrm{cs}}$ should be large[33] compared with a very small number, the ratio of thickness h to the typical radius of curvature R. We shall not restrict ourselves to the membrane approximation outlined here: later on in this chapter, we derive the missing expression for the bending energy.

We shall also assume small displacements. This approximation has been defined earlier in equation (2.32a), and assumes that the gradients of the displacement field remain small. Then, the slope of the plate profile $w(x, y)$ is everywhere small:

$$\left|\frac{\partial w}{\partial x}\right| \ll 1, \quad \left|\frac{\partial w}{\partial y}\right| \ll 1. \tag{6.55}$$

This rules out the possibility of rotations by a finite angle. It is possible to derive more general plate equations so as to relax any small displacements approximation. The latter can be useful for practical applications and numerical simulations.

6.4.2   2D stress–strain relations

Let (Oxy) be the plane of the reference configuration of the plate, and z the transverse direction, see Fig. 6.9. The deformed shape of the plate is characterized by a displacement

---

[32] A membrane is defined as an elastic plate (or shell) that can sustain tangential stress but no flexural moments: it has a vanishing bending modulus.

[33] We assumed earlier that the strain remained small everywhere, $|\epsilon_{\alpha\beta}| \ll 1$ (Hookean elasticity); this is indeed compatible with the assumption in equation (6.54), as the thickness h is much smaller than R by assumption.

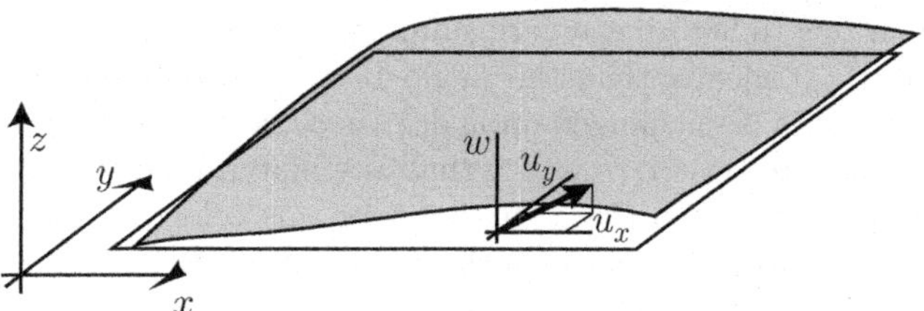

Fig. 6.9 Deformation of a thin plate. The displacement is decomposed into its in-plane components, $u_x(x,y)$ and $u_y(x,y)$ and its (out-of-plane) deflection $w(x,y)$.

field defined over the centre plane, whose in-plane components are denoted[34] $u_x(x,y)$ and $u_y(x,y)$ while the out-of-plane displacement is denoted $w(x,y)$. This out-of-plane displacement is also called the *deflection*. The equations for planar membranes derived below are based on a reduction to two dimensions of the equations for three-dimensional elasticity derived in Chapter 2.

This reduction starts from the stress-free condition on the upper and lower edges of the plate:

$$\sigma_{xz}(x,y,\pm h/2) = \sigma_{yz}(x,y,\pm h/2) = \sigma_{zz}(x,y,\pm h/2) = 0. \tag{6.56}$$

The derivatives of the various components of the stress tensor are related by the condition of mechanical equilibrium $\partial\sigma_{ij}/\partial x_j = 0$ and are all of the same order of magnitude. Because the plate dimension in the transverse direction, h, is much smaller than any other length scale, the stress components appearing in equation (6.56) cannot depart much from their value on the free edges, which is zero. This means that these particular components of the stress tensor are much smaller than all the other ones, something that we write:

$$\sigma_{xz}(x,y,z) = \sigma_{yz}(x,y,z) = \sigma_{zz}(x,y,z) = 0 \tag{6.57}$$

in a first approximation. This is consistent with the previous remark that any dependence of strain and stress with respect to the transverse coordinate z can be neglected as long as only stretching effects are considered. We shall see in the next section and in Appendix D how this statement must be refined so as to account for bending effects.

When using the Cauchy–Poisson relations of Hookean elasticity, given earlier in equation (2.61), the equation (6.57) can be turned into conditions for the strain tensor:

$$\epsilon_{xz} = 0 \qquad \epsilon_{yz} = 0, \tag{6.58a}$$

$$\epsilon_{zz} = -\frac{\nu}{1-\nu}\left(\epsilon_{xx} + \epsilon_{yy}\right). \tag{6.58b}$$

[34] In the present analysis of elastic plates, we rename some of the quantities that were introduced in the geometrical section at the beginning of this chapter. This will not introduce any conflict of notation. The in-plane components of the displacement, called $u(x,y)$ and $v(x,y)$ earlier, are now written u_x and u_y, while the deflection is denoted $w(x,y)$. This new notation is intended to single out the deflection that plays a specific role in the approximation of small displacements. Also, the coordinates (x,y) are sometimes referred to as x_α for $\alpha = x,y$ to take advantage of Einstein summation. All these notations are consistent with the general conventions listed at the beginning of this book.

The first set of equations (6.58a) has a very simple interpretation if one recalls the analysis of Section 2.2.5: $\epsilon_{xz} = 0$ means that the angle between small material vectors initially parallel to the x and z axes is unchanged upon deformation: this remains a right angle. Using the argument with $\epsilon_{yz} = 0$ too, we conclude that any material curve initially perpendicular to the centre surface (parallel to the z direction in this reference configuration) remains perpendicular to the deformed centre surface after transformation. In simpler words, the material lines transverse to the centre surface remain transverse: there is no out-of-plane shear.[35] This condition of orthogonality is known as the Kirchhoff kinematical hypothesis.[36]

When plugged into the remaining constitutive relations (2.61), the equations (6.58) yield effective constitutive relations for a plate, which link the in-plane strain to the in-plane stress:[37]

$$\sigma_{xx}(x,y) = \frac{E}{1-\nu^2}\left(\epsilon_{xx}(x,y) + \nu\,\epsilon_{yy}(x,y)\right)$$

$$\sigma_{yy}(x,y) = \frac{E}{1-\nu^2}\left(\epsilon_{yy}(x,y) + \nu\,\epsilon_{xx}(x,y)\right) \tag{6.59}$$

$$\sigma_{xy}(x,y) = \frac{E}{1+\nu}\,\epsilon_{xy}(x,y).$$

Out-of-plane components, such as the transverse strain ϵ_{zz}, have been eliminated. All quantities are evaluated along the centre surface, which, at the dominant order, amounts to averaging over the plate thickness.[38]

To obtain a complete set of equations we still need to express the strain in terms of the displacement (Section 6.4.3) and then to write the conditions of mechanical equilibrium (Section 6.4.4).

6.4.3 Keeping nonlinearity connected to the deflection only

The definition of the strain tensor given in Chapter 2, equation (2.9), yields the following expression for the in-plane components $\epsilon_{\alpha\beta}$ needed in equation (6.59) above:

$$\epsilon_{\alpha\beta}(x,y) = \frac{1}{2}\left(\frac{\partial u_\alpha}{\partial x_\beta} + \frac{\partial u_\beta}{\partial x_\alpha}\right) + \frac{1}{2}\left(\frac{\partial u_x}{\partial x_\alpha}\frac{\partial u_x}{\partial x_\beta} + \frac{\partial u_y}{\partial x_\alpha}\frac{\partial u_y}{\partial x_\beta} + \frac{\partial w}{\partial x_\alpha}\frac{\partial w}{\partial x_\beta}\right), \tag{6.60}$$

where the Greek indices run over the directions x and y tangent to the plate (also noted x_α for $\alpha = x$ or y). All the quantities in this equation refer to the centre surface, and are functions of x and y only (the strain $\epsilon_{\alpha\beta}(x,y)$ was noted $\epsilon_{\alpha\beta}^{\mathrm{cs}}$ earlier and we shall now simplify the notation as there is no ambiguity).

---

[35] Shear deformations remain possible in the plane of the mean surface, that is with $\epsilon_{xy} \neq 0$. Note that the out-of-plane shear has to be zero for a thin plate (or shell) only.

[36] The word *hypothesis* comes from the fact that this assumption is sometimes used as a starting point to derive the equations for elastic plates or shells.

[37] The transverse displacement w will nevertheless be kept in the final equations, through the non-linear terms in the definition of strain: this membrane theory is not just a theory of elasticity in two dimensions.

[38] This averaging is consistent with the former qualitative analysis of stretching, which involves a uniform deformation of the plate throughout its thickness. No such averaging will be performed when we consider bending later on.

The above definition of the strain includes all non-linear terms and is equally valid for small or finite displacements. Here, we assume that the displacement is small; we could therefore attempt to drop all non-linear terms in the right-hand side of equation (6.60), see Section 2.2.11. This would lead to the so-called Love–Kirchhoff equations for plates, which indeed are linear equations. Here, we make a slightly less restrictive approximation: we discard the first two non-linear terms but keep the third one[39] that depends on the deflection $w(x, y)$. The result reads

$$\epsilon_{\alpha\beta}(x, y) = \frac{1}{2} \left(\frac{\partial u_\alpha}{\partial x_\beta} + \frac{\partial u_\beta}{\partial x_\alpha} \right) + \frac{1}{2} \frac{\partial w}{\partial x_\alpha} \frac{\partial w}{\partial x_\beta}. \tag{6.61}$$

The reason for dropping the first two non-linear terms in the full definition (6.60), is that they are quadratic functions of the gradients $\partial u_\gamma / \partial x_\alpha$ of the *in-plane* components of the displacement. The same gradients define the linearized strain $\epsilon_{\alpha\beta}^{\text{lin}} = (1/2) \left(\partial u_\alpha / \partial x_\beta + \partial u_\beta / \partial x_\alpha \right)$, and so are retained at the linear order anyway. These gradients being small by assumption, the quadratic terms are always dominated by the linear ones and can safely be neglected.

The situation is different with the other non-linear terms, that is those involving the deflection w. Since we are only interested in computing the in-plane strain components, this deflection does not appear at all at the linear order. As a result, these non-linear terms may well not be dominated by another term, and it is safer to retain it.

Our previous analysis of isometric deformations of a surface can be used to illustrate in a concrete manner the specificity of the non-linear term involving $w(x, y)$. In equation (6.10) we noted that the in-plane displacement u_α are small, third-order quantities in the vicinity of a developable surface with a horizontal tangent plane. denoting r as the distance to the point with a horizontal tangent plane, we write $u_\gamma \sim r^3$. We also noted that the deflection w is second order, $w \sim r^2$. Going to the gradients, we find $u_{\gamma,\alpha} \sim r^2$ while $w_{,\alpha} \sim r$ (where commas in the subscript are for derivation). In this particular example, the gradients of the different components of the displacement are of different orders of magnitude. More accurately the gradients of the in-plane components are second order, hence are much smaller than the gradients of the deflection. As a result, we find that the linear term $u_{\alpha,\beta}$ is actually of the same order of magnitude as the non-linear term $(w_{,\alpha} w_{,\beta})$. Then, it perfectly makes sense to retain a single non-linear term, as we did in equation (6.61).

The explicit expression of the simplified definition (6.61) of the strain reads:

$$\epsilon_{xx}(x, y) = \frac{\partial u_r(x, y)}{\partial x} + \frac{1}{2} \left(\frac{\partial w(x, y)}{\partial x} \right)^2 \tag{6.62a}$$

$$\epsilon_{xy}(x, y) = \frac{1}{2} \left(\frac{\partial u_x(x, y)}{\partial x} + \frac{\partial u_y(x, y)}{\partial y} \right) + \frac{1}{2} \frac{\partial w(x, y)}{\partial x} \frac{\partial w(x, y)}{\partial y} \tag{6.62b}$$

$$\epsilon_{yy}(x, y) = \frac{\partial u_y(x, y)}{\partial y} + \frac{1}{2} \left(\frac{\partial w(x, y)}{\partial y} \right)^2. \tag{6.62c}$$

[39] We are doing an approximation of small displacement, but only a partial one, that is for the in-plane components of the displacement only. By doing so, we loose the invariance of the equations by rotations with a finite angle (slopes are assumed to be small) but at the same we can account rather accurately for isometric deformations. This point of the derivation of the equations for elastic plates is a rather subtle one. It probably explains why it took a century and a half to find the equations for thin plates, while the equations for thin rods go back to Euler.

Note that we introduced the same approximate definition of strain earlier in our analysis of developable surfaces, see equation (6.11): the present derivation of the equations for plates is tightly connected to the geometrical problem of developable surfaces.

6.4.4 Mechanical equilibrium

So far, we have written the linear constitutive equations for the plate in (6.59) and defined the strain in (6.62). When combined, these equations allow one to compute the stress tensor associated with an arbitrary displacement of the centre surface. It remains to write the condition of mechanical equilibrium. These equations were derived in Chapter 2, equation (2.77), from the condition that the energy is stationary. They are non-linear in general. We emphasized at the end of Section 2.5 that this nonlinearity in the equation of equilibrium derives from that found in the definition of strain, by variation of energy. As a result, when the linearized definition of the strain is used (approximation of small displacements), one should use the linearized equations of equilibrium—conversely, when the exact, non-linear strain is used, the non-linear terms in the equations of equilibrium should be retained. Here, we use a hybrid definition of strain, which was linearized with respect to u_x and u_y but not with respect to w. By minimization of the elastic energy, one gets the linear version (2.51) of the equations of mechanical equilibrium along the in-plane directions:

$$\frac{\partial \sigma_{\alpha i}}{\partial x_i} + \rho\, g_\alpha = 0 \qquad \text{for } \alpha = x, y, \tag{6.63a}$$

and the non-linear version (2.76) along the out-of-plane direction, z:

$$\frac{\partial\left[(\mathbf{e}'_i \cdot \mathbf{e}_z)\, \sigma_{ij}\right]}{\partial x_j} + \rho\, g_z = 0,$$

which can be put in a more explicit form using the definition (2.71) for the deformed material frame $\mathbf{e}'_i$:

$$\frac{\partial}{\partial x_j}\left[\frac{\partial(z + u_z)}{\partial x_i}\, \sigma_{ij}\right] + \rho\, g_z = 0. \tag{6.63b}$$

To avoid cumbersome notations, we have dropped the letters PK (for Piola–Kirchhoff) labelling the stress tensor in the context of finite displacements. The source term $\rho\, g_i$ denotes a volumic density[40] of external force applied on the material of the plate. Since we consider small strain, the volumic mass is close to its value in the reference configuration, $\rho \approx \rho^0$, and can be considered to be a known quantity.

As noted earlier, the stress components $\sigma_{zk} = \sigma_{kz}$ with $k = x, y, z$ having at least one index z, are negligible compared with the in-plane components $\sigma_{\alpha\beta}$, because of the free boundary conditions on the edges (see also Appendix D). Then, the index i in equation (6.63a) can be restricted to the in-plane directions x and y, and the same holds for the indices i and j in equation (6.63b). Following our conventions, these indices will all be replaced with Greek letters below.

---

[40] Since stress has been averaged over the thickness, $g_i(x, y)$ is actually the average over z of the density of force by unit mass. To account for the weight of the plate, for instance, one should set $g_i = -g\, \delta_{iz}$, assuming that z denotes the vertical axis oriented upwards.

Using the fact that $\partial z/\partial x_i = 0$ when $i = \alpha$ is an in-plane index, equation (6.63b) simplifies to:

$$\frac{\partial}{\partial x_\beta}\left[\frac{\partial w}{\partial x_\alpha}\,\sigma_{\alpha\beta}\right] + \rho\,g_z = 0,$$

where $w(x,y) = u_z(x,y,z=0)$ denotes the deflection of the centre surface. When the derivative of the product in the left-hand side is expanded, one of the terms, namely $(\partial w/\partial x_\alpha)\,(\partial\sigma_{\alpha\beta}/\partial x_\beta)$, is zero by the in-plane equation (6.63a)—recall that its index i is in fact restricted to in-plane directions, $i = \alpha$. This yields the equations of equilibrium in their final form:

$$\frac{\partial\sigma_{\alpha\beta}}{\partial x_\beta} + \rho\,g_\alpha = 0 \tag{6.64a}$$

$$\frac{\partial^2 w}{\partial x_\alpha\,\partial x_\beta}\,\sigma_{\alpha\beta} + \rho\,g_z = 0. \tag{6.64b}$$

The non-linear term in the second equation represents the net out-of-plane force obtained by coupling curvature and in-plane stress, an effect that has already been interpreted in Figure 2.11.

These equations (6.64), together with (6.59) and (6.62), form a complete set that describes elastic plates in the membrane approximation. In the next section, we cast them to a simpler form by means of algebraic transformations.

6.4.5　Stress balance and the Airy potential

We shall now assume that the external distributed force is purely normal: its in-plane components g_α (with a Greek index α) vanish

$$g_\alpha(x,y) = 0.$$

This assumption makes it possible to rewrite the F.–von K. equations in condensed form, see equation (6.88). This involves introducing the Airy potential χ. If the external force is not purely normal ($g_\alpha \neq 0$), this rewriting is not possible, and one should retain the displacement as the primary unknown, as we do for instance in Section 7.3, equation (7.1).

Under the assumption $g_\alpha(x,y) = 0$, the equations for in-plane equilibrium (6.64a) can be solved generically[41] in terms of an auxiliary function $\chi(x,y)$, called the Airy potential:

$$\sigma_{xx} = \frac{\partial^2\chi}{\partial y^2}, \qquad \sigma_{xy} = -\frac{\partial^2\chi}{\partial x\,\partial y}, \qquad \sigma_{yy} = \frac{\partial^2\chi}{\partial x^2}. \tag{6.65}$$

Using the Airy potential, it is possible to eliminate the three in-plane components of the stress tensor, $\sigma_{\alpha\beta}$, subject to the two conditions (6.64a), in favour of a single scalar function, $\chi(x,y)$. A similar trick has been used to compute the twist energy of a rod, see equation (3.38).

One can furthermore eliminate the in-plane components of the displacement u_x and u_y as we did earlier in section 6.2.1: in this section, we used exactly the same definition for

---

[41] The equations (6.64a) are the equation of equilibrium for any elasticity problem in two dimensions. The Airy potential is therefore not only useful for plate problems, but for any 3D elasticity problem that is invariant along a particular direction.

the in-plane strain as we do now in equation (6.62). The result of this elimination is the geometric compatibility condition (6.12):

$$2\frac{\partial^2 \epsilon_{xy}}{\partial x \partial y} - \frac{\partial^2 \epsilon_{xx}}{\partial y^2} - \frac{\partial^2 \epsilon_{yy}}{\partial x^2} = \left\{ \frac{\partial^2 w}{\partial x^2} \frac{\partial^2 w}{\partial y^2} - \left(\frac{\partial^2 w}{\partial x \partial y} \right)^2 \right\},$$

where the quantity on the right-hand side is the Gauss curvature.

The left-hand side of this equation can be expressed in term of the in-plane components of the stress using the constitutive relations (6.59):

$$2\frac{\partial^2 \epsilon_{xy}}{\partial x \partial y} - \frac{\partial^2 \epsilon_{xx}}{\partial y^2} - \frac{\partial^2 \epsilon_{yy}}{\partial x^2} = -\frac{1}{E} \left(\frac{\partial^4 \chi}{\partial y^4} + \frac{\partial^4 \chi}{\partial x^4} + 2\frac{\partial^4 \chi}{\partial x^2 \partial y^2} \right)$$

$$= -\frac{\Delta^2 \chi}{E},$$

where we have introduced the two-dimensional harmonic operator as

$$\Delta = \frac{\partial^2}{\partial x^2} + \frac{\partial^2}{\partial y^2}.$$

The biharmonic operator Δ^2 is defined by composition: $\Delta^2 f = \Delta(\Delta f)$.

Combining the two equations above we find the equation satisfied by χ, which is a compatibility condition for the in-plane strain:

$$\Delta^2 \chi + E \left\{ \frac{\partial^2 w}{\partial x^2} \frac{\partial^2 w}{\partial y^2} - \left(\frac{\partial^2 w}{\partial x \partial y} \right)^2 \right\} = 0. \tag{6.66}$$

This equation is biharmonic, a characteristic feature of the theory of elasticity. The source term in equation (6.66), proportional to Young's modulus, is the Gauss curvature, introduced in equation (6.13).

The other equilibrium condition (6.64b), along z, can be rewritten in terms of the Airy potential:

$$-h \left(\frac{\partial^2 \chi}{\partial x^2} \frac{\partial^2 w}{\partial y^2} + \frac{\partial^2 \chi}{\partial y^2} \frac{\partial^2 w}{\partial x^2} - 2\frac{\partial^2 \chi}{\partial x \partial y} \frac{\partial^2 w}{\partial x \partial y} \right) = h\,\rho\,g_z, \tag{6.67}$$

where we have included an overall factor h to anticipate the generalization to the F.–von K. equations. Equations (6.66) and (6.67) are the equations for plates in the membrane approximation. The two unknown functions are the deflection $w(x,y)$ and the Airy potential for the in-plane stress, $\chi(x,y)$. The question of boundary conditions is discussed below. These equations were derived[42] by Föppl in 1907, see reference (L. Föppl, 1907).

6.4.6   A word of caution about wrinkles

The above equations for membranes must be used with care as their solutions are often not unique, and may also not be smooth.

---

[42] In the preface of the textbook *Zwang und Drang* by Föppl father and son (A. Föppl and L. Föppl, 1924), Oravas mentions that the same problem of large displacements of a thin elastic plate in which the extension of the plate is not neglected had also been studied by Bubnov (1872–1919).

Under large enough compressive stress, a thin elastic plate might buckle, a phenomenon that we shall study in detail in the forthcoming chapters. This buckling arises from a balance between compression—the stretching energy associated with compressive in-plane stress—and bending. It is typically described using the F.–von K. equations, derived at the end of the present chapter. Now, the membrane equations above can be viewed as the F.–von K. equations in the formal limit $D \to 0$, that is when the bending modulus goes to zero. In this limit, we shall see that the buckling thresholds go to zero, which means that any compressive stress, no matter how small, will lead to buckling with the membrane equations. As it does not take into account the bending energy, the membrane theory cannot correctly predict the buckled shape. Many shapes are possible, including some with undulations having an arbitrarily small wavelength. This phenomenon, called wrinkling, is typical in the presence of the large compressive loads in F.–von K. equations, as shown for instance in Chapter 10, Section 10.2. We refer to Le Dret and Raoult's work (H. Le Dret and A. Raoult, 1995) for a mathematically rigorous presentation of wrinkling in membranes, and Libai and Simmonds' book (A. Libai and J. G. Simmonds, 1998, §V.T.6) for a review on wrinkling in shells. Wrinkling can be illustrated by the following example.

Consider a square membrane with size $L \times L$, whose reference configuration is defined by $0 \le x \le L$ and $0 \le y \le L$. Assume that the edge $x = 0$ is clamped, that is $u_x(0, y) = 0$, $u_y(0, y) = 0$ and $w(0, y) = 0$. The opposite edge $x = L$ is also clamped, and moved in such a way that the membrane is compressed along the x direction: $u_x(L, y) = -\alpha L$, $u_y(L, y) = 0$ and $w(L, y) = 0$, where $\alpha > 0$ is the average rate of contraction imposed on the plate. To be consistent with the small displacement approximation, we assume $\alpha \ll 1$. One solution of the membrane equations with these boundary conditions is the simple contraction, defined by $u_x(x, y) = -\alpha x$, $u_y(x, y) = 0$, $w(x, y) = 0$. This solution corresponds to a uniform uniaxial strain $\epsilon_{xx} = -\alpha$. The resulting stress field can easily be obtained by using the effective 2D stress–strain relations for a plate; being uniform in the plate, it satisfies the equations of equilibrium. The membrane energy of this solution then reads $E\,h\,L^2\,\alpha^2$, up to a numerical factor of order one that depends on Poisson's ratio.

For the same problem, one can build solutions of lower energy by allowing out-of-plane displacements. Such solutions will be isometric to the reference configuration, and so have a vanishing membrane energy. For any function $f(s)$ with $0 \le s \le 1$ such that $f(0) = 0$, $f(1) = 0$ and with the normalization condition $\int_0^1 f'^2(s)\,\mathrm{d}s = 1$, we can build an embedding with cylindrical symmetry:

$$w(x, y) = \sqrt{2\,\alpha}\,L\,f\left(\frac{x}{L}\right), \qquad u_x(x, y) = -\alpha\,L \int_0^{\frac{x}{L}} f'^2(s)\,\mathrm{d}s, \qquad u_y(x, y) = 0.$$

The proposed profile w has cylindrical symmetry and is given by f, up to a rescaling. The in-plane displacement has been chosen such that $\epsilon_{xx} = \partial u_x/\partial x + (\partial w/\partial x)^2/2$ vanishes. The other strain components ϵ_{xy} and ϵ_{yy} also vanish by cylindrical symmetry, the displacement being a function of x only: this displacement defines an embedding, that is a deformation without in-plane strain, $\epsilon_{\alpha\beta} = 0$. As a result, the stretching energy is zero. It can easily be checked that both the imposed boundary conditions and the equilibrium equations are satisfied: this is a solution to our problem. Obviously, many choices of f are possible and the solution of the membrane equations is not unique. The possible lack of smoothness of the solutions is illustrated by noticing that, in this example, we can take $f(s)$ to be piecewise linear but not $\mathcal{C}^1$ (in physical terms, sharp creases have a vanishing energy in the membrane approximation): Take for instance f to a sawtooth function, that is $f' = \pm 1$, with alternating signs on small intervals $s \in [2i/(2n), 2i + 1/(2n)]$.

This simple example illustrates the potential difficulties associated with the membrane equations. To avoid these difficulties, one can use the F.–von K. equations instead, and take advantage of the regularization provided by the bending term that smoothes out the wrinkles. One can still use the membrane equations provided that the solution exhibits tensile stress everywhere. Membrane solutions displaying compressive stress are a sign of inconsistency as this stress can be removed, and the energy lowered, by making wrinkles in the transverse direction.[43]

We shall not discuss membrane equations further in the rest of this book: one can merely view them as a intermediate step in the derivation of the F.–von K. equations, which are the equations relevant to the particular plate problems that we shall study in this book.

6.5 Equilibrium: the Föppl–von Kármán equations

There are a number of practical situations where the bending energy plays a role and the membrane approximation is unsuited. When a piece of paper is bent into a cylinder for instance, its neutral plane deforms isometrically, and the membrane theory yields a vanishing elastic energy,[44] which makes it inappropriate: the equilibrium configuration in fact derives from the minimization of the *bending* energy between all possible *isometric* deformations. A similar although more subtle situation arises during the crumpling of a piece of paper: the neutral plane deforms isometrically everywhere, except near conical or ridge-like singularities. In these regions, stretching of the neutral plane remains very small in magnitude, but the corresponding stretching energy is of the *same* order of magnitude as the bending energy. As a result, the total elastic energy (bending plus stretching) has to be minimized. These examples, studied in the following chapters, illustrate the limitations of the membrane theory and motivate the derivation of the F.–von K. equations below: they account in a consistent way for both stretching and bending.

In the present section we relax the main assumption of the membrane theory, given in equation (6.54): we no longer assume that the bending energy is negligible in front of the stretching energy. We shall nevertheless remain in the framework of Hookean elasticity (small strain). This requires in particular that the two-dimensional strain along the centre surface remains small, and that the radii of curvature of the plate are much larger[45] than the thickness, h. Furthermore, we keep the assumption of small displacement (small slope), as expressed by equation (6.55). As argued above, this could be relaxed at the price of a greater computational burden.

---

[43] The net outcome of these wrinkles is to replace the constitutive relation with an effective one that is identical to the original one for tensile stress, but yields vanishing stress when the strain is negative (contraction), see reference (H. Le Dret and A. Raoult, 1995). This is connected with a standard method in non-convex analysis, called quasi-convexification.

[44] The Gauss curvature that appears in the second term of equation (6.66). The solution of the equations is then $\chi = 0$. The membrane stress is zero everywhere, hence a vanishing stretching energy.

[45] From equation (6.3), the strains induced by bending are of the same order as the ratio of the thickness to the typical radius of curvature of the plate. Even near singularities of a crumpled sheet of paper, the smallest radius of curvature of the centre surface may remain much larger than h because it is of order h divided by the small angular deflection of the plate near the singularity, see Chapter 9.

6.5.1 Short track

The aim of the following sections, 6.5.2 to 6.5.5, is to compute the net force (this force is normal to the plate and counted per unit area) due to bending. It is given in its final form in equation (6.82) as $(-D\,\Delta^2 w)$.

The most direct approach to derive this bending force is by energetic arguments; we shall do this much later in this chapter in Section 6.6.3; see equation (6.102) in particular. This is also the approach we shall use for shells, see Section 12.4.2. The energetic argument boils down to the following: the density of bending energy is essentially the square of the mean curvature, $(\Delta w)^2$; after two integrations by parts its variation can be written as the integral of $2(\Delta^2 w)\,\delta w$ plus boundary terms. In this variation of the energy, any coefficient in front of the infinitesimal deflection δw can be interpreted as force normal to the plate. Thus $\Delta^2 w$ is our bending force per unit area, up to a sign and a coefficient that can be worked out easily.

Even though the energetic derivation of the bending force is concise and mathematically elegant, it does not provide a clear mechanical picture of the origin of the bending force. For this reason, we present an alternative derivation below; it explicitly relates the bending force to the actual 3D stress in the plate.

Readers satisfied with a formal derivation of the bending force can skip directly to Section 6.5.6, relying on Section 6.6.3 for an energetic derivation of the bending force.

A fully detailed derivation of the equations for elastic plates is given in Appendix D, based upon 3D elasticity and on a formal expansion with respect to the small aspect ratio of the plate, following Ciarlet's work (Ph. G. Ciarlet, 1980). This derivation also explains where the Kirchhoff hypothesis comes from.

6.5.2 Dependence of strain on the transverse coordinate

The main assumption that we shall use to derive the bending energy is that the Kirchhoff hypothesis correctly predicts the in-plane stress when the thickness is sufficiently small. Now, recall the interpretation of the bending energy given in Fig. 6.1(b): all filaments perpendicular to the centre surface are parallel in the reference configuration, but they do not remain so when the centre surface is curved, because of the Kirchhoff kinematical hypothesis: they are convergent on one side of the curved centre surface, and divergent on the other side. Geometrically, the distances between points lying on two neighbouring filaments are affected by curvature of the centre surface: these points come closer on one side of the centre surface, and go farther apart on the other side. This leads to a non-zero elastic energy even though the strain is zero on average through the thickness. This is the signature of bending effects.

We can now turn this qualitative argument into an explicit calculation of the bending energy. In terms of the displacement, Kirchhoff's kinematical hypothesis,[46] $\epsilon_{\alpha z} = 0$, takes the form:

$$\frac{1}{2}\left(\frac{\partial u_x}{\partial z} + \frac{\partial u_z}{\partial x}\right) = 0.$$

---

[46] In Appendix D, we do not use Kirchhoff's kinematical hypothesis as a starting point but instead recover it on the basis of a perturbative approach. Therefore, the word 'hypothesis', used for historical reasons, can sometimes be inappropriate. In fact, the Kirchhoff hypothesis is not verified at higher orders in this expansion.

At leading order, we do not need to keep track of the dependence of u_z on z, and so we write $u_z(x, y, z) \approx u_z(x, y, 0) = w(x, y)$. The equation above yields $\partial u_x / \partial z = -w_{,x}(x, y)$ and, by integration with respect to z:

$$u_x(x, y, z) = u_x(x, y, 0) - z \, \frac{\partial w(x, y)}{\partial x}. \tag{6.68}$$

A similar equation for u_y can be derived. These two equations for u_x and u_y can be written in compact form:

$$u_\alpha(x, y, z) = u_\alpha(x, y, 0) - z \, w_{,\alpha}(x, y).$$

The in-plane strain is then computed by taking in-plane gradients and evaluating its symmetric part with respect to the indices α and β:

$$\epsilon_{\alpha\beta}(x, y, z) = \epsilon_{\alpha\beta}(x, y, z = 0) - z \, w_{,\alpha\beta}(x, y), \tag{6.69}$$

where we neglect higher-order terms, in a sense that will be explained later on. When deriving the equation above, we have implicitly assumed[47] that the non-linear terms in the definition of the strain do not need to be considered when one seeks the dependence of $\epsilon_{\alpha\beta}$ with respect to the coordinate z. We have just established in equation (6.69) that the curvature of the centre surface, given by the second derivatives $w_{,\alpha\beta}(x, y)$, introduces a variation of the strain across the thickness, as was announced at the beginning of this section. This is the fundamental kinematical relation for bending.

We shall now derive the in-plane stress $\sigma_{\alpha\beta}(x, y, z)$ associated with the strain $\epsilon_{\alpha\beta}(x, y, z)$. To do so, we use[48] the effective two-dimensional constitutive relations derived earlier in equation (6.59) to obtain:

$$\sigma_{\alpha\beta}(x, y, z) = \frac{E}{1 + \nu} \left(\epsilon_{\alpha\beta} + \frac{\nu}{1 - \nu} \, \epsilon_{\gamma\gamma} \, \delta_{\alpha\beta} \right)$$

$$= \sigma_{\alpha\beta}(z = 0) - z \left[\frac{E}{1 + \nu} \left(w_{,\alpha\beta} + \frac{\nu}{1 - \nu} \, w_{,\gamma\gamma} \, \delta_{\alpha\beta} \right) \right]. \tag{6.70}$$

Note that the term in square brackets depends on the curvature of the centre surface. The aim of the coming sections is to derive the net force, called the bending force, that arises out of the correction to the stress that appears in square bracket in equation above.

6.5.3   Bending is balanced by normal shear force

The in-plane balance of the 3D stress tensor written in equation (6.63a) is now used to calculate the stress component $\sigma_{\alpha z}$. When the terms in the divergence are suitably grouped, this equation yields an equation for $\sigma_{\alpha z}$ in terms of the in-plane stress $\sigma_{\alpha\beta}$:

$$\frac{\partial \sigma_{\alpha z}}{\partial z} = - \frac{\partial \sigma_{\alpha\beta}(x, y, z)}{\partial x_\beta}. \tag{6.71}$$

To make the exposition simpler we assumed that the applied force (if there is any) is normal, $g_\alpha = 0$. An arbitrary normal force $g_\alpha = 0$ can be restored in the equations of equilibrium at the end.

---

[47] This turns out to be true at the dominant order, as explained in Appendix D.

[48] Some care should be taken with these relations, which are based on the Kirchhoff hypothesis and may not be true at all orders. See Appendix D for a complete justification.

Plugging equation (6.70) into the right-hand side, we obtain

$$\frac{\partial \sigma_{\alpha z}}{\partial z} = -\frac{\partial \sigma_{\alpha \beta}(x, y, 0)}{\partial x_\beta} + z \left[\frac{E}{1 - \nu^2} \, w_{,\beta\beta\alpha} \right] \tag{6.72}$$

after suitable renaming of the dummy index γ.

The left-hand side is an exact derivative. Its integral with respect to z yields the variation $\sigma_{\alpha z}(x, y, h/2) - \sigma_{\alpha z}(x, y, -h/2)$, a quantity that cancels by the boundary conditions (6.56) on the upper and lower edges of the plates, $\sigma_{\alpha z}(x, y, z = \pm h/2) = 0$. This implies that the integral of the right-hand side cancels too. Using $\int_{-h/2}^{+h/2} z \, \mathrm{d}z = 0$, this yields

$$h \frac{\partial \sigma_{\alpha \beta}(x, y, 0)}{\partial x_\beta} = 0. \tag{6.73}$$

We have just rederived equation for in-plane equilibrium (6.64a), written here in the particular case when $g_\alpha = 0$ (no applied in-plane force): this equation has the same expression as in the membrane theory, except that the in-plane stress is replaced by its average value (which is also, at leading order with respect to the small parameter, its value on the centre surface $z = 0$).

We have just shown that the first term in right-hand side of equation (6.72) is zero. Solving this differential equation for $\sigma_{\alpha z}$ with respect to the variable z, one finds:

$$\sigma_{\alpha z} = \frac{E \, w_{,\beta\beta\alpha}}{2 \left(1 - \nu^2\right)} \left(z^2 - \left(\frac{h}{2}\right)^2 \right), \tag{6.74}$$

where the constant of integration has been determined from the boundary condition $\sigma_{\alpha z}(x, y, \pm h/2) = 0$.

Note that we have gone one step beyond Kirchhoff hypothesis here: the small but non-zero value of the stress $\sigma_{\alpha z}$ has been computed—this quantity is of order h^2 here, although it is zero in the membrane theory.

We shall now define the normal shear force q_α by

$$q_\alpha(x, y) = \int_{-h/2}^{+h/2} \sigma_{\alpha z} \, \mathrm{d}z. \tag{6.75}$$

By definition, q_x represents a normal shear force in the plane (xz) and q_y a normal shear force in the plane (yz). As we shall see shortly, this shear force, which arises out of bending effects, leads to a net transverse force in the equations of equilibrium.

Plugging equation (6.74) in the integral, we obtain an explicit expression for the shear force due to bending:

$$q_\alpha = -\frac{E \, h^3 \, w_{,\beta\beta\alpha}}{12 \left(1 - \nu^2\right)}. \tag{6.76}$$

6.5.4   Net bending force

Let us return to the equation of mechanical equilibrium. The equation for equilibrium along the transverse direction z become, once integrated over the thickness:

$$\int_{-h/2}^{+h/2} \frac{\partial}{\partial x_l} \left(\sigma_{lj} \frac{\partial(z + u_z)}{\partial x_j} \right) \mathrm{d}z + \int_{-h/2}^{+h/2} \rho \, g_z \, \mathrm{d}z = 0, \tag{6.77}$$

where we recall that the mass density ρ can be replaced by the known quantity ρ^0, the mass density in the reference configuration, since we assume the strain to be small. In the last term, g_z is the density of the applied force in the transverse direction, per unit mass of the body. The integral yields the applied force per unit area, which we call f_z; it is related to the applied force per unit mass by the obvious definition:

$$f_z(x,y) = \int_{-h/2}^{h/2} \rho\, g_z(x,y,z)\,\mathrm{d}z. \tag{6.78}$$

In the first integral of equation (6.77), the term corresponding to $l = z$ yields an exact derivative; as earlier, the integral of this term cancels by the boundary conditions on the upper and lower edges of the plate. Therefore, the index l can be restricted to in-plane directions and we set $l = \alpha$. In addition, one can approximate $\partial(z + u_z)/\partial z \approx 1$ when $j = z$ since $\partial u_z/\partial z$ is small under the assumption of small displacement—note that this approximation does not make sense for $j \neq z$, because the other term $\partial z/\partial x_\alpha$ is zero. This leads to

$$\int_{-h/2}^{h/2} \frac{\partial}{\partial x_\alpha}\left(\sigma_{\alpha\beta} \frac{\partial u_z}{\partial x_\beta}\right)\,\mathrm{d}z + \int_{-h/2}^{h/2} \frac{\partial \sigma_{\alpha z}}{\partial x_\alpha}\,\mathrm{d}z + \int_{-h/2}^{h/2} \rho\, g_z\,\mathrm{d}z = 0. \tag{6.79}$$

The first term was already present in the membrane theory: we have already interpreted the fact that the combination of in-plane stress $\sigma_{\alpha\beta}$ and curvature of the centre surface can give rise to a net out-of-plane force. This term is non-zero in the membrane theory and its small correction due to bending effects can be neglected: we can approximate it by its value at the centre surface $z = 0$:

$$\int_{-h/2}^{h/2} \frac{\partial}{\partial x_\alpha}\left(\sigma_{\alpha\beta} \frac{\partial u_z}{\partial x_\beta}\right)\,\mathrm{d}z \approx h\,\frac{\partial}{\partial x_\alpha}\left(\sigma_{\alpha\beta}(x,y,0)\,\frac{\partial w(x,y)}{\partial x_\beta}\right).$$

Expanding the outermost derivative and again using the condition for in-plane equilibrium (6.73) we obtain

$$\int_{-h/2}^{h/2} \frac{\partial}{\partial x_\alpha}\left(\sigma_{\alpha\beta} \frac{\partial u_z}{\partial x_\beta}\right)\,\mathrm{d}z = h\,\sigma_{\alpha\beta}\,\frac{\partial^2 w}{\partial x_\alpha\,\partial x_\beta},$$

where the transverse displacement at leading order $w(x,y)$ is the displacement of the centre surface $w(x,y) = u_z(x,y,0)$, a quantity that is independent of z by definition.

The second integral in equation (6.79) can be identified as the 2D divergence of the shear force q_α, and the last term as the density of applied normal force f_z per unit *area* of the plate. We can rewrite this equation as

$$h\,\sigma_{\alpha\beta}\,\frac{\partial^2 w}{\partial x_\alpha\,\partial x_\beta} + \frac{\partial q_\alpha}{\partial x_\alpha} + f_z = 0, \tag{6.80}$$

where all quantities are independent on z and can be seen as defined on the centre surface.

Comparison of equations (6.80) and (6.64b) reveals that the only change brought about by bending in the membrane theory is a new term, the second term in equation (6.80). This term represents a net normal force per unit area due to bending and is

$$f_{\mathrm{b}}(x,y) = \frac{\partial q_\alpha}{\partial x_\alpha}. \tag{6.81}$$

With the help of equation (6.75), this bending force can be calculated as

$$f_{\rm b} = -\frac{E\,h^3\,w_{,\alpha\alpha\beta\beta}}{12\,(1-\nu^2)} = -D\,\Delta^2 w, \tag{6.82}$$

where we have introduced the bending modulus of the plate

$$D = \frac{E\,h^3}{12\,(1-\nu^2)}. \tag{6.83}$$

This expression is compared with a similar one obtained for the bending of a rod of flat cross-section and the occurrence of the factor $(1-\nu^2)$ in the denominator is discussed in details in Section 6.7.1.

Although it appears with a very small prefactor,[49] that scales like the third power h^3 of the thickness, this bending force $f_z^{\rm b}$ is not always negligible in front of the stretching term—in the first term of equation (6.80), the in-plane stress gets multiplied by a second derivative of the deflection, a small quantity by assumption.

6.5.5   Comparison with the bending of rods

Our derivation of the bending force $f_{\rm b}$ for plates is in fact closely related to the Kirchhoff equation for rods: the bending force for a plate can be interpreted by looking at the bending of planar elastic curves (elastic curves are twistless rods).

Consider an elastic curve straight whose reference configuration is along the x axis, and current configuration deviates weakly from this axis. Call $w(x)$ this small deflection along the z axis. Because the deviation is small as well as its derivatives, the tangent to the curve is $\mathbf{d}_3 = \mathbf{e}_x + w'(x)\,\mathbf{e}_z$, and the constitutive relation of rods yields the internal moment as

$$\mathbf{M} = EI\,\mathbf{d}_3 \times \mathbf{d}_3' = -EI\,w''\,\mathbf{e}_y,$$

using the equations established in Chapter 3. Projecting the equation for equilibrium of moments $\mathbf{M}' + \mathbf{d}_3 \times \mathbf{F} = \mathbf{0}$ along the y axis, we find

$$F_z(x) = -EI\,w'''(x), \tag{6.84}$$

which shows that a normal shear force $F_z = \mathbf{F} \cdot \mathbf{e}_z$ is required to balance any variation in the internal moment. This F_z is to rods what the normal shear q_α is to plates, and the equation above is a variant of equation (6.76) for plates.

Finally, the equilibrium of forces of the rod, $\mathbf{F}' + \mathbf{p} = \mathbf{0}$, yields, when projected in the z direction

$$F_z' + p_z = 0$$

where $\mathbf{p}$ is the distributed (external) force and $p_z = \mathbf{p} \cdot \mathbf{e}_z$ is similar to the external loading f_z for plates. This shows that bending results in an equivalent normal force per unit length

$$f_{\rm b}^{\rm rod}(x) = F_z'(x)$$

in elastic rods. This equation is similar to equation (6.81) for plates.

---

[49] The bending force arises out of the normal shear stress $\sigma_{z\alpha}$, which is much smaller than in-plane stress.

Combining these equations, the bending force appears to be related to the fourth derivative of the deflection

$$f_{\mathrm{b}}^{\mathrm{rod}} = -EI\, w''''(x),\qquad(6.85)$$

which is similar to the bending term (6.82) for plates.

6.5.6   The Föppl–von Kármán equations

Let us now return to equation (6.80). Its first term describes stretching effects and is similar to that found in the membrane theory. Its second term, representing bending forces, is formally a second-order correction, which has just been worked out explicitly in equation (6.82). By collecting both terms, we obtain the first F.–von K. equation:

$$D\,\Delta^2 w - h\left(\frac{\partial^2\chi}{\partial x^2}\frac{\partial^2 w}{\partial y^2} + \frac{\partial^2\chi}{\partial y^2}\frac{\partial^2 w}{\partial x^2} - 2\,\frac{\partial^2\chi}{\partial x\partial y}\frac{\partial^2 w}{\partial x\,\partial y}\right) - f_z = 0.\qquad(6.86a)$$

As explained at the start of Section 6.4.5, this formulation of the plate equations based on the Airy potential assumes that the in-plane force $f_\alpha = h\,g_\alpha$ is zero. In the following, we shall assume in addition that the normal component f_z is zero:

$$f_z(x,y) = 0.$$

The inhomogeneous term f_z can easily be restored in the equations for transverse equilibrium whenever needed.

The second F.–von K. equation is just the compatibility condition for the in-plane strain introduced earlier in (6.66):

$$\Delta^2\chi + E\left\{\frac{\partial^2 w}{\partial x^2}\frac{\partial^2 w}{\partial y^2} - \left(\frac{\partial^2 w}{\partial x\,\partial y}\right)^2\right\} = 0.\qquad(6.86b)$$

Indeed, the equations of equilibrium in the in-plane directions are the same as in the membrane theory: equation (6.64a) has been rederived in equation (6.73). It is therefore again possible to solve them formally using the Airy potential, defined in equation (6.65). This Airy potential satisfies the same equation as before, which expresses in-plane equilibrium.

Let us use the differential operator $[.,.]$ introduced earlier in equation (1.13) as follows:[50]

$$[U,V] = \left(\frac{\partial^2 U}{\partial x^2}\frac{\partial^2 V}{\partial y^2} + \frac{\partial^2 V}{\partial x^2}\frac{\partial^2 U}{\partial y^2} - 2\frac{\partial^2 U}{\partial x\,\partial y}\frac{\partial^2 V}{\partial x\,\partial y}\right),\qquad(6.87)$$

where $U(x,y)$ and $V(x,y)$ are two arbitrary functions. One can then rewrite the Föppl–von Kármán equations in a more condensed form:

$$D\,\Delta^2 w - h\,[w,\chi] = 0\qquad(6.88a)$$

$$\Delta^2\chi + \frac{E}{2}\,[w,w] = 0.\qquad(6.88b)$$

---

[50] This operator is invariant under rotation in the (x,y) plane, being the derivative with respect to λ at $\lambda = 1$ of the determinant of the Hessian matrix of the function $(x,y) \mapsto (U + \lambda V)$. Recall that, by definition, the Hessian matrix is built up using the values of the second derivatives of a function as entries.

This set of equations describes an elastic plate under the above-stated assumptions. They are partial differential equations for two unknown functions, the deflection $w(x, y)$ and the Airy potential $\chi(x, y)$. Their coefficients are the thickness of the plate h, its bending modulus D defined in equation (6.83) and its Young's modulus E. These equations are non-linear: the non-linear term in the second equation is nothing but the Gauss curvature $[w, w]/2$. It reflects the underlying geometrical structure of the problem, as discussed in Section 6.8.3. The non-linear term of the first equation, i.e. $[w, \chi]$, couples the curvature of the centre surface with in-plane stress in the balance of vertical forces. These equations were derived by von Kármán in reference (T. von Kármán, 1910). He did this by combining Föppl's membrane theory with Kirchhoff's theory for the bending of plates—although he did not refer to Föppl in his published work.

The F.–von K. equations have been written in terms of the Airy potential χ above. Sometimes it is better to avoid introducing the Airy potential by keeping the in-plane displacement (u, v) as unknown. In this case, the following formulation of the equilibrium is appropriate:

$$\frac{\partial \sigma_{\alpha\beta}}{\partial x_\beta} = 0 \tag{6.89a}$$

$$D\,\Delta^2 w - h\,\frac{\partial^2 w}{\partial x_\alpha\,\partial x_\beta}\,\sigma_{\alpha\beta} = 0, \tag{6.89b}$$

where the first equation is for in-plane equilibrium, see equation (6.64a), and the second one is just a rewriting of equation (6.86a) for the equilibrium in the normal direction. In the presence of applied forces, the right-hand sides of the above equations should be changed to $(-\rho\,g_\alpha)$ and $(-h\,\rho\,g_z)$ respectively. In this formulation, the in-plane stress $\sigma_{\alpha\beta}$ must be seen as an auxiliary variable depending on the main unknown (u, v, w) through the definition (6.61) of in-plane strain $\epsilon_{\alpha\beta}$ and the constitutive relations (6.59).

6.6 Elastic energy

In the previous section and in Appendix D, we derive the F.–von K. equations by solving the equations of equilibrium order by order. In the current section, we want to emphasize that these equations can be derived by minimizing the relevant energy—these equations being derived through a consistent schema from the general equations of elasticity, it is not surprising that the variational structure carries over. This plate energy is derived below. It is made up of a stretching and a bending term. It has the same physical content as the F.–von K. equations, and so is conceptually equivalent to them. However, there are a number of situations when it is more convenient to use the energetic approach than the F.–von K. equations. This includes computing approximate numerical solutions of the F.–von K. equations, for instance, as illustrated in the following chapters.

6.6.1 Derivation of the plate energy

To derive the elastic energy of a plate, we begin with the *volumic* density of elastic energy $e_{\rm el}$ given in equation (2.66) for an isotropic material with linear elastic response:

$$e_{\rm el} = \frac{1}{2}\sigma_{ij}\,\epsilon_{ij} = \frac{1}{2}(\sigma_{\alpha\beta}\,\epsilon_{\alpha\beta} + \sigma_{\alpha z}\,\epsilon_{\alpha z} + \sigma_{zz}\,\epsilon_{zz}). \tag{6.90}$$

In the particular case of a plate, the normal shear stress $\sigma_{\alpha z}$ and the transverse stress σ_{zz} are dominated[51] by in-plane stress $\sigma_{\alpha\beta}$, and so we retain only the first term, with in-plane indices α and β in the equation above:

$$e_{\mathrm{el}} = \frac{\sigma_{\alpha\beta}\,\epsilon_{\alpha\beta}}{2} = \frac{\sigma_{xx}\,\epsilon_{xx} + \sigma_{yy}\,\epsilon_{yy} + 2\,\sigma_{xy}\,\epsilon_{xy}}{2}.$$

In this expression, we can use the effective 2D constitutive relations (6.59) to eliminate the stress in favour of the strain:

$$e_{\mathrm{el}} = \frac{E}{2\,(1-\nu^2)} \left((\epsilon_{xx} + \nu\,\epsilon_{yy})\,\epsilon_{xx} + (\epsilon_{yy} + \nu\,\epsilon_{xx})\,\epsilon_{yy} + 2\,(1-\nu)\epsilon_{xy}{}^2 \right)$$

$$= \frac{E}{2\,(1-\nu^2)} \left((\epsilon_{xx} + \epsilon_{yy})^2 - 2(1-\nu)\,(\epsilon_{xx}\,\epsilon_{yy} - \epsilon_{xy}{}^2) \right). \tag{6.91a}$$

By using the inverse constitutive relations instead, one could alternatively eliminate the strain in favour of the stress, which leads to:

$$e_{\mathrm{el}} = \frac{1}{2\,E} \left((\sigma_{xx} + \sigma_{yy})^2 - 2(1+\nu)\,(\sigma_{xx}\,\sigma_{yy} - \sigma_{xy}{}^2) \right). \tag{6.91b}$$

The next step will be to integrate this last form of the energy density with respect to the transverse direction z. We shall take advantage of the very specific dependence of this energy on z, and carry out the integration in a formal way first: given a quadratic form Q that acts on a vector $\underline{\epsilon}$, that is $Q : \underline{\epsilon} \mapsto (\underline{\epsilon} \cdot \underline{\underline{Q}} \cdot \underline{\epsilon})$, where $\underline{\underline{Q}}$ is a symmetric matrix, we shall compute the integral of this quadratic form

$$\int_{-h/2}^{h/2} \mathrm{d}z\, \underline{\epsilon}(z) \cdot \underline{\underline{Q}} \cdot \underline{\epsilon}(z)$$

over a domain $z \in [-h/2, h/2]$ when $\underline{\epsilon}(z)$ is an affine function of z:

$$\underline{\epsilon}(z) = \underline{\epsilon}_0 + z\,\underline{\epsilon}_1.$$

By expanding $\underline{\epsilon}(z)$ in the integral and noticing that the cross-terms $\underline{\epsilon}_0 \cdot \underline{\underline{Q}} \cdot \underline{\epsilon}_1$ that are linear with respect to z cancel upon integration by symmetry of the domain, we obtain:

$$\int_{-h/2}^{h/2} \mathrm{d}z\, \underline{\epsilon}(z) \cdot \underline{\underline{Q}} \cdot \underline{\epsilon}(z) = \int_{-h/2}^{h/2} \mathrm{d}z\, (\underline{\epsilon}_0 + z\,\underline{\epsilon}_1) \cdot \underline{\underline{Q}} \cdot (\underline{\epsilon}_0 + z\,\underline{\epsilon}_1)$$

$$= \underline{\epsilon}_0 \cdot \underline{\underline{Q}} \cdot \underline{\epsilon}_0 \int_{-h/2}^{h/2} \mathrm{d}z + \underline{\epsilon}_1 \cdot \underline{\underline{Q}} \cdot \underline{\epsilon}_1 \int_{-h/2}^{h/2} z^2 \mathrm{d}z$$

$$= h\,\underline{\epsilon}_0 \cdot \underline{\underline{Q}} \cdot \underline{\epsilon}_0 + \frac{h^3}{12}\,\underline{\epsilon}_1 \cdot \underline{\underline{Q}} \cdot \underline{\epsilon}_1. \tag{6.92}$$

---

[51] This argument can be made perfectly accurate based on the expansion given in Appendix D. Indeed, $\epsilon_{\alpha\beta}$ and $\sigma_{\alpha\beta}$ are found there to be of order ϵ^2, and so the first term in the right-hand side of equation (6.90) is of order ϵ^4. This is to be compared with $\epsilon_{\alpha z}$ and $\sigma_{\alpha z} = \mathcal{O}(\epsilon^3)$ for the second term, which is therefore of order ϵ^6, and with $\epsilon_{zz} = \mathcal{O}(\epsilon^3)$ (we do not know if the effective 2D constitutive relations are valid beyond order three) and $\sigma_{zz} = \mathcal{O}(\epsilon^4)$ for the last term, which is therefore of order ϵ^7 or less.

These two terms depend on the average value of $\underline{\epsilon}$ on the one hand, and on its linear part with z on the other. There is no cross-term involving the product of $\underline{\epsilon}_0$ and $\underline{\epsilon}_1$— this explains why the stretching and bending energies of a plate will ultimately appear uncoupled.

We can apply this calculation to compute the integral of the elastic energy in equation (6.91a) with respect to z. For this purpose, we set $\underline{\epsilon}(z)$ to be the vector[52] whose entries are the in-plane strain components, $\underline{\epsilon} = \{\epsilon_{xx}, \epsilon_{yy}, \epsilon_{xy}\}$. By equation (6.69), this vector has indeed an affine dependence on z. The constant term $\underline{\epsilon}_0 = \{\epsilon_{xx}, \epsilon_{yy}, \epsilon_{xy}\}_{z=0}$ contains the strain along the centre surface $z = 0$, while the linear one $\underline{\epsilon}_1 = \{-w_{,xx}, -w_{,yy}, -w_{,xy}\}$ has the opposites of the curvatures for components—recall that we write $w(x, y) = u_z(x, y, 0)$ for the deflection of the centre surface. One should further identify $\underline{\underline{Q}}$ to be the quadratic form that maps the in-plane strain $\underline{\epsilon}$ to the energy density given in the right-hand side of equation (6.91b). Applying formula (6.92), we obtain two terms:

$$\int_{-h/2}^{h/2} e_{\mathrm{el}} \, \mathrm{d}z = \frac{E\,h}{2\,(1-\nu^2)} \left[(\epsilon_{xx} + \epsilon_{yy})^2 - 2\,(1-\nu)\,(\epsilon_{xx}\,\epsilon_{yy} - \epsilon_{xy}{}^2)\right]_{z=0} \cdots$$

$$+ \frac{E\,h^3}{24\,(1-\nu^2)} \left[\left(\frac{\partial^2 w}{\partial x^2} + \frac{\partial^2 w}{\partial y^2}\right)^2 - 2(1-\nu)\left(\frac{\partial^2 w}{\partial x^2}\frac{\partial^2 w}{\partial y^2} - \left(\frac{\partial^2 w}{\partial x\,\partial y}\right)^2\right)\right]. \qquad (6.93)$$

In this equation, all quantities in the right-hand side have to be evaluated along the centre surface. Since any explicit dependence on the variable z has now be removed, we shall implicitly evaluate all quantities at the centre surface, $z = 0$ in the following of this section.

Upon integration with respect to the in-plane variables x and y, we obtain the elastic energy $\mathcal{E}_{\mathrm{el}}$ of a plate as the sum of two terms, which we call the stretching and bending energies:

$$\mathcal{E}_{\mathrm{el}} = \mathcal{E}_{\mathrm{s}} + \mathcal{E}_{\mathrm{b}}. \qquad (6.94)$$

By identification with equation (6.93), these two energies are

$$\mathcal{E}_{\mathrm{s}} = \frac{E\,h}{2\,(1-\nu^2)} \iint \mathrm{d}x\,\mathrm{d}y \left[(\epsilon_{xx} + \epsilon_{yy})^2 - 2(1-\nu)\,(\epsilon_{xx}\,\epsilon_{yy} - \epsilon_{xy}{}^2)\right] \qquad (6.95a)$$

and

$$\mathcal{E}_{\mathrm{b}} = \frac{D}{2} \iint \mathrm{d}x\,\mathrm{d}y \left((\Delta w)^2 - 2\,(1-\nu')\,\frac{[w, w]}{2}\right). \qquad (6.95b)$$

In the equation above, we have used the definition (6.83) of the bending modulus D and have expressed the Gauss curvature $K = [w, w]/2$ in terms of the bracket operator to make the writing more concise. Recall that all quantities are implicitly evaluated along the centre surface, $z = 0$.

By minimizing the sum of the stretching and bending energies given above in equations (6.95a) and (6.95b) with respect to the unknown displacement $u_x(x, y, 0)$, $u_y(x, y, 0)$ and $w(x, y)$ of the centre surface, one can derive the F.–von K. equations. Of course, this

---

[52] What we call $\underline{\epsilon}$ is not a vector strictly speaking, as it collects the components of the physical strain tensor $\underline{\underline{\epsilon}}$, which is of rank two. In a particular frame, this $\underline{\epsilon}$ is just the argument of the quadratic form Q.

involves using the definition (6.61) to compute the strain in terms of the displacement first.

6.6.2   Plate energy as a function of Airy potential

There are two possible approaches for deriving the F.–von K. by variational principles. The first one involves minimizing the plate energy expressed with respect to all components of the displacement (both the in-plane and transverse components); the second approach involves cancelling the variation of the plate energy with respect to the Airy potential $\chi(x, y)$ and the transverse displacement $w(x, y)$. In the following, we shall demonstrate only the second approach. This involves rewriting the plate energy in terms of the Airy potential first, which is the aim of the present section; the actual variation is carried out in the next section.

Note that the variational principle relevant to the χ-w formulation is a condition of stationary energy (existence of a saddle point) and not a condition of energy minimum. The other variational principle based on in-plane and transverse displacement, however, is a genuine principle of minimum energy, as expected.

To rewrite $\mathcal{E}_\mathrm{s}$ in terms of χ, we shall first recall that the density of elastic energy has been expressed as a function of the strain in equation (6.91a) and as a function of the stress in equation (6.91b). The integrand for the stretching energy is just the same quantity, the density of elastic energy, evaluated along the centre surface, and so equation (6.91b) provides an expression for $\mathcal{E}_\mathrm{s}$ in terms of the stress

$$\mathcal{E}_\mathrm{s} = \frac{h}{2E} \iint \mathrm{d}x\,\mathrm{d}y \left[(\sigma_{xx} + \sigma_{yy})^2 - 2(1 + \nu)\left(\sigma_{xx}\,\sigma_{yy} - \sigma_{xy}{}^2\right) \right]. \tag{6.96}$$

We can then use the definition (6.65) of the Airy potential to rewrite this as a function of χ:

$$\mathcal{E}_\mathrm{s} = \frac{h}{2E} \iint \mathrm{d}x\,\mathrm{d}y \left((\Delta\chi)^2 - 2(1 + \nu)\frac{[\chi,\chi]}{2} \right). \tag{6.97a}$$

For reference, we recall the definition of the bending energy:

$$\mathcal{E}_\mathrm{b} = \frac{D}{2} \iint \mathrm{d}x\,\mathrm{d}y \left((\Delta w)^2 - 2(1 - \nu)\frac{[w,w]}{2} \right). \tag{6.97b}$$

6.6.3   Variational structure of the F.–von K. equations

In the rest of this section, we shall show that one can recover the F.–von K. equations by cancelling the variation in the sum of the stretching and bending energies given in equation (6.97), plus any potential energy representing the applied forces. It is important to note that the unknowns $w(x, y)$ and $\chi(x, y)$ are not independent (G.W. Hunt, G. J. Lord, and A.R. Champneys, 1999). They are related by the compatibility condition (6.66), which comes from the elimination of in-plane displacement in the definition of strain. We shall recall this compatibility condition here:

$$\Delta^2\chi + \frac{E}{2}\,[w,w] = 0. \tag{6.98}$$

Since the unknowns χ and w are subject to a constraint, the condition of stationarity of the energy will involve Lagrange multipliers, as explained in Section 1.3.5. The constraint (6.98) is not a single constraint: it has to be satisfied for all x and y, and so there is a two-parameter family of associated Lagrange multipliers $\lambda(x, y)$. From equation (1.31), the condition for the stationarity of the energy reads, as in constrained minimization problems:

$$\delta\mathcal{E}_{\mathrm{s}} + \delta\mathcal{E}_{\mathrm{b}} + \iint \mathrm{d}x\,\mathrm{d}y\,\lambda(x, y)\,\delta\left(\Delta^2\chi + \frac{E}{2}\,[w, w]\right) = 0. \tag{6.99}$$

In this expression, δJ stands for the linear variation of the function J caused by the infinitesimal changes $\delta\chi(x, y)$ and $\delta w(x, y)$ of the unknown functions. For instance, with $J = \iint \mathrm{d}x\,\mathrm{d}y\,w^2(x, y)$, the variation would read

$$\delta J = 2\iint \mathrm{d}x\,\mathrm{d}y\,w(x, y)\,\delta w(x, y)$$

as this is how J is changed with the replacement $w \to w + \delta w$. The notations δ and Δ should not be confused, as δ means the variation for a small change δw and $\delta\chi$ of the unknowns, while Δ is the harmonic operator.

To work out the variations in equation (6.99) above, we shall need a few identities involving the bracket operator. The first identity reads, for $f(x, y)$, $g(x, y)$ and $k(x, y)$ arbitrary smooth functions:

$$[f, g]\,k = f\,[g, k] + \frac{\mathrm{d}}{\mathrm{d}x}\left((f_{,x}\,g_{,yy} - f_{,y}\,g_{,xy})\,k - f\,g_{,yy}\,k_{,x} + f\,g_{,xy}\,k_{,y}\right)$$
$$+ \frac{\mathrm{d}}{\mathrm{d}y}\left((f_{,y}\,g_{,xx} - f_{,x}\,g_{,xy})\,k - f\,g_{,xx}\,k_{,y} + f\,g_{,xy}\,k_{,x}\right). \tag{6.100}$$

This identity can be verified by a direct calculation. It implies that expressions like $(\iint [f, g]\,k\,\mathrm{d}x\,\mathrm{d}y)$ are invariant under permutations of the functions $f(x, y)$, $g(x, y)$ and $k(x, y)$, up to boundary terms, something that we write as

$$\iint \mathrm{d}x\,\mathrm{d}y\,k\,[f, g] = \iint \mathrm{d}x\,\mathrm{d}y\,f\,[g, k] + \mathrm{BT}, \tag{6.101}$$

where the term BT stands for some boundary terms coming from integration by parts, which are not given here.

We shall now proceed to the calculation of the variation in equation (6.99). Let us start with the last term in the stretching energy (6.97a):

$$\delta\left(\iint [\chi, \chi]\right) = 2\iint [\chi, \delta\chi],$$

by bilinearity and symmetry of the bracket operator. Taking k to be the constant function $k(x, y) = 1$, $f = \chi$ and $g = \delta\chi$, we find using equation (6.101) that the variation in the equation above reduces to boundary terms,[53] since $[1, w] = 0$:

$$\delta\left(\iint [\chi, \chi]\right) = \iint 0\,\mathrm{d}x\,\mathrm{d}y + \mathrm{BT}.$$

---

[53] By a similar calculation, the integral of the Gauss curvature $K = [w, w]/2$ can be reduced to boundary terms. This is related to the Gauss–Bonnet theorem, which states that this integral depends only on the topology of the manifold, and on the configuration of the boundary.

By a similar calculation, the variation of the last term $\iint [w, w]$ in equation (6.95b) can be reduced to boundary terms. For the last term in equation (6.99), we compute

$$\frac{E}{2} \iint \lambda \, \delta \left([w, w]\right) = E \iint \lambda \, [w, \delta w] = E \iint [w, \lambda] \, \delta w + \mathrm{BT}.$$

Collecting all these results, we rewrite the condition of stationarity of the energy in equation (6.100) as:

$$\iint \left(\frac{h}{2E} \delta \left[(\Delta \chi)^2\right] + \frac{D}{2} \delta \left[(\Delta w)^2\right] + E \, [w, \lambda] \, \delta w + \lambda \, \delta \left[\Delta^2 \chi\right] \right) + \mathrm{BT} = 0.$$

In the expression above, the first term comes from the stretching energy, the second term from the bending energy, and the last two terms from the constraint.

It remains to expand the first order variations of the squared quantities, and to carry out integration by parts so as to extract the factors $\delta w(x, y)$ and $\delta \chi(x, y)$, as explained in Appendix A. This yields

$$\iint \left(\left\{ \frac{h}{E} \Delta^2 \chi + \Delta^2 \lambda \right\} \delta \chi + \left\{ D \, \Delta^2 w + E \, [w, \lambda] \right\} \delta w \right) + \mathrm{BT}. \tag{6.102}$$

According to the general principles of constrained minimization, the two factors in braces must vanish at the equilibrium, and the Lagrange multiplier $\lambda(x, y)$ must be such that the functions $\chi(x, y)$ and $w(x, y)$ satisfy the constraint (6.98). Note that we make no attempt to compute the boundary terms, which are related to the loads applied on the lateral edge of the plate. The corresponding boundary conditions will be derived in the following chapters, whenever needed.

By canceling the factor in front of $\delta \chi$ in equation (6.102), we find that

$$\lambda(x, y) = -\frac{h}{E} \chi(x, y) + \mu(x, y),$$

where $\mu(x, y)$ is a solution of $\Delta^2 \mu = 0$. This indeterminacy in the function λ is related to the fact that the Airy potential $\chi(x, y)$ is also the solution of a partial differential equation $\Delta^2 \chi + h/2 \, [w, w] = 0$, and so is determined up to a solution of the homogeneous equation $\Delta^2 \chi = 0$ as well. This homogeneous solution must be determined by the boundary conditions. Therefore, we can forget about the term $\mu(x, y)$ in the equation above, and write simply:

$$\lambda(x, y) = -\frac{h}{E} \chi(x, y). \tag{6.103}$$

By putting this expression back into the second factor in braces in equation (6.102), we arrive at the following condition for the existence of stationary point:

$$D \, \Delta^2 w - h \, [w, \chi] = 0.$$

We have just recovered the first F.–von K. equation (6.88a) as the condition of stationarity of the energy, for the plate energy defined in equations (6.94) and (6.97). The other F.–von K. equation is the kinematical constraint (6.98).

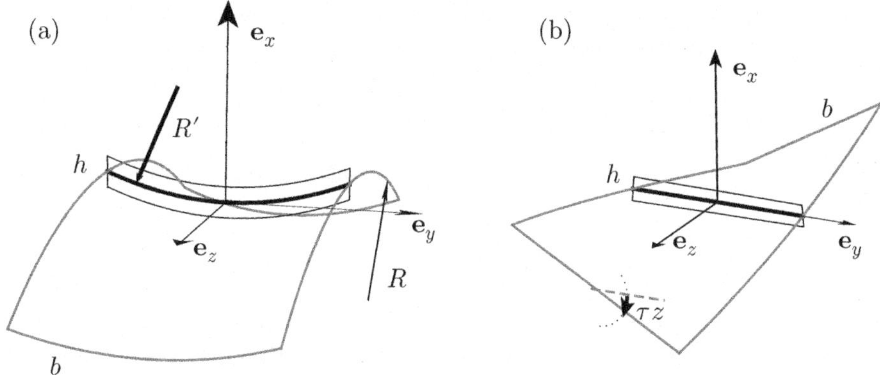

Fig. 6.10 Bending and twisting of a narrow band, with thickness $h \ll b$.

6.7 Narrow plates: consistency with the theory of rods

In this section, we consider the case of a narrow plate whose centre surface has a width b that is much smaller than its length. Let R be the typical length along the long dimension of the plate, $R \gg b$. We assume that b is still much larger than the thickness h, so that the plate theory applies. This object can be viewed both as a narrow plate and as a flat rod: its rectangular cross-section, of dimensions $h \times b$, is much smaller than the typical axial length, R, and so the main assumption underlying the theory of thin rods is satisfied.

In this section, we show that the plate and the rod description are indeed equivalent for such narrow bands: the plate equations in the limit of a narrow domain $(b \ll R)$, and the rod equations in the limit of a flat cross-section $(b \gg h)$ both yield the same result, provided b is not too large. The two descriptions agree for b much larger than h but also much smaller than $b^* \sim (a\,R)^{1/2}$, as we show below. When b becomes of order b^* or larger, the rod theory does not apply.

We first consider the flexion of a narrow plate, in Section 6.7.1, and then the twist of a narrow plate, in Section 6.7.2.

6.7.1 Flexion of a narrow plate

For the analysis of flexion, we use the geometry in Fig. 6.10: the origin of the Euclidean coordinates is taken to be the intersection of the center line with a particular cross-section, taken as the origin for the curvilinear coordinate, $s = 0$. The x axis is perpendicular to the centre surface there, the y axis is along the width of the plate, the z axis being along the centre line. According to the rod theory, the deflection of the cross section $s = 0$ is given by equation (3.17a):

$$X(x, y) = -\frac{\nu}{2R}(x^2 - y^2),$$

where R is the radius of curvature imposed to the centre line. The general formula has been simplified using the fact that the centre line passes through the origin of the coordinates: then, $x_0 = 0$, $y_0 = 0$, and so $\langle x \rangle = 0$ and $\langle y \rangle = 0$ by equation (3.20), and $\eta = 0$

by equation (3.19). The deflection of the curve defined as the intersection of the centre surface ($x = 0$) with the cross-section $s = 0$ reads:

$$X(0, y) = \frac{\nu\, y^2}{2\,R}. \tag{6.104}$$

Equation (3.9) then yields the deflection of an arbitrary point lying along the centre surface as

$$w(y, s) = \mathbf{r}(0, y, s) \cdot \mathbf{e}_x = R\left(\cos\frac{s}{R} - 1\right) + X(0, y)\cos\frac{s}{R} \approx -\frac{s^2}{2\,R} + \frac{\nu\, y^2}{2\,R},$$

where $|y| \le b/2$. This approximation made here is accurate in the vicinity of the origin, for $|s| \ll R$. This warrants that the slopes remain small, a condition for the F.–von K. equations to be valid. Over this region, one can further approximate s with z:

$$s \approx z.$$

The calculation of the bending energy turns out to be simpler near the origin $|s| \ll R$ as there is no need to take into account geometric nonlinearity there. Since the band is bent uniformly, the density of elastic energy per unit length is uniform along the whole band.

Now viewing the strip as a plate, we shall compute the bending energy associated with this deflection of the centre line. To do so, we first need to compute the second derivatives[54] of the deflection:

$$w_{,yy} = \frac{\nu}{R}, \quad w_{,ss} = -\frac{1}{R}, \quad w_{,ys} = 0.$$

The centre surface appears to be curved with a curvature along the transverse direction s that is comparable in magnitude to the curvature $1/R$ of the centre line (unless $\nu = 0$, which is not a generic case anyway). For materials with a positive Poisson's ratio, $\nu > 0$, these two curvatures appear to be pointing in opposite directions and so the centre surface has negative Gauss curvature

$$K = \frac{[w, w]}{2} = -\frac{\nu}{R^2}, \tag{6.105}$$

something that is sometimes called *anticlastic* curvature in the field of rod theory. From equation (6.95b), the bending energy per unit length of this plate reads

$$\frac{\mathcal{E}_b}{L} = \frac{D}{2}\int_{-b/2}^{b/2} \mathrm{d}y\left(\left(\frac{1}{R} - \frac{\nu}{R}\right)^2 - 2\,(1 - \nu)\,K\right) = \frac{b\,D\,[(1 - \nu)^2 - 2\,(1 - \nu)\,(-\nu)]}{2\,R^2}.$$

Using the definition of the bending modulus for a plate

$$D = \frac{E\,h^3}{12\,(1 - \nu^2)},$$

we find, after simplifications

$$\frac{\mathcal{E}_b}{L} = \frac{1}{2}\frac{E\,h^3\,b}{12}\frac{1}{R^2}.$$

---

[54] Note that the in-plane directions are y and $s \sim z$ here, while they used to be x and y at the beginning of this chapter. The variables in the previous expressions are changed accordingly.

This is identical to the bending energy from the rod theory, given in equation (3.25) as

$$\frac{\mathcal{E}_{\mathrm{rod}}}{L} = \frac{E\,I}{2}\,\frac{1}{R^2},$$

the moment of inertia being $I = h^3\,b/12$ by equation (3.27).

We have just shown that the bending energy of a rod with a flat cross-section is equal to the bending energy of the plate having the same centre surface. Therefore, the rod and the plate descriptions are consistent for a narrow plate, provided the stretching energy in the plate remains negligible. To estimate this stretching energy, we first have to estimate the in-plane strain. The latter cannot vanish everywhere (when $\nu \neq 0$) as the centre surface has a non-zero Gauss curvature K and is therefore non-developable. The F.–von K. equation expressing the compatibility of in-plane displacements, in equation (6.88b), yields

$$\frac{\sigma_{\alpha\beta}}{b^2} \sim \frac{E}{R^2}, \tag{6.106}$$

since $\sigma_{\alpha\beta} \sim \chi/b^2$, the dominant contribution to partial in-plane derivatives coming from $\partial/\partial_y \sim 1/b$. Note that we do not keep track of Poisson's ratio ν, of order 1, in this scaling argument and we assume $\nu \neq 0$. Using Hooke's law, $\sigma_{\alpha\beta} \sim E\,\epsilon_{\alpha\beta}$, we get an estimate for the in-plane strain:

$$\epsilon_{\alpha\beta} \sim \left(\frac{b}{R}\right)^2.$$

Therefore, the stretching energy per unit length of the band is of order

$$\frac{\mathcal{E}_{\mathrm{s}}}{L} \sim E\,h\,b\left(\frac{b}{R}\right)^4.$$

Comparison with the bending energy, $\mathcal{E}_{\mathrm{b}}/L \sim E\,h^3\,b/R^2$, reveals that the stretching energy is indeed negligible as long as

$$b \ll b^*, \quad \text{with } b^* \sim (h\,R)^{1/2}.$$

This means that Kirchhoff's theory is applicable to rods that have a flat cross-section, $b \gg h$ (but only as long as $b \ll b^*$).

This raises the question of what happens for wider bands, such that $b \sim b^*$. Then, the plate's stretching and the bending energies are comparable. H. Lamb solved this problem as long ago as 1890: he derived an analytic solution for the profile of the centre line $X(0, y)$, for arbitrary values of the rescaled width b/b^*, see Fig. 6.11. In the limit $b/b^* \to 0$, Kirchhoff's theory for a rod with a flat cross-section is recovered from his solution. In the opposite limit of a very wide band $b/b^* \to \infty$, but still with $b \ll R$, he finds that the cross-section becomes almost flat, and the centre surface goes to a cylinder, which is a developable surface. This is not surprising, as any stretching energy becomes very penalizing in this limit. We refer to Horace Lamb's original reference (H. Lamb, 1891), to Love's presentation (A. E. H Love, 1927, §335E) and to another reference using modern notations (R. T. Shield, 1992) for the details of the calculation. The outcome of this theory is that a band has an effective bending modulus in the form

$$\frac{E\,h^3\,b}{12}\left(1 + \frac{\nu^2}{1-\nu^2}\,Q\left(\frac{b}{b^*}\right)\right), \tag{6.107}$$

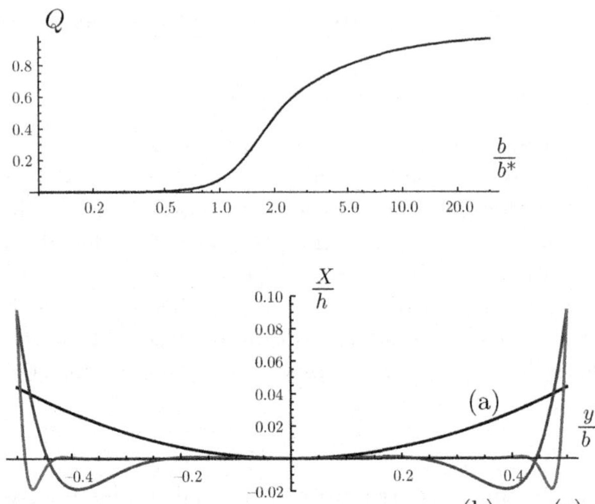

Fig. 6.11 H. Lamb's analysis of the flexion of a band with intermediate width, $b \sim b^*$. *Top*: universal function Q giving the effective bending modulus of the band as a function of rescaled width, b/b^*, see equation (6.107). *Bottom*: rescaled profiles of the cross-sections for various rescaled width with $\nu = .3$: (a) $b/b^* = 1$, (b) $b/b^* = 10$, (c) $b/b^* = 40$. This profile is circular with curvature ν/R for small b/b^*, as implied by equation (6.104), and flat for large b/b^* (except in a small boundary layer along the long edges): the centre surface is then close to a cylinder, which is a developable surface.

where Q is a universal[55] function of the rescaled width b/b^*, and b^* is defined as

$$b^* = \frac{\sqrt{2}\,(h\,R)^{1/2}}{(3\,(1-\nu^2))^{1/4}}.$$

When its argument goes to zero ($b \ll b^*$), Q tends to 0, and Kirchhoff's rod theory is recovered: the bending modulus in equation (6.107) reads $Eh^3b/12$. In the other limit, that is when its argument goes to infinity ($b \gg b^*$), Q tends to 1 and one recovers the bending modulus of a plate, namely

$$\frac{E\,h^3\,b}{12\,(1-\nu^2)},$$

with the usual factor $(1-\nu^2)$ in the denominator.

6.7.2   Twist of a narrow plate

According to the theory of thin elastic rods, the twist stiffness of a rod with a very flat cross-section $h \times b$ with $h \ll b$ can be determined by solving a harmonic equation (3.39a) for a potential $\bar{\chi}$, with the boundary condition (3.39b). The following solution

---

[55] This function depends however on the shape of the cross-section: rectangular, elliptic. . .

$$\bar{\chi}(y, z) = -\frac{x^2}{2} + \frac{h^2}{8}$$

satisfies both the equation in the interior of the domain, for $|x| < h/2$ and $|y| < b/2$, and the boundary condition along the long edges, $x = \pm h/2$, $|y| < b/2$. However, it does not satisfy the boundary condition along the short edges $|x| < h/2$, $y = \pm b/2$: it is not the exact solution but is nevertheless close to it when $b \gg h$, except in two small neighbourhoods of the short edges. For a large aspect ratio $b \gg h$, these end effects are negligible and the twist stiffness is close to

$$\mu\,J = \frac{E}{2\,(1+\nu)}\, 4 \iint \left(\bar{\chi}_{,x}^2 + \bar{\chi}_{,y}^2 \right) \mathrm{d}x\,\mathrm{d}y$$

$$= \frac{2\,E}{(1+\nu)}\, b \int_{-\frac{h}{2}}^{\frac{h}{2}} x^2\,\mathrm{d}x = \frac{2\,E}{(1+\nu)}\,\frac{b\,h^3}{12}, \tag{6.108}$$

where we have used the expression (3.41) of J as a function of the potential $\bar{\chi}$.

Using the geometrical conventions in Fig. 6.10(b), which are the same as for the analysis of bending, we compute the deflection of the centre surface of a flat, twisted rod with respect to its tangent plane (Oyz) using the general parameterisation (3.34):

$$w(y, z) = \mathbf{r}(x = 0, y, z) \cdot \mathbf{e}_x = -y\,\sin(\tau\,z) \approx -\tau\,y\,z,$$

where we have used $x_0 = 0$ and $y_0 = 0$ since the origin of the coordinates lies on the centre line. Note that we are interested in the deflection of the centre surface only, whose equation is $x = 0$. The linearization of the sine function is accurate in a region of length that is small compared with $1/\tau$, to which we restrict the analysis. The curvature of this centre surface is given by

$$w_{,yy} = 0, \quad w_{,zz} = 0, \quad w_{,yz} = -\tau,$$

showing that the twisted band has negative Gauss curvature:

$$K = -\tau^2. \tag{6.109}$$

Plugging the above expressions in the definition (6.95b) of the bending energy, we obtain, as earlier:

$$\frac{\mathcal{E}_\mathrm{b}}{L} = \frac{D}{2} \int_{-b/2}^{b/2} -2\,(1-\nu)\,(-\tau^2)\,\mathrm{d}y = \frac{1}{2}\,\frac{2\,E\,b\,h^3}{12\,(1+\nu)}\,\tau^2.$$

This bending energy of plate theory is equal to the twist energy $\frac{1}{2}\mu\,J\,\tau^2$ of rod theory when one uses for the twist modulus $\mu\,J$ the value given above in equation (6.108).

As with bending, the plate and the rod theories are consistent with each other provided the stretching energy in the plate is negligible. In order to determine the range of widths over which the two theories are consistent, one has to compare the stretching energy to the bending energy. The estimate of the stretching energy of a twisted band is very similar to that for a bent band. By equations (6.105) and (6.109), the Gauss curvature is $-\nu/R^2$ for bending and $-\tau^2$ for twisting. Repeating the same argument as before, replacing $1/R$ with τ, one estimates the stretching energy of a twisted band as:

$$\frac{\mathcal{E}_\mathrm{s}}{L} \sim E\,h\,b\,(\tau\,b)^4.$$

Therefore, the rod and plate descriptions agree for

$$b \ll b^*, \quad \text{where } b^* \sim \left(\frac{h}{\tau}\right)^{1/2}.$$

When b becomes as large as b^*, the rod theory is no longer valid, as the stretching energy is not negligible.

6.8   Discussion

In this section, we discuss a few general features of the Föppl–von Kármán equations. This section can be skipped in a first reading: the different points below will be referred to whenever needed in the following chapters.

6.8.1   Range of validity

In several places in this chapter, we stressed the assumptions underlying the equations for elastic plates. Let us summarize them here. First of all, the thickness h must be much smaller than the other dimensions of the plate and the strain was assumed small (Hookean elasticity). Therefore the curvature of the centre surface should remain much smaller than the inverse thickness:

$$\left|\frac{\partial^2 w}{\partial x_\alpha \, \partial x_\beta}\right| \ll \frac{1}{h} \qquad \text{for } \alpha, \beta = x, y. \tag{6.110a}$$

We also assumed that the displacements were small:

$$\left|\frac{\partial u_\alpha}{\partial x_\beta}\right| \ll 1 \quad \text{and} \quad \left|\frac{\partial w}{\partial x_\beta}\right| \ll 1. \tag{6.110b}$$

This approximation means that the F.–von K. equations are not invariant by finite rotations in the 3D Euclidean space: finite rotations of plates are not solutions of the F.–von K. equations, although they describe physically rigid-body motions. This assumption is nevertheless very useful as it greatly simplifies the formalism. As explained, different plate equations have to be used whenever one wants to deal with finite rotations.

One of the trickiest points in the derivation of the plate equations was to allow that the deflection, although small, is much larger than the in-plane displacements:

$$\frac{\partial w}{\partial x_\alpha} = \mathcal{O}\left(\sqrt{\left|\frac{\partial u_\beta}{\partial x_\gamma}\right|}\right). \tag{6.110c}$$

As already explained, this points to the fact that the in-plane and out-of-plane displacements are generically of different orders of magnitude, something that ultimately points to the fact that the reference, planar configuration is invariant upon up–down mirror symmetry. As a result, the change in in-plane lengths caused by a purely transverse displacement is *second order* (it has to be unchanged should the sign of the transverse displacement be changed), while that of a in-plane displacement are *first order*.

The order of magnitude of the various terms in the F.–von K. equations can now be estimated as follows. By the definition (6.83), the bending modulus D is of order $E\,h^3$ since the Poisson coefficient, ν, is a number of order one. By taking all derivatives with respect to x or y to be of order $1/L$, L being the typical dimension of the plate, and by balancing all terms in the F.–von K. equations (6.88), we get, in order of magnitude:

$$E\,h^3\,\frac{w}{L^4} \sim h\,\frac{\chi\,w}{L^4} \quad \text{and} \quad \frac{\chi}{L^4} \sim \frac{E\,w^2}{L^4}.$$

This yields, by elimination:

$$w \sim h \quad \text{and} \quad \chi \sim E\,h^2.$$

The deflection, the in-plane stress and strain are therefore of order:

$$w \sim h, \qquad \sigma_{\alpha\beta} \sim \frac{\chi}{L^2} \sim E\left(\frac{h}{L}\right)^2, \qquad \epsilon_{\alpha\beta} \sim \frac{\sigma_{\alpha\beta}}{E} \sim \left(\frac{h}{L}\right)^2. \tag{6.111}$$

Using these estimates, the conditions of validity (6.110) of the F.–von K. equations reduce to:

$$h \ll L.$$

This is consistent, as the assumption of a small aspect ratio $h/L \ll 1$ was indeed at the heart of our derivation of the equations for elastic plates.

Note that equation (6.111) implies that the deflection w should be proportional to the thickness h. In practice, this deflection can be many times larger than the thickness h, as long as it remains much smaller than the dimension of the plate, L. Therefore, the strongly non-linear buckling regime (postbuckling) can be approached using the F.–von K. equations, as we do in Chapter 8.

6.8.2 Reduction to the Elastica in one dimension

In the case of deformations with cylindrical symmetry, that is when the deformation is invariant with respect to, say, the y coordinate, the F.–von K. equations take a similar form as the equations for the Elastica (a twistless, planar elastic rod). This is discussed in detail in Section 8.5.3—however, it should be noted that the F.–von K. equations, not being geometrically exact, have a more limited range of validity than the equations for the Elastica, which can handle finite rotations.

6.8.3 Underlying geometrical structure

There is a rather tight connection between the proof that we gave of the *Theorema egregium* in the first part of this chapter, and the structure of the F.–von K. equations. In fact, the equations for elastic plates appear as an extension of Gauss' theorem when the in-plane strain, instead of being merely set to zero, is such that it minimizes the elastic energy. The strong connection between the geometry of surfaces and the elasticity of plates has ultimately to do with the fact that isometric deformations of the centre plane involve bending energy only, and therefore cost much less energy than a generic deformation.

This remark explains why the Gauss curvature appears in the F.–von K. equations, in the bracket term in the second equation (6.86b), later rewritten as $[w, w]$ in equation (6.88b). The Gauss curvature, which measures by how much the centre surface departs from a developable configuration, is the source term in the equation for the Airy potential, χ, that governs in-plane stress. As a result, no stretching will be present in the plate if the centre surface can deform isometrically.

We emphasize that the relation between in-plane stress (or the Airy potential) on the one hand, and the Gauss curvature on the other is *non local*: given the profile of the Gauss curvature $K(x, y)$ of the centre surface, the Airy potential must be determined by solving a biharmonic equation after taking into account the proper boundary conditions, discussed in the next chapters. The value of the Airy potential at any point is then a function of the value of the Gauss curvature at all other points of the plate, and of the boundary conditions.[56] In other words, the elastic energy of an elastic plate cannot be written as the integral of some local elastic energy, function of the deflection $w(x, y)$ and its derivatives only.

6.8.4   Singular limit of small thickness

In the F.–von K. equations, the highest power of the small parameter, h, comes in front of the highest derivative of one of the unknown functions, $w(x, y)$—this happens in the bending term $D \, \Delta^2 w$, where $D = Eh^3/[12(1 - \nu^2)]$. As a result, the nature of the differential system changes when one takes the limit $h \to 0$: it passes from fourth order with respect to w to second order. In this case, the solutions of the differential system typically vary abruptly when $h \to 0$. Such a limit is called singular.[57] As a result of this singular limit, the behaviour of the solutions of the F.–von K. equations when the thickness h goes to zero is often not simple. In Section 6.4.6, we showed that the solutions with $D = 0$ (membrane equations) can be non-unique. Very often, the physically relevant solution of the problem can only be determined by considering bending effects. This is what we do in Chapter 9 to study the conical points and the folds in crumpled paper, both being singular solutions of the equations when h goes to zero.

While this singular structure makes the equations for plates more difficult to approach, it explains why elastic plates can display complex (and interesting) behaviours. The example of crumpled paper is a good one: the membrane theory does a rather poor job, as it yields many solutions with a vanishing energy, which furthermore exhibit singularities; a correct understanding of these singularities and of the overall mechanical response of a crumpled piece of paper involves a boundary layer theory based on the full F.–von K. equations.

6.9   Conclusion

The F.–von K. equations for elastic plates have a rather rich and complex mathematical structure, being non-linear partial differential equations of fourth order with respect to both the deflection $w(x, y)$ and the Airy potential $\chi(x, y)$. Their nonlinearity has a geometrical origin, connected with Gauss' *Theorema egregium* for the isometric deformations of surfaces. This makes them universal in some sense: much like the Navier–Stokes equations for fluids, no additional material parameters enters the equations when a geometrically non-linear

---

[56] This is best demonstrated by a simple example, inspired from Chapter 9. Imagine that we try to apply a circular cone whose angle at the tip, α, is different from $\pi/2$ onto a plane. It is clear that this will involve substantial stretching energy distributed over the whole surface of the cone. On the other hand, the Gauss curvature is zero everywhere, except in the vicinity of the tip, since both the cone and the plane are developable surfaces.

[57] See Appendix B on boundary layers for an example of such singular limits.

system goes from a linear to a non-linear behaviour. This is qualitatively different from the constitutive nonlinearity encountered in the presence of finite strain, which involves several, and often many specific parameters. In contrast, geometric nonlinearity makes it possible to characterize the response of a slender elastic system using very few reduced parameters, something that makes them well suited to analytical approaches. While the 'universal' instabilities of fluids has been extensively studied for several decades now, thin elastic systems have received comparatively less attention. In the following chapters, we study solutions of the F.–von K. equations in various geometries. We shall often rely on the geometrical underlying structure of these equations to devise analytical solutions or consistent approximation scheme.

References

Ph. G. Ciarlet. A justification of the von Kármán equations. *Archive for Rational Mechanics and Analysis*, 73(4):349–389, 1980.

H. Le Dret and A. Raoult. The nonlinear membrane model as variational limit of three-dimensional nonlinear elasticity. *Journal de Mathématiques Pures et Appliquées*, 75:551–580, 1995.

A. Föppl and L. Föppl. *Drang und Zwang: eine höhere festigkeitslehre für Ingenieure*. R. Oldenbourg, München, 1924.

L. Föppl. *Vorlesungen über technische Mechanik*, volume 5. B. G. Teubner, Leipzig, Germany, 1907.

G.W. Hunt, G. J. Lord, and A.R. Champneys. Homoclinic and heteroclinic orbits underlying the post-buckling of axially-compressed cylindrical shells. *Computer Methods in Applied Mechanics and Engineering*, 170:239–251, 1999.

H. Lamb. Sur la flexion d'un ressort élastique plat. *Philosophical Magazine*, 31(5):182–188, 1891.

A. E. H Love. *A Treatise on the Mathematical Theory of Elasticity*. (Reprinted by Dover, New York, 1944). Cambridge University Press, 1927.

A. Libai and J. G. Simmonds. *The Nonlinear Theory of Elastic Shells*. Cambridge University Press, 2nd edition, 1998.

R. T. Shield. Bending of a beam or wide strip. *Quarterly Journal of Mechanics and Applied Mathematics*, 45(4):567–573, 1992.

M. Spivak. *A Comprehensive Introduction to Differential Geometry*, volume 5. Publish or Perish, Inc., Houston (TX), 2nd edition, 1979.

T. von Kármán. Festigkeitsprobleme in Maschinenbau. In F. Klein and C. Müller, editors, *Encyklopädie der Mathematischen Wissenschaften*, volume 4, pages 311–385. Teubner, 1910.

7
End effects in plate buckling

This chapter and the next cover some aspects of the buckling of thin plates. In the present chapter, the loading is uniaxial and applied along the long dimension of the plate, but in the next chapter we deal with more general biaxial loadings. This chapter is concerned with end effects in the weakly non-linear regime, i.e. how the buckling depends on the finite (but very long) length of the plate. The next chapter addresses the elastic post-buckling far above the buckling threshold but without end effects.

7.1   A historical background on end effects

In the early 1980s, end effects in 1D patterns was an active topic of research involving various groups. The main motivation was related to Rayleigh–Bénard convection, an instability that is present in fluids heated from below. More precisely, the goal was to understand the observed changes of wavelength of Rayleigh–Bénard convection rolls when the temperature difference is increased above the instability threshold. In the general problem of the influence of end effects on patterns, the Rayleigh–Bénard problem in fact turned out to be rather non-generic: Rayleigh–Bénard rolls tend (S. Zaleski, Y. Pomeau, and A. Pumir, 1984) to merge perpendicular to a lateral wall, although the basic assumption in the study of end effects was that rolls are *parallel* to the lateral boundary, an *unstable* situation. In Rayleigh–Bénard convection, the best studied situation, particularly by Koschmieder (E. L. Koschmieder, 1974), was that of concentric rolls in a liquid filling a small vertical gap between two large horizontal discs. The very regular target pattern found experimentally was misleading because it did not reveal the influence of the lateral boundary conditions, but was instead the result of a horizontal temperature gradient in the radial direction. In such a circular geometry the wavelength of the rolls in super-critical conditions is uniquely fixed (Y. Pomeau, S. Zaleski, and P. Manneville, 1985) by a constraint of neutrality against bending, independent of any effect of the lateral boundary conditions.

Therefore—and this was not so well understood at this time—the prime physical example of selection by end effects was in the buckling of a long but finite elastic plate under compression, a suggestion made in (Y. Pomeau, 1981) and later tested experimentally by Wesfreid *et al.* (M. Clément, E. Guyon, and J. E. Wesfreid, 1981; M. Boucif, J. E. Wesfreid, and E. Guyon, 1991). The general problem of buckling instability was perhaps not very familiar to most scientists working in the field of non-equilibrium systems, with the exception of an early attempt to apply amplitude equations to a model for buckling (C. G. Lange and A. C. Newell, 1971). Compared with fluid mechanical problems, those in elasticity have in some sense a richer mathematical structure because of their variational formulation: the equilibrium configuration is a minimum of the elastic energy. This is significant for two reasons. First there exists something like an optimal structure of a buckled plate. This yields an optimal wavelength, for instance, a concept that does not exist for Rayleigh–Bénard convective rolls. Furthermore, exact Noether invariants in elasticity problems allow one to relate the slowly varying buckling parameters in the bulk, and close to the end,[1] as shown in Section 7.6.2.

The derivation of the end effects in (Y. Pomeau and S. Zaleski, 1981) relied upon a discussion of the topology of 'trajectories' representing the roll system that is similar to that presented in Section 4.4.2 when deriving the oscillations of the wavelength of the solution as the control parameter changes. A more analytical approach was followed in (M. C. Cross *et al.*, 1980; P. G. Daniels, 1981; M. C. Cross *et al.*, 1983), with a detailed discussion of the matching of the solution between the neighbourhood of the end and the bulk solution. In addition to the end effects in plate buckling discussed (Y. Pomeau, 1981) by using the Noether invariant, a discussion of the same problem by the method of amplitude equations was presented in (M. Potier-Ferry, 1983). The experiments showed (M. Boucif, J. E. Wesfreid, and E. Guyon, 1991) a pattern broadly consistent with the theoretical picture. Discrepancies were attributed to the difference between the simplified boundary conditions used in theoretical works and the real life situation. In any case, it was shown unambiguously that the changes of wavenumber occurring above threshold depend crucially on the anchoring conditions at the short ends of the plate, a fact that supports the theory presented next.

A last point deserves consideration: the end effects in patterns are an extreme case of a more general class of bifurcations with non-uniform parameters in space. The 'end effect' we consider is caused by an abrupt 'discontinuity' in the underlying equations in the form of boundary conditions. The other extreme of the same general class of problems is when the bifurcation parameters changes very slowly over the entire domain where the instability takes place. In plates, this could be realized with slowly varying width or thickness, for instance. When the bifurcation parameter changes very slowly as a function of the longitudinal coordinate, from values below threshold to above threshold, the wavelength of the bifurcated pattern above threshold is uniquely determined by using an adiabatic method (Y. Pomeau and S Zaleski, 1983). As usual in this kind of situation, this simply means that the wavelength is within an exponentially small range in the adiabatic limit, which becomes finite where the change of parameter becomes a sharp jump, as in the case of boundary conditions at a short end of a plate.

7.2 Geometry

The loading geometry is sketched in Fig. 7.1. The plate is a rectangular and has a large-aspect ratio. It is freely supported along its long sides, which are bound to a frame, shown in light grey in the figure: the displacement of the lateral edges of the plate are imposed by the frame, although the plate can freely rotate there. The joint between the plate and the frame acts as a hinge, with axis along the y axis, that is along the long sides of the plates. The lateral frame is then compressed longitudinally,[2] as shown by the arrows in the figure.

---

[1] In non-variational problems, as for the rolls in the Rayleigh–Bénard convection, the exact Noether invariant can be replaced by a kind of adiabatic invariant (Y. Pomeau and S. Zaleski, 1981; M. C. Cross *et al.*, 1980). This invariant exists near the instability threshold only. Its derivation is far more difficult than that based on Noether invariants.

[2] Buckling of the frame itself is prevented by means of appropriate vertical guides.

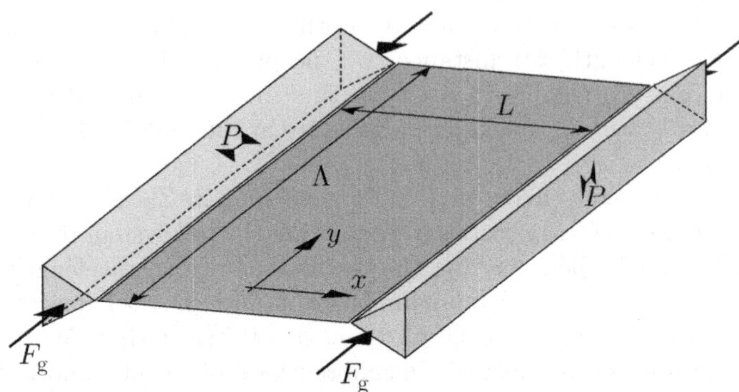

Fig. 7.1 Geometry of the buckling problem: a long rectangular plate of width L and length $\Lambda \gg L$ is attached to lateral guides. These guides are under simple compression with a longitudinal force F_{g}, and impose a longitudinal contraction rate P along the edges of the plate. We study the resulting buckling patterns in the plate.

The lateral frame is assumed to be much stiffer than the plate overall and so its shape is unaffected by the deformations of the plate. The frame imposes on the plate a uniform longitudinal contraction along its long edges.[3] Then, the plate is longitudinally compressed, and its planar configuration may become unstable: above a loading threshold, out-of-plane oscillations develop, which allow the plate to relax some of the imposed compression—at the price of some bending energy. This is a buckling bifurcation, similar to that introduced in Appendix A, Section A.5 for the Elastica.

The experimental realization of such a loading geometry is discussed in Section 7.7 and in more details in reference (M. Boucif, J. E. Wesfreid, and E. Guyon, 1991). It can be achieved either by compressing the plate longitudinally, as reported in Chapter 8, or by thermal[4] effects. These experiments confirmed the existence of post-buckling transitions with changes of wavenumber, as discussed theoretically in the following sections.

Below, we shall compute the onset of bifurcation from unbuckled to buckled state, i.e. the values of the frame contraction such that the planar solution is no longer the only solution of the F.–von K. equations. Above threshold, but close to it, methods of bifurcation theory can be used to find the pattern with the lowest elastic energy. The general question of the so-called 'preferred solution' of the bifurcation problem, loosely defined as the solution that is most likely to be observed in experiments, is actually not so simple to answer. We shall analyse in some depth how the buckled solution is *selected* by the finite length of the plate. The boundary conditions along the short sides of the plate play an important role in this selection process.

The outline of this chapter is as follows. In Section 7.3, we derive the governing equations for the plate, which serve as a starting point for the subsequent sections. In Section 7.4, we

---

[3] Different buckling geometries are studied in the following chapters, resulting in different buckling patterns. For instance, we will be concerned in Chapter 8 with the case of *clamped* boundary conditions along the long sides, and of a *biaxial* imposed stress.

[4] Such thermal effects are induced, for instance, by heating a plate held by a frame with a lower thermal expansion coefficient. As temperature increases, the plate attempts to expands relative to the frame, and so becomes compressed isotropically.

study the linear stability of an infinitely long plate and show that, above a critical value P_c of the loading P, the planar solution, defined in Section 7.3.2, is no longer the solution with the lowest energy. For $P > P_c$, this planar solution loses its stability while stable, buckled solutions appear. The selection of the wavelength of the buckling pattern by energy minimization is addressed in Section 7.5. This analysis of an infinitely long plate is rather classical. The case of a large but finite length, considered next, is not so. In this case, it is not so easy to predict what structure will actually appear in experiments, because some patterns may be metastable and the actual choice of the physical solution may depend on the history of the loading. This question, addressed theoretically in reference (Y. Pomeau, 1981) for general models, is reviewed in Section 7.6 below. Buckling experiments (M. Boucif, J. E. Wesfreid, and E. Guyon, 1991) on plates in a geometry similar to the one considered here are briefly summarized in Section 7.7.

7.3 Governing equations

In this section, the equations governing the elasticity of the strip are derived from the F.–von K. equations, and written in a dimensionless form. Boundary conditions on the long sides are specified as well.

7.3.1 F.–von K. equations

We recall the form of the F.–von K. equations derived in Chapter 6. We call x and y the coordinates parallel to the plane of the plate, x being in the transverse direction and y in the longitudinal one. The long edges of the plates are $x = \pm L/2$, where L is the plate width. The reference configuration is the planar, stress-free configuration of the strip, and the components of the displacement are written $(u(x,y), v(x,y), w(x,y))$, or equivalently $(u_x(x,y), u_y(x,y), w(x,y))$. We use the notation $(u_x, u_y) = (u, v)$ for the in-plane displacement, and w for the deflection. Writing D the bending modulus of the plate, defined in equation (6.83), h its thickness, E and ν the values of Young's modulus and Poisson's ratio, the equations for a plate at equilibrium take the following form, in terms of the displacement:

$$\epsilon_{\alpha\beta} = \frac{u_{\alpha,\beta} + u_{\beta,\alpha}}{2} + \frac{w_{,\alpha}\, w_{,\beta}}{2} \tag{7.1a}$$

$$\sigma_{xx} = \frac{E}{1-\nu^2}\left(\epsilon_{xx} + \nu\,\epsilon_{yy}\right), \quad \sigma_{yy} = \frac{E}{1-\nu^2}\left(\nu\,\epsilon_{xx} + \epsilon_{yy}\right), \quad \sigma_{xy} = \frac{E}{1+\nu}\,\epsilon_{xy}, \tag{7.1b}$$

$$\sigma_{\alpha\beta,\beta} = 0 \tag{7.1c}$$

$$D\,\Delta^2 w - h\,\sigma_{\alpha\beta}\, w_{,\alpha\beta} = 0. \tag{7.1d}$$

We use again a compact notation where commas in subscripts are for partial derivatives. The first equation is the definition of strain, linearized with respect to the in-plane displacement, but not with respect to the deflection $w(x,y)$, as introduced earlier in equation (6.62). The second set of equations is the effective constitutive relations (6.59). The third one is the condition of in-plane equilibrium (6.64a) in the absence of distributed applied force. The last one is for the out-of-plane equilibrium given in equation (6.86a); it includes the flexural term.

To describe the deformations of the plate, one has to solve the above set of equations together with the following boundary conditions, relevant for the particular geometry under consideration:

$$u_x\left(\pm\frac{L}{2},y\right)=0,\qquad u_y\left(\pm\frac{L}{2},y\right)=-Py,\quad w\left(\pm\frac{L}{2},y\right)=0,\qquad(7.2a)$$

$$\frac{\partial^2 w}{\partial x^2}\left(\pm\frac{L}{2},y\right)=0.\qquad(7.2b)$$

The boundary conditions (7.2a) on the displacement express the fact that the edge of the plate are bound to the lateral guides. Indeed, each guide is compressed with an applied longitudinal force $F_{\mathrm{g}}>0$. The guides being much stiffer than the plate, they deform as if their edges were free of stress. This loading geometry, relevant for the guides, is called the simple compression (applied axial force with free boundary conditions on the lateral edges) and has been analysed in Section 2.4.3. According to the results of this section, the guides under simple compression contract by a factor:

$$-\epsilon^{\mathrm{g}}_{yy}=P,\ \text{where}\ P=\frac{F_{\mathrm{g}}}{S_{\mathrm{g}}\,E_{\mathrm{g}}},\qquad(7.3)$$

under the applied loading F_{g}, where S_{g} is the sectional area of each guide, and E_{g} its Young's modulus. The first two boundary conditions in equation (7.2a) then ensure that the displacements, hence the longitudinal strain, are continuous at the long edges of the plate.

The two remaining boundary conditions for w written above are those that hold for simple support: no out-of-plane deformation is possible. However, the plate can freely rotate about the y axis at its edge, and so the y component of the bending moment should be zero there. Much as with a rod, a case already studied[5] in Section 3.6.4, this bending moment is proportional to the curvature, that is to the second derivative of the deflection, hence equation (7.2b).

We have now written down all the equations that we are going to solve in the rest of this chapter—except for the boundary conditions along the *short* edges, considered later. Before proceeding to the solution of these equations, we recall an alternative formulation of the equations for plates, in terms of the Airy potential $\chi(x,y)$:

$$\Delta^2\chi+\frac{E}{2}\,[w,w]=0\qquad(7.4a)$$

$$D\,\Delta^2 w-h\,[\chi,w]=0\qquad(7.4b)$$

$$\sigma_{xx}=\chi_{,yy},\quad\sigma_{yy}=\chi_{,xx},\quad\sigma_{xy}=-\chi_{,xy}.\qquad(7.4c)$$

This set of equations is equivalent to the set (7.1), with the in-plane displacement eliminated in favour of the Airy potential. We recall that the first equation in the group above is the compatibility condition that warrants the existence of a displacement field, that the second equation is again the condition for out-of-plane equilibrium, while the third equation is the definition of Airy potential, which yields the in-plane stress in terms of the unknown χ. For

---

[5] In equation (4.9), we derived the boundary conditions at the free edge of an elastic rod (a hair strand). One condition takes the form $\theta_{,s}=0$. This is similar to the present condition $w_{,xx}=0$, since the angle θ is equivalent to the slope $w_{,x}$ the derivative $\theta_{,s}$ of the deflection in the limit of small displacements.

the current problem, it is not convenient to remove the in-plane displacement altogether, as the boundary conditions are naturally expressed in terms of the displacement, and are quite cumbersome in terms of χ. However, the energy is more compactly expressed as a function of χ, and so we shall use both formulations.

7.3.2 Unbuckled solution

There exists a simple solution to the above equations, which describes the planar, unbuckled state. This solution has uniform strain and stress, and the deflection is zero. Using a nought in the superscript to label this particular solution, we have:

$$u_x^0(x,y) = 0, \quad u_y^0(x,y) = -P\,y, \quad w^0(x,y) = 0$$

$$\epsilon_{xx}^0(x,y) = 0, \quad \epsilon_{yy}^0(x,y) = -P, \quad \epsilon_{xy}^0(x,y) = 0$$

$$\sigma_{xx}^0(x,y) = -\frac{\nu\,E}{1-\nu^2}\,P, \quad \sigma_{yy}^0(x,y) = -\frac{E}{1-\nu^2}\,P, \quad \sigma_{xy}^0(x,y) = 0,$$

$$\chi^0(x,y) = -\frac{E\,P}{2\,(1-\nu^2)}\,\left(x^2 + \nu\,y^2\right).$$

This is always a solution of the problem, that is of equations (7.1) and (7.2), but this solution is not always stable. Above a critical value P_{c} of the loading calculated below, that is for $P > P_{\mathrm{c}}$, non-planar solutions appear. Very much as in the analysis of the Elastica made in Section A.5, these solutions have fewer symmetries than the planar solutions. This is called a buckling bifurcation.

7.3.3 Rescaled F.–von K. equations with prestress

We shall now transform the plate equations, first by using the unbuckled solution as the reference configuration instead of the stress-free configuration, and second by introducing dimensionless quantities, following the method given in Section 1.2.4 for the generic case of the equations for a pendulum. According to our general conventions, a star denotes the magnitude that is used for the rescaling, while a bar is for the rescaled variable. For instance, an obvious choice for rescaling the in-plane coordinates is $\overline{x} = x/x^*$ and $\overline{y} = y/y^*$ with $x^* = y^* = L$. The long edges, defined formerly by their Cartesian equation $x = \pm L/2$, are now defined by $\overline{x} = \pm 1/2$. The new rescaled quantities, $\overline{u}_\alpha$, $\overline{w}$, etc., which are functions of the rescaled variables $(\overline{x}, \overline{y})$ by convention, are defined by:

$$u_\alpha(x,y) = u_\alpha^0(x,y) + u^*\,\overline{u}_\alpha(\overline{x},\overline{y}) \tag{7.5a}$$

$$w(x,y) = w^*\,\overline{w}(\overline{x},\overline{y}) \tag{7.5b}$$

$$\epsilon_{\alpha\beta}(x,y) = \epsilon_{\alpha\beta}^0(x,y) + \epsilon^*\,\overline{\epsilon}_{\alpha\beta}(\overline{x},\overline{y}) \tag{7.5c}$$

$$\sigma_{\alpha\beta}(x,y) = \sigma_{\alpha\beta}^0(x,y) + \sigma^*\,\overline{\sigma}_{\alpha\beta}(\overline{x},\overline{y}) \tag{7.5d}$$

$$\chi(x,y) = \chi^0(x,y) + \chi^*\,\overline{\chi}(\overline{x},\overline{y}) \tag{7.5e}$$

$$P = \epsilon^*\,\overline{P}. \tag{7.5f}$$

In these equations, the constant terms describe the unbuckled solution, which now takes the form $\overline{u}_\alpha(\overline{x}, \overline{y}) = 0$ and $\overline{w}(\overline{x}, \overline{y}) = 0$. The scales u^*, w^*, etc. can be chosen arbitrarily. However, they should be chosen in such a way that the final equations have as few parameters as possible. Following the method presented in Section 1.2.4, one fixes these scales so as to balance the various terms of the equations we are trying to solve. For instance, the definition of the strain (7.1a) yields $\epsilon^* = u^*/L$ and $\epsilon^* = (w^*/L)^2$, since $x^* = y^* = L$. Doing the same with the other equations, one is led to suggest the following rescalings:[6]

$$x^* = y^* = L, \qquad u^* = \frac{D}{E\,h\,L}, \qquad w^* = \sqrt{\frac{D}{E\,h}}, \qquad (7.5\text{g})$$

$$\epsilon^* = \frac{D}{E\,h\,L^2}, \qquad \sigma^* = \frac{D}{h\,L^2}, \qquad \chi^* = \frac{D}{h}. \qquad (7.5\text{h})$$

Once again, these scales are only a matter of convention and other choices are possible—although different conventions might not be as convenient as this one. Up to numerical factors of order one, these scales are just those underlying the F.–von K. equations, derived earlier in equation (6.111).

The typical strain ϵ^* introduced here gives the correct order of magnitude of the critical strain P_c for buckling. Since D was defined as $D = Eh^3/(12(1 - \nu^2))$, this ϵ^* is equal, up to a numerical factor of order one, to the squared aspect ratio of the plate, $(h/L)^2$, a number that is very small by assumption. This means that the buckling parameter $\overline{P}$ can take values of order one (which it does at the onset of buckling), although the physical strain remains small, as assumed in our derivation of the equations for plates.

We can now plug the rescalings (7.5) into the F.–von K. equations (7.1), to obtain the equations in their dimensionless form:

$$\overline{\epsilon}_{\alpha\beta} = \frac{\overline{u}_{\alpha,\beta} + \overline{u}_{\beta,\alpha}}{2} + \frac{\overline{w}_{,\alpha}\,\overline{w}_{,\beta}}{2} \qquad (7.6\text{a})$$

$$\overline{\sigma}_{xx} = \frac{\overline{\epsilon}_{xx} + \nu\,\overline{\epsilon}_{yy}}{1 - \nu^2}, \qquad \overline{\sigma}_{yy} = \frac{\nu\,\overline{\epsilon}_{xx} + \overline{\epsilon}_{yy}}{1 - \nu^2}, \qquad \overline{\sigma}_{xy} = \frac{\overline{\epsilon}_{xy}}{1 + \nu}, \qquad (7.6\text{b})$$

$$\overline{\sigma}_{\alpha\beta,\beta} = 0 \qquad (7.6\text{c})$$

$$\Delta^2\overline{w} + \frac{\overline{P}}{1 - \nu^2}\,(\nu\,\overline{w}_{,\overline{x}\,\overline{x}} + \overline{w}_{,\overline{y}\,\overline{y}}) - \overline{\sigma}_{\alpha\beta}\,\overline{w}_{,\alpha\beta} = 0. \qquad (7.6\text{d})$$

All quantities are now defined in terms of the rescaled space variables $(\overline{x}, \overline{y})$. As a result, all differential operators, such as the harmonic operator Δ and the bracket operator $[,]$ used below, implicitly refer to *rescaled* coordinates. For instance, Δ now[7] denotes $\partial^2/\partial\overline{x}^2 + \partial^2/\partial\overline{y}^2$, which is L^2 times the original harmonic operator $\partial^2/\partial x^2 + \partial^2/\partial y^2$. Similarly, $\overline{\sigma}_{\alpha\beta,\beta}$ now denotes $\partial\overline{\sigma}_{\alpha\beta}/\partial\overline{x}_\beta$.

Compared with the initial set of equations (7.1), the new one above (7.6) has got rid of the coefficients such as D, h and E. Another difference, due to the change of the reference

---

[6] The present rescalings are different from those introduced in Section 6.8.1. The reference deflection w^* reads $D = \sqrt{D/(Eh)}$ in the present chapter, as opposed to $w^* = h$ formerly. As shown by plugging in the definition (6.83) of D, the two rescalings differ only by a numerical factor of order one, depending on Poisson's ratio.

[7] We could have introduced a bar notation for the rescaled operators, but this leads to cumbersome notations.

configuration, is the appearance of the two terms in the middle, both proportional[8] to $\overline{P}$, in the out-of-plane equilibrium (7.6d). They come from the cross-terms in the expansion of the non-linear term $\sigma_{\alpha\beta}\, w_{,\alpha\beta}$ in the original equation, when the stress takes its non-zero value in the unbuckled configuration, $\sigma_{\alpha\beta} = \sigma^0_{\alpha\beta} \neq 0$ for $\alpha = \beta$. These new terms reveal the existence of a stress proportional to $\overline{P}$ in the new reference configuration—something called residual stress—and we shall see that they tend to destabilize the planar solution when $\overline{P}$ increases.

The boundary conditions (7.2) take the following form, in terms of the rescaled quantities:

$$\overline{u}_x\left(\pm\frac{1}{2},\overline{y}\right) = 0, \qquad \overline{u}_y\left(\pm\frac{1}{2},\overline{y}\right) = 0, \quad \overline{w}\left(\pm\frac{1}{2},\overline{y}\right) = 0, \tag{7.7a}$$

$$\overline{w}_{,\overline{x}\,\overline{x}}\left(\pm\frac{1}{2},\overline{y}\right) = 0. \tag{7.7b}$$

Since the new reference configuration satisfies the boundary conditions, the inhomogeneous terms in the right-hand side have disappeared.

Finally, when expressed in terms of the rescaled Airy potential $\overline{\chi}(\overline{x},\overline{y})$, the plate equations (7.6) can be rewritten as:

$$\Delta^2\overline{\chi} + \frac{[\overline{w},\overline{w}]}{2} = 0 \tag{7.8a}$$

$$\Delta^2\overline{w} + \frac{\overline{P}}{1-\nu^2}\left(\nu\,\overline{w}_{,\overline{x}\,\overline{x}} + \overline{w}_{,\overline{y}\,\overline{y}}\right) - [\overline{\chi},\overline{w}] = 0 \tag{7.8b}$$

$$\overline{\sigma}_{xx} = \overline{\chi}_{,\overline{y}\,\overline{y}}, \quad \overline{\sigma}_{yy} = \overline{\chi}_{,\overline{x}\,\overline{x}}, \quad \overline{\sigma}_{xy} = -\overline{\chi}_{,\overline{x}\,\overline{y}}. \tag{7.8c}$$

In the rest of this chapter, we shall exclusively deal with rescaled symbols, except when stated otherwise. To keep the notation simple, *we shall therefore omit the bars* and write the rescaled quantities such as $\overline{w}$ and $\overline{\chi}$ as w and χ.

7.4 Linear stability analysis

We shall first consider the case of a plate that is infinitely long in the y direction. Above a critical value P_c of the loading P, the planar solution, $\overline{u}_\alpha(\overline{x},\overline{y}) = 0$ and $\overline{w}(\overline{x},\overline{y}) = 0$, bifurcates to a buckled configuration. This transition is continuous:[9] the amplitude of the bifurcated solution departs smoothly from zero above the buckling threshold. When the loading is exactly the critical loading, $P = P_c$, the unbuckled solution coexists with buckled solutions of vanishingly small amplitude. This provides a way of computing the critical loading P_c: it is such that the *linearized* F.–von K. equations, together with the relevant boundary conditions, have a non-trivial solution—that is a solution different[10] from $u_\alpha = 0$ and $w = 0$ for some x and y. This solution of the linearized equations, denoted $(u_\alpha^{(1)}, w^{(1)})$, is

---

[8] More accurately, the transverse prestress σ^0_{xx} is proportional to (νP), while the longitudinal prestress σ^0_{yy} is directly proportional to P. This explains the form of the two new terms in equation (7.6d).

[9] The first instability of a planar elastic plate is always a super-critical bifurcation, as will be shown in Section 8.4.2.

[10] Recall that, from now on, the bars above rescaled quantities are implicit.

sometimes called a marginal mode.[11] The equations to be satisfied by the small perturbation $(u_\alpha^{(1)}, w^{(1)})$, are derived from the original set of equations (7.6) by linearization near the flat solution, $(u_\alpha = 0, w = 0)$:

$$\epsilon_{\alpha\beta} = \frac{u_{\alpha,\beta} + u_{\beta,\alpha}}{2} \tag{7.9a}$$

$$\sigma_{xx} = \frac{\epsilon_{xx} + \nu\,\epsilon_{yy}}{1 - \nu^2}, \qquad \sigma_{yy} = \frac{\nu\,\epsilon_{xx} + \epsilon_{yy}}{1 - \nu^2}, \qquad \sigma_{xy} = \frac{\epsilon_{xy}}{1 + \nu}, \tag{7.9b}$$

$$\sigma_{\alpha\beta,\beta} = 0 \tag{7.9c}$$

$$\Delta^2 w + \frac{P}{1 - \nu^2}\left(\nu\,w_{,xx} + w_{,yy}\right) = 0. \tag{7.9d}$$

The boundary conditions (7.7), which are already linear, are unmodified.

An important simplification takes place here, as the equations of the problem happen to split into a first set of equations for the in-plane unknowns u_α, $\epsilon_{\alpha\beta}$, $\sigma_{\alpha\beta}$ on the one hand, and a second set for the out-of-plane unknown $w(x, y)$ on the other. There is no coupling between in-plane and out-of-plane unknowns at the linear order. The first set contains equations (7.9a), (7.9b), (7.9c) and the two first boundary conditions in (7.2a); the second set contains equation (7.9d) and the remaining boundary conditions. Now, the equations for the in-plane variables do not depend on the loading parameter P, and so have no other solution than $u_\alpha(x, y) = 0$, no matter what the value of P is:[12]

$$u_\alpha^{(1)}(x, y) = 0, \quad \epsilon_{\alpha\beta}^{(1)}(x, y) = 0, \quad \sigma_{\alpha\beta}^{(1)}(x, y) = 0, \quad \chi^{(1)}(x, y) = 0. \tag{7.10}$$

We are left with the linearized equations for the deflection:

$$\Delta^2 w^{(1)} + \frac{P}{1 - \nu^2}\left(\nu\,\frac{\partial^2 w^{(1)}}{\partial x^2} + \frac{\partial^2 w^{(1)}}{\partial y^2}\right) = 0 \tag{7.11a}$$

$$w^{(1)}\left(\pm\frac{1}{2}, y\right) = 0, \qquad \frac{\partial^2 w^{(1)}}{\partial x^2}\left(\pm\frac{1}{2}, y\right) = 0. \tag{7.11b}$$

In this linearized version of the F.–von K. equations, stretching is not accounted for. These linear equations for plates are known as the Kirchhoff equations for plates and, from a historical perspective, were derived before the full F.–von K. equations.

For some specific values of P, we expect non-trivial solutions $w^{(1)}(x, y)$ of these equations to exist. These solutions are sought after[13] in the form:

$$w^{(1)}(x, y) = A_1 \sin(a\,y) \cos(b\,x), \tag{7.12}$$

---

[11] Let us note by anticipation that such a marginal mode appears as the vertical tangent in the bifurcation diagram in Fig. 8.7, right.

[12] The vanishing of $\chi^{(1)}$, that is of $\sigma_{\alpha\beta}^{(1)}$ can also be obtained directly by symmetry considerations, see Section 8.4.2.

[13] The particular form of $w^{(1)}$ in equation (7.12) can be justified as follows. First, note that $w^{(1)}$ can be decomposed in Fourier modes along the longitudinal direction y. The function $w^{(1)}$ being the solution of a linear system of differential equations, all its modes, which are of the form $A(a) \sin(a\,y) f_a(x)$ or $A(a) \cos(a\,y) f_a(x)$, must satisfy the same equation. This yields an *ordinary* differential equation for $f_a(x)$, which is integrable. After taking the boundary conditions into account, one obtains f_a in the form given in equation (7.12). A similar argument will be used in Section 8.4.2 for a different buckling problem.

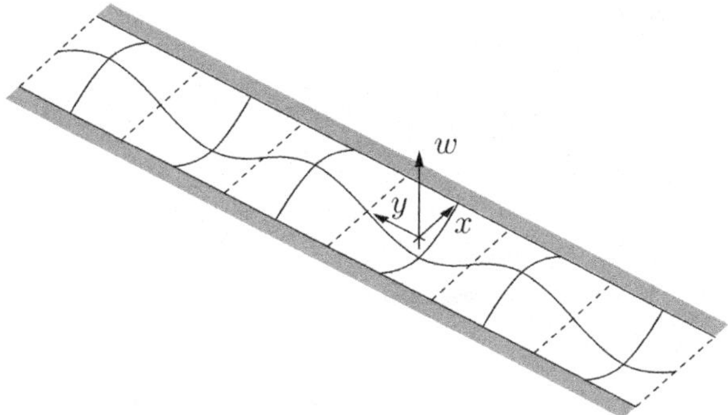

Fig. 7.2 First unstable pattern in the case of an infinitely long strip, used in the analysis of linear stability in equation (7.12).

where A_1 is the buckling amplitude, to be determined later. The following choice for b allows the boundary conditions (7.11b) to be satisfied:

$$b = \pi. \tag{7.13}$$

Other values like $b = 3\pi$ are also compatible with these boundary conditions, but they would appear for higher loads, and so are discarded (see Section 8.4.2).

Both the critical load P_{c} and the longitudinal wavenumber a remain to be computed. By plugging the form of $w^{(1)}$ given above into the linearized equation (7.11a), one obtains:

$$A_1 \left[(-a^2 - \pi^2)^2 + \frac{P}{1 - \nu^2} \left(-\nu\,\pi^2 - a^2 \right) \right] \left(\sin(a\,y)\,\cos(\pi\,x) \right) = 0, \tag{7.14}$$

an equation that must hold for all x and y such that $|x| \leq 1/2$. The only solution is when the factor in square brackets vanishes:

$$P = P(a) = (1 - \nu^2) \frac{(a^2 + \pi^2)^2}{a^2 + \nu\,\pi^2}. \tag{7.15}$$

This equation is represented graphically in Fig. 7.3. Any set of values of P and a that satisfy this relation yields a solution of the linearized F.–von K. equations. The lowest possible value of P in equation (7.15) yields the critical value of the longitudinal wavenumber, a_{c}, and the critical strain P_{c}. This is expressed by the set of equations

$$P'(a_{\mathrm{c}}) = 0, \qquad P_{\mathrm{c}} = P(a_{\mathrm{c}}).$$

The solution of the above system, found by computing the derivative of P explicitly, reads:

$$P_{\mathrm{c}} = 4\,\pi^2\,(1 - \nu)^2\,(1 + \nu), \qquad a_{\mathrm{c}} = \pi\,\sqrt{1 - 2\,\nu}.$$

To obtain the critical load in physical units, we restore the implicit bar on top of P_{c}, and multiply by ϵ^* which has been scaled out. This yields

$$\epsilon^*\,\overline{P}_{\mathrm{c}} = \frac{D}{E\,h\,L^2} \left[4\,\pi^2\,(1 - \nu)^2\,(1 + \nu) \right] = \frac{\pi^2\,(1 - \nu)}{3} \left(\frac{h}{L} \right)^2. \tag{7.16}$$

As anticipated based on dimensional arguments, this buckling strain is a very small number in the limit $h \ll L$ that we consider.

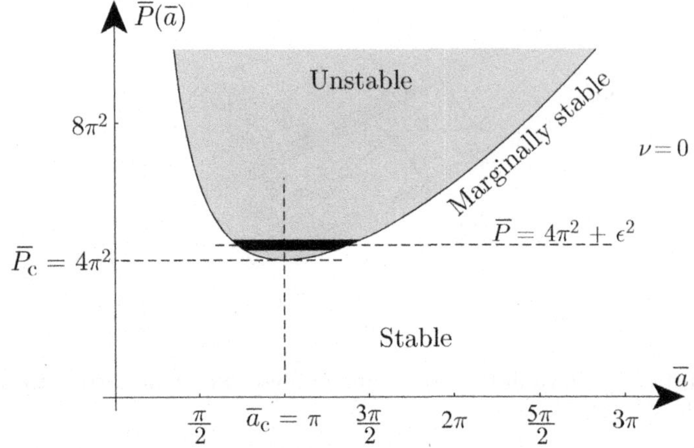

Fig. 7.3 Diagram of linear stability of the infinitely long strip, for $\nu = 0$: intensity of loading, $\overline{P}$, versus longitudinal wavenumber, $\overline{a}$, for marginal stability, as given by equation (7.15). The minimum of the curve yields the bifurcation threshold. Below the curve, the strip is stable. Above, it is unstable and there is a continuum of wavenumbers a that make the strip unstable (thick segment); the selection of the wavenumber is done by the minimization of the elastic energy, see Section 7.5.

In the rest of this chapter, we shall restrict our analysis to the case of a plate with zero Poisson's ratio, and set

$$\nu = 0. \tag{7.17}$$

Most materials have a Poisson's ratio around $\nu = 0.3$, and so the case $\nu = 0$ is not very representative. However, this leads to the simplest mathematical expressions. The extension of the following analysis to the case of an arbitrary Poisson's ratio does not raise any fundamental difficulty but involves significantly more intricate algebra.

For $\nu = 0$, the rescaled critical load and the marginal modes given earlier become:

$$P_c = 4\pi^2, \qquad a_c = \pi. \tag{7.18}$$

The critical longitudinal wavelength $2\pi/a_c$ is then twice the strip width. Having determined the critical value of the loading and the buckling mode of infinitesimally small amplitude at threshold, we are now done with the analysis of linear stability.

So far, the amplitude A_1 has remained arbitrary since the equations for $w^{(1)}$ are linear by construction. In the next section, we show how this buckling amplitude A_1 can be computed. This involves pushing further a systematic expansion whose first step is the linear stability analysis just carried out.

7.5 Buckling amplitude near threshold

We analyse in this section the solutions of the full F.–von K. equations when the amplitude of the perturbation is small but non-zero. In particular, we compute the amplitude of the buckling pattern, which has remained undetermined from the analysis of linear stability.

7.5.1 Outline of the perturbation scheme

Slightly above the bifurcation threshold, we seek the unknowns u_α and w by expansion in powers of a small parameter, ϵ, defined below:

$$u_\alpha(x,y) = \epsilon\, u_\alpha^{(1)}(x,y) + \epsilon^2\, u_\alpha^{(2)}(x,y) + \cdots \tag{7.19a}$$

$$w(x,y) = \epsilon\, w^{(1)}(x,y) + \epsilon^2\, w^{(2)}(x,y) + \cdots \tag{7.19b}$$

By this expansion, the buckling amplitude above threshold appears to be proportional to the expansion parameter ϵ at dominant order. One could attempt to define ϵ as the difference between the actual loading strain P and its critical value P_{c}, since we are doing an expansion that is valid just above threshold. However, such an expansion would be inconsistent, due to the invariance of the system by mirror symmetry with respect to the (x,y) plane. The consequence of this symmetry is that, if a buckling profile $\epsilon\, w^{(1)}(x,y)$ is a solution of the equations, then the symmetric one, $-\epsilon\, w^{(1)}(x,y)$, will be a solution too, for the same value of P. The buckling amplitude, proportional to ϵ, has an undetermined sign. This is related to the classical fact that the buckling amplitude above threshold in a continuous bifurcation is actually proportional to the *square root* of the distance to threshold, $\epsilon \propto \pm\sqrt{P - P_{\mathrm{c}}}$. Therefore, we define ϵ as:

$$\begin{aligned} P &= P_{\mathrm{c}} + \epsilon^2 \\ &= 4\,\pi^2 + \epsilon^2, \end{aligned} \tag{7.20}$$

since we have found that $P_{\mathrm{c}} = 4\pi^2$ for $\nu = 0$. The indeterminacy in the sign of ϵ then appears as the consequence of the fact that ϵ^2 is defined implicitly through its square, ϵ^2, in equation (7.20) above.

When plugged into the equations of the problem, the present expansion leads, at the dominant order (1), to the linear stability analysis of the previous section. Therefore, $u_\alpha^{(1)}(x,y) = 0$ and we use for $w^{(1)}(x,y)$ the expression derived previously.

Proceeding order by order, the next step is to pull this linear solution back into the governing equations and compute the correction due to nonlinearity. This correction is small because the amplitude is small, exactly for the same reason that ϵ^2 is small compared with ϵ when ϵ tends to zero. This calculation of the second-order correction is done in Section 7.5.2 and is fairly straightforward.

In Section 7.5.3, the expansion is pushed one step further. Corrections to the solution that are cubic in the small amplitude A_1 are computed. At this order, a resonance effect is found. To keep a well-behaved expansion, one has to impose a certain relation between the cubic term, the parameters of the problem and the distance to threshold. This additional relation, which makes it possible to continue the expansion beyond third order, is called a *solubility condition*. It ultimately yields the amplitude of the buckled solution, so far undetermined, as a function of the control parameters of the problem, namely the loading and the wave number in the y direction.

In Section 7.5.4, we show that the minimization of the elastic energy yields a unique wavenumber for the bifurcated structure, in the case of an infinitely long strip.

7.5.2   Expansion to second order

Let us try to push further the expansion of the buckled solution written so far at order 1. We take the linear solution (7.12), and pick an arbitrary value of the longitudinal wavenumber a within the unstable band shown by a bold line in Fig. 7.3. The selection of the value of a will be discussed below in Section 7.5.4—note that a has to be equal to a_c at threshold only, and we are now investigating what happens above threshold.

We mentioned above that the first-order cancellation of $u_\alpha^{(1)}(x,y) = 0$ follows from symmetry. The detailed argument is that changing the sign of ϵ in the expansion (7.19) must correspond to the mirror symmetry $u_\alpha \mapsto u_\alpha$ and $w \mapsto (-w)$. Stated differently, $\epsilon\, u_\alpha^{(1)} + \epsilon^2\, u_\alpha^{(2)} + \cdots$ has to be an even function of ϵ, while $\epsilon\, w^{(1)} + \epsilon^2\, w^{(2)} + \cdots$ has to be an odd function of ϵ. Therefore, all functions $u_\alpha^{(k)}(x,y)$ have to cancel for odd values of the order k, and so do all functions $w^{(k)}(x,y)$ for even values of k. In particular, we get, at order 2:

$$w^{(2)}(x,y) = 0. \tag{7.21}$$

The out-of-plane equilibrium (7.6d) is then automatically satisfied at this order. The remaining conditions yield:

$$\epsilon_{\alpha\beta}^{(2)} = \frac{u_{\alpha,\beta}^{(2)} + u_{\beta,\alpha}^{(2)}}{2} + \left[\frac{w_{,\alpha}^{(1)}\, w_{,\beta}^{(1)}}{2}\right] \tag{7.22a}$$

$$\sigma_{xx}^{(2)} = \epsilon_{xx}^{(2)}, \quad \sigma_{yy}^{(2)} = \epsilon_{yy}^{(2)}, \quad \sigma_{xy}^{(2)} = \epsilon_{xy}^{(2)}, \tag{7.22b}$$

$$\overline{\sigma}_{\alpha\beta,\beta} = 0 \tag{7.22c}$$

together with the boundary condition

$$u_\alpha^{(2)}\left(\pm\frac{1}{2}, y\right) = 0. \tag{7.22d}$$

Note the very particular form of the constitutive relations (7.22b) for the case $\nu = 0$ that we are considering: the rescaled 2D strain and stress tensors are equal. If we were considering arbitrary values of ν, the original effective 2D stress–strain relations (7.9b) would have to be used instead.

The set of equations (7.22) is a linear system of partial differential equations for the functions $u_\alpha^{(2)}$, and we shall now proceed to solve it by a standard method, taking advantage of the translational invariance in the y direction. The set of PDEs (7.22) has a single type of source term, given by the square bracket in the right-hand side of equation (7.22a). The function $w^{(1)}$ has been determined earlier, and so these source terms are known. To take advantage of the linearity of the problem at second order, we rewrite these source terms by expanding the products of trigonometric functions as:

$$\left[\frac{(w_{,x}^{(1)})^2}{2}\right] = \frac{A_1^{\,2}\,\pi^2}{8}\,(1 - \cos(2\pi\,x)) - \frac{A_1^{\,2}\,\pi^2}{8}\,(1 - \cos(2\pi\,x))\,\cos(2a\,y)$$

$$\left[\frac{w_{,x}^{(1)}\, w_{,y}^{(1)}}{2}\right] = \qquad\qquad - \frac{A_1^{\,2}\,a\,\pi}{8}\,\sin(2\pi\,x)\,\sin(2a\,y)$$

$$\left[\frac{(w_{,y}^{(1)})^2}{2}\right] = \frac{A_1^{\,2}\,a^2}{8}\,(1 + \cos(2\pi\,x)) + \frac{A_1^{\,2}\,a^2}{8}\,(1 + \cos(2\pi\,x))\,\cos(2a\,y).$$

We have effectively performed a Fourier decomposition of these source terms with respect to the longitudinal variable y. In the right-hand sides above, the first column lists the source terms that are 'constant' with respect to y, while the second column lists the modes with wavevector $2a$ along y. By linearity, one can solve the original linear set of PDEs mode by mode, that is by superposition of two solutions, one that is independent of y, and one that is a pure Fourier mode with respect to y, and has a wavevector $2a$. No other modes are present in the source terms. For each of these pure modes, the linear set of PDEs becomes a set of linear *ordinary* differential equations. We shall skip over the details of the resolution of this linear ODEs with constant coefficients, which has the unique solution:

$$u_x^{(2)}(x,y) = A_1^{\,2}\,\frac{\pi}{16}\,\left[\sin(2\pi\,x) - \sin(2\pi\,x)\,\cos(2a\,y)\right] \tag{7.23a}$$

$$u_y^{(2)}(x,y) = A_1^{\,2}\,\frac{a}{16}\,\left[-(1 + \cos(2\pi\,x))\,\sin(2a\,y)\right]. \tag{7.23b}$$

The associated in-plane stress can be computed by plugging this solution back into equations (7.22a) and (7.22b). It can be checked that these stress components derive from the following Airy potential:

$$\chi^{(2)}(x,y) = \frac{A_1^{\,2}}{8}\,\left[\frac{a^2\,x^2 + \pi^2\,y^2}{2} + \frac{-a^4\,\cos(2\pi\,x) + \pi^4\,\cos(2a\,y)}{4\,a^2\,\pi^2}\right]. \tag{7.24}$$

We have fully computed the solution at second order, and we can now proceed to the next order.

7.5.3 Cubic order: solubility condition

At the next, cubic, order one finds a solubility condition for the expansion in powers of ϵ, which yields at the end the sought after amplitude A_1 as a function of the 'control' parameters a and P. The third order solution $w^{(3)}$ is given by reading equation (7.6d) at third order:

$$\left\{\left(\Delta^2 + P\,\frac{\partial^2}{\partial y^2}\right) w\right\}^{(3)} = [\chi^{(2)}, w^{(1)}], \tag{7.25}$$

where the right-hand side is computed, as earlier, by inserting the known values of $\chi^{(2)}$ and $w^{(1)}$:

$$[\chi^{(2)}, w^{(1)}] = -\frac{3\,(a^4 + \pi^4)\,A_1^{\,3}}{16}\,\cos(\pi\,x)\,\sin(a\,y)$$

$$+ \frac{A_1^{\,3}}{16}\,\left[\pi^4\,\cos(\pi\,x)\,\sin(3a\,y) - a^4\,\cos(3\pi\,x)\,\sin(a\,y)\right]. \tag{7.26}$$

The boundary conditions are derived directly from equation (7.11b). In the equation above, the right-hand side gives the Fourier modes present in the decomposition of the source term $[\chi^{(2)}, w^{(1)}]$, with respect to both x and y. Noticing that the equation for $w^{(3)}$ is linear, like that at second order, we seek a solution by linear superposition:

$$w^{(3)}(x,y) = w^{(3:A)}(x,y) + w^{(3:B)}(x,y) + w^{(3:C)}(x,y) \tag{7.27a}$$

of three pure Fourier modes, the possible values of the wavevectors being imposed by the inhomogeneous terms:

$$w^{(3:A)}(x,y) = W_A^{(3)} \, \cos(\pi\,x)\,\sin(a\,y) \tag{7.27b}$$

$$w^{(3:B)}(x,y) = W_B^{(3)} \, \cos(\pi\,x)\,\sin(3a\,y) \tag{7.27c}$$

$$w^{(3:C)}(x,y) = W_C^{(3)} \, \cos(3\pi\,x)\,\sin(a\,y). \tag{7.27d}$$

The three constants $W_A^{(3)}$, $W_B^{(3)}$ and $W_C^{(3)}$ introduced here should be determined by plugging these expressions back into equation (7.25).

The differential operator in the left-hand side of equation (7.25) acts on each Fourier component $w^{(3:I)}(x,y)$, with $I = A, B, C$ as a multiplication by a constant, each of these components being a pure Fourier mode. With obvious notations, these scalars read:

$$\left(\Delta^2 + P\,\frac{\partial^2}{\partial y^2}\right)_A = (-\pi^2 - a^2)^2 - P\,a^2$$

$$\left(\Delta^2 + P\,\frac{\partial^2}{\partial y^2}\right)_B = (-\pi^2 - 9\,a^2)^2 - 9\,P\,a^2$$

$$\left(\Delta^2 + P\,\frac{\partial^2}{\partial y^2}\right)_C = (-9\,\pi^2 - a^2)^2 - P\,a^2.$$

Inserting into these expressions the expansion (7.20) for P, as well as the following expansion for a which, again, is expected to vary above threshold:

$$a = \pi + \delta\,\epsilon + \ldots, \tag{7.28}$$

we obtain:

$$\left(\Delta^2 + P\,\frac{\partial^2}{\partial y^2}\right)_A = 0 \qquad - \epsilon^2\,\pi^2\,(1 - 4\,\delta^2) + \cdots \tag{7.29a}$$

$$\left(\Delta^2 + P\,\frac{\partial^2}{\partial y^2}\right)_B = 64\,\pi^4 + \cdots \tag{7.29b}$$

$$\left(\Delta^2 + P\,\frac{\partial^2}{\partial y^2}\right)_C = 96\,\pi^4 + \cdots, \tag{7.29c}$$

where dots denote higher order terms. In the generic form of the expansion (7.28) for a, we have introduced a coefficient $\delta = a^{(1)}$ that remains to be determined.

The set of equations (7.29) reveals that the differential operator acts as a second-order quantity when applied on the Fourier mode labelled A, depending on x and y like $\cos(\pi\,x)\,\sin(a\,y)$, although it acts like a non-zero constant on the two other modes.

Now plugging the expansion (7.29) for the differential operator and that for $w^{(3)}$ given in equation (7.27a) into equation (7.25), and identifying the coefficients on both sides Fourier mode by Fourier mode, we obtain:

$$0 \times W_A^{(3)} = -\frac{3\,\pi^4\,A_1{}^3}{8} \tag{7.30a}$$

$$64\,\pi^4 \times W_B^{(3)} = \frac{\pi^4\,A_1{}^3}{16} \tag{7.30b}$$

$$96\,\pi^4 \times W_B^{(3)} = -\frac{\pi^4\,A_1{}^3}{16}. \tag{7.30c}$$

Everything works well for the last two Fourier modes, whose amplitude read:

$$W_B^{(3)} = \frac{A_1{}^3}{1024}$$

$$W_C^{(3)} = -\frac{A_1{}^3}{1536}.$$

However, the solution of equation (7.30a) involves a division by zero. This zero comes from the vanishing of the differential operator at the onset of bifurcation, that is for $\epsilon = 0$, when applied on this particular mode. This vanishing is by no means a coincidence: the third-order Fourier mode labelled A was already present in $w^{(1)}$ at first order, in equation (7.12); the condition for the onset of buckling, written thereafter in equation (7.14), is precisely that the same[14] differential operator, applied on the same Fourier mode, vanishes.

Formally, $W_A^{(3)}$ should be chosen as infinite, and this suggests that the solution $w^{(3)}$, and more accurately the term $w^{(3:A)}$ is not oscillatory but instead grows linearly with y. This is what happens generically when a system is forced at its frequency, something called a resonance[15] phenomenon.

Such a resonant forcing is not possible here, as the amplitude of buckling must remain everywhere small. The paradox of the division by zero in equation (7.30a) is solved by noticing that the left-hand side has been written incorrectly: the A Fourier mode in the left-hand side of the master equation (7.25) does not come from the zeroth order of the operator, which is zero, times the third order of w, but instead from the second-order of the operator, times the *first* order of w. The left-hand side of equation (7.30a) is corrected accordingly:

$$-\pi^2\,(1 - 4\,\delta^2) \times A_1 = -\frac{3\,\pi^4\,A_1{}^3}{8}. \tag{7.32}$$

This equation is called a solubility condition. It does not allow one to compute the coefficient $W_A^{(3)}$ of the solution at third order as we expected. Instead, it gives a condition involving the first-order amplitude A_1, which makes it possible to continue the expansion at third order and beyond. This solubility condition fixes A_1 as:

$$A_1{}^2 = \frac{8\,(1 - 4\,\delta^2)}{3\,\pi^2}. \tag{7.33}$$

---

[14] The only difference in these operators is that the former expression (7.14) deals with arbitrary values of ν, while we are now considering $\nu = 0$ only.

[15] A similar resonance phenomenon can be observed in the equation of the harmonic oscillator, $X_{,tt} + X = A\sin(q\,t)$, whose general solution is $X(t) = A\sin(qt)/(1 - q^2) + b\sin(t + \phi)$, with b and ϕ arbitrary constants. When $q \neq \pm 1$ this is the general solution; however, if $q = \pm 1$, that is when forcing is applied at the natural frequency of the system, this solution is no longer valid and a resonance occurs. Then, the equation still has solutions, but growing linearly in time: $X(t) = -A\,t\cos t + b\sin(t + \phi)$. In the (weakly) non-linear case, this growing solution is replaced by a periodic solution with a finite (but large) amplitude.

This weakly non-linear solution only exists—no complex numbers here—in the domain $4\delta^2 < 1$. Multiplying both sides of the inequality by ϵ^2, and identifying $(\delta\,\epsilon) \approx a - a_c$ and $\epsilon^2 = P - P_c$, we rewrite this condition as $4\,(a - a_c)^2 < P - P_c$. On the other hand, the expansion of the local Cartesian representation of the instability curve $P = P(a)$ given by equation (7.15) with $\nu = 0$ for small ϵ yields $P - P_c \approx 4\,(a - a_c)^2$. Therefore, a real root A_1 of equation (7.33) exists *above* the instability curve only, which is consistent.

To sum up, the third order solution reads

$$w^{(3)}(x,y) = W_A^{(3)} \cos(\pi\,x)\,\sin(a\,y)$$

$$+ \frac{A_1{}^3}{1024}\,\cos(\pi\,x)\,\sin(3a\,y) - \frac{A_1{}^3}{1536}\,\cos(3\pi\,x)\,\sin(a\,y). \quad (7.34)$$

The coefficient $W_A^{(3)}$ could be determined by pushing the expansion further, much like A_1 required solving the expansion at third order. The amplitude A_1 is given by the solubility condition (7.33) in terms of the parameter δ. This parameter δ can be chosen freely, and so we have constructed so far a one-parameter family of weakly non-linear solutions. In the rest of this chapter, we investigate the selection of the longitudinal wavevector δ. The selection by energy minimization, relevant for the case of an infinitely long strip, is investigated first.

7.5.4   Selection of longitudinal wavenumber

Since we have constructed a one-parameter family of weakly non-linear solutions, it is natural to study which solution has the lowest energy among this family. This provides a reasonable selection mechanism for the parameter $\delta = a^{(1)}$ when the strip is infinitely long. To carry out this calculation, one has to plug back the solution constructed so far in perturbations into the general expression (6.97) for the elastic energy of a plate, carry out the integrations formally, and minimize the density of elastic energy with respect to the free parameter δ. Like any other quantity, this energy has to be computed order by order with respect to ϵ. By symmetry, the energy is a even function of ϵ. At order zero, the energy is just the energy of the unbuckled solution, which does not depend on δ. The second order correction to the energy near the threshold is zero, as this is precisely the condition for the onset of instability. This means that the lowest order at which δ enters into the energy is the fourth order. The calculation of this energy to such a high order is very cumbersome. We give a simple argument below, which shortcuts this lengthy calculation, showing that the selection of δ by energy minimization requires $\delta = 0$.

Consider a typical continuous bifurcation for an order parameter w, taking place when a control parameter P reaches a value P_c. The canonical form of the energy describing this instability reads:

$$\mathcal{E}(P,w) = w^4 - (P - P_c)\,w^2.$$

Let ϵ be the square-root of the distance to threshold, $\epsilon^2 = P - P_c$. Since we expect a square-root behaviour of w with respect to $P - P_c$ near threshold, we rescale w by ϵ as earlier, and define A_1 by $w = \epsilon\,A_1$. Then, the energy takes the form

$$\mathcal{E} = \epsilon^4\,(A_1{}^4 - A_1{}^2).$$

Minimization with respect to A_1 shows that the amplitude satisfies $4\,A_1{}^3 - 2\,A_1 = 0$, that is $A_1{}^2 = 1/2$.

We shall modify this classical description to account for the fact that a one-parameter family of bifurcated solutions appears at the onset of the instability. Let δ be this parameter, as earlier. This is taken into account by replacing $P - P_c = \epsilon^2$ by $\epsilon^2 f(\delta)$, where $f(\delta) > 0$ is a positive function of δ whose precise form depends on the details of the system under consideration. The modified energy reads $\mathcal{E} = \epsilon^4 \left(A_1{}^4 - f(\delta) A_1{}^2 \right)$. With this energy, all the modes labelled by δ become simultaneously unstable for $P > P_c$, as required. Repeating the previous analysis, we find that the amplitude of a particular mode is given by $A_1{}^2 = f(\delta)/2$. Plugging back into the energy, we derive the energy associated with a particular unstable mode above threshold:

$$\mathcal{E} = \epsilon^4 \left(\frac{f^2}{4} - f\,\frac{f}{2} \right) = -\frac{\epsilon^4 f^2}{4} = -\frac{1}{4}\epsilon^4 A_1{}^4 = -\frac{1}{4}\,w^4.$$

Therefore, the buckled solution that has the lowest energy among the family is just one that maximizes $f(\delta)$, or equivalently the one with the largest squared amplitude $A_1{}^2$ or w^2.

Extrapolating these results to the original system, an infinitely long strip, we infer that energy minimization selects the value of δ that makes $A_1{}^2$ maximum. The dependence of A_1 on δ was given earlier in equation (7.33). It is obvious that the wavenumber making $A_1{}^2$ the largest is just

$$\delta = 0. \tag{7.35}$$

Near threshold and for an infinitely long plate, the 'optimal' wavenumber is the one that first becomes unstable at threshold. However, for a large but finite plate, the experimentally observed wavenumber is not necessarily the one that yields the lowest energy. As we shall see in the next section, there may exist other local minima of the energy that are linearly stable, and may therefore be observed. This comes from the selection of the wavenumber a by end effects.

7.6 Wavenumber selection by end effects

As shown in the previous section, minimization of the energy provides one possible way to select the wavenumber above the instability threshold—this selection mechanism applies when the plate is infinitely long. In different geometries, different mechanisms may apply. In the present section, we first consider the case of a half-infinite plate, and afterwards the case of a finite length along y. We show that the wavenumber is selected by end effects and not only by a minimization of the elastic energy in the bulk.

7.6.1 Outline

The idea of selection by end effects requires some introduction. It follows from the remark that, because of the translational invariance of the F.–von K. equations in the y direction, there is an associated Noether invariant. This is a function of the solution that can be computed explicitly for any value of y and that turns out to be independent of y. The explicit expression of this invariant is derived in Section 7.6.2.

The next three sections are devoted to an explicit calculation of this invariant near the buckling threshold. Section 7.6.3 deals with the simple case of a periodic bifurcated buckling pattern, relevant for computing the invariant in the bulk, that is far from the short edge. The derivation of the invariant at the short edge is done in two steps. Section 7.6.4 is devoted to a derivation of the so-called amplitude equation. Near threshold, it yields a solution for the merging of the solution in the bulk and near the short edge—this merging takes place on distances far larger than the wavelength of the pattern. In Section 7.6.5, building upon this amplitude solution, one derives the expression of the Noether invariant at the short edge. This yields the selection of the wavelength for the case of a semi-infinite strip: the wavenumber must lie in a narrow interval, much narrower than that permitted for an infinitely extended pattern. Lastly, in Section 7.6.6, we explain how this picture is to be changed for the realistic case of a long but finite strip. There the pattern is obtained by gluing two half-infinite solutions, the matching of the solutions in the middle giving a condition of quantization of the wavenumber within the narrow band already mentioned.

7.6.2 A buckling invariant

The selection mechanism is based on the variational structure of the equations, and on the fact that they are invariant by translation along the long direction y. Under these conditions, a Noether invariant[16] exists. This invariant takes the form of a scalar quantity that, for any equilibrium solution of the F.–von K. equations, takes a constant value along the strip length. This invariant relates the profile of the solution along any transverse section of the strip at a coordinate y to that at any other coordinate y'. This property can be used to match the solution in the bulk of the strip to its profile near its short ends, something that is at the heart of the selection mechanism. This selection via the end effects operates in non-variational systems (Y. Pomeau and S. Zaleski, 1981) as well, where an adiabatic invariant replaces the exact Noether invariant–see Section 4.4.3 of this book about adiabatic invariants.

A general theory yields the expression of the Noether invariant from the energy of the system for any particular symmetry. In the specific case at hand, this leads one to introduce the quantity $M(y)$ as:

$$M(y) = \int_{-1/2}^{1/2} \mathrm{d}x \; \left(- [\partial_\beta(\sigma_{\alpha\beta})] \, u_{\alpha,y} + [\Delta^2 w + P\, w_{,yy} - \partial_\beta(\sigma_{\alpha\beta}\, w_{,\alpha})] \, w_{,y} \right). \tag{7.36}$$

Here, we have introduced the shorthand notation ∂_i for the partial derivative with respect to x_i. We shall not justify the form of $M(y)$ proposed above, and refer the reader to textbooks for the theory of Noether invariants.[17] In the integrand, the first quantity in square brackets is just the left-hand side of the equation (7.6c) for in-plane equilibrium, and it vanishes for equilibrium solutions. The second quantity in square brackets can be written as $\Delta^2 w + P\, w_{,yy} - \sigma_{\alpha\beta}\, w_{,\alpha\beta} - \sigma_{\alpha\beta,\beta}\, w_{,\alpha}$. This is the sum of the left-hand sides of the two equations of equilibrium, written earlier in equations (7.6c) and (7.6d), in the particular case $\nu = 0$

---

[16] Noether invariants are conserved quantities found in physical systems, that reflect the presence of continuous symmetries in the equations, and their variational structure. Classical examples of Noether invariants include the linear momentum in Lagrangian mechanics, related to the invariance under space translation, and the conservation of energy, related to the invariance under time translation.

[17] To give more details on this derivation, the form of the integrand in equation (7.36) reads $(\delta\mathcal{F}/\delta u_\alpha)\, u_{\alpha,y} + (\delta\mathcal{F}/\delta w)\, w_{,y}$, according to the theory of Noether invariants. Here, $(\delta\mathcal{F}/\delta u_\alpha)$ and $(\delta\mathcal{F}/\delta w)$ denote the variation of the elastic energy (6.95) associated with the F.–von K. equations, with respect to the unknown functions $u_\alpha(x,y)$ and $w(x,y)$. We have also made use of the rescalings and considered the change of reference configuration (see Section 7.3.3) to obtain equation (7.36).

considered. As a result, this second square bracket vanishes as well for any equilibrium solution. As a result,

$$M(y) = 0 \tag{7.37}$$

for any solution (u_x, u_y, w) of the equations for the strip.

So far, we have constructed a quantity $M(y)$ that vanishes everywhere, for any equilibrium solution. This may look rather uninteresting. However, the remarkable property of this quantity $M(y)$ is that it is an exact derivative: there exists a second quantity $N(y)$, called the Noether invariant, such that $M(y) = \mathrm{d}N(y)/\mathrm{d}y$. By equation (7.37), this new quantity $N(y)$ will be independent of y, for any solution of the F.–von K. equations:

$$\frac{\mathrm{d}N(y)}{\mathrm{d}y} = 0. \tag{7.38}$$

In the rest of this section, we perform a series of algebraic manipulations, to make $M(y)$ explicitly appear as an exact derivative, and provide the expression of $N(y)$. This result is applied in the following sections, where we evaluate the invariant $N(y)$ for various equilibrium profiles of the strip.

Let us first introduce an intermediate result. For any smooth function $f(x,y)$, the following identity can be established by expanding all derivatives:

$$f_{,y}\,\Delta^2 f = -\frac{\mathrm{d}(Bf)}{\mathrm{d}x} + \frac{1}{2}\frac{\mathrm{d}(Cf)}{\mathrm{d}y}, \tag{7.39a}$$

where we have introduced the non-linear differential operators B and C:

$$(Bf) = f_{,x^2}\,f_{,xy} - f_{,x^3}\,f_{,y} - 2\,f_{,y}\,f_{,xy^2} \tag{7.39b}$$

$$(Cf) = (f_{,x^2})^2 - 2\,(f_{,xy})^2 + 2\,f_{,y}\,f_{,y^3} - (f_{,y^2})^2. \tag{7.39c}$$

Equation (7.39a) allows one to integrate by parts the biharmonic term in equation (7.36) as:

$$M_1 = \int_{-1/2}^{1/2} \mathrm{d}x\,\left([\Delta^2 w]\,w_{,y}\right) = \frac{1}{2}\frac{\mathrm{d}}{\mathrm{d}y}\left(\int_{-1/2}^{1/2}\mathrm{d}x\,(Cw)\right) - \Big[(Bw)\Big]_{x=-1/2}^{+1/2}. \tag{7.40a}$$

The last term in this equation is a boundary term[18] denoting the variation, for any particular y, of the quantity in brackets from $x = -1/2$ to $x = +1/2$. Now, on the edges $x = \pm 1/2$, the boundary conditions (7.7) impose $w = 0$ and $w_{,xx} = 0$; as a result, $(Bw) = 0$ there, and the boundary term vanishes. The remaining term in the right-hand side is an exact derivative, which yields the first contribution to the Noether invariant $N(y)$.

We now proceed to the other terms in equation (7.36). The easiest one, proportional to the loading P, can be directly rewritten as a derivative:

$$M_2 = \int_{-1/2}^{1/2} \mathrm{d}x\,P\,w_{,yy}\,w_{,y} = \frac{1}{2}\frac{\mathrm{d}}{\mathrm{d}y}\left(\int_{-1/2}^{1/2}\mathrm{d}x\,P\,{w_{,y}}^2\right). \tag{7.40b}$$

---

[18] There is no conflict of notation between the square brackets for boundary terms, which bear subscripts and superscripts, those for the Monge–Ampère operator, like in the Gauss curvature, $[w,w]/2$, which bear a comma in-between, and those for regular parenthesis as in the integrand of (7.36), which are not decorated.

Finally, we consider the two remaining terms in the integrand defining $M(y)$. They are first integrated by parts:

$$M_3 = \int_{-1/2}^{+1/2} \mathrm{d}x \; (-\partial_\beta(\sigma_{\alpha\beta})\, u_{\alpha,y} - \partial_\beta(\sigma_{\alpha\beta}\, w_{,\alpha})\, w_{,y})$$

$$= -\int_{-1/2}^{+1/2} \mathrm{d}x \; \frac{\partial[\sigma_{\alpha\beta}\,(u_{\alpha,y} + w_{,\alpha}\, w_{,y})]}{\partial x_\beta} + \int_{-1/2}^{+1/2} \mathrm{d}x \; \sigma_{\alpha\beta}\,(u_{\alpha,y\beta} + w_{,\alpha}\, w_{,\beta y}).$$

When $\beta = x$, the integrand of the first term in the right-hand side is an exact derivative and the integral evaluates to a boundary term $[\sigma_{\alpha x}\,(u_{\alpha,y} + w_{,\alpha}\, w_{,y})]_{x=-1/2}^{+1/2}$. This boundary term is zero since u_α and w vanish for all y along the edges $x = \pm 1/2$ by the boundary conditions (7.7), and so do their derivatives along these edges, $u_{\alpha,y}$ and $w_{,y}$. Therefore, we take $\beta = y$ in this term. Permuting the derivative and the integration, one obtains

$$M_3 = \frac{\mathrm{d}}{\mathrm{d}y}\left(-\int_{-1/2}^{+1/2} \mathrm{d}x \; (\sigma_{\alpha y}\,(u_{\alpha,y} + w_{,\alpha}\, w_{,y}))\right) + \int_{-1/2}^{+1/2} \mathrm{d}x \; \sigma_{\alpha\beta}\,(u_{\alpha,y\beta} + w_{,\alpha}\, w_{,\beta y}).$$

To make the second integral in M_3 appear as an exact derivative, we use the trick given in the end of Section 1.2.2; since $\sigma_{\alpha\beta}$ is a symmetric tensor that gets contracted with the tensor $(u_{\alpha,y\beta} + w_{,\alpha}\, w_{,\beta y})$, we can replace the latter by its symmetric part:[19]

$$\sigma_{\alpha\beta}\,(u_{\alpha,y\beta} + w_{,\alpha}\, w_{,\beta y}) = \sigma_{\alpha\beta}\left(\frac{u_{\alpha,y\beta} + u_{\beta,y\alpha}}{2} + \frac{w_{,\alpha}\, w_{,\beta y} + w_{,\beta}\, w_{,\alpha y}}{2}\right)$$

$$= \sigma_{\alpha\beta}\,\frac{\partial}{\partial y}\left(\frac{u_{\alpha,\beta} + u_{\beta,\alpha}}{2} + \frac{1}{2}\, w_{,\alpha}\, w_{,\beta}\right) = \sigma_{\alpha\beta}\,\frac{\partial \epsilon_{\alpha\beta}}{\partial y}$$

$$= \frac{1}{2}\,\frac{\partial(\sigma_{\alpha\beta}\, \epsilon_{\alpha\beta})}{\partial y}.$$

To write the last equality, we have used a result established at the end of Section 2.4.2, namely that $\delta(\sigma_{\alpha\beta}\, \epsilon_{\alpha\beta})/2 = \sigma_{\alpha\beta}\, \delta\epsilon_{\alpha\beta}$, where δa stands for any infinitesimal variation of a quantity a. The equality holds in particular when δ is replaced with ∂_y.

We can now rewrite the last piece M_3 as:

$$M_3 = \frac{\mathrm{d}}{\mathrm{d}y}\left(\int_{-1/2}^{+1/2} \mathrm{d}x \; \left(-\sigma_{\alpha y}\,(u_{\alpha,y} + w_{,\alpha}\, w_{,y}) + \frac{\sigma_{\alpha\beta}\, \epsilon_{\alpha\beta}}{2}\right)\right).$$

To clean up this expression, we symmetrize the tensor doubly contracted with $\sigma_{\alpha\beta}$, and expand the second double contraction. Some terms cancel out, leading to:

$$M_3 = \frac{1}{2}\frac{\mathrm{d}}{\mathrm{d}y}\left(\int_{-1/2}^{+1/2} \mathrm{d}x \; (\sigma_{xx}\, \epsilon_{xx} - \sigma_{yy}\, \epsilon_{yy} - \sigma_{\alpha y}\, w_{,\alpha}\, w_{,y})\right). \tag{7.40c}$$

---

[19] The symmetric part of a tensor a_{ij} is $(a_{ij} + a_{ji})/2$.

Each of the pieces forming $M = M_1 + M_2 + M_3$ has now been rewritten as an exact derivative. Collecting everything, we obtain an explicit formula for the Noether invariant as:

$$N(y) = \frac{1}{2} \int_{x=-1/2}^{+1/2} \mathrm{d}x \, \left((Cw) + P \, w_{,y}{}^2 + \sigma_{xx} \, \epsilon_{xx} - \sigma_{yy} \, \epsilon_{yy} - \sigma_{\alpha y} \, w_{,\alpha} \, w_{,y} \right), \qquad (7.41)$$

where the operator (Cw) is defined in equation (7.39c).

We have established that the quantity $N(y)$ defined above is the anti-derivative of $M(y)$, provided the displacement (u_x, u_y, w) satisfies the boundary conditions. Moreover, when this triple is a solution of the F.–von K. equations for the strip, $M(y) = 0$ and $N(y)$ is independent of y.

7.6.3 Calculation of the invariant in the bulk

Let us consider a strip that is clamped along one of its short ends (or both, see Fig. 7.4). The solutions of the F.–von K. equations far from the edges is the same as that studied above in Section 7.4 for an infinite length. This solution has to merge smoothly with the ends. This merging takes place under the constraint that the value of the invariant $N(y)$ is the same near the ends and in the bulk. This brings in two questions: how to compute N in the bulk on the one hand (which is relatively easy), and near the two short edges on the other (which is more difficult)? The invariant in the bulk is calculated in the present Section; the solution near the end is derived in section 7.6.4 and the invariant there is computed in Section 7.6.5.

The invariant in the bulk can be found by plugging into the explicit expression of N the dominant terms of the weakly non-linear solution for u_α and w computed in Section 7.5.

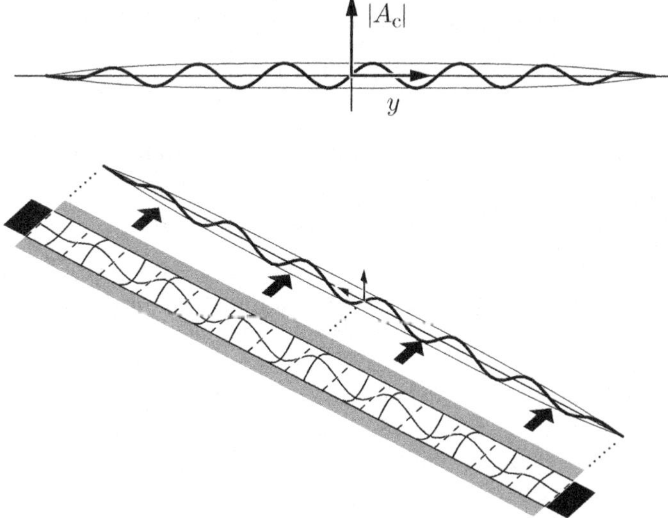

Fig. 7.4 Merging of a periodic pattern of bumps with the ends. The boundary conditions are free support along the lateral edges (grey rectangles), and clamping along the short edge (black rectangles). The envelope is a hyperbolic tangent near each end, as studied in Section 7.6.5. Matching these solutions with the solution of the bulk is constrained by conservation of the Noether invariant N, see Section 7.6.6.

Then, it is only a matter of elementary (but lengthy) calculations to compute the expansion for $N(y)$. It turns out that the dominant contribution to $N(y)$ is third order in ϵ, and reads:

$$N(y; \epsilon, \delta) = -\frac{16\,\pi}{3}\left(1 - 4\,\delta^2\right)\delta\,\epsilon^3 + \cdots \tag{7.42}$$

As expected, for any solution parameterised by ϵ and δ, the invariant $N(y)$ is in fact independent of y.

To obtain this expression, we have used the expansions (7.20) for the loading P and (7.28) for wavenumber a. the solubility condition (7.33) has been used to evaluate the squared amplitude $A_1{}^2$ whenever needed, hence the factor $(1 - 4\,\delta^2)$.

As we shall see later, the Noether invariant is a fourth-order quantity with respect to ϵ near the clamped end. To match it with its value in the bulk, we need to determine the latter one order further. This requires us to first push the expansion of the wavenumber a further:

$$a = \pi + \delta\,\epsilon + \delta'\,\epsilon^2 + \ldots, \tag{7.43}$$

where we have introduced $\delta' = a^{(2)}$. We are investigating the selection of the longitudinal wavelength, which implies finding the values of δ and δ' chosen by the system. The third-order calculation (7.42) of the invariant in the bulk is complemented with the next order as follows:

$$N(y; \epsilon, \delta, \delta') = -\frac{16\,\pi}{3}\left(1 - 4\,\delta^2\right)\delta\,\epsilon^3 + \frac{1}{3}\left(1 - 4\,\delta^2\right)\left(1 - 28\,\delta^2 - 16\pi\,\delta'\right)\epsilon^4 + \cdots \tag{7.44}$$

A particular case that will turn out to be important is when $\delta = a^{(1)} = 0$. Then, the above expression for the Noether invariant far from the edges simplifies to:

$$N(y; \epsilon, \delta = 0, \delta') = \frac{1 - 16\pi\,\delta'}{3}\,\epsilon^4 + \cdots \tag{7.45}$$

In the next two sections, we compute the value of this invariant near the short edges.

7.6.4   Solution near a short edge: amplitude equation

It remains to find the invariant N at a short edge that is clamped. The periodic solution of constant amplitude, derived above by expansion in powers of A_1, is not valid there as it does not satisfy the clamped boundary condition. Therefore one has to devise a strategy to merge the solution in the bulk and close to a short edge. This is done in this section, by allowing the amplitude of the solution to vary slowly. We derive an amplitude equation that is the limit of the original F.–von K. equations for small gradients of the amplitude. This amplitude theory has applications in many problems in mechanics and in physics. It was developed by Segel and by Newell and Whitehead (L. A. Segel, 1969; A. C. Newell and J. A. Whitehead, 1969) for Rayleigh–Bénard convection between parallel plates with free boundary condition.[20] In the present case, the general amplitude equation can be derived straightforwardly. This is first because there is no dynamics in the problem at hand and second because the variation occurs only in one space direction, along the coordinate y.

---

[20] It was pointed out later (E. D. Siggia and A. Zippelius, 1981) that, for this specific physical situation, the amplitude equation is more complicated than usual because of the possibility of large scale flows.

The periodic solutions of the F.–von K. equations in series of powers of A_1 near threshold depend on two quantities, the amplitude A_1 and the phase of the circular function: the factor $\sin(a\,y)$ in $w(x,y)$ can be changed to $\sin(a\,y + \phi)$, where ϕ is a constant arbitrary phase. These solutions therefore belong to a doubly continuous family, parameterised by an amplitude and a phase. These parameters have no reason to remain constant, and one may wonder what happens when the real amplitude A_1 and the phase ϕ depend weakly, in a sense to be made precise later, on the space[21] variable y. This slow dependence gives rise to a new class of post-buckling solutions, which includes the one relevant for the merging with the lateral edge: the slow variation of the amplitude allows the solution that is periodic in y far from the short edge to decay to zero so as to satisfy the boundary condition on the short edge. The separation of scale between the wavelength of the solution, $2\pi/b = 2$ in our dimensionless units, and the typical distance of variation of the amplitude, of order $\epsilon^{-1} \gg 2$, is crucial in this analysis.

Let us first assume that the solution of the F.–von K. equations oscillates with a slowly varying complex amplitude. This leads one to replace the assumed lowest order solution $A(x,y) = \epsilon\, A_1 \sin(a\,y) \cos(\pi\,x)$, where A_1 was a real constant, by

$$A(x,y) = \epsilon\,\frac{A_{\mathrm{c}}(y)\,e^{i\pi y} + A_{\mathrm{c}}^*(y)\,e^{-i\pi y}}{2}\,\cos(\pi\,x), \tag{7.46}$$

where $A_{\mathrm{c}}(y)$ is a slowly varying complex function of y, and $A_{\mathrm{c}}^*(y)$ its complex conjugate. The wavenumber along y is now included both in the rapid phase of $e^{\pm i\pi y}$, where it takes the constant value π, and in an eventual slow variation of the complex amplitude. In particular, by taking $A_{\mathrm{c}}(y) = A_1 e^{i\,\epsilon\,\delta\,y}$, where δ is an arbitrary number, this wavenumber becomes $\pi + \epsilon\,\delta$, as earlier in the analysis of the bulk solution.

The derivation of the equation for $A_{\mathrm{c}}(y)$ from the full original F.–von K. equations is very similar to the calculation of the amplitude A_1 as a function of ϵ and δ presented in Section 7.5. This amplitude was found in equation (7.32), which we rewrite in the following form, after multiplying both sides by ϵ^2:

$$\epsilon^2 A_1 + 4\,(i\,\epsilon\,\delta)^2 A_1 = \frac{3\,\pi^2\,\epsilon^2}{8}\,A_1{}^3,$$

where $i = \sqrt{-1}$. The derivation of the equation for A_{c} can be shortened by rewriting this algebraic equation for A_1 into a differential equation for $A_{\mathrm{c}}(y)$. One substitutes $A_{\mathrm{c}}(y)$ for A_1 in the linear part of this equation and replaces $(i\,\epsilon\,\delta)$ by the derivation operator $\mathrm{d}/\mathrm{d}y$ with respect to the slow variable y. Finally the cubic term $A_1{}^3$ is replaced[22] by $(|A_{\mathrm{c}}|^2 A_{\mathrm{c}})$. The result is another solubility condition, now in the form of an amplitude (differential) equation:[23]

$$\epsilon^2 A_{\mathrm{c}} + 4\,\frac{\mathrm{d}^2 A_{\mathrm{c}}}{\mathrm{d}y^2} = \frac{3\,\pi^2\,\epsilon^2}{8}\,|A_{\mathrm{c}}|^2 A_{\mathrm{c}}. \tag{7.47}$$

---

[21] In time dependent problems, as for patterns in fluids, for instance, a slow *time dependence* of the complex amplitude should also be considered.

[22] The cubic term cannot be written in the form $A_{\mathrm{c}}{}^3$, as this would break the invariance of the equations with respect to an arbitrary phase, $A_{\mathrm{c}} \mapsto A_{\mathrm{c}} \exp(i\,\alpha'')$.

[23] The amplitude equation (7.47) has an interesting structure, being the Euler–Lagrange condition for the extremum of the functional

$$\mathcal{F}(A) = \int \mathrm{d}y \left(-\epsilon^2 |A_{\mathrm{c}}|^2 + 4\left|\frac{\mathrm{d}A_{\mathrm{c}}}{\mathrm{d}y}\right|^2 + \frac{3\pi^2\,\epsilon^2 |A_{\mathrm{c}}|^4}{16} \right).$$

By construction, the solutions with constant amplitude A_1 studied earlier are particular solutions of this amplitude equation: the algebraic relation (7.33) between A_1, ϵ and δ is recovered by putting $A_c(y) = A_1 e^{i \epsilon \delta y}$ in the amplitude equation (7.47).

The equation (7.47) can be solved in general by means of elliptic integrals. We shall only need its simplest non-trivial solution, the hyperbolic tangent:

$$A_c(y) = \frac{2\sqrt{2}}{\pi \sqrt{3}} \tanh\left(\frac{\epsilon y}{2\sqrt{2}}\right) \exp\left(i \alpha''\right). \tag{7.48}$$

Here, we have introduced a constant arbitrary phase α'', which will be useful below. By the definition (7.46) of A_c, the associated deflection reads

$$w(y) = \frac{2\sqrt{2}\,\epsilon}{\pi \sqrt{3}} \tanh\left(\frac{\epsilon y}{2\sqrt{2}}\right) \cos(\pi y + \alpha'') \cos(\pi x). \tag{7.49}$$

This solution interpolates between a vanishing amplitude at $y = 0$ and a non-zero constant value far from the origin. The typical length scale for the variation of the amplitude is ϵ^{-1} which, for small ϵ, is much larger than the wavelength of the buckling pattern, 2 in our dimensionless units. This makes the derivation of the amplitude equation consistent: the space dependence of A_c brings a small perturbation to the equation, as do the non-linear effects, although the fast dependence remains the same everywhere.

In the following section, we extend the solution (7.49) just derived, which is valid near the clamped edge, to represent a half-infinite strip.

7.6.5   Half-infinite strip

We shall now study the case of a half-infinite strip, whose short edge is clamped. We show precisely how these clamped boundary conditions can be accounted for. It turns out that a continuum of half-infinite solutions exists near threshold, with a wavenumber inside a narrow band. The case of a *finite* rectangular plate is studied in the following section, and yields a discrete set of solutions, as expected.

To build the solution for a half-infinite strip, we make use of the solution (7.49) above, which tends to a periodic pattern for large y, and has a vanishing amplitude $A_c(y)$ near the clamped edge $y = 0$. This condition $A_c(y) \to 0$ for small y is necessary but not sufficient for the clamped boundary conditions to be satisfied: by working out these boundary conditions, we derive below relations between the parameters of the hyperbolic tangent solution, which leaves at the end one free parameter, an angle called α'.

The clamped boundary conditions applied on the short edge at $y = 0$ read:

$$w = 0 \qquad \text{and} \qquad \frac{\partial w}{\partial y} = 0. \tag{7.50}$$

These equations express the fact that both the deflection w and the slope $w_{,y}$ of the plate measured perpendicular to the edge vanish along a clamped short edge. There are also boundary conditions for the in-plane displacement, namely $u_x = 0$ and $u_y = 0$. However, these boundary conditions need not be considered, as the in-plane displacement is always very small near the clamped edge.[24]

---

[24] The in-plane displacement, which appears at second order in the expansion for symmetry reasons, is everywhere of order the square of the deflection in our dimensionless units. Close to the clamped edge,

The particular solution (7.49), based on the hyperbolic tangent, is not the generic solution of the amplitude equation. We need to write the most general solution in the vicinity of the clamped edge in order to derive the solution in the half-strip. To do this, we return to the amplitude equation (7.47). Focusing on the vicinity of the edge, where A_c is very small, we neglect the non-linear term in the right-hand side. The two remaining terms in the left-hand side balanced each other for $y \sim \epsilon^{-1}$. We consider a neighborhood of the clamped edge of width much smaller than ϵ^{-1}, which allows us to drop the term $\epsilon^2 A_c$ as well. We are left with the following equation

$$2 \frac{\mathrm{d}^2 A_c}{\mathrm{d}y^2} = 0. \tag{7.51}$$

Its solutions are affine functions of y with complex coefficients, which we write in polar form as

$$A_c = i\,\gamma\,e^{i\,\alpha} + \gamma'\,e^{i\,\alpha'}\,y,$$

where γ, α, γ' and α' are four real parameters. The extra factor i in front of γ is a purely a matter of convenience and could be incorporated into the phase α. In terms of the deflection, this solution reads

$$w(x,y) = [\gamma \sin(\pi y + \alpha) + \gamma' y \cos(\pi y + \alpha')]\,\cos(\pi x). \tag{7.52}$$

The boundary condition (7.50) imposes

$$\alpha = 0 \qquad \text{and} \qquad \gamma = -\frac{\gamma' \cos\alpha'}{\pi},$$

which allows one to rewrite the generic solution (7.52) as a function of two real parameters, the angle α' and the amplitude γ for instance:

$$w(x,y) = \gamma' \left[y \cos(\pi y + \alpha') - \frac{\cos\alpha'}{\pi} \sin(\pi y) \right] \cos(\pi x). \tag{7.53}$$

Because of the approximations underlying equation (7.51), this solution is not valid for arbitrary values of y. This is obvious from the presence of a *secular* term[25] proportional to y, which becomes unbounded for large values of y: the linearization used to derive this solution breaks down when the amplitude of this secular term becomes too large.

This defines a matching problem:[26] equation (7.53) defines the so-called inner solution, which varies asymptotically as a power law of the variable in general, this power law being here a simple linear law. For large y, the secular term is indeed dominant in equation (7.53):

$$w(x,y) \approx y\,\gamma' \cos(\pi y + \alpha') \cos(\pi x). \tag{7.54}$$

This equation is valid for $y \gg 1$ but not too far away from the edge either, as the secular term cannot become arbitrarily large.

the deflection is very small compared with its value in the bulk. As a result, the in-plane displacement is negligible near the clamped edge, and the conditions $u_x = 0$ and $u_y = 0$ there are automatically satisfied.

[25] Such a term is called *secular*, after Lagrange, because in perturbations of the Keplerian orbit of a planet due to other planets, the effect of this type of term would be noticeable after centuries, *saeculae* in Latin.

[26] In Appendix B, matching problems are discussed in the context of boundary layer theory.

The outer solution, valid not too close from the edge, is given by the hyperbolic tangent solution in equation (7.49); when its argument is small, that is for $y \ll \epsilon^{-1}$, this hyperbolic tangent can be replaced by its argument which yields, from equation (7.49):

$$w(y) \approx \frac{\epsilon^2 \, y}{\pi \sqrt{3}} \, \cos(\pi \, y + \alpha'') \, \cos(\pi \, x), \quad \text{for } 1 \ll y \ll \epsilon^{-1}. \tag{7.55}$$

This is the expression for the outer solution in the intermediate range $1 \ll y \ll \epsilon^{-1}$, i.e. at a distance from the edge that is large compared with the wavelength, but small compared with the length scale ϵ^{-1}.

We arrive at the matching problem sketched in Fig. 7.5. The inner and outer solutions given in equations (7.54) and (7.55), although derived by different approximation schemes, represent the same physical function. They should match in the so-called intermediate region, $1 \ll y \ll \epsilon^{-1}$, where they are both valid. The matching is possible only if both the phase of the oscillations and the coefficient of their linear envelope are equal, see Fig. 7.5. This leads to two new relations:

$$\gamma' = \frac{\epsilon^2}{\pi \sqrt{3}} \quad \text{and} \quad \alpha' = \alpha'', \tag{7.56}$$

which are the matching conditions for the inner and outer solutions.

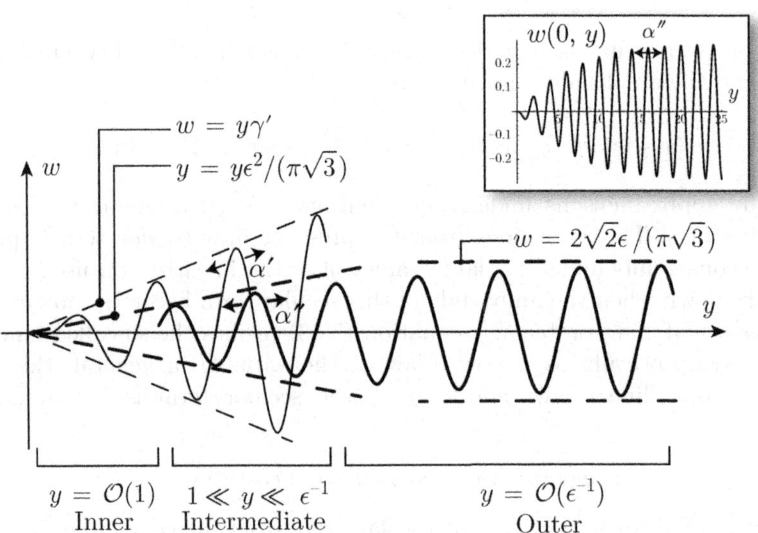

Fig. 7.5 Matching problem near the clamped short edge at $y = 0$. The outer solution, shown with a thick line, has a hyperbolic tangent as envelope, while the inner solution is given by equation (7.53). At dominant order, a correct matching requires one to adjust both the relative phase of the solutions and the slope of their linear envelope. The matched solution is shown in the inset for $\epsilon = 0.3$ and $\alpha'' = 0$. Matching allows one to compute the phase of the solution in the bulk, as fixed by the clamped boundary conditions.

Using the matching conditions (7.56), one can rewrite the inner and outer solutions as functions of the loading ϵ and of the free parameter α'':

$$w(x,y) = \cos(\pi x) \times \begin{cases} \dfrac{\epsilon^2}{\pi\sqrt{3}} \left(y\cos(\pi y + \alpha'') - \dfrac{\cos\alpha''}{\pi}\sin(\pi y) \right) & \text{for } y \ll \epsilon^{-1}, \\[2ex] \dfrac{2\sqrt{2}\,\epsilon}{\pi\sqrt{3}} \tanh\left(\dfrac{\epsilon\,y}{2\sqrt{2}}\right) \cos(\pi y + \alpha'') & \text{for } y \gg 1. \end{cases} \tag{7.57}$$

By construction, these functions match in the region of overlap, $1 \ll y \ll \epsilon^{-1}$ (intermediate region), in the limit of small ϵ.

The asymptotic behaviour of this matched solution far from the edge can be obtained by taking the limit $y \gg \epsilon^{-1}$ in the second part of equation (7.57). This yields:

$$w(x,y) \approx \frac{2\sqrt{2}\,\epsilon}{\pi\sqrt{3}} \cos(\pi y + \alpha'')\cos(\pi x) \qquad \text{for } y \gg \epsilon^{-1}. \tag{7.58}$$

Let us compare this with the solution in the bulk, derived in equations (7.12) and (7.33) by a weakly non-linear analysis:

$$w(x,y) = \frac{2\sqrt{2}\,\epsilon}{\pi\sqrt{3}} \sqrt{1 - 4\delta^2}\,\cos(a\,y + \alpha''')\cos(\pi x). \tag{7.59}$$

In this equation, we have replaced the original factor $\sin(a\,y)$ with the more general form $\cos(a\,y + \alpha''')$ by introducing an arbitrary phase α'''. This is made possible by the invariance by translation along y.

One expects to recover the bulk behaviour by taking the limit of large y in our matched solution representing the neighbourhood of the edge $y = 0$. However, comparing equations (7.58) and (7.59) reveals two discrepancies, which one has to get rid of. First, the factor $\sqrt{1 - 4\delta^2}$ is present in the second equation only. This discrepancy is suppressed by requiring

$$\delta = 0, \tag{7.60}$$

where we recall that $\delta = a^{(1)}$ is the first-order correction to the longitudinal wavenumber a. The second difference is in the phase, which reads $(\pi y + \alpha'')$ in the first equation and $(a\,y + \alpha''')$ in the second one. This discrepancy looks more serious, as it is impossible to tune two phases with different wavenumbers, π and a, over the whole bulk of the strip: the phases drift with respect to each other by about $(a - \pi)\Lambda$ over the strip length. Although $(a - \pi)$ is small, the length of the plate Λ is arbitrarily large, and the drift cumulated over the plate length may be significant.

This second discrepancy is removed by noticing that the phase drift at position y is of order $(a - \pi)\,y \approx \epsilon^2\,y\,\delta'$ since the expansion of a reads $a = \pi + \epsilon^2\,\delta' + \cdots$ when $\delta = 0$. This phase drift remains negligible as long as $y \ll \epsilon^{-2}$. The solution (7.57) is not valid very far from the edge, but only within the range $y \ll \epsilon^{-2}$. This solution has to be matched with equation (7.59) further away from the edge, where the perturbation $\delta'\epsilon^2$ to a becomes significant. This requires $\alpha''' = \alpha''$, and the solution very far from the edge takes the form:

$$w(x,y) = \frac{2\sqrt{2}\,\epsilon}{\pi\sqrt{3}} \cos((\pi + \epsilon^2\,\delta')\,y + \alpha'')\cos(\pi x) \qquad \text{for } y \sim \epsilon^{-2} \text{ or larger.} \tag{7.61}$$

This construction of the function $w(x, y)$ with two matchings given by equations (7.57) and (7.61) can be summarized in a concise way:

$$
w(x, y) = \cos(\pi x) \times
\begin{cases}
\frac{\epsilon^2}{\pi \sqrt{3}} \left(y \cos(\pi y + \alpha'') - \frac{\cos \alpha''}{\pi} \sin(\pi y) \right) & \text{for } y \ll \epsilon^{-1}, \\
\frac{2\sqrt{2}\,\epsilon}{\pi \sqrt{3}} \tanh\left(\frac{\epsilon y}{2\sqrt{2}} \right) \cos((\pi + \epsilon^2 \delta')\,y + \alpha'') & \text{for } y \gg 1.
\end{cases}
\tag{7.62}
$$

We have simply corrected the wavelength in the second part of equation (7.57) to make it consistent with equation (7.61). As we showed, this correction extends the range of validity of our solution to very large values of y, namely $y = \mathcal{O}(\epsilon^{-2})$ and beyond.

To complete this construction of the solution $w(x, y)$, it remains to express the perturbation δ' to the wavenumber in terms of the free parameter α''. This number δ' appears in the outer solution, given by the second part of equation (7.62), but is absent from the inner solution, given by the upper block. Now, there is no reason to introduce an additional degree of freedom in the process of matching and one should be able to express this δ' as a function of the parameters of the inner solution, α'' and ϵ. This is indeed possible in principle by pushing this boundary layer analysis to the next order in ϵ. In practice, this is most cumbersome.

To circumvent the calculation of the solution to next order, we use the conservation of the Noether invariant and impose that $N(y)$ takes the same value in the inner and outer regions. This leads to an additional relation, which ultimately yields δ' as a function of α''. The invariant can be computed near the clamped edge by plugging the inner solution, given by the upper block in equation (7.62) into the definition (7.41) of the invariant.[27] After integration of trigonometric functions, we obtain:

$$
N = -\frac{2\,\epsilon^4 \sin^2 \alpha''}{3}
$$

in the inner region and at dominant order in ϵ.

The calculation of this invariant in the outer region has already been done in equation (7.45):

$$
N = \frac{1 - 16\pi\, \delta'}{3}\, \epsilon^4.
$$

By identification of the two values of the invariant $N(y)$, we deduce:

$$
\delta' = \frac{(1 + 2 \sin^2 \alpha'')}{16\pi}.
\tag{7.63}
$$

As anticipated, δ' depends directly on the angle α''. This equation completes our construction of the solution $w(x, y)$ near a clamped edge. Together with (7.62), it yields the post-buckled solution for a half-infinite strip as a function of the applied loading ϵ and a free parameter,[28] α'', which can be interpreted as the free phase in the bulk.

---

[27] When we studied the clamped boundary conditions, we mentioned that the in-plane displacement is formally negligible near the clamped edge, where the amplitude is very small; see footnote 24. For the same reason, the last terms in the integrand defining $N(y)$ in equation (7.41), which scale like the in-plane strain and stress or their square, are negligible in front on the first two terms, which depend only on w, close to the clamped edge. In the calculation of $N(y)$ near the clamped edge, only the first two terms have to be considered in the integrand.

[28] In the case of a half-infinite strip, it is possible in principle to show which value of α'' yields the minimum of elastic energy, by expanding the energy at the right order. However, we shall not attempt to

Equation (7.63) restricts the range of wavenumber that can be found in a half-infinite strip to a bandwidth of order ϵ^2:

$$\pi + \epsilon^2\,\delta'_- \le a \le \pi + \epsilon^2\,\delta_+, \tag{7.64a}$$

where

$$\delta'_- = \frac{1}{16\,\pi} \quad \text{and} \quad \delta'_+ = \frac{3}{16\,\pi}. \tag{7.64b}$$

This range is much narrower than that for an infinite strip, which is of order $\epsilon = \sqrt{P - P_{\mathrm{c}}}$ when $\delta \ne 0$, as revealed by the parabolic profile in Fig. 7.3.

In the following section, the case of a long but finite plate is studied.

7.6.6   Strip of large but finite length

Let us now consider a plate of finite but large length Λ, whose short ends located at $y = 0$ and $y = \Lambda$ are clamped. The solution for such a plate is found by gluing together two half-infinite solutions of the previous section, each one merging with one short edge of the plate. In the bulk, the global solution must have a unique wavenumber along y, that is $a \approx \pi + \epsilon^2\,\delta'$. Besides, there is another condition imposing that the two expressions of the solution in the bulk have the same phase.

Let α''_0 and α''_Λ be the phases of the half-infinite solutions built in equation (7.62) at the ends at $y = 0$ and at $y = \Lambda$ respectively. The uniqueness of the wavenumber in the bulk, which allows one to match these two solutions, yields:

$$\delta' = \frac{(1 + 2\sin^2\alpha''_0)}{16\pi} = \frac{(1 + 2\sin^2\alpha''_\Lambda)}{16\pi}, \tag{7.65}$$

and so $(\sin\alpha''_0)$ and $(\sin\alpha''_\Lambda)$ must be equal or opposite. Calling $\mu(\delta')$ one particular root of this equation for α'':

$$\mu(\delta') = \arcsin\left(\sqrt{\frac{16\pi\,\delta' - 1}{2}}\right) \in \left[0, \frac{\pi}{2}\right], \tag{7.66a}$$

the general solutions for α''_0 and α''_Λ read:

$$\alpha''_0(\delta') = \pm\mu(\delta') + k_0\,\pi \quad \text{and} \quad \alpha''_\Lambda(\delta') = \pm\mu(\delta') + k_\Lambda\,\pi, \tag{7.66b}$$

where the general solution for α''_0 and α''_Λ is expressed using two arbitrary integers k_0 and k_Λ and two arbitrary signs. These signs in front of $\mu(\delta')$ can be chosen independently for α''_0 and for α''_Λ.

Furthermore there is a condition of compatibility of the phase between the two ends. The pattern constructed from the end at $y = 0$ is given by equations (7.62) with $\alpha'' = \alpha''_0$, while the one constructed from the other end is given by the same equation but with $\alpha'' = \alpha''_\Lambda$ and y replaced by $(\Lambda - y)$. This yields two formulae for the deflection as functions of y, which are identical in the bulk where the hyperbolic tangent is equal to one, up to an oscillatory term which reads:

$$\cos[(\pi + \epsilon^2\,\delta')\,y + \alpha''_0] \quad \text{and} \quad \cos[(\pi + \epsilon^2\,\delta')\,(\Lambda - y) + \alpha''_\Lambda]$$

do this here. The case of a strip of finite length is fully treated in Section 7.6.6: then, the selection of the wavenumber in the bulk does not follow from a principle of energy minimization, which would require one to carry the ϵ expansion at next order.

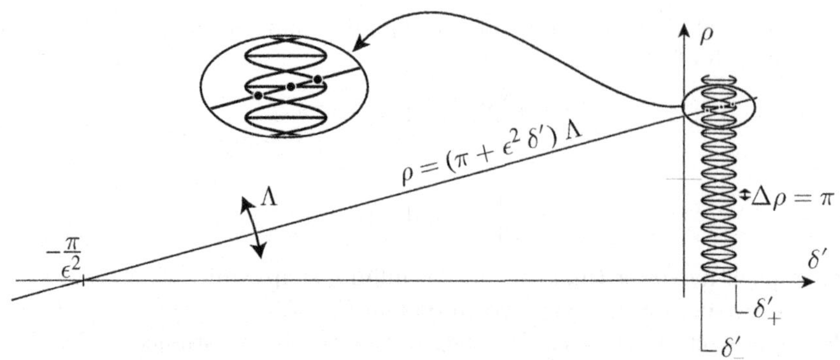

Fig. 7.6 Graphical solution of the quantization condition in equation (7.67a) for given values of the plate length Λ and loading ϵ. This implicit equation for δ' is represented by the line, the ladder structure on the right-hand side being obtained by combining equations (7.66a) and (7.67b) to express ρ as a function of δ'. Intersections of the line and the ladder, shown by the small discs, yield the values of δ' that allow a smooth matching of the solutions constructed from either end.

in either solution. For these two factors to be equal, the arguments of the cosine function have to be equal everywhere, or opposite everywhere, modulo 2π. They cannot be equal, as they vary in opposite directions with y, and so must be opposite:

$$(\pi + \epsilon^2\,\delta')\,y + \alpha''_0 = -[(\pi + \epsilon^2\,\delta')\,(\Lambda - y) + \alpha''_\Lambda].$$

This equality is *a priori* valid modulo 2π, but it can be turned into a regular equality by a proper choice of k_0 and k_Λ. From the equation above, we deduce the following quantization condition:[29]

$$(\pi + \epsilon^2\,\delta')\,\Lambda = \rho(\delta') \tag{7.67a}$$

where

$$\rho(\delta') = -(\alpha''_0(\delta') + \alpha''_\Lambda(\delta')).$$

Using equation (7.66b), this phase ρ can be rewritten as:

$$\rho(\delta') = k\,\pi + \begin{cases} -2\,\mu(\delta') \\ 0 \\ +2\,\mu(\delta') \end{cases}. \tag{7.67b}$$

In this formula, the brace refers to the three possible values of the sum $(\pm\mu \pm \mu)$ arising by addition of α''_0 and α''_Λ, while $k = -(k_0 + k_\Lambda)$ is an integer. Like $\mu(\delta')$, defined in equation (7.66a), the quantized phase ρ is a function of δ'—and of an additional integer k.

The selection of the wavenumber δ' and the matching of the solutions constructed from the two short ends ultimately involve solving equation (7.67a) for δ', with prescribed values of the plate length Λ and loading ϵ. This is done graphically in Fig. 7.6.

---

[29] Equation (7.67a) is called a quantization condition as it imposes the value of the phase variation ($a\,\Lambda$) along the length of the plate, as happens in similar mathematical problems arising in quantum mechanics.

Now, when the strip length Λ varies with a fixed loading ϵ, the slope of the line changes in this figure. By following the solutions δ', one is then lead to the diagram in Fig. 7.7. The function $\delta'(\Lambda)$ is multivalued, as there are multiple intersections in Fig. 7.6.

On the other hand, when the strip length Λ is fixed with a varying loading ϵ, one obtains the diagram in Fig. 7.8. The oscillating curves for $P(a)$ were found by elimination of μ in the previous equations:

$$P - P_{\mathrm{c}} = \epsilon^2 = \frac{16\pi\,\delta'}{1 + 2\,\sin^2\mu}\,\epsilon^2 = \frac{16\pi\,(a - \pi)}{2 - \cos(2\,\mu)}$$

$$= \frac{16\pi\,(a - \pi)}{2 - \cos(a\,\Lambda - k\,\pi)} = \frac{16\pi}{2 \pm \cos(a\,\Lambda)}\,(a - \pi). \tag{7.68}$$

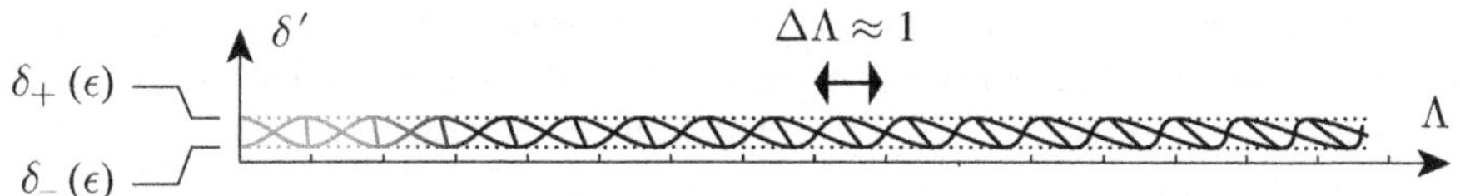

Fig. 7.7 Possible values of the rescaled wavenumber δ', for fixed loading ϵ, when the strip length Λ is changed. The curves are obtained by tracking the solutions δ' in Fig. 7.6 when the slope of the line is changed. The analysis is only valid when the strip is much longer that the size of the boundary regions near the clamped end, that is for $\Lambda \gg \epsilon^{-1}$, that is where the curves are in dark grey. The pattern is almost periodic for increments of Λ by one in rescaled units (that is by the plate width L in physical units), corresponding to the insertion of one bump—that is half a wavelength—in the bulk of the plate.

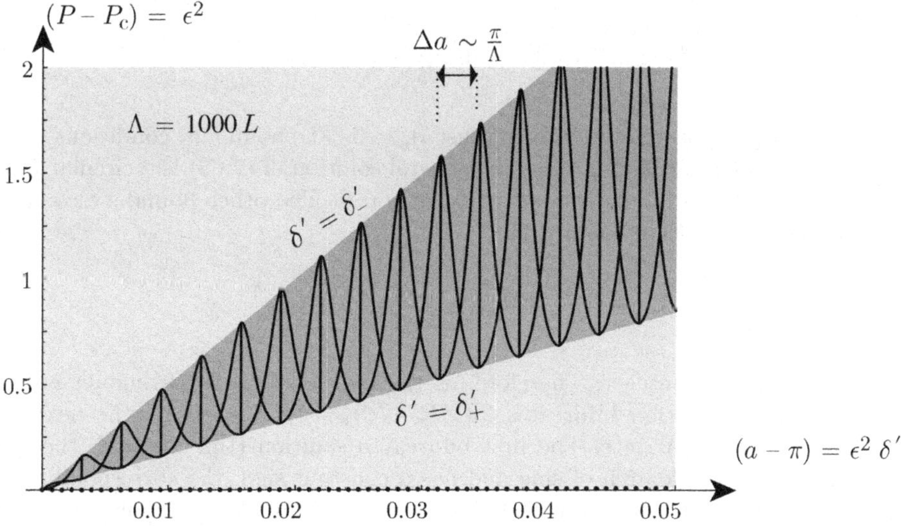

Fig. 7.8 Diagram of solutions for a fixed strip length (here $\Lambda = 10^3$), showing the possible wavenumber a as a function of the applied loading P. The instability region for an infinite strip is shown in light grey, its parabolic lower boundary (dotted curve very close to the axis) being given by equation (7.15). The angular sector shown in dark grey represents the allowed range of values for δ', from equation (7.64a).

The above series of equalities can be derived using equations (7.20), (7.66a), (7.43), and (7.67) sequentially. For large Λ and small $(a - \pi)$, this yields two fast oscillating curves $P(a)$ in the figure. The envelope of the curves are the two lines $P - P_c = 16\pi\,(a - \pi)/3$ and $P - P_c = 16\pi\,(a - \pi)$, which correspond to $\delta' = \delta'_+$ and $\delta' = \delta'_-$ respectively. Besides this oscillating curve, the solutions $\rho = k\,\pi$ in equation (7.67b), lead to $a = k\,\pi/\Lambda$. They are represented by the series of vertical segments on the diagram.

Recall that for an infinite strip, we found an unstable region delimited by a parabola (light grey region in Fig. 7.8), while for a half-infinite strip this region narrows down to an angular sector (dark grey triangle) as δ' stays within an interval, $\delta' \in [\delta'_-, \delta'_+]$. For a finite strip, the quantization condition (7.67a) further restricts the solutions to a discrete set of values of δ', which can be read off from Fig. 7.7 or 7.8. The curves in Fig 7.8 undergo fast oscillations for large Λ (that is for large aspect ratio Λ/L in physical units). The period of these oscillations for the variable a is very small, of order π/Λ. Therefore, when Λ is large, this discrete set of allowed values for the wavenumber a becomes denser and denser, and yields the continuous sector $\delta \in [\delta_-, \delta_+]$ in the limit Λ infinite.

7.6.7   Solutions with an oscillating amplitude

In this section, we show that the amplitude equation for a long strip has solutions other than the ones constructed before above threshold. These solutions are such that the amplitude A_c oscillates in the bulk. We show that such solutions are linearly unstable, which is not surprising.

This new type of solution can be constructed as follows. Consider the amplitude equation (7.47) for a long but finite plate of length Λ. It can be seen as a bifurcation problem in itself. The amplitude equation linearized near $A_c = 0$ takes the simple form:

$$\epsilon^2\, A_c^{\mathrm{lin}} + 4\,\frac{\mathrm{d}^2 A_c^{\mathrm{lin}}}{\mathrm{d}y^2} = 0, \tag{7.69}$$

where A_c^{lin} is small, the base solution being just $A_c = 0$. The boundary conditions for a long plate are $A_c^{\mathrm{lin}} = 0$ for $y = 0$ and $y = \Lambda$. The general solution of (7.69) is a circular function, $(\sin \epsilon\, y/2)$. It satisfies the boundary condition at $y = 0$. The other boundary condition at $y = \Lambda$ yields the quantization condition:

$$\frac{\epsilon\,\Lambda}{2} = k\,\pi,$$

where k is any non-zero integer. Therefore, as the reduced loading parameter ϵ increases above zero, the first solution bifurcates for $(P - P_c)_1 = \epsilon_1{}^2 = 4\,\pi^2/\Lambda^2$, the next one for $(P - P_c)_2 = \epsilon_2{}^2 = 16\,\pi^2/\Lambda^2$, etc. The first bifurcated solution (the one with the smallest $\epsilon = \epsilon_1$), is made of a single arch of sine and has a constant sign; the corresponding profile has no node. Tracking this solution while increasing Λ yields a wavenumber in the range $[\pi + \epsilon^2\,\delta'_+, \pi + \epsilon^2\,\delta'_-]$: it is the solution constructed before, which has an almost constant amplitude in the bulk. In contrast, the next solutions, corresponding to $\epsilon = \epsilon_2$, ϵ_3 etc. yields envelopes $A_c(y)$ with more and more nodes in the bulk. They are all linearly unstable, because a perturbation concentrated near such an internal node lowers the elastic energy, when conveniently tailored.

7.6.8 Symmetry of the solutions

The three types of solutions corresponding to the three possible choices for ρ in equation (7.67b) have different symmetries with respect to the centre of the strip, $y = \Lambda/2$. A pattern is symmetric with respect to this centre if the phase there, $a\,\Lambda/2 + \alpha_0''$ is zero modulo π, that is if it gives $a\,\Lambda/2 + \alpha_0'' = j\,\pi/2$, where j is an even integer. The pattern is antisymmetric if this phase is $\pi/2$ modulo π, that is if it gives $a\,\Lambda/2 + \alpha_0'' = j\,\pi/2$, where j is an odd integer. Now, $(a\,\Lambda)$ is the left-hand side of the quantization condition (7.67a), whose right-hand side is equal to $-(\alpha_0'' + \alpha_\Lambda'')$. Combining these equations, we find that the pattern is symmetric when $(\alpha_0'' - \alpha_\Lambda'')/\pi$ is an even integer, antisymmetric if this is an odd integer, and has no particular symmetry otherwise. For the pattern to be symmetric, the terms $\pm\mu$ have to cancel each other in the difference of α_0'' and α_Λ'' given by equation (7.66b), and so the signs in front of μ must be the same in this equation. This corresponds to $\rho = k\,\pi \pm 2\,\mu$ in equation (7.67b), i.e. to the two winding branches in the ladder structure in Fig. 7.6. In contrast, the horizontal steps in this ladder correspond to the choice $\rho = k\,\pi + 0$, i.e. to solutions having no particular symmetry.

7.7 Experiments

Boucif *et al.* (M. Boucif, J. E. Wesfreid, and E. Guyon, 1991) carried a careful experimental study of the non-linear selection of the wavelengths of post-buckled configurations. This showed well the changes of wavenumber above the critical threshold for long elastic plates. However the theory just presented applies only in general terms to the experiments for reasons we are going to discuss, and that are well explained in the original paper.

In our notation, the relation between P and a that is relevant for the *experiments* is:

$$-0.018\,(P - P_c) < (a - \pi) < 0.107\,(P - P_c). \tag{7.70}$$

The numbers -0.018 and 0.107 reflect the value of Poisson's ratio of the film used in the experiments. The relations derived in equation (7.64) *under the approximation of a vanishing Poisson's ratio* read:

$$\frac{1}{16\pi} < \delta' < \frac{3}{16\pi} \quad \text{that is} \quad 0.020\,(P - P_c) < (a - \pi) < 0.060\,(P - P_c).$$

There is some discrepancy in the values of these bounds, which can be attributed to the fact that we do not use the actual value of ν in the theory, and that the boundary conditions considered in the theory are difficult to realize experimentally (the influence of the boundary conditions along the small edges on the buckling is studied in reference (M. Potier-Ferry, 1983) with a simplified model).

The coefficient in the lower bound in equation (7.70) is negative, although it was expected to be positive from the theory: experimentally, the value $a = \pi$ seems to be permitted above threshold, that is $\delta'_- < 0$. This difference is significant because it changes the aspect of the diagram in Fig. 7.8, which goes as shown in Fig. 7.9 when $\delta'_- < 0$. This diagram is obtained by replacing equation (7.68) with:

$$P - P_c = \frac{I}{1 \pm J\,\cos(a\,\Lambda)}\,(a - \pi), \tag{7.71}$$

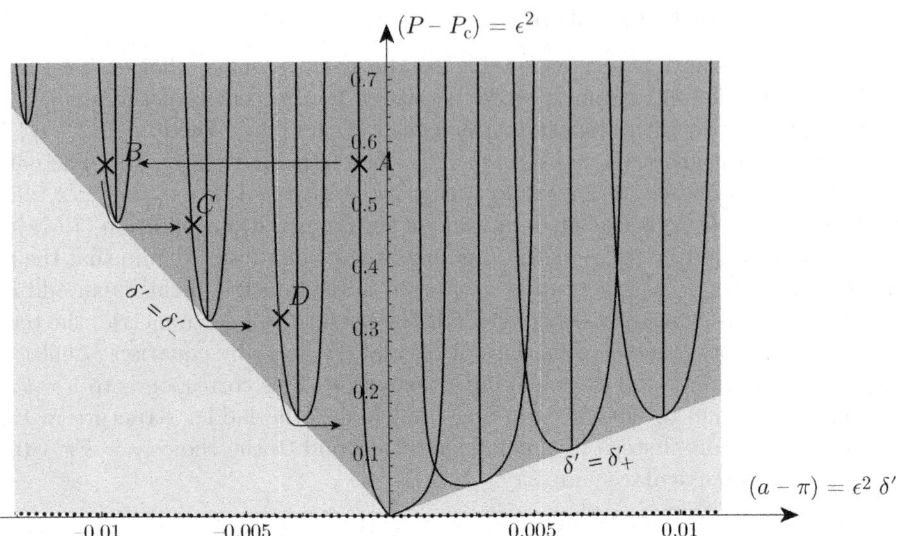

Fig. 7.9 Variation of the wavenumber with the loading for a strip of finite length, when $\delta'_- < 0$. This is otherwise the same diagram as in Fig. 7.8. It is based on equation (7.71) with the values of I and J given in the main text. The arrows show that the solution can be forced to vary discontinuously (jumps) when the loading is decreased.

where I and J are two positive numerical constants that allow one to recover the experimental coefficients in the bounds for $(P - P_{\rm c})/(a - \pi)$ given in equation (7.70). They are $I = 22.5$ and $J = 1.40$. Because J is larger than 1, $(P - P_{\rm c})$, as computed by equation (7.71), can become arbitrarily large when the denominator vanishes, that is for $\cos(a\,\Lambda) = \pm 1/J$. This leads to the vertical asymptotes in the diagram in Fig. 7.9.

When ϵ increases from zero, the solution can follows one of the two branches starting from the 'origin' $(a - \pi, P - P_{\rm c}) = (0, 0)$ and arrives at, say, point A in Fig. 7.9. During this increase in the loading, the wavevector a remains almost constant:[30] as ϵ grows, the solution follows the asymptotic branch and the wavenumber tends smoothly toward the value that makes zero the denominator in equation (7.71). Indeed, Boucif *et al.* (M. Boucif, J. E. Wesfreid, and E. Guyon, 1991) found that the wavenumber does not change much upon loading.

By various tricks they managed to change the wavenumber of the solution above threshold (point B in Fig. 7.9). Decreasing the loading later, they observed bifurcations that can be explained very simply. Suppose the solution has just landed on a branch that is not the one that goes all the way down to the linear bifurcation threshold (point B). By lowering the loading ϵ, one has to follow a trajectory of solutions in the $(a - \pi, P - P_{\rm c})$ plane that lands on one of the edges of the angular sector drawn by the limit values of the wavenumber consistent with the boundary conditions at the short ends (dark grey sector in the diagram).

---

[30] The situation is quite different in the case $\delta'_- > 0$, as revealed by Fig. 7.8. Then, no branch of the solution has a vertical asymptote $(P - P_{\rm c}) \to \infty$. The solution has to jump to another branch if the loading is increased indefinitely. This jump takes place when δ' reaches δ'_- if ϵ increases (and when it reaches δ'_+ if ϵ decreases, hence a hysteresis loop).

When this limit is reached, a discontinuous transition from one buckling mode to another was observed in the experiments (see figure 8 in reference (M. Boucif, J. E. Wesfreid, and E. Guyon, 1991)). This transition was well defined because under decreasing load the wavenumber stayed roughly constant (downward vertical path in our diagram), and it jumped suddenly (horizontal path in our diagram, since the loading ϵ does not change much during the quick jump). Physically the decrease or increase of the wavenumber a amounted to adding or subtracting half a wavelength to the structure, although sometimes a full wavelength was suppressed or added. As the loading decreased even more, another transition took place, of the same kind, down to the critical linear threshold where the buckling disappeared.

These experiments show a nice agreement with the idea of selection of the wavenumbers by end effects. In particular the range of possible wavenumbers was far smaller than anything derived from an analysis of stability of an infinitely long structure: the band of wavenumbers was found to be much smaller than that of linearly unstable modes, and also less than the band of so-called Eckhaus-stable modes, which lies inside the band of linear stability.

7.8   Conclusion

The analysis above followed the usual pattern of perturbation theory, and it does show what happens near the onset of instability threshold: the plate buckles and reaches a state of lowest energy that is periodic in the y direction. We extensively discussed the case of a strip of finite length with clamped boundary conditions at the short ends, and showed that the longitudinal wavelength is fixed by a quantization condition. We found that this wavelength can have only finitely many values above threshold. Furthermore, these values were found in a narrow band close to $2\,L$ in physical units, where L is the width of the strip. The width of this band was found to be much smaller than in the case of an infinite strip.

This chapter was concerned with a weakly non-linear regime, where the analytic calculations can be carried rather completely, at least for a uniaxial loading. In the coming chapter, Chapter 8, we shall examine what happens in a similar geometry, i.e. for long rectangular plate under biaxial loading but at a finite distance distance from threshold.

Part of the theoretical analysis presented in this chapter was published in Reference (Y. Pomeau, 1981) in a preliminary form. The experiments discussed in the text were presented in reference (M. Boucif, J. E. Wesfreid, and E. Guyon, 1991).

References

M. Boucif, J. E. Wesfreid, and E. Guyon. Experimental study of wavelength selection in the elastic buckling instability of thin plates. *European Journal of Mechanics. A. Solids*, 10(6):641–661, 1991.

M. C. Cross, P. G. Daniels, P. C. Hohenberg, and E. D. Siggia. Effect of distant sidewalls on wave-number selection in Rayleigh–Bénard convection. *Physical Review Letters*, 45(11):898–901, Sep. 1980.

M. C. Cross, P. G. Daniels, P. C. Hohenberg, and E. D. Siggia. Phase-winding solutions in a finite container above the convective threshold. *Journal of Fluid Mechanics*, 127:155–183, 1983.

M. Clément, E. Guyon, and J. E. Wesfreid. Multiplicité de modes de déformation d'une plaque sous compression. *Comptes Rendus de l'Académie des Sciences–Series II–Mécanique*, 293:87–89, 1981.

P. G. Daniels. The effect of distant side-walls on the evolution and stability of finite-amplitude Rayleigh–Bénard convection. *Proceedings of the Royal Society of London. Series A, Mathematical and Physical Sciences*, 378(1775):539–566, 1981.

E. L. Koschmieder. Bénard convection. *Advances in Chemical Physics*, 26:177–212, 1974.

C. G. Lange and A. C. Newell. The post-buckling problem for thin elastic shells. *SIAM Journal on Applied Mathematics*, 21(4):605–629, 1971.

A. C. Newell and J. A. Whitehead. Finite bandwidth, finite amplitude convection. *Journal of Fluid Mechanics*, 38:279–303, 1969.

M. Potier-Ferry. Amplitude modulation, phase modulation and localization of buckling patterns. In J. M. T. Thompson and G. W. Hunt, editors, *Collapse: The Buckling of Structures in Theory and Practice*, pages 149–159. Cambridge University Press, 1983.

Y. Pomeau. Non linear pattern selection in a problem of elasticity. *Journal de Physique, Lettres*, 42:L1, 1981.

Y. Pomeau and S. Zaleski. Wavelength selection in one dimensional cellular structures. *Journal de Physique*, 42:515, 1981.

Y. Pomeau and S. Zaleski. Pattern selection in a slowly varying environment. *Journal de Physique, Lettres*, 44(4):L135–141, 1983.

Y. Pomeau, S. Zaleski, and P. Manneville. Axisymmetric cellular structures revisited. *Zeitschrift für angewandte Mathematik und Physik (ZAMP)*, 36(3):367–394, 1985.

L. A. Segel. Distant side-walls cause slow amplitude modulation of cellular convection. *Journal of Fluid Mechanics*, 38(1):203–224, 1969.

E. D. Siggia and A. Zippelius. Pattern selection in Rayleigh–Bénard convection near threshold. *Physical Review Letters*, 47(12):835–838, Sep 1981.

S. Zaleski, Y. Pomeau, and A. Pumir. Optimal merging of rolls near a plane boundary. *Physical Review A*, 29(1):366–370, Jan 1984.

8
Finite amplitude buckling of a strip

In this chapter, we study the elastic buckling and postbuckling of a long rectangular plate submitted to in-plane anisotropic compression, as shown in Fig. 8.1. This study illustrates how the F.–von K. equations can characterize the elastic response of a plate in a concrete situation. The geometry is simple enough to allow explicit calculations. The behaviour of a plate under stress turns out to be quite complex nonetheless, as a consequence of geometric nonlinearity in the equations. We consider applied loads that are typically a few times the critical stress for the initial buckling; the limit of buckling under large applied stress, the so-called 'strongly non-linear buckling' will be discussed elsewhere, in the second half of Chapter 10 of this book. We also consider the plate to be infinite in one direction. This is a fair assumption for representing the experiments done with plates having a large length to width ratio; the influence of the finite length of the buckling of a strip has been studied in details in Chapter 7.

The motivation for studying an infinitely long, rectangular plate geometry is twofold. This geometry is a natural extension of Euler's *Elastica* problem, where the one-dimensional rod is replaced by a two-dimensional plate held along its sides. Because the F.–von K. equations couple displacements in the two in-plane directions, the strip displays a very rich behaviour, many buckling patterns being genuinely two dimensional. A second motivation for the present analysis was to explain the worm-like delamination of compressed thin

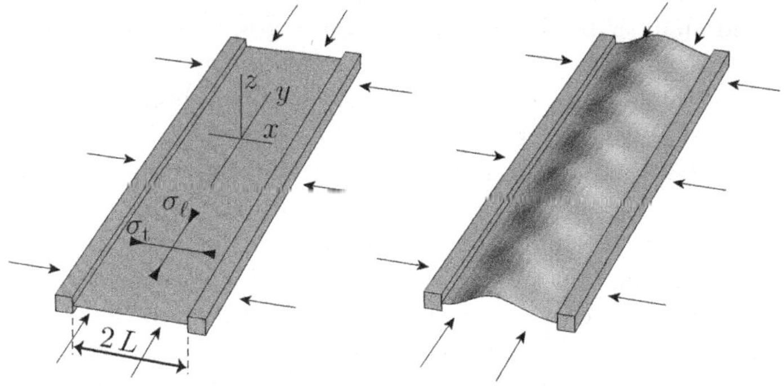

Fig. 8.1 The system is a long elastic strip clamped along both sides. A biaxial in-plane compression (σ_t, σ_ℓ) is induced in the strip and the resulting buckling and post-buckling is studied. Buckling amplitude is exaggerated to help visualization. Reproduced from reference (B. Audoly, B. Roman, and A. Pocheau, 2002), with kind permission of *The European Physical Journal* (*EPJ*), © EDP Sciences, Società Italiana di Fisica, Springer-Verlag 2002.

elastic films, as explained in Section 8.10: the infinite rectangular strip provides a simplified geometry for the analysis of patterns in thin-film delamination.

A variety of methods can be used to study the buckling of a plate, ranging from linear stability analysis and (numerical) Ritz procedure, to the analysis in the strongly non-linear limit done in Chapter 10. Using a combination of such methods we propose a rather exhaustive analysis, both analytical and numerical, of the buckling of an infinite elastic strip. We report on experiments as well, and compare them against theoretical predictions.

The chapter is organized as follows. In Section 8.1, we present a very short review on plate buckling. In Section 8.2, we present the experiments done on a long rectangular plate subjected to biaxial compression. The equations relevant for buckling are derived in Section 8.3 by including prestress in the F.–von K. equations (the notion of prestress is explained later). The analysis starts in Section 8.4 with the stability of the unbuckled configuration. In Section 8.6, we study the vicinity of a special point in the plane of loading parameters, where the wavelength of the primary unstable mode becomes infinite. In Section 8.7, the stability of an exact solution, the Euler column, in investigated. The yet unexplored regions of the phase diagram are studied in Section 8.8 using a semi-analytical Ritz procedure. In Section 8.9, we return to the experiments and compare them with the predictions of the theory. Some patterns observed in thin film delamination are discussed in Section 8.10. Finally, we mention in Section 8.11 the limitations and extensions of the analysis.

8.1 A short review on buckling

Buckling is a very classical subject in elasticity. In particular, there is a vast literature concerning the buckling of elastic plates and it is difficult, if not impossible, to make a complete review of this topic. In this section, we propose a short review, concerned mostly with the particular geometry that we study.

The study of finite amplitude effects in buckling problems was triggered by the experimental observations and analysis by Stein of mode jumping in elastic plates under compressive stress (M. Stein, 1959a, b). The interest in this problem came from the aerospace industry where the body of various flying machines, made by assembling large plates with small curvature, contributed to the overall rigidity. One topic that was actively investigated was the relationship between the large scale stress and the eventual change of shape of the metallic cover under strain. A classical question by that time was the analysis of the initial buckling of structure. Stein experimentally studied the post-buckling behaviour, with the aim of determining the effective rigidity beyond the buckling threshold. He himself proposed an explanation of the secondary bifurcation based upon the intuition that higher-order modes of instability of the unbuckled configuration may show up as a secondary bifurcation. This idea was presented later in various ways. In particular, an influential review paper by Bauer *et al.* (L. Bauer, H. B. Keller, and E. L. Reiss, 1975) claimed on the basis of general arguments that a secondary bifurcation is to be expected whenever the initial buckling is degenerate and gives several linearly unstable modes for the same loading—this situation is not generic and usually requires the parameters of the system, such as the aspect ratio, to have a special value. This motivated many subsequent papers, most of them theoretical, some including theorems and their full proof, although a few experiments were reported. Since most papers considered the buckling of rectangular plates, the basic theoretical method was to expand the energy on a finite basis of functions for the displacement, and to find the state of lowest elastic energy. In the case of plates with finite length, it was recognized early on (E. J. Holder and D. Schaeffer, 1984; D. Schaeffer and M. Golubitsky, 1979) that the occurrence of secondary bifurcations depends sensitively on the boundary conditions. Despite this, many authors use unphysical boundary conditions, such as boundary conditions for the value of the Airy potential instead of its physically meaningful second derivatives.

We are aware of only one paper, by Nakamura and collaborators (T. Nakamura and K. Uetani, 1979) dealing with a geometry close to that considered here, namely a long rectangular plate. They consider a set of possible values for the in-plane aspect ratio of the plate (length to width ratio), but they never formally take the limit of an infinitely large aspect ratio. In this limit, which we consider, the longitudinal wavenumber becomes a continuous parameter; the presence of a continuous wavenumber makes it possible to grab the post-buckling behaviour analytically, and allows the phase diagram to be derived without relying on a purely numerical approach. Another advantage of dealing with an infinitely long plate is that an exact, post-buckled solution of the non-linear equations is available, and it is natural to study its secondary bifurcations. In contrast, for a plate having a finite aspect ratio, the primary unstable modes are available close to initial threshold only, although the secondary bifurcations take place at a finite distance from this threshold. The use of trial functions to study secondary bifurcations is often poorly controlled in these conditions. Nakamura (T. Nakamura and K. Uetani, 1979) uses a clever method of analysis by looking at the neighbourhood of a co-dimension two bifurcation where two modes become unstable together simultaneously, something that can be obtained by changing both the load and the aspect ratio of the plate. The analysis of the bifurcation near this point of co-dimension two gives access to a quantitatively exact theory, although in a very small domain in the space of control parameters.

Buckling problems have been looked at by applied mathematicians. Everall and Hunt study the possible resonances between the primary unstable mode and the secondary one, when the ratio of their wavelengths is a rational number (P. R. Everall and G. W. Hunt, 1999; G. W. Hunt and P. R. Everall, 1999)

Some authors put their work under the umbrella of buckling problems, although the equations they study have a loose relation with plate elasticity (C.-S. Chien, 1989). Most papers (with the exception of (R. Maaskant and J. Roorda, 1992)) are concerned with uniaxial compression, and so do not address the complex structure of primary and secondary bifurcation that we unfold in this chapter for the case of an arbitrary external stress. We end this quick (and partial) discussion with a list of papers, in addition to those already explicitly mentioned, on post-buckling phenomena in plates (M. Uemura and O-Il Byon, 1977, 1978; R. Maaskant and J. Roorda, 1992; J.-J. Gervais, O. Abderrahmann, and R. Pierre, 1997; P. R. Everall and G. W. Hunt, 2000; C.-S. Chien, S.-Y. Gong, and Z. Mei, 2000).

8.2   The experiments

We first describe experiments (B. Audoly, B. Roman, and A. Pocheau, 2002) designed to achieve the loading geometry shown in Fig. 8.1 and performed by Benoît Roman and Alain Pocheau (IRPHE, Université d'Aix-Marseille, France). The set-up is shown in Fig. 8.2.

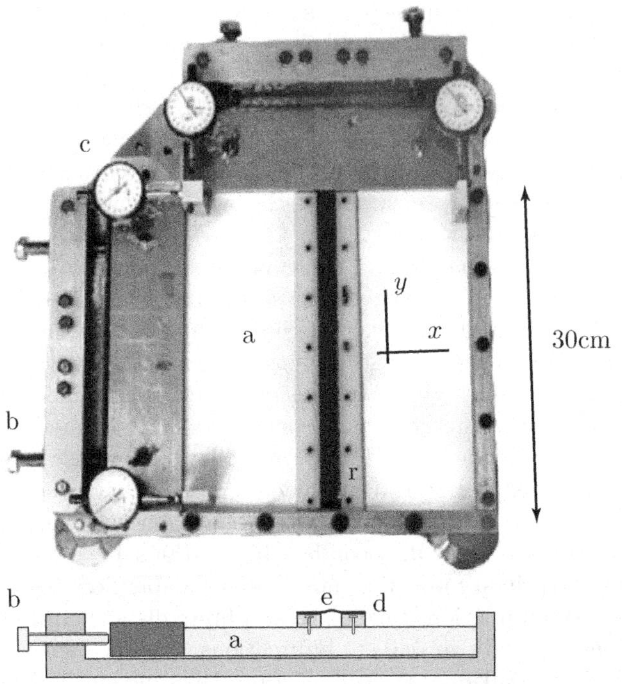

Fig. 8.2 Experimental set-up made up of a 30×30 cm^2 wide substratum (a) compressed along orthogonal directions using four screws (b), of four dial indicators (c) showing the imposed strain and of two rulers (d) screwed on the substratum, on to which the strip (e) is glued. The substratum, being thick, imposes the displacement along the edges of the film. Designed by B. Roman and A. Pocheau, image reproduced from (B. Audoly, B. Roman, and A. Pocheau, 2002) with kind permission of *The European Physical Journal (EPJ)*, © EDP Sciences, Società Italiana di Fisica, Springer-Verlag 2002.

The elastic film is a long, rectangular piece of polyvinylcarbonate (PVC) clamped along its long edges. A strip of this material is biaxially compressed by binding it to a $30 \times 30 \times 1$ cm^3 PVC block (substratum) thick enough to be unaffected by the film deformation. The substratum does not buckle during the experiment and is sucked downwards to prevent lifting off the base at large in-plane compressions. It is squeezed along two orthogonal directions and the amounts of imposed shortening in the transverse and longitudinal direction with respect to the film can be tuned independently using a set of screws. The imposed biaxial strain is controlled by four dial indicators. The resulting stress is denoted σ_t in the transverse direction and σ_ℓ in the longitudinal one. Note that σ_ℓ plays a similar role to the longitudinal loading parameter P from the previous chapter (although σ_ℓ is a residual *stress* while P used to be a differential *strain*).

The strip is made of a polycarbonate strip that is 30 cm long, 2 cm wide (in-plane aspect ratio $15 : 1$) and 0.1 mm thick. It is not bound directly to the substratum but instead glued to small PVC rulers that are themselves screwed on to the substratum (Fig. 8.2). These rulers transmit the compression from the substratum to the strip while keeping it at a distance from the substratum, hereby allowing it to buckle freely.

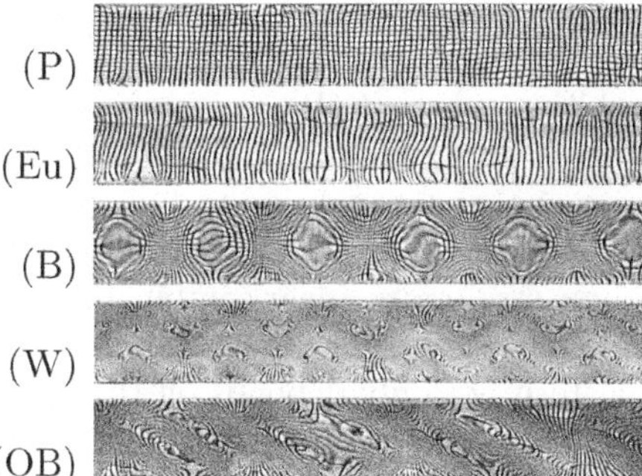

Fig. 8.3 Buckling patterns identified from the distortion of a grid reflected on the strip surface (observed strip size is 20×150 mm^2), reproduced from reference (B. Audoly, B. Roman, and A. Pocheau, 2002): (P) initial level of distortion in the absence of buckling, (Eu) cylindrical profile (Euler column) revealed by the disappearance of almost all the longitudinal stripes, (B) bumps revealed by a symmetric distortion, (W) worms, (OB) oblique bumps. With kind permission of *The European Physical Journal (EPJ)*, © EDP Sciences, Società Italiana di Fisica, Springer-Verlag 2002.

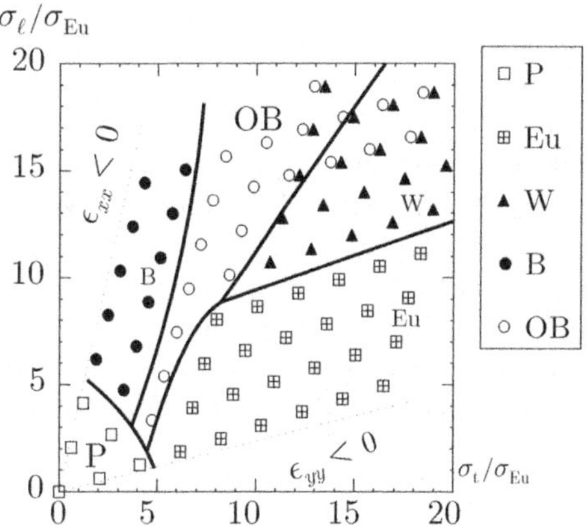

Fig. 8.4 Experimental phase diagram showing the domains of existence of various buckling patterns in Fig. 8.3, from (B. Audoly, B. Roman, and A. Pocheau, 2002). Note the bistability region with both oblique bumps and worms. The lines give the approximate experimental regions of stability of the patterns. This diagram will be compared with the theoretical one in Fig. 8.29. With kind permission of *The European Physical Journal (EPJ)*, © EDP Sciences, Società Italiana di Fisica, Springer-Verlag 2002.

Upon compression of the substrate, the buckling pattern of the strip was determined from the distortions of the image of a grid reflected on the strip surface (see snapshots in Fig. 8.3). The corresponding phase diagram in the (σ_t, σ_ℓ) stress space is shown in Fig. 8.4. In the range of compressions that was investigated, four different buckling modes were observed, besides the initial planar configuration (P):[1] a cylindrical profile known as the Euler column (Eu) at dominant transverse compressions, a bump mode (B) at dominant longitudinal compressions, a mode with oblique bumps (OB) and a worm like pattern (W) for transverse and longitudinal compressions of about the same magnitude. Observations were limited to an angular sector in the stress space[2] (σ_t, σ_ℓ) due to Poisson's ratio of the film: the set-up allows only positive strain components ϵ_{xx} and ϵ_{yy} to be imposed to the film, and by Hooke's law (6.59) for 2D elasticity these strains are linear combinations of σ_t and σ_ℓ depending on ν. For some values of the loading parameters, either the bumps or oblique bumps patterns can be observed: the observed pattern not only depends on the present loading conditions, but also on the history of loading (multistability). Gaps between domains correspond to buckling amplitudes that are too small for the patterns to be identified, as happens near the transition lines.

8.3   Equations for the compressed strip

With the aim of accounting for these experimental results, we solve the F.–von K. equations and analyse the buckling behaviour of a infinitely rectangular plate under biaxial compression.

8.3.1   F.–von K. equations with initial compression

We write the F.–von K. equations with residual compression (σ_t, σ_ℓ), i.e. we write the equations satisfied by the deflection $w(x, y)$ and the *change* of the Airy potential $\chi(x, y)$ taking the unbuckled, compressed state (sketched in Fig. 8.1 left) as a reference. Following the method of Section 1.2.4, we write these equations in dimensionless form.

For that purpose, we introduced rescaled quantities. Natural scales for the various variables follow from the analysis of Section 6.8.1, which has already been applied to plate equations to yield the scales summarized in equation (6.111): in-plane lengths are made dimensionless using the half-width L of the plate (note that the half-width was written $L/2$ in the previous chapter), deflection using h, in-plane stress using $E\,(h/L)^2$, and therefore χ using $E\,h^2$. Denoting dimensionless variables with an overline, this leads to the following rescalings:

$$(x, y) = (L\,\overline{x}, L\,\overline{y}) \tag{8.1a}$$

$$\Delta^2 = \frac{1}{L^4}\overline{\Delta}^2 \tag{8.1b}$$

$$(\sigma_t, \sigma_\ell) = \left(\frac{E\,h^2}{L^2}\,\overline{\sigma}_t, \frac{E\,h^2}{L^2}\,\overline{\sigma}_\ell\right) \tag{8.1c}$$

$$w(x, y) = h\,\overline{w}(\overline{x}, \overline{y}) \tag{8.1d}$$

$$\chi(x, y) = E\,h^2\,\overline{\chi}(\overline{x}, \overline{y}) + \left(-\sigma_t\frac{x^2}{2} - \sigma_\ell\frac{y^2}{2}\right) \tag{8.1e}$$

---

[1] The distortions in the planar state are due to various imperfections of the set-up. A perfect set-up would yield a perfectly rectangular pattern.

[2] In this chapter, to make the notations more obvious, we call σ_t, σ_ℓ the imposed stress components, σ_{xx}, σ_{yy}, that are the transverse and longitudinal stress in the unbuckled configuration.

$$u_x(x,y) = E\,L\,h^2\overline{u_x}(L\,\overline{x},L\,\overline{y}) + \frac{(-\sigma_t + \nu\,\sigma_\ell)\,x}{E} \tag{8.1d}$$

$$u_y(x,y) = E\,L\,h^2\overline{u_y}(L\,\overline{x},L\,\overline{y}) + \frac{(\nu\,\sigma_t - \sigma_\ell)\,y}{E}. \tag{8.1e}$$

There are many other possible ways of introducing dimensionless variables, but the present choice will turn out to be particularly convenient. In equation (8.1e), the imposed compression (σ_t,σ_ℓ), has been incorporated in such a way that $\overline{\chi} \equiv 0$, $\overline{u_x} \equiv 0$, $\overline{u_y} \equiv 0$, $\overline{w} \equiv 0$ describes the state of reference with biaxial compression $\sigma_{\alpha\beta} = -\sigma_t\delta_{\alpha x}\delta_{\beta x} - \sigma_\ell\delta_{\alpha y}\delta_{\beta y};$. The stress (σ_t,σ_ℓ) in the reference configuration is called prestress, or residual stress. The quantity $\overline{\chi}$ appears as an *increment* of in-plane stress with respect to prestress. Note that, according to our conventions, *positive* values of σ_t and σ_ℓ correspond to *negative* stress, i.e. to residual *compression*. The opposite case, namely tensile residual stress, is stable and does not lead to buckling.

Putting these definitions into the F.–von K. equations (6.86), we obtain the dimensionless F.–von K. equations with prestress:

$$\frac{1}{12(1-\nu^2)}\,\overline{\Delta}^2\overline{w} + \overline{\sigma}_t\,\frac{\partial^2\overline{w}}{\partial\overline{x}^2} + \overline{\sigma}_\ell\,\frac{\partial^2\overline{w}}{\partial\overline{y}^2} - [\overline{\chi},\overline{w}] = 0 \tag{8.2a}$$

$$\overline{\Delta}^2\overline{\chi} + \frac{1}{2}[\overline{w},\overline{w}] = 0. \tag{8.2b}$$

This set of equations is similar to equations (7.8a) and (7.8b) derived in the previous chapter, although with slightly different rescalings.

Here, we have used the formulation of the plate equations in terms of the Airy potential χ. This formulation is quite popular, especially in the mathematical literature, as the equations can be expressed in a very compact form. However, it is not very well suited for dealing with realistic boundary conditions, such as the ones considered here. The other formulation of the plate equations, based on the in-plane displacements, u and v, was used in the previous chapter and would lead here again to simpler calculations. We shall nevertheless base the following analysis on the Airy potential, with the aim of comparing the advantages and drawbacks of the two methods. The result is that introducing the Airy potential is useful only in the rare cases where the geometry is simple enough that there exists an obvious solution for the Airy potential.

8.3.2   Clamped boundary conditions

We investigate the deformation of a strip held along its long sides, as in Fig. 8.1. These so-called *clamped* boundary conditions have to be imposed in addition to the F.–von K. equations above, which describe deformation of the strip away from the edges. Because the in-plane deformations do not appear explicitly in the present formulation of the F.–von K. equations, based on the Airy potential χ, some additional work is needed for us to enforce these clamping conditions.

The reader should not be mislead by the use of the variables σ_t and σ_ℓ for characterizing the applied loading. Under clamped boundary conditions, we are actually imposing *strain* and not stress in the film. The stress components σ_t and σ_ℓ are those that *would* be measured in the unbuckled configuration; in the buckled configuration, physical stress is different. To be more explicit, we could have used the (differential) strain ϵ_t and ϵ_ℓ as control parameters. This strain is slightly more physical as it is conserved (on average in the sample) upon buckling. In any case, these differential strains are related to residual stress in the unbuckled configuration by the stress–strain relations for plane-stress elasticity, and it is mathematically equivalent to choosing $(\epsilon_t, \epsilon_\ell)$ or (σ_t, σ_ℓ) for characterizing the applied loading. It should be recalled that, by convention, these quantities refer to the (possibly unstable) unbuckled state.

Now, in the presence of clamped edges, the displacement (u_x, u_y, w) as well as the slope $\partial w/\partial x$ are imposed along the edges (we recall that w is the deflection, or the vertical component of the displacement):

$$w(x = \pm L, y) = 0, \qquad\qquad \frac{\partial w}{\partial x}(x = \pm L, y) = 0, \tag{8.3a}$$

$$u_x(x = \pm L, y) = \pm\frac{(-\sigma_t + \nu\,\sigma_\ell)\,L}{E}, \qquad u_y(x = \pm L, y) = \frac{(\nu\,\sigma_t - \sigma_\ell)\,y}{E}, \tag{8.3b}$$

where L is still the half-width of the plate (see Fig. 8.1). Equation (8.3b) is written in terms of the displacements (u_x, u_y, w), with the natural (stress-free) state of the strip as reference. In these equations, the right-hand terms express that the free plate has been forced into a compressed state along its edges. These terms disappear when we make use of the quantities $\overline{w}, \overline{u_x}, \overline{u_y}$, because the latter take the *biaxially stressed* state as reference state:

$$\overline{w}(\overline{x} = \pm 1, \overline{y}) = 0, \qquad\qquad \frac{\partial \overline{w}}{\partial \overline{x}}(\overline{x} = \pm 1, \overline{y}) = 0, \tag{8.4}$$

$$\overline{u_x}(\overline{x} = \pm 1, \overline{y}) = 0, \qquad\qquad \overline{u_y}(\overline{x} = \pm 1, \overline{y}) = 0. \tag{8.5}$$

The boundary conditions (8.3b) involve the in-plane displacements u_x and u_y, which have been eliminated from the F.–von K. equations in favour of the Airy potential. In order to express them in terms of the explicit unknowns of the F.–von K. equations, w and χ, one needs the following relations, derived by combining Hooke's equations (6.59) for plane stress elasticity with the non-linear stress–strain relations (6.62) :

$$\left[\frac{\partial u_x}{\partial x}\right] = -\frac{\nu}{E}\frac{\partial^2\chi}{\partial x^2} + \frac{1}{E}\frac{\partial^2\chi}{\partial y^2} - \frac{1}{2}\left(\frac{\partial w}{\partial x}\right)^2 \tag{8.6a}$$

$$\left[\frac{\partial u_y}{\partial y}\right] = \frac{1}{E}\frac{\partial^2\chi}{\partial x^2} - \frac{\nu}{E}\frac{\partial^2\chi}{\partial y^2} - \frac{1}{2}\left(\frac{\partial w}{\partial y}\right)^2 \tag{8.6b}$$

$$\left[\frac{\partial u_x}{\partial y} + \frac{\partial u_y}{\partial x}\right] = -\frac{2(1+\nu)}{E}\frac{\partial^2\chi}{\partial x\partial y} - \left(\frac{\partial w}{\partial x}\frac{\partial w}{\partial y}\right). \tag{8.6c}$$

Square bracket are used in equation (8.6a) and in the following to allow one to recognize these quantities easily when they appear. The relations above are valid everywhere on the plate.

They allow one to express the boundary conditions for u_x and u_y in terms of w and χ. Along the edges $(x = \pm L, y)$, an obvious identity, combined with equation (8.3b), yields:

$$\frac{\partial^2 u_x}{\partial y^2} = \frac{\partial}{\partial y}\left[\frac{\partial u_x}{\partial y} + \frac{\partial u_y}{\partial x}\right] - \frac{\partial}{\partial x}\left[\frac{\partial u_y}{\partial y}\right] = -\frac{(2+\nu)}{E}\frac{\partial^3 \chi}{\partial x \partial y^2} - \frac{1}{E}\frac{\partial^3 \chi}{\partial x^3} = 0, \qquad (8.7a)$$

$$\left[\frac{\partial u_y}{\partial y}\right] = \frac{1}{E}\frac{\partial^2 \chi}{\partial x^2} - \frac{\nu}{E}\frac{\partial^2 \chi}{\partial y^2} = \frac{(\nu\,\sigma_{\mathrm{t}} - \sigma_\ell)}{E}, \qquad (8.7b)$$

where terms as $\partial w/\partial x_{(\pm L, y)}$ were discarded thanks to equation (8.3a).

These two boundary conditions for $\partial^2 u_x/\partial y^2$ and $\partial u_y/\partial y$ in (8.7) are weaker than the original ones in equation (8.3b) for $u_x(\pm L, y)$ and $u_y(\pm L, y)$, because they were obtained by derivation with respect to y. For example, they allow the edges to slide apart from each other:

$$u_x(\pm L, y) = \pm\left(\frac{(-\sigma_{\mathrm{t}} + \nu\,\sigma_\ell)\,L}{E} + C\right) \quad u_y(\pm L, y) = 0,$$

where C is an arbitrary constant, a situation which is clearly incompatible with the original clamping condition. In order to prevent such a sliding of the edges, one has to impose specifically, using equation (8.3b), that the distance between the film edges remains constant:

$$\frac{u_x(+L, 0) - u_x(-L, 0)}{2\,L} = \frac{(-\sigma_{\mathrm{t}} + \nu\,\sigma_\ell)}{E}.$$

The left-hand term can be written as $1/2L \int_{x'=-L}^{L}\left[\partial u_x/\partial x(x', 0)\right]\,\mathrm{d}x'$. Hence by equation (8.6a):

$$\frac{1}{2\,L}\int_{x'=-L}^{L}\left(-\frac{\nu}{E}\frac{\partial^2 \chi}{\partial x^2} + \frac{1}{E}\frac{\partial^2 \chi}{\partial y^2} - \frac{1}{2}\left(\frac{\partial w}{\partial x}\right)^2\right)\,\mathrm{d}x' = \frac{(-\sigma_{\mathrm{t}} + \nu\,\sigma_\ell)}{E}. \qquad (8.8)$$

This expression no longer makes use of the in-plane displacement (u, v). Note that this condition needs only be imposed at, say, $y = 0$: imposing it for any y would be redundant with equation (8.7a), which can be recovered by derivation of the equation above with respect to y.

Two other similar conditions are needed to prevent any shearing of the boundaries in the form

$$\delta u_x(\pm L, y) = 0 \quad \delta u_y(\pm L, y) = \pm C,$$

or combined rotation of these boundaries in the form

$$\delta u_x(\pm L, y) = \mp C\,y \quad \delta u_y(\pm L, y) = \pm C\,x.$$

These two conditions will be automatically satisfied in the following because of symmetries in the deformations that we shall impose. For this reason, we do not write these additional boundary conditions explicitly.

To sum up, the clamping conditions yield, in terms of the dimensionless functions $\overline{w}$ and $\overline{\chi}$, four boundary conditions along both edges $(\overline{x} = \pm 1, \overline{y})$, for all $\overline{y}$:

$$\overline{w} = 0, \qquad\qquad\qquad \frac{\partial \overline{w}}{\partial \overline{x}} = 0, \qquad\qquad \text{by (8.4)} \qquad (8.9a)$$

$$-(2+\nu)\frac{\partial^3 \overline{\chi}}{\partial \overline{x}\,\partial \overline{y}^2} - \frac{\partial^3 \overline{\chi}}{\partial \overline{x}^3} = 0 \qquad \frac{\partial^2 \overline{\chi}}{\partial \overline{x}^2} - \nu\frac{\partial^2 \overline{\chi}}{\partial \overline{y}^2} = 0, \qquad \text{by (8.7)} \qquad (8.9b)$$

plus one additional constraint to be imposed at $\overline{y} = 0$ only:

$$\int_{\overline{x}'=-1}^{1} \left(-\nu\frac{\partial^2 \overline{\chi}}{\partial \overline{x}^2} + \frac{\partial^2 \overline{\chi}}{\partial \overline{y}^2}\right)_{(\overline{x}',0)} \mathrm{d}x' - \frac{1}{2}\int_{\overline{x}'=-1}^{1} \left(\frac{\partial \overline{w}}{\partial \overline{x}}(\overline{x}',0)\right)^2 \mathrm{d}x' = 0. \qquad (8.9c)$$

8.4   Linear stability

One of the few analytical solutions to the F.–von K. equations (8.2) is the 1D Euler column introduced below. For more general geometries, there are no exact solutions to these non-linear equations, and a standard approach is to seek solutions of small amplitude, i.e. to restrict oneself to a *weakly non-linear* regime. This was done in Chapter 7 for uniaxial residual stress, including the discussion of finite size effects in the long direction. In Chapter 10, another possible approach is discussed in the opposite limit, namely the strongly non-linear regime.

The weakly non-linear analysis is studied in the present section. This approach is first justified rigorously in Section 8.4.1 as we show that the initial buckling[3] is a super-critical (also called continuous) bifurcation. As a result, this buckling can be studied using the *linearized* equations. In a sub-critical bifurcation the stationary amplitude of the new mode would change according to the same square root law but would be linearly unstable; then, the stable solution always has a finite amplitude and is out of reach of a perturbative approach. This perturbative approach is the aim of Section 8.4.2 (linear stability analysis).

8.4.1   Initial buckling is always super-critical

In this subsection, we prove that the initial buckling of the strip is super-critical (continuous): the amplitude of the deflection due to buckling varies above the threshold with a square root law,[4] typical of a super-critical bifurcation. This follows from the structure of the F.–von K. equations, and applies to any plate geometry as well as to other types of boundary conditions. It is however specific to the *initial* buckling bifurcation. Later, in Section 8.8, we shall indeed encounter sub-critical (discontinuous) bifurcations in our study of secondary bifurcations.

Technically, we proceed as follows: we assume that there exists a buckled configuration $(w_{\mathrm{b}}, \chi_{\mathrm{b}}) \neq (0,0)$ satisfying the F.–von K. equations for given values of the imposed stress $(\sigma_{\mathrm{t}}, \sigma_{\ell})$. This equilibrium configuration $(w_{\mathrm{b}}, \chi_{\mathrm{b}})$ can be stable, metastable or even unstable (here 'stability' refers to the value of the elastic energy: a state is stable if no other state exists with a lower energy. It is linearly stable if no state with a lower energy exists nearby,

---

[3] By *initial* buckling, we mean the *first* bifurcation affecting the planar configuration, as opposed to *secondary* buckling studied later.

[4] Note that, as a consequence of this, the deflection is a continuous function of the imposed stress, hence the term 'continuous'.

and it is metastable if it is linearly stable although another state with a lower energy exists, but not in its vicinity). Then, we prove that the unbuckled state $(w, \chi) \equiv (0,0)$ is *linearly unstable* with respect to small perturbations of the form $(\epsilon\, w_{\mathrm{b}}, \epsilon^2\, \chi_{\mathrm{b}})$. As a result, the planar configuration is either the absolute minimum of energy or is unstable, and the initial buckling cannot be sub-critical.

Suppose that the buckled state $(w_{\mathrm{b}}, \chi_{\mathrm{b}}) \not\equiv (0,0)$ is given. Let us consider the following one-parameter family of plate configurations, indexed by a real number v (having eliminated the in-plane displacement from the equations, there should be no ambiguity with the new parameter v, to be used in the current section only):

$$w_v(x,y) = v\, w_{\mathrm{b}}(x,y) \qquad \chi_v(x,y) = v^2\, \chi_{\mathrm{b}}(x,y). \tag{8.10}$$

This family is used to interpolate between the buckled configuration and the planar one: $v = 1$ corresponds to the buckled state $(w_{\mathrm{b}}, \chi_{\mathrm{b}})$, $v = 0$ to the planar one, and $v = -1$ to the reflection of the buckled one $(w_{\mathrm{b}}, \chi_{\mathrm{b}})$ by horizontal mirror symmetry.

A nice property of this one-parameter family is that all configurations (w_v, χ_v) satisfy the clamping boundary conditions (8.9). This is in fact due to the structure of the stress–strain relations (8.6) and can easily be extended to all boundary conditions that can be expressed as linear combinations of the displacements and their gradients, as freely supported edges for example.

For a similar reason, all configurations (w_v, χ_v) satisfy the in-plane equilibrium condition (8.2b), which is linear, as can be checked straightforwardly. However they do not satisfy the *out-of-plane* equilibrium condition (8.2a) as it is non-linear.

Because the configurations (8.10) both satisfy the in-plane equilibrium and the boundary conditions for an arbitrary v, it makes sense to plot the plate energy $\mathcal{E}_{\mathrm{FvK}} = \mathcal{E}_{\mathrm{s}} + \mathcal{E}_{\mathrm{b}}$ derived earlier in Section 6.6.2 as a function of the parameter v. Plugging equation (8.10) into the definition of $\mathcal{E}_{\mathrm{s}}$ and $\mathcal{E}_{\mathrm{b}}$ in equation (6.97), one finds that $\mathcal{E}_{\mathrm{s}} \propto \chi^2$ and $\mathcal{E}_{\mathrm{b}} \propto w^2$ are even polynomials of the variable v, of degree four and two respectively. As a result, the total energy $\mathcal{E}_{\mathrm{FvK}}$ is a polynomial of fourth degree that involves only even powers of v. If $\mathrm{d}^2\mathcal{E}_{\mathrm{FvK}}(0)/\mathrm{d}v^2 \geq 0$, the derivative $\mathrm{d}\mathcal{E}_{\mathrm{FvK}}(0)/\mathrm{d}v(v)$ would only vanish at $v = 0$, which would contradict our assumption that $v = 1$ is also an equilibrium configuration. Therefore, $\mathcal{E}''_{\mathrm{FvK}}(v = 0) < 0$, and the planar configuration is indeed linearly unstable with respect to small perturbations of the form $(\epsilon\, w_{\mathrm{b}}, \epsilon^2\, \chi_{\mathrm{b}})$, for small ϵ. As a result, the planar strip can become unstable through a super-critical bifurcation only (see Fig. 8.5).

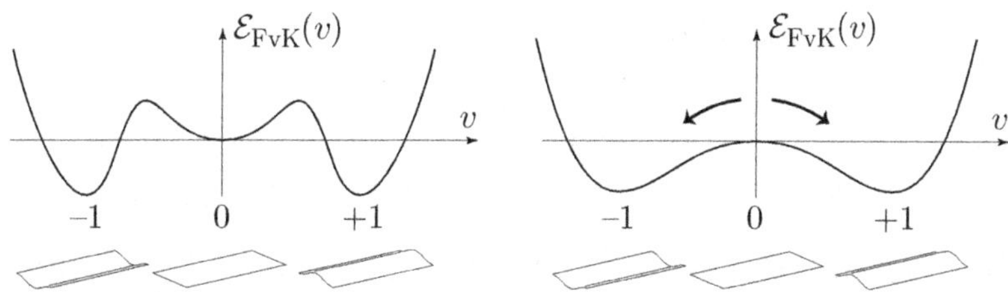

Fig. 8.5 Because the reduced F.–von K. energy $\mathcal{F}(v)$ is an even polynomial of v of fourth degree, sub-critical bifurcations (left) are not possible: as soon there exists a buckled equilibrium configuration, the planar one becomes linearly unstable. As a result, the initial bifurcation can only be super-critical (right).

8.4.2   Linear stability of the unbuckled configuration

In this subsection, we seek solutions of infinitesimally small amplitude, i.e. we study the *linearized* F.–von K. equations for the strip. According to the results of Section 8.4.1, the existence of such solutions can indeed be used to track the onset of super-critical buckling of the strip.

It can be seen from equations (8.2b) and (8.9c) that $\overline{w}$ and $\overline{\chi}$ are not of the same order of smallness in the limit of small amplitude: if $\overline{w}$ is of order, say, ϵ then $\overline{\chi}$ will be of order ϵ^2. This is consistent with equation (8.10). Linearization removes any quantity of order ϵ^2, and the linearized F.–von K. equations for the strip boil down to $\overline{\chi}^{[1]} \equiv 0$ and

$$\frac{1}{12(1-\nu^2)}\,\overline{\Delta}^2\overline{w}^{[1]} + \overline{\sigma}_t\,\frac{\partial^2\overline{w}^{[1]}}{\partial\overline{x}^2} + \overline{\sigma}_\ell\,\frac{\partial^2\overline{w}^{[1]}}{\partial\overline{y}^2} = 0 \quad \text{from (8.2)} \tag{8.11a}$$

$$\overline{w}^{[1]}(\pm 1,\overline{y}) = 0, \quad \frac{\partial\overline{w}^{[1]}}{\partial\overline{x}}(\pm 1,\overline{y}) = 0 \quad \text{along the edges,} \tag{8.11b}$$

where the unknowns have been linearized using the notations

$$\overline{\chi}(\overline{x},\overline{y}) = \overline{\chi}^{[1]}(\overline{x},\overline{y}) + \cdots \quad \text{and} \quad \overline{w}(\overline{x},\overline{y}) = \overline{w}^{[1]}(\overline{x},\overline{y}) + \cdots$$

The equations above are similar to equations (7.11), although with different boundary conditions and scaling conventions.

Note that the boundary condition (8.9c) does not play any role at the linear order. When doing the linear stability analysis below, we drop overlines and superscript '[1]' for convenience.

Because the strip is assumed to be infinitely long and invariant under translation in the y direction and because of the linearity of the equations of linear stability, any Fourier component of a solution w remains itself a solution. Therefore, one can seek solutions in the form of pure Fourier modes:

$$w(x,y) = w_k(x)\,\cos(k\,y). \tag{8.12}$$

The subscript k for the Fourier mode should not be confused with the implicit superscript [1] used to denote quantities of order ϵ.

For any such function $w_k(x)$, equation (8.11a) becomes the ordinary differential equation with respect to the transverse coordinate:

$$\frac{d^4 w_k(x)}{dx^4} - \left(2\,k^2 - 12\,(1-\nu^2)\,\sigma_t\right)\frac{d^2 w_k(x)}{dx^2} + k^2\left(k^2 - 12\,(1-\nu^2)\,\sigma_\ell\right)w_k(x) = 0$$

with boundary conditions $w_k(\pm 1) = w_k'(\pm 1) = 0$. \hfill (8.13)

We are interested in finding the values of k and the set of critical compressions (σ_t,σ_ℓ) such that equation (8.13) admits non-trivial (non-zero) solutions.

Let us study the case $k = 0$ first. The above equation reads, for $k = 0$,

$$\frac{d^4 w_0(x)}{dx^4} + 12\,(1-\nu^2)\,\sigma_t\,\frac{d^2 w_0(x)}{dx^2} = 0 \quad \text{with b. c. } w_0(\pm 1) = w_0'(\pm 1) = 0. \tag{8.14}$$

Its solutions can be obtained by straightforward analysis. Given the symmetry $x \mapsto (-x)$ of this equation, one can look for solutions $w_0(x)$ that are either even or odd functions of x. Even solutions are of the form $w_0(x) = \alpha\,\cos([12\,(1-\nu^2)\,\sigma_t]^{1/2}\,x) + \beta$ where α and β

are yet unknown constants. Boundary conditions at $x = \pm 1$ impose $\sigma_t = n^2\,\pi^2/12\,(1-\nu^2)$ and $\beta = (-1)^{n+1}\,\alpha$ for some strictly positive integer n. Only the lowest critical value ($n = 1$) of the stress σ_t is physical, because higher ones give marginal modes near the planar configuration when the latter is already unstable. For $n = 1$, the marginal mode of stability reads:

$$w_{\mathrm{Eu}}(x, y) = \frac{1 + \cos(\pi\,x)}{2}. \tag{8.15a}$$

This is a solution of the linearized equation when

$$\sigma_t = \sigma_{\mathrm{Eu}} \stackrel{\mathrm{def}}{=} \frac{4\pi^2\,D}{h\,(2L)^2} = \frac{E\,\pi^2}{3(1-\nu^2)}\left(\frac{h}{2\,L}\right)^2, \tag{8.15b}$$

where, for later reference, we restored the physical dimensions in the critical stress using equation (8.1c). Recall that the width of the plate is $(2L)$ and not L. This explains why we have made the quantity $(2L)$ appear in the above. This mode is called the Euler column (hence the subscript Eu) because it is cylindrical (invariant in the y direction) and has an exact equivalent in Euler's theory of the planar *Elastica*. The factor $1/2$ in w_{Eu} is conventional since the buckling amplitude cannot be determined by a linear stability analysis; this coefficient has been chosen in such a way that the maximum deflection, at $x = 0$, is one unit in rescaled variables. Note that the linear buckling threshold for this particular mode, given by equation (8.15b), depends only on σ_t and not on σ_ℓ.

A similar analysis can be carried out for odd solutions $w_0(x)$ of the variable x. The corresponding critical stress σ_t is found to satisfy the equation $\sin\tilde{\sigma}_t/\tilde{\sigma}_t = \cos\tilde{\sigma}_t$ with $\tilde{\sigma}_t = \sigma_t/\sigma_{\mathrm{Eu}} > 0$, the smallest root of which is 4.49341, a number significantly larger than 1. The critical stress for antisymmetric buckling is therefore larger than σ_{Eu}, and will be discarded. From Sturm–Liouville theory, the number of nodes of the solutions increases with the 'energy'[5] and so we shall not need to examine solutions $w_0(x)$ that have a higher number of nodes.

We now look at modes that are not necessarily cylindrical by dropping the restriction $k = 0$. The characteristic polynomial for the ordinary differential equation (8.13), corresponding to solutions in the form $w_k(x) = \exp(q\,x)$, reads:

$$\frac{q^4}{12(1-\nu^2)} - q^2\left(\frac{2\,k^2}{12(1-\nu^2)} - \sigma_t\right) + \left(\frac{k^4}{12(1-\nu^2)} - \sigma_\ell\right) = 0.$$

This second degree polynomial in (q^2) has the determinant:

$$\delta = \sigma_t{}^2 + \frac{4\,k^2}{12(1-\nu^2)}\left(\sigma_\ell - \sigma_t\right),$$

and roots:

$$\lambda_\pm = \frac{12(1-\nu^2)}{2}\left(\frac{2\,k^2}{12(1-\nu^2)} - \sigma_t \pm \sqrt{\delta}\right). \tag{8.16}$$

For simplicity, we shall assume $\delta \geq 0$ in the following. The case $\delta < 0$ can be treated similarly and does not lead to any physical bifurcation, as tedious calculations show.

---

[5] This is similar to what happens with the one-dimensional Schrödinger equation on a finite interval, the dimensionless critical stress $\sigma_t/\sigma_{\mathrm{Eu}}$ playing the role of the energy of an eigenstate.

As earlier, solutions $w_k(x)$ can be assumed to be either even or odd functions of x but the argument above suggests that odd w_k's are unphysical (and this can also be checked by explicit calculations). The roots (8.16) of the characteristic polynomial yield the generic symmetric solution of equation (8.13) as:

$$w_k(x) = \alpha_+ \cosh(\sqrt{\lambda_+}\, x) + \alpha_- \cosh(\sqrt{\lambda_-}\, x) = 0,$$

where, by convention, $\cosh(\sqrt{\lambda_\pm}\, x)$ stands for $\cos(\sqrt{-\lambda_\pm}\, x)$ when $\lambda_\pm$ is negative.

The coefficients $\alpha_\pm$ must be imposed by the boundary conditions $w_k(\pm 1) = w'_k(\pm 1) = 0$, which yield two linear equations:

$$\begin{cases} \alpha_+ \cosh(\sqrt{\lambda_+}) + \alpha_- \cosh(\sqrt{\lambda_-}) = 0 \\ \alpha_+ \sqrt{\lambda_+} \sinh(\sqrt{\lambda_+}) + \alpha_- \sqrt{\lambda_-} \sinh(\sqrt{\lambda_-}) = 0, \end{cases}$$

with a similar convention: $\sqrt{\lambda_\pm} \sinh(\sqrt{\lambda_\pm}) = -\sqrt{-\lambda_\pm} \sin(\sqrt{-\lambda_\pm})$ when $\lambda_\pm < 0$. There exist non-trivial solutions (α_+, α_-), provided the determinant of the system vanishes:

$$\sqrt{\lambda_+} \tanh(\lambda_+) = \sqrt{\lambda_-} \tanh(\lambda_-), \tag{8.17}$$

with $\lambda_\pm(\nu, k, \sigma_{\rm t}, \sigma_\ell)$ defined in (8.16). Again, for negative $\lambda_\pm$, $[\sqrt{\lambda_\pm} \tanh(\sqrt{\lambda_\pm})]$ is a shorthand for $[-\sqrt{-\lambda_\pm} \tan(\sqrt{-\lambda_\pm})]$.

Equation (8.17) is the equation for the neutral modes of the strip near the planar configuration. These modes are also called the marginally stable modes. A bifurcation takes place at the lowest value of the loading where neutral modes appear. We shall therefore introduce:

$$\tilde\sigma_{\rm t}(\nu, k, \sigma_\ell) = \min\left\{\sigma_{\rm t} \mid (\nu, k, \sigma_{\rm t}, \sigma_\ell) \text{ is a root of } (8.17)\right\}, \tag{8.18}$$

a function that is plotted numerically in Fig. 8.6 versus k, for three different values of σ_ℓ.

In the phase diagram that we are building, the label P stands for the region where the planar configuration is stable. Let us note as ∂P the edge of this region, that is the set of loading parameters for which the planar configuration looses stability. The parametric equation of the curve ∂P in the loading plane $(\sigma_{\rm t}, \sigma_{\rm t})$ can be obtained as:

$$\sigma_{\rm t}^{\partial{\rm P}}(\nu, \sigma_\ell) = \min_k \tilde\sigma_{\rm t}(\nu, k, \sigma_\ell). \tag{8.19}$$

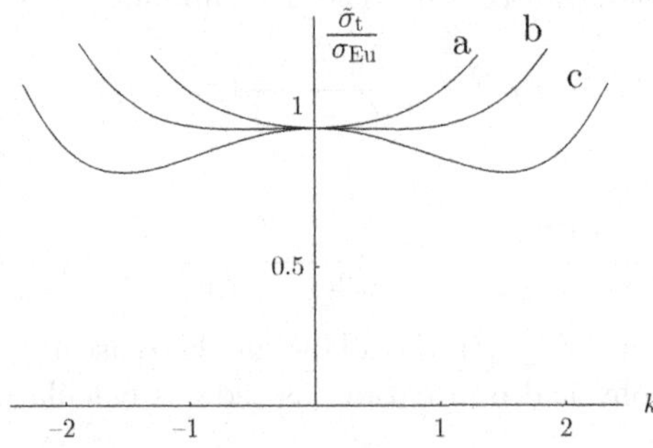

Fig. 8.6 Plots of $\tilde\sigma_{\rm t}(\nu = 0.3, k, \sigma_\ell)$ defined in (8.18) for (a) $\sigma_\ell = \sigma_{\rm Eu}$, (b) $\sigma_\ell = 2\sigma_{\rm Eu}/3$, (c) $\sigma_\ell = (0.3\,\sigma_{\rm Eu})$.

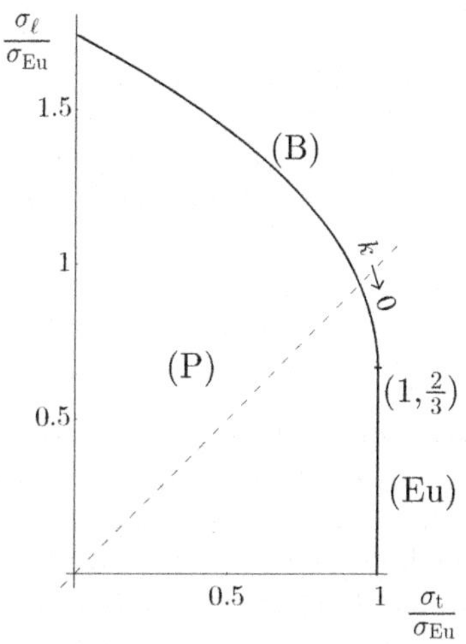
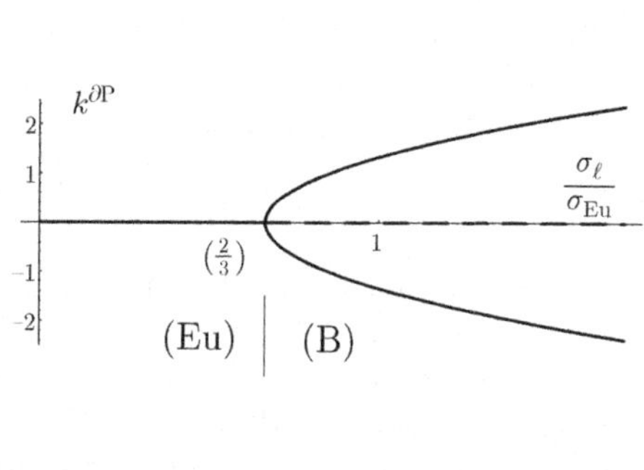

Fig. 8.7 *Left*: diagram of linear stability of the planar strip in the (σ_t, σ_ℓ) stress space. Marginal modes are the Euler column (Eu), with $k = 0$, for $\sigma_\ell < 2/3\,\sigma_{Eu}$ or a bump mode (B), with $k \neq 0$, for $\sigma_\ell > 2/3\,\sigma_{Eu}$. The dashed line corresponds to isotropic (equi-biaxial) loading, $\sigma_t = \sigma_\ell$. *Right*: plot of the wavenumber $k^{\partial P}$ of the linearly unstable bump modes along the curve of marginal stability of the planar region. Note the transition to zero of $k^{\partial P}$ at $\sigma_\ell/\sigma_{Eu} = 2/3$.

This parametric equation has been used to draw the domain of linear stability of the planar strip for $\nu = 0.3$ in Fig. 8.7, left. The value $k^{\partial P}(\nu, \sigma_t)$ of the wavenumber k along this boundary is the wavenumber of the marginal mode. It is plotted in Fig. 8.7, right.

Figure 8.6 shows that the value $k^{\partial P}$ of k where the minimum is reached in equation (8.19) bifurcates for a critical value of σ_ℓ which can be calculated as $(2\,\sigma_{Eu}/3)$. Indeed, for $\sigma_\ell < 2\,\sigma_{Eu}/3$, $k^{\partial P} = 0$ and the unstable mode is the Euler column (Eu) while for $\sigma_\ell > 2/3\,\sigma_{Eu}$ it is a bump mode B having a harmonic dependence on the longitudinal coordinate y. These results, summarized in Fig. 8.7, left, can already account for the relative layout of the planar (P), Euler (Eu) and bumps (B) regions in the experimental diagram in Fig. 8.4.

8.5 The Euler column, an exact solution

In this section, we show that the Euler column, encountered earlier as a marginal mode i.e. a solution of the *linearized* plate equations for $\sigma_t = \sigma_{Eu}$, is in fact an exact solution of the non-linear F.–von K. equations. To do so, we seek solutions of the F.–von K. equations that are invariant along the longitudinal direction y.

8.5.1 Derivation

Solution that are invariant along the longitudinal direction y have cylindrical symmetry— by cylinder, we mean surfaces invariant by translation in a given direction, which do not necessarily have a circular cross-section. Cylindrical profiles are developable and therefore have a vanishing Gauss curvature: at any point of the surface, the direction y is a principal direction of zero curvature (see Section 6.2.2). Therefore, the equation for in-plane equilibrium reduces to:

$$\Delta^2 \chi = \frac{\partial^4 \chi}{\partial x^4} + 2\,\frac{\partial^4 \chi}{\partial x^2 \partial y^2} + \frac{\partial^4 \chi}{\partial y^4} = 0.$$

Now, $\partial^4 \chi / \partial y^4 = \partial^2 \sigma_{xx} / \partial y^2$ vanishes because stress is assumed to be independent of y. Note that the Airy potential itself, being not directly observable, might depend on y. For the same reason,

$$\frac{\partial^4 \chi}{\partial x^2 \partial y^2} = \frac{\partial^2 \sigma_{yy}}{\partial y^2}$$

vanishes as well. Therefore the equation for in-plane equilibrium reads $\partial^4 \chi / \partial x^4 = 0$ and is readily integrated as: $\chi(x,y) = \chi_3\, x^3/6 + \chi_2\, x^2/2 + \chi_1\, x + \tilde{\chi}(y)$, where χ_3, χ_2 and χ_1 are integration constants and $\tilde{\chi}(y)$ an arbitrary smooth function of y. Using now the boundary conditions (8.9b) for χ and dropping the unphysical terms (those that are constant or linear in x or y and yield no contribution to the stress upon double derivation), one obtains the potential associated with the stress *change* due to buckling as

$$\chi(x,y) = \chi_2 \left(\nu\,\frac{x^2}{2} + \frac{y^2}{2} \right). \tag{8.20}$$

The cylindrical solution that we are deriving, the Euler column, has uniform, biaxial in-plane stress. The constant χ_2 is determined below.

Because the in-plane stress is constant, the equation for w is an ordinary differential equation (ODE) with constant coefficients:

$$\frac{1}{12(1-\nu^2)}\,\frac{d^4 w(x)}{dx^4} + (\sigma_t - \chi_2)\,\frac{d^2 w(x)}{dx^2} = 0 \tag{8.21}$$

$$\text{with the b.c. (8.9a)} \quad w(\pm 1) = w'(\pm 1) = 0.$$

This equation is very similar to that studied earlier for linear stability: by replacing σ_t with $(\sigma_t - \chi_2)$ in equation (8.14) one obtains the equation just above.[6] Reusing the previous analysis, we find a solution of the non-linear F.–von K. equations in the form:

$$\begin{cases} w(x,y) = w_{\mathrm{m}}\,\dfrac{1+\cos(\pi x)}{2} & \chi_2 = \sigma_t - \sigma_{\mathrm{Eu}} & \text{(Euler buckling for } \sigma_t \ge \sigma_{\mathrm{Eu}}) \\[2mm] w(x,y) = 0 & \chi_2 = 0 & \text{(planar solution for } \sigma_t < \sigma_{\mathrm{Eu}}) \end{cases}. \tag{8.22}$$

As earlier, oscillating profiles $w(x)$ can be discarded because they represent higher energy configurations. The constant χ_2 has been found in terms of the transverse compression σ_t. In contrast to the linear stability analysis, the amplitude w_{m} is not a free parameter, and it is set by the condition (8.9c), which has not yet been used. Indeed, plugging the expressions (8.20) for χ and (8.22) for w into equation (8.9c), one obtains a non-linear equation for the amplitude of deformation:

$$w_{\mathrm{m}}^2 = \frac{4}{3} \left(\frac{\sigma_t}{\sigma_{\mathrm{Eu}}} - 1 \right). \tag{8.23}$$

---

[6] This is all consistent as χ_2, being a quantity of second order, was taken to be zero in the analysis of linear stability.

To summarize, the Euler column appears to be an exact solution of the F.–von K. equations for $\sigma_t > \sigma_{Eu}$, with a deflection:

$$w(x,y) = w_m \frac{1 + \cos \frac{\pi x}{L}}{2} \qquad \text{where } w_m = \pm \frac{2\,h}{\sqrt{3}} \left(\frac{\sigma_t}{\sigma_{Eu}} - 1 \right)^{1/2}, \tag{8.24a}$$

and uniform stress, given by (8.1e) and (8.20):

$$\sigma_{xx} = -\sigma_{Eu} \quad \sigma_{yy} = -(\sigma_\ell - \nu\,(\sigma_t - \sigma_{Eu})) \quad \sigma_{xy} = 0. \tag{8.24b}$$

In these expressions, we have restored the physical dimensions such as E, h, and used the planar, stress-free configuration as the reference configuration: the above expressions yield the physical stress in the plate upon buckling. This buckling bifurcation is super-critical and the maximum amplitude indeed changes like the square root of the buckling parameter $(\sigma_t/\sigma_{Eu} - 1)$ above threshold, the generic behaviour for super-critical bifurcations.

Note that the non-linear terms in the F.–von K. equations have been dropped from the beginning of the analysis by taking advantage of the vanishing of Gauss curvature for cylindrical profiles. The Euler bifurcation results from another nonlinearity, coming from the boundary condition (8.9c) used to derive the amplitude of buckling.

8.5.2 Range of validity

We shall now look at the validity of this Euler column solution, given that the F.–von K. equations were derived under a small displacement approximation: for consistency, we have to investigate the range of compressions where the Euler solution is indeed compatible with the approximations underlying the F.–von K. equations. The *stability* (both linear and non-linear) of this solution is a different matter, investigated in Section 8.7.

It could be thought that the Euler solution is valid only in a weakly non-linear regime, i.e. when the dimensionless buckling parameter satisfies $[(\sigma_t/\sigma_{Eu}) - 1] \ll 1$. As we show below, the domain of validity of the F.–von K. equations is in fact much wider, and their solutions can be used even well above the buckling threshold, when the buckling parameter is *significantly larger than unity* The reason lies in the geometrical origin of the nonlinearity in the F.–von K. equations, which makes them *exact* in some sense.

Let us come back to the assumptions used to derive the F.–von K. equations earlier in Chapter 6. The condition (6.110b) that the slope be small reads, above the buckling threshold $\sigma_t > \sigma_{Eu}$:

$$\frac{\partial w}{\partial x_\alpha} \sim \frac{1}{L}\, h \left(\frac{\sigma_t}{\sigma_{Eu}} - 1 \right)^{1/2} \ll 1. \tag{8.25a}$$

The condition (6.110a) of curvature small compared to $1/h$ imposes:

$$h\,\frac{\partial^2 w}{\partial x_\alpha^2} \sim \frac{h^2}{L^2} \left(\frac{\sigma_t}{\sigma_{Eu}} - 1 \right)^{1/2} \ll 1 \tag{8.25b}$$

Now, the release of in-plane compression $\sigma_{\alpha\beta}$ is at most of order $\sigma_{\mathrm{Eu}} \sim E(h/L)^2$, and the typical squared slope is of order

$$\left(\frac{\partial w}{\partial x}\right)^2 \sim \left(\frac{h}{L}\right)^2 \left(\frac{\sigma_{\mathrm{t}}}{\sigma_{\mathrm{Eu}}}\right).$$

Therefore, all the terms in equation (8.6) for the in-plane strain have the same order of magnitude, and condition (6.110b) restricts the use of the F.–von K. equations to compressions such that:

$$\left|\frac{\partial u_\alpha}{\partial x_\beta}\right| \sim \left(\frac{h}{L}\right)^2 \left(\frac{\sigma_{\mathrm{t}}}{\sigma_{\mathrm{Eu}}} - 1\right) \ll 1. \tag{8.25c}$$

In this expression, $h/L \ll 1$.

Comparing all these conditions (8.25) we find that, well above the buckling threshold (i.e. in the limit of large dimensionless buckling parameter $\sigma_{\mathrm{t}} \gg \sigma_{\mathrm{Eu}}$) and for[7] $h \ll L$, the hardest one to fulfill is equation (8.25a). To summarize the previous conditions, the F.–von K. equations can be used provided the imposed compression is not too large:

$$\left(\frac{\sigma_{\mathrm{t}}}{\sigma_{\mathrm{Eu}}}\right)^{1/2} \ll \frac{L}{h}. \tag{8.26}$$

Since the right-hand in this inequality is a large number, the F.–von K. equations break down for large values of the buckling parameter $\sigma_{\mathrm{t}}/\sigma_{\mathrm{Eu}}$, which is well in the non-linear post-buckled regime.

Using the definition (8.19) of σ_{Eu}, the condition of validity of the Euler solution (8.26) can be rewritten in the form:

$$\left(\frac{\sigma_{\mathrm{t}}}{E}\right)^{1/2} \ll 1. \tag{8.27}$$

By Hooke's law, the ratio σ_{t}/E is of the order of the differential strain, up to a coefficient of order one depending on Poisson's ratio. The condition of validity of the F.–von K. is that this typical strain remains small; in this particular geometry, the domain of validity of the plate equations is just that strain remain small, which is the assumption underlying Hookean elasticity. This is because, again, in this particular geometry with clamped lateral boundaries, finite rotations cannot take place unless strain becomes finite too—this is not always what happens in plate or rod elasticity, see Section 2.2.11.

8.5.3 Comparison with the theory of rods

The Euler column is essentially one-dimensional (its profile is cylindrical) and indeed has an equivalent in the theory of rods, which we investigate in this section. We consider the buckling of a rod in a geometry similar to that of the strip (Fig. 8.8). The rod is taken inextensible and its energy only includes a bending term proportional to the square of the curvature.

---

[7] $h \ll L$ is precisely the condition under which the plate equations were derived. In the experiments reported at the beginning, $h/L \sim 1/100$.

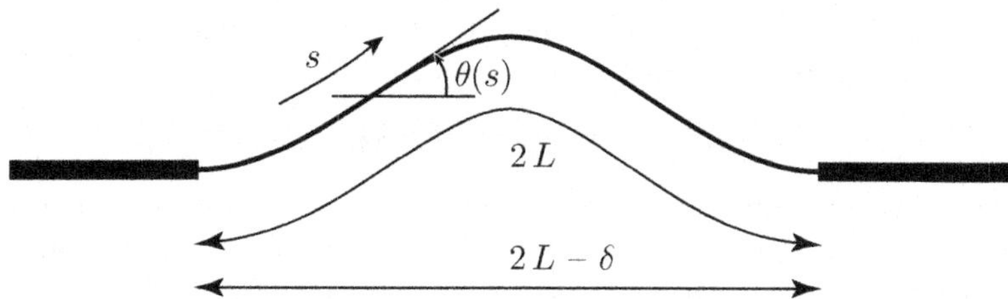

Fig. 8.8 Buckling of an inextensible bar, in a geometry similar to that of the strip. We use geometrically exact equations, which permit to assess the validity of the Euler solution of the F.–von K. equations.

Equilibrium equations for an inextensible rod

The equations for rods are given in several places in this book, in Appendix A for instance. They are briefly recalled below. Let $s \in [-L, L]$ be the curvilinear coordinate along the rod, $\theta(s)$ the local direction of the tangent, and EI the bending stiffness of the rod (we use the same notation as in Part I of this book, Chapter 3 in particular, except that the rod extends now from $s = -L$ to $+L$ although it used to go from 0 to L; L is now the half-length of the rod to make the notation consistent with that for the Euler column). The bending energy of the rod reads:

$$\mathcal{E}_{\text{rod}} = \frac{EI}{2} \int_{L}^{L} \dot{\theta}^2(s) \, ds, \tag{8.28}$$

In equation (8.28), $\dot{\theta}(s) = d\theta/ds$. One has to minimize the energy $\mathcal{E}_{\text{rod}}$ with the constraint that the ends are fixed: the quantities

$$x(L) - x(-L) = \int_{-L}^{L} \cos[\theta(s)] \, ds, \tag{8.29a}$$

$$y(L) - y(-L) = \int_{-L}^{L} \sin[\theta(s)] \, ds \tag{8.29b}$$

are prescribed.

Following the method presented in Section 1.3.5, this constrained minimization problem is addressed using two Lagrange multipliers, F_x and F_y, one for each constraint (8.29). As usual, these Lagrange multipliers, which are conjugate quantities of a displacement, can be interpreted as the force applied by the support at the end. In order to obtain the equilibrium equations, one has to compute the variation of:

$$\delta \left(\mathcal{E}_{\text{rod}} + F_x \int_{-L}^{L} \cos[\theta(s)] \, ds + F_y \int_{-L}^{L} \sin[\theta(s)] \, ds \right)$$

$$= \left[EI \, \dot{\theta}(s) \, \delta\theta(s) \right]_{s=-L}^{L} + \int_{-L}^{L} \left(-EI \, \ddot{\theta}(s) - F_x \sin[\theta(s)] + F_y \cos[\theta(s)] \right) \delta\theta(s) \, ds. \tag{8.30}$$

This calculation is similar to that in Appendix A. As usual, some integration by parts must be carried out to bring an integrand proportional to the variation itself and not to its derivatives.

For this energy to be stationary with respect to arbitrary changes $\delta\theta(s)$ of $\theta(s)$ satisfying the constraints (8.29), the integrand has to vanish for all s:

$$EI\,\ddot{\theta}(s) + F_x\,\sin[\theta(s)] - F_y\,\cos[\theta(s)] = 0. \tag{8.31}$$

The boundary term in equation (8.30) vanishes automatically since $\delta\theta(\pm L) = 0$ with clamped edges. As usual for constrained minimization problems, the Lagrange multipliers F_x and F_y must be chosen such that the variables given in equations (8.29) take their imposed values.

The buckled bar solution

In the following, we consider only *symmetric* buckling geometries, i.e. odd functions $\theta(s)$, as in Fig. 8.8. Then, the integral condition $\int_{-L}^{L} \sin[\theta(s)]\,\mathrm{d}s = 0$ is automatically satisfied, and the corresponding Lagrange multiplier F_y can be discarded. Since $F_y = 0$, we can now omit the subscript in the force and write $F = F_x$. Equations (8.31) then take the form:

$$EI\,\ddot{\theta}(s) + F\,\sin[\theta(s)] = 0 \qquad \text{with b.c. } \theta(\pm L) = 0,$$

$$\text{with the constraint:} \qquad \int_{s=-L}^{+L} \cos[\theta(s)]\,\mathrm{d}s = 2\,L - \delta,$$

where δ denotes the imposed axial shortening of the bar. Let us introduce the square root of the compression rate, a convenient dimensionless parameter for this problem:

$$\varepsilon = \left(\frac{\delta}{2\,L}\right)^{1/2}. \tag{8.32}$$

The problem is made dimensionless by using $s^* = \delta^* = L$ as the scale for all lengths, and $F^* = EI/L^2$ as the scale for the compression force. For the sake of readability, we omit the bars on top of dimensionless quantities, such as $\bar{s} = s/s^*$; the dimensionless form of the equations reads:

$$\ddot{\theta}(s) + F\,\sin[\theta(s)] = 0 \qquad \text{with b.c. } \theta(\pm 1) = 0, \tag{8.33a}$$

$$\text{and } \frac{1}{2}\int_{s=-1}^{+1} \cos\theta(s)\,\mathrm{d}s = 1 - \varepsilon^2. \tag{8.33b}$$

In equation (8.33a), one recognizes the equation for a physical[8] pendulum of finite amplitude: θ is the angle with respect to the vertical direction, and the variable s plays the role of time. Solutions to this non-linear equation are known in the form of elliptic integrals, see Appendix A. However, such functions are not very convenient to handle, and we shall seek $\theta(s)$ as an expansion in ε instead.

The problem is invariant under mirror reflection $(x, y) \mapsto (-x, y)$, which changes θ into $(-\theta)$ and leaves F unchanged. One should notice that the sign of ε is also somewhat arbitrary due to the presence of a square root in its definition. This suggests that θ_ε and F_ε should be sought in the form of series in odd powers of ε, and even powers respectively:

---

[8] By physical pendulum, we mean a pendulum satisfying the equation $\ddot{\theta} + \omega^2\,\sin\theta = 0$ rather than the linearized (simple pendulum) version $\ddot{\theta} + \omega^2\,\theta = 0$.

$$\theta_\varepsilon(s) = \varepsilon\,\theta^{[1]}(s) + \varepsilon^3\,\theta^{[3]}(s) + \mathcal{O}(\varepsilon^5), \tag{8.34a}$$

$$F_\varepsilon = F^{[0]} + \varepsilon^2\,F^{[2]} + \mathcal{O}(\varepsilon^4). \tag{8.34b}$$

Plugging these expansions (8.34) into the equilibrium equations (8.33a), one obtains at order linear in ε:

$$\ddot{\theta}^{[1]}(s) + F^{[0]}\,\theta^{[1]}(s) = 0 \qquad \text{with boundary conditions } \theta^{[1]}(\pm1) = 0. \tag{8.35}$$

The same equation was encountered in the analysis of the Euler column. It has the following solution:

$$F^{[0]} = \pi^2 \qquad \theta^{[1]}(s) = -2\,\sin(\pi\,x), \tag{8.36}$$

where the prefactor (-2) in $\theta^{[1]}$ has been chosen such that equation (8.33b) is satisfied up to order (ϵ^2); the minus sign represents *upwards* buckling. By integration of the kinematical equation $y'(s) = \sin\theta(s)$, we find the deflection in the form

$$y(s) = \frac{4\,L\,\epsilon}{\pi}\,\frac{1 + \cos\frac{\pi\,x}{L}}{2} + \dots, \tag{8.37}$$

where we have restored the physical dimensions that were implicit in the calculation, and omitted higher-order terms.

At next order, the equilibrium equations (8.33a) yields, for $\theta^{[3]}$:

$$\left(\frac{\mathrm{d}^2}{\mathrm{d}s^2} + \pi^2\right)\theta^{[3]}(s) = \frac{\pi^2}{6}\left(\theta^{[1]}(s)\right)^3 - F^{[2]}\,\theta^{[1]}(s) \qquad \text{with } \theta^{[3]}(\pm1) = 0, \tag{8.38a}$$

while the equation (8.33b) for the amplitude writes, at order ε^4:

$$\frac{1}{2}\int_{-1}^{+1}\left[\theta^{[3]}\,\theta^{[1]} - \frac{1}{24}\left(\theta^{[1]}\right)^4\right]\mathrm{d}s = 0. \tag{8.38b}$$

These equations can be solved to yield:

$$F^{[2]} = \frac{\pi^2}{4} \qquad \theta^{[3]}(s) = -\frac{1}{24}\,(7 + 2\,\cos(2\pi\,s))\,\sin(2\pi\,s), \tag{8.39}$$

which gives the buckling of an inextensible bar at second order in $\delta/L = \varepsilon^2$. This expansion can be continued at higher orders.

Comparison with the F.–von K. equations

We shall now stress the similarities and point out some differences between the buckling of plate, which is extensible longitudinally, and the buckling of an inextensible rod.

Let A be the length of the plate in the invariant direction. Comparison of the bending energies reveals that the coefficient $(D\,A)$ for the plate model should be identified with the bending modulus EI of the rod: a plate that deforms with cylindrical symmetry has an effective bending modulus that is proportional to its width.

Since the rod is inextensible, it buckles under an imposed displacement δ with a vanishing threshold $\delta_{\mathrm{c}} = 0$, as shown by equation (8.32). In contrast, the plate, which is compressible in its plane, has a non-zero threshold. It is remarkable, however, that the thresholds are identical when the buckling is force controlled. A plate buckles into the Euler pattern for $\sigma_{\mathrm{t}} = \sigma_{\mathrm{Eu}}$. The magnitude of the total force applied by the support along either clamped edge reads $(-\sigma_{xx}) A h = 4\pi^2 DA/(2 L)^2$ by equation (8.15b). For the rod, the critical force reads $F^* F^{[0]} = EI\pi^2/L^2$, which is indeed the same results when EI is identified with DA.

We should also note that the deflections of the plate and the rod have the same sinusoidal profile near threshold, that is the same dependence on the variables s or x; compare equations (8.24a) and (8.37). Note that the amplitudes of the deflection, however, cannot be identified as this amplitude is of the order of the thickness h in the plate model although the parameter h is absent from the rod model.

An important difference between the two models has to do with geometric nonlinearity. The F.–von K. equations for the plate assume small rotations while the rod model is geometrically exact. This introduces a discrepancy between the two models. We shall consider more specifically loadings that are well above the buckling threshold; there, the other fundamental difference between the two models (inextensible rod on the one hand, longitudinally stretchable plate on the other) becomes irrelevant.[9] In these conditions, comparison of the two models will allow us to recover the range of validity of the F.–von K. equations, by comparison with the rod model, which is exact even for large deflections, provided self-intersections of the bar are allowed. The solutions $F^{[0]}$, $F^{[2]}$, $\theta^{[1]}$ and $\theta^{[3]}$ above can be used to get the first non-zero contributions to the elastic energy of the bent bar:

$$\mathcal{E}_{\mathrm{rod}}(\varepsilon) = e^{[2]}\,\varepsilon^2 + e^{[4]}\,\varepsilon^4 + \mathcal{O}(\varepsilon^4), \tag{8.40}$$

where the coefficients can be calculated as:

$$e^{[2]} = \frac{1}{2}\int_{-1}^{+1}\left(\dot{\theta}^{[1]}\right)^2 \mathrm{d}s = 2\pi^2 \qquad e^{[4]} = \frac{1}{2}\int_{-1}^{+1}\left(2\,\dot{\theta}^{[3]}\,\dot{\theta}^{[1]}\right)\mathrm{d}s = \frac{\pi^2}{2}. \tag{8.41}$$

The approximations underlying the F.–von K. theory amount to truncating this energy to the first term:

$$\mathcal{E}_{\mathrm{FvK}}(\varepsilon) = e^{[2]}\,\varepsilon^2.$$

The plate theory is therefore suited under the condition that the quartic term in (8.40) remains negligible compared with the quadratic one:

$$\frac{e^{[4]}\,\varepsilon^4}{e^{[2]}\,\varepsilon^2} = \frac{\varepsilon^2}{4} < \rho \quad \text{hence } \varepsilon^2 < 4\,\rho,$$

where ρ, small and dimensionless, is the relative error allowed in the energy. In terms of plate variables, and using Hooke's law, we have $\epsilon^2 = \delta/(2L) \approx \sigma_{\mathrm{t}}/E$. The above condition rewrites

$$\frac{\sigma_{\mathrm{t}}}{E} \approx \frac{\sigma_{\mathrm{t}}}{\sigma_{\mathrm{Eu}}}\left(\frac{h}{L}\right)^2 < 4\,\rho. \tag{8.42}$$

The condition derived previously in equation (8.26) is recovered. In the experiments, the aspect ratio of the film is $h/L = 0.01$. Accepting an inaccuracy of $\rho = 1\%$ on the energy, one can safely use the F.–von K. equations up to dimensionless buckling parameters $\sigma_{\mathrm{t}}/\sigma_{\mathrm{Eu}}$ as large as 400, i.e. for compressions 400 times the threshold value.

---

[9] Recall that the buckling bifurcation results from a competition between stretching and bending effects. Well above threshold, the strain is penalizing (it costs a lot of energy) and the plate minimizes its energy by almost completely relaxing this strain, behaving effectively like an inextensible plate.

8.6   Transition from finite to infinite wavelengths

In Section 8.4, the part of the phase diagram describing the initial buckling was derived (Fig. 8.7). It comprises a special point:

$$\sigma_t^{cr} = \sigma_{Eu} \qquad \sigma_\ell^{cr} = \frac{2}{3}\sigma_{Eu}. \tag{8.43}$$

There, the marginal mode made of bumps (B on the diagram) has a vanishing longitudinal wavenumber $k \to 0$, hence an infinite wavelength $\Lambda = 2\pi/k$: nearby $(\sigma_t^{cr}, \sigma_\ell^{cr})$, there is a smooth transition from bump patterns to the Euler column.

In the present section, we explore the vicinity of this critical point using Poincaré's method of normal forms. Taking advantage of the fact that wavelengths are large, we derive the elastic energy of the strip as a function of a reduced set of parameters, based on mere symmetry considerations. This same reasoning is classically used to study Lifshitz' point of magnetic materials,[10] hence our use of the denomination of Lifshitz' point later on.

These simple analytical arguments will allow us to predict multistable states of the strip on a qualitative basis, and therefore to anticipate and interpret the more detailed numerical results of Section 8.7.

8.6.1   Derivation of a reduced energy

So far, two buckling modes have been found in the stability analysis of the planar strip: the Euler column and the symmetric bumps with a finite non-zero longitudinal wavenumber k, given earlier in equation (8.12). In the following, we consider states of the strip that are linear combinations of these two modes, with amplitudes[11] e and s:

$$w(x,y) = e\,\frac{1 + \cos\frac{\pi x}{L}}{2} + s\,\cos(k\,y)\,\frac{1 + \cos\frac{\pi x}{L}}{2}.$$

We seek an expression of the elastic energy $\mathcal{E}$ of the strip as a function of e, s, k and of the imposed compressions σ_t and σ_ℓ. We focus on the vicinity of Lifshitz' point and introduce the small parameters:

$$\hat{\sigma}_t = \frac{\sigma_t - \sigma_{Eu}}{\sigma_{Eu}} \quad \text{and} \quad \hat{\sigma}_\ell = \frac{\sigma_\ell - \frac{2}{3}\sigma_{Eu}}{\sigma_{Eu}}. \tag{8.44}$$

This change of coordinate is sketched in Fig. 8.9.

In the vicinity of Lifshitz' point, bifurcations are toward small amplitude solutions and so both e and s are small. The energy being invariant under the symmetries $e \mapsto (-e)$ (up/down invariance), $s \mapsto (-s)$ (shift of π in the phase of the bumps), $k \mapsto (-k)$ (orientation of the y axis), the expansion of the energy with respect to these quantities should only contain *even* powers.

---

[10] Usually, ferromagnets have a spontaneous magnetization pointing in the same direction everywhere. However it may happen, for instance in some garnets, that the direction of magnetization changes periodically in space. Such materials are called helimagnets, the extreme case is Néel's anti-ferromagnets where magnetization is in opposite directions from one lattice site to the next. Lifschitz' point is a point in the parameter space where the wavelength of the modulation becomes infinite (I. M. Lifshitz, 1969; I. M. Lifshitz and A. Y. Grosberg, 1974).

[11] The amplitude s of the symmetric mode should not be confused with the curvilinear coordinate of the rod used in the previous section.

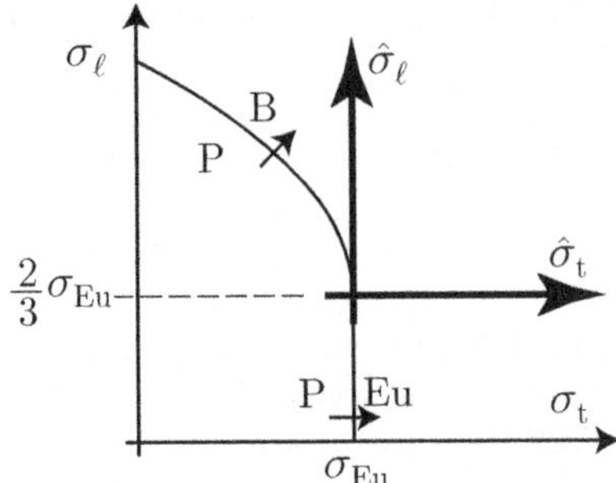

Fig. 8.9 Change of coordinate allowing one to study the vicinity of Lifshitz' point.

The contribution to the energy of the Euler column is simply written as:

$$\mathcal{E}_{\mathrm{Eu}}(e) = \mathcal{E}(e, s = 0) = -\frac{1}{2}\hat{\sigma}_t\, e^2 + \frac{1}{4}\, e^4.$$

This is indeed the simplest form of the energy that yields a bifurcation for the parameter $e \neq 0$ when $\hat{\sigma}_t > 0$, and satisfies the symmetry $e \mapsto (-e)$. The choice of coefficients is somewhat arbitrary thanks to rescalings, as explained below.

In order to write down the contribution of the bumps, which depends on both s and k, one can replace e by s in the above expression and try to restore the dependence of the coefficients on k by making some guess. Again, k is small and one seeks expansions in powers of k^2. The coefficient of the quartic term, $1/4$, is finite in the energy of the Euler column ($k = 0$). For small $k \neq 0$, it will slightly depart from this value $1/4$ but this does not induce any bifurcation: we disregard the dependence of the quartic term on k, and keep the $1/4$ coefficient unchanged. The same does not hold for the quadratic coefficient, $(-\hat{\sigma}_t/2)$, which changes sign around $\hat{\sigma}_t = 0$, even for $k = 0$. In order to find its dependence on k, we note that:

- it should reduce to $(-\mu\,\hat{\sigma}_t/2)$ for $k = 0$ (μ positive constant), an expression which is similar[12] to that of the Euler column, $(-\hat{\sigma}_t/2)$, given earlier;
- the coefficient $\hat{\sigma}_\ell$ is associated with a second derivative in the y direction in the F.– von K. equations, hence with k^2 here;
- according to the stability analysis of the planar state, the wavenumber k of the optimal pattern undergoes a super-critical bifurcation near $\hat{\sigma}_\ell = 0$: it goes to zero when $\hat{\sigma}_\ell$ approaches zero from above, and vanishes for $\hat{\sigma}_\ell < 0$.

The simplest form of the energy that satisfies all these constraints is:

$$\mathcal{E}_{\mathrm{B}}(s, k) = \mathcal{E}(e = 0, s, k) = -\frac{1}{2}\left[\mu\,\hat{\sigma}_t + \frac{k^2}{2}\,\hat{\sigma}_\ell - \frac{k^4}{4}\right] s^2 + \frac{1}{4}\, s^4,$$

as will be checked below, in Section 8.6.2. Owing to the term in square brackets, the optimal k indeed bifurcates from zero to finite values above $\hat{\sigma}_\ell = 0$: the planar state becomes unstable towards bumps rather than Euler column.

---

[12] Here, one has to introduce an additional parameter, μ, different from 1 because the limit $k \to 0$ is singular for the elastic energy of the bumps, see Fig. 8.24 for instance.

It remains to couple the bumps and Euler column through a cross-term. We have so far considered expansion of the energy up to fourth order in the deflections e and s. At this order only one cross-term is compatible with the symmetries: $(e^2 s^2)$. Collecting all these contributions, one obtains the reduced energy of the strip:

$$\mathcal{E}(e, s, k) = -\frac{1}{2}\hat{\sigma}_t\, e^2 + \frac{1}{4}\, e^4 - \frac{1}{2}\left[\mu\,\hat{\sigma}_t + \frac{k^2}{2}\,\hat{\sigma}_\ell - \frac{k^4}{4}\right] s^2 + \frac{1}{4}\, s^4 + \frac{\kappa}{2}\, e^2\, s^2. \tag{8.45}$$

In this expression, most of the coefficients have been set to arbitrary and convenient values by rescaling the variables: rescaling e [resp. $\hat{\sigma}_t$, s, k, $\hat{\sigma}_\ell$] allows one to set the coefficient of e^4 [resp. $(\frac{1}{2}\hat{\sigma}_t\, e^2)$, s^4, $(k^4 s^2)$, $(\hat{\sigma}_\ell k^2 s^2)$] to the value proposed above. The two remaining coefficients, κ and μ, cannot be chosen arbitrarily and must instead be computed by the method of Section 8.8:

$$\mu = 0.408248 \text{ and } \kappa = \frac{1}{\mu} = 2.44949 \qquad \text{for a Poisson's ratio } \nu = 0.3. \tag{8.46}$$

They satisfy the following inequalities:

$$0 < \mu = \frac{1}{\kappa} < \kappa, \tag{8.47}$$

which is in fact all we need to know about these coefficients.

In the rest of this section, we derive the phase diagram of the strip near Lifshitz' point $(\sigma_t = \sigma_{\mathrm{Eu}}, \sigma_\ell = 2/3\,\sigma_{\mathrm{Eu}})$ by studying the minima of the energy function $\mathcal{E}(e, s, k)$ as a function of the applied loading $(\hat{\sigma}_t, \hat{\sigma}_\ell)$.

8.6.2 Recovering the initial buckling

Let us first investigate the domain of the $(\hat{\sigma}_t, \hat{\sigma}_\ell)$ plane where the unbuckled configuration (P) is stable. Stability with respect to Euler buckling is:

$$\left.\frac{\partial^2 \mathcal{E}}{\partial e^2}\right|_{(e,s)=(0,0)} = -\hat{\sigma}_t > 0,$$

and that with respect to bumps:

$$\left.\frac{\partial^2 \mathcal{E}}{\partial s^2}\right|_{(e,s)=(0,0)} = -\left[\mu\,\hat{\sigma}_t + \frac{k^2}{2}\,\hat{\sigma}_\ell - \frac{k^4}{4}\right] > 0 \quad \text{for all } k.$$

Therefore the domain of linear stability of the unbuckled (planar) state is given by the conditions:

$$\hat{\sigma}_t < 0 \text{ and } \max_k \left[\mu\,\hat{\sigma}_t + \frac{k^2}{2}\,\hat{\sigma}_\ell - \frac{k^4}{4}\right] < 0, \tag{8.48}$$

or more explicitly:

$$\hat{\sigma}_t < 0 \text{ and } \begin{cases} \mu\,\hat{\sigma}_t < 0 & \text{if } \hat{\sigma}_\ell < 0, \\ \mu\,\hat{\sigma}_t + \frac{\hat{\sigma}_\ell^2}{4} < 0 & \text{if } \hat{\sigma}_\ell > 0. \end{cases}$$

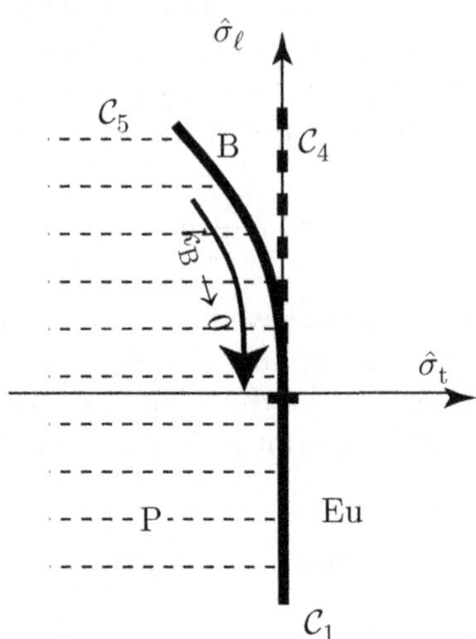

Fig. 8.10 Domain of stability of the planar configuration near the Lifshitz' point, consistent with the results of Section 8.4. The labels are P for the planar configuration ($e = s = 0$), (Eu) for the Euler column ($e \neq 0$, $s = 0$), B for the bumps ($s \neq 0$, $e = 0$).

This domain is bounded by two curves:

$$\mathcal{C}_1 : \qquad\qquad \hat{\sigma}_t = 0 \text{ and } \hat{\sigma}_\ell < 0 \qquad\qquad (8.49)$$

$$\mathcal{C}_5 : \qquad\qquad \hat{\sigma}_t = -\frac{1}{4\,\mu}\hat{\sigma}_\ell^2 = 0 \text{ and } \hat{\sigma}_\ell > 0. \qquad\qquad (8.50)$$

This domain is shown in Fig. 8.10. By construction, it is the same as in the phase diagram in Fig. 8.7 up to a translation, which is due to the change of coordinates in this plane. The strip becomes unstable against the Euler column across $\mathcal{C}_1$, and against a bump pattern across $\mathcal{C}_5$. For later reference, we also include the curve $\mathcal{C}_4$ defined by

$$\mathcal{C}_4 : \qquad \hat{\sigma}_t = 0 \text{ and } \hat{\sigma}_\ell > 0. \qquad\qquad (8.51)$$

Along this curve, the branch corresponding to planar solutions bifurcates into a Euler column. This transition is not observable physically since the planar solutions are already unstable against bump patterns there, but it is included for later reference.

In the planar configuration, the reduced energy of the strip is simply:

$$\mathcal{E}_{\mathrm{P}} = \mathcal{E}(e = 0, s = 0, k) = 0. \qquad\qquad (8.52)$$

The wavenumber k of the most unstable bump pattern along $\mathcal{C}_5$ is the one that achieves the maximum in equation (8.48) above:

$$k = \pm\sqrt{\hat{\sigma}_\ell} \quad \text{i.e.} \quad \Lambda = \frac{2\,\pi}{k} \propto \frac{1}{\left(\sigma_\ell - \frac{2}{3}\sigma_{\mathrm{Eu}}\right)^{1/2}} \sim \frac{1}{\left(\sigma_t - \sigma_{\mathrm{Eu}}\right)^{1/4}}. \qquad (8.53)$$

This wavelength becomes infinite when Lifshitz' point $(\hat{\sigma}_t, \hat{\sigma}_\ell) = (0, 0)$ is approached from above ($\hat{\sigma}_\ell > 0$).

8.6.3 Stability of the Euler column

Similarly, one can investigate the stability of the Euler column with respect to a varicose pattern, i.e. a pattern obtained by superposition[13] of symmetric bumps onto the Euler column solution: both $e \neq 0$ and $s \neq 0$ are non-zero. The Euler column, which is the base state of the present stability analysis, is characterized by:

$$\left. \frac{\partial \mathcal{E}}{\partial e} \right|_{s=0} = e \left(e^2 - \hat{\sigma}_{\rm t} \right) = 0 \qquad \text{hence} \qquad e = \pm \sqrt{\hat{\sigma}_{\rm t}} \text{ provided } \hat{\sigma}_{\rm t} > 0,$$

and has an elastic energy:

$$\mathcal{E}_{\rm Eu} = -\frac{1}{4} \hat{\sigma}_{\rm t}^2. \tag{8.54}$$

Stability with respect to a perturbation made of bumps requires:

$$\left. \frac{\partial^2 \mathcal{E}}{\partial s^2} \right|_{(e,s)=(\pm\sqrt{\hat{\sigma}_{\rm t}},0)} = -\left[\mu\, \hat{\sigma}_{\rm t} + \frac{k^2}{2} \hat{\sigma}_\ell - \frac{k^4}{4} \right] + \kappa\, \hat{\sigma}_{\rm t} > 0 \quad \text{for all } k.$$

The domain of stability of the Euler column is therefore given by the conditions:

$$\hat{\sigma}_{\rm t} > 0 \quad \text{and} \quad \max_k \left[-(\kappa - \mu)\, \hat{\sigma}_{\rm t} + \frac{k^2}{2} \hat{\sigma}_\ell - \frac{k^4}{4} \right] < 0,$$

which can be rewritten as:

$$\hat{\sigma}_{\rm t} > 0 \quad \text{and} \quad \begin{cases} -(\kappa - \mu)\, \hat{\sigma}_{\rm t} < 0 & \text{if } \hat{\sigma}_\ell < 0, \\ -(\kappa - \mu)\, \hat{\sigma}_{\rm t} + \frac{\hat{\sigma}_\ell^2}{4} < 0 & \text{if } \hat{\sigma}_\ell > 0. \end{cases}$$

In the $(\overline{\sigma}_\ell, \overline{\sigma}_{\rm t})$ plane this domain is bounded by the line $\mathcal{C}_1$ defined earlier, and by a new curve:

$$\mathcal{C}_3 : \qquad \hat{\sigma}_{\rm t} = \frac{1}{4(\kappa - \mu)} \hat{\sigma}_\ell^2 \text{ and } \hat{\sigma}_\ell > 0. \tag{8.55}$$

This domain of stability of the Euler column is shown in Fig. 8.11. Along the borderline $\mathcal{C}_1$, common with the planar region, the Euler column flattens, $e \to 0$, by a smooth transition to the planar configuration, while along $\mathcal{C}_3$, it becomes unstable with respect to varicose patterns (s becomes non-zero). The wavelength of the most unstable varicose pattern is given by equation (8.53) again.

8.6.4 Bump pattern

The analysis of the bump pattern ($e = 0$, $s \neq 0$) is similar. We skip the details of the calculations and simply state the final results. The amplitude of the bump configuration is given by:

$$e = 0, \quad s = \pm \left(\mu\, \hat{\sigma}_{\rm t} + \frac{1}{4} \hat{\sigma}_\ell^2 \right)^{1/2}, \quad k_{\rm B} = \pm \sqrt{\hat{\sigma}_\ell}, \tag{8.56}$$

---

[13] This is called 'varicose' because if, for instance, the Euler column is upwards, every bump makes the column look wider and taller (upwards bumps) or narrower and lower (downward bumps) alternatively.

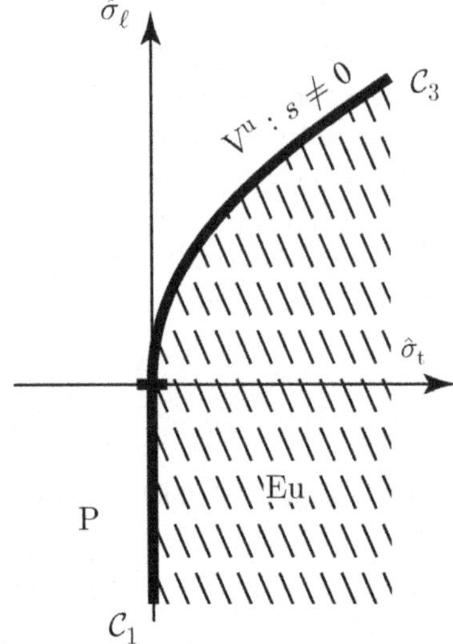

Fig. 8.11 Domain of stability of the Euler column near the critical point, which extends the diagram derived earlier in Section 8.4.

and has the elastic energy:

$$\mathcal{E}_{\mathrm{B}} = -\frac{1}{4}\left(\mu\,\hat{\sigma}_t + \frac{1}{4}\hat{\sigma}_\ell^2\right)^2 . \tag{8.57}$$

A stability analysis yields the domain where the bump patterns can be observed, as shown in Fig. 8.12. Along the curve $\mathcal{C}_5$ defined earlier, the amplitude s of the bumps goes to zero,

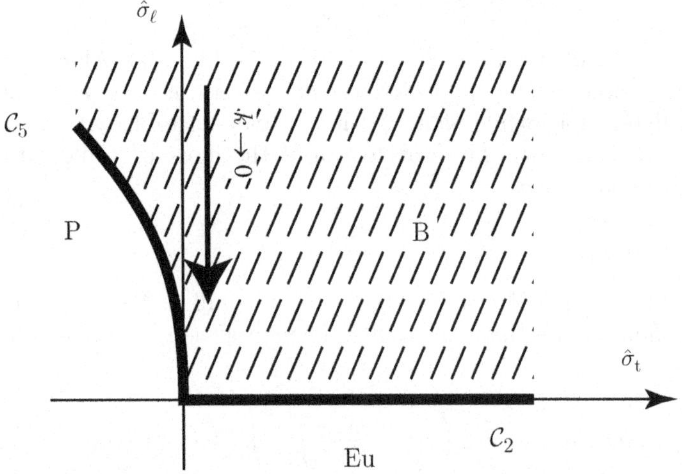

Fig. 8.12 Domain of stability of the bump pattern $(e = 0,\ s \neq 0)$ near Lifshitz' point, which extends the diagram derived earlier in Fig. 8.7, Section 8.4.

and there is a smooth transition back to the planar state. The bumps region B is bounded from below by an additional curve:

$$\mathcal{C}_2: \qquad \hat{\sigma}_\ell = 0 \text{ and } \hat{\sigma}_t > 0. \tag{8.58}$$

Along this curve, bumps become unstable against varicose patterns, as e bifurcates to a non-zero value—as we shall see below, these varicose patterns are themselves unstable and a discontinuous bifurcations brings the strip to the Euler column configuration.

8.6.5 Varicose pattern

A similar analysis shows that the varicose patterns (corresponding to both e and s non-zero, see Fig. 8.25) make the energy stationary in the region where Eu and B overlap, i.e. in between $\mathcal{C}_2$ and $\mathcal{C}_3$, but these varicose patterns are always linearly unstable—at least in the vicinity of Lifshitz' point, where the concept of reduced energy holds. This pattern is labelled V^u in the figures.

For these unstable patterns,

$$e = \pm \left(\frac{(\kappa\mu - 1)\,\hat{\sigma}_t + \kappa\hat{\sigma}_\ell^2/4}{\kappa^2 - 1} \right)^{1/2}, \quad s = \pm \left(\frac{(\kappa - \mu)\,\hat{\sigma}_t - \hat{\sigma}_\ell^2/4}{\kappa^2 - 1} \right)^{1/2}, \tag{8.59}$$

with a wavenumber given by (8.53). The energy of this pattern is:

$$\mathcal{E}_V = \frac{1}{4} \frac{\left((\kappa - \mu)\,\hat{\sigma}_t - \hat{\sigma}_\ell^2/4\right)^2}{\kappa^2 - 1} - \frac{1}{4}\hat{\sigma}_t^2. \tag{8.60}$$

The condition that the square roots in the equation above have positive arguments yields the domain of existence of this unstable branch, which is in between $\mathcal{C}_2$ and $\mathcal{C}_3$, as shown in Fig. 8.13. Although physically unobservable, this branch of unstable solutions will be useful for building up a geometric picture of the phase diagram later on.

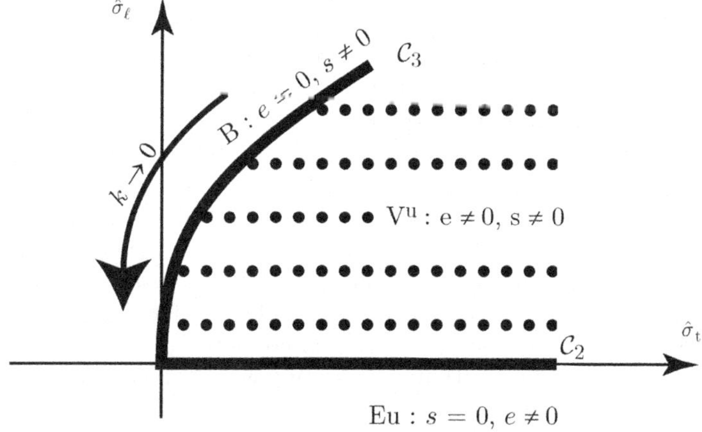

Fig. 8.13 Domain of existence of an unstable varicose pattern (both e and s non-zero).

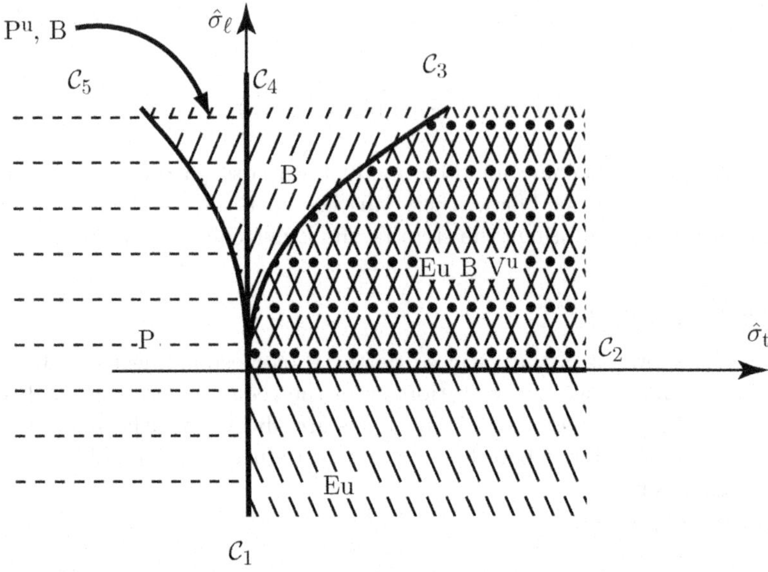

Fig. 8.14 Phase diagram of the strip in the vicinity of the Lifshitz' point. In the region bounded by C_2 and C_3, the strip is multistable: both bumps (B) or an Euler column (Eu) are stable. In this region, there also exists an unstable pattern (V$^\mathrm{u}$, varicose).

8.6.6 Conclusion

The relative position of the curves C_i $(1 \leq i \leq 5)$ encountered previously can be derived from the inequalities (8.47). Figure 8.14 combines all these diagrams. The final diagram is consistent with the experimental phase diagram presented earlier in Fig. 8.4. The relative position of the P, Eu and B regions is similar. Multistability has indeed been observed experimentally in a region located in between the Eu and B regions.

In Fig. 8.15, we propose a geometrical picture of the phase diagram. In the plane of the loading $(\hat{\sigma}_\ell, \hat{\sigma}_\mathrm{t})$, a small loop enclosing Lifshitz' point is followed (see insert). The angle $\theta = [\tan^{-1}(\hat{\sigma}_\ell/\hat{\sigma}_\mathrm{t}) - \pi]$ is an azimuthal angle measuring the type of loading. Starting at the intersection $\theta = 0$ with the half-line $\hat{\sigma}_\mathrm{t} < 0$, $\hat{\sigma}_\ell = 0$ and turning counter-clockwise (increasing θ), the loading is first such that only the planar configuration is stable: in the space of order parameters (top diagram), the axis $e = s = 0$ is followed. The energy remains zero (bottom diagram), according to equation (8.52).

When the curve C_1 is crossed, the planar configuration becomes unstable (dashed lines), and the strip adopts the Euler configuration, which has a lower energy according to (8.54). In the top diagram the amplitude e becomes non-zero, following a continuous (super-critical) pitchfork bifurcation. This branch remains stable until it merges with the unstable[14] configuration V$^\mathrm{u}$ (unstable varicose pattern) upon crossing C_3, and becomes unstable. The top diagram shows that this bifurcation is discontinuous: at the end of the Eu branch, the

---

[14] The exponent u in the labels is used for unstable patterns.

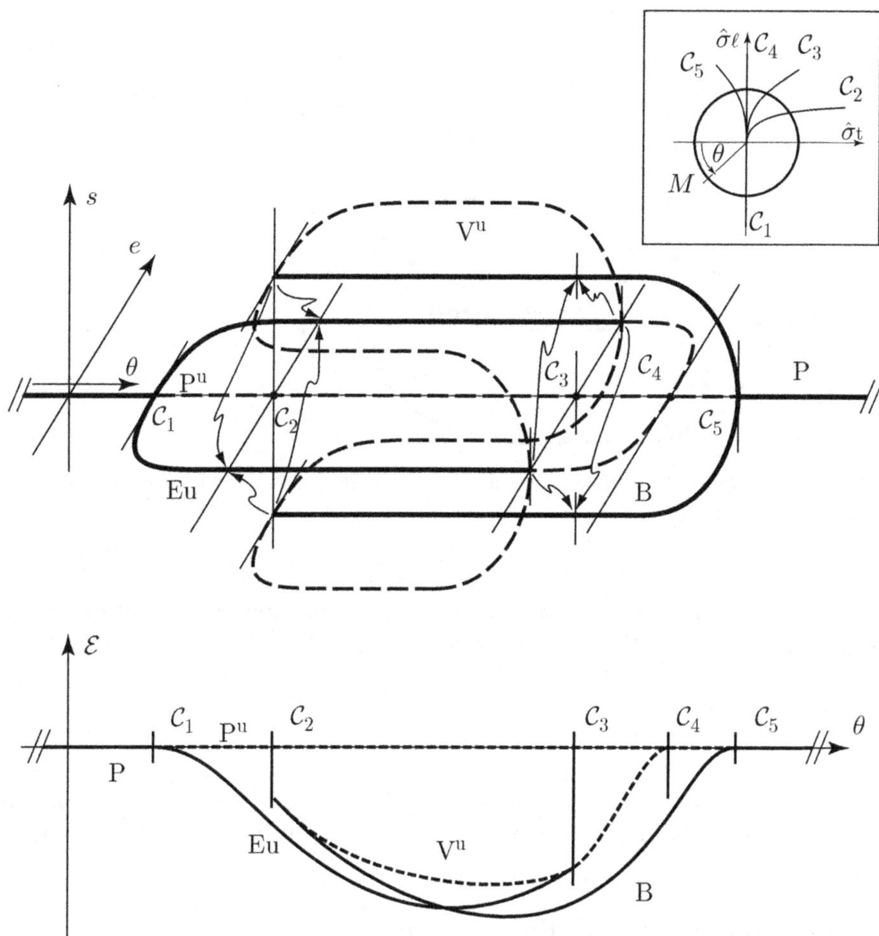

Fig. 8.15 Schematic bifurcation diagram of the strip with Poisson's ratio $\nu = 0.3$, obtained by following a small loop around the critical point (see insert). The control parameter θ is the curvilinear coordinate along the loop in the plane of the loading parameters. In the top diagram, the state of the strip is shown using the order parameters, e and s. The jagged arrows represent discontinuous bifurcations. In the bottom diagram, the reduced elastic energy along the different branches is plotted. Dashed lines denote unstable states.

system has to jump to the remote B branch, as indicated by jagged arrows. This is called a discontinuous (sub-critical) bifurcation.

A similar bifurcation takes place when moving the other way around, i.e. when θ is decreased: across the curve $\mathcal{C}_2$, a super-critical pitchfork bifurcation makes s non-zero, and the bumps branch disappears following a sub-critical pitchfork bifurcation as it merges with the other side of the unstable varicose branch.

In the region between $\mathcal{C}_2$ and $\mathcal{C}_3$ where both bumps and Euler column are stable, the system is multistable: even though a single configuration, say Eu, achieves the absolute minimum of energy for a given value of the loading parameters (σ_t, σ_ℓ), the transition from

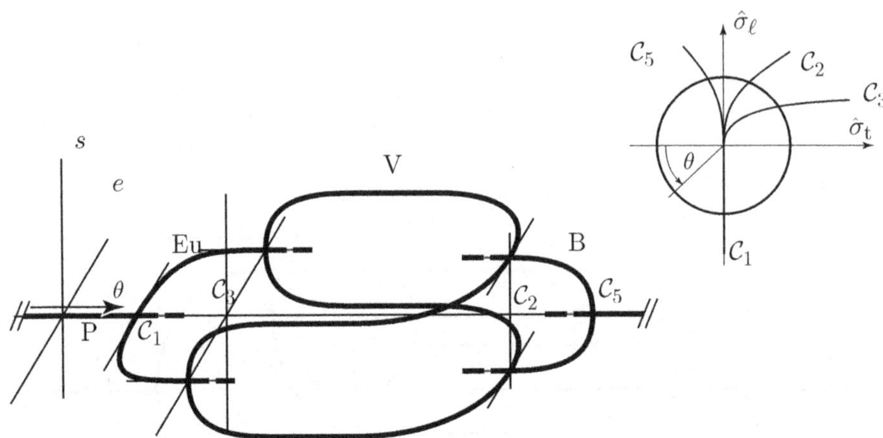

Fig. 8.16 Simple diagram that yields the actual diagram in Fig. 8.15 by folding the varicose branch (V) and making it unstable ($\mathcal{C}_3$ and $\mathcal{C}_2$ swapped).

the local energy minimum, B for instance, to the absolute minimum does not usually take place in the experiments, as this involves crossing an energy barrier. In this region, the state of the strip depends on the history of loading.

To conclude this topic, we note that the diagram in Fig. 8.15 can be thought of as deriving from the 'folding' of the varicose branch, starting from the simpler diagram shown in Fig. 8.16 where the curves $\mathcal{C}_2$ and $\mathcal{C}_3$ have been swapped, making the varicose branch stable. The latter diagram may correspond to a strip with a Poisson's ratio different from 0.3 (κ and μ depend on ν) but it is merely given for the purpose of visualization.

8.7 Linear stability of the Euler column

Since we obtained the Euler column as an *exact* solution of the F.–von K. equations, it is possible to investigate its linear stability by a method similar to that used in Section 8.4.2 for the planar configuration. Such an analysis allows one to extend the phase diagram of the strip by including the possible secondary buckling of the Euler column.

8.7.1 Equations for the linear stability analysis

The in-plane stress in the Euler column is uniform, and has the value given above in equation (8.24b):

$$\sigma_{xx} = -\sigma_{\mathrm{Eu}} \quad \sigma_{yy} = -(\sigma_\ell - \nu\,(\sigma_{\mathrm{t}} - \sigma_{\mathrm{Eu}})) \quad \sigma_{xy} = 0.$$

At large dimensionless buckling parameter ($\sigma_{\mathrm{t}}\,\&\,\sigma_\ell \gg \sigma_{\mathrm{Eu}}$), the transverse compressive stress ($-\sigma_{xx}$) is well released as it goes from σ_{t} to $\sigma_{\mathrm{Eu}} \ll \sigma_{\mathrm{t}}$ upon buckling, but the longitudinal compression ($-\sigma_{yy}$) remains large, close to ($\sigma_\ell - \nu\,\sigma_{\mathrm{t}}$), see Fig. 8.17. This compressive residual stress may induce a secondary buckling of the Euler column, addressed here by studying its linear stability. We shall continue this analysis numerically

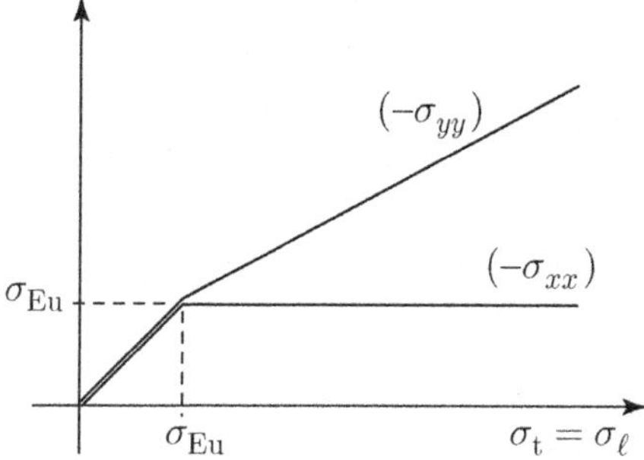

Fig. 8.17 Stresses in the Euler column for initial isotropic compression ($\sigma_{\mathrm{t}} = \sigma_{\ell}$). The transverse applied compression is efficiently released as $(-\sigma_{xx})$ saturates to σ_{Eu} even for large initial stress, but the longitudinal compression is not, as it continues to increase with a slope $(1 - \nu) > 1/2$. This induces a secondary bifurcations of the column, investigated below.

towards the end of this chapter, in order to catch finite amplitude effects (discontinuous bifurcations).

Let $(\overline{w}_{\mathrm{lin}}, \overline{\chi}_{\mathrm{lin}})$ be a small perturbation of the state of the strip near the Euler column. We consider perturbed states of the plate of the form

$$\overline{w}(\overline{x}, \overline{y}) = \overline{w}_{\mathrm{Eu}}(\overline{x}, \overline{y}) + \overline{w}_{\mathrm{lin}}(\overline{x}, \overline{y}), \qquad \overline{\chi}(\overline{x}, \overline{y}) = \overline{\chi}_{\mathrm{Eu}}(\overline{x}, \overline{y}) + \overline{\chi}_{\mathrm{lin}}(\overline{x}, \overline{y}),$$

where quantities with label Eu refer to the Euler column solution, and bars refer to the dimensionless variables introduced earlier in equation (8.1). The linear stability is addressed by studying the existence of such perturbations $(\overline{w}_{\mathrm{lin}}, \overline{\chi}_{\mathrm{lin}})$ that leave the film in mechanical equilibrium. These marginally unstable solutions reveal super-critical[15] secondary bifurcations of the Euler column. After plugging into the equilibrium equations and expanding to first order in $\overline{w}_{\mathrm{lin}}$ and $\overline{\chi}_{\mathrm{lin}}$, one obtains the F.–von K. equations (8.2) linearized near the Euler column solution (8.24):

$$\frac{\Delta^2 \overline{w}_{\mathrm{lin}}}{12\,(1 - \nu^2)} + \overline{\sigma}_{\mathrm{Eu}} \frac{\partial^2 \overline{w}_{\mathrm{lin}}}{\partial \overline{x}^2} + (\overline{\sigma}_{\ell} - \nu\,(\overline{\sigma}_{\mathrm{t}} - \overline{\sigma}_{\mathrm{Eu}})) \frac{\partial^2 \overline{w}_{\mathrm{lin}}}{\partial \overline{y}^2} \cdots$$

$$+ \frac{\pi^2\,\overline{w}_{\mathrm{m}}}{2} \cos(\pi\,\overline{x}) \frac{\partial^2 \overline{\chi}_{\mathrm{lin}}}{\partial \overline{y}^2} = 0$$

$$\overline{\Delta}^2 \overline{\chi}_{\mathrm{lin}} - \frac{\pi^2\,\overline{w}_{\mathrm{m}}}{2} \cos(\pi\,\overline{x}) \frac{\partial^2 \overline{w}_{\mathrm{lin}}}{\partial \overline{y}^2} = 0, \tag{8.61}$$

where $\overline{w}_{\mathrm{m}}(\sigma_{\mathrm{t}}, \nu)$ is the maximum deflection of the Euler column defined in equation (8.24a), in dimensionless form.

---

[15] Since the results of Section 8.4.1 only apply to the planar state, the Euler column could eventually loose its stability through a *sub-critical* bifurcation. In that case, the secondary buckling threshold estimated from the presence of marginal modes would be overestimated. That the bifurcation is super-critical will in fact be checked numerically later.

The boundary conditions are derived from equation (8.9) as:

$$\overline{w}_{\text{lin}}(\pm 1) = 0, \qquad \frac{\partial \overline{w}_{\text{lin}}}{\partial \overline{x}}\bigg|_{\overline{x}=\pm 1} = 0, \tag{8.62a}$$

$$\left[\frac{\partial^2 \overline{\chi}_{\text{lin}}}{\partial \overline{x}^2} - \nu \frac{\partial^2 \overline{\chi}_{\text{lin}}}{\partial \overline{y}^2}\right]_{\overline{x}=\pm 1} = 0, \qquad \left[(2+\nu)\frac{\partial^3 \overline{\chi}_{\text{lin}}}{\partial \overline{x}\,\partial \overline{y}^2} + \frac{\partial^3 \overline{\chi}_{\text{lin}}}{\partial \overline{x}^3}\right]_{\overline{x}=\pm 1} = 0. \tag{8.62b}$$

The third boundary condition (8.9c) is always satisfied with perturbations that have a harmonic dependence on the coordinate $\overline{y}$, as its role is to prevent the edges from sliding away from each other through a global motion of the lateral clamps. With the aim of keeping notations simple, we drop bars on top of perturbed quantities and omit the subscript 'lin' for the perturbation in the following.

8.7.2   Symmetries of the marginal modes

From the symmetries of the base solution in this analysis of stability—the Euler column is invariant by translation along the y direction—one can restrict the stability analysis to harmonic perturbations: w and χ can be chosen in the form $w = [\cos(q\,y)\,w_q(x)]$ and $\chi = [\cos(q\,y)\,\chi_q(x)]$ for a single given q. By repeating the argument given in Section 8.4.2, we consider perturbations $(\overline{w}_q, \overline{\chi}_q)$ that are pure Fourier modes with a given wavenumber q in the y direction. This greatly simplifies the analysis as the set of partial differential equations (8.61) is turned into a set of *ordinary* differential equations.

The reflection invariance of the Euler column $x \mapsto (-x)$ can be exploited in a similar manner to impose a symmetry for $w_q(x)$ and $\chi_q(x)$: both $w_q(x)$ and $\chi_q(x)$ can be assumed to be even *or* odd functions of x. This can be shown by the same argument as earlier, considering the symmetric part of the perturbation $(\mathcal{S}w, \mathcal{S}\chi)$ and the antisymmetric[16] one $(\mathcal{A}w, \mathcal{A}\chi)$ defined as:

$$[\mathcal{S}f](x,y) = \frac{f(x,y) + f(-x,y)}{2} \qquad [\mathcal{A}f](x,y) = \frac{f(x,y) - f(-x,y)}{2}$$

instead of the Fourier components: by similar arguments, if (w, χ) is a solution, then both $(\mathcal{S}w, \mathcal{S}\chi)$ and $(\mathcal{A}w, \mathcal{A}\chi)$ are solutions too.

To sum up, we can consider only perturbations in the form:

$$w(x,y) = w_q(x)\cos(q\,y) \quad \text{and} \quad \chi(x,y) = \chi_q(x)\cos(q\,y), \tag{8.63}$$

in a range of values of q, where $w_q(x)$ and $\chi_q(x)$ are either both even (stability with respect to symmetric perturbations), or both odd (antisymmetric perturbations) functions of x.

---

[16] Note that the perturbations do not necessarily have the *same* symmetry as the Euler column: they can be antisymmetric while the Euler column is symmetric. In fact, one has to consider perturbations that are *eigenfunctions* for the symmetry operator that leave the equations invariant, such as the symmetry with respect to the plane (Oyz). The corresponding eigenvalues can eventually differ from 1: here, it is (-1) for antisymmetric perturbations.

Inserting the particular form of the perturbation above into the linearized F.–von K. equations (8.61), one obtains the equations for marginal modes:

$$
\frac{1}{12\,(1-\nu^2)}\frac{d^4 w_q(x)}{dx^4} + \left(\sigma_{\mathrm{Eu}} - \frac{2\,q^2}{12\,(1-\nu^2)}\right)\frac{d^2 w_q(x)}{dx^2} + \left(\frac{q^4}{12\,(1-\nu^2)}\cdots\right.
$$

$$
\left. - q^2\,(\sigma_\ell - \nu(\sigma_{\mathrm{t}} - \sigma_{\mathrm{Eu}}))\right) w_q(x) - \frac{q^2\,\pi^2\,w_{\mathrm{m}}}{2}\cos(\pi x)\,\chi_q(x) = 0 \tag{8.64a}
$$

$$
\frac{d^4\chi_q(x)}{dx^4} - 2\,q^2\,\frac{d^2\chi_q(x)}{dx^2} + q^4\,\chi_q(x) + \frac{q^2\,\pi^2\,w_{\mathrm{m}}}{2}\cos(\pi x)\,w_q(x) = 0
$$

with boundary conditions:

$$
w_q(x = \pm 1) = 0, \qquad w_q'(x = \pm 1) = 0,
$$

$$
\left[\chi_q'' + q^2\,\nu\,\chi_q\right]_{x=\pm 1} = 0, \qquad \left[\chi_q^{(3)} - (2+\nu)\,q^2\,\chi_q'\right]_{x=\pm 1} = 0. \tag{8.64b}
$$

This set of equations (8.64) defines the linearized problem near the Euler column configuration. The equations were first derived by Jensen (H. M. Jensen, 1993). A detailed analysis of their solutions was given in (B. Audoly, 1999; B. Audoly, B. Roman, and A. Pocheau, 2002).

This linearized problem is mathematically well-posed since there are eight homogeneous boundary conditions for a homogeneous differential system of order 4 with two unknown functions w_q and $\chi_q(x)$. As a result, marginal modes only exist for specific values of the parameters σ_{t}, σ_ℓ, q and ν—more accurately, for discrete (quantized) values of σ_{t} if σ_ℓ, q and ν are fixed. These specific values correspond to the loss of linear stability of the Euler column.

8.7.3 Principle of the numerical implementation

We proceed to the solution of the linearized F.–von K. equations (8.64) near the Euler column configuration. The shooting method, used in the numerics, is outlined in the present section. The results of the analysis are given next, in Section 8.7.4, and a detailed computer implementation of the method is proposed in Section 8.7.5.

Let us introduce a vector $W(x)$ defining a perturbation that is a solution of the linear system (8.64a) at a point x:

$$
W(x) = \left\{ w_q(x), w_q'(x), \frac{d^2 w_q(x)}{dx^2}, \frac{d^3 w_q(x)}{dx^3}, \chi_q(x), \chi_q'(x), \frac{d^2\chi_q(x)}{dx^2}, \frac{d^3\chi_q(x)}{dx^3} \right\}, \tag{8.65}
$$

and consider a particular initial condition at $x = 0$, for which the j-th component of $W(0)$ is 1, while all the others vanish (j is an integer, $1 \le j \le 8$). Let us label as '(j)' the corresponding solution of (8.64a). By definition, $[W^{(j)}(0)]_i = \delta_{ij}$, where δ is the Kronecker symbol. These functions $W^{(j)}(x)$ span all possible solutions of the differential equation (8.64a): by linear superposition, the solution with arbitrary initial conditions $[W(0)]_i = w_i$ at $x = 0$ can be written:

$$
W(x) = \Sigma_{i=1}^{8} w_i\, W^{(i)}(x). \tag{8.66}
$$

The matrix made up of the column vectors $W^{(0)}(a)$, $W^{(1)}(a)$, ... arranged into columns is the *shooting* matrix T_0^a from $x = 0$ to $x = a$ for the differential equation (8.64a): by definition, the shooting matrix is an 8 by 8 matrix whose entries are related to the components of the vector $W(x = a)$ by

$$[T_0^a]_{ij} = [W^{(j)}(a)]_i \qquad 1 \leq i, j \leq 8. \tag{8.67}$$

Notice that for $a = 0$, the shooting matrix is by definition the identity matrix. When a is arbitrary, the shooting matrix T_0^a can be computed by integrating numerically the differential equation with the eight fundamental initial conditions $W^{(i)}(0)$, and collecting the final values of w_q and χ_q and their derivatives in a table. The shooting matrix contains all the necessary information about the system of differential equations and depends on its parameters σ_t, σ_ℓ, q and ν.

Once the shooting matrix has been calculated, it remains to enforce the boundary conditions (8.64b). Given the form of the boundary conditions, one introduces a rectangular 8×4 matrix B:

$$B = \begin{pmatrix} 1 & 0 & 0 & 0 & 0 & 0 & 0 & 0 \\ 0 & 1 & 0 & 0 & 0 & 0 & 0 & 0 \\ 0 & 0 & 0 & 0 & \nu q^2 & 0 & 1 & 0 \\ 0 & 0 & 0 & 0 & 0 & -q^2(2+\nu) & 0 & 1 \end{pmatrix}, \tag{8.68}$$

which is such that $[B.W(\pm 1)]$ yields the left-hand sides of the boundary conditions (8.64b) at $x = \pm 1$.

For marginal modes w, these boundary conditions must be satisfied: $B.W(\pm 1) = 0$, which can be written as:

$$\forall c \in \{1, \dots 4\} \qquad 0 = B_{cj}\, W(\pm 1)_j \tag{8.69}$$

$$= B_{cj}\, w_i\, W^{[i]}(\pm 1)_j \qquad \text{by (8.66)} \tag{8.70}$$

$$= B_{cj}\, [T_0^{\pm 1}]_{ji}\, w_i \qquad \text{by (8.67),} \tag{8.71}$$

where an arbitrary marginal mode is represented by the state $w = W(0)$ of the differential equation at $x = 0$.

The latter equation allows us to rewrite the problem of existence of marginal modes in a matrix form:

$$\exists w \neq 0_{\mathbb{R}^8} \qquad \left(B.T_0^{+1}\right).w = 0 \text{ and } \left(B.T_0^{-1}\right).w = 0.$$

Owing to the reflection invariance $x \mapsto (-x)$, it is not necessary to compute both T_0^{+1} and T_0^{-1}: it is sufficient to impose that the perturbation be either symmetric ($w'(0) = w'''(0) = \chi'(0) = \chi'''(0) = 0$, i.e. $w_{2j+2} = 0$) *or* antisymmetric ($w_{2j+1} = 0$), and that the boundary conditions be satisfied at $x = +1$ only. This remark allows one to put the condition of marginal stability in the following matrix form:

$$\det\left(B(\nu, q).T_0^1(\sigma_t, \sigma_\ell, \nu, q).^t P^\pm\right) = 0, \tag{8.72}$$

where $\det(.)$ is the determinant of the 4×4 matrix in parenthesis, P^+ (resp. P^-) is the 4×8 matrix projecting vectors onto the subspace of symmetric (resp. antisymmetric) perturbations for $W(0)$: $P_{ij}^+ = \delta_{2(i-1),(j-1)}$ for $1 \leq i \leq 4$ and $1 \leq j \leq 8$ (resp. $P_{ij}^- = \delta_{2(i-1),(j-2)}$).

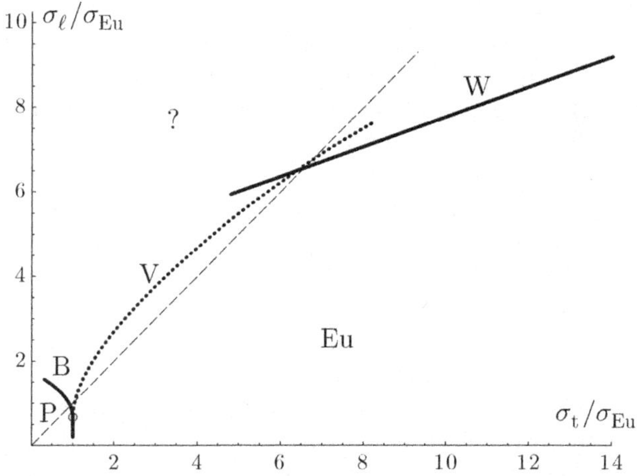

Fig. 8.18 Phase diagram for the long rectangular strip complemented by the analysis of linear stability of the Euler column, for a Poisson's ratio $\nu = .3$. Dashed line represents isotropic loading, $\sigma_{\mathrm{t}} = \sigma_\ell$. See Section 8.7.4 for an explanation of symbols V and W; the other ones are as in Fig. 8.7.

To sum up, the Euler column is marginally unstable when the compressions $(\sigma_{\mathrm{t}}, \sigma_\ell)$ are roots of equation (8.72), where $B(\nu, q)$ is the matrix (8.68) expressing boundary conditions (8.64b), and $T_0^1(\sigma_{\mathrm{t}}, \sigma_\ell, \nu)$ is the shooting matrix from $x = 0$ to 1 for the system of ordinary differential equations (8.64a). Loss of stability is against a symmetric perturbation when $P^\pm = P^+$, or to an antisymmetric one for $P^\pm = P^-$. For the present problem, the differential equations cannot be integrated analytically, and the shooting matrix T_0^1 is found by numerical integration.

8.7.4   Numerical results

Critical stress for secondary buckling

Figure 8.18 shows the phase diagram of a strip complemented by the analysis of linear stability of the Euler configuration. In this diagram, Poisson's ratio of the strip has been set to a value $\nu = 0.3$, which is typical of many materials (with the noticeable exception of glass, rubber, sand...); the influence of ν is studied in detail below. Two curves bounding the Euler region (Eu) have been added to the earlier diagram in Fig. 8.7. Each one of these curves is defined as follows: for any value of $(\sigma_{\mathrm{t}}, \sigma_\ell)$ below the curve, the determinant in equation (8.72) has a constant sign when the parameter q is varied. Above the curves, this determinant changes sign over a range of q. By choosing either $P^\pm = P^+$ or P^-, one studies the stability of the Euler column either with respect to varicose modes, denoted V or to worm-like patterns,[17] denoted W. The detailed numerical procedure used to compute these curves is presented in Section 8.7.5.

---

[17] Worm-like patterns, labelled (W) in Fig. 8.3, are similar to a Euler column with the crest snaking from one side to the other. They are described mathematically by the Euler column solution, combined with an antisymmetric perturbation to the deflection (amplitude a) that is periodic along the column.

Selection of the secondary pattern

Symmetric or antisymmetric patterns are in competition, and the one with the lowest critical compression appears first. At least not too far from instability thresholds this also gives the pattern with the lowest energy, although it is well possible that this does remain true far above threshold. The curves of marginal stability of the Euler column are physically meaningful only if the column is linearly stable there against any other perturbation. For this reason, the plot of these curves is not extended far beyond their intersection point, at $\sigma_t/\sigma_{\mathrm{Eu}} \sim 6.5$ (for $\nu = 0.3$). The column will become unstable for sufficiently large σ_ℓ with respect to worm-like patterns (W) if $\sigma_t/\sigma_{\mathrm{Eu}} > 6.5$, and with respect to varicose patterns (V) if $\sigma_t/\sigma_{\mathrm{Eu}} < 6.5$, as shown in the diagram in Fig. 8.18.

We emphasize that the present section deals with linear stability only, although applied to a non-planar basic state of the plate. As such, it yields the domain of stability of the Euler column but cannot predict the shape of the strip much beyond the (secondary) buckling threshold. There remains a region in the parameter space that still resists analysis, and will be addressed by a different approach later in Section 8.8.

Since there are two possible secondary patterns (symmetric and antisymmetric), an interesting question arises: can they 'cooperate', i.e. can the energy of the strip be better relaxed by a superposition of *both* patterns? In fact, this can only happen in the region of the diagram that is still unstudied, because the symmetry arguments in Section 8.7.2 have shown that secondary patterns appear one at a time. In fact, we shall see in Section 8.8, using specific tools, that cooperation between symmetric and antisymmetric modes indeed takes place in some regions of the parameter space. This will lead us to a new pattern of 'oblique bumps'.

Longitudinal wavelength

Figure 8.19 shows the wavelength of the secondary marginal mode, for both the varicose and worm-like patterns. We emphasize that only one of the patterns is selected at the onset of the secondary bifurcation, as explained above, even though both curves have been plotted over the full range of $\sigma_t/\sigma_{\mathrm{Eu}}$ for later reference. The critical wavelengths are given

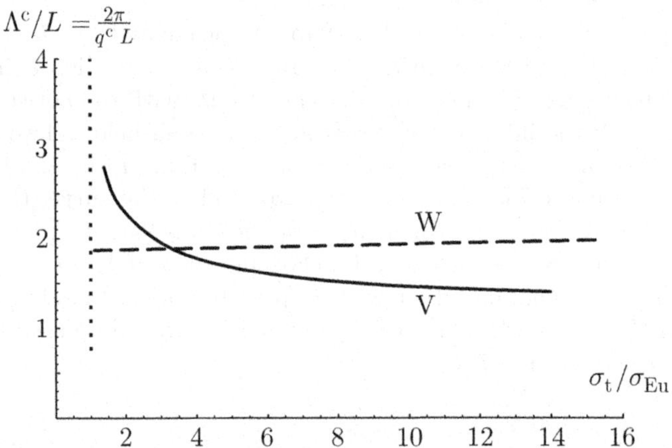

Fig. 8.19 Wavelength of the secondary patterns at the onset of the bifurcation, for $\nu = 0.3$, expressed in units of half the strip width, L.

by $\Lambda^{\mathrm{cr}} = 2\,\pi\,L/q^{\mathrm{cr}}$, where q^{cr} is the critical wavenumber, i.e. the value of q that is the root of the determinant in equation (8.72)—recall that, by definition, this determinant has exactly one root q at any regular point of the curve of marginal stability.

Longitudinal wavelengths of the secondary patterns are found to be of the order of magnitude of the strip width (this is easy to understand, because L is the only in-plane length at hand that can be used to form Λ with the correct physical dimension). The wavelength of the worm-like pattern is in fact remarkably constant, and very close to $2\,L$. The curve for the varicose pattern is discussed below.

The secondary modes

The shape of the film above the secondary buckling threshold is represented in Fig. (8.20), where an Euler column superimposed on either a symmetric or antisymmetric pattern has been plotted. To allow visualization, the amplitude of the symmetric or antisymmetric modes has been set to a finite value in this figure, although it is in fact vanishingly small at the onset of the bifurcation. The amplitude of the Euler column, however, is finite and continuous across the secondary bifurcation.

The varicose pattern amounts to a longitudinal modulation of the deflection that is uniform over the width of the strip, while the worm-like one involves a transverse, periodic deformation of the column.

Vicinity of Lifshitz' point

In Section 8.6, we studied the vicinity of Lifshitz' point $(\sigma_{\mathrm{t}}, \sigma_\ell) = (\sigma_{\mathrm{Eu}}, 2/3\,\sigma_{\mathrm{Eu}})$. On the basis of symmetry considerations, we predicted that the curve of marginal stability of the Euler column corresponding to varicose secondary patterns reached the critical point at its end (infinite wavelengths) with a vertical tangent. This is indeed verified by numerical computations (see phase diagram above, in Fig. 8.18).

It was furthermore predicted that all longitudinal wavelengths should scale as in equation (8.53) near this Lifshitz' point. In particular, for Λ_{V}:

$$\Lambda_{\mathrm{V}} \propto \frac{L\,\sigma_{\mathrm{Eu}}}{(\sigma_{\mathrm{t}} - \sigma_{\mathrm{Eu}})^{1/4}} \quad \text{for } \sigma_{\mathrm{t}} \text{ slightly above } \sigma_{\mathrm{Eu}}. \tag{8.73}$$

This accounts well for the divergence of the wavelength Λ along the varicose branch in Fig. 8.19 at $\sigma_{\mathrm{t}}/\sigma_{\mathrm{Eu}} = 1$.

Note that the analysis in Section 8.6 applies to *symmetric* bump patterns but not to *antisymmetric* ones. This analysis was indeed based on the remark that there is a continuous transition from the *symmetric* bump pattern to the Euler column across Lifshitz' point, where the bump wavelength diverges. This is why no divergence of the worm-like branch is obtained in Fig. 8.19 near Lifshitz' point, in contrast to the varicose branch.

Dependence on Poisson's ratio

Numerical results reported so far concern a specific (and very common) value of Poisson's ratio, $\nu = 0.3$. In this section, we investigate the influence of this parameter, and let it vary. To allow visualization of the results in 2D diagrams, we concentrate on the case of an isotropic[18] initial compression, imposing $\sigma_{\mathrm{t}} = \sigma_\ell$.

---

[18] By isotropic, we mean that the residual stress is isotropic *in the plane of the film*. It is not isotropic in all space directions, begin zero in the direction perpendicular to this plane. This is usually called *equi-biaxial* stress, but we refer to it as 'isotropic' in the absence of ambiguity.

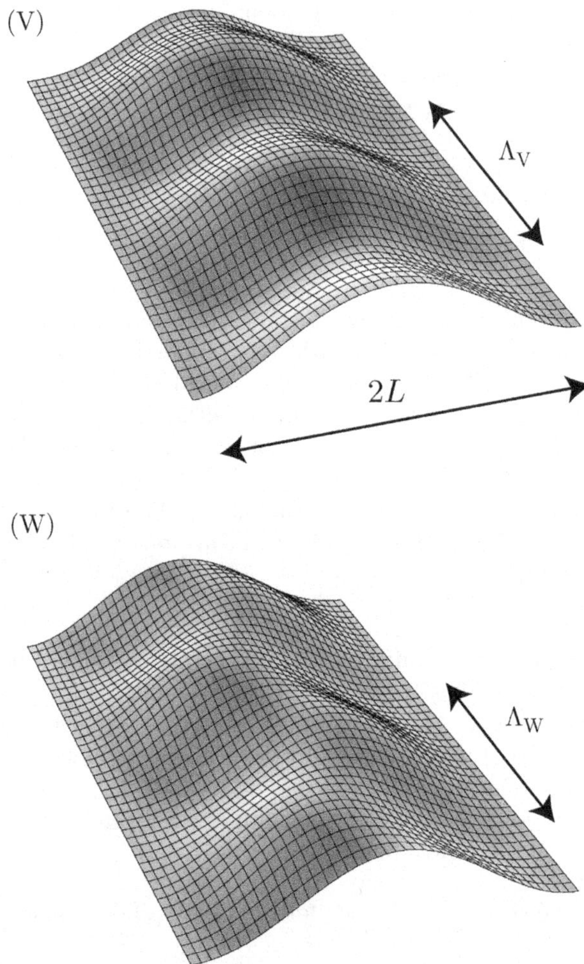

Fig. 8.20 Film profiles obtained by secondary buckling of the Euler column under the effect of the residual longitudinal compression (B. Audoly, 1999). The amplitude of the deflection is arbitrary. Depending on the applied compression, the most stable secondary pattern is either antisymmetric (worm like pattern, W) or symmetric (varicose, V).

Figure 8.21 shows the dependence of the critical (isotropic) compression leading to destabilization of the Euler column, as a function of Poisson's ratio ν. Values along the vertical axis correspond to abscissae of the intersection of the diagonal $\sigma_t = \sigma_\ell$ (isotropic loading) with either the V or W-curve in Fig. 8.18.

For a critical value ν_c of Poisson's ratio, the curves corresponding to the varicose and worm-like patterns cross when $\sigma_t = \sigma_\ell$. For small values of Poisson's ratio ($\nu < \nu_c$) and under isotropic loading, the Euler column becomes first linearly unstable with respect to varicose patterns when the loading is gradually increased. By contrast, for larger values of Poisson's ratio ($\nu > \nu_c$), and still for isotropic loading, the column is first unstable against worm-like patterns. Since ν_c is determined from the intersection of two curves that themselves do not

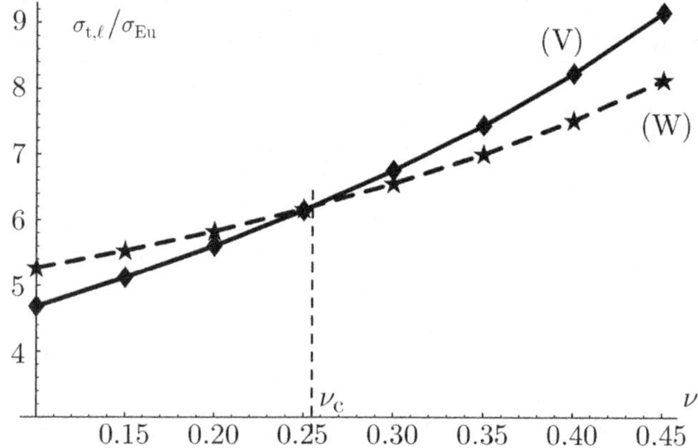

Fig. 8.21 Critical isotropic compression $\sigma_{\mathrm{t}} = \sigma_{\ell}$ destabilizing the Euler column, by symmetric (V) or antisymmetric (W) secondary buckling (B. Audoly, 1999), as a function of the Poisson ratio ν of the film.

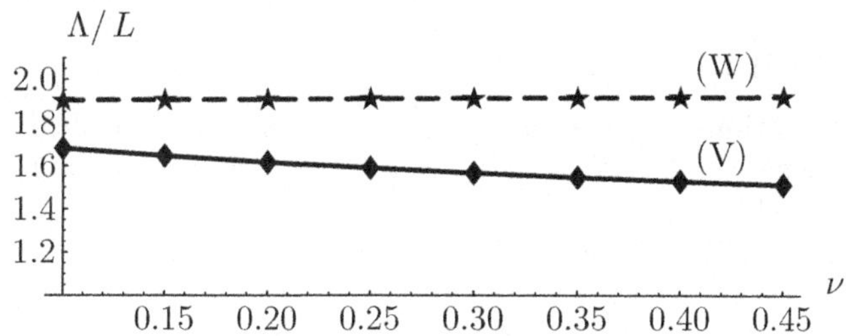

Fig. 8.22 Dependence of the longitudinal wavelength Λ of the secondary buckling patterns on Poisson's ratio ν of the film (B. Audoly, 1999).

depend on any parameter, its value is universal:

$$\nu_{\mathrm{c}} = 0.255 \pm 0.001.$$

This dependence of the unstable pattern on ν can be interpreted as follows. The stretching elastic energy of a plate can be written, from equation (6.95a), as:

$$\mathcal{E}_{\mathrm{s}} = \frac{E}{1 - \nu^2} \iint \left(\epsilon_{xx}^2 + \epsilon_{yy}^2 + 2\nu\, \epsilon_{xx}\epsilon_{yy} + 2(1 - \nu)\, \epsilon_{xy}^2 \right) \, \mathrm{d}x \, \mathrm{d}y.$$

This expression shows that the relative energetic cost of shear ($\epsilon_{xy}{}^2$ term) versus compression (ϵ_{xx}^2 and ϵ_{yy}^2 terms) increases as ν decreases. This explains that the varicose secondary pattern is preferred at lower values of ν: being symmetric it involves less shear than the worm-like (antisymmetric) one.

The dependence of the longitudinal wavelength on Poisson's ratio of the strip can also be plotted (Fig. 8.22). These wavelengths depend very little on Poisson's ratio of the film, ν.

8.7.5   Full commented Mathematica® code

In this section, we present and comment on the full code used for the numerical computations. It is based on version 6 of Mathematica software (Wolfram Research, 2007) from

Wolfram Research, Inc. This illustrates how a symbolic calculation language, rather than the more commonly used finite-element implementations, or lower-level languages such as C, can be used to implement these computations in no more than a few pages of code. This section can be skipped in a first reading.

The following source code examples were typeset using proprietary fonts ©Wolfram Research, Inc., with kind permission of Wolfram Research.

Using our system of dimensionless quantities for the numerical calculations (omitting the bars for readability), we first specify the definition of the critical stress $\overline{\sigma}_{\mathrm{Eu}} = \sigma_{\mathrm{Eu}}/(E\,h^2/L^2)$ for Euler buckling which reads, from equation (8.15b):

```
In[1]:=  σE = π² / (12 (1 - ν²));
```

as well as dimensionless buckling amplitude w_{m}, which reads, from equation (8.23):

```
In[2]:=  wm = If[σt > σE, (2/√3) √(σt/σE - 1) , 0];
```

We consider perturbations that have a harmonic dependence on the longitudinal variable:

```
In[3]:=  χ = (χq[#1] Cos[q #2]) &;
         w = (wq[#1] Cos[q #2]) &;
```

In Mathematica, functions f such as $f(x,y) = x + y^2$ are specified with the syntax $\mathtt{f = (\#1 + \#2^2)\&}$, where #1 and #2 stands for the first and second arguments of the function, when followed by the & sign.

The definition for the biharmonic operator reads:

```
In[5]:=  BiLaplacian[f_] := Function[{x, y},
             Derivative[4, 0][f][x, y] + 2 Derivative[2, 2][f][x, y]
               + Derivative[0, 4][f][x, y]];
```

The plate equations linearized near the Euler column solution have been given in equation (8.61):

```
In[6]:=  FvKLin = Simplify[{

             (1 / (12 (1 - ν²))) BiLaplacian[w][x, y] + σE ∂{x,2} w[x, y]

             + (σ1 - ν (σt - σE)) ∂{y,2} w[x, y] + (π² wm / 2) Cos[π x] ∂{y,2} χ[x, y],

             BiLaplacian[χ][x, y] - (π² wm / 2) Cos[π x] ∂{y,2} w[x, y]} /. y → 0];
```

with the associated boundary conditions (8.62):

```
In[7]:=  BndrCnd =
           Simplify[Flatten[
             {w[x, y] , ∂x w[x, y],
               ∂{x,2} χ[x, y] - ν ∂{y,2} χ[x, y] ,
               - (2 + ν) ∂x,{y,2} χ[x, y] - ∂{x,3} χ[x, y]}
             /. y → 0]];
```

The unknowns of the linear stability problem are the Fourier modes $\mathtt{wq(x)}$ and $\chi\mathtt{q}(x)$. In order to implement the collection of vectors $W^{(i)}$ introduced in equation (8.66), we take advantage of the multidimensional numerical integration capabilities of Mathematica 6, and use the convention that $\mathtt{wq}(\mathtt{x})$ and $\chi\mathtt{q}(x)$ are *vector valued* functions. Each component of these vectors corresponds to a particular initial condition, labelled (i) earlier. In the numerics, it is more convenient to use a different ordering convention than that used earlier for the indices i and for the components of the vector W in equation (8.65):

```
In[8]:= InitCnd =
     Transpose[{
        {wq[0], wq''[0], χq[0], χq''[0],
         wq'[0], wq'''[0], χq'[0], χq'''[0]},
        IdentityMatrix[8]
       }
      ] //
      Map[#[[1]] == #[[2]] &, #] &;
   InitCnd // MatrixForm
```

```
Out[9]//MatrixForm=
```

$$
\begin{pmatrix}
\mathtt{wq}[0] == \{1,\ 0,\ 0,\ 0,\ 0,\ 0,\ 0,\ 0\} \\
\mathtt{wq}''[0] == \{0,\ 1,\ 0,\ 0,\ 0,\ 0,\ 0,\ 0\} \\
\chi\mathtt{q}[0] == \{0,\ 0,\ 1,\ 0,\ 0,\ 0,\ 0,\ 0\} \\
\chi\mathtt{q}''[0] == \{0,\ 0,\ 0,\ 1,\ 0,\ 0,\ 0,\ 0\} \\
\mathtt{wq}'[0] == \{0,\ 0,\ 0,\ 0,\ 1,\ 0,\ 0,\ 0\} \\
\mathtt{wq}^{(3)}[0] == \{0,\ 0,\ 0,\ 0,\ 0,\ 1,\ 0,\ 0\} \\
\chi\mathtt{q}'[0] == \{0,\ 0,\ 0,\ 0,\ 0,\ 0,\ 1,\ 0\} \\
\chi\mathtt{q}^{(3)}[0] == \{0,\ 0,\ 0,\ 0,\ 0,\ 0,\ 0,\ 1\}
\end{pmatrix}
$$

where the first four components in $\mathtt{wq}$ and $\chi\mathtt{q}$ now correspond to symmetric solutions, and the last four to antisymmetric ones. The list $\mathtt{InitCnd}$ above yields the initial condition of the linearized differential system. With this conventions, the fifth index in the vector valued functions $\mathtt{wq}(\mathtt{x})$ and $\chi\mathtt{q}(x)$, for instance, yields the solution of the linearized equations with the initial condition $\mathtt{wq}(0) = 0$, $\mathtt{wq}'(0) = 1$, $\mathtt{wq}''(0) = 0, \ldots, \chi\mathtt{q}'''(0) = 0$, as shown by reading along the fifth column in the right-hand side of the output $\mathtt{Out}[9]$.

We arrive at the heart of the calculation, namely the integration of the linearized equations. We define a function, $\mathtt{ShootMatrices}$, that performs then numerical integration and returns the two matrices $(B \cdot T_0^1 \cdot P^{\pm})$ corresponding to symmetric $(+)$ and antisymmetric $(-)$ modes. The $\mathtt{Block}$ statement avoids name conflicts, and allows us to deal with the global variables ν, $\sigma\mathtt{t}$, ... as local variables within the statement. This procedure is implemented below and works as follows. At label $\mathtt{SM1}$, we collect the linearized equations and the initial conditions, and request numerical integration from $x = 0$ to 1 with the built-in function $\mathtt{NDSolve}$, retaining only the $\mathtt{First}$ (and only) solution in the list of solutions returned. At label $\mathtt{SM2}$, using a rather compact notation, we evaluate the values of the four boundary conditions at $x = 1$ with each one of the eight solutions of the linearized equations just found. The resulting matrix has dimensions 4×8 (rows correspond to boundary conditions and columns to initial conditions). We have just computed the matrix $(B \cdot T_0^1)$. The shooting matrices are obtained at label $\mathtt{SM3}$ by splitting this matrix into two square 4×4 matrices as this amounts, with our simple ordering convention, to performing the projections P^+ for the first four columns and P^- for the last four ones.

```
In[10]:= ShootMatrices[vArg_, σtArg_, σlArg_, qArg_] :=
      Block[{v = vArg, σt = σtArg, σl = σlArg, q = qArg},
        NDSolve[
           Join[Map[# == 0 &, FvKLin] (*SM1*), InitCnd],
           {wq, χq},
           {x, 0, 1}] // First // (*SM2*) ((BndrCnd /. x → 1) /. #) & //
         (*SM3*) {Take[#, All, {1, 4}], Take[#, All, {5, 8}]} &
        ];
```

As an example, we perform the numerical integration and compute this matrices $(B \cdot T_0^1 \cdot P^{\pm})$ with Poisson's ratio $\nu = 0.3$, isotropic compression $\overline{\sigma}_t = 3.$, $\overline{\sigma}_\ell = 3$ and wavenumber $\overline{q} = 3.14$:

In[11]:= `ShootMatrices[.3, 3., 3., 3.14] // Map[MatrixForm, #] & // Column`

Out[11]=

$$\begin{pmatrix} 11.3282 & 1.39664 & 45.9045 & 0.308567 \\ 56.2836 & 5.91425 & 218.088 & -0.653106 \\ -106.256 & 1.24998 & -124.981 & 35.9398 \\ -284.643 & -20.1483 & -711.978 & 56.1003 \end{pmatrix}$$

$$\begin{pmatrix} 2.80897 & 0.313262 & 5.26568 & -0.00740715 \\ 11.7631 & 1.39664 & 23.9408 & -0.453809 \\ -9.48765 & 0.466823 & -31.1148 & 8.86191 \\ -54.8107 & -4.12013 & -34.4348 & 11.7176 \end{pmatrix}$$

The first matrix corresponds to symmetric perturbations, and the second matrix to anti-symmetric ones.

It remains to scan over the parameters $(q, \overline{\sigma}_t, \overline{\sigma}_\ell)$ and detect the vanishing of the determinant of either matrix. These determinants, which we call the marginal stability indicator, are computed by:

In[12]:=
```
MarginalStabIndicator[
    vArg_Real, σtArg_Real, σlArg_Real, qArg_Real,
    symm_ /; (symm == "+" || symm == "-")
] :=
    ShootMatrices[vArg, σtArg, σlArg, qArg][[If[symm == "+", 1, 2]]] // Det;
```

Scanning with respect to two out of these three parameters, $\overline{q}$ and $\overline{\sigma}_\ell$ for instance, can be represented graphically in a plane. Below, we show a density plot that reveals the parabola-like curve along which the determinant of the symmetric (left) or antisymmetric (right) matrix vanishes in the $(\overline{q}, \overline{\sigma}_\ell)$ plane, for $\nu = 0.3$ and $\overline{\sigma}_\ell = 3$.

In[13]:=
```
Table[
    ContourPlot[MarginalStabIndicator[.3, 3., σl, q, symm],
    {q, 0., 6.}, {σl, 0., 10.},
    PlotRange → {-1. 10^3, 1. 10^3},
    Contours → 15, ContourStyle → None,
    ColorFunction →
      (If[# > .5, GrayLevel[.7 + .3 (# - .5) / .5], GrayLevel[.3 * # / .5]] &),
    ClippingStyle → {Black, White},
    Frame → False, Axes → True, AxesLabel → {"q/L^-1", "σl"},
    PlotLabel → "Symmetry" <> symm <> " , ν=.3, σt=3."],
    {symm, {"+", "-"}}] // GraphicsRow
```

Out[13]=

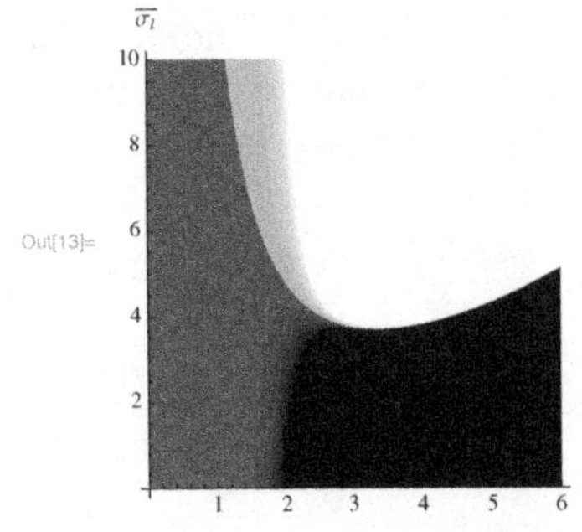

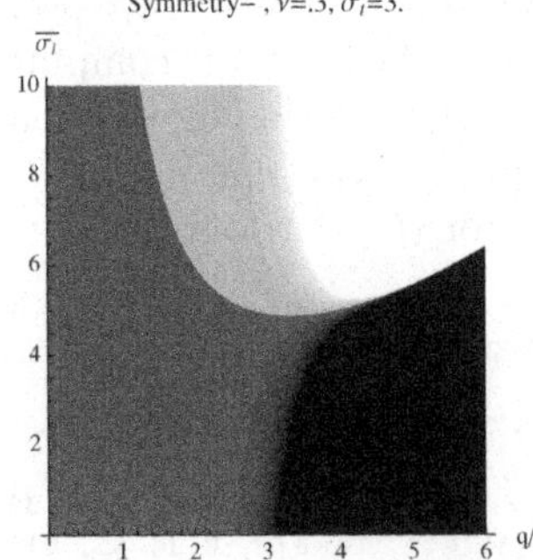

Note the use of a custom grey levels to emphasize the change of sign of the determinant. The minima of these curves correspond to marginal stability of the plate for this particular value of $\overline{\sigma}_t$. It remains to find a procedure to compute these minima automatically, for arbitrary values of $\overline{\sigma}_t$.

To do this, we use a constrained minimization procedure that works best with an estimate and a bracket of the optimal values of q and $\overline{\sigma}_\ell$ as a function of σ_t. These estimates and brackets do not need to be very accurate; they only need to be good enough for the numerical minimization procedure to converge. The values below were determined by trial and error:

```
In[14]:= q0[σtrans_, "+"] := 2.5 * (σtrans - (σE /. ν → .3))^(1/4);
         qMin[σtrans_, "+"] := 2. * (σtrans - (σE /. ν → .3))^(1/4);
         qMax[σtrans_, "+"] := 3.5 * (σtrans - (σE /. ν → .3))^(1/4);
         σl0[σtrans_, "+"] := 1. + .7 * σtrans;
         σlMin[σtrans_, "+"] := .6 * σtrans;
         σlMax[σtrans_, "+"] := 1.6 + σtrans;
```

and for the antisymmetric one

```
In[20]:= q0[σtrans_, "-"] := 3.5;
         qMin[σtrans_, "-"] := 3.1;
         qMax[σtrans_, "-"] := 4.;
         σl0[σtrans_, "-"] := σtrans;
         σlMin[σtrans_, "-"] := 3. + .2 * σtrans;
         σlMax[σtrans_, "-"] := 4. + .6 * σtrans;
```

The divergence of q for σ_t close to σ_{Eu} for the symmetric perturbation has been inferred from equation (8.53).

For given values of σ_t and ν, we are seeking the minimum value of σ_ℓ lying on the curve when the determinant, called the indicator, vanishes. Constrained minimization is very easy to implement in recent version of Mathematica, using the syntax `FindMinimum[{var, cstr}, ...]`, where *var* is the variable to be minimized and *cstr* the constraint, here in the form of an equality:

```
In[26]:= ECBoundary[nu_, σtrans_, symm_] :=
           FindMinimum[
             {σl, MarginalStabIndicator[nu, σtrans, σl, q, symm] == 0},
             {{q, q0[σtrans, symm], qMin[σtrans, symm], qMax[σtrans, symm]},
              {σl, σl0[σtrans, symm], σlMin[σtrans, symm], σlMax[σtrans, symm]}},
             PrecisionGoal → 4, MaxIterations → 200
           ] // Last // {σt → σtrans, σl → (σl /. #), q → (q /. #), ν → nu} &
```

As an illustration, we can recover the position of the minimum in the graph Out[13] given above, for the symmetric mode:

```
In[27]:= ECBoundary[.3, 3., "+"]
```

```
Out[27]= {σt → 3., σl → 3.71034, q → 3.34159, ν → 0.3}
```

and for the antisymmetric mode:

```
In[28]:= ECBoundary[.3, 3., "-"]
```

```
Out[28]= {σt → 3., σl → 4.88347, q → 3.3285, ν → 0.3}
```

Scanning over the values of σ_t, here with $\nu = .3$, is achieved by:

```
In[29]:= ECBoundarySampling["+"] = Table[ECBoundary[.3, σtr, "+"], {σtr, 1.1, 14., .2}];
         ECBoundarySampling["-"] = Table[ECBoundary[.3, σtr, "-"], {σtr, 1.1, 14., .2}];
```

The two curves to the stability at the edge of the Euler region in the diagram are plotted using the command

```
In[31]:= ListPlot[{σt, σl} /. {ECBoundarySampling["+"], ECBoundarySampling["-"]},
    AxesOrigin → {0, 0},
    Joined → True,
    PlotStyle → {Black, Directive[Black, Dashed]},
    AxesLabel → {"σt", "σl"}
    ]
```

and the wavelength of the instability, given for reference, by

```
In[32]:= ListPlot[{σt, 2 π / q} /. {ECBoundarySampling["+"], ECBoundarySampling["-"]},
    AxesOrigin → {0, 0},
    Joined → True,
    PlotStyle → {Black, Directive[Black, Dashed]},
    AxesLabel → {"σt", "2π/(q*L)"}
    ]
```

In this section, we have used different units for the loading compared with previous plots. In Fig. 8.18, for instance, the bifurcation diagram has been presented in the plane $(\overline{\sigma}_t/\overline{\sigma}_{\mathrm{Eu}}, \overline{\sigma}_\ell/\overline{\sigma}_{\mathrm{Eu}})$. The plots in the present section are based on $(\overline{\sigma}_t, \overline{\sigma}_\ell)$ instead. The values on the axes therefore differ by a factor $\overline{\sigma}_{\mathrm{Eu}}$ equal to 0.90 when $\nu = 0.3$, which must not be overlooked.

8.8   Extension of the diagram

In the preceding sections, we started to build up the phase diagram of an infinite elastic strip, by first addressing the initial buckling bifurcation (Section 8.4), by solving the non-linear F.–von K. equations for the Euler column solution (Section 8.5) and by studying the linear stability of the latter (Section 8.7). This led us to the phase diagram shown in Fig. 8.18.

The upper part of the diagram has remained unexplored so far. We have nevertheless collected a few hints on the way:

- [bump mode]   Just above the upper (curved) boundary of the P region in the diagram, the planar strip takes a configuration made up of bumps (B). These bumps are symmetrical with respect to the middle of the strip. The analysis of linear stability showed that the planar strip is marginally unstable with respect to a bump mode along this boundary, and the arguments of Section 8.4.1 established the super-critical nature of the initial buckling bifurcation.

- [Lifshitz' point]   In the vicinity of Lifshitz' point, there is potentially instability against two modes, the bumps and the Euler column. In the space of energy, the corresponding branches connect to the undisturbed planar state either way around the Lifshitz' point and overlap, as shown in Fig. 8.14. They are connected through an unstable varicose branch (see Fig. 8.15). Therefore, in the vicinity of the Lifshitz' point, destabilization of the Euler column, as studied in Section 8.7, takes place through a *sub-critical* bifurcation: the V curve in the phase diagram in Fig. 8.18 marks the *end* of the Euler branch. There must exist another branch corresponding to another stable configuration, on to which the strip will jump when moving too far in the space of parameters.

- [oblique bumps]   As noticed earlier, a stable pattern of superposed symmetric and antisymmetric bump modes is to be expected in the upper part of the diagram because both modes are available as neutral modes there as shown by the analysis of linear stability. In order to address this issue, an energy including a possible coupling between the two instability modes must be considered. This is done below.

Below, we shall extend the ideas introduced in Section 8.6 when studying the vicinity of Lifshitz' point. This method will allow us to finally completely fill in the phase diagram of the instabilities of the strip.

8.8.1   Principle of the method

At this point, we have exhausted all possible analytically exact approaches of the problem: the planar solution, the Euler column and their infinitesimal perturbations are the only exact solutions of the F.–von K. equations at hand. Any further progress on the buckling of the strip requires some approximations and/or direct numerical computations.

In the rest of this section, we present an approximation scheme that extends in a very natural way the preceding analysis. It builds on an extension of the ideas introduced in Section 8.6: limiting the possible configurations of the buckled strip to a superposition of a few modes (three) of deformations, we minimize the strip energy over a subspace of the full (infinite dimensional) space of configurations.

This approximation scheme, called a Ritz method[19], is widely used to find minima of energy functions in infinite dimensional space (here, the set of configurations of the plate). The idea is to minimize the energy over a well-chosen, finite-dimensional subspace. We shall indeed restrict our attention to specific configurations of the strip. Being based on

---

[19] In the context of differential equations, Ritz methods are known as Galerkin methods and are at the basis of the numerical method of finite elements, for instance.

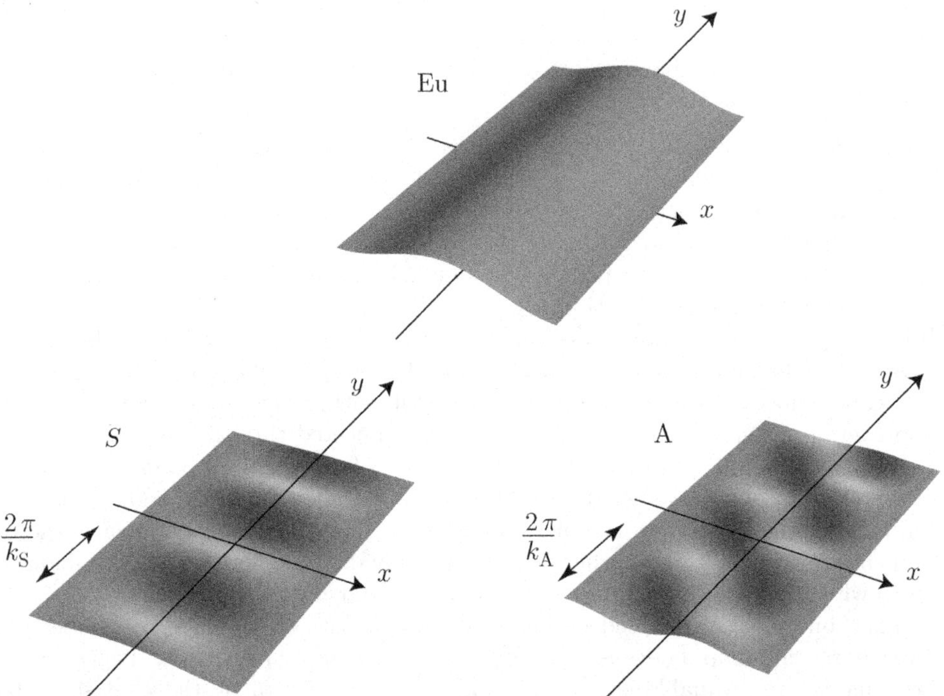

Fig. 8.23 The three fundamental modes of deformation used in Ritz procedure. The trial configurations of the plate are obtained as arbitrary linear superposition of these modes. Two additional parameters, k_{S} and k_{A}, allow tuning the wavelength of the bump and varicose modes independently.

an approximation, the method does not yield an exact solution of the equations. However, the finite-dimensional subspace is built from available analytical results near instability threshold, and the approximation scheme provides a relatively simple method to investigate the behaviour of the strip at a finite distance from threshold, consistent with the properties of the linear stability. Although the Ritz method is often implemented through a numerical minimization, in the present case almost all steps of the calculation can be carried out exactly. Doing so offers a few important advantages, as discussed below.

Trial form of the deflection

The first step in the Ritz method is to choose the allowed modes of deformations. In an attempt to be the least arbitrary as possible, we allow the plate to deform according to all the modes of deformation encountered so far: the Euler column (Eu), a symmetric bump mode (B), and an antisymmetric one (W). These modes are represented in Fig. 8.23. Therefore we restrict our analysis, somewhat arbitrarily, to deflections of the form:

$$w(x,y) = e\,w^{\mathrm{Eu}}(x) + s\,w^{S}(x)\,\cos(k_{\mathrm{S}}\,y) + a\,w^{A}(x)\,\cos(k_{\mathrm{A}}\,y + \Phi_{\mathrm{A}}). \qquad (8.74)$$

The amplitudes $\{e, s, a\}$ are adjustable parameters, as well as the two wavenumbers $\{k_{\mathrm{S}}, k_{\mathrm{A}}\}$. Note that s induces the bump (B) pattern, and a the worm-like (W) pattern. These five parameters are called the trial parameters. They span the five-dimensional subspace of configurations over which minimization of the energy will be performed. In the case of resonances, that is when k_S and k_A are commensurate, there is a sixth trial parameter, the relative phase Φ_A between the symmetric and antisymmetric perturbations.

The strip being infinite, the phase shift Φ_A is irrelevant whenever the wavenumbers k_S and k_A are incommensurate, i.e. when the ratio k_S/k_A is irrational.[20] Shifting the origin of the longitudinal coordinate y by $(2\pi n/k_S)$, where n is a signed integer, amounts to a change of the *relative* phase Φ_A by:

$$\Phi_A{}' = \Phi_A + 2\pi \left(n' + n \left(\frac{k_A}{k_S} - 1 \right) \right), \tag{8.75}$$

where n' is a second signed integer expressing that Φ_A is defined modulo 2π only. When k_A/k_S is irrational, the new phase $\Phi_A{}'$ can be made arbitrarily close to any given real value, such as zero, by choosing the integers n and n' appropriately. In that case, Φ_A has no physical meaning and the energy cannot depend on it. In contrast, in *resonant* cases, such as $k_S = k_A$ or $k_S = 2\,k_A$ for example, the value of Φ_A is physically meaningful, as we shall see in the following discussion of resonances.

In order to fully characterize the possible modes of deformations, one has to specify the behaviour of the trial functions $w^{\mathrm{Eu}}(x)$, $w^S(x)$ and $w^A(x)$ with respect to their variable x. The quality of the approximation depends largely on this choice. For the cylindrical mode, the profile of the non-linear Euler column is obviously the best candidate:

$$w^{\mathrm{Eu}}(x) = 1 + \cos(\pi x). \tag{8.76}$$

Recall that dimensionless quantities are used implicitly. In particular, the strip edges are located at $x = \pm 1$. The function $w^{\mathrm{Eu}}(x)$ has an overall arbitrary amplitude that is incorporated in the adjustable parameter e, and the normalization for this amplitude differs by a factor of 2 from what is was earlier. The form of w^{Eu} above is borrowed from (8.24a), which ensures that the procedure is *exact* in the Euler region of the diagram.

For the two other transverse profiles w^S and w^A, a natural choice would be to use the neutral modes present along the curves of marginal stability of the Euler column. Unfortunately, they have no closed analytical form. For this reason, we choose the following simple functions, that are either symmetric or antisymmetric:

$$w^S(x) = 1 + \cos(\pi x) \qquad w^A(x) = \sin(\pi x) + \frac{\sin(2\,\pi x)}{2}, \tag{8.77}$$

and satisfy the boundary conditions (8.9a):

$$w(\pm 1, y) = 0 \qquad \left. \frac{\partial w(x, y)}{\partial x} \right|_{x = \pm 1} = 0.$$

We emphasize that the two longitudinal wavenumbers k_S and k_A are tunable, and are *a priori* independent—cooperative effects between the bump modes S and A can in fact lead to $k_S = k_A$ but this is then a feature of the physical system and is not imposed *a priori*.

---

[20] In practice, the essential property of the number k_A/k_S is to be close to a fraction whose numerator and denominator are both small, e.g. $1/1$, $1/2$, $2/1$, etc. rather than being rational or not. In the experiments, the plate has a finite length and only a limited number of wavelengths are observable. In these circumstances, only simple ratios, such as $1/1$, $1/2$, $2/1$, etc. are likely to yield significant coupling between the symmetric and antisymmetric modes.

The dependence of the trial functions on the longitudinal variable y for the S and A bump modes has been chosen to be harmonic. This form is exact at the onset of the secondary buckling patterns, as shown in equation (8.63). It is therefore the most natural candidate for the approximation scheme.

Solving for χ

We could likewise impose a trial form of the Airy potential $\chi(x, y)$, parameterised by a small set of additional tunable parameters. There exists a better strategy that involves no further approximation:[21] the plates equations can be solved analytically for the Airy potential in a rectangular domain such as the one we consider. Indeed, the second F.–von K. equation (8.2b) for χ:

$$\overline{\Delta}^2 \overline{\chi} = -\frac{\partial^2 \overline{w}}{\partial \overline{x}^2} \frac{\partial^2 \overline{w}}{\partial \overline{y}^2} + \left(\frac{\partial^2 \overline{w}}{\partial \overline{x} \partial \overline{y}} \right)^2 , \tag{8.78}$$

is subject to the boundary conditions (8.9b) and (8.9c) and can be solved when the trial form (8.74) of $w(x, y)$ is inserted. This leads to an explicit, although lengthy, expression for $\chi(x, y)$ in terms of the five tunable parameters $\{e, s, a, k_S, k_A\}$ (plus Φ_A in the resonant case), given in equation (8.85) below.

Since χ is a solution of a linear partial differential equation (PDE) with linear boundary conditions (8.9b) and (8.9c), it depends linearly on the inhomogeneous terms in both the PDE and the boundary conditions. As we shall show, χ is in fact a quadratic function of the amplitudes $\{e, s, a\}$, the coefficients of which are functions of $\{k_S, k_A\}$ that are calculated below in Section 8.8.2.

Elastic energy in the trial subspace

Although we have solved for χ the second F.–von K. equation, which expresses the compatibility of strain, is does not make any sense[22] to attempt to solve the first F.–von K. equation (8.2a). This equation expresses the equilibrium, i.e. the minimization of energy in the original, full space of configurations. Taking advantage of the variational structure of the F.–von K. equations, we replace equation (8.2a) by the condition of energy minimum within the finite-dimensional subspace of allowable configurations.

This energy is found by applying the rescalings (8.1) to the plate energy defined in equations (6.94) and (6.95). This yields, in terms of rescaled variables:

$$\mathcal{E}_{\text{Ritz}} = \iint \left\{ \frac{(\Delta w)^2}{24 (1 - \nu^2)} - \frac{1}{12 (1 + \nu)} \frac{[w, w]}{2} \dots \right.$$
$$+ \frac{\sigma_{xx}^2}{2} + \frac{\sigma_{yy}^2}{2} - \nu \, \sigma_{xx} \, \sigma_{yy} + (1 + \nu) \, \sigma_{xy}^2 \dots$$
$$\left. + (-\sigma_{xx} + \nu \, \sigma_{yy}) \, \sigma_{\text{t}} + (-\sigma_{yy} + \nu \, \sigma_{xx}) \, \sigma_{\ell} \right\} \mathrm{d}x \, \mathrm{d}y. \tag{8.79}$$

---

[21] The advantage of solving the second F.–von K. equations exactly instead of using a trial form for χ is that it avoids shrinking the subspace of admissible configurations any further. Specifying the form or the Airy potential would amount to impose additional constraints on the allowable configurations of the strip.

[22] One would be led to an overdetermined problem, since there are an 'infinite' number of equations, one for any value of x in the PDE, although there are only finitely many adjustable parameters.

The last line contains the forcing term that come from the residual compression in the planar state, as expressed by the inhomogeneous terms in equations (8.1). The second term on the first line of the integrand may be omitted as its integral can be transformed into a boundary term by the Gauss–Bonnet theorem, and this boundary term vanishes with clamped boundary conditions. Before carrying out the minimization of the energy in equation (8.79), the in-plane stress must be expressed in terms of the Airy potential using its definition: $\sigma_{xx} = \chi_{,yy}$, $\sigma_{yy} = \chi_{,xx}$ and $\sigma_{xy} = -\chi_{,xy}$, where the commas in subscripts denote partial derivation; the Airy potential itself has to be expressed as a function of the trial parameters $\{e, s, a, k_{\mathrm{S}}, k_{\mathrm{A}}\}$ (and eventually Φ_{A}) using the formal solution of the second F.–von K. equation determined previously. In the end, the problems is the minimization of a function $\mathcal{E}_{\mathrm{Ritz}}$ of the trial parameters, for a given value of the loading $(\sigma_{\mathrm{t}}, \sigma_{\ell})$.

Anticipating the forthcoming calculation, we remark that the energy has a simple dependence on some of the trial parameters. This remark allows one to easily solve the minimization problem with respect to these parameters. Indeed, Ritz energy is a biquadratic[23] function of the amplitudes of the trial modes $\{e, s, a\}$, as shown by the first terms in equation (8.86), plus two forcing terms that are the product of either applied compression (σ_{t} and σ_{ℓ}) times a quadratic form of $\{e, s, a\}$. All coefficients depend on Poisson's ratio ν, on integrals of products of the transverse functions $w^{\mathrm{Eu}}(x)$, $w^{\mathrm{S}}(x)$, $w^{\mathrm{A}}(x)$ and their derivatives, and on the wavenumbers k_{S} and k_{A}.

8.8.2 Outline of the calculation

We shall now carry out the computation outlined above. By minimizing the elastic energy we shall derive the full phase diagram of the strip.

Resonances between the bump modes

In order to solve the F.–von K. equation (8.78) without yet specifying the numerical values of $\{e, s, a\}$, one has to expand in Fourier modes both the unknown Airy potential:

$$\chi(x, y) = \sum_{q} \cos(q\,y)\,\chi_q(x) + \sum_{q \neq 0} \sin(q\,y)\,\tilde{\chi}_q(x) + \frac{\bar{\chi}\,y^2}{2} \tag{8.80}$$

and the right-hand side of the same equation (8.78):

$$\frac{\partial^2 w}{\partial x^2}\frac{\partial^2 w}{\partial y^2} - \left(\frac{\partial^2 w}{\partial x\,\partial y}\right)^2 = \sum_{q} \cos(q\,y)\,K_q(x) + \sum_{q \neq 0} \sin(q\,y)\,\tilde{K}_q(x). \tag{8.81}$$

The last term in equation (8.80), $(\bar{\chi}\,y^2/2)$, does not belong to the Fourier expansion. It has to be included as second derivatives of χ only are physical: $\bar{\chi}$ corresponds to the coefficient of the mode $q = 0$ in the (standard) Fourier expansion of the physical stress $\sigma_{xx} = \partial^2\chi/\partial y^2$.

In equation (8.81), the Fourier components of Gauss curvature, $K_q(x)$ and $\tilde{K}_q(x)$, are functions of all the trial parameters that define $w(x, y)$. They are found by plugging the trial

---

[23] By biquadratic we mean a quadratic function of the *squared* trial parameters, i.e. a quartic function with only even powers, such as $a + b\,X^2 + c\,X^4$ in the case of a single variable X or as in (8.86) more generally.

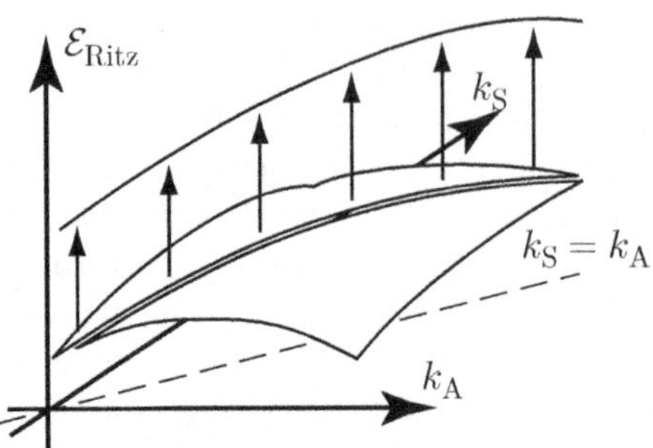

Fig. 8.24 Line of discontinuity of the energy due to resonances, when this energy is plotted as a function of k_{S} and k_{A}. Only the resonance $k_{\mathrm{S}} = k_{\mathrm{A}}$ is shown, but a similar discontinuity occurs for all the values of k_{S} and k_{A} listed in (8.84).

form (8.74) of w into the left-hand side of equation (8.81), decomposing in Fourier modes, and identifying the coefficients K_q and $\tilde{K}_q$ in the right-hand side. Once these functions $K_q(x)$ and $\tilde{K}_q(x)$ are known, each Fourier component of $\chi(x,y)$ can be computed by integrating along the transverse (x) direction the *ordinary* differential equation for each Fourier mode:

$$q^4 \frac{\mathrm{d}^4 \chi_q(x)}{\mathrm{d}x^4} - 2\,q^2 \frac{\mathrm{d}^2 \chi_q(x)}{\mathrm{d}x^2} + \chi_q(x) = -K_q(x), \qquad (8.82)$$

which is derived from the PDE (8.78). There are similar equations for $\tilde{\chi}_q(x)$. The boundary conditions (8.9b) and (8.9c) provide boundary conditions for $\chi_q(x)$; for brevity, they are not given here.

Note that the Gauss curvature has only a finite number of Fourier modes on the right-hand side of (8.78), and so does the Airy potential χ: the Gauss curvature is quadratic in $w(x,y)$. As a result, sums over the Fourier wavenumber q run over the values:

$$q \in \{0, k_{\mathrm{S}}, k_{\mathrm{A}}, 2\,k_{\mathrm{S}}, 2\,k_{\mathrm{A}}, k_{\mathrm{S}} \pm k_{\mathrm{A}}\}, \qquad (8.83)$$

where terms with a $\cos((k_{\mathrm{S}} - k_{\mathrm{A}})\,y)$ dependence, for example, arise from the decomposition of trigonometric functions:

$$\cos(k_{\mathrm{S}}\,y)\cos(k_{\mathrm{A}}\,y) = \frac{1}{2}\left(\cos((k_{\mathrm{S}} + k_{\mathrm{A}})\,y) + \cos((k_{\mathrm{S}} - k_{\mathrm{A}})\,y)\right).$$

When $k_{\mathrm{S}} = k_{\mathrm{A}}$, for example, the Fourier modes $q = 0$ and $q = k_{\mathrm{A}} - k_{\mathrm{S}}$ merge into a single one. Consequently, the coefficients $K_0(x)$, as well as $\chi_0(x)$, take on a special form when $k_{\mathrm{S}} = k_{\mathrm{A}}$. Some terms are present only when the two wavelengths are strictly equal. As a result, the elastic energy, seen as a function of all the trial parameters including the wavenumbers k_{S} and k_{A}, is also discontinuous when $k_{\mathrm{S}} = k_{\mathrm{A}}$. This 'resonance effect' is illustrated in Fig. 8.24. The case $k_{\mathrm{S}} = k_{\mathrm{A}}$ is not the only case of resonance.

Indeed, the elastic energy (8.79) can be expressed formally as a linear combination of the products (w^2), χ^2 and of χ. The function χ itself is linear in w^2 as shown by equation (8.78). As a result, the elastic energy includes a fourth power of w. Because the trial form of $w(x,y)$ involves three Fourier modes $q = 0, k_{\mathrm{S}}, k_{\mathrm{A}}$, the Fourier decomposition of the elastic energy will involve wavenumbers of the form $(m\,k_S \pm n\,k_A)$ with $m \geq 0$, $n \geq 0$ and $m + n \leq 4$. As a result, the possible resonances are:

$$k_A = \pm k_S, \quad k_A = \pm 2\,k_S, \quad k_A = \pm\frac{1}{2}k_S, \quad k_A = \pm 3\,k_S, \quad k_A = \pm\frac{1}{3}\,k_S. \tag{8.84}$$

When computing the elastic energy, each of these resonances must be considered separately.

   This resonance phenomenon is important: resonant patterns may release the elastic energy more efficiently than non-resonant ones. We shall indeed see later that they are present in a significant region of the phase diagram of the strip, especially towards large initial stress.

Outline of the calculation

We shall skip many details of the calculations and focus on the discussion of the final diagram. To give some flesh to this outline, we shall list the first Fourier components of the Airy potential $\chi(x, y)$ in the non-resonant case, i.e. when none of the equalities in equation (8.84) holds:

$$\overline{\chi} = -\frac{a^2(8\pi^2 + 5k_A{}^2\nu) + 4(2e^2\pi^2 + s^2(\pi^2 + 3k_S{}^2\nu))}{64(-1 + \nu^2)} \tag{8.85a}$$

$$
\begin{aligned}
\chi^{q=0}(x) = {}&+\frac{1}{64}x^2\left(5a^2k_A{}^2 + 12k_S{}^2s^2\right.\\[2mm]
&\left.+\frac{\nu}{1-\nu^2}\left(a^2(8\pi^2 + 5k_A{}^2\nu) + 4(2e^2\pi^2 + s^2(\pi^2 + 3k_S{}^2\nu))\right)\right)\\[2mm]
&-\left(\frac{1}{8\,\pi^4}a^2k_A{}^2\pi^2 + \frac{1}{2\,\pi^4}k_S{}^2\pi^2s^2\right)\cos(\pi x)\\[2mm]
&-\frac{1}{32\,\pi^4}\left(-a^2k_A{}^2\pi^2 + k_S{}^2\pi^2s^2\right)\cos(2\pi x)\\[2mm]
&+\frac{a^2k_A{}^2\cos(3\pi x)}{72\pi^2} + \frac{a^2k_A{}^2\cos(4\pi x)}{512\pi^2}
\end{aligned}
\tag{8.85b}
$$

$$\chi^{q=k_A}(x) = \cdots$$

These expressions are obtained by analytical integration of (8.82) with suitable boundary conditions. Similar expressions for the six remaining Fourier components of $\chi(x)$ can be derived. This yields the Airy potential in the non-resonant case. Resonances are then considered, each individually, each time yielding alternative forms of $\chi(x, y)$.

   The next step is to plug the above expression for $\chi(x, y)$ and the trial form of $w(x, y)$ into the definition for the density of elastic energy of the strip, which is the integrand in equation (8.79). This density of elastic energy has then to be integrated over the surface of the plate. Integration with respect to the transverse coordinate $-L \leq x \leq L$ is carried out symbolically. The plate being infinitely long, one should not attempt to integrate over the longitudinal variable $-\infty < y < \infty$. Instead, we compute the average density of elastic energy per unit length of the plate, by retaining the terms with a vanishing wavenumber $q = 0$ only; all other terms in the energy are zero on average. In the following, we may refer to this energy density per unit length of the plate simply as the 'energy' of the plate.

It takes the form:

$$\mathcal{E}_{\mathrm{Ritz}}(e,s,a,k_{\mathrm{S}},k_{\mathrm{A}},\nu) = c_{400}(k_{\mathrm{S}},k_{\mathrm{A}},\nu)\,e^4 + c_{200}(k_{\mathrm{S}},k_{\mathrm{A}},\nu)\,e^2$$

$$+ c_{040}(k_{\mathrm{S}},k_{\mathrm{A}},\nu)\,s^4 + \cdots + c_{004}(k_{\mathrm{S}},k_{\mathrm{A}},\nu)\,a^4 + \ldots$$

$$+ c_{220}(k_{\mathrm{S}}\cdots)\,e^2\,s^2 + c_{202}(k_{\mathrm{S}}\cdots)\,e^2\,a^2 + c_{022}(k_{\mathrm{S}}\cdots)\,s^2\,a^2$$

$$- \sigma_{\mathrm{t}}\left[c'_{200}(k_{\mathrm{S}}\cdots)\,e^2 + c'_{020}(k_{\mathrm{S}}\cdots)\,s^2 + c'_{002}(k_{\mathrm{S}}\cdots)\,a^2 \right]$$

$$- \sigma_{\ell}\left[c''_{020}(k_{\mathrm{S}}\cdots)\,s^2 + c''_{002}(k_{\mathrm{S}}\cdots)\,a^2 \right]. \tag{8.86}$$

The functions $c_{ijk}(k_{\mathrm{S}},k_{\mathrm{A}},\nu)$ are not all listed here. A typical coefficient reads, in the non-resonant case:

$$c_{004}(k_{\mathrm{S}},k_{\mathrm{A}},\nu) = -\,k_{\mathrm{A}}{}^2\,\pi^2\,\sinh(2\,k_{\mathrm{A}}) \tag{8.87}$$

$$\times \left(\frac{9\,\pi^6\,(9\,\nu\,\pi^2)\,(2\,k_{\mathrm{A}}(1+\nu)\cosh(2\,k_{\mathrm{A}}) + \cdots)}{8\,k_{\mathrm{A}}{}^2\,(4\,k_{\mathrm{A}}{}^2 + \pi^2)^2 \cdots} + \cdots \right) + \cdots$$

This expansion for c_{004}, the coefficient of a^4 in $\mathcal{E}_{\mathrm{Ritz}}$, contains about 500 terms.

From this explicit expression (8.86) for Ritz energy one can derive by minimization the equilibrium configuration of the strip, for arbitrary values of the applied compression not too far from threshold.

8.8.3   Analytical minimization with fixed longitudinal wavelengths

As anticipated, the strip energy (8.86) is a biquadratic form of the amplitudes $\{e,s,a\}$ (first three lines of this equation) plus either σ_{t} or σ_{ℓ} times a quadratic form. In the present section, we take advantage of this simple dependence to solve formally the minimization problem with respect to the amplitudes $\{e,s,a\}$. Minimization with respect to the wavenumbers k_{S} and k_{A} can be done numerically only, and is left to the next section. In the present section, we consider k_{S} and k_{A} as fixed.

To explain the method, we shall work out a simpler but similar example, which reads:

$$\mathcal{E}(e,s) = \frac{1}{4}\,e^4 + \frac{\alpha(\sigma_{\mathrm{t}},\sigma_{\ell})}{2}\,e^2 + \frac{1}{4}\,s^4 + \frac{\beta(\sigma_{\mathrm{t}},\sigma_{\ell})}{2}\,s^2 + \frac{\gamma}{2}\,e^2\,s^2. \tag{8.88}$$

Compared with the actual energy (8.86), we have discarded the antisymmetric bump mode ($a = 0$) and normalized e and s in such a way that the coefficients of e^4 and s^4 are $1/4$. This will be sufficient for explaining the method.

Comparison with (8.86) shows that, for any fixed k_{S} and k_{A}, the functions $\alpha(\sigma_{\mathrm{t}},\sigma_{\ell})$ and $\beta(\sigma_{\mathrm{t}},\sigma_{\ell})$ depend linearly on σ_{t} and σ_{ℓ} and γ is a numerical constant.

In a first step, we minimize $\mathcal{E}(e,s)$ with respect to the amplitude e. Given that the dependence of $\mathcal{E}$ on e is biquadratic: $\mathcal{E} = e^4/4 + (\alpha + \gamma\,s^2)\,e^2/2 + \cdots$, minimization is straightforward. Two cases must be considered:

$$C_0 : \text{ if } \alpha + \gamma\,s^2 > 0, e = 0 \text{ and } \mathcal{E} = \frac{s^4}{4} + \frac{\beta\,s^2}{2}, \tag{8.89}$$

$$C_1 : \text{ if } \alpha + \gamma\,s^2 < 0, e = \pm\sqrt{-\alpha - \gamma\,s^2} \text{ and } \mathcal{E} = \frac{1-\gamma^2}{4}\,s^4 + \frac{\beta - \alpha\,\gamma}{2}\,s^2 - \frac{\alpha^2}{4}.$$

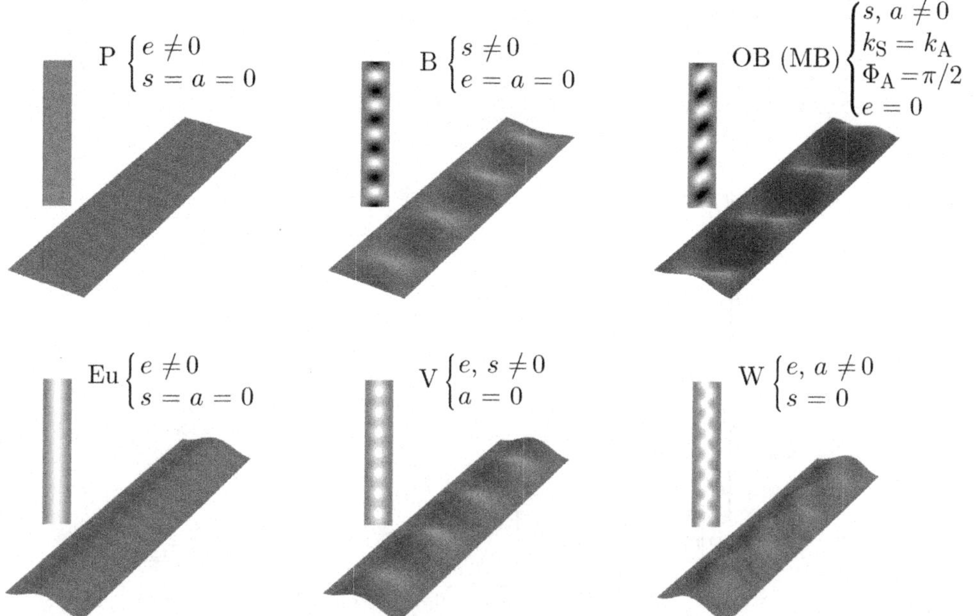

Fig. 8.25 Visualization of the buckling modes that are present in the diagram. The oblique bump (OB) mode is a resonant variant of the mixed bump (MB) mode.

Minimization with respect to s is carried out in the same way. For each case in equation (8.89), one has to consider two sub-cases depending on the sign of the coefficient of s^2 in the relevant expression of the energy:

$$C_{00} : \text{ if } (\alpha + \gamma s^2 > 0 \,\&\, \beta > 0), \qquad e = 0,\, s = 0 \text{ and } \mathcal{E} = 0,$$

$$C_{01} : \text{ if } (\alpha + \gamma s^2 > 0 \,\&\, \beta < 0), \qquad e = 0,\, s = \pm\sqrt{-\beta} \text{ and } \mathcal{E} = -\frac{\beta^2}{4},$$

$$C_{10} : \text{ if } (\alpha + \gamma s^2 < 0 \,\&\, \beta - \gamma\alpha > 0),\, e = \pm\sqrt{-\alpha - \gamma s^2},\, s = 0 \text{ and } \mathcal{E} = -\frac{\alpha^2}{4}, \qquad (8.90)$$

$$C_{11} : \text{ if } (\alpha + \gamma s^2 < 0 \,\&\, \beta - \gamma\alpha < 0),\, e = \pm\sqrt{-\alpha - \gamma s^2},\, s = \pm\sqrt{-\beta + \gamma\alpha}$$

$$\text{and } \mathcal{E} = -\frac{(\beta - \gamma/.\alpha)^2}{4(1-\gamma^2)} - \frac{\alpha^2}{4}.$$

The inequalities can be made fully explicit by plugging in the relevant values of e and s just determined:

$$C_{00} : (\alpha > 0 \,\&\, \beta > 0) \qquad e = 0,\, s = 0$$

$$C_{01} : (\alpha - \gamma\beta > 0 \,\&\, \beta < 0) \qquad e = 0,\, s = \pm\sqrt{-\beta}$$

$$C_{10} : (\alpha > 0 \,\&\, \beta - \gamma\alpha < 0) \qquad e = \pm\sqrt{-\alpha},\, s = 0 \qquad (8.91)$$

$$C_{11} : \left(\frac{\alpha - \gamma\beta}{1-\gamma^2} < 0 \,\&\, \beta - \gamma\alpha < 0\right)\, e = \pm\sqrt{\frac{-\alpha + \gamma\beta}{1-\gamma^2}},\, s = \pm\sqrt{\frac{-\beta + \gamma\alpha}{1-\gamma^2}}.$$

This solves the problem of minimizing our simplified energy $\mathcal{E}(e, s)$ with respect to e and s: we have determined all local minima of the energy depending on the control parameters α, β, γ. They are located along four distinct *branches*, C_{ij}, which correspond to two bifurcations associated with e (either $e = 0$ or $\pm\sqrt{\cdots}$) and s (either $s = 0$ or $s = \pm\sqrt{\cdots}$).

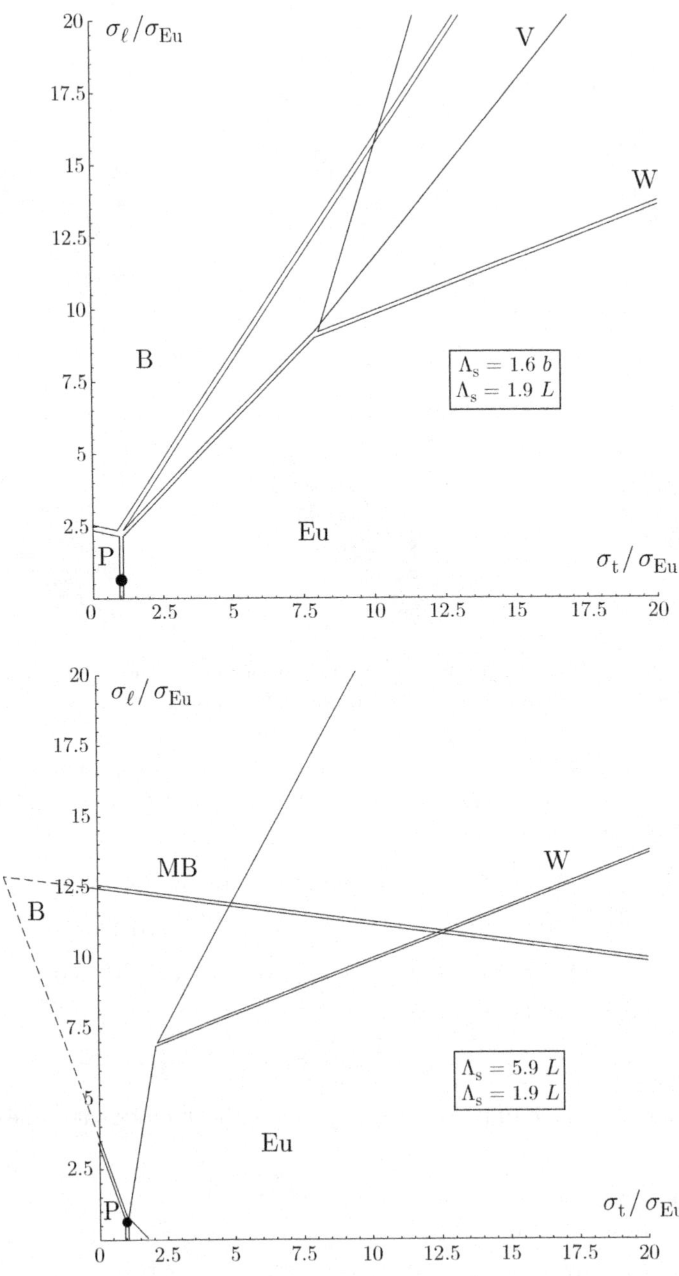

Fig. 8.26 Intermediate phase diagram of the strip, for $\nu = 0.3$. Configurations that are a local minimum of energy are shown; they were obtained by symbolic minimization of the strip energy with respect to the amplitudes $\{e, s, a\}$ only. Longitudinal wavelengths Λ_S and Λ_A are prescribed to two different arbitrary sets of values (top and bottom). Further minimization with respect to these wavelengths will yield the final phase diagram. The dot indicates the location of Lifshitz' point.

Table 8.1 Summary of the possible buckling modes with the trial form of the displacement (8.74). Each one corresponds to a branch in the plot of the energy versus the two control parameters $(\sigma_\mathrm{t}, \sigma_\ell)$

e	0	0	0	0	1	1	1	1
s	0	0	1	1	0	0	1	1
a	0	1	0	1	0	1	0	1
symbol	P		B	MB	Eu	W	V	AM
name	planar	×	sym. bumps	mixed bumps	Euler	worm	varicose	all modes

One could think that these branches are mutually exclusive, in the sense that for any given value of α, β and γ, only one set of inequalities C_{ij} can be satisfied at a time: either the coefficient of e^2 (resp. of s^2) is positive *or* it is negative, the energy minimum being accordingly either $e = 0$ or $e = \pm\sqrt{\dots}$. In fact, these possibilities are not mutually exclusive as the coefficient of e^2 in (8.89) does not only depend on (α, β, γ) but also on s. As a result, e and s circularly depend on each other and there can be more than one local equilibrium at a time, in some range of the parameters (α, β, γ). This is obvious from the equations above, where the inequalities defining different C_{ij}'s involve different functions of α, β and γ. In other words, the various branches can overlap, which makes the strip multistable, as shown in the diagrams in Fig. 8.26.

Minimization of the actual strip energy with respect to $\{e, s, a\}$ is carried out similarly. There is one more amplitude, a, than in the simplified model, hence one more possible bifurcation, and *eight* branches of equilibrium solutions. These branches are summarized in Table 8.1. In the upper part of the table, 0 (resp. 1) indicates that the corresponding trial amplitude has not (resp. has indeed) bifurcated, i.e. that its value is 0 (resp. non-zero). In the lower part of the table, the shorthand and full name of the branches are given, the symbol × marking those that are unused because they do not appear in the range of stress investigated here.

These branches are shown in Fig. 8.26. They were plotted using the expression (8.86) for Ritz energy. The two diagrams correspond to two different sets of values for $\{k_\mathrm{S}, k_\mathrm{A}\}$, which are fixed until the next section. The upper one uses the typical wavelengths obtained in the analysis of stability of the Euler column (see Fig. 8.19). The lower one explores a larger wavelength $2\pi/k_\mathrm{S}$, which the analysis of Section 8.6 showed to be more appropriate near Lifshitz' point.

A few remarks can be made before we proceed to numerical minimization over k_S and k_A. The first is that, because α and β are affine functions of σ_t and σ_ℓ, as noted above, and because of the affine dependence of the inequalities in (8.91) on α and β, the branches in the phase diagram have straight boundaries. The domains in Fig. 8.26 are indeed polygonal. However, the curved boundaries are curved in the final diagram, as k_S or k_A are allowed to vary there.

Second, the upper diagram in Fig. 8.26 can be understood from the results of the previous sections: the relative positioning of the varicose, worm-like and Euler regions agrees well with the linear stability analysis of the Euler column in Section 8.7. This indicates that the choice of the trial form (8.74) is appropriate: the strip has not been 'shaken up' too much by the drastic reduction of the space of configuration. More quantitative assessment of the approximation involved in Ritz procedure is given later.

The lower diagram is for a longer wavelength, $\Lambda_S = 2\pi/k_S \approx 3\,L$. It is relevant for the vicinity of Lifshitz' point (dot in the diagrams), where the wavelength of the symmetric bumps is known to diverge. The relative layout of the planar, Euler and (symmetric) bumps regions, and the overlap (multistability) between the B and Eu branches are all consistent with the results based on the expansion of the energy in normal form—this is not very surprising as the present numerical analysis is just a refined version of the arguments given in Section 8.6.

We now proceed to minimize the energy with respect to the two remaining trial parameters, k_S and k_A. This will allow us to bring together the previous and somehow partial pieces of information on the phase diagram into a broader and consistent picture.

8.8.4 Full minimization

We have derived an analytical expression of Ritz strip energy along either bifurcation branch, and inequalities characterizing the (infinitesimal) stability of the corresponding configurations. So far, these quantities are all given as functions of two 'internal' parameters, k_S and k_A, and one single dimensionless[24] material parameter, ν. By numerical minimization over k_S and k_A we can now optimize these internal parameters. This leads to the final phase diagram of the strip shown in Figs. 8.27–8.29 for $\nu = 0.3$.

Figure 8.27 shows the domains of existence of the various buckling patterns: in each domain, the pattern shown has the lowest energy. This diagram is an extension of Fig. 8.26 with further numerical minimization over k_S and k_A. The relative layout and the overlap of the domains in Fig. 8.27 can be understood using the exploded view in Fig. 8.28. In the latter, curved arrows indicate super-critical bifurcations. For example, transverse compression applied on the planar (P) state leads to the Euler column (Eu) through a bifurcation involving the amplitude e (lower left-hand side arrow in Fig. 8.28), as investigated earlier in Section 8.5. If longitudinal compression is increased subsequently, the Euler column can eventually bifurcate super-critically towards a varicose pattern (V region, curved arrow marked s), or towards a worm-like pattern (W region, arrow marked a)—see the analysis of Section 8.7. Note that the varicose pattern is stable in a very narrow window in the space of loading parameters only—this has been confirmed by detailed numerical simulations (M.-W. Moon, 2004; E. A. Jagla, 2007).

These diagrams also reveal sub-critical bifurcations, i.e. manifolds of solutions in the configuration space that are *not* smoothly connected to another manifold, and end abruptly. Across such a bifurcation line, one configuration of the strip stops and the strip has to jump discontinuously to a different state: the trial amplitudes vary by a finite amount while the applied stress is changed infinitesimally. This was discussed in Section 8.6.6 in the vicinity of Lifshitz' point (see Fig. 8.15 in particular). Such sub-critical bifurcations are denoted by dashed curves parallel to a stability boundary in Fig. 8.28. Sub-critical bifurcations may end up in an unstable pattern, which is then indicated with a superscript 'u'.

Figure 8.29 shows the configuration of the strip that is the absolute energy minimum for a given loading. The solid curve separating the OB (oblique bumps) and W (worms) regions, for instance, is the locus where these two patterns have the same energy, in the area where they are both stable.

---

[24] Other strip parameters, such as E, L are present implicitly as they are required to define dimensionless quantities.

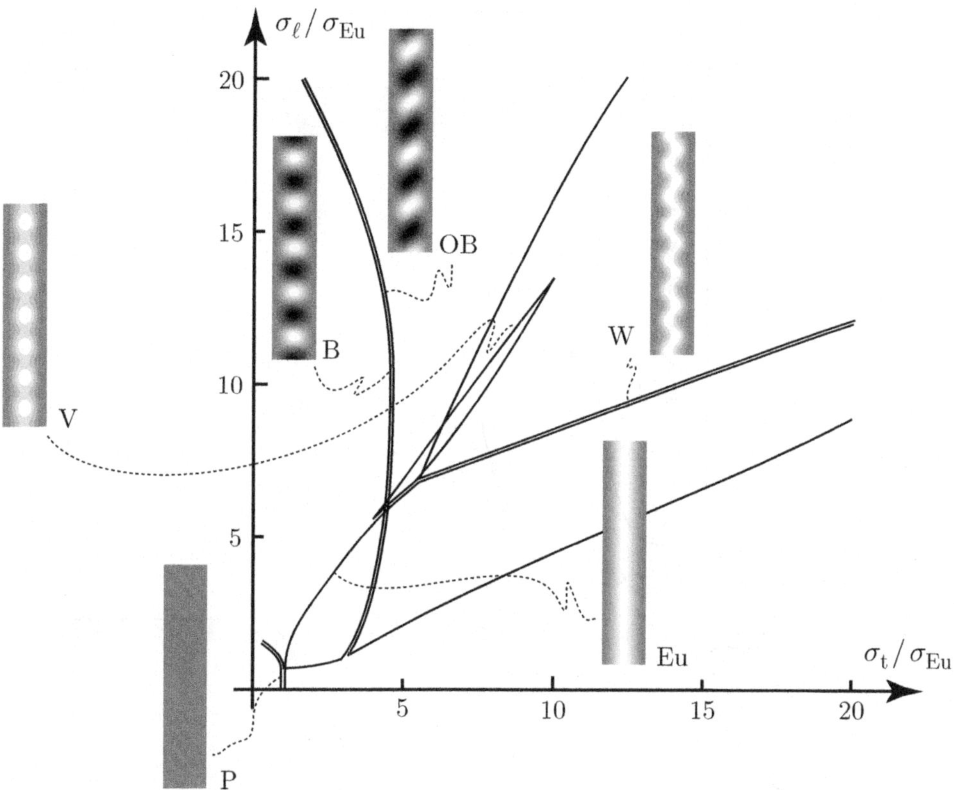

Fig. 8.27 Theoretical phase diagram of a strip with Poisson's ratio $\nu = 0.3$ showing all the energy branches, including the metastable ones. The dashed curves show the pattern associated with a given boundary.

In these diagrams, a new mode, made of 'oblique bumps' (OB), has appeared. It is a resonant variant of the 'mixed bumps' described in Table 8.1: it corresponds to $e = 0$ and has both s and a non-zero with the resonant condition

$$k_{\mathrm{E}} = k_{\Lambda} \quad \text{and} \quad \Phi_{\mathrm{A}} = \frac{\pi}{2} \quad \text{in equation (8.74).} \tag{8.92}$$

As explained in Section 8.8.2, some coupling effects between the symmetric and antisymmetric modes are present only when the wavenumbers k_{S} and k_{A} are equal. Oblique bumps are obtained as it turns out that the optimum phase shift between the symmetric and antisymmetric mode is[25] $\Phi_{\mathrm{A}} = \pi/2$. Note that the amplitudes s and a vary independently when the loading is changed; for instance, along the left edge of the OB region in Fig. 8.28,

[25] That the optimum phase shift Φ_{A} between symmetric and antisymmetric bumps be a multiple of $\pi/2$ can be understood as follows. Equation (8.74) can be rewritten $w(x,y) = e\, w^{\mathrm{Eu}}(x) + s\, w^{\mathrm{S}}(x)\, \cos(k_{\mathrm{S}}\, y) + a'\, w_{\mathrm{A}}(x)\, \cos(k_{\mathrm{A}}\, y) + a''\, w_{\mathrm{A}}(x)\, \sin(k_{\mathrm{A}}\, y)$, where the antisymmetric mode is now described using the amplitudes $\{a', a''\}$, related to the quantities $\{a, \Phi_{\mathrm{A}}\}$ used earlier by $a' + i\, a'' = a\, \exp(i\, \Phi_{\mathrm{A}})$. In this form, the energy is still a biquadratic function of $\{e, s, a', a''\}$. In the case of oblique bumps, one is above the bifurcation threshold for a'' $(a'' = \pm\sqrt{\cdots})$ but not for a' $(a' = 0)$. Hence $\Phi_{\mathrm{A}} = \pi/2$.

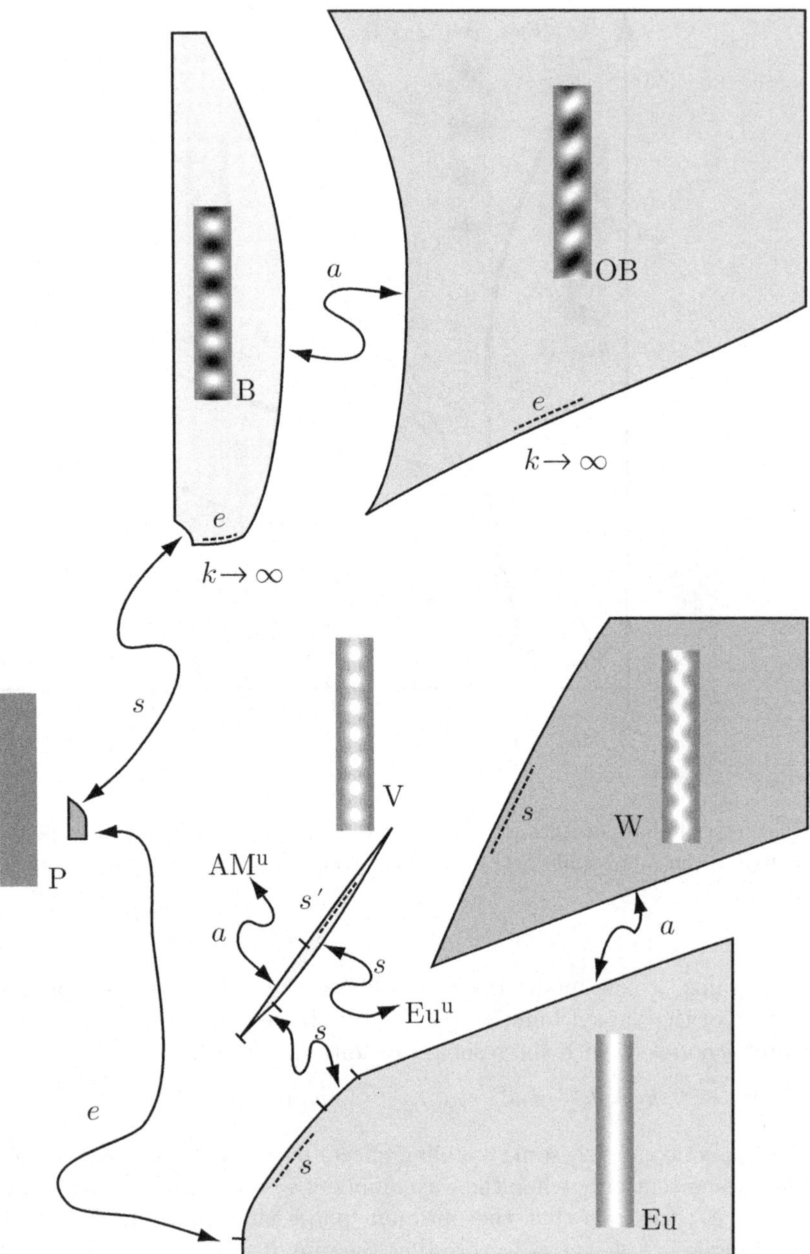

Fig. 8.28 Exploded view of the phase diagram for a strip with Poisson's ratio $\nu = 0.3$. Supercritical (pitchfork) bifurcations are indicated by arrows connecting branches, with the parameter undergoing a bifurcation written along the arrow. Dashed lines denote sub-critical (discontinuous) bifurcation. Patterns that are unstable are indicated with a superscript 'u'.

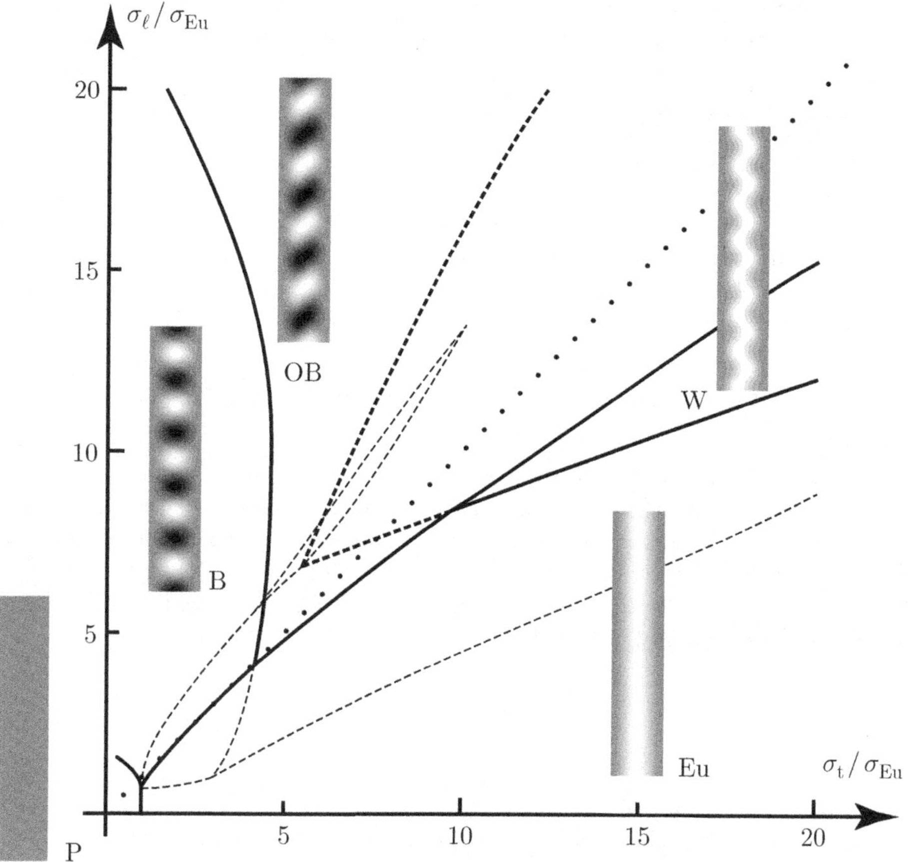

Fig. 8.29 Theoretical phase diagram of a strip with Poisson's ratio $\nu = 0.3$ showing the lowest energy configuration. The edges of the metastable (linearly stable) regions are shown with dashed curves, and in thick dashed lines for the edge of the worm pattern. Isotropic loading is shown by the dotted line ($\sigma_{\mathrm{t}} = \sigma_\ell$).

the amplitude u is vanishingly small while s remains finite. As a result, the oblique bumps are oriented almost perpendicular to the axis of the strip in the neighbourhood of the B to OB transition ($|s| \gg |a|$). They become more and more oblique as one moves away from this bifurcation curve in the parameter space ($|s| \sim |a|$).

8.9 Comparison with buckling experiments

We shall first come back to the buckling experiments presented at the beginning of this chapter, in Section 8.2. The overall shapes of the experimental and theoretical phase diagrams agree well, as show by comparison of Figs. 8.4 and 8.29. The theory perfectly accounts for the various patterns observed experimentally and for the relative layout of the corresponding regions in the phase diagram. However, the exact locations of the bifurcations curves slightly differ from theory to experiment. The sources of discrepancy are analysed below.

8.9.1   Finite length effects

By restricting the unstable wavenumbers k_S and k_A to a discrete set of possible values, the finite length of the strip stabilizes the planar state, and contributes to widening the P domain. This could be accounted for by the theory, by performing a minimization over a discrete set of values $k_A, k_S = 2\pi/(nL)$, where L is the finite length of the strip and n is an integer, instead of considering k_S and k_A as continuous. Finite size effects near the onset of bifurcation have been discussed in detail in Chapter 7.

8.9.2   Sensitivity to experimental imperfections

The main difficulty in the experiment comes from the presence of a small parameter, the film aspect ratio $h/(2L) = 5 \times 10^{-3}$, which makes the measurements sensitive to the imperfections of the setup (N. Yamaki, 1959). Somewhat paradoxically, the existence of this small parameter makes it possible to use a framework for the theoretical analysis, namely the plate theory, that is simpler than the full equations of 3D elasticity.

The sensitivity to imperfections in the experiments can be explained as follows. In-plane displacements of the film edge as small as

$$2\,L\,\frac{\sigma_{\mathrm{Eu}}}{E} \approx 2\,L\,\frac{(h/L)^2}{12} \approx 2\ \mu\mathrm{m}$$

induce film strain of order σ_{Eu}/E by Hooke's law, and perturb the residual stress by about σ_{Eu}: they are large enough to make the film buckle. This makes the set-up extremely sensitive to mechanical gaps. Such gaps are present at any mechanical binding in the set-up, for instance between base and substrate. It also makes it sensitive to the quality of the ruler–film binding. Only after many efforts was a satisfactory initial state of the film achieved in the experiments: gluing the strip edges to the rulers without inducing spurious stress in the film turned out to be a critical step. This unwanted stress is mainly due to the chemical action of the glue used to bind the strip to the rulers. Any method of binding the strip to the rulers, by gluing or by clamping in a metallic jaw, induces unwanted stress.[26]

The order of magnitude of this spurious residual stress can be estimated from the slightly deformed image of the grid in the absence of any applied compression—this refers to the imperfect reflection of the rectangular grid in unbuckled (P) state in Fig. 8.3. In this snapshot of the imperfectly planar state, the width of the strip is $2L = 2$ cm and the grid pattern typically deviates by an apparent distance $\delta x = 5$ mm. In Fig. 8.30, elementary geometric considerations show that the typical, small initial slope of the strip can be estimated from δx by: $\delta\theta = \delta x/d$, where d is the distance of observation. By equation (8.23), this amounts to a typical inhomogeneity in the residual stress:

---

[26] In many applications it is desirable to bind tightly two pieces of matter with as little spurious stress as possible, something that is not easy.

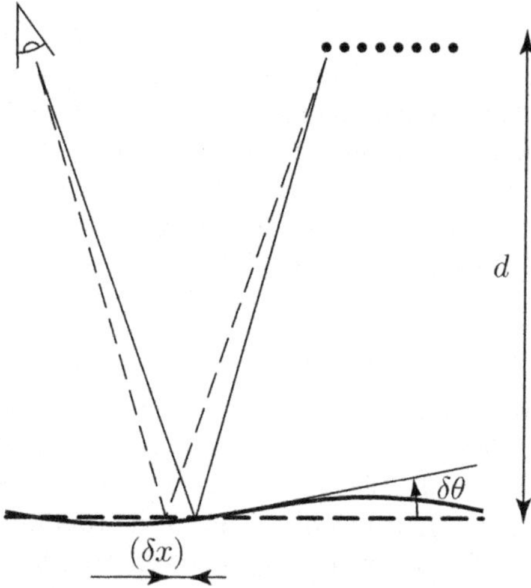

Fig. 8.30 Estimation of the initial non-planarity of the strip in the experiment, based on the apparent distortions in the image of the grid in the planar state.

$$(\delta\sigma_{t,\ell}) = \sigma_{\mathrm{Eu}}\,\frac{3^2}{4^2}\left(\frac{w_{\mathrm{m}}}{h}\right)^2$$

$$\sim \left(E\,(\delta\theta)\frac{L}{h}\right)^2$$

$$\sim E\left(\frac{(\delta x)}{d}\frac{L}{h}\right)^2.$$

$$\sim \sigma_{\mathrm{Eu}}, \tag{8.93}$$

using the above numerical values and $d \sim 1\,\mathrm{m}$ in Fig. 8.30. As a result, there is an uncertainty of order one unit in the experimental location of the bifurcation lines along either the σ_t and σ_ℓ axis in Fig. 8.4.

Temperature variations of ΔT in the room hosting the experiment induce thermal stress of order:

$$E\,(\Delta T)\,(\alpha_{\mathrm{s}} - \alpha_{\mathrm{f}}) \sim \frac{\sigma_{\mathrm{Eu}}}{(h/L)^2}(\Delta T)\,(\alpha_{\mathrm{s}} - \alpha_{\mathrm{f}}), \tag{8.94}$$

where $\alpha_{\mathrm{s,f}}$ are the thermal expansion coefficients for the film and for the metallic frame supporting the experiment. Using numerical values, a change of only $\Delta T = 2°\mathrm{C}$ induces a stress of order σ_{Eu}, enough to make the film buckle. Here again, the sensitivity is enhanced by the presence of a small parameter, h/L. For better control, the set-up had to be put in a temperature-controlled box.

Given the high sensitivity to imperfections, observations can only be expected to match the theory whenever both the buckling amplitude and the compressions are not too small. Mechanical gaps in the set-up amount to a change of origin in the phase diagram and can in fact account for the overestimation of the size of the P domain in Fig. 8.4 as compared with Fig. 8.29.

8.9.3 Multistability

Multistability makes comparison between theory and experiments more difficult as experimental patterns can depend on the loading history (see Fig. 8.4). Multistability also explains that the experimental W domain in Fig. 8.4 is wider than could be expected by considering the pattern with lowest energy only, shown by solid lines in Fig. 8.29: the experimental worm region extends past the domain where its energy is the lowest, but it lies well within the region of linear stability (thick dashed lines in Fig. 8.29).

8.9.4 Limitations of the model

Another source of discrepancy comes from the fact that the theory is based on an approximation, as a small set of (well-chosen) trial functions is used for the deflection. Comparison with the available exact results in Fig. 8.35 suggests that the approximation involves an additional error of order σ_{Eu} on the thresholds, that is of order one unit in rescaled variables. As explained in Section 8.11, this uncertainty can be removed by running detailed numerical simulations.

8.10 Application: interpretation of delamination patterns

The original motivation for the study of buckling patterns of a rectangular elastic strip, analyzed in detail in this chapter, and for the buckling experiments of Section 8.2 is related to the delamination of compressed thin films, a topic that we shall review shortly.

Coating a substrate with a thin solid film is done in a number of technological applications. Thin films may for instance provide thermal or electrical insulation of a substrate, play the role of a chemical barrier between the substrate and its surroundings, help harden or improve the optical properties of optical polymers. We refer to (G. Gioia and M. Ortiz, 1997) for a review on this topic.

Residual stress is often present in thin coated films, and arises by a variety of mechanisms; for a detailed review, see the book by Freund and Suresh (L. B. Freund and S. Suresh, 2003). Often, residual stress appears by chemical effects during the deposition process itself. Another source of residual stress comes from the difference in thermal expansion coefficients between film and substrate, when the film is obtained at high temperature, by vapour deposition for instance, and subsequently cooled at room temperature. This residual stress is compressive when the film has a lower thermal expansion coefficient than the substrate (the substrate would spontaneously contract more than the film). Here, we focus on the case of compressive residual stress in the film. When this residual stress is large enough, the film may buckle. This instability is similar to the buckling instability of a free plate studied in Section 8.5, except that the substrate plays the role of an elastic foundation. The driving force of the instability comes from the stretching energy of the film; it is in competition with its bending energy and with the substrate's elastic energy. The instability takes place when the substrate is compliant enough (H. G. Allen, 1969).

When the interface between film and substrate is weak, delamination can take place in addition: a crack propagates along the interface, allowing the film to lift off the substrate along the debonded area. Crack propagation is driven by film buckling (A. G. Evans and J. W. Hutchinson, 1984) and leads to blisters. Complex and diverse blister

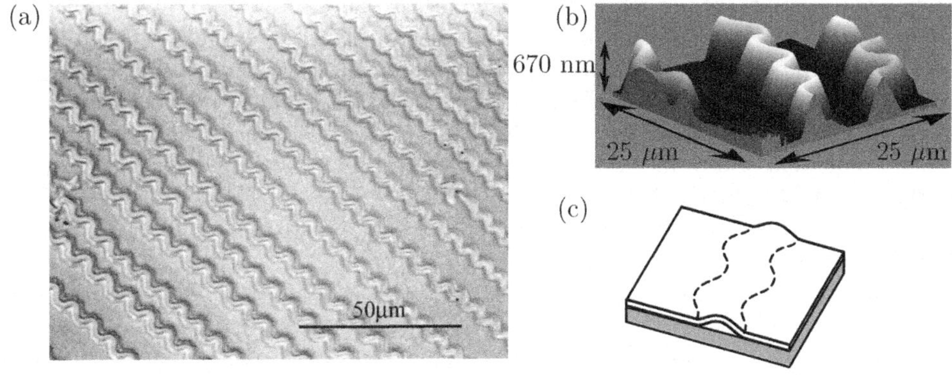

Fig. 8.31 Telephone cord delamination. Patterns in amorphous carbon films obtained by ion beam deposition on sapphire wafers observed by (a) laser microscopy and (b) atomic force microscopy. (c) Sketch of the pattern, which is formed when the film lifts off the substrate along an undulating region and buckles. Images (a) and (b) courtesy of X. D. Zhu, reproduced from reference (X. D. Zhu, K. Narumi, and H. Naramoto, 2007) with kind permission of IOP Publishing Ltd.

morphologies can be obtained, depending on the experimental conditions (D. Nir, 1984; A. A. Volinsky, 2003), such as long straight-sided blisters, circular blisters with or without wavy edges, worm-like (or telephone cord-like) blisters (G. Gille and B. Rau, 1984; J. W. Hutchinson, M. D. Thouless, and E. G. Liniger, 1992), and even circular dots aligned along a line (A. A. Volinsky, 2003; G. Parry *et al.*, 2004). Circular blisters have been analysed in (J. W. Hutchinson, M. D. Thouless, and E. G. Liniger, 1992). Here, we focus mainly on elongated morphologies, namely the straight-sided, the worm-like and the dotted ones.

8.10.1 Telephone cord blister morphologies

Worm-like delamination, also called telephone cord delamination, is probably the most common morphology in thin-film delamination and shows up under diverse experimental conditions. Such patterns are shown in Fig. 8.31: the film lifts off the substrate along an elongated region with wavy boundaries.

Two main difficulties arise in the analysis of complex morphologies such as worm-like patterns. Formulation of the problem couples film buckling and propagation of the interface crack between film and substrate. Indeed, the film can be modelled as a free plate with clamped edges, the blister being much larger than the film's thickness. It is clamped along a curve that is the edge of the debonded region; as a result, its profile depends on the geometry of the debonded region. Conversely, the film profile determines the loading on the interfacial crack (J. W. Hutchinson and Z. Suo, 1992) and influences its propagation. Delamination results from a coupling between two non-linear processes: film buckling and crack propagation.

In worm-like delamination, buckling takes place in a domain whose geometry is not elementary and there are no analytical solutions to the buckling problem. It was first suggested by Jensen (H. M. Jensen, 1993) that this wavy pattern could be understood as resulting from an instability affecting a straight-sided pattern: in his pioneering paper, he studied the stability of the lateral cracks along the sides of an infinitely long, straight-sided blister. He found that worm-like patterns can be formed as a result of this instability, at least in a limited range of values of the parameters. His analysis suggests too that varicose patterns[27] should most often be formed, even though those patterns are not observed in experiments under isotropic loading. There is strong experimental evidence that the propagation of interface cracks cannot be described merely by energetic arguments as interfacial toughness depends on a mode mixity parameter (J. W. Hutchinson and Z. Suo, 1992); varicose pattern is probably inhibited by this mode mixity (B. Audoly, 2000; M.-W. Moon *et al.*, 2002).

To circumvent the difficulties associated with the interplay of buckling and lateral crack propagation, we consider the buckling of a long rectangular strip with prescribed, straight lateral boundaries,[28] thereby following the simplified approach of references (B. Audoly, 1999; B. Audoly, B. Roman, and A. Pocheau, 2002). We are brought back to the problem addressed at the beginning of this chapter, and we shall now discuss how it can account qualitatively for the delamination morphologies.

In most experiments, residual stress is isotropic.[29] In the theoretical diagram in Fig. 8.27 and in the experimental one in Fig. 8.4, two patterns appear to be in competition for large isotropic stress: worms and oblique bumps. In the latter, the bumps are alternatively upwards and downwards. This is incompatible with the presence of the substrate, which has not been taken into account in our analysis. We conjecture that this mode becomes a bubble mode in the presence of a substrate, the downward bumps being replaced by a region of contact with the substrate. The resulting bubble mode is more constrained than the initial oblique bump mode and, therefore, can be expected to have a significantly higher energy. As shown by the detailed numerical calculations presented in Section 8.11.2 and Fig. 8.37, the worm-like pattern has the lowest energy for large isotropic stress when the substrate is considered. To sum up, the buckling analysis of a rectangular plate undertaken in the present chapter qualitatively explains the predominance of worm-like delamination patterns, which appear to be the pattern of lowest energy compatible with the presence of substrate.

Even in the absence of substrate blocking downwards buckling, a rectangular strip buckles most of the time into a worm-like pattern under isotropic residual stress. This is demonstrated by a simple classroom experiment, shown in Fig. 8.32, whereby a film made of plastic, a transparency used for overhead projection, is clamped along a rectangular metallic frame at room temperature, and then heated using a hairdryer. Thermal expansion of the film is much larger than that of the metallic frame and, as a result, compressive residual stresses are induced in the film upon heating.

[27] In retrospect, it appears that symmetric perturbations leading to varicose patterns are obtained in the analysis of reference (H. M. Jensen, 1993) mainly because the primary buckling pattern, which is the base state of the stability analysis, has been forced to be a Euler mode, although it is in fact a bump mode under isotropic compression, see Section 8.4.2. In view of this, the presence of a symmetric linearly unstable mode does not point to linear instability affecting the Euler mode, but merely to the fact that the strip tries to switch from the imposed Euler profile to its fundamental mode made of bumps.

[28] Doing so, we disregard another possible source of instability, namely that interface propagation is easier when the crack front is curved. This phenomenon has been shown to be relevant for the destabilization of a circular blister (J. W. Hutchinson, M. D. Thouless, and E. G. Liniger, 1992).

[29] Again, by 'isotropic', we mean what is usually called 'equi-biaxial'.

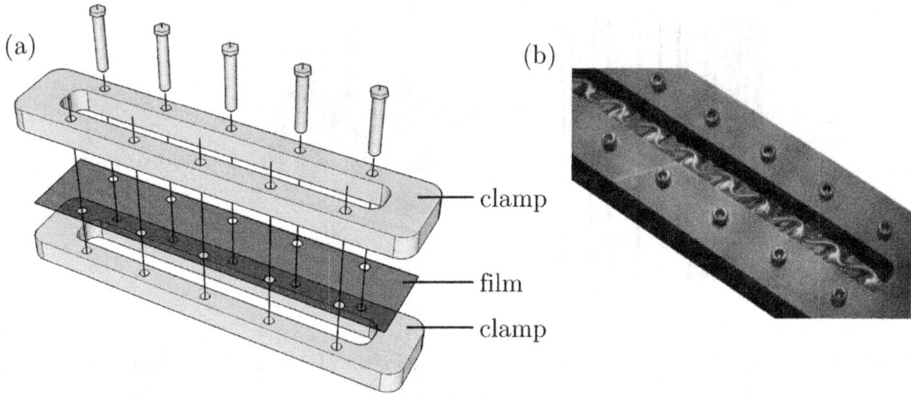

Fig. 8.32 A classroom experiments, designed by B. Roman and A. Pocheau, demonstrating worm-like buckling of a long rectangular film. (a) The film is clamped at room temperature in a rectangular frame. (b) Compressive residual stress is obtained by heating the film with a hairdryer, the thermal expansion coefficient of the film, made of plastic material, being much larger than that of the metallic frame.

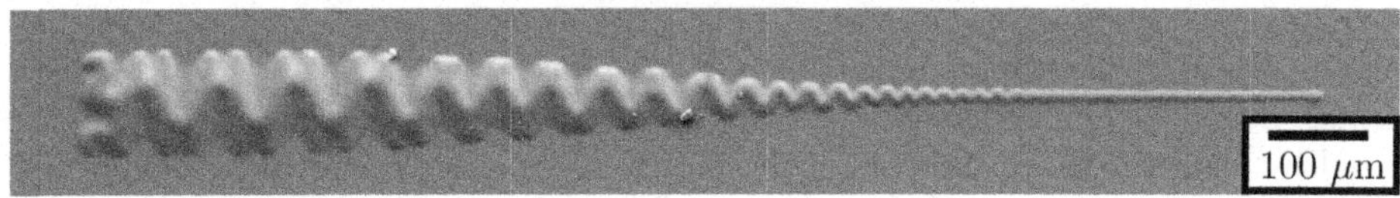

Fig. 8.33 Buckle delamination in a diamond-like carbon film on top of a silicon substrate. Delamination occurs in a tapered strip obtained in a controlled manner by lithography: inside this strip, the interface adhesion between film and substrate is weaker. Note the coexistence of worm-like delamination, straight blister, and a region without delamination (the patterned strip extends past the arrest point of delamination on the right-hand side of the picture). Image courtesy of M.-W. Moon, reproduced from reference (M.-W. Moon, 2004) with kind permission of Elsevier.

The idea that the worm-like pattern in thin film delamination reflects an elastic instability of the debonded region, and that crack propagation can be neglected in a first approach, is confirmed by another experiment (M.-W. Moon, 2004) at small scale, shown in Fig. 8.33. In this elegant set-up, lithographic techniques are used to produce a region of low adhesion between film and substrate. This region is shaped as a strip with a slowly varying width. Under experimental conditions such that delamination is restricted to the low adhesion region, as in Fig. 8.33, one can observe a combination of buckling patterns changing with the width of the strip; worm-like where the strip is wide, cylindrical Euler buckling for intermediate width, and no buckling at all beyond the 'tip' on the right-hand side of Fig. 8.33, where the low adhesion region is narrow (this unbuckled region with low adhesion is not visible in the picture but it does extend past the tip). This is perfectly consistent with the fact that the dimensionless buckling number $\sigma_{t,\ell}/\sigma_{\mathrm{Eu}}$ varies, for fixed residual stress $\sigma_{t,\ell}$, in inverse proportion to the critical stress $\sigma_{\mathrm{Eu}} \propto 1/(2L)^2$, where $(2L)$ is the width of the pattern. As a result, the buckling number decreases when the width of the strip decreases,

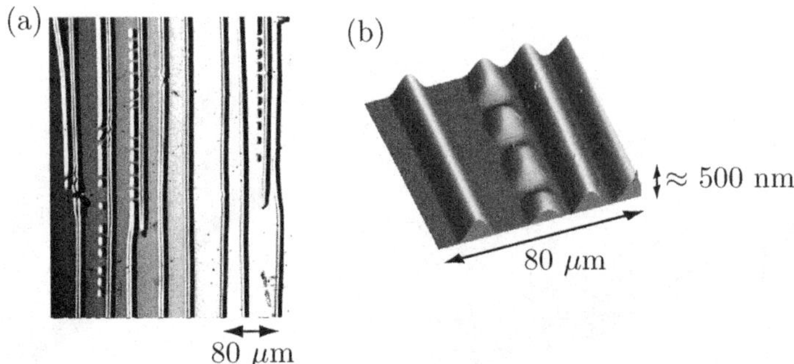

Fig. 8.34 Bubble delamination. A 240 nm thick nickel thin film is deposited onto a polycarbonate substrate using ion-beam sputtering. Straight delamination blisters (Eu) are induced by the application of a transverse compression. When the transverse compression is subsequently removed, a longitudinal compression is effectively induced along the straight blisters (G. Parry *et al.*, 2004), and dots are progressively formed (M. George *et al.*, 2002). Here, they are observed 15 days after release of the uniaxial compression (a) by optical microscopy and (b) by atomic force microscopy. Images courtesy of M. George, C. Coupeau, J. Colin, F. Cleymand and F. Grilhé (PHYMAT-CNRS, Poitiers, France) reprinted from (M. George *et al.*, 2002) with kind permission from Taylor & Francis, Ltd., http://www.informaworld.com.

and one moves from the upper-right corner in the phase diagram in Fig. 8.27 to the lower-left corner. Doing so, one successively crosses the worm region, the Euler region (bumps are penalized by the substrate) and finally the planar region. This corresponds exactly to the sequence of patterns shown in Fig. 8.33, going from left to right. It is remarkable that the the worm-like pattern can be reproduced even though the lateral boundaries are effectively frozen. This confirms that an elastic instability is the primary cause for such delamination patterns.

8.10.2   Other patterns

Straight-sided blisters, corresponding to the cylindrical Euler buckling pattern, are very seldom observed in delamination experiments under isotropic residual stress, $\sigma_t = \sigma_\ell$. However, they have been reported in a few experiments under anisotropic[30] residual stress, when the stress is predominantly transverse with respect to the blister's main orientation, $\sigma_t > \sigma_\ell$. This corresponds to moving downwards in the phase diagram in Fig. 8.27 or 8.29, where Euler buckling indeed becomes the dominant pattern.

Under residual stress that is predominantly longitudinal, $\sigma_t < \sigma_\ell$, a delamination pattern made of bubbles has been observed (M. George *et al.*, 2002; A. A. Volinsky, 2003), see Fig. 8.34. Here again, this is consistent with the bumps pattern obtained in the upper part of the theoretical phase diagrams earlier (by the same argument, the alternating upwards and downwards bumps give rise to a series of bubble in the presence of a substrate).

8.11   Limitations and extensions of the model

In the buckling analysis of a rectangular strip presented in Sections 8.4 to 8.8, we have tried to push the analytical calculation of the phase diagram as far as possible, avoiding

---

[30] By anisotropic, we mean biaxial but not equi-biaxial, that is $\sigma_t \neq \sigma_\ell$.

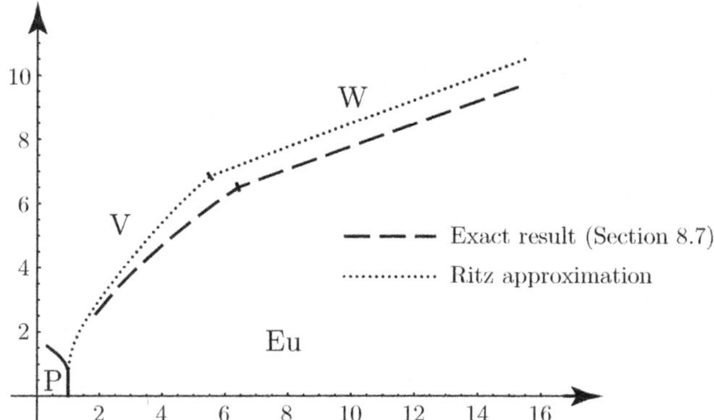

Fig. 8.35 Comparison of the exact (Section 8.7) and Ritz (Section 8.8) predictions concerning the secondary bifurcation from the Euler pattern. Ritz method, by effectively making available fewer unstable modes, overestimates the secondary buckling threshold. This overestimation, of the order 10%, could be improved by extending the basis of trial functions in equation (8.74). The overall shape of the edge of the planar region and the transition between varicose and worm-like unstable patterns are correctly captured by a Ritz method.

relying on fully numerical simulations. There are some approximations and limitations in this analysis, but it provides good insights into the mechanics of buckling, and makes it possible to track the various bifurcations, super-critical as well as sub-critical,[31] easily. Even though it is based on approximations, the predictions of this approximate analysis turn out to be remarkably close to the exact results.

Whenever high accuracy is desirable, or when the substrate has to be taken into account, this analysis can be complemented by direct numerical simulations. In this section, we comment briefly on the limitations of the analysis, and report on the numerical simulations that extend it.

8.11.1   Accuracy of Ritz procedure

This derivation of the phase diagram was based on an approximate method, called a Ritz method. By our choice of trial functions for the deflection, the solution is warranted to be accurate in the lower part of the diagram, that is in the P and Eu regions, and slightly above them; see Figs. 8.35 and 8.36. Far away from these regions, the predictions should not be expected to be accurate.[32] Being approximate, our solutions do not satisfy[33] the equations of equilibrium except, by construction, for the planar and Euler patterns. The most direct

[31] Being able to track sub-critical (discontinuous) energy branches is very valuable. The plate is indeed a macroscopic object that does not necessarily adopt the lowest energy configuration but instead tends to follow a given local energy minimum as long as it exists.

[32] Even the secondary bifurcation from the Euler column to worms or varicose modes, computed exactly in Section 8.7, are recovered only approximately by a Ritz method, whose trial functions w^S and w^A do not exactly match the secondary unstable modes; see Figs. 8.35 and 8.36.

[33] This provides a simple mean of assessing the error involved in our approximation, given by the magnitude of the unbalanced transverse force in the left-hand side of the equation of equilibrium (8.2a) in the smooth setting.

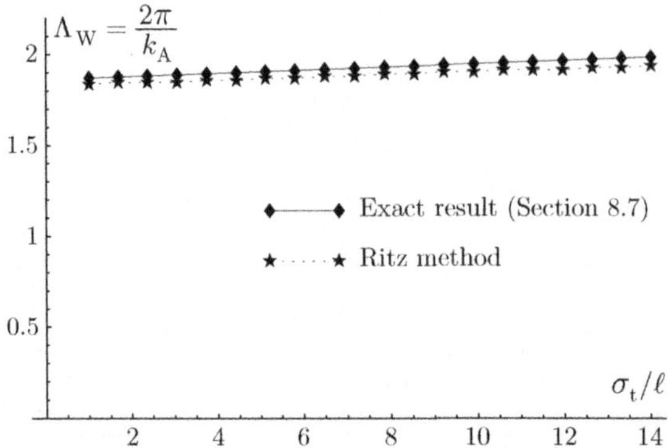

Fig. 8.36 Comparison of Ritz and exact prediction concerning the wavelength Λ_A of the worm-like pattern along the curve of marginal stability of the Euler configuration with respect to worms.

approach[34] for improving and better controlling the accuracy of the phase diagram would be to undertake fully numerical simulations. Such simulations have been done, based on the finite-elements method (M.-W. Moon, 2004) or on triangular meshes of springs (E. A. Jagla, 2007), for instance.

8.11.2 Presence of substrate

The presence of the substrate prevents downward buckling. The resulting buckling patterns may be in partial contact with the substrate, and cannot be easily incorporated into our analysis. In reference (G. Parry *et al.*, 2006), the buckling of a long, rectangular elastic strip has been investigated numerically, taking into account the non-penetration condition with the substrate (unilateral buckling). The results, shown in Fig. 8.37, are remarkably similar to those of the phase diagram presented earlier in Fig. 8.27, once the patterns involving downward buckling, such as the bumps patterns, have been removed or replaced with similar bubble patterns in partial contact with substrate.

8.11.3 Coupling with crack propagation

As explained in the introduction to the present section, the analysis of delamination is made difficult as it couples two non-linear phenomena: film buckling and crack propagation. This makes numerical simulations of delamination difficult too. Recently, an important step forward has been achieved, with numerical simulations by Jagla (E. A. Jagla, 2007) taking into account both film buckling over an arbitrary domain, and interface crack propagation. The results are impressive in that they reproduce a variety of typical delamination morphologies observed in experiments.

---

[34] Another possibility of improving the accuracy of the Ritz method would be to include more parameters in the trial form of the deflection, but the symbolic calculations quickly become untractable, even with the aid of a computer, when more terms are added.

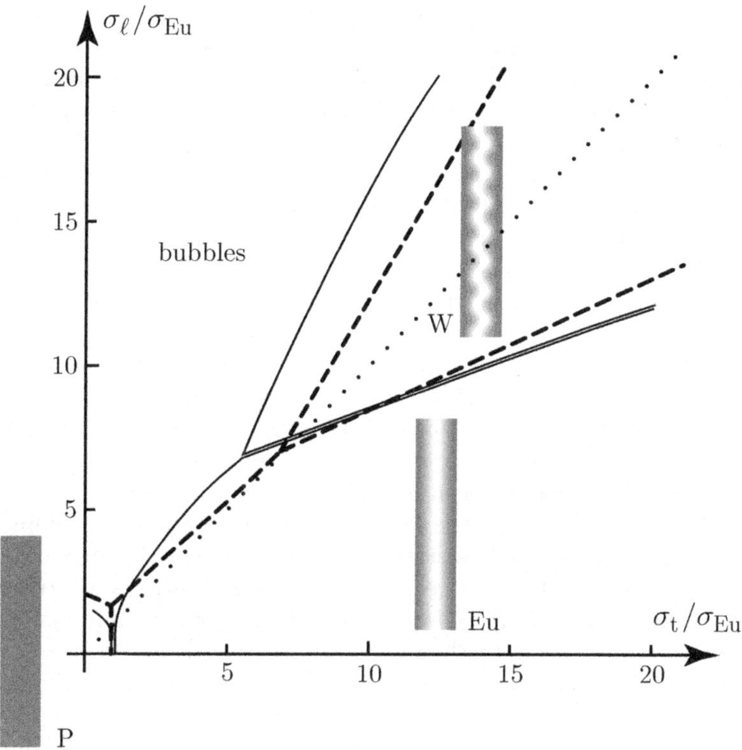

Fig. 8.37 Comparison of the results of Section 8.8 with finite-element simulations taking into account the presence of the substrate (unilateral buckling). Solid curves are the boundaries of some of the branches already shown in Fig. 8.27—only the branches compatible with the non-penetration condition are shown: planar, Euler column and worm. Dashed curves are the numerical phase diagram for the unilateral buckling problem, from reference (G. Parry *et al.*, 2006), showing the lowest energy configuration; in the upper part of this numerical diagram, a pattern made of a series of bubbles is obtained, which is an avatar of the bump pattern modified by the presence of the substrate. The dotted line denotes isotropic loading ($\sigma_{\mathrm{t}} = \sigma_\ell$).

8.12 Conclusion

In this chapter, we studied the buckling and post-buckling behaviour of a long, rectangular plate loaded under biaxial in-plane compression and clamped along its long edges. Our aim was to give a concrete example of the application of the F.–von K. equations. A rich experimental phase diagram was obtained, which was later interpreted using a combination of theoretical approaches: linear stability analysis of the planar state, expansion of the energy near a critical point using normal modes, linear stability of bifurcated states, and a quasi-analytical Ritz procedure. Quite interestingly, this combination of theoretical approaches was needed to reach a complete and consistent picture of the post-buckling. Application to thin-film delamination has been presented: the phase diagram of a rectangular strip with fixed, straight boundaries provides a good qualitative account of the various elongated delamination patterns observed in experiments.

This chapter provides yet another example of a seemingly simple system, a rectangular elastic plate described by Hookean elasticity, whose behaviour turns out to be quite complex. Although the geometric ingredient in the theory ultimately amounts to an equation as simple as $\mathrm{d}\ell^2 = \mathrm{d}x^2 + \mathrm{d}y^2 + \cdots$, and the elastic response is assumed to be linear, by Hooke's relations, the intertwining of these two leads to a rich phase diagram. The almost inexhaustible richness of buckling problems will show up in the next chapters, devoted to the limit of large loads, i.e. well above threshold. There, completely new methods of analysis will be necessary and a striking self-similar pattern of polyhedral folds will be shown to give the state of lowest elastic energy. Before we investigate this problem, we shall need to introduce new geometrical objects in the field of elasticity, the folds and the d-cones.

In addition to references to papers by other authors, given earlier, we shall mention that the linear stability analysis of Section 8.7 has been published in (B. Audoly, 1999) in the case of equi-biaxial compression; the experiments of Section 8.2 and the final phase diagrams of Section 8.8 have appeared in (B. Audoly, B. Roman, and A. Pocheau, 2002), but the details of the analysis have not yet been published. The buckling analysis presented in this chapter has been applied recently to the case of a stiff film bonded to a compliant substrate and subjected to compressive residual stress. This problem arises for instance in polymeric substrate coated with a metallic interface. Because of the high stiffness contrast between film and substrate, buckling can take place even in the absence of delamination. A combination of analytical approaches, similar to the ones presented in this chapter, and ranging from linear stability analysis to asymptotic solutions in the large load limit, have been used to derive a very rich phase diagram (B. Audoly and A. Boudaoud, 2008,a,b,c), illustrating the power of exact or approximate analytical methods for the analysis of buckling.

References

B. Audoly and A. Boudaoud. Buckling of a thin film bound to a compliant substrate (part 1). Formulation, linear stability of cylindrical patterns, secondary bifurcations. *Journal of the Mechanics and Physics of Solids*, 56(7):2401–2421, 2008.

B. Audoly and A. Boudaoud. Buckling of a thin film bound to a compliant substrate (part 2). A global scenario for the formation of herringbone pattern. *Journal of the Mechanics and Physics of Solids*, 56(7):2422–2443, 2008.

B. Audoly and A. Boudaoud. Buckling of a thin film bound to a compliant substrate (part 3). Herringbone solutions at large buckling parameter. *Journal of the Mechanics and Physics of Solids*, 56(7):2444–2458, 2008.

H. G. Allen. *Analysis and Design of Structural Sandwich Panels*. Pergamon Press, New York, 1969.

B. Audoly, B. Roman, and A. Pocheau. Secondary buckling patterns of a thin plate under in-plane compression. *The European Physical Journal B*, 27:7–10, 2002.

B. Audoly. Stability of straight delamination blisters. *Physical Review Letters*, 83(20):4124–4127, Nov 1999.

B. Audoly. Mode-dependent toughness and the delamination of compressed thin films. *Journal of the Mechanics and Physics of Solids*, 48(11):2315–2332, 2000.

L. Bauer, H. B. Keller, and E. L. Reiss. Multiple eigenvalues lead to secondary bifurcation. *SIAM Review*, 17(1):101–122, 1975.

C.-S. Chien, S.-Y. Gong, and Z. Mei. Mode jumping in the von Kármán equations. *SIAM Journal of Scientific Computing*, 22(4):1354–1385, 2000.

C.-S. Chien. Secondary bifurcations in the buckling problem. *Journal of Computational and Applied Mathematics*, 25(3):277–287, 1989.

A. G. Evans and J. W. Hutchinson. On the mechanics of delamination and spalling in compressed films. *International Journal of Solids and Structures*, 20(5):455–466, 1984.

P. R. Everall and G. W. Hunt. Arnold tongue predictions of secondary buckling in thin elastic plates. *Journal of the Mechanics and Physics of Solids*, 47(10):2187–2206, 1999.

P. R. Everall and G. W. Hunt. Mode jumping in the buckling of struts and plates: a comparative study. *International Journal of Non-Linear Mechanics*, 35:1067–1079, 2000.

L. B. Freund and S. Suresh. *Thin Film Materials: Stress, Defect Formation and Surface Evolution.* Cambridge University Press, 2003.

J.-J. Gervais, O. Abderrahmann, and R. Pierre. Finite element analysis of the buckling and mode jumping of a rectangular plate. *Dynamics and Stability of Systems*, 12:161–185, 1997.

M. George, C. Coupeau, J. Colin, F. Cleymand, and F. Grilhé. Delamination of metal thin films on polymer substrates: from straight-sided blisters to varicose structures. *Philosophical Magazine A*, 82(3):633–641, 2002.

G. Gioia and M. Ortiz. Delamination of compressed thin films. *Advances in Applied Mechanics*, 33:119–192, 1997.

G. Gille and B. Rau. Buckling instability and adhesion of carbon layers. *Thin Solid Films*, 120(2):109–121, 1984.

G. W. Hunt and P. R. Everall. Arnold tongues and mode-jumping in the supercritical post-buckling of an archetypal elastic structure. *Proceedings of the Royal Society A: Mathematical, Physical and Engineering Sciences*, 455(1981):125–140, 1999.

E. J. Holder and D. Schaeffer. Boundary conditions and mode jumping in the von Kármán equations. *SIAM Journal on Mathematical Analysis*, 15(3):446–458, 1984.

J. W. Hutchinson and Z. Suo. Mixed mode cracking in layered materials. *Advances in Applied Mechanics*, 29:63–191, 1992.

J. W. Hutchinson, M. D. Thouless, and E. G. Liniger. Growth and configurational stability of circular, buckling-driven film delaminations. *Acta Metallurgica et Materialia*, 40(2):295–308, 1992.

E. A. Jagla. Modeling the buckling and delamination of thin films. *Physical Review B (Condensed Matter and Materials Physics)*, 75(8):085405, 2007.

H. M. Jensen. Energy release rates and stability of straight-sided, thin-film delaminations. *Acta Metallurgica et Materialia*, 41(2):601–607, 1993.

I. M. Lifshitz and A. Y. Grosberg. State diagram of a polymer globule and the problem of self-organization of its spatial structure. *Soviet Physics JETP*, 38(6):1198–1208, 1974.

I. M. Lifshitz. Some problems of the statistical theory of biopolymers. *Soviet Physics JETP*, 28(6):1280–1286, 1969.

M.-W. Moon, H. M. Jensen, J. W. Hutchinson, K. H. Oh, and A. G. Evans. The characterization of telephone cord buckling of compressed thin films on substrates. *Journal of the Mechanics and Physics of Solids*, 50(11):2355–2377, 2002.

M.-W. Moon, K.-R. Lee, K. H. Oh, and J. W. Hutchinson. Buckle delamination on patterned substrates. *Acta Materialia*, 52(10):3151–3159, 2004.

R. Maaskant and J. Roorda. Mode jumping in biaxially compressed plates. *International Journal of Solids and Structures*, 29(10):1209–1219, 1992.

D. Nir. Stress relief forms of diamond-like carbon thin films under internal compressive stress. *Thin Solid Films*, 112(1):41–50, 1984.

T. Nakamura and K. Uetani. The secondary buckling and post-secondary-buckling behaviours of rectangular plates. *International Journal of Mechanical Sciences*, 21(5):265–286, 1979.

G. Parry, C. Coupeau, J. Colin, A. Cimetière, and J. Grilhé. Buckling and post-buckling of stressed straight-sided wrinkles: experimental AFM observations of bubbles formation and finite element simulations. *Acta Materialia*, 52(13):3959–3966, 2004.

G. Parry, A. Cimetiere, C. Coupeau, J. Colin, and J. Grilhe. Stability diagram of unilateral buckling patterns of strip-delaminated films. *Physical Review E (Statistical, Nonlinear and Soft Matter Physics)*, 74(6):066601, 2006.

D. Schaeffer and M. Golubitsky. Boundary conditions and mode jumping in the buckling of a rectangular plate. *Communications in Mathematical Physics*, V69(3):209–236, 1979.

M. Stein. Loads and deformations of buckled rectangular plates. Technical report R40, NASA, 1959.

M. Stein. The phenomenon of change in buckle pattern in elastic structures. Technical Report R39, NASA, 1959.

M. Uemura and O-Il Byon. Secondary buckling of a flat plate under uniaxial compression. Part 1: theoretical analysis of simply supported flat plate. *International Journal of Non-Linear Mechanics*, 12(6):355–370, 1977.

M. Uemura and O-Il Byon. Secondary buckling of a flat plate under uniaxial compression. Part 2: analysis of clamped plate by F.E.M. and comparison with experiments. *International Journal of Non-Linear Mechanics*, 13(1):1–14, 1978.

A. A. Volinsky. Experiments with in-situ thin film telephone cord buckling delamination propagation. *Materials Research Society Symposium Proceedings*, 749:W10.7.1–W.10.7.6, 2003.

Wolfram Research, Inc. Mathematica edition: Version 6.0. Champaign, IL (USA), 2007.

N. Yamaki. Postbuckling behavior of rectangular plates with small initial curvature loaded in edge compression. *Journal of Applied Mechanics*, 26(3):407–414, 1959.

X. D. Zhu, K. Narumi, and H. Naramoto. Buckling instability in amorphous carbon films. *Journal of Physics: Condensed Matter*, 19(23):236227, 2007.

9

Crumpled paper

9.1   Introduction

The cover of this book displays the image of a piece of paper that was crumpled and then flattened, a very simple experiment that yields a rather complex pattern. A polygonal network is visible in the picture, with straight edges connecting vertices. This two-dimensional pattern reveals the fact that, upon crumpling, a piece of paper deforms into a complicated polyhedral surface whose flat faces merge along straight edges. The key property of this polyhedral surface is that it has the same intrinsic geometry as the initial planar sheet of paper: distances measured along the surface are unaffected by the crumpling process.[1] This simple observation suggests that the crumpling patterns are connected to the geometrical constraint of conserving the lengths along the sheet. It also suggests that the equations for elastic plates, which favour isometric deformations of the centre surface are a suitable framework for studying crumpling (M. Ben Amar and Y. Pomeau, 1997).

The present chapter analyses in some detail the structure of crumpled paper based on the general equations for elastic plates, as derived in Chapter 6. This structure is shown to be built of two elementary bricks, namely developable cones and ridges. Conical singularities are the subject of Section 9.2, while ridges are studied in Section 9.3.

Upon unfolding a piece of crumpled paper or, better, a sheet of the material used for making overhead transparencies,[2] permanent and sharply localized marks are revealed. These marks reflect the focusing of stress at some points where the limit yield of the material is exceeded, leading to irreversible plastic deformations. Such a remark has more than an anecdotical character as it is also relevant for a car crash where the body (sheet-metal) is often strongly deformed at localized points. The focusing of stress and energy during paper crumpling has fascinated physicists for more than a decade—for a recent review on crumpling, we refer to reference (T. A. Witten, 2007). The ridge singularity was first characterized based on scaling laws (A. Lobkovsky et al., 1995). Later, a simple ridge geometry was analysed based on a boundary layer equations (A. E. Lobkovsky, 1996), for which we propose a solution in the present chapter; this ridge geometry has been solved numerically in (A. E. Lobkovsky and T. A. Witten, 1997). The extension of energy focusing to the case of elastic manifolds in higher dimensions has been investigated, e.g., in (S. C. Venkataramani et al., 2000). The second type of singularity, describing the conical singularities, was studied simultaneously (M. Ben Amar and Y. Pomeau, 1997); experimentally, this singularity has been studied by forcing a thin film into the circular opening using a tip (S. Chaïeb, F. Melo,

---

[1] As explained in the following, these distances are unchanged only as long as the small width of the edges connecting the vertices is disregarded. This small width arises out of bending effects, and is small in the limit of thin sheets.

[2] Overhead transparencies are better suited to experiments than ordinary paper, as they remain in the elastic regime in a wider range of stress conditions than sheets of regular paper.

Fig. 9.1 Crumpling of a piece of aluminium foil reveals a network of polygonal ridges. Ridges meet at vertices which are tips of developable cones. Image courtesy of Télémaque Audoly.

and J.-C. Géminard, 1998; E. Cerda and L. Mahadevan, 1998), as explained in Section 9.2.3 of this chapter. More complex geometries comprising several developable cones moving in response to an increasing loading have been studied in (A. Boudaoud *et al.*, 2000).

9.2 Conical singularities

The formation of conical singularities in an initially planar elastic plate can be qualitatively explained in the following way (M. Ben Amar and Y. Pomeau, 1997). Let us try to find the state of minimum elastic energy of a plate that is forced to pass through a given non-planar curve. Whenever possible, this state will be such that the centre surface of the plate remains developable, i.e. such that the lengths measured along the centre surface remain unaffected by the deformation. Indeed, developability makes the stretching energy of the plate vanish. Depending on the curve, a developable configuration consistent with the prescribed non-planar curve may or may not exist. Even when it exists mathematically, it may describe a surface intersecting itself and be physically irrelevant. In the absence of physically relevant, smooth solution having a vanishing stretching energy, one may try to extend the class of solutions to surfaces comprising singularities. In this section, we consider a particular type of singularity, the developable cones, called 'd-cones'. The d-cones are subjected to a kinematical constraint of developability, in a way that is explained below. The experiment reported in Section 9.2.3 shows that such a d-cone configuration can be obtained by pushing a plate through a ring using a sharp object (E. Cerda and L. Mahadevan, 1998).

Before discussing this experiment, we shall first introduce in Section 9.2.1 the geometry of a developable cone. An arbitrary conical surface has a zero Gauss curvature $K = 0$: its curvature is zero along the generatrices, which are also principal directions—see equation (6.25). However, a cone is only developable locally and cannot be globally developed on to a plane in general as a circle drawn with its centre at the tip has a perimeter that may be different from 2π times its radius. In fact, the Gauss curvature is not everywhere zero: it is undefined at the tip of the cone, obviously a singular point for any quantity, such as the

Gauss curvature, related to second derivatives of the surface parameterisation. A specific condition has to be imposed for a cone to be developable around its tip.

This condition will appear when one considers the mechanics of a plate deformed into a cone: unless it is satisfied, the perimeters of some circles drawn on the centre surface differ from their value in the natural configuration, and as a result a large stretching energy is stored in the deformed plate. In contrast, when the developability condition is satisfied, the elastic energy has no stretching contribution (except in the vicinity of the tip of a d-cone) but only a (formally small) bending contribution.

Once the necessary geometrical tools have been introduced, we discuss the experimental realization of a d-cone using a specific loading geometry. This requires us to first consider the contact of a d-cone with a cylinder, a rather complex issue that is first looked at in Section 9.2.2 for the simpler one-dimensional case of an elastic rod (or Elastica) pushed on to a plane. In this simple geometry, we derive the condition for the take-off of a rod from a plane with which it makes contact. A very similar condition applies for the merging of the cone with the ring on which it is pushed, in the experimental realisation of the d-cone discussed next in Section 9.2.3.

Lastly, in Section 9.2.4 we come to the point that was alluded to at the beginning of this chapter: there is stress focusing at the tip of a d-cone. This focusing is studied in two steps. First, the F.–von K. equations are solved far away from the tip by expansion in powers of (h/r), h being the thickness of the plate and r the distance to the tip. The size of the tip area is of order h times a large parameter, and there a boundary layer method yields a parameterless (or numerical) form of the F.–von K. equations. Based on this result, one can show that near the tip the stress is of order E (Young's modulus) times a small geometrical parameter. This explains the marks left on a piece of crumpled paper because such a stress can exceed the elasticity limit of the material of the plate.

9.2.1   Geometry of developable cones

We focus here on the geometry of developable cones, or d-cones. The condition that circles with their centre at the tip have a perimeter $2\pi\rho$, ρ being the radius measured along the surface, is transformed into an integral condition on the function $g(\theta)$ that defines the cone in cylindrical coordinates. Then we show that exactly the same condition is to be satisfied to minimize the stretching energy of a plate deformed into a cone. The explicit form of this condition is intricate in general (that is for $g(.)$ arbitrary) but becomes simple when the cone deviates slightly from a plane, which happens when the function $g(.)$ is small with respect to 1. It turns out that this approximation is relevant here as it corresponds to the range of validity of the F.–von K. equations: besides the condition that curvatures must be smaller than the inverse of h, there is an additional requirement that the displacements from the reference configuration involves local rotations with an angle β such that $|\beta| \ll 1$. Our parameterisation $g(\theta)$ of the d-cone is such that $\beta = \tan^{-1}(g)$ (see below), and so the F.–von K. equations are applicable provided $|g| \ll 1$.

In cylindrical coordinates, a general conical surface with its tip at the origin of the coordinate system has equation $z = r\,g(\theta)$ in cylindrical coordinates, r and θ being the polar coordinates in the horizontal plane, and z the vertical coordinate. The distance to the tip of a point on the cone is $\rho = r\sqrt{1 + g^2(\theta)}$. By expanding for small $g(.)$, we find the parametric equation for the material curve obtained by deforming a circle of radius ρ drawn

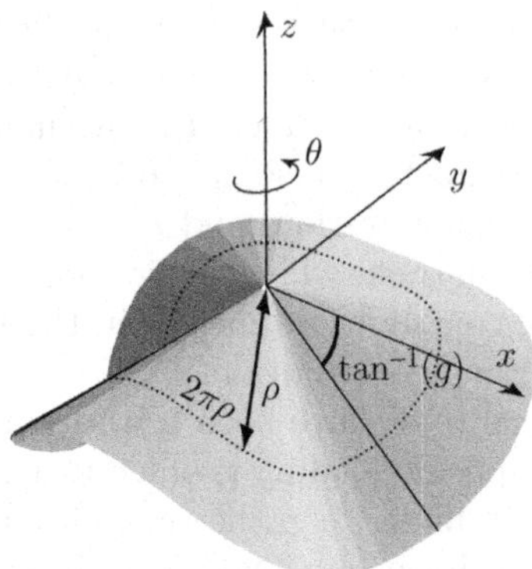

Fig. 9.2 Geometry of a developable cone (d-cone). To prevent large stretching stress away from the cone, a geometrical constraint must be satisfied: the perimeter of circles of radius ρ centred at the tip should remain equal to $2\pi\rho$ upon deformation.

on the undeformed plate:

$$r(\theta) \approx \rho \left(1 - \frac{1}{2} g^2(\theta) \right),$$

and

$$z(\theta) = r(\theta)\, g(\theta) \approx \rho\, g(\theta).$$

This gives a parametric equation for the deformed circle, with parameter θ; note that the deformed circle is defined as the locus of points lying on the cone, at a fixed distance ρ from the tip. The parameter is θ. The square of the length element along this circle is

$$\mathrm{d}s^2 = \mathrm{d}z^2 + \mathrm{d}r^2 + r^2\mathrm{d}\theta^2 = \rho^2 \left(g'^2 + g^2 g'^2 + (1 - g^2) \right) \mathrm{d}\theta^2.$$

In this expression and thereafter, g' stands for $\mathrm{d}g/\mathrm{d}\theta$, g'' for the second derivative with respect to θ, etc. In the limit of a small deflection g, this becomes:

$$\mathrm{d}s \approx \rho\, \mathrm{d}\theta \left(1 + \frac{g'^2 - g^2}{2} \right).$$

At the first non-trivial order in $g(.)$ the condition that the perimeter of this circle is $2\pi\rho$ therefore reads:

$$I = \frac{1}{2} \int_0^{2\pi} \mathrm{d}\theta \left(g'^2 - g^2 \right) = 0. \tag{9.1}$$

This equation may be seen as an extension of the Gauss–Bonnet theorem to surfaces whose Gauss curvature has a point singularity. For smooth surfaces, the Gauss–Bonnet theorem states that the integral of the Gauss curvature K over a piece of smooth surface with the topology of a disc is 2π minus the contour integral, taken along the boundary of this surface of the intrinsic curvature of this boundary.[3] When the surface is smooth and developable, the Gauss curvature is zero and this theorem expresses that the integral of the

---

[3] The intrinsic curvature of a curve drawn on a surface measures its signed curvature, as seen from within this surface. It can be defined, up to a sign, as the inverse radius of the circle drawn in the tangent plane that

intrinsic curvature of an arbitrary closed curve is 2π times an integer. The intrinsic curvature of such a curve being invariant by isometry (M. Spivak, 1979), we get, after developing the surface onto a plane, that the integral of the signed curvature of any smooth closed planar curve is a multiple of 2π, something that can indeed be checked directly.[4] By reversing the argument, the Gauss–Bonnet theorem is seen to imply the *Theorema egregium*, that is the conservation of Gauss curvature by isometries. A cone is singular at the tip and the Gauss–Bonnet theorem is not applicable; however equation (9.1) provides a developability condition for the neighbourhood of the tip in the form of a contour integral, very much like the Gauss–Bonnet theorem.

The same condition, $I = 0$, is relevant to the deformation of an elastic plate into a cone, as we shall show now: it makes a large contribution to the stretching energy vanish. This condition is another example of the tight connection between the Gauss curvature and the stretching energy of a plate, as discussed in Chapter 6.

Let us start from the F.–von K. equations given in equations (6.88):

$$\frac{h^3 E}{12(1 - \nu^2)} \Delta^2 w - h[w, \chi] = 0 \tag{9.2a}$$

$$\Delta^2 \chi + \frac{E}{2}[w, w] = 0, \tag{9.2b}$$

where the notation is that used throughout this text: $\chi(x, y)$ is the Airy potential, $w(x, y)$ the small deflection of the plate out of the $z = 0$ plane, h the plate thickness and (E, ν) are respectively Young's modulus and Poisson's ratio. The bracket $[U, V]$ is for the Monge–Ampère operator, defined earlier in equation (1.13):

$$[U, V] = \frac{\partial^2 U}{\partial x^2}\frac{\partial^2 V}{\partial y^2} + \frac{\partial^2 V}{\partial x^2}\frac{\partial^2 U}{\partial y^2} - 2\frac{\partial^2 U}{\partial x \partial y}\frac{\partial^2 V}{\partial x \partial y}. \tag{9.3}$$

The components of the stress tensor are given by second derivatives of the Airy potential, see definition (6.65). We shall now seek a solution of the F.–von K. equations such that the deflection $w(x, y)$ describes a cone far from the tip, in a sense to be made precise. The conical surface we shall consider has the same equation as before:

$$w(r, \theta) = r\,g(\theta). \tag{9.4}$$

A constant function $g(\theta)$ would give a regular cone with half-angle $\phi = \tan^{-1}(1/g)$. Unless $g - 0$ (or equivalently $\phi = \pi/2$), this solution is singular as it is not $\mathcal{C}^1$ at the tip $r = 0$: it is then continuous but not differentiable. This singularity appears in the bending terms in the F.–von K. equation, i.e. the fourth derivative $\Delta^2 w$ in equation (9.2a), which depends on r like r^{-3} when computed with the function w in (9.4).

As shown in Chapter 6 the F.–von K. equations can be derived from a variational principle. The functional to be minimized is the elastic energy stored in the deformed plate. The stretching contribution to the elastic energy was given in equation (6.97a) as:

fits best the curve around a given point. For a more general definition, see for instance reference (M. Spivak, 1979).

[4] The tangent $\mathbf{t}$ to a planar curve can be parameterised by an angle θ, $\mathbf{t} = (\cos\theta, \sin\theta)$ in Cartesian coordinates. The curvature is then $\mathrm{d}\theta/\mathrm{d}s$, where s denotes the curvilinear abscissa. For a simply connected close curve, we indeed get that $\int \mathrm{d}s\,\mathrm{d}\theta/\mathrm{d}s = 2k\pi$, where k is the number of turns of the tangent.

$$\mathcal{E}_\mathrm{s} = \frac{h}{2\,E} \iint \mathrm{d}x\,\mathrm{d}y \left((\Delta\chi)^2 - 2(1+\nu)\,\frac{[\chi,\chi]}{2} \right)$$

$$= \frac{h}{2\,E} \iint \mathrm{d}x\,\mathrm{d}y \left(\chi_{,xx}{}^2 + \chi_{,yy}{}^2 - 2\,\nu\,\chi_{,xx}\,\chi_{,yy} + 2\,(1+\nu)\,\chi_{,xy}{}^2 \right).$$

We shall first derive a lower bound for this stretching energy—the aim is to show later that this lower bound becomes very large unless the condition $I = 0$ is satisfied. To derive this lower bound, we discard the last term in the second integrand, which is always positive and we use the inequality $a^2 + b^2 - 2\,\nu\,a\,b \geq q(\nu)\,(a^2 + b^2)$, valid for all real numbers a and b, where $q(\nu) > 0$ is the smallest number among $(1 + \nu)$ or $(1 - \nu)$—this inequality comes from the fact that the eigenvalues of the positive quadratic form $(a^2 + b^2 - 2\,\nu\,a\,b)$ are $(1 \pm \nu)$, which are two non-zero, positive numbers given the physically admissible range of Poisson's ratios, $-1 < \nu \leq 1/2$. This yields:

$$\mathcal{E}_\mathrm{s} \geq \frac{q(\nu)\,h}{2\,E} \iint \mathrm{d}x\,\mathrm{d}y\,(\Delta\chi)^2.$$

We are now ready to show that this lower bound is very large unless the developability condition is satisfied. The asymptotic behaviour of the Airy potential χ itself depends on the function $g(.)$. It is found by formally solving equation (9.2b) for $\Delta\chi$, assuming the Gauss curvature $K = [w,w]/2$ as given. One obtains after one application of Green's theorem:[5]

$$\Delta\chi(\mathbf{r}) = -\frac{E}{2\pi} \iint \mathrm{d}^2\mathbf{r}'\,K(\mathbf{r}')\,\ln|\mathbf{r}' - \mathbf{r}|, \qquad (9.5)$$

[5] For solving the Poisson equation $\Delta f = h$ in two dimensions for f, one classically introduces the so-called Green's function $G(\mathbf{p}, \mathbf{q}) = \ln|\mathbf{q} - \mathbf{p}|/2\pi$. As can be checked by explicit calculation the harmonic operator with respect to the variable $\mathbf{p}$ (for a fixed value of $\mathbf{q}$) onto this Green function, $\Delta_\mathbf{p} G(\mathbf{p}, \mathbf{q})$, yields zero whenever $\mathbf{p} \neq \mathbf{q}$. This function $\Delta_\mathbf{p} G(\mathbf{p}, \mathbf{q})$ is nonetheless singular, and in fact has a Dirac contribution at $\mathbf{p} = \mathbf{q}$. This Dirac contribution can be computed by applying Green's second identity, $\iint_D (u\,\Delta v - v\,\Delta u) = \int_{\partial D}(u\,v_{,\mathrm{n}} - v\,u_{,\mathrm{n}})$, on the domain D defined as the disc of unit radius centred at $\mathbf{q}$, the contour ∂D then being the circle of unit radius centred at $\mathbf{q}$, with $u = 1$ and $v = K_\mathbf{q}$. Here, $u_{,\mathrm{n}}$ denotes the outer normal derivative of the function u along the contour. This yields:

$$\Delta_\mathbf{p} G(\mathbf{p}, \mathbf{q}) = \delta(\mathbf{p} - \mathbf{q}).$$

The family of functions $K_\mathbf{q}(\mathbf{p})$ yields the solution of the harmonic equation when the source term is a Dirac function centred on an arbitrary point $\mathbf{q}$. This is precisely the definition of a Green's function. A solution to the Poisson equation with an arbitrary source term such as equation (9.2b):

$$\Delta(\Delta\chi)(\mathbf{p}) = -E\,K(\mathbf{p})$$

with $\mathbf{p} = (x, y)$ can then be found by using the following decomposition of the source term into Dirac contributions:

$$\Delta(\Delta\chi)(\mathbf{p}) = \iint \mathrm{d}\mathbf{q}\,(-E\,K(\mathbf{q}))\,\delta(\mathbf{p} - \mathbf{q}).$$

By linear superposition, a solution is just:

$$\Delta\chi(\mathbf{p}) = \iint \mathrm{d}\mathbf{q}\,(-E\,K(\mathbf{q}))\,\frac{\ln|\mathbf{p} - \mathbf{q}|}{2\pi}.$$

This leads to equation (9.5).

where Δ is the usual 2D harmonic operator. The kernel of the harmonic operator chosen to write equation (9.5) assumes that the boundary conditions are such that the Gauss curvature decays to zero at infinity. Therefore the stress near the tip is dominated by the contribution arising from the local curvature. No matter whether the cone is developable at the tip (d-cone) or not, its Gauss curvature is zero everywhere else. Therefore the elastic energy of the elastic plate, which tends by assumption to a cone far from the tip, is dominated by the following contribution far from the tip, that is for large $r = |\mathbf{r}|$:

$$\Delta\chi(\mathbf{r}) \approx -\frac{E}{2\pi} \ln(r) \int d^2\mathbf{r}' \, K(\mathbf{r}'). \tag{9.6}$$

The integral in the right-hand side is dominated by contributions coming from the core of the cone—this core is defined as the small region near the tip where bending effects are important. Equation (9.6) means that, for the calculation of $\Delta\chi$ in the far field, the tip appears as a point-like source with a weight given by the integral in the right-hand side.

At this point, one has to recall that the stretching energy in the plate is proportional to the integral of $[\Delta\chi(\mathbf{r})]^2$, or larger. Therefore, unless the coefficient of the logarithm in equation (9.6) is zero, the elastic energy of the cone is at least proportional to $r^2 \ln^2(r)$ for large r, a very large quantity for a plate of large dimensions, that grows even faster than the total area. Clearly, this is incompatible with the elastic energy being minimum and the integral on the right-hand side of (9.6) has to vanish:

$$\int d^2\mathbf{r}' \, K(\mathbf{r}') = 0.$$

This condition can be re-expressed as an integral condition for $g(.)$, which will be shown to be identical to the geometrical condition given earlier in equation (9.1). To get this integral condition, let us compute the geometrical integral

$$I' = \int d^2\mathbf{r}' \, K(\mathbf{r}') = \int_0^R dr \, r \int_0^{2\pi} d\theta \, K(\mathbf{r}').$$

In the integral over the modulus r we introduce an upper bound R to get rid of possible divergences at large distances.

We use the property that the Gauss curvature $K(x, y) = \frac{1}{2}[w, w]$ is the divergence of a 2D vector field $\mathbf{P}(x, y)$:

$$K - \frac{1}{2}[w, w] = \frac{\partial^2 w}{\partial x^2}\frac{\partial^2 w}{\partial y^2} - \left(\frac{\partial^2 w}{\partial x \partial y}\right)^2 = -\frac{1}{2}\operatorname{div}\mathbf{P}, \tag{9.7}$$

where

$$\mathbf{P} = \left[\frac{\partial w}{\partial y}\frac{\partial^2 w}{\partial x \partial y} - \frac{\partial w}{\partial x}\frac{\partial^2 w}{\partial y^2}\right]\mathbf{e}_x + \left[\frac{\partial w}{\partial x}\frac{\partial^2 w}{\partial x \partial y} - \frac{\partial w}{\partial y}\frac{\partial^2 w}{\partial x^2}\right]\mathbf{e}_y,$$

where $\mathbf{e}_x$ and $\mathbf{e}_y$ are two orthonormal unit vectors in the horizontal plane. By the divergence theorem,[6] the integral $I' = \iint d^2\mathbf{r}' \, K(\mathbf{r}')$, when computed inside a large circle surrounding

---

[6] We are applying the divergence theorem to the centre surface of the plate. This assumes that, when the bending effects are taken into account, the Gauss curvature remains smooth everywhere, including near the tip. This condition is reasonable, as it prevents the bending energy from being infinite. We shall indeed prove later on that the curvature remains finite at the tip when bending is taken into account.

the tip of the cone, is the flux of $-\mathbf{P}/2$ across the boundary of the domain. This integral of the flux across a large circle tends to a constant as the radius of the circle becomes large, and this constant is precisely the value of I' we are trying to obtain (this is because the difference between the flux across two large circles is the integral of the Gauss curvature over the area between them; the Gauss curvature vanishes far from the tip and this integral tends to zero when the two circles bounding the ring-shaped domain become both very large).[7] Therefore, the integral I is equal to the flux of $-1/2\mathbf{P}$ across a large circle, in the range where the Cartesian equation of the surface can be approximated by the simple equation of a cone. This yields a simple expression for the flux:

$$I' = -\frac{1}{2}\int \mathbf{P}(\rho,\theta).\mathbf{n}(\theta)\,\rho\,\mathrm{d}\theta.$$

Here (ρ,θ) are the polar coordinates in the plate, centred at the tip and $\mathbf{n}$ is the outer normal to the circle. By noticing that $\rho\,\mathbf{n}(\theta) = \mathbf{r}$ is just the vector joining the tip to the current point on the circle, we get:

$$I' = -\frac{1}{2}\int_{-\pi}^{\pi} \mathrm{d}\theta\,\mathbf{r}.\mathbf{P}$$

$$= -\frac{1}{2}\int_{-\pi}^{\pi} \mathrm{d}\theta\left[x\left(\frac{\partial w}{\partial y}\frac{\partial^2 w}{\partial x\partial y} - \frac{\partial w}{\partial x}\frac{\partial^2 w}{\partial y^2}\right) + y\left(\frac{\partial w}{\partial x}\frac{\partial^2 w}{\partial x\partial y} - \frac{\partial w}{\partial y}\frac{\partial^2 w}{\partial x^2}\right)\right].$$

It is a matter of tedious but straightforward algebra now to get the value of this expression for a conical deflection $w = r\,g(\theta)$. The final result is:

$$I' = -\frac{1}{2}\int_{-\pi}^{\pi} \mathrm{d}\theta\,g(\theta)\left(g(\theta) + \frac{\mathrm{d}^2 g(\theta)}{\mathrm{d}\theta^2}\right) \tag{9.8}$$

This shows, as expected, that I' does not depend on the radius ρ of the domain of integration if it is large enough. This integral must be zero to make the large contribution to the stretching energy of the cone vanish. By integration by parts of the second term, the condition $I' = 0$ above appears to be identical to the geometrical condition derived before in equation (9.1). We have therefore shown that the energy minimization of an elastic plate forced into a conical shape selects particular conical configurations, namely *developable cones* (d-cones).

In Section 9.2.4, we shall explain how the asymptotic behaviour of the Airy potential and of $w(x,y)$, as derived above, can be matched with the near-tip solution.

9.2.2   The Elastica pushed on to a plane

In the coming three sections, we shall deal with various questions related to an experimental realization of a d-cone. In the present section, we depart a bit from the main line of Part II of this book, as we do not deal here with plates but with rods. A rod pushed on to a

---

[7] This assumes implicitly, something that has to be checked *a posteriori*, that the next order in the large distance behaviour of $w(x,y)$ beyond the simple cone does not bring in finite contributions to the large distance behaviour of I. This follows from the estimates given below for these next order corrections, which are at least of order $1/r$ times the dominant contribution, and so yield a contribution to the curvature along the generatrices of the cone of order $1/r^2$ (this curvature is zero at the dominant order for large r). Therefore the Gauss curvature decays like $1/r^3$ at large r: a product of the regular curvature of the cone of order $1/r$ perpendicular to the generatrices times the one of order $1/r^2$ along the generatrices.

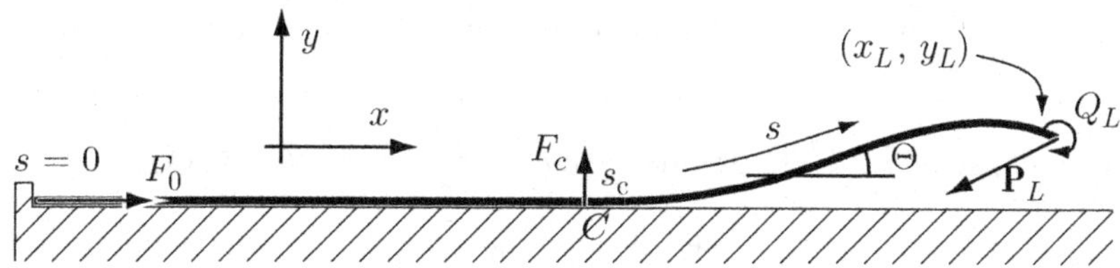

Fig. 9.3 The Elastica pushed on to a plane provides a simple geometry for deriving the boundary condition at the moving endpoint of a contact region.

plane provides a simple geometry where the conditions holding at the endpoint of a contact region can be easily analysed. By free contact region, we mean that the contact region is not fixed in advance but instead results from a mechanical condition that follows from the condition of energy minimum. The slightly more difficult case of a contact between a cone and a ring on which it is pushed is considered in the following sections.

Let us consider a planar elastic curve of length L pushed on to a plane as shown in Fig. 9.3. The end at $s = 0$ is clamped while the other end at $s = L$ has a prescribed position (x_L, y_L). Let $\mathbf{P}_L$ and $Q_L\,\mathbf{e}_z$ be the force and moment applied on the rod at $s = L$. We study configurations such that contact with the plane occurs along a part of the rod only, with arc length $0 \leq s \leq s_{\mathrm{c}}$. Here, s_{c} denotes the arc length of the point C in Fig. 9.3 where the rod lifts off the plane. The Cartesian coordinates (x, y) are such that the plane has equation $y = 0$, and $\Theta(s)$ denotes the orientation of the tangent[8] $\mathbf{t}(s)$ to the rod with respect to the x axis.

The boundary condition holding at the endpoint C, derived below, accounts for the contact forces near this point. The underlying physics is the short-range interaction between the two solids but we do not here introduce a microscopic description of these interactions. We shall only impose that the plane is impenetrable, which is expressed by the condition $y(s) \geq 0$, and minimize the total energy of the system with this constraint.

We rederive below the equations for the equilibrium of a planar Elastica by minimization of its elastic energy. Many equivalent derivations can be proposed, see (C. Majidi, 2007) for a discussion of the various possible approaches. In the presence of a free region of contact, the Euler–Lagrange procedure provides an additional equation that fixes the position of C. Minimization of the energy for an elastic rod has already been presented in Section 4.2.1 when the rod is unconstrained in space. We adapt this derivation to the case of contact with the plane, skipping the details of the calculation that have been already presented and focusing on the specificities of the present geometry. The energy of the Elastica is the integral $\mathcal{E}_{\mathrm{rod}} = EI/2 \int_0^L \Theta'^2 \, \mathrm{d}s$, where the curvature Θ' is again defined as the derivative of the tangent orientation $\Theta(s)$ with respect to s. The Cartesian coordinates of an arbitrary point on the centre line read $x = \int_0^s \cos\Theta \, \mathrm{d}s$, $y = \int_0^s \sin\Theta \, \mathrm{d}s$, taking the clamped endpoint at $s = 0$ as the origin of the coordinate system. The optimal configuration is found by minimization of the energy. At first order, the variation of elastic energy of the Elastica due to an arbitrary perturbation $\delta\Theta(s)$ in its profile has been expressed as follows,

---

[8] This is the same quantity as lowercase $\theta(s)$ in our former derivation of the equations for the Elastica in Section 4.2.1, but we changed to an uppercase notation here because θ is now the second cylindrical coordinate.

using integration by parts—see equations (3.52) in the general 3D case and (4.7) for the 2D hair:

$$\delta\mathcal{E}_{\text{rod}} = \int_0^L EI\,\Theta'\,\delta\Theta'\,\mathrm{d}s = [EI\,\Theta'\,\delta\Theta]_0^L - \int_0^L EI\,\Theta''\,\delta\Theta\,\mathrm{d}s, \tag{9.9}$$

where the squared brackets $[EI\,\Theta'\,\delta\Theta]$ denote the boundary terms coming from the integration by parts. Special care must be taken in the present geometry: integration by parts makes the second derivative Θ'' appear, a quantity that might not be well defined at $s = s_{\text{c}}$ where smoothness of the solution is not warranted. The integrand in the above equation might not be well defined at s_{c}. To avoid this difficulty, we shall rewrite the above equation, by first splitting the elastic energy in two contributions[9] corresponding to the two regions on either sides of the point s_{c} and integrating by parts separately:

$$\delta\mathcal{E}_{\text{rod}} = \int_0^{s_{\text{c}}^-} EI\,\Theta'\,\delta\Theta'\,\mathrm{d}s + \int_0^{s_{\text{c}}^+} EI\,\Theta'\,\delta\Theta'\,\mathrm{d}s$$

$$= [EI\,\Theta'\,\delta\Theta]_0^{s_{\text{c}}^-} - \int_0^{s_{\text{c}}^-} EI\,\Theta''\,\delta\Theta\,\mathrm{d}s + [EI\,\Theta'\,\delta\Theta]_{s_{\text{c}}^+}^L - \int_{s_{\text{c}}^+}^L EI\,\Theta''\,\delta\Theta\,\mathrm{d}s.$$

Here, we write $s_{\text{c}}^{\pm}$ for the limit of a possibly discontinuous function when s_{c} is approached from above or from below. Writing f as the quantity inside the square brackets denoting the boundary terms, we rearrange them as

$$[f]_0^{s_{\text{c}}^-} + [f]_{s_{\text{c}}^+}^L = f(s_{\text{c}}^-) - f(0) + f(L) - f(s_{\text{c}}^+)$$

$$= (f(L) - f(0)) - (f(s_{\text{c}}^+) - f(s_{\text{c}}^-)) = [f]_0^L - [\![f]\!]_{s_{\text{c}}},$$

where the double square brackets $[\![f]\!]_{s_{\text{c}}}$ denote the discontinuity of the argument at s_{c}: $[\![f]\!]_{s_{\text{c}}} = f(s_{\text{c}}^+) - f(s_{\text{c}}^-)$. This allows one to rewrite the variation of the elastic energy as:

$$\delta\mathcal{E}_{\text{rod}} = [EI\,\Theta'\,\delta\Theta]_0^L - [\![EI\,\Theta']\!]_{s_{\text{c}}}\,\delta\Theta(s_{\text{c}}) - \int_0^{s_{\text{c}}^-} EI\,\Theta''\,\delta\Theta\,\mathrm{d}s - \int_{s_{\text{c}}^+}^L EI\,\Theta''\,\delta\Theta\,\mathrm{d}s. \tag{9.10}$$

Comparison with the smooth case in equation (9.9) reveals an important difference: a new term $(-[\![EI\,\Theta']\!]_{s_{\text{c}}}\,\delta\Theta(s_{\text{c}}))$ is present in the presence of a discontinuity—note that we have factored $\delta\Theta$ out of the double square bracket in this new term as this function is smooth at s_{c} by assumption.

The remaining derivation of the equations of equilibrium is unaffected by the free region of contact. The applied force and moment at the endpoint $s = L$ are taken into account by considering the variation of the total energy, which is the sum of the elastic energy written above and of a potential energy:

$$\delta\mathcal{E}_{\text{rod}} - \mathbf{P}_L \cdot \delta\mathbf{r}(L) - Q_L\,\delta\Theta(L), \tag{9.11}$$

where $\mathbf{r} = (x, y)$ is the position of a point along the centre line.

Contact with the impenetrable plane is taken care of by the method of Lagrange multipliers, see Section 1.3.5. We assume that there is no friction. Then, the equilibrium results

---

[9] We assume that the curvature remains bounded, an assumption to be checked at the end. As a result, the contribution to the elastic energy coming from the neighbourhood of s_{c} remains negligible.

from minimization of the total (elastic plus potential) energy subjected to the constraint. The constraint is $y(s) \geq 0$ for all s. We assume that the contact region is $0 \leq s \leq s_\mathrm{c}$, and so the equality is reached in the region of contact, $y(s) = 0$ for $0 \leq s \leq s_\mathrm{c}$, while the inequality is strict elsewhere, $y(s) > 0$ for $s > s_\mathrm{c}$. A single constraint c would be associated with a single Lagrange multiplier λ. Here, we have a family of constraints $y(s) = 0$ indexed by s in the range $0 \leq s \leq s_\mathrm{c}$. The associated Lagrange multiplier $p(s)$ then becomes a *function* of the parameter s. According to the general method of Lagrange multipliers, we add to the variation of the physical function to be minimized a term $(-\lambda\,\delta c)$ coming from the constraint, see Section 1.3.5. Here the constraints are indexed by s and this term reads:

$$-\int_0^{s_\mathrm{c}} p(s)\,\delta y(s)\,\mathrm{d}s.$$

Introducing the vector quantity $\mathbf{p}(s) = p(s)\,\mathbf{e}_y$, we rewrite the constraint term as

$$-\int_0^{s_\mathrm{c}} \mathbf{p}(s) \cdot \delta\mathbf{r}(s)\,\mathrm{d}s. \tag{9.12}$$

Given that the above term is homogeneous to an energy, the quantity $\mathbf{p}(s)$ can be interpreted as a density of external force per unit length of the rod, along the direction perpendicular to the plane. Therefore, the Lagrange multiplier $p(s)$ can be interpreted as the contact pressure (strictly speaking it is actually a force per unit *length*, given that the Elastica is one-dimensional) between the plane and the rod. The last term above to be included in the variation can then be interpreted loosely as the potential energy associated with this contact pressure.

Adding now the constraint term (9.12) to the physical energy (9.10) and (9.11), we consider the total variation

$$\delta\mathcal{E}_{\mathrm{tot}} = [EI\,\Theta'\,\delta\Theta]_0^L - [\![EI\,\Theta']\!]_{s_\mathrm{c}}\,\delta\Theta(s_\mathrm{c}) - \int_0^{s_\mathrm{c}^-} EI\,\Theta''\,\delta\Theta\,\mathrm{d}s - \int_{s_\mathrm{c}^+}^L EI\,\Theta''\,\delta\Theta\,\mathrm{d}s$$

$$- \int_0^L \mathbf{p}(s) \cdot \delta\mathbf{r}(s)\,\mathrm{d}s - \mathbf{P}_L \cdot \delta\mathbf{r}(L) - Q_L\,\delta\Theta(L).$$

Note that we have extended the range of integration of the contact pressure term using the convention that $p(s) = 0$ for $s > s_\mathrm{c}$.

Before working out the above variation, we note that the perturbation of the *position* of a current point along the centre line, $\delta\mathbf{r}(s)$, required for the calculation of the work of forces, depends indirectly on the perturbation $\delta\Theta$ to the *orientation* of the tangent to the centre line. This difficulty has been solved in Chapter 3, equation (3.58), by introducing an auxiliary function, interpreted as the internal force in the rod:

$$\mathbf{F}(s) = \int_s^L \mathbf{p}(s')\,\mathrm{d}s' + \mathbf{P}_L, \tag{9.13}$$

where the term $\mathbf{P}_L$ could be incorporated in the integrand as a Dirac contribution localized on the edge $s = L$.

This auxiliary function allows the total variation $\delta\mathcal{E}_{\text{tot}}$ to be rewritten using integration by parts, as explained in Section 3.6, see equation (3.59):

$$\delta\mathcal{E}_{\text{tot}} = (EI\,\Theta'(L) - Q_L)\,\delta\Theta(L) - [\![EI\,\Theta']\!]_{s_c}\,\delta\Theta(s_c)$$

$$-\int_0^{s_c^-}(EI\,\Theta'' + F_n)\,\delta\Theta\,\mathrm{d}s - \int_{s_c^+}^{L}(EI\,\Theta'' + F_n)\,\delta\Theta\,\mathrm{d}s, \quad (9.14)$$

where F_n is the normal component of the internal force,

$$F_n = F_y\,\cos\Theta - F_x\,\sin\Theta.$$

Other terms are present in equation (3.59) but they vanish here with the boundary conditions $\delta\mathbf{r}(0) = \mathbf{0}$ and $\delta\Theta(0) = 0$ at the clamped endpoint $s = 0$.

Equilibrium is reached when the above variation is zero for an arbitrary perturbation $\delta\Theta(s)$. As expected, the first term in equation (9.14) yields the boundary condition associated with the applied moment, $EI\,\Theta'(L) = Q_L$, see Section 3.6.4, and the two integrals yield the equations of equilibrium for the planar Elastica, $EI\,\Theta'' + F_n = 0$ for $s < s_c$ (region of contact) and for $s > s_c$ (free region). These equations are the 2D version of the Kirchhoff equation expressing the balance of moments for a three-dimensional rod, derived in equation (3.63b)—the other Kirchhoff equation (3.63a) has been introduced in an integrated form above in equation (9.13).

More interesting is the remaining (second) term in the variation (9.14), which leads to the condition

$$[\![EI\,\Theta']\!]_{s_c} = 0, \text{ that is } EI\,\Theta'(s_c^-) = EI\,\Theta'(s_c^+). \quad (9.15)$$

This equation imposes the continuity of the internal moment $EI\,\Theta'$ across the endpoint of the contact region, C. It follows from the condition of energy minimum and is far from obvious, as we shall shortly see: the internal force, unlike the internal moment, is discontinuous at C. This condition (9.15) is specific to a free contact problem, and is exactly the boundary condition that we were seeking: the degree of freedom associated with the unknown position of the endpoint C can be fixed using this additional equation. A simple illustration of this boundary condition is shown in Fig. 9.4.

Having derived the equations of equilibrium for the Elastica pushed on to a plane, see Fig. 9.3, we can now outline the solution of these equations. In the contact region, $s \leq s_c$, the tangent has constant orientation, $\Theta = 0$, and so $\Theta' = 0$ and $\Theta'' = 0$. From the balance of moments, $EI\,\Theta'' + F_n = 0$, we conclude that $F_n = F_y$ vanishes in the contact region, that is for $s < s_c$: the internal force is purely longitudinal there. Given $EI\,\Theta'(s_c^-) = 0$, the continuity condition (9.15) yields

$$EI\,\Theta'(s_c^+) = 0.$$

For the sake of completeness, we give here the expression of the contact pressure $p(s)$, which has a rather unexpected structure. We start by computing the internal force $\mathbf{F}(s)$, given by equation (9.13) as:

$$F_x(s) = P_L^x, \qquad F_y(s) = \int_s^L p(s')\,\mathrm{d}s' + P_L^y,$$

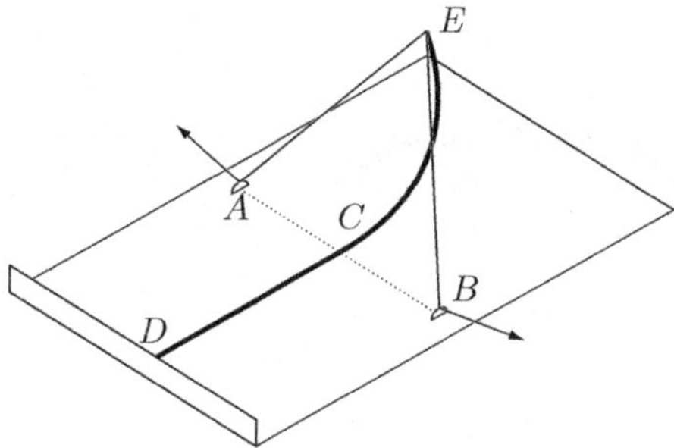

Fig. 9.4 Simple illustration of the boundary condition (9.15). A rod is laid down on a plane, clamped at one end D, and then bent by tightening a cord fastened at the other end E, passing through two small rings A and B. Upon tightening, part of the rod bends while the other part of the rod is pushed on to the plane. No matter what tension is applied, the lift-off point, C, is fixed and aligned with A and B. By equation (9.15), the internal moment applied across point C is zero, as it is zero in the flat region. The global balance of torques on the bent part of the rod then imposes that the moment applied by the cord with respect to an axis passing through C is zero. This requires C to lie within the plane AEB.

where P_L^x and P_L^y) are the components of the applied force $\mathbf{P}_L$. In the absence of friction, we find that the component of the internal force in the direction parallel to the plane of contact is uniform: $F_x = P_L^x$ everywhere (it will eventually be balanced by the reaction of the support at $s = 0$). For the other component, we find $F_y = P_L^y$ in the free part ($s > s_c^+$) since $p(s') = 0$ there does not contribute to the integral for F_y; moreover, we have just shown $F_y = 0$ for $s < s_c$. Plugging these values back for $F_y(s)$ into the equation above, we obtain the contact pressure in the form:

$$p(s) = (-P_L^y)\, \delta_D(s - s_c). \tag{9.16}$$

The applied force has to be downward ($P_L^y < 0$) for our assumption regarding the geometry of contact to be consistent ($p \geq 0$). Then, the contact force exerted by the plane is simply a concentrated force localized at the point C that exactly balances the normal component of the applied force $\mathbf{P}_L$, as sketched in Fig. 9.3. This point reaction force is represented by a Dirac contribution[10] $\delta_D(s - s_c)$ in the contact pressure. The contact pressure vanishes everywhere else.

The main purpose of this section was to show that boundary conditions at the free contact point can be derived from the variational formulation of the problem with appropriate Lagrange multipliers. In the Elastica equation, which is second order, the non-trivial boundary condition is the continuity of the first derivative of Θ at the merging with the plane, as derived in equation (9.15). For the contact of a cone with a circle, the underlying equation is fourth order, and the relevant condition of contact will similarly involve the cancellation of the second derivative of a displacement. Note that a numerical investigation

---

[10] The Dirac function is written δ_D, where the subscript D is to avoid confusion with infinitesimal variations, such as in $\delta\Theta$ or $\delta\Theta'$. This singularity is a mathematical singularity: for a rod of finite diameter, there will be a smoothing of this singular force over distances of the order of the diameter of the rod. It is well possible too that if the interaction between the rod and the solid surface is attracting in some range, this could show up in a transition region, which is certainly an interesting issue.

Fig. 9.5 Experimental d-cone obtained by pushing an elastic film through a glass container with a pencil, after Cerda and Mahadevan (E. Cerda and L. Mahadevan, 1998). Photograph courtesy of Carmine Audoly.

of the equilibria of an Elastica pushed on to a plane is proposed in (R. H. Plaut *et al.*, 1999), based on similar equations to those derived here.

9.2.3 Elastic sheet pushed into a cylinder

This section presents a simple experiment where a d-cone is formed on an elastic plate. This experiment, shown in Fig. 9.5, was first proposed in reference (S. Chaïeb, F. Melo, and J.-C. Géminard, 1998). The conical solution far from the tip was first published in (E. Cerda and L. Mahadevan, 1998) and in more detail in (E. Cerda and L. Mahadevan, 2005). Numerical simulations are presented in (T. Liang and T. A. Witten, 2005). In this simple geometry, an elastic sheet is pushed through a cylinder using a sharp tip. In response to this applied force, the sheet takes the form of a d-cone with its tip at the point of application of the force. Interestingly, the sheet is in contact with the cylinder along part of its boundary, and lifts off elsewhere. This overall shape is quite easy to understand: a partial contact with the cylinder makes possible the invagination of the sheet into the cylinder without violating the developability condition given earlier in equation (9.1). Below we shall derive the shape of the plate far from the tip, which is conical—this is an outer solution in the language of boundary layers. We proceed in a manner very similar to the analysis of the Elastica pushed on to a plane: the elastic energy of a conical sheet is minimized, subjected to the condition of non-penetration of the walls of the cylinder.

We shall assume, and check at the end, that the sheet deforms without significant stretching, at least far from the tip. The elastic energy of the cone is then dominated by its bending contribution:

$$\mathcal{E}_\mathrm{b} = 2\pi \, \frac{Eh^3}{24(1 - \nu^2)} \int_0^L \mathrm{d}r \, r \int_{-\pi}^{\pi} \mathrm{d}\theta \, (\Delta w)^2 , \tag{9.17}$$

where L denotes the radius of the circular sheet bent into a cone. The parametric equation of a cone is $w = r\,g(\theta)$, where g is a function of θ to be determined. The quantity Δw then takes the form $\Delta w = \frac{1}{r}\,(g'' + g)$. Therefore,

$$\int_0^L \mathrm{d}r\, r \int_{-\pi}^{\pi} \mathrm{d}\theta\, (\Delta w)^2 = \left(\int_0^L \frac{\mathrm{d}r}{r} \right) \int_{-\pi}^{\pi} \mathrm{d}\theta\, (g'' + g)^2 .$$

The integral over r diverges logarithmically for r tending to zero. This divergence is unphysical, because close to the tip, the surface is no longer a geometrical cone, as shown below in Section 9.2.4. The surface differs significantly from a cone at a distance of the tip of order h/ϵ, where ϵ is the small magnitude of the dimensionless function $g(\theta)$. This can be taken as a lower bound for the integral over r, which leads one to replace the diverging integral $\left(\int_0^L \mathrm{d}r/r \right)$ by the logarithm $\ln(L\epsilon/h)$. Later on, this logarithm will be seen to appear as an overall multiplicative constant that can be set arbitrarily to one by rescaling the energy. Therefore the reduced dimensionless energy of bending reads:

$$\mathcal{E}_{\mathrm{b}}^\dagger = \frac{1}{2} \int_{-\pi}^{\pi} \mathrm{d}\theta\, (g'' + g)^2 . \tag{9.18}$$

Minimization of this functional with respect to the unknown function $g(\theta)$ gives in principle the profile of the cone. This minimization is subjected to two constraints: the extended Gauss–Bonnet condition (9.1) and the fact that the cone cannot penetrate into the cylinder. The latter condition writes as $g(\theta) \geq \epsilon$, where ϵ is the angle of the geometrical cone that has its tip at the end of the pushing tip and its base on the circle at the edge of the cylinder:

$$\epsilon = \frac{d}{r}, \tag{9.19}$$

where d is the penetration of the pushing tip into the cylinder, and r the radius of the cylinder, see Fig. 9.6. This number ϵ has to be small for the assumption underlying the F.–von K. equations to be valid: the sheet undergoes small rotations from its planar configuration. We are now going to derive the equation to be satisfied by the d-cone at equilibrium under these constraints. The method closely follows that of the previous Section 9.2.2. In particular, the free contact with the cylinder is associated with a specific boundary conditions at the endpoints of the contact region.

We consider an arbitrary perturbation of the cone profile $\delta g(\theta)$. The variation of the reduced bending energy then reads

$$\delta \mathcal{E}_{\mathrm{b}}^\dagger = \int_{-\pi}^{+\pi} \mathrm{d}\theta\, (g'' + g)\,(\delta g'' + \delta g).$$

We would like to integrate by parts the terms in the integrand that are proportional to $(\delta g'')$ but this makes higher-order derivatives of δg appear, which are not necessary well-defined at the edges of the contact regions. As earlier, this difficulty is avoided by first splitting the domain of integration into two parts, corresponding to the region of contact $\mathcal{D}_{\mathrm{c}}$ and to the free region (without contact) $\mathcal{D}_{\mathrm{f}}$. For the sake of simplicity, we shall not write the equations in full generality. We focus on a simple geometry observed in typical experiments, assuming that there is a single region of contact, a single bump where the cone lifts off from the cylinder and that the solution is symmetric—other geometries are possible and can be

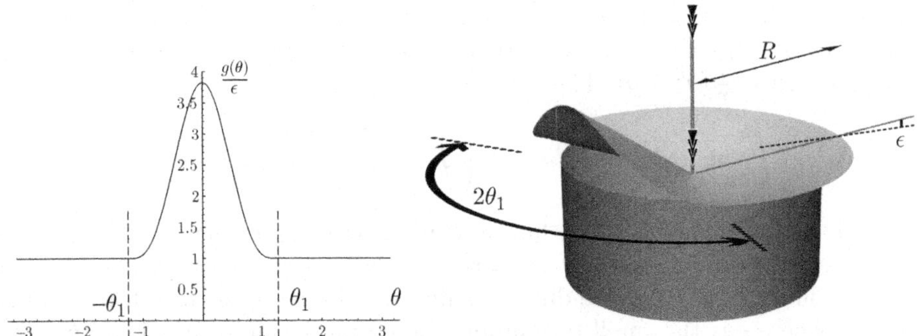

Fig. 9.6 (Left) Rescaled numerical solution $g(\theta)$ for the profile of the d-cone. (Right) Three dimensional representation of the d-cone pushed into a cylinder: compare with the experiment in Fig. 9.5.

treated similarly (E. Cerda and L. Mahadevan, 2005). Then, the coordinate system can be chosen such that that the lift-off takes place in the range

$$\mathcal{D}_f: \qquad -\theta_1 < \theta < \theta_1, \tag{9.20a}$$

where θ_1 is half the angular size of the free region, a number that will be determined below. The complementary region is in contact, and is written as

$$\mathcal{D}_c: \qquad \theta_1 < \theta < 2\pi - \theta_1. \tag{9.20b}$$

The integration by parts leads to:

$$\delta\mathcal{E}_b^\dagger = \left[(g + g'')\,\delta g' - (g' + g''')\,\delta g\right]_{-\theta_1}^{\theta_1} + \left[\cdots\right]_{\theta_1}^{2\pi-\theta_1}$$
$$+ \int_{-\theta_1}^{\theta_1} (g + 2\,g'' + g^{(4)})\,\delta g\;\mathrm{d}\theta + \int_{\theta_1}^{2\pi-\theta_1} \cdots\;\mathrm{d}\theta,$$

where the dots stand for terms that are repeated over the second domain. Note that the coefficient of δg in the last explicit integral includes a fourth derivative $g^{(4)} = \mathrm{d}^4 g/\mathrm{d}\theta^4$, although the same variation in the Elastica problem yielded a second derivative. This is because, for the Elastica problem, the unknown is the slope, although the unknown function, $g(\theta)$, is now directly (i.e. without derivatives) related to the displacement.

As before, we shall reorder the boundary terms produced by the integration by parts to make discontinuities appear at the edges of the contact region, denoted by double square brackets:

$$\delta\mathcal{E}_b^\dagger = -[\![(g + g'')\,\delta g']\!]_{(-\theta_1)} - [\![(g + g'')\,\delta g']\!]_{(+\theta_1)}$$
$$+ [\![(g' + g''')\,\delta g]\!]_{(-\theta_1)} + [\![(g' + g''')\,\delta g]\!]_{(+\theta_1)}$$
$$+ \int_{\mathcal{D}_f} (g + 2\,g'' + g^{(4)})\,\delta g\;\mathrm{d}\theta + \int_{\mathcal{D}_c} (g + 2\,g'' + g^{(4)})\,\delta g\;\mathrm{d}\theta.$$

These boundary terms can be simplified by making some assumptions regarding the degree of smoothness of the solution at the points $\pm\theta_1$. Obviously, the function $g(\theta)$ has to be continuous there, as the cone is not allowed to tear apart. The derivative $\delta g'(\theta)$ has to be continuous too: a discontinuity in $g'(\theta)$ would be associated with a sharp fold that physically has an infinite bending energy. Therefore, we assume g and g', as well as δg and $\delta g'$, to be continuous. Then, the terms associated with discontinuities can be simplified as:

$$\delta\mathcal{E}^{\dagger}_{\mathrm{bend}} = -[\![g'']\!]_{(-\theta_1)}\,\delta g'(-\theta_1) - [\![g'']\!]_{(+\theta_1)}\,\delta g'(\theta_1) + [\![g''']\!]_{(-\theta_1)}\,\delta g(-\theta_1)$$

$$+[\![g''']\!]_{(+\theta_1)}\,\delta g(\theta_1) + \int_{\mathcal{D}_{\mathrm{f}}} (g + 2\,g'' + g^{(4)})\,\delta g\,\mathrm{d}\theta + \int_{\mathcal{D}_{\mathrm{c}}} (g + 2\,g'' + g^{(4)})\,\delta g\,\mathrm{d}\theta. \qquad (9.21)$$

The variation associated to the extended Gauss–Bonnet condition (9.1) reads:

$$\delta I = \int_{-\pi}^{\pi} (g'\,\delta g' - g\,\delta g)\,\mathrm{d}\theta = -\int_{-\pi}^{\pi} (g + g'')\,\delta g\,\mathrm{d}\theta. \qquad (9.22)$$

The boundary terms $[g'\,\delta g]$ from the integration by parts are zero since the function inside the bracket is periodic and smooth at the endpoints $\pm\theta_1$ of the contact region: it contains no derivative higher than first order. We shall multiply this variation δI by a quantity, denoted λ, which is the Lagrange multiplier associated with the Gauss–Bonnet condition[11] $I = 0$.

Finally the free region of contact is taken care of using a continuous set of Lagrange multipliers, as in the Elastica problem:

$$-\int_{-\pi}^{\pi} p(\theta)\,\delta g(\theta)\,\mathrm{d}\theta.$$

In this equation, $p(\theta)$ is the normalized contact pressure with the cone, a function of θ that can only be non-zero for $g(\theta) = \epsilon$, that is wherever the d-cone is actually in contact with the cylinder. Building up from the previous analysis of the Elastica in contact, we anticipate the structure of the pressure $p(\theta)$ and decompose it to a smooth part $p_{\mathrm{s}}(\theta)$ which is non-zero only inside the contact region, and two punctual forces with magnitudes $p_D^{\pm}$ at the edges of the contact region, represented by Dirac contributions δ_D to the pressure:

$$p(\theta) = p_{\mathrm{s}}(\theta) + p_D^{-}\,\delta_D(\theta + \theta_1) + p_D^{+}\,\delta_D(\theta - \theta_1), \qquad (9.23)$$

with $p_{\mathrm{s}} = 0$ on $\mathcal{D}_{\mathrm{f}}$. The contribution to the variation coming from the contact condition can be rewritten as:

$$-p_D^{-}\,\delta g(-\theta_1) - p_D^{+}\,\delta g(\theta_1) - \int_{-\pi}^{\pi} p_{\mathrm{s}}(\theta)\,\delta g(\theta)\,\mathrm{d}\theta. \qquad (9.24)$$

Collecting the terms given in equations (9.22), (9.23) and (9.24), we write the variation of the augmented energy, which has to be zero at equilibrium for an arbitrary $\delta g(\theta)$:

$$\delta\mathcal{E}^{\dagger}_{\mathrm{b}} - \lambda\,\delta I - \int_{-\pi}^{\pi} p(\theta)\,\delta g(\theta)\,\mathrm{d}s =$$

$$-[\![g'']\!]_{(-\theta_1)}\,\delta g'(-\theta_1) - [\![g'']\!]_{(+\theta_1)}\,\delta g'(\theta_1)$$

$$+ ([\![g''']\!]_{(-\theta_1)} - p_D^{-})\,\delta g(-\theta_1) + ([\![g''']\!]_{(+\theta_1)} - p_D^{+})\,\delta g(\theta_1)$$

$$+ \int_{\mathcal{D}_{\mathrm{c}}\cup\mathcal{D}_{\mathrm{f}}} \delta g\left\{(1 + \lambda)\,g + (2 + \lambda)\,g'' + g^{(4)} - p_{\mathrm{s}}(\theta)\right\}\mathrm{d}s. \qquad (9.25)$$

---

[11] This Lagrange multiplier serves to enforce the conservation of the perimeters of circles whose centre is at the tip of the d-cone, as we noted above. This multiplier is therefore related to the so-called hoop strain (orthoradial component of the strain tensor in our cylindrical coordinates).

In this variation we set to zero the coefficients in front of the arbitrary perturbation $\delta g(\theta)$ to obtain the following equations. The coefficient of $\delta g'(\pm\theta_1)$ yields the boundary condition associated with the free contact,

$$[\![g'']\!]_{\pm\theta_1} = 0, \tag{9.26a}$$

which is very similar to equation (9.15) for the Elastica pushed on a plane—recall that we can identify $\delta\Theta$ in the Elastica with $\delta g'$ here, and the internal moment in the Elastica with g'' here. The coefficient of $\delta g(\pm\theta_1)$ in equation (9.25) yields the balance of normal forces at the lift-off points, which will be used to compute the amplitude $p_D^{\pm}$ of the point reaction forces of the support there:

$$p_D^{\pm} = [\![g''']\!]_{\pm\theta_1}. \tag{9.26b}$$

The integral term gives the equation for the equilibrium profile of the cone in either domain $\mathcal{D}_\mathrm{f}$ and $\mathcal{D}_\mathrm{c}$:

$$(1+\lambda)\,g + (2+\lambda)\,g'' + g^{(4)} = p_\mathrm{s}(\theta). \tag{9.26c}$$

We have written down the equations for the equilibrium, including all relevant conditions at the lift-off points, and we can proceed to solve them. The easiest part is for the contact region $\mathcal{D}_\mathrm{c}$, for which $g(\theta) = \epsilon$. This implies $g'' = 0$ and $g^{(4)} = 0$. Plugging back into equation (9.26c), we compute the smooth part of the contact pressure

$$p_\mathrm{s}(\theta) = (1+\lambda)\,\epsilon \quad \text{for } \theta \in \mathcal{D}_\mathrm{c}. \tag{9.27}$$

In the bump-like region $\mathcal{D}_f$, that is for $|\theta| < \theta_1$, we have to solve equation (9.26c) with $p_\mathrm{s} = 0$:

$$(1+\lambda)\,g + (2+\lambda)\,g'' + g^{(4)} = 0. \tag{9.28}$$

This equation comes with four boundary conditions expressing the fact that g and g' are smooth at the lift-off points:

$$g(\pm\theta_1) = \epsilon, \qquad g'(\pm\theta_1) = 0. \tag{9.29a}$$

Two other boundary conditions derive from equation (9.26a):

$$g''(\pm\theta_1) = 0, \tag{9.29b}$$

since $g'' = 0$ in the contact region.

The differential equation (9.28) in the free part is linear with constant coefficients. Its characteristic polynomial, $X^4 + (2+\lambda)\,X^2 + (1+\lambda)$, has roots $X = \pm i\,a$ and $X = \pm i$. Here, a is a shorthand for:

$$a = \sqrt{1+\lambda}. \tag{9.30}$$

We deduce that the general solution of equation (9.28) is a linear combination of $\cos(a\,\theta)$, $\sin(a\,\theta)$ and $\cos(\theta)$, $\sin(\theta)$. The solution of this fourth-order differential equation that satisfies the four boundary conditions (9.29a) can be found explicitly:

$$g(\theta) = \epsilon \left(\frac{\sin\theta_1}{H} \cos(a\,\theta) - \frac{a\,\sin(a\,\theta_1)}{H} \cos(\theta) \right), \tag{9.31}$$

where

$$H = \sin(\theta_1)\cos(a\,\theta_1) - a\,\sin(a\,\theta_1)\cos(\theta_1). \tag{9.32}$$

As usual in constrained minimization problems, the actual value of the Lagrange multiplier λ is fixed by requiring that the associated constraint, the extended Gauss–Bonnet condition $I = 0$, is satisfied:

$$\frac{2\pi - 2\theta_1}{2}\,(0 - \epsilon^2) - \frac{1}{2}\int_{-\theta_1}^{\theta_1}(g'^2 - g^2)\,\mathrm{d}\theta = 0.$$

This yields, after using the explicit form (9.31) of g in the free part:

$$\theta_1 H + a\sin(\theta_1)\sin(a\theta_1) - \pi H = \frac{1 - a^2}{2H}\sin^2(\theta_1)\left(\theta_1 + \frac{\sin(a\theta_1)\cos(a\theta_1)}{a}\right). \tag{9.33}$$

This yields a first equation for λ (or $a(\lambda)$) and θ_1. A second equation for these two unknowns is provided by equation (9.29b), which we have not yet used:

$$a\tan\theta_1 = \tan(a\,\theta_1). \tag{9.34}$$

Solving these two equations numerically for λ and $\theta_1{}^{12}$ yields the two remaining unknowns of the problem:

$$a = 3.8045, \qquad \theta_1 = 1.2129. \tag{9.35}$$

The angular size of the free region is therefore (E. Cerda and L. Mahadevan, 1998):

$$2\,\theta_1 \approx 139°.$$

Note that this angular size is independent of the penetration of the tip d into the cylinder, that is of the parameter ϵ, at least for small values[13] of ϵ. The corresponding numerical solution $g(\theta)$ is plotted in Fig. 9.6, together with a three-dimensional representation of the d-cone.

As in the Elastica problem of Section 9.2.2, the contact pressure $p(\theta)$ is singular at the edges of the contact region, $(\pm\theta_1)$. This singular contribution is given by equation (9.26b), which expresses the balance of normal forces at these edges. Using the above solution for $g(\theta)$, we find the amplitude of each of these point forces:

$$p_D^{\pm} = 38.698\,\epsilon.$$

This force intensity has no absolute meaning since it derives from a reduced energy $\mathcal{E}_b^{\dagger}$: for consistency, we should have written $\lambda^{\dagger}$ and $p^{\dagger}$ as the quantities that scale with the undetermined coefficient that has been factored out of the energy as it depends on the core size. However, this force can be compared with the contribution to the reaction force coming from the (smooth) part of the pressure, in the interior of the contact region:

$$\int_{\theta_1}^{2\pi-\theta_1} p_s(\theta)\,\mathrm{d}\theta = (2\pi - 2\theta_1)(1 + \lambda)\,\epsilon = 55.833\,\epsilon.$$

---

[12] The quantities a and H are auxiliary variables that are given explicitly in terms of λ and θ_1 by equations (9.30) and (9.32).

[13] For values of the rescaled penetration ϵ comparable to one, the size of the free region, $2\theta_1$ becomes a function of ϵ.

Finally, we shall briefly discuss the range of validity of this solution. A first requirement, already discussed above, is that the tip penetrates into the cylinder by a distance d much smaller than the cylinder radius r. Then, the typical angle ϵ remains small, see equation (9.19), and so does the slope of the plate, as assumed for the derivation of the F.–von K. equations.[14] For our d-cone solution to be valid, the penetration d must therefore not be too large, but it should not be too small either. If it is too small, the deformation of the sheet will be localized near the point of application of the force, and its shape will remain a surface of revolution.[15] The d-cone solution studied above can be seen as describing the strongly post-buckled regime of this initially axisymmetric deformation. We now estimate the critical stress corresponding to this instability, which will provides a lower bound on the penetration for our d-cone solution to be relevant. In the configuration of revolution, a typical deflection d corresponds to slopes of order d/r. These slopes vary themselves over the characteristic in-plane distance r, yielding curvatures of order d/r^2 far from the tip.[16] From equation (9.8) with $g \sim d/r$, in-plane strain will be of order $(d/r)^2$. Buckling results from a competition of bending and stretching effects, and by balancing the typical density of bending energy, which is Eh^3 times the square of the typical curvature, $(d/r^2)^2$, with the typical density of stretching energy, which is Eh times the squared in-plane strain $(d^2/r^2)^2$, we find that the penetration d is of order of the thickness h at the buckling threshold. The d-cone solution is valid well above this threshold only, when the stretching energy becomes negligible in front of the bending one:

$$d \gg h. \tag{9.36}$$

The thickness h of the plate is assumed to be much smaller than the radius r of the cylinder and this lower bound on the penetration d is well compatible with the other assumption, $d \ll r$.

9.2.4 Conical singularities: stress focusing

We have not yet considered what happens near the tip, except to notice that the geometry there cannot be that of a perfect cone (this would make the bending energy diverge). This is a typical situation where boundary layers are formed: far from the tip, the solution is the conical one just described. The minimization of the bending energy yields a well defined function $g(\theta)$, all that is needed to define the shape of the plate far from the tip. This does not explain how the surface looks like near the tip, since by plugging this conical solution into the total elastic energy, one gets a divergent result. This divergence is removed by considering bending effects, which are of the same order of magnitude as stretching in the vicinity of the tip. Once the divergence is removed, the true physical solution is simple to describe in words: far from the tip, in a sense to be given shortly, the surface is a developable cone with almost no stretching and its energy is dominated by bending

[14] This assumption can be dropped by using a formulation of the equations valid for finite displacements (E. Cerda and L. Mahadevan, 2005), something that does not change the results in any essential way.

[15] The corresponding shape can be computed by solving the F.–von K. equations, which in the presence of cylindrical symmetry become ordinary differential equations.

[16] Curvatures can be much higher at the tip, but the curvatures that are relevant for studying the buckling of the configuration of revolution are those far from the tip.

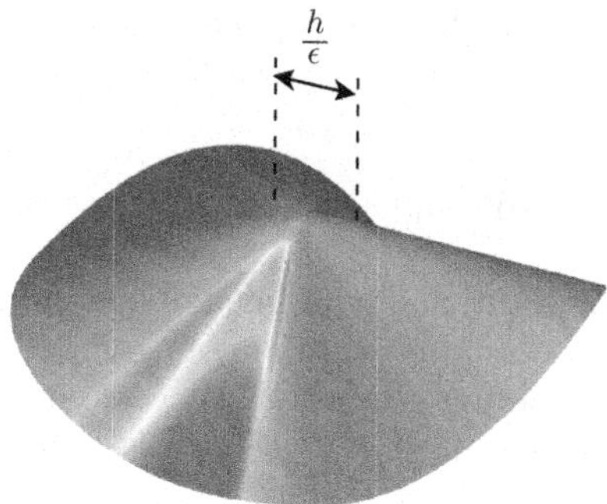

Fig. 9.7 In a small region with typical dimension h/ϵ near the tip, the deformation of the d-cone is regularized by bending effects. This assumes that this region, although much larger than h, remains much smaller than the outer radius R: $h \ll \epsilon R$.

terms. Close to the tip, stretching becomes important and regularizes the deformation, see Fig. 9.7. The maximum curvature at the tip will be estimated below based on these ideas. The inner solution, relevant near the tip, is not known to date. It could depend on the way the external forces are applied. We shall simply present scaling arguments relevant to this inner solution, and outline a consistent procedure for solving the outer problem by expansion.

As in the previous section, we use cylindrical coordinates and take the parametric equation of the cone in the far field as:

$$w(r, \theta) = r\, g(\theta). \tag{9.37}$$

In equation (9.37), as before, r and θ are the radius and the polar angle in the (x, y) plane, the origin being at the tip of the cone. The function $g(\theta)$ defines the conical profile of the surface in the far field, as computed previously. The singularity in the bending term in the F.–von K. equation is apparent, as the fourth derivative $h^3 \Delta^2 w$ in equation (9.2a) depends formally on r like $1/r^3$ when $w(r, \theta)$ is given by equation (9.4). Therefore, at small scale, the bending effects represented by this biharmonic operator can no longer be neglected and a theory valid near the tip has to be devised, where bending remains finite. This solution of the inner equation must also tend asymptotically (at large r) toward the cone solution (9.37). Because there is only one length scale in the problem, h, the perfectly conical solution becomes invalid at distances from the tip of order h.

This can be the basis of a formal solution of the F.–von K. equations, in which the far field solution derived above is only the beginning of an expansion in inverse powers of the distance.[17] The solutions in the far field are given by an expansion:

$$w = w_{(0)} + w_{(1)} + \cdots \qquad \text{and} \qquad \chi = \chi_{(0)} + \chi_{(1)} + \cdots \tag{9.38}$$

These functions are defined by a different expansion in the inner region but, according to the general principles of boundary layer theory given in Appendix B, their far-field and near-field expressions must coincide in the so-called intermediate (or matching) region.

---

[17] In fact, it is simpler but equivalent to see it as an expansion in powers of h^2 because of the structure of the F.–von K. equations.

At dominant order in the far field, we drop the bending term in the F.–von K. equations (9.2a) and (9.2b) because it involves a power of the small thickness h larger than all other terms:

$$[w_{(0)}, \chi_{(0)}] = 0 \quad \text{and} \quad \Delta^2 \chi_{(0)} + \frac{E}{2}[w_{(0)}, w_{(0)}] = 0. \tag{9.39}$$

This set of equations has many solutions. As argued above, solutions such that $\Delta \chi$ does not converge to zero at infinity have a large stretching energy, proportional to the area of the plate. Such solutions describe loading geometries that are not relevant here, such as when the plate is stretched along its boundaries. The cone-like solution is preferred as it involves no stretching energy:

$$w_{(0)} = r\, g(\theta), \qquad \chi_{(0)} = 0.$$

In the present expansion of solutions of the F.–von K. equations in inverse powers of the distance to the tip, the dominant contribution is precisely the cone-like solution studied in the previous section.

The next order solution, $(w_{(1)}, \chi_{(1)})$ involves a power of $1/r$ larger than the dominant term, for a reason explained below. It is given by a set of equations similar to equation (9.39), but derived from the F.–von K. equations by retaining all terms at next order:

$$\frac{h^3 E}{12(1 - \nu^2)}\, \Delta^2 w_{(0)} - h\,[w_{(0)}, \chi_{(1)}] = 0 \tag{9.40a}$$

and

$$\Delta^2 \chi_{(1)} + E\,[w_{(0)}, w_{(1)}] = 0. \tag{9.40b}$$

Recall that the expansion (9.38) was assumed to be with increasing powers of h^2: $w_{(0)}$ is of order h^0 while $\chi_{(1)}$ is of order h^2, etc. This is why, for instance, $h^3 \Delta^2 w_{(0)}$ is of the same order as $h[w_{(0)}, \chi_{(1)}]$, both terms being present in the equation above.

In polar coordinates, the various quantities present in the linearized F.–von K. equations take the form:

$$\Delta^2(r\, g(\theta)) = \frac{g(\theta) + 2g''(\theta) + g^{(4)}(\theta)}{r^3}, \quad [r\, g(\theta), \chi_{(1)}] = \frac{g(\theta) + g''(\theta)}{r}\,\frac{\partial^2 \chi_{(1)}}{\partial r^2}. \tag{9.41}$$

The second equality can be derived from the definition of the bracket (or Monge–Ampère) operator in Cartesian coordinates given in equation (6.87). With the help of these expressions, one writes explicitly the equation (9.40a) as:

$$\frac{h^3 E}{12(1 - \nu^2)}\,\frac{1}{r^3}(g(\theta) + 2g'' + g^{(4)}) = \frac{h}{r^2}\left[r^2(g + g'')\frac{\partial^2 \chi_{(1)}}{\partial r^2}\right]. \tag{9.42}$$

This equation for the Airy potential $\chi_{(1)}$ calls for a couple of remarks. First it is actually an *ordinary* differential equation for the function $\chi_{(1)}(r, \theta)$, where θ is considered as a parameter, since no derivative of this function with respect to θ is present. Therefore, the solution of equation (9.42) can be written straightforwardly as:

$$\chi_{(1)}(r, \theta) = -\frac{h^2 E}{12(1 - \nu^2)}\,\frac{g(\theta) + 2g''(\theta) + g^{(4)}(\theta)}{g(\theta) + g''(\theta)}\,\ln(r) + r\,C(\theta) + B(\theta). \tag{9.43}$$

In equation (9.43), the functions $C(\theta)$ and $B(\theta)$ are arbitrary constants of integration. Because the stress is a second spatial derivative of the Airy potential, the stress coming from the contribution $rC(\theta)$ decays as $1/r$ at large distances from the tip of the cone. Once integrated along a circle whose perimeter is $2\pi r$, this stress gives a constant contribution that can be interpreted as a force exerted on the cone at infinity. Here we assume that the cone is free of forces applied at infinity and this force has to vanish, $C(\theta) = 0$.

The function $B(\theta)$ can always be incorporated into the argument of the logarithm. Indeed, changing the scale for measuring the distances amounts to adding to $B(\theta)$ a contribution proportional to the coefficient of $\ln(r)$ on the right-hand side of equation (9.43). Therefore the constant of integration $B(\theta)$ has no meaning by itself. It should be considered together with the precise definition of the argument of the logarithm of r in this equation. The calculation of $B(\theta)$ requires the present expansion to be carried out in full detail up to the next order, something that is beyond the scope of the present chapter. Therefore, we shall continue as if $B(\theta)$ and $C(\theta)$ were zero: let us write $\chi_{(1)}(r,\theta) = E\,\hat{\chi}_{(1)}(\theta)\,\ln(r)$ the function obtained by taking $C = B = 0$ in equation (9.43). This gives:[18]

$$\hat{\chi}_{(1)}(\theta) = -\frac{h^2}{12(1-\nu^2)}\frac{g + 2g'' + g^{(4)}}{g + g''}.$$

Another difficulty with the solution (9.43) arises when one tries to apply it to the particular geometry of Section 9.2.3, an elastic sheet pushed into a cylinder. In this case, the function $g(\theta)$ is not continuous beyond its third derivative. Its fourth derivative $g^{(4)}(\theta)$ has a finite jump across the values $\pm\theta_1$ of the angle θ where the cone lifts off the cylinder. At higher orders in the expansion, things get worse because derivatives of $g(\theta)$ even higher than the fourth derivative are required. As a result, two boundary layers take place in the angular domains near these values $\pm\theta_1$ of the angle θ, which smooth out the jumps of the fourth derivative of $g(\theta)$. These boundary layers should introduce higher-order effects in the bending energy, namely effects that appear at higher order in the expansion of this bending energy with respect to the thickness of the plate. We shall assume that this can be done and continue as if the function $g(\theta)$ were infinitely smooth. Note that this lack of smoothness of $g^{(4)}(\theta)$ is specific to the geometry of a plate pushed into a cylinder, and does not occur, for instance, in the d-cones that make up a crumpled sheet of paper.

It remains to compute the first-order correction of the deflection. The equation (9.40b) for the quantity $w_{(1)}(r,\theta)$ is rewritten by expressing the bracket operator in cylindrical coordinates, as in equation (9.41):

$$[w_{(0)}, w_{(1)}] = \frac{g(\theta) + g''(\theta)}{r}\frac{\partial^2 w_{(1)}}{\partial r^2} \tag{9.44}$$

and by expressing the biharmonic operator in the left-hand side of this equation in cylindrical coordinates, as was done earlier in equation (9.41):

$$\Delta^2\left(\hat{\chi}_{(1)}(\theta)\,\ln r\right) = \frac{4(\ln r - 1)}{r^4}\,\hat{\chi}_{(1)}''(\theta) + \frac{\ln r}{r^4}\,\hat{\chi}_{(1)}''''(\theta).$$

This yields the equation for the deflection $w_{(1)}(r,\theta)$ in an explicit form:

$$\frac{g(\theta) + g''(\theta)}{r}\frac{\partial^2 w_{(1)}}{\partial r^2} = -(4\,\hat{\chi}_{(1)}''(\theta) + \hat{\chi}_{(1)}''''(\theta))\frac{\ln r}{r^4} + 4\,\hat{\chi}_{(1)}''(\theta)\frac{1}{r^4}. \tag{9.45}$$

---

[18] We assume that the given profile of the cone is such that the equation $g + g'' = 0$ has no root. If it had a root, the present expansion in powers of $1/r$ would have to be changed in the vicinity of these roots and a boundary layer analysis be carried out.

This equation has exactly the same structure as equation (9.42) for $\chi_{(1)}$: it is an *ordinary* differential equation with respect to r with a parameter θ. Integration with respect to r is straightforward, and yields:

$$w_{(1)}(r,\theta) = M(\theta) + r\,N(\theta) + \hat{w}_{(1)}(\theta)\,\frac{h}{r} + \bar{w}_{(1)}(\theta)\,\frac{h\ln r}{r}, \qquad (9.46)$$

where $M(\theta)$ and $N(\theta)$ are constants of integration, while $\hat{w}_{(1)}(\theta)$ and $\bar{w}_{(1)}(\theta)$ are shorthands for two long expressions depending on $\hat{\chi}_{(1)}$ and g that can be found explicitly by identification with equation (9.45). The first constant of integration $M(\theta)$ should be zero as it makes the outer solution discontinuous near the tip $r \to 0$, which makes the matching with an inner solution impossible in the vicinity of the tip. The second term describes a conical solution, $w_{(1)} \sim r\,N(\theta)$, and can therefore be incorporated into the dominant order $w_{(0)} = r\,g(\theta)$ by redefining $g(\theta)$. The two other terms are specific to the expansion to this order and yield the expected result, namely that, up to logarithms, the solution of this equation depends on r like $1/r$:

$$w_{(1)} \sim h^2 \frac{\hat{w}_{(1)}}{r},$$

where $\hat{w}_{(1)}$ is a known function of θ, although its full expression is not given here. Being the solution of an auxiliary numerical problem without any parameter, this function takes values of order one. This can be written as well as $w_{(1)} \sim r\,h^2/r^2$. The dominant contribution to the deflection being $w_{(0)} = rg(\theta)$, this shows that we have consistently derived the next term in an expansion for large r whose small parameter is the dimensionless ratio $(h/r)^2$.

We outlined the principle of an expansion that allows one to compute the d-cone solution in the far field, that is for $r \gg h$. This expansion breaks down when r becomes comparable to h. More accurately, the large distance expansion breaks down for r of the order of or less than h/ϵ. This can be seen for instance in equation (9.45) for $w_{(1)}$, which shows that this quantity is of order h divided by a function of θ proportional to g and its derivatives. Therefore $w_{(1)}$ is of order $h^2/\epsilon r$ and the true expansion parameter (at least for ϵ small) is $h/\epsilon r$. Moreover this makes the expansion compatible with the application of the F.–von K. equations all the way from r very large to the tip of the cone, because the radius of curvature of the plate is everywhere much larger than $1/h$. The angle of the cone is small, of order ϵ, and the curvature near the tip is of order ϵ/h, much less than $1/h$. The maximum stress is where the stretching is maximum, i.e. near the tip of the cone. The estimation of this maximum stress would require one to describe the vicinity of the tip, that is to derive an inner solution.

To obtain this inner solution and ultimately the order of magnitude of the maximum stress we notice first that the far field does not introduce any length scale in the problem: a cone is invariant under a dilation with its centre at the tip. Therefore it is natural to expect that, by taking h as unit length, the only length scale in the formulation of the problem,[19] one will obtain a set of parameterless equations valid both near and far from the tip. However this is not fully sufficient for our purpose, since we need to take into account the other small parameter, the typical slope ϵ of the cone. This leads one to define

---

[19] The situation here is different from that of plate buckling, where the boundary condition on the edges of the plate introduce a length scale, the width of the plate.

a *scaled* function $g(.)$, say $\bar{g}(.)$ of order 1 such that $g(\theta) = \epsilon\,\bar{g}(\theta)$, the number ϵ being small. Now we attempt to rewrite the full F.–von K. equations in a manner that is consistent with the asymptotic behaviour $w \approx \epsilon\,\bar{g}(\theta)\,r$ for large r but without any small or large parameter. We call this a purely numerical problem. This rewriting can be achieved by introducing:

$$\bar{r} = \frac{\epsilon\,r}{h}, \quad \bar{w}(\bar{r},\theta) = \frac{w}{h}, \quad \bar{\chi}(\bar{r},\theta) = \frac{\chi}{Eh^2}. \tag{9.47}$$

Note that this scaling is consistent with the large distance behaviour of the solution, since we have both $w(r,\theta) = r\,g(\theta)$ and $\bar{w}(\bar{r},\theta) = \bar{r}\,\bar{g}(\theta)$ for r and $\bar{r}$ large. In terms of the new quantities, the F.–von K. equations become:

$$\frac{1}{12(1-\nu^2)}\overline{\Delta}^2 \bar{w} - [\bar{w},\bar{\chi}] = 0, \tag{9.48a}$$

and

$$\overline{\Delta}^2\bar{\chi} + \frac{1}{2}[\bar{w},\bar{w}] = 0. \tag{9.48b}$$

In equations (9.48a) and (9.48b), the space derivatives are with respect to the overlined variable: $\overline{\Delta}$ is now

$$\frac{\partial^2}{\partial\bar{r}^2} + \frac{1}{\bar{r}}\frac{\partial}{\partial\bar{r}} + \frac{1}{\bar{r}^2}\frac{\partial^2}{\partial\theta^2}$$

and so on.

The boundary conditions are $\overline{\Delta}\,\bar{\chi}$ tends to zero as $\bar{r}$ tends to infinity (far field), although $\bar{w}$ tends to $\bar{r}\,\bar{g}(\theta)$. In the asymptotic conditions as well as in the equations (9.48a) and (9.48b) any small or large parameter has now disappeared. Therefore, one expects that, for a given $\bar{g}(.)$ and for finite values of $\bar{r}$, the solution will take numerical values of order 1.

These rescalings yield the order of magnitude of the various mechanical quantities. Let us estimate the physical stress σ_{yy} for instance. It is related to χ by $\sigma_{yy} = \partial^2\chi/\partial x^2$. Turning now to the dimensionless quantities just introduced, one obtains:

$$\sigma_{yy} = E\epsilon^2\frac{\partial^2\bar{\chi}}{\partial\bar{x}^2}.$$

The quantity $\partial^2\bar{\chi}/\partial\bar{x}^2$ being determined by a purely numerical problem, it is a number of order 1. This shows that the stress in the tip area is formally small, as it should be, namely of order $E\epsilon^2$ (recall that $g \sim \epsilon$ is the typical angular deviation of the cone generatrices far from the tip, a quantity that vanishes in the reference, planar configuration). Nevertheless this stress can be larger than the limit yield of the material. An angular deviation in the range 0.05 is quite small (0.5 mm over 1 cm) and it yields already a stress of order 2.5×10^{-3} times Young's modulus (usually very large). This can be a way of testing materials for very large stress while keeping large deformations confined in a very small region.

To summarize, the cone solution was build in two steps: far from the tip, the asymptotic conical shape has been found by minimizing the bending energy under the various constraints, including the generalized Gauss–Bonnet condition and the continuity of the curvature at the endpoints of the region of contact with the cylinder. Then, the scaling laws in the tip area have been found by balancing the various terms in the F.–von K. equations to make them consistent with the cone-like behaviour at large distances and with the balance of bending and stretching in the core.

As apparent from the above discussion, the inner solution for the cone is still an unsettled question. Even the scaling for the size of the core region is still being debated. There has been some experimental and numerical observations suggesting that the core size could scale differently from what was assumed above. A fractional power $h^{1/3} R^{2/3}$ of the thickness h and the radius R of the plate (E. Cerda, 1999; T. Liang and T. A. Witten, 2005) has been suggested for this core size, at least in some range of the parameters. However it has been pointed out that such a fractional scaling of the core size cannot be valid asymptotically for small h (T. A. Witten, 2007).

9.3 Ridge singularities

The rolling of a plane into a cylinder is a very familiar example of a smooth[20] isometric deformation: it leaves the distances along the surface unchanged whilst the tangent plane changes continuously with the position. Non-smooth deformations can leave the distances unchanged as well: this is illustrated by taking a sheet of paper and folding it along a straight line. The result is two flat half-planes merging along a straight fold.

The cover of this book shows that similar folds can appear spontaneously when a piece of paper is crumpled. At large scales such fold can be viewed as a discontinuity for the tangent plane that jumps abruptly from one orientation to another. However, no real discontinuity can occur along an elastic sheet of finite thickness as it would yield an infinite curvature and so an infinite bending energy. Therefore a real elastic plate has a finite radius of curvature along the fold, which depends on the thickness of the plate and, most probably, tend to zero as this thickness tends to zero.

Folds are in fact commonly observed in thin elastic plates subject to large deformations. In many cases, the classical Hookean elasticity does not apply: along folds, the material is often strained beyond the limit of reversible elasticity, making the standard Hookean approach invalid. Indeed, a piece of crumpled paper does not usually return to its planar shape once left free. Nevertheless the calculation of the fold structure in the elastic limit has enough interest in itself and may apply to carefully chosen situations.

As explained in reference (A. E. Lobkovsky, 1996), a fold similar to the ones on the cover of this book can be formed by applying an elastic plate on to a frame with two sharp angles. This geometry is shown in Fig. 9.8. The plate deformation can be analysed using the F.–von K. equations, which require that the normal to the surface changes only by a small angle from one side of the fold to the other. This assumption is not crucial, but saves a lot of effort by allowing one to use the F.–von K. equations instead of their covariant generalization.[21] For a reason that will soon become clear, we shall assume that the fold has a finite length, denoted L. Since the computation is far from straightforward, we shall first estimate the orders of magnitude of the solution and use them in a second step to devise a scheme of approximation for solving the F.–von K. equations in this geometry.

The coming developments have an interest in themselves but they are also relevant for the building of solutions of the F.–von K. equations in the large load limit, the purpose of

---

[20] The displacement associated with the rolling of a plane on to a cylinder is typically at least $\mathcal{C}^2$: it is continuous with continuous first and second derivatives.

[21] In the original paper by Lobkovsky and collaborators (A. E. Lobkovsky, 1996), this assumption of small rotations is not used: a covariant formulation of the equations for plates is used.

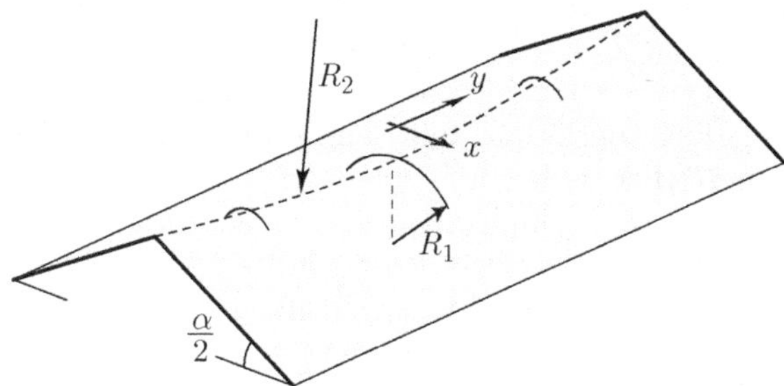

Fig. 9.8 Fold created by applying an elastic plate on to a rigid frame, shown with bold lines, having sharp angles (A. E. Lobkovsky, 1996). The dashed line is the crest of the ridge, with radius of curvature R_2 along the ridge and R_1 in the perpendicular direction.

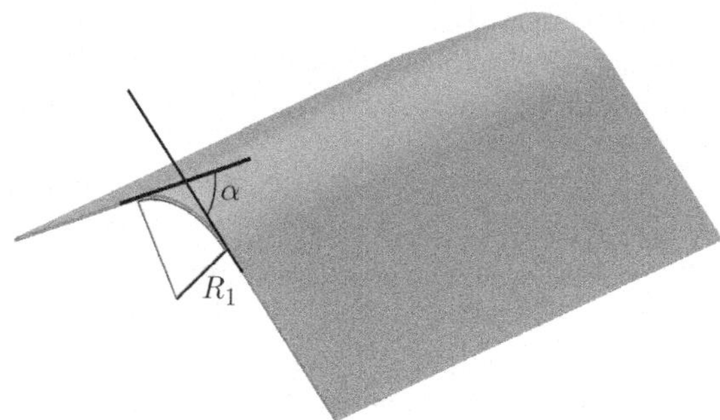

Fig. 9.9 Failure of the fold construction based on a cylindrical profile: for a fixed fold angle α, the bending energy diverges as the fold width goes to zero. This points to the fact that the actual fold structure is more complex than a cylinder, and that the fold properties vary along its axis.

Chapter 10. In this limit, a compressed plate tends to a polyhedral surface with straight edges that are folds from the point of view of elasticity theory. The minimization of the energy of this polyhedral structure requires one to know the energy of a single fold as a function of its length and angle, something that is precisely the aim of the present chapter.

9.3.1   Order of magnitude estimates

Let us first try to construct the fold according to what will appear to be a naive—although instructive—approach. We consider a cylindrical surface[22] made up of two half-planes making a small angle α, connected by a cylindrical region (fold), which is another cylinder with typical radius of curvature R_1, see Fig. 9.9. This cylindrical shape is developable, and so cancels any stretching energy. The only remaining contribution to the elastic energy comes from bending and can be estimated in orders of magnitude using the F.–von K. theory:

---

[22] Note that cylinders are defined as 3D surfaces spanned by the set of parallel lines, called the generatrices, resting on a given 3D curve. Their cross-section is not necessarily circular.

the bending energy per unit length along the fold is found by integration *across* the fold of the bending energy density $Eh^3/12(1 - \nu^2)(\Delta w)^2$. The order of magnitude of the mean curvature, Δw, is $\Delta w \sim 1/R_1$. The width of the fold (perpendicular to its axis) is of order $R_1 \alpha$. Therefore the bending energy per unit length of the fold is $Eh^3(1/R_1)^2 \alpha R_1$. Since this energy depends on R_1 like $1/R_1$, it is lower when R_1 is larger, i.e. when the fold is wider. It becomes clear that our cylindrical profile is far from optimal when a sharp angle is imposed on the edges of the plate ($R_1 \to 0$), as in Fig. 9.8.

To minimize its bending energy, the fold would like to be as wide as possible, although its width is constrained to be zero at its endpoints by the boundary conditions. Moreover the profile should locally be close to a developable one to lower the stretching energy: the fold width is therefore expected to vary slowly along the fold. Given all these constraints, we conclude that the actual fold is not cylindrical: its width goes to zero near its endpoints but increases in the middle of the fold. This is clearly better for the bending energy. The price to pay in order to decrease the bending energy involves some stretching energy, as the resulting profile is not developable—this price is cheap if the fold width varies slowly along the crest. As we show below, the consequence is that the maximum fold width depends on the length L of the fold, something that is rather counter-intuitive. It is essential to consider a finite fold length L in order to be able to build consistently the fold solution.

Let us now consider the correct geometry, sketched in Fig. 9.8, and derive the scaling relations for the ridge width, curvature and energy as a function of the ridge length and of the plate thickness. The bending energy is the lineic density found above, multiplied by the length of the fold $\mathcal{E}_\mathrm{b} \sim Eh^3 \alpha L/R_1$. Here, R_1 denotes as before the typical radius of curvature perpendicular to the fold, which remains to be determined. The stretching energy $\mathcal{E}_\mathrm{s}$ is obtained by integration of the energy density $h/E\,(\Delta\chi)^2$ in the fold region with an area of order $(\alpha R_1 \times L)$. The Airy potential χ is a solution of the second F.–von K. equation:

$$\Delta^2\chi + \frac{E}{2}\frac{1}{R_1 R_2} = 0.$$

Here, R_2 is the radius of curvature in the direction of the fold, see Fig. 9.8. This radius is found later from a balance of the bending and stretching energies. Since any quantity changes much faster across the fold than parallel to it (in other terms we assume $\alpha R_1 \ll L$ and $R_1 \ll R_2$, an assumption to be checked at the end), the harmonic operator can be estimated as $\Delta \sim (1/\alpha R_1)^2$. Combining all those estimates, one gets the stretching energy:

$$\mathcal{E}_\mathrm{s} \sim E\,h\,L\,\frac{R_1^3 \alpha^5}{R_2^2}.$$

The two radii of curvature R_1 and R_2 are related by a geometrical relation. This relation is found by expressing in two different manners the vertical deflection δ of the midpoint of the fold, called the sag: δ expresses by how much this midpoint is lowered upon folding (the crest becomes curved in deformed configuration). The typical slope of the ridge (dashed line in Fig. 9.8) is δ/L. Since this slope changes sign over the width length L, this yields the curvature $1/R_2 \sim \delta/L^2$, hence $\delta \sim L^2/R_2$. A similar calculation can be done in the direction perpendicular to the fold. Then, one has to replace R_2 by R_1 and the L by the fold width αR_1. This yields $\delta \sim (\alpha R_1)^2/R_1 \sim \alpha^2 R_1$. By equating the two expressions for δ, we find the geometrical relation:

$$R_2 \sim \frac{L^2}{\alpha^2 \, R_1}.$$

Written in the form $R_2 = (L/(\alpha \, R_1)) \, (1/\alpha) \, L$, where the two factors in front are large numbers, the radius R_2 appears to be much larger than L, as expected.

The estimate for R_2 allows one to rewrite the stretching energy as:

$$\mathcal{E}_\mathrm{s} \sim \frac{E \, h}{L^3} \, R_1^5 \, \alpha^9.$$

Comparing now $\mathcal{E}_\mathrm{s}$ and $\mathcal{E}_\mathrm{b}$, one can see that by decreasing R_1, $\mathcal{E}_\mathrm{s}$ will decrease (for $R_1 \to 0$, the profile is developable as $R_2 \to \infty$ and the stretching energy vanishes) whilst, as already noticed, $\mathcal{E}_\mathrm{b}$ increases. Increasing R_1 changes the balance the other way. The optimal value for the sum of the two energies, at least as far as the orders of magnitude are concerned, is when they are equal, which gives:

$$R_1 \sim h^{1/3} L^{2/3} \alpha^{-4/3}. \tag{9.49}$$

Note that the angle α cannot be too small, as the calculation assumed that the width of the fold, i.e. αR_1 is much smaller than L, which requires $\alpha \gg (h/L)$.

The final (and useful) result is that the energy of the fold is of order

$$\mathcal{E}_\mathrm{fold} \sim E \, h^{8/3} \, \alpha^{7/3} \, L^{1/3}. \tag{9.50}$$

This scaling has been originally proposed by Lobkovsky (A. Lobkovsky *et al.*, 1995; A. E. Lobkovsky, 1996); it was proved rigorously by Conti and Maggi (S. Conti and F. Maggi, 2008). Note the fancy exponents in this order of magnitude estimate, which are rarely seen in plate mechanics. They arise from the separation of scale in the bending and stretching energies ($E \, h^3$ versus $E \, h$ prefactors) and from geometrical relations.

9.3.2   Rescaled parabolic equations

In this section, we shall reduce the F.–von K. equations to a slightly simpler form relevant for the fold geometry: a boundary layer analysis of these equations is used to find their solution relevant for this geometry. Later on, we shall further reduce these equations and describe in some details what happens near the endpoints of the fold.

The F.–von K. equations read in general, see equation (6.88):

$$\frac{h^3 E}{12(1 - \nu^2)} \Delta^2 w - h[w, \chi] = 0 \tag{9.51a}$$

$$\Delta^2 \chi + \frac{E}{2}[w, w] = 0, \tag{9.51b}$$

where all symbols χ, $h, \ldots$ have the usual meaning. The reduction we shall now make belongs to the general category of the parabolic approximations (the so-called lubrication approximation in fluid mechanics, a word that sounds strange in the present context) valid when every quantity changes far more rapidly as a function of one coordinate than of the other. In the present case, we shall denote as y the coordinate along the fold, and x the one perpendicular to it. Since we expect a much faster variation in the x direction than in the y direction, we may approximate the biharmonic operator by its term with the highest derivative with respect to the fast variable x: $\Delta^2 \approx (\partial^2/\partial x^2)^2$. This yields the 'parabolic' version of the F.–von K. equations:[23]

$$\frac{h^3 E}{12(1 - \nu^2)} \frac{\partial^4 w}{\partial x^4} - h[w, \chi] = 0 \tag{9.52a}$$

$$\frac{\partial^4 \chi}{\partial x^4} + \frac{E}{2}[w, w] = 0. \tag{9.52b}$$

Let us consider the typical scales of the various quantities entering in these equations. The length of the fold L has been introduced earlier. It yields the typical scale $y^* = L$ for the coordinate y along the fold. Let x^* be the typical width of the fold, which provides the natural scale in the transverse direction x. In Section 9.3.1, we showed that this width is related to the transverse curvature of the fold by $x^* \sim \alpha R_1$. The typical scale w^* for the deflection w is obtained by noticing that, far from the fold, the Cartesian equation of the plate reads $w = \pm x\,\alpha/2$: in the matching region $(x \sim x^*)$, the order of magnitude w^* of the deflection is given by $w^* \sim \alpha\,x^*$. From equation (9.51a), χ scales like $\chi^* = Eh^2 L^2/x^{*2}$. Putting this into the second parabolic F.–von K. equation, one obtains that, to have there all terms of the same order of magnitude, x^* should scale as $x^* \sim \alpha^{-1/3}\,h^{1/3}\,L^{2/3}$, the same result as obtained previously in equation (9.49) since $x^* \sim R_1\alpha$.

Scaling transformations (sometimes called 'stretching transformations', a wording we shall avoid because of possible confusion with the mechanical phenomenon of stretching) based on the above estimates allow one to write the parabolic F.–von K. equations in a completely parameterless form, thereby leading to a purely *numerical* problem:

$$\frac{\partial^4 \overline{w}}{\partial \overline{x}^4} - [\overline{w}, \overline{\chi}] = 0 \tag{9.53a}$$

$$\frac{\partial^4 \overline{\chi}}{\partial \overline{x}^4} + \frac{1}{2}[\overline{w}, \overline{w}] = 0. \tag{9.53b}$$

In these equations, we have defined the rescaled coordinates and functions by

$$\overline{y} = \frac{y}{y^*}, \qquad \overline{x} = \frac{x}{x^*}, \qquad \overline{w} = \frac{w}{w^*}, \qquad \overline{\chi} = \frac{\chi}{\chi^*},$$

where the typical scales are :

$$y^* = L, \qquad x^* = \frac{h^{1/3}\,L^{2/3}}{[12\,(1 - \nu^2)]^{1/6}\,\alpha^{1/3}}, \qquad w^* = \alpha\,x^*, \qquad \chi^* = \frac{E\,w^{*2}\,x^{*2}}{L^2}. \tag{9.54}$$

These scales are those derived by the scaling analysis just above, with additional conventional numerical factors chosen so as to cancel those present in the F.–von K. equations. The derivatives that enter into $[\overline{w}, \overline{\chi}]$ and $[\overline{w}, \overline{w}]$ are implicitly taken with respect to the *rescaled* coordinates $\overline{x}$ and $\overline{y}$.

That the equations (9.53) define a purely numerical problem is obvious because, first, they do not include any parameter but only numbers of order 1. Second, they are subject to the boundary conditions that $\overline{\chi}$ tends to a constant as $\overline{x}$ tends to infinity (this constant may

---

[23] Notice that this parabolic approximation cannot be used to simplify the other terms $[w, \chi]$ and $[w, w]$ in the F.–von K. equations because each one of these quantities involves the same order of derivation with respect to each variable: any term in these bracket operators involves two derivations with respect to y and two derivations with respect to x.

be taken as zero, because $\overline{\chi}$ only appears through its derivatives in the parabolic F.–von K. equations) although $\overline{w}$ tends to $\pm\overline{x}$ as $\overline{x}$ tends to $\pm\infty$.

After this reduction of the F.–von K. equations to their dimensionless parabolic form (9.53), we still have a set of *partial* differential equations. In contrast, the usual outcome of a boundary layer analysis is to get rid of at least one variable. With a special choice of conditions, these equations can be reduced further to a set of coupled *ordinary* differential equations, as we show in the coming section (the merging of this parabolic form of the F.–von K. equations with the end points is a rather difficult issue that is considered later).

9.3.3 Reduction to a set of coupled ordinary differential equations

Lobkovsky (A. E. Lobkovsky, 1996) has found a way of reducing the parabolic F.–von K. equations to a set of simpler non-linear ordinary differential equations (ODEs). We shall follow this derivation in its main lines in the rest of this section. The key step is a change of variable assuming that $w(x, y)$ and $\chi(x, y)$ are functions of a scaled value of x, denoted

$$\xi = \frac{x}{q(y)}; \tag{9.55}$$

the length scale $q(y)$ is a function of y and it varies along the fold—this $q(y)$ can be thought of as the local width of the ridge, which will be determined later. In equations, the assumption reads:[24]

$$w(x, y) = -q(y)\, p_1(\xi), \qquad \chi(x, y) = q(y)\, p_2(\xi). \tag{9.56}$$

In this equation and until the end of this section, we drop the overlines for the sake of readability, although rescaled quantities are implied.

Note that the same scaling function $q(y)$ is used for both the Airy potential and the deviation w. The miraculous effect of this change of variables is that it transforms the Monge–Ampère operator in a very simple way. As shown by a direct calculation,

$$[q(y)\, p_i(\xi), q(y)\, p_j(\xi)] = \frac{\ddot{q}}{q}\left\{ (p_i - \xi\, p_i')\, p_j'' + (p_j - \xi\, p_j')\, p_i'' \right\}. \tag{9.57}$$

In the right-hand side of this expression, the function q and its derivatives $\dot{q}$ and $\ddot{q}$ are implicitly functions of the variable y, although the functions p_i and their derivatives p_i' and p_i'' are implicitly functions of the variable ξ. To make obvious the difference between derivatives with respect to the longitudinal variable y and to the rescaled transverse variable ξ, we use dots for the first ones, and primes for the second.

Applying equation (9.57) with $i = j = 1$, we find that the Gauss curvature transforms in a very compact manner under this change of variables:

$$K = \frac{1}{2}[w, w] = \frac{\ddot{q}}{q}\,(p_1 - \xi\, p_1')\, p_1''. \tag{9.58}$$

---

[24] By convention, we have put a additional minus sign in the definition (9.56), absent in reference (A. E. Lobkovsky, 1996). This is not essential, due to the symmetry $p_1 \rightarrow -p_1$ in the equations. This is simply to have the boundary condition $p_1' \rightarrow 1/2$ for $\xi \rightarrow +\infty$ later on, while the slope $w_{,x}$ is asymptotically negative on the side $x > 0$ with the conventions in Fig. 9.8. Note too that the subscripts 1 and 2 in $p_1(\xi)$ and $p_2(\xi)$ have no relationship with the labelling of coordinates.

Furthermore the parabolic F.–von K. equations (9.53) take the following form:

$$-\frac{p_1^{(4)}}{q^3} + \frac{\ddot{q}}{q}\left((p_1 - \xi\,p_1')\,p_2'' + (p_2 - \xi\,p_2')\,p_1''\right) = 0,$$

$$\frac{p_2^{(4)}}{q^3} + \frac{\ddot{q}}{q}\,(p_1 - \xi\,p_1')\,p_1'' = 0,$$

where $p_i^{(4)}$ stands for the fourth derivative of $p_i(\xi)$ with respect to its argument ξ. Multiplying both sides of the equations by $q^3(y)$, we obtain

$$-p_1^{(4)} + \{\ddot{q}\,q^2\}\left((p_1 - \xi\,p_1')\,p_2'' + (p_2 - \xi\,p_2')\,p_1''\right) = 0, \tag{9.59a}$$

$$p_2^{(4)} + \{\ddot{q}\,q^2\}\,(p_1 - \xi\,p_1')\,p_1'' = 0. \tag{9.59b}$$

Even though the arguments y for p and ξ for p_i are omitted for the sake of readability, the above equations define a set of *partial* differential equations in the plane (ξ, y). However, the dependence on y is limited to the two identical terms $\{\ddot{q}\,q^2\}$ in curly braces. If we assume that these terms are independent of y, something that we can try to make happen since we have the freedom to define the function $q(y)$ as we like, the above set of equations become a set of *ordinary* differential equations for the variable ξ only. Calling $(-A)$ the constant value of $\{\ddot{q}\,q^2\}$, we end up with the following equation for the local fold width $q(y)$:

$$\frac{\mathrm{d}^2 q}{\mathrm{d}y^2} + A\,\frac{1}{q^2(y)} = 0. \tag{9.60}$$

The constant A can be set to one, as we show below. When the origin $y = 0$ is chosen to be the midpoint of the fold, the solution $q(y)$ has to be even by symmetry. At the endpoints of the fold, $y = \pm 1/2$ in rescaled variables, a sharp angle is imposed by the boundary conditions and the width of the fold has to vanishes: the condition $q(\pm 1/2) = 0$ holds. These conditions, together with equation (9.60), completely specify the amplitude function $q(y)$, as we now show.

9.3.4   Local fold width

Taking advantage of the separation of variables introduced in the previous section, we first solve for the fold width $q(y)$, that is we consider the longitudinal variable y first. The solution involving the transverse variable ξ is more difficult and will be considered next.

The normalization of the amplitude function $q(y)$ is arbitrary. Indeed, all the observable quantities like the deflection w or the stress given by the derivatives of χ are unchanged in equation (9.56) when $q(y)$ is divided by an arbitrary numerical factor, while both p_1 and p_2 are multiplied by the same factor. As a result, we can choose $A = 1$ in equation (9.60).

The equation (9.60) for $q(y)$ with $A = 1$ can be solved by elementary methods. In a first step, it can be transformed into a first-order equation upon multiplication of equation (9.60) by $\dot{q}$ and integration:

$$\frac{1}{2}\dot{q}^2 = \frac{1}{q} - \frac{1}{q_0}. \tag{9.61}$$

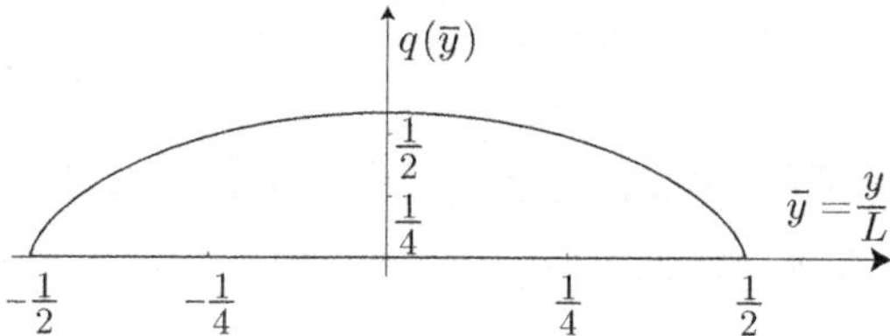

Fig. 9.10 Transverse rescaled width of the fold, $q(\bar{y})$, as a function of rescaled longitudinal coordinate $\bar{y} = y/L$. Note that the width $q(\bar{y})$ goes to zero at both endpoints of the fold $y = \pm L/2$, where the curvature is indeed infinite due to the boundary conditions imposed by the angular support.

Here, we have introduced a constant of integration q_0 that remains to be fixed. This q_0 that can be interpreted as the maximum fold width q (in rescaled coordinates) at the midpoint of the fold where $y = 0$ and $\dot{q} = 0$ by symmetry.

After extracting the square root for $\dot{q}$ in equation (9.61) the variables q and y separate. One more integration yields an implicit relation between $q(y)$ and y:

$$\int_{q_0}^{q(y)} \frac{\mathrm{d}q'}{\sqrt{1/q' - 1/q_0}} = \pm\sqrt{2}\, y\,.$$

The integral in the left-hand side can be computed by changing the variable of integration and introducing ϕ such that $|q| = q_0 \cos^2 \phi$. This yields:

$$-\left(\cos^{-1}\left(\sqrt{\frac{q}{q_0}} \right) + \sqrt{\frac{q}{q_0}\left(1 - \frac{q}{q_0} \right)} \right) = \pm\sqrt{\frac{2}{q_0^3}}\, y\,,$$

where $\cos^{-1}$ denotes the arc cosine function. The boundary condition $q(\pm 1/2) = 0$ at the endpoints of the fold has not yet been used. When inserted into the equation above, it gives the value of q_0:

$$q_0 = \left(\frac{2}{\pi^2} \right)^{1/3}, \tag{9.62}$$

which itself allows one to rewrite the implicit equation for the solution $q(y)$ in its final form:

$$\cos^{-1}\left(\sqrt{\frac{q(y)}{q_0}} \right) + \sqrt{\frac{q(y)}{q_0}\left(1 - \frac{q(y)}{q_0} \right)} = \pi\,|y|\,. \tag{9.63}$$

The local fold width $q(y)$ has been determined completely. This function is plotted in Fig. 9.10—note that the overline in $\bar{y}$, which marks rescaled quantities, is omitted in the main text but has been reintroduced in the plot for reference.

9.3.5　Equations for transverse fold profile

Our particular choice of the function $q(y)$ amounts to taking $A = 1$ in equation (9.60). As a result, the factor $\{\ddot{q}\, q^2\}$ is equal to (-1) in equations (9.59), and the latter can be rewritten as a set of fourth-order ordinary differential equations[25] for the functions $p_1(\xi)$ and $p_2(\xi)$:

$$p_1'''' + p_2''\,(p_1 - \xi\,p_1') + p_1''\,(p_2 - \xi\,p_2') = 0, \tag{9.64a}$$

$$p_2'''' - p_1''\,(p_1 - \xi\,p_1') = 0. \tag{9.64b}$$

In order to solve these equations and find the fold profile, one must specify the boundary conditions for $\xi \to \pm\infty$, which is what we do now.

Owing to the invariance of the problem upon reflection with respect to the fold axis, $x \mapsto -x$, that is $\xi \mapsto -\xi$, one only needs to solve the equations over the interval $0 \le \xi < +\infty$. By symmetry, the derivatives p_1', p_1''', p_2' and p_2''' whose order is odd must cancel at the origin, $\xi = 0$. This yields four boundary conditions at the centre line of the fold ($\xi = 0$).

Other boundary conditions must be imposed for $\xi \to +\infty$ as the generic behaviour of a solution (p_1, p_2) of the system (9.64) diverges at infinity, leading to infinite stress far away from the ridge. Physically the stress is expected to decay as the plate recovers a planar configuration, and this must be imposed by asymptotic conditions on the equations. Note that not all stress components have the same order of magnitude: lengths perpendicular to the ridge were rescaled using a typical length x^* much smaller than that in the longitudinal direction, $y^* = L$. Therefore $\sigma_{yy} = \partial^2\chi/\partial x^2$ is formally much larger than any other component of the stress. The set of differential equations (9.64) must be understood as the leading order in a systematic expansion with respect to the small parameter h/L. At dominant order, only the formally dominant component of the stress, $\sigma_{yy} = \chi_{,xx} = p_2''(\xi)/q(y)$, has to tend to zero far away from the ridge:

$$p_2''(\xi) \to 0 \qquad \text{for } \xi \to \infty.$$

This behaviour has to be compared with the generic asymptotic behaviour of p_2, which can easily be guessed. Far away from the ridge, the plate recovers a planar shape and so the transverse curvature $w_{,xx} = -p_1''(\xi)/q(y)$ must go to zero (recall that commas in indices denote partial derivatives). As a result, the generic asymptotic behaviour of solutions of equation (9.64b) is given by $p_2'''' = 0$: it is a polynomial of degree three or less. The condition $p_2'' \to 0$ written above imposes that the coefficients of the two higher-order terms in this polynomial, ξ^2 and ξ^3, cancel. This counts as two boundary conditions for $\xi \to +\infty$, which we write as $p_2''' \to 0$ and $p_2'' \to 0$.

Far away from the ridge, the plate has a slope $-\alpha/2$ on the side $x \to +\infty$ and $+\alpha/2$ on the other side, see Fig. 9.8. This leads to the asymptotic condition $w_{,x} \to -\alpha/2$ for $x \to \infty$. The other condition, $w_{,x} \to +\alpha/2$ for $x \to -\infty$ follows by symmetry: $w(x, y)$ is an even function of x and so $w_{,x}(x, y)$ is odd. Equations (9.54) and (9.56) can be used to write this condition in terms of the unknown p_1:

$$p_1' \to +\frac{1}{2} \qquad \text{for } \xi \to \infty.$$

---

[25] The differential system in equation (9.64) is of total order eight but it can be transformed into a fourth-order system by noticing that there are two constants of integration and by using the Legendre transformations of $p_1(\xi)$ and $p_2(\xi)$ as unknown functions instead of the functions $p_1(\xi)$ and $p_2(\xi)$ themselves, see reference (A. E. Lobkovsky, 1996). However, the boundary conditions, when expressed in terms of the Legendre transformations, become significantly more difficult to write. Therefore, we shall not use this transformation.

So far, we have obtained seven boundary or asymptotic conditions while the differential system (9.64) is of order eight.

The last condition reflects the geometry of the loading. We have already shown that the asymptotic behaviour of $p_2(\xi)$ is given by a polynomial of degree three at most, but its cubic and quadratic coefficients vanish with the two asymptotic conditions introduced above: asymptotically, $p_2(\xi)$ is an affine function, and so p_2' has a finite limit for $\xi \to \infty$. This limit, value which we write $p_2'(+\infty)$, is in fact related to the axial force of compression F_y applied on the fold. This force can be computed by measuring the integrated internal stress transmitted across an imaginary cut along the x axis:

$$F_y = \int_{-\infty}^{+\infty} \left(-\sigma_{yy} \right) \mathrm{d}x = \int_{-\infty}^{+\infty} \left(-\chi_{,xx} \right) \mathrm{d}x = -\int_{-\infty}^{+\infty} \frac{p_2''(\xi)}{q(y)} \, \mathrm{d}x$$

$$= -\int_{-\infty}^{+\infty} p_2''(\xi) \, \mathrm{d}\xi = -\left(p_2'(+\infty) - p_2'(-\infty) \right) = -2\, p_2'(+\infty). \tag{9.65}$$

In this formula, we have continued to omit the bar for rescaled quantities: F_y as a shorthand for the axial force rescaled by the typical force $F^* = h\, \chi^*/x^*$, which is built from equation (9.54). The integrals in the definition of F_y are taken along any line drawn on the plate perpendicular to the fold (constant y), this force being independent of y as expected.

In a loading geometry such that the fold is subject to longitudinal compression $F_y > 0$, as in reference (A. E. Lobkovsky and T. A. Witten, 1997), the boundary condition $p_2' \to -F_y/2$ must be taken into account. Here, we consider the simplest loading geometry, whereby the two triangles that support the edges of the plate (called the triangular supports thereafter) can move freely away from each other. As a result, the axial force F_y is zero, and the boundary condition $p_2'(\xi) \to 0$ for $\xi \to \infty$ holds. The function $p_2(\xi)$ therefore tends to a constant at infinity.

To sum up, the boundary conditions for the differential system (9.64) are, near the origin:

$$p_1'(0) = 0 \quad p_1'''(0) = 0 \quad p_2'(0) = 0 \quad p_2'''(0) = 0, \tag{9.66a}$$

while, far away from ridge, for $\xi \to +\infty$:

$$p_2''' \to 0 \quad p_2'' \to 0 \quad p_2' \to 0 \quad p_1' \to \frac{1}{2}. \tag{9.66b}$$

The system (9.64) being of order eight, its integration subject to these eight boundary conditions is a well-posed problem, whose numerical solution is studied in the next section.

9.3.6 Numerical solution for the transverse fold profile

The set of ODEs (9.64) cannot simply be integrated by progressing from an initial value at a starting point (the so-called Cauchy problem) as conditions are imposed at the two 'endpoints' of the interval, $\xi = 0$ and $\xi \to +\infty$. This defines a so-called two-point boundary value problem. Shooting from either endpoint is a possibility: this involves guessing the missing initial conditions, integrating, and finally adjusting the guessed initial conditions so as to satisfy the conditions on the other end. Such a shooting method is used, for instance, in Chapter 14 of this book to obtain a numerical solution for the circular fold on an elastic ball. However, as happens in Chapter 13 for the analysis of the elastic torus, the shooting method

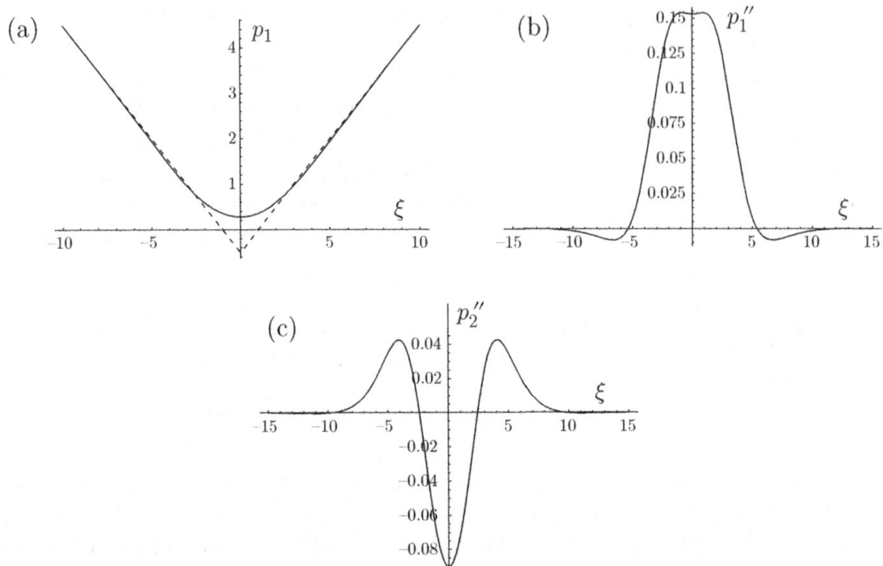

Fig. 9.11 Numerical solution of the fold equations (9.64) with boundary conditions (9.66). The function $p_1(\xi)$ is connected to the fold profile by equation (9.56), while p_1'' yields the transverse curvature $w_{,xx}$ up to a rescaling, $p_1'' = -q(y)\,w_{,xx}/w^*/x^{*2}$, and p_2'' yields the dominant (longitudinal) stress component, σ_{yy}, again up to a rescaling $p_2'' = q(y)\,\sigma_{yy}/\chi^*/x^{*2}$. Note the affine asymptote to $p_1(\xi)$ with slope $\pm 1/2$, as imposed by the condition (9.66b).

is doomed to fail in practice whenever the generic solution of the equation diverges at infinity. This is what happens here and we need to resort to a more advanced integration scheme, called the relaxation method (?)Ch. 17]PressTeukolskyVetterling-Numerical-recipes-in-C-2002. Its principle is explained in Section 13.6.2.

The problem is posed over an infinite domain, $\xi \in [0, +\infty[$. It is possible to make a change of variable mapping this infinite domain to the segment $[0, 1[$, but it is much simpler, although slightly less accurate, to solve the differential system over a finite arbitrary interval $[0, \xi_m]$, applying the asymptotic boundary conditions at the large but finite limit value ξ_m, and check the numerical convergence of the solution when ξ_m is made large.

This relaxation method yields a solution of differential equations (9.64) with the conditions (9.66). This solution is shown in Fig. 9.11. The fold profile, reconstructed using the equation $w(x, y) = -q(y)\,p_1(\xi)$, is shown in the form of slices in Fig. 9.12, while a three-dimensional visualization is proposed in Fig. 9.13. The computed values of the unknown functions and their even derivatives at the origin read:

$$p_1(0) = 0.278, \qquad p_1''(0) = 0.153, \qquad p_2(0) = 0.131, \qquad p_2''(0) = -0.090, \tag{9.67}$$

the odd derivatives being zero by symmetry. Although the boundary layer equations (9.64) for a fold have already been proposed in reference (A. E. Lobkovsky, 1996), the associated boundary and asymptotic conditions have not been discussed and this seems to be the first published solution of these equations.

The sag in the middle of the fold is clearly visible in Fig. 9.13. It can indeed be observed experimentally by looking at a sheet pinched in two places (the endpoints of the fold).

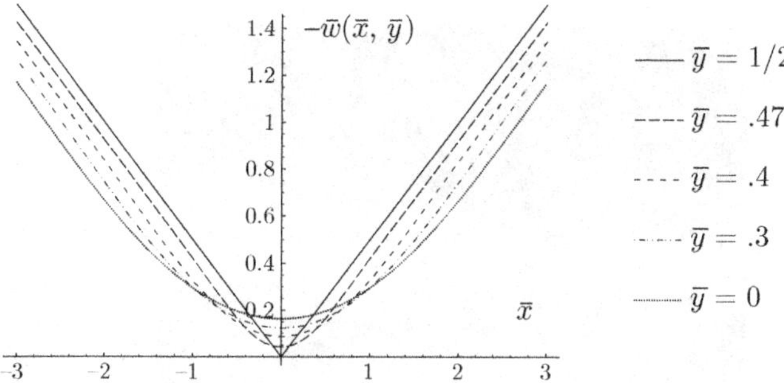

Fig. 9.12 Transverse slices of the fold profile $w(x,y)$ determined from the numerical solution of the fold equations (9.64). Slices are shown for various values of the coordinate $\overline{y} = y/L$ along the fold. Rescaled quantities are shown: $\overline{x} = x/x^*$ and $\overline{w} = ww^*$, see equation (9.54).

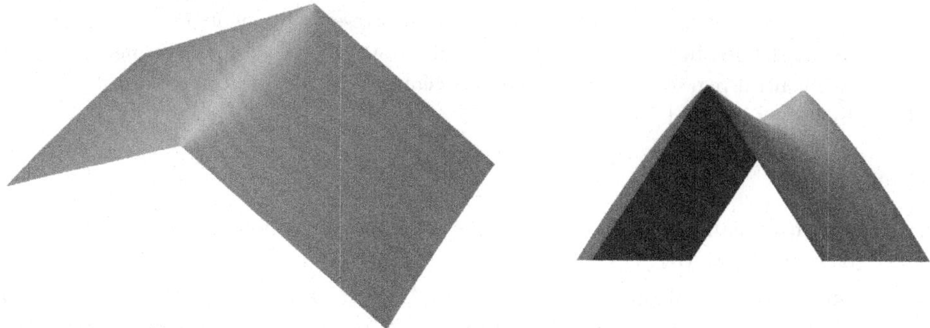

Fig. 9.13 Three-dimensional visualization of the fold profile reconstructed from the numerical solution of the boundary layer equations (9.64). Variation of the fold width, as described by the function $q(y)$, is visible from the reflection on the general view (left), while the quasi-axial view (right) clearly reveals the sag.

It has been observed in reference (A. E. Lobkovsky, 1996) by direct minimization of the plate energy with the boundary conditions shown corresponding to Fig. 9.8. It is due to the increase of the radius of curvature when one moves from the endpoints towards the midpoint of the fold. Our boundary layer solution allows one to find the exact expression of the sag S, which we define as the displacement of the midpoint of the ridge during folding (when the angle α increases from 0 to its final value) assuming that its endpoints are fixed:

$$S = -w(0,0) = -\overline{w}(0,0)\, w^* = q(0)\, p_1(0)\, w^* = \frac{0.163\, h^{1/3}\, L^{2/3}\, \alpha^{2/3}}{[12\,(1-\nu^2)]^{1/6}}, \tag{9.68}$$

where the numerical prefactor comes from the numerical solution, $q(0)\, p_1(0) = 0.163$.

The transverse curvature profile plotted in Fig. 9.11(b) is based on our solution of the boundary layer equation; this profile matches very well with the profile obtained by direct numerical simulations of the plate equations, in the limit of small thickness—compare with Fig. 3 in reference (B. A. DiDonna and T. A. Witten, 2001).

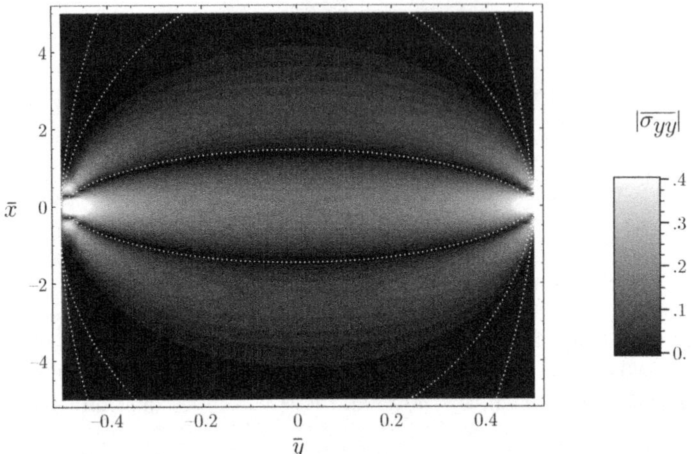

Fig. 9.14 Level map for the dominant (longitudinal) stress component, $\overline{\sigma_{yy}} = \sigma_{,yy}/(\sigma_{yy})^*$, the scale $(\sigma_{yy})^*$ being defined in the main text. The white dotted lines show the change of sign of this component, the central region being under compression, $\sigma_{yy} < 0$. Note the concentration of stress near the endpoints of the fold.

The stress tensor can be visualized too. Its leading component is the longitudinal one, σ_{yy}. It is given by $\sigma_{yy} = (\sigma_{yy})^* \overline{\chi}_{,\bar{x}\bar{x}} = (\sigma_{yy})^* p_2''(\xi)/q(y)$, where the scale for the longitudinal stress comes from equation (9.54): $(\sigma_{yy})^* = \chi^*/x^{*2} = E\,h^{2/3}\,\alpha^{4/3}/[12\,(1-\nu^2)]^{1/3}\,L^{2/3}$. Note that the stress diverges near the endpoints. This is not surprising as the imposed transverse curvature $w_{,xx}$ becomes infinite there. This stress concentration near the endpoints leads to irreversible deformations (scars) very much as near the tip of a d-cone. The transverse variation of σ_{yy} is given by the function $p_2''(\xi)$, plotted in Fig. 9.11. Note its alternating sign, related to the constraint $F_y = -\int_{-\infty}^{+\infty} d\xi\, p_2'' = 0$ in equation (9.65): the longitudinal stress is compressive ($p_2'' < 0$) in the central region defined by $|\xi| < 2.394$ (that is for $|x| < 2.394\, q(|y|)$), extensional for $2.394 < |\xi| < 9.587$, again compressive for $9.587 < |\xi| < 17.464$, etc. These numbers are the roots of p_2''. A level map of $|\sigma_{yy}|$ is given in Fig. 9.14.

The moment M required to fold the plate by an angle α is probably one of the most interesting quantities that can be obtained from the present solution: it characterizes the global response of the plate and could be measured experimentally. To compute this moment, we start by calculating the torque transmitted across the plane of symmetry of the fold, $x = 0$. This torque is measured with respect to the y-axis passing through the endpoints of the fold. By convention, we compute the torque applied by the side $x > 0$ of the fold on to the side $x < 0$ across the crest. Let us consider an element dy along this crest, centred at a coordinate y. The transverse curvature at this point $(x = 0, y)$ of the centre line is $w_{,xx}(0, y) = -p_1''(0)/q(y)$. The internal moment transmitted across this centre line element is therefore $(-D\,w_{,xx}\,dy)$. A transverse force resultant $h\,\sigma_{xx}(0, y)\,\mathbf{e}_x\,dy$ is transmitted as well. This force contributes to the moment M proportionally to its arm with respect to the y-axis, see Fig. 9.15; this arm can be expressed as $w(0, y)$. This yields the following expression for the bending moment:

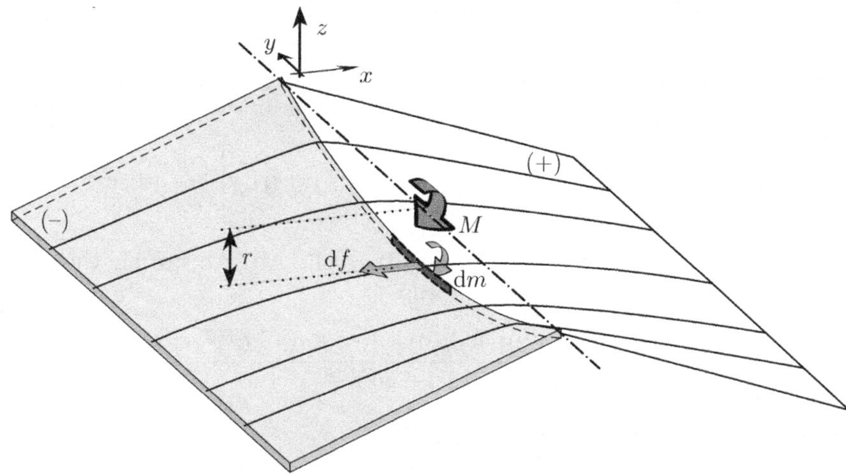

Fig. 9.15 In equations (9.69) and (9.70), the moment applied from the (+) half of the plate, defined by $x > 0$, on to the other half (−), defined by $x < 0$, is computed. This moment M is defined with respect to the y axis of the coordinate system (dashed dotted line). Every surface element along the imaginary cut $x = 0$ brings two contributions to M, expressed by the two terms in the integrand of equation (9.69), which are of comparable magnitude: one is the internal moment $dm \propto (-D\,w_{,xx})$, measured by definition with respect to the *centre of mass of the surface element*; the other one comes from the resultant force on this element, $df \propto (h\sigma_{xx})$, multiplied by the arm $r \propto (-w)$ with respect to the y axis.

$$M = \int_{-L/2}^{+L/2} (-D\,w_{,xx} + h\,w\,\sigma_{xx})\ \mathrm{d}y. \tag{9.69}$$

Indeed, when $w_{,xx} < 0$ as in Fig. 9.8, the contribution to the moment M applied from the side $x > 0$ is positive with our conventions. When $\sigma_{xx} < 0$ and $w < 0$, as in this figure, the transverse force yields another contribution. The integral in the equation above has to be taken along the fold crest, $x = 0$. Factoring the scale M^* computed by dimensional analysis:

$$M^* = \frac{E\,h^{8/3}\,L^{1/3}\,\alpha^{4/3}}{[12(1-\nu^2)]^{5/6}},$$

one computes the non-dimensional moment $\overline{M}$ as:

$$\overline{M} = \frac{M}{M^*} = \int_{1/2}^{-1/2} (-\overline{w}_{,\bar{x}\bar{x}} + w\,\chi_{,yy})\ \mathrm{d}\overline{y}.$$

Using the definition (9.56) of w and χ in terms of p_1 and p_2 and omitting the overbars, one gets:

$$\overline{M} = \int_{-1/2}^{+1/2} \left(\frac{1}{q(y)}\,p_1''(0) - q(y)\,p_1(0)\,\ddot{q}(y)\,p_2(0) \right)\ \mathrm{d}y$$

$$= \left[\int_{-1/2}^{+1/2} \frac{\mathrm{d}y}{q(y)} \right] p_1''(0) - \left[\int_{-1/2}^{+1/2} q(y)\,\ddot{q}(y)\,\mathrm{d}y \right] p_1(0)\,p_2(0).$$

In this expression, the factors in brackets are known because the function $q(y)$ has been calculated in equation (9.63), and the other factors involving p_1 and p_2 are given by equation (9.67). Skipping the details of the calculation of the integrals (the function $q(y)$ is known implicitly and so these integrals must be rewritten in terms of the variable q):

$$\left[\int_{-1/2}^{+1/2} \frac{\mathrm{d}y}{q(y)}\right] = (2\pi)^{2/3} \qquad \left[\int_{-1/2}^{+1/2} q(y)\,\ddot{q}(y)\,\mathrm{d}y\right] = -(2\pi)^{2/3}.$$

This yields the value $\overline{M} = (2\pi)^{2/3}[p_1''(0) + p_1(0)\,p_2(0)]$. After restoring the dimensional factor M^*, the physical moment takes the form:

$$\begin{aligned}
M &= \frac{(2\pi)^{2/3}[p_1''(0) + p_1(0)\,p_2(0)]\;E\,h^{8/3}\,L^{1/3}\,\alpha^{4/3}}{[12(1-\nu^2)]^{5/6}} \\[2mm]
&= \frac{0.645\;E\,h^{8/3}\,L^{1/3}\,\alpha^{4/3}}{[12(1-\nu^2)]^{5/6}}.
\end{aligned}$$

(9.70)

By balancing the moments applied on the side $x > 0$ of the fold, it appears that the moment applied by the triangular support balances exactly the moment applied to the other side of the fold through the crest. This shows that, as expected, M given above is also the moment applied on the triangular supports to fold the plate.

The elastic energy of the fold can now be calculated by progressively bending the fold from the flat configuration. The element of work done by the operator is $M\,\mathrm{d}\alpha$ as he or she has to apply the moment M on the articulated triangular supports when they rotate by an angle $\mathrm{d}\alpha$. If this is done quasi-statically, the final elastic energy in the plate is equal to the total work done by an external operator:

$$\begin{aligned}
\mathcal{E}_{\text{fold}} &= \int M\,\mathrm{d}\alpha \\[2mm]
&= \frac{7}{3}\frac{(2\pi)^{2/3}\,[p_1'' + p_1\,p_2]_{\xi=0}}{[12(1-\nu^2)]^{5/6}}\;E\,h^{8/3}\,L^{1/3}\,\alpha^{7/3} \\[2mm]
&= \frac{1.505}{[12(1-\nu^2)]^{5/6}}\;E\,h^{8/3}\,L^{1/3}\,\alpha^{7/3}.
\end{aligned}$$

(9.71)

Note that the dimensional analysis has yielded with little effort the correct form of the energy, see equation (9.50), but without the numerical prefactor. To the best of our knowledge, this is the first time that the energy of a fold is calculated with the numerical prefactor. Note that the numerical constants in equations (9.68), (9.70) and (9.71) are universal, as the boundary layer equations (9.64) for the fold have no parameter.

We make two remarks before closing this discussion. The first is that the second term in equation (9.64b) can be recognized as the Gauss curvature, see equation (9.58). Therefore, the integral of the Gauss curvature along any line with constant coordinate y (perpendicular to the fold) is zero

$$\int_{-\infty}^{+\infty} K\,\mathrm{d}x = \int p_2''''(\xi)\,q(y)\,\mathrm{d}\xi = 0,$$

as $p_2''' \to 0$ on both sides for $\xi \to \pm\infty$. This means that the integral of the Gauss curvature over the full plate area $|y| < L/2$ is zero. This condition is similar to that encountered earlier for the cone problem: it was then interpreted as a condition for the stretching energy to remain finite far from the singularity.

The second remark is about the asymptotic profile far from the fold. To zeroth order, we have imposed $w \to \pm\alpha\, x/2$ on either side, $x \to \pm\infty$. This is the Cartesian equation for two half-planes making an angle α. To next order in a boundary layer expansion we get a first correction $w(x,y) \approx -q(y)\, p_1(x/q(y))$. For large $\xi = x/q(y)$, p_1 has an affine asymptote, $p_1(\xi) \approx \pm\xi/2 + p_1^\infty$, see Fig. 9.11. The constant p_1^∞ is given by the analysis of the numerical solution as $p_1^\infty = -0.507$. For the deflection this yields $w(x,y) \approx -\alpha\,|x|/2 + w^*\left(-p_1^\infty\right) q(y)$ for $x \to \pm\infty$. This shows that, on either side, the fold goes asymptotically to an almost flat half-infinite cylinder whose axis is tilted by an angle $\alpha/2$ and whose profile is given by equation $z = w^*\left(-p_1^\infty\right) q(y)$. Up to a multiplicative constant $-p_1^\infty\, w^*$, the intersection of the fold with a vertical plane $x = x_0$ far away from the ridge ($x_0 \to \pm\infty$) yields a curve that is just the graph of the function $q(y)$ given in Fig. 9.10. This can indeed be seen in Fig. 9.13, right. The presence of vertical tangents along the sides of the domain points is inconsistent with the assumption of small slope underlying the F.–von K. equations, and has to be resolved.

To complete the story, we also need to look at the endpoints of the ridge. It happens that, there, the parabolic approximation of the F.–von K. equations breaks down because the derivative of w (for instance) along the fold direction becomes of the order of or bigger than the derivatives perpendicular to it, contrary to what is assumed in the parabolic approximation. So another boundary layer sets in there, which has to be matched with the fold solution just found. This is the topic of the next section.

9.3.7   Boundary condition near conical points at the ends of a fold

We have just identified two regions where the ridge solution is inconsistent: along the long edges of the rectangular domain, where the slope becomes infinite, and at the endpoints of the ridge. This was noticed by Conti and Maggi (S. Conti and F. Maggi, 2008), who considered a ridge formed in a diamond-shaped plate and not in a rectangular plate, thereby removing the region with infinite slope from the domain; in addition, they relax any boundary condition along the edge near the endpoints of the ridge. With these modified domain and boundary conditions, they were able to prove Lobkovsky's scaling; this indicates that the inconsistencies found in the previous section are minor, and can probably be cured by taking into consideration some additional boundary layers. A scaling analysis of the boundary layer arising near the endpoints of the ridge is proposed below.

For the solution derived in the previous section, the parabolic approximation underlying equation (9.53) becomes invalid near the endpoints of the fold. To begin with, we derive the behaviour of the function $q(y)$ near $y = 1/2$, one of the endpoints of the fold—the properties at the other end are found by symmetry. This behaviour is derived straightforwardly from the original ODE, equation (9.60). Assuming it behaves like a power, one finds that $q(y) \sim (y - \frac{1}{2})^{2/3}$. The numerical prefactor is found next by inserting this power law back into the equation. This gives

$$q(y) \approx \left(\frac{9}{2}\right)^{1/3} \left(y - \frac{1}{2}\right)^{2/3}. \tag{9.72}$$

Alternatively, this expansion can be found directly by expanding the implicit solution given in (9.63)

The deflection $w(x, y)$ is given by $w(x, y) = -q(y) p_1(x/q(y))$ and so its derivatives with respect to x and y are:

$$\frac{\partial w}{\partial x} = -p_1' \quad \text{and} \quad \frac{\partial w}{\partial y} = -\dot{q}(y)\,(p_1 - \xi\,p_1'),$$

where, as before, $\xi = x/q(y)$. Therefore the derivative of w with respect to x remains finite near the end point of the fold, although the derivative $w_{,y}$ diverges like $(y - 1/2)^{-1/3}$. This makes invalid the parabolic approximation, that was based upon the assumption that any quantity, like $w(x, y)$, changes much more quickly as a function of x than as a function of y. Therefore there is a neighbourhood of the end points where this parabolic approximation fails and has to be replaced by another one. It turns out that the diverging solution given by the parabolic approximation yields the large distance behaviour of the d-cone solution and so it has to be matched with the near-tip solution. There is however an added twist, so to speak, to the case considered in Section 9.2. The large distance behaviour of a d-cone was assumed to minimize the bending energy only, because it was assumed, somewhat implicitly, that the shape of the cone was smooth. Here we face a different situation: the d-cone is not smooth, since it includes a dihedral angle along the fold (see Fig. 9.8). If one follows the derivation in Section 9.2, the contribution to the energy of the far field is limited to the pure bending energy. This cannot be valid for the present case because one would have to assume at infinity a cone with a dihedral angle, which would lead to an infinite bending energy. The smoothing of this sharpness in the fold structure is precisely what was considered in the previous section. Therefore the asymptotic behaviour of the d-cone at the end of a fold depends on the direction when this d-cone is at the end of a fold: in almost all directions it tends to a smooth cone, except along the directions where two different plane orientations meet, making a fold in the sense of the previous section. This matching of the fold solution and the d-cone is actually possible. This depends on the range of values of y and x where the parabolic approximation for the fold becomes invalid. Coming back to the original unscaled quantities, the derivative $\partial w/\partial x$ scales like α, the dihedral angle of the fold, although $\partial w/\partial y$ scales like $\alpha\{l/L\,((y - 1/2)/L)^{-1/3}$, where L is the length of the fold and $l = h^{1/3}L^{2/3}\alpha^{-1/3}$ is its width at the centre. Therefore the parabolic approximation breaks down when the two derivatives become of the same order of magnitude, which happens when $(y - 1/2)$ becomes of order h/α or smaller. The fold has non-trivial (x, y) dependence in the domain where ξ is of order 1. From its definition, $\xi = x/l\,(L/(y - 1/2))^{2/3}$. Therefore, the transition region is where $(y - 1/2) \sim h/\alpha$ as we have just seen and where ξ is of order 1, that is where $x \sim h\alpha^{-1}$, x being the distance top the tip. The two scalings are consistent with the fact that $x \sim y \sim h/\alpha$ is the scaling for the size of the tip of the d-cone in general. This completes the picture of the fold, showing that it ends like a d-cone with a condition at infinity slightly different from the usual one for a smooth cone.

9.4 Conclusion

This chapter has explained how stress focuses in crumpled elastic structures. The large-scale geometry is fairly simple to describe: to minimize the stretching energy, a constrained plate that is planar at rest tends to buckle and to become a polyhedral surface. In the limit of a negligible thickness, this surface is an isometric deformation of the plane. It is only because of the bending effects that the d-cones and the folds become non-singular and have a finite size. This chapter analysed how this width can be found by solving the F.–von K. equations. Somehow this leaves pending the question of the geometry of this polyhedral surface in the embedding 3D space. There is indeed no general answer to that question, because it depends on the way the external loading is applied. See reference (B. Audoly and A. Boudaoud, 2008) for an application of the fold geometry to the analysis of a buckling problem in the limit of a large buckling parameter.

References

B. Audoly and A. Boudaoud. Buckling of a thin film bound to a compliant substrate (part 3). Herringbone solutions at large buckling parameter. *Journal of the Mechanics and Physics of Solids*, 56(7):2444–2458, 2008.

M. Ben Amar and Y. Pomeau. Crumpled paper. *Proceedings of the Royal Society A: Mathematical, Physical and Engineering Sciences*, 453(1959):729–755, 1997.

A. Boudaoud, P. Patrício, Y. Couder, and M. Ben Amar. Dynamics of singularities in a constrained elastic plate. *Nature*, 407:718–720, 2000.

E. Cerda, S. Chaieb, F. Melo, and L. Mahadevan. Conical dislocations in crumpling. *Nature*, 401(6748):46–49, 1999.

E. Cerda and L. Mahadevan. Conical surfaces and crescent singularities in crumpled sheets. *Physical Review Letters*, 80(11):2358–2361, Mar. 1998.

E. Cerda and L. Mahadevan. Confined developable elastic surfaces: cylinders, cones and the Elastica. *Proceedings of the Royal Society A: Mathematical, Physical and Engineering Sciences*, 461(2055):671–700, 2005.

S. Conti and F. Maggi. Confining thin elastic sheets and folding paper. *Archive for Rational Mechanics and Analysis*, 187:1–48, 2008.

S. Chaïeb, F. Melo, and J.-C. Géminard. Experimental study of developable cones. *Physical Review Letters*, 80(11):2354–2357, Mar. 1998.

B. A. DiDonna and T. A. Witten. Anomalous strength of membranes with elastic ridges. *Physical Review Letters*, 87(20):206105, Oct. 2001.

A. Lobkovsky, S. Gentges, H. Li, D. Morse, and T. A. Witten. Scaling properties of stretching ridges in a crumpled elastic sheet. *Science*, 270(5241):1482–1485, 1995.

A. E. Lobkovsky. Boundary layer analysis of the ridge singularity in a thin plate. *Physical Review E (Statistical, Nonlinear and Soft Matter Physics)*, 53(4):3750–3759, Apr. 1996.

A. E. Lobkovsky and T. A. Witten. Properties of ridges in elastic membranes. *Physical Review E (Statistical, Nonlinear and Soft Matter Physics)*, 55(2):1577–1589, Feb. 1997.

T. Liang and T. A. Witten. Crescent singularities in crumpled sheets. *Physical Review E (Statistical, Nonlinear and Soft Matter Physics)*, 71(1):016612, 2005.

C. Majidi. Remarks on formulating an adhesion problem using Euler's elastica. *Mechanics Research Communications*, 34(1):85–90, 2007.

R. H. Plaut, S. Suherman, D. A. Dillard, B. E. Williams, and L. T. Watson. Deflections and buckling of a bent Elastica in contact with a flat surface. *International Journal of Solids and Structures*, 36(8):1209–1229, Mar. 1999.

W. H. Press, S. A. Teukolsky, W. T. Vetterling, and B. P. Flannery. *Numerical Recipes in C*. Cambridge University Press, 2002.

M. Spivak. *A Comprehensive Introduction to Differential Geometry*, volume 5. Publish or Perish, Inc., Houston (TX), 2nd edition, 1979.

S. C. Venkataramani, T. A. Witten, E. M. Kramer, and R. P. Geroch. Limitations on the smooth confinement of an unstretchable manifold. *Journal of Mathematical Physics*, 41(7):5107–5128, 2000.

T. A. Witten. Stress focusing in elastic sheets. *Reviews of Modern Physics*, 79:643, 2007.

10
Fractal buckling near edges

In this chapter, the buckling of thin elastic sheets leading to fractal patterns[1] is studied. Some of these fractal buckling patterns have already been shown in Fig. 5.1, Chapter 5, but they have not yet been explained. They comprise a cascade of folds at smaller and smaller scales and can be observed near the edge of torn plastic sheets used to make garbage bags, as well as near the edge of some plant leaves. As we argued, this rippling phenomenon can be accounted for using the model of an elastic plate having a stretched edge. Stretching near the edge results from irreversible deformations during the tearing process when the edge is stretched beyond the elastic limit (in plastic bags), or from enhanced biological growth near the edge (in the leaves or flowers of some species of plants). Besides this irreversible process, the material response can be assumed elastic. In the following we explain these fractal structures by the theory of elastic plates.

In the second half of the present chapter, we study another situation that leads to fractal patterns, namely when an elastic plate is compressed well above the threshold of instability. The near threshold situation has been studied in Chapter 7; what happens at finite distance from threshold has been the topic of Chapter 8. Below we show that far above threshold the buckling pattern may become fractal.

These two types of fractal structures (in a free sheet with a stretched edge, and in clamped plates loaded far above threshold of buckling) can be observed in experiments. The fractal pattern of a torn elastic sheet can be observed by looking closely at the cut left by the tearing of a ductile sheet, such as a plastic sheet used for wrapping goods or for garbage bags, see Fig. 10.1. Over several length scales, tiny wrinkles near the edge are observed to merge into larger wrinkles farther from the edge. Seen globally this yields a sort of 'fractal' structure. The structures at smaller scales close to the edge are typically in the millimetre range, although the 'large scale' ones are usually centimetric. We shall not make use of the general terminology and theory of fractals. Practically the number of steps in the cascade is restricted to three or four in the experiments, five at most in the numerics. Therefore, speaking of fractal dimension would not make much sense, because of the limited range of available length scales. Nevertheless, we put in evidence a few non-trivial and unexpected facts. The first one is the relation between this problem of a set of cascading ripples and the general question of the embedding of a 2D surface that has a prescribed metric in the ordinary 3D Euclidean space, a famous problem of differential geometry about which

---

[1] Fractals have been a popular topic of study in the last thirty years. A good reference is the first book by Mandelbrot (B. Mandelbrot, 1977). In a nutshell a fractal is a structure that looks the same when seen after a change of scale. Usually fractals exist in a limited range of scales, since their property of self-similarity does not hold at very large scales because of the finiteness of the object under study and at very small scales where the continuous medium approximation breaks down. For instance the fractal patterns seen in elasticity stop at large scales defined by the lateral width of the buckled plate although the small scale is given by the thickness of the plate.

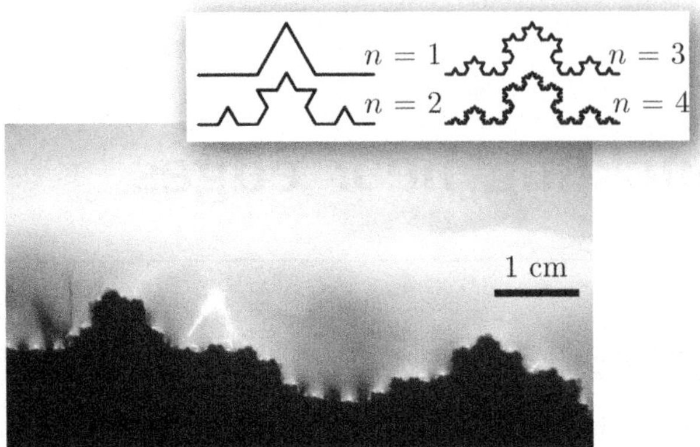

Fig. 10.1 Close-up of the edge of a piece of thick, ductile plastic sheet that has been torn by hand. The fractal undulations are very similar to Koch snowflake (inset), obtained by iteration over an integer n of a planar geometric construction (H. von Koch, 1904). Very similar fractal undulations can be observed in various other systems that have a stretched boundary, such as plant leaves, see Fig. 5.1(e).

little is known in general. As we shall show, there seems to exist many, if not infinitely many, solutions of this embedding problem; we shall investigate how the bending energy selects a particular pattern out of these many embeddings, which turns out to be fractal, see Section 10.1 below.

In the other geometry, that of strongly post-buckled plate with clamped edges, fractal buckling patterns can also be observed in elementary experiments. As seen in Chapter 8, a flat elastic plate with clamped lateral boundaries buckles when pushed along its own plane so as to release longitudinal compression. Near threshold the buckling pattern is smooth and depends on the details of the particular boundary conditions used in the experiment—for a long rectangular plate, it is periodic along the long direction (see Fig. 8.3, for instance). Here, we specifically consider the limit when the plate is stressed far above the buckling threshold.[2] This makes a rather interesting problem of the limit of bifurcation much above threshold, although the usual range of interest is close to threshold. In the geometry too, a kind of fractal structure shows up. At large scales the plate is pleated like an accordion. Such an accordion does not satisfy the boundary conditions. To accommodate for the presence of the boundary, a new set of folds at a smaller scale is inserted near the boundary; this process is repeated in a cascade. Its smallest scale can be derived from an analysis of the F.–von K. equations. This makes a rather sharp difference with the first item of this chapter, since the geometry here is simple, although the calculation of the energy rests on the outcome of a rather complex boundary layer analysis. This is in some sense an extreme case of a strong nonlinearity.

The general lesson is after all not so surprising. It tells that non-trivial effects may happen near a boundary as happens in any situation where boundary layers show up. However, the

---

[2] As seen in many places in this book, the assumption that the applied stress is much larger than the critical stress for buckling (large buckling parameter) is not incompatible with the assumptions underlying the F.–von K. equations (at least up to a certain limit, obviously).

present situation is a little bit special and the standard boundary layer theory does not apply: a cascade takes place over an intermediate scale and allows the large scale solution to adapt itself to the conditions imposed on the edge.

10.1   Case of residual stress near a free edge

The problem considered below has already been introduced in Chapter 5, where a simplified analysis based on a rod model was presented: in this preliminary analysis, we almost entirely forgot the elastic sheet, except for a very thin ribbon near the stretched edge. By solving the Kirchhoff equation for this geometry, we were able to reproduce the undulations found in surfaces which we called godet-like structures. This 1D simplification of the problem does not explain how the wrinkles occurring near the edge can merge with the flat part of the plate in the interior of the sheet.

The present section addresses the fractal pattern resulting from the quasi-static tearing of a thick (ductile) plastic bag, as described in (E. Sharon *et al.*, 2002), with a cutting speed typically in the range of one centimetre per second. The fundamental phenomenon responsible for buckling is simply the divergence of the stress near the tip of a (moving) crack in 2D. Indeed, the stress diverges like the inverse squared distance to the tip of the crack by an effect mathematically similar to the point effect in electrostatics. Therefore, the plastic sheet is stretched well beyond its limit of elasticity in the vicinity of the crack and reaches a state a permanent deformation, which is stronger and stronger closer to the edge on the path followed by the crack. Once the sheet has been torn into two pieces, it relaxes toward equilibrium and the subsequent deformations are reasonably well described by Hooke's law for linear elasticity, but with a permanent stretching near the edge. By a classical buckling effect, the plate minimizes its elastic energy by wrinkling near this edge, the third physical dimension being used to store the length in excess. In the present section, we show that this elastic relaxation of the elastic plate can even produce a self-similar pattern, even though the loading is modelled by a very simple and smooth residual stress profile near the edge.

As explained earlier, the salad leaf features enhanced growth near its edge and is mechanically similar to a torn plastic sheet. In both systems a cascade of folds with a smaller and smaller wavelength towards the edge can be observed. Compared with other 'classical' self-similar structures, such as snowflakes, see Fig. 10.1, the elasticity problem has a specific flavour: it does not depend too much on the way the structure is being grown, being the solution of a minimization problem, without obvious dynamical underpinning. Somewhat similar fractal structures near edges or boundaries have been described in minimization problems arising in the theory of ferromagnetic materials (L. D. Landau, 1938, 1943) and martensitic phase transitions (S. Conti, 2000).

10.1.1   Forward

In the following sections, we adapt the equations for elastic plates to account for the presence of a stretched edge. This stretching can be specified as a residual stress in the planar configuration—this residual stress is sometimes called a *prestress*, and has been encountered in several places already, see for instance Section 7.3.3. After deriving the governing equations, we present a very efficient numerical procedure allowing one to find the minimizers, which are the profiles with minimal elastic energy. These minimizers are found to be self-similar profiles with up to five generations. Unlike in Chapter 8, we do not present the numerical scheme for solving this plate problem in full detail; we simply outline the numerical method and discuss the results. The difference with the rod model presented earlier in Chapter 5 is that we now consider the full problem, not restricted to the close neighbourhood of the edge. It is probable that the present work will shed light on self-similarity in elastic systems (Y. Pomeau and S. Rica, 1997; W. Jin and P. Sternberg, 2002; H. Ben Belgacem, 2002), as well as in other systems where the patterns are determined by energy minimization. It could be interesting to see if similar self-similar geometrical structures could show up in problems with a variational underpinning and with no obvious intrinsic length scale, such as in the general theory of relativity.

10.1.2 Geometry of a surface with a stretched edge

In this section we start by defining the geometrical problem of embedding a surface that has a prescribed metric describing a stretched edge. Building on these geometrical tools, we shall formulate in the following sections the associated problem of minimizing the elastic energy of a plate, and implement it numerically.

Let us consider a semi-infinite elastic sheet, whose reference configuration is chosen to be planar and contained in the (x, y) plane. Let $y = 0$ be the edge in this reference configuration, the sheet occupying the half-plane $y > 0$, x being arbitrary. This reference configuration is chosen by convention, and solely for the purpose of tagging the material points along the sheet: it is not assumed to be stress-free. The pairs (x, y) are used as Lagrangian coordinates along the centre surface. The presence of a stretched edge is imposed mathematically by specifying a natural, non-Euclidean metric in these Lagrangian variables. Neighbouring points (x, y) and $(x + \mathrm{d}x, y + \mathrm{d}y)$ would like to be separated by a distance, called the *natural* distance, which is described by a non-Euclidean metric and is different (being actually larger) from the (Euclidean) distance $\sqrt{\mathrm{d}x^2 + \mathrm{d}y^2}$ in this somewhat arbitrary planar configuration. One solution for them to achieve their natural distance is to have the surface deform in three dimensions—what we have here is a purely geometrical model for buckling.

Let us call $\mathrm{d}s$ the natural distance between neighbouring points, defined as the *relaxed* length of a small ribbon-like piece of material cut along the shortest path joining these two points. We ask how $\mathrm{d}s$ can depend on x, y, $\mathrm{d}x$ and $\mathrm{d}y$. Since $\mathrm{d}s$ is fundamentally a distance, its square $\mathrm{d}s^2$ has to depend quadratically on $\mathrm{d}x$ and $\mathrm{d}y$. Moreover, we shall assume that the physical process causing the residual stress (tearing or differential growth) is invariant along the direction x parallel to the edge. This means that the coefficients of the quadratic form $\mathrm{d}s^2$ of $\mathrm{d}x$ and $\mathrm{d}y$ may depend only on y. Therefore, the most general expression for this natural distance[3] is

$$\mathrm{d}s^2 = G(y)\,\mathrm{d}x^2 + H(y)\,\mathrm{d}x\,\mathrm{d}y + I(y)\,\mathrm{d}y^2, \tag{10.1}$$

where G, H and I are three functions that describe the irreversible deformations or the growth near the edge. The case of no residual stress, that is of a stress-free planar configuration, corresponds to $G(y) = 1$, $H(y) = 0$ and $I(y) = 1$ and the metric is then

---

[3] When a spring is stretched beyond its elastic limit, its natural length is effectively reset to a different (larger) value. The non-Euclidean metric considered here generalizes this effect of irreversible stretching to a 2D medium.

Euclidean; in the general case, these conditions should be recovered asymptotically in the interior, that is for $y \to \infty$, since the residual strain is assumed to be localized near the edge. Physically, G represents stretching along the boundary, I stretching perpendicular to the boundary, while H represents some shear deformation caused by the tearing process (this H can be expected to be identically zero in the case of tissue growth by symmetry). These three functions could be computed if a detailed model of ductile fracture or biological growth was given; in the present analysis, they are considered to be given as parameters, and are assumed to satisfy some reasonable assumptions, detailed later.

A useful remark that simplifies the problem is that the shear and the transverse residual strain, H and I, can be relaxed by means of a suitable in-plane deformation of the surface. Indeed, consider the x-invariant transformation such that the point (x, y) moves to $(x + u(y), y + v(y))$; this is essentially a sliding motion of the region near the edge since any material line parallel to the edge $y = 0$ moves by an in-plane, rigid-body translation. Using the two fundamental formulae (2.8) and (2.9) for the strain, we find that the distance between neighbouring material points are redefined by this sliding transformation as:

$$\mathrm{d}s^2 = \mathrm{d}x^2 + 2u'(y)\,\mathrm{d}x\,\mathrm{d}y + \left(1 + 2v'(y) + u'^2(y) + v'^2(y)\right)\mathrm{d}y^2.$$

Provided H and I are not too different from 0 and 1 respectively (in the following, we shall consider only the case of infinitesimal residual strain anyway), one can choose the functions u and v such that $2u' = H$ and $2v' + v'^2 = I - 1 - u'^2$. Therefore, a simple in-plane sliding transformation of this kind can be used to redefine the shear and transverse parts of the metric in a perfectly arbitrary manner. The initial problem of finding a deformation, called an embedding, of surface leading to the metric (10.1) can be simplified: by sliding the edge in its own plane first, we are brought to the simpler problem of finding embedding with a prescribed metric of the particular form:

$$\mathrm{d}s^2 = (1 + g(y))^2\,\mathrm{d}x^2 + \mathrm{d}y^2, \tag{10.2}$$

where we note $G(y) = (1 + g(y))^2$, which is convenient since we assume a small residual strain, that is $|g| \ll 1$. The H and I contributions to the residual strain are irrelevant to the buckling problem and have been effectively removed by means of an in-plane transformation. This is not the case of the longitudinal stretching, described by the function $g(y)$, which is coupled to the out-of-plane deformations.

An interesting exercise is to compute the Gauss curvature associated with the metric (10.2). The Gauss curvature has been defined in Chapter 6 for a surface in the 3D Euclidean space, and is given by the product of the principal curvatures at any particular point. One may wonder how the Gauss curvature can be computed from the target metric only, without knowing the embedding, that is the final shape of the surface. What makes it possible is that the Gauss curvature is invariant by isometric deformations, and so will be the same for any embedding. Amusingly, it can even be computed without using any embedding at all, for instance by studying the perimeters of circles, see Fig. 6.4. Another possibility, left as an exercise to the interested reader, is to study the convergence or divergence of two close geodesics[4] that are initially parallel. The Gauss curvature can then be calculated by

---

[4] A geodesic is defined to be the shortest path between any pair of its points (at least when the points are not too far apart). In Euclidean geometry, geodesics are straight lines. The equation for the geodesics can be obtained from Euler–Lagrange equations, see Appendix A, by imposing that the curvilinear distance

analogy with what happens on a sphere, where the geodesics are the great circles. For yet another approach to this problem, see footnote 6.

As mentioned above, we assume that the stretching profile $g(y)$ is a given, positive[5] function of y, such that $g(y) \to 0$ for $y \to +\infty$ (there is no stretching in the interior of the plate). This function $g(y)$ is also assumed to be peaked near the edge $y = 0$, where the growth or the residual strain are maximum. However, we assume small residual strains (and in fact show that small strains can very well account for fractal patterns) and so $|g(y)| \ll 1$ everywhere. The function $g(y)$ changes over a well-defined length scale, assumed to be much larger than the thickness of the plate. An example of such a function $g(y)$ is given in Fig. 5.3.

The geometrical structure underlying the plate equations has been presented in Chapter 6: since the bending energy of a plate is formally much smaller than its stretching energy in the limit of a small thickness, the equilibrium configurations are such that the stretching energy is minimized in the first place. This brings us to the question of whether this stretching energy can be made exactly zero; this is exactly the problem of embeddings that we have mentioned before, since the stretching energy measures the difference of the metric in actual configuration to the natural metric. Interestingly, such embeddings may of may not exist depending on the choice of the function $g(y)$, may or may not be smooth, and their existence (or not) for a given arbitrary function $g(y)$ remains an open question (M. Spivak, 1979)—this problem is studied in detail in Chapter 11. Again, this geometrical problem is crucial for predicting the equilibrium shape of a plate. If smooth embeddings exist, the stretching energy will be minimized by having the centre surface of the plate come close to an embedding; in addition, embeddings can be selected by minimizing the bending energy if they are not unique. In contrast, if smooth embeddings do not exist, the actual equilibrium solution will either try to approach an embedding that has singularities (see Chapter 9) or the shape that is the non-zero, smooth minimum of the stretching energy.

In an attempt to build an explicit example of an embedding for our problem, consider for instance the rippled surface given by its Cartesian equation

$$z(x,y) = \epsilon \cos(kx)\, A(y),$$

where ϵ is a small number, allowing us to use the simple framework of small displacement for evaluating the strain, k is the wavelength of the pattern, and $A(y)$ is the envelope of the ripples, which we can try to adjust later. Equation (6.42) yields the Gauss curvature of such a surface as:

$$K(x,y) = \frac{\partial^2 z}{\partial x^2}\frac{\partial^2 z}{\partial y^2} - \left(\frac{\partial^2 z}{\partial x \partial y}\right)^2$$

$$= -\epsilon^2 k^2 \left(\cos^2(kx)\, A(y)A''(y) + \sin^2(kx)A'^2(y)\right),$$

where we have neglected $\partial z/\partial x$ and $\partial z/\partial y$, which are small compared with 1 in the denominator since ϵ is small. For this surface to be a solution, the metric should be invariant along the edge, and so the Gauss curvature should not depend on x. This happens for

---

along the geodesics is minimal and so does not change to first order for an arbitrary perturbation of the curve.

[5] The case $g(y) > 0$, which we consider corresponds to compressive stress $\sigma_{yy} > 0$ along the boundary in the planar configuration, see inset in Fig. 10.2; therefore, for large enough residual strain, this planar configuration may be an *unstable* equilibrium.

instance when the envelope is of the form $A(y) = \exp(\rho y)$, as can be checked directly: this gives a solution to our embedding problem when the residual strain $g(y)$ is itself an exponential function. However, our guess for $z(x, y)$ does not work in the general case when $g(y)$ is not exponential.

In the following, embeddings are determined numerically by minimization of an elastic energy that penalizes stretching, much like in the F.–von K. equations.

10.1.3   Accounting for residual strain

We now proceed to the derivation of our model. Let $u(x, y)$, $v(x, y)$ and $w(x, y)$ be the components of the displacement, along the x, y and z axes respectively. We follow the derivation of the F.–von K. equations given earlier in Chapter 6, Section 6.4, and extend it to account for the residual strain along the edge. The first step is to compute the strain; this can be done under the approximation of small slope and small in-plane displacement. Since the reference (planar) configuration is no longer stress-free, one can define several strain tensors, which must be carefully distinguished. The first one, denoted $\epsilon^{p/a}$, expresses the change in the metric between the planar reference configuration and the actual configuration. The second one, denoted $\epsilon^{n/a}$ expresses the difference between the natural metric and the metric in the actual configuration. The geometric calculation of the strain given earlier in equation (6.61) now holds for $\epsilon^{p/a}$. However, the effective bidimensional constitutive law for the plate, given earlier in equation (6.59), applies to the physical strain $\epsilon^{n/a}$, since by definition the in-plane stress would vanish in a natural configuration, which is such that $\epsilon^{n/a} = 0$. We need to connect these two strains, and this can be done by the chain rule $\epsilon^{p/a} = \epsilon^{p/n} + \epsilon^{n/a}$, where $\epsilon^{p/n}$ expresses the difference of the metric in the planar case and the natural metric: identifying with equation (10.2) we obtain $\epsilon_{xx}^{p/n} = g(y)$, $\epsilon_{xy}^{p/n} = 0$ and $\epsilon_{yy}^{p/n} = 0$: recall that we assume $|g(y)| \ll 1$. Combining these equations, we arrive at the following definition for the physical strain:

$$\epsilon_{xx} = u_{,x} + \frac{w_{,x}^{\,2}}{2} - g(y) \tag{10.3a}$$

$$\epsilon_{yy} = v_{,y} + \frac{w_{,y}^{\,2}}{2} \tag{10.3b}$$

$$\epsilon_{xy} = \frac{1}{2}\left(u_{,y} + v_{,x} + w_{,x}\, w_{,y}\right). \tag{10.3c}$$

We again use the notation $u_{,x} = \partial u / \partial x$ etc. In this equation and in the following, $\epsilon_{\alpha\beta}$ stands for the physical in-plane strain, that is $\epsilon_{\alpha\beta} = \epsilon_{\alpha\beta}^{n/a}$. The other, non-physical strain tensors will not be needed in the following.

The residual strain along the edge appears in the equations through the inhomogeneous term $-g(y)$ in the right-hand side of ϵ_{xx}. When residual strain arises from an irreversible *stretching* of the free edge, e.g. by plastic flow near the crack tip, $g(y) > 0$ and the longitudinal strain ϵ_{xx} in the actual configuration can be made less negative by balancing $g(y)$ with the term $w_{,x}^{\,2}/2$, roughly speaking. This makes the strain less negative, so it is decreased in absolute value and so is the stretching energy. Non-zero values of $w_{,x}^{\,2}/2$ can typically be obtained by having the sheet make ripples along the edge, something that breaks the original invariance of the problem along this direction. This argument explains that the residual stress describing irreversible stretching along the edge is able to drive a buckling instability.

In the strain definitions above, residual strain is included. The rest of the derivation of the plate equations is identical to what has been done earlier in Chapter 6 in the absence of residual strain. Therefore, a complete set of equations for the current problem of a plate buckling under residual stress is composed of (i) the strain definition, just given in equation (10.3), (ii) the effective constitutive relations (from Hooke's law) for a plate, given in equation (6.59), and (iii) the F.–von K. equations (6.89) for the equilibrium. For the problem at hand, these equations have no analytical solution—recall that even the simpler geometrical problem of finding embeddings, that is solving for $\epsilon_{\alpha\beta} = 0$ has no analytical solution.[6] Therefore, we resort to a numerical solution of the problem. Unfortunately traditional methods for solving the F.–von K. equations numerically are inappropriate for the problem at hand: the equations are non-linear; we need to resolve wrinkles over a wide range of length scales, hence a numerical mesh having a very large number of elements is necessary; we would like to be able to follow equilibrium solutions when the parameters are changed. Using a standard numerical scheme such as finite differences, solution tracking is completely impractical as the convergence is much too slow. For this reason, we introduce a *toy* buckling problem that is mathematically similar to the original one but is amenable to very efficient numerical simulations, and turns out to feature self-similar solutions. A similar approach has been used recently to investigate the herringbone buckling of stiff films bound to a compliant substrate (B. Audoly and A. Boudaoud, 2008) using a simplified plate model. Even though the model introduced in the following is simplified, great care is taken to preserve the geometrical structure underlying the equations, which we claim is at the heart of this fractal buckling. Using this toy model, we do not claim to make any quantitative prediction regarding the fractal wrinkles; instead, our emphasis is on understanding the basic mechanism responsible for these wrinkles.

10.1.4 Plate energy

Since our aim is to set up a simplified buckling problem, it is natural to work on the energy rather than on the equations of equilibrium. Let us first rewrite the energy for a plate given in equation (6.95) as:

$$\mathcal{E} = \iint \mathrm{d}x\,\mathrm{d}y \left(\frac{E\,h}{2(1-\nu^2)} \left[\nu\left(\epsilon_{xx} + \epsilon_{yy}\right)^2 + (1-\nu)\left(\epsilon_{xx}^2 + 2\epsilon_{xy}^2 + \epsilon_{yy}^2\right) \right] \right.$$
$$\left. + \frac{E\,h^3}{24(1-\nu^2)}(\Delta w)^2 \right), \tag{10.4}$$

where E and ν are Young's modulus and Poisson's ratio of the plate. Furthermore the components of the strain ϵ_{xx}, etc. are related to the displacement by the equations (10.3) and so they include a contribution arising from residual stress. The first part of the integrand,

---

[6] Eliminating by derivation u and v from the equations $\epsilon_{xx} = \epsilon_{yy} = \epsilon_{xy} = 0$, one finds an equation for $w(x,y)$ which reads $w_{,xx}\,w_{,yy} - w_{,xy}^2 = -g''(y)$. The left-hand sides can be identified as the Gauss curvature in the actual configuration (under the approximation of small deflection from a planar configuration). Indeed, this calculation is similar to that done earlier in Section 6.2.1, and we have just shown that the Gauss curvature associated with the natural metric is $(-g''(y))$. The mathematical form of this equation explains why it is difficult to find embeddings analytically: it is a non-linear partial differential equation for the unknown function $w(x,y)$.

proportional to h, is the stretching energy and the second one, proportional to h^3, is the bending energy. In the bending energy, we have omitted a term proportional to the Gauss curvature $K = [w, w]/2$ in the integrand.[7]. The energy functional $\mathcal{E}$ encodes the usual Hookean constitutive relations as well as the differential equations of equilibrium, which can all be rederived from it by minimization with respect to small variations δu, δv and δw of the displacement, see Section 6.6.

The problem can be formulated in terms of dimensionless quantities. One of the advantages of doing so is that numerical round-off errors are minimized by dealing with numerical values that are neither very small nor very large in absolute value. The out-of-plane deviation is measured in units of the thickness h, as is usual in plate buckling problems. The in-plane length scale relevant for the coordinates x and along y is the typical distance of variation of $g(y)$, and is noted Λ. As usual in this text, the dimensionless quantities are written with overlines, see Section 1.2.4. They are such that $\overline{w} = w/h$, $\overline{x} = x/\Lambda$, $\overline{y} = y/\Lambda$, $u = \overline{u}\, h^2/\Lambda$, $v = \overline{v}\, h^2/\Lambda$. Note that the deflection w is *a priori* much larger, by a factor Λ/h, than the in-plane displacements. This classical rescaling in plate theory follows from a balance of the linear and non-linear terms in the definition of strain, as explained earlier in Section 7.3.3 for instance. The residual strain g is already dimensionless but is further rescaled to make all terms in the definition of ϵ_{xx} formally of the same order of magnitude. This leads to a new, overlined g such that $\overline{g} = \Lambda^2 g/h^2$. The assumption behind this definition is that $\overline{g}$ is of order 1. This implies that the physical residual strain g is of order h^2/Λ^2, a number that is small compared with 1. Since this scaling has been found by balancing all the terms relevant for the buckling analysis of the plate, this typical residual strain h^2/Λ^2 is in fact an order of magnitude estimate of the critical residual strain for buckling, and we shall assume that this also gives the general order of magnitude of the residual strain, that is the amplitude of the function $g(.)$. For consistency, we rescale the energy according to $\mathcal{E} = \overline{\mathcal{E}} E h^5/((1 - \nu^2)\,\Lambda^4)$.

Using these rescalings, the energy functional reads

$$\overline{\mathcal{E}} = \frac{1}{2}\iint \mathrm{d}\overline{x}\,\mathrm{d}\overline{y}\,\left(\left[\nu\,(\overline{\epsilon}_{xx} + \overline{\epsilon}_{yy})^2 + (1 - \nu)\,(\overline{\epsilon}_{xx}^2 + 2\overline{\epsilon}_{xy}^2 + \overline{\epsilon}_{yy}^2) \right] + \frac{1}{12}(\overline{\Delta w})^2 \right). \qquad (10.5)$$

The plate's bending modulus has been effectively set to the value $1/12$.

In the expression above, the overlined components of the strain tensor have the same relation to the displacement and to the residual strain as the non-overlined quantities, as they appear in equation (10.3). For instance, the (xx) component of the strain is:

$$\overline{\epsilon}_{xx} = \overline{u}_{,\overline{x}} + \frac{\overline{w}_{,\overline{x}}^{\,2}}{2} - \overline{g}(\overline{y}).$$

As we have done many times, we shall drop the bars on top of the rescaled quantities from now on, as this makes the notation easier to read: we deal with rescaled quantities unless specified otherwise.

10.1.5 Periodicity along the edge, Fourier analysis

We restrict our investigations to solutions that are periodic in the direction x parallel to the edge—note we do not assume a harmonic (sinusoidal) dependence of w on x, however.

---

[7] By Gauss–Bonnet theorem, this omitted term can be re-expressed as a curvilinear integral along the edge $y = 0$ of the plate. Therefore, by omitting this term, we do not modify the equation of equilibrium in the interior of the plate, but effectively replace the boundary conditions along the edge $y = 0$ by non-physical ones. We shall do several other approximations like this in the following.

Let k be the fundamental wavenumber in the x direction, defined as 2π divided by the wavelength. This wavenumber will later be found by energy minimization. Taking advantage of the periodicity, we make a Fourier decomposition of the various quantities with respect to the variable x. By convention, the Fourier components carry an integer superscript $[q]$, corresponding to wavenumber $(q\,k)$. A strain component $\epsilon_{\alpha\beta}(x,y)$ for instance is decomposed as

$$\epsilon_{\alpha\beta}(x,y) = \sum_q \epsilon_{\alpha\beta}^{[q]}(y)\sin(q\,k\,y).$$

In this equation, the Fourier coefficients $\epsilon_{\alpha\beta}^{[q]}(y)$ are functions of y to be determined.

The stretching energy, given by the first terms in the integrand of equation (10.5), is a quadratic form of the strain components. Using the Parseval formula[8] one can write the corresponding integral over x as a quadratic combination of the Fourier coefficients of the strain. This leaves a quadratic functional of the Fourier coefficients $\epsilon_{\alpha\beta}^{[q]}(y)$, that is physically the energy per unit length in the x direction. It reads:

$$\mathcal{E} = \frac{1}{2}\int \mathrm{d}y \sum_{qq'\alpha\alpha'\beta\beta'} \epsilon_{\alpha\beta}^{[q]}(y)\, A_{\alpha\beta,\alpha'\beta'}^{q,q'}\, \epsilon_{\alpha'\beta'}^{[q']}(y). \tag{10.6}$$

The numerical coefficients $A_{\alpha\beta,\alpha'\beta'}^{q,q'}$ can be computed from Eq. (10.4). So far everything is exact, the only assumption being that the pattern is periodic with respect to x. Note that the integral with respect to the periodic variable x has been replaced by a discrete sum, by Fourier analysis.

10.1.6 A simplified stretching energy

We introduce an approximation that makes it possible to eliminate in-plane displacements (u, v) from the energy functional. This approximation preserves the underlying mathematical structure of the problem and the special role played by embeddings. Other approximations of the plate equations have been suggested in the literature, which do not preserve the geometry; as a result, the energy may be overestimated by a very large factor (W. Jin and P. Sternberg, 2002). With the present approximation, this difficulty is avoided as the embeddings continue to have a vanishing stretching energy (see discussion below). Equation (10.6) above defines a quadratic form in the Fourier components $\epsilon_{\alpha\beta}^{[q]}(y)$. Instead of minimizing the full quadratic form (10.6), we set to zero some suitably chosen components $\epsilon_{\alpha\beta}^{[q]}(y)$ and minimize with respect to the remaining ones. We choose to impose $\epsilon_{xx}^{[q]} = 0$ and $\epsilon_{xy}^{[q]} = 0$ for $q > 0$; these additional kinematical constraints are somewhat arbitrary; nevertheless, they are extremely practical as they allow the in-plane displacement u and v to be computed explicitly as a function of $w(x,y)$, leaving us with a single unknown w. Again, we do not claim that the predictions of this model will be exact but simply that they are qualitatively correct and that they suggest a convincing mechanism for buckling in a fractal manner.

---

[8] This standard result in Fourier analysis relates the integral of the square of a periodic function of x over one period to the infinite sum of the squared moduli of its Fourier components.

In the framework of this approximation one obtains after some algebra the stretching energy in terms of $w(x, y)$ only:

$$\mathcal{E}_{\mathrm{s}} = \frac{1}{2} \int \mathrm{d}y \left[\left(\left\langle \frac{w_{,x}^2}{2} \right\rangle - g(y) \right)^2 + \frac{1}{2} \sum_{q>0} \left(\frac{\{w_{,xx}\, w_{,yy} - w_{,xy}^2\}^{[q]}}{k^2\, q^2} \right)^2 \right], \qquad (10.7)$$

where brackets $\langle f \rangle = f^{[0]}$ stand for average with respect to x. The details of the calculation are similar to those published in (B. Audoly and A. Boudaoud, 2008) in a different geometry.

All the terms in the stretching energy $\mathcal{E}_{\mathrm{s}}$ have a simple geometric interpretation. Gauss' *Theorema egregium* states that Gauss curvature K is conserved by embeddings. The Gauss curvature of the abstract manifold with metric (10.2) is $K = -g''(y)$ in the limit of a weakly stretched edge (when the non-scaled, physical residual strain satisfies $|g| \ll 1$); the Gauss curvature associated with a profile $w(x, y)$ reads $K = w_{,xx}\, w_{,yy} - w_{,xy}^2$. Therefore, the condition of zero stretching corresponds to:

$$w_{,xx}\, w_{,yy} - w_{,xy}^2 = -g''(y) \quad \leftrightarrow \quad \mathcal{E}_{\mathrm{s}} = 0. \qquad (10.8\mathrm{a})$$

As outlined in footnote 6, this equation is derived from the condition that the in-plane strain is identically zero. By averaging equation (10.8a) above over x, and integrating twice with respect to y, one obtains the following equation, which holds for any value of y and for any embedding:

$$\left\langle \frac{w_{,x}^2}{2} \right\rangle = g(y). \qquad (10.8\mathrm{b})$$

Indeed, $\langle w_{,xx}\, w_{,yy} - w_{,xy}^2 \rangle = -\langle w_{,x}\, w_{,xyy} + w_{,xy}^2 \rangle = -\langle (w_{,x}\, w_{,xy})_{,y} \rangle = -1/2 \langle (w_{,x})^2_{,yy} \rangle$, where we have used integration by parts, and neglected the boundary terms by averaging over many periods. This equation can also be derived directly by averaging equation (10.3a) for ϵ_{xx} which is zero for an embedding, using $\langle u_x \rangle = 0$; the last equality $\langle u_x \rangle = 0$ can itself be established by noticing that $\langle u_x \rangle$ is the average of a derivative and so is bounded by the amplitude of variation of u, a finite number since the plate does not deform far away from edge for $y \to \infty$, divided by the length of the plate along the x axis, which is assumed to be infinite.

Similarly, by taking any non-zero Fourier component q of equation (10.8a), we find

$$\{w_{,xx}\, w_{,yy} - w_{,xy}^2\}^{[q]} = 0 \qquad (10.8\mathrm{c})$$

for any embedding.

In retrospect, the stretching energy (10.7) appears to be built as a sum of terms that penalize any deviation from a strainless embedding, that is any deviation from the equalities (10.8b) and (10.8c). Our simplified stretching energy is positive, and vanishes if and only if the deflection $w(x, y)$ describes such an embedding. Our schema of approximation respects the geometrical underpinning of the F.–von K. equations.

10.1.7 Numerical results and interpretation

In the following, we minimize the total energy, that is the sum of the stretching energy (10.7) and the bending energy as given by the last term in (10.4), with respect to the fundamental wavenumber k and with respect to the Fourier components $w^{[q]}(y)$ of the unknown deflection

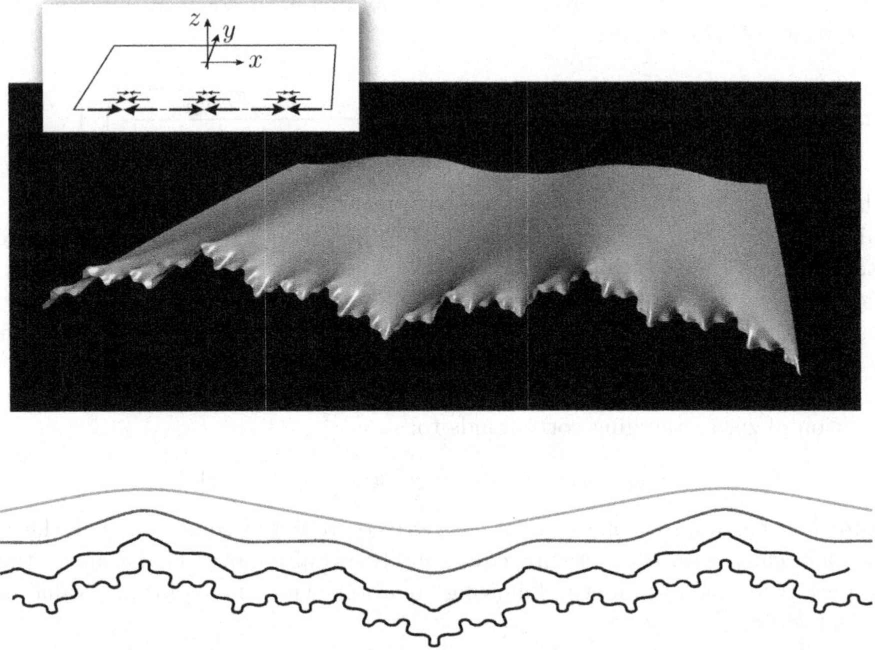

Fig. 10.2 Numerical solution of the equilibrium of a plate with a stretched edge (insert) buckling into a self-similar pattern. Five generations of wrinkles can be observed. Simulation parameters are plate thickness $h = 2.9 \times 10^{-5}$, width 1, metric $g(y) = 1/y$ for $y > 0.008$ with affine continuation for $y < 0.008$. The optimum fundamental wavelength is $2\pi/k = 0.54$. *Top*: 3D visualization of the numerical solution. *Bottom*: Cuts of the centre surface along vertical planes parallel to the edge at various distances from the edge, revealing the formation of new wrinkles by period tripling towards the edge. Amplitude of the deflection is enhanced to aid visualization.

($q = 0, 1, 2, \cdots$). The numerical minimization is based on a 1D finite element representation of the functions $w^{[q]}(y)$ ($q = 0 \leq q \leq q_{\max}$). The particular form of our stretching energy (10.7) leads to a very efficient implementation allowing interactive calculations. Real-time user feedback on the minimizer is particularly valuable in the presence of many metastable configurations; such a feedback also made possible the tracking of solution (Figs 10.4–10.5).

By minimization, we obtained self-similar solutions with up to five generations of wrinkles as shown in Fig. 10.2. This particular one was obtained with the choice[9] $g(y) = 1/y$ for $y > 0.008$ and an affine continuation for $0 < y < 0.008$. Experimenting with various profiles for $g(y)$, we found that the absolute minima of energy were self-similar solutions when $g(y)$ is sufficiently peaked near $y = 0$.

The cascade of wrinkles is generated by period tripling starting with the fundamental wavenumbers k, and the successive harmonics $3\,k$, $9\,k$, $27\,k$, $81\,k$. This period tripling can

---

[9] It may look surprising that our function $g(y)$ takes on values that are of order 100 near the origin, although the rescaled values of g were assumed to be of order one. In fact, the absolute magnitude of g is irrelevant due to the scaling $g \sim w^2$ in our equations. In other words, if we divide g by 100 and w by 10 for any numerical solution of our problem, we have another solution.

be understood by mere symmetry considerations. Indeed, when we write the equations of equilibrium for the deflection $w(x,y)$ and Fourier transform along direction x, we get a set of coupled equations for the Fourier components $w^{[q]}(y)$. Because of the underlying symmetry, the equations for the odd Fourier components $q = 1, 3, 5 \ldots$ are uncoupled to the even components. As a result, the similarity factor α describing the cascade near the edge[10] must map odd integers onto odd integers, in such a way that the equations for the associated Fourier components remain independent upon the scaling transformation: this leaves odd integers (3, 5, etc.) as eligible values for the magnification factor α in the space of wavenumbers. In our simulation, $\alpha = 3$ was found to be the most favourable one energetically, and $\alpha = 5$ yields solutions having an energy slightly higher only. This compares well with the experimental factor (3.2) reported in reference (E. Sharon *et al.*, 2002). The slight discrepancy can be attributed to experimental conditions where the nonlinearity is of order one, although our theory assumes it to be small (here, we assume $|g| \ll 1$ and small slopes). In other sets of experiments, a factor close to five has sometimes been observed,which further confirms our prediction.

A detailed analysis of the stretching energy (10.7) shows that self-similar solutions are invariant under the scaling transformation:

$$x' \to 3\,x \quad y' \to 3\,y \quad q' \to q/3 \quad \frac{q'\,w'_{q'}(y')}{\sqrt{g(y')}} \to \frac{q\,w_q(y)}{\sqrt{g(y)}}, \tag{10.9}$$

where we restrict our analysis to a magnification factor 3. This invariance is confirmed by the collapse of numerical functions $w^{[q]}(y)$ shown in Fig. 10.3.

The order of magnitude of the fundamental wavevector k is given by the macroscopic extent of the plate, i.e. the inverse width of the range of y where $g(y)$ differs significantly from zero. This is the width of our simulation domain. This gives the longest wavelength of the solution. Its smallest wavelength is determined by a mechanism analysed below. The cascade takes place over all intermediate scales by successive period tripling.

Bending regularizes the wrinkles at small scales because it sharply increases the energy of the short scale oscillations. Let λ be the smallest wavelength of the wrinkling, and $\ell = g(0)/g'(0)$ be the *small* length scale induced by $g(y)$ near the edge. The cutoff λ is found by balancing stretching and bending energies at scale λ. For the metric used in the simulations of Fig. 10.4, this yields

$$\lambda \sim h^{2/5}\ell^{3/5}. \tag{10.10}$$

These exponents are explained as follows. One assumes that the length scale for the smallest wrinkle near the edge is the same in the x direction and in the y direction, say λ_m. This is the shortest wavelength near the edge. With our particular metric profile, the dilation factor, a dimensionless quantity grows like ℓ/y near the edge, where ℓ is a typical length scale, used here to make $g(y)$ dimensionless. Note that with other power laws for $g(y)$ the

---

[10] By similarity factor α we understand here the change in wavenumber related to the change in distance to the edge. It is difficult to define such notions accurately because a wavenumber along x can be mapped only loosely to the distance y to the edge: since the functions under consideration depend continuously on y. Nevertheless the trend is clear: a component of the deviation with a certain wavenumber along x has noticeable values in a narrow range of values of y, which get closer and closer to the edge as the index q of the Fourier component increases. The details of the scaling factors, etc. are to be found in equations (10.9).

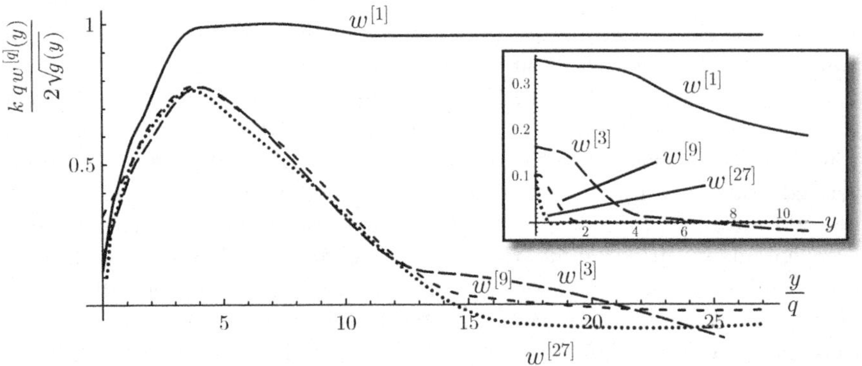

Fig. 10.3 Fourier components $w^{[q]}(y)$ ($q = 1, 3, 9, 27$) of the deflection for a self-similar solution with four generations (insert) and their collapse using transform (10.9). The fundamental mode $w^{[1]}(y)$ need not collapse for intermediate and large values of y that correspond to points far away from the distended edge, where there is no cascade. Note that fixed-point function is not universal in the sense that it depends on the metric. This fixed point function is defined 'experimentally' by the one ensuring the best the collapse of the numerical data. Same numerical parameters as in Fig. 10.2 except for $h = 2.9 \times 10^{-4}$. Reproduced from reference (B. Audoly and A. Boudaoud, 2003) with kind permission of the American Physical Society.

final exponents would be different. The stretching $g(y)$ felt by the last wrinkle is of order ℓ/λ_m. It has to be balanced, at least in order of magnitude, by the out-of-plane tilting angle induced by the wrinkle, which is associated with a stretching of order $(\partial w/\partial x)^2$. Since x there is of order λ_m, the balance between $g(y)$ and $(\partial w/\partial x)^2$ yields

$$w^2 \sim \ell \lambda_m$$

The other relation follows from the mechanical equilibrium. The bending energy per unit area is $Eh^3(\Delta w)^2$. In order of magnitude this gives $Eh^3(\Delta w)^2 \sim Eh^3 w^2/\lambda_m^4$. The order of magnitude of the energy of stretching is less easy to estimate. This energy turns out to be dominated by the large-scale deformation (this can be seen also as due to the stress generated by the large-scale strain). The energy density associated with a vertical deviation w over a horizontal distance ℓ is in general of order $E h \left(w^4/\ell^4\right)$. Therefore the balance of stretching and bending energy yields $w^2 \sim h^2\ell^4/\lambda_m^4$. Taking now into account the relation between $g(.)$ and w, one finds $\lambda \sim h^{2/5}\ell^{3/5}$, in good agreement with the numerics. The scaling law given in equation (10.10) is checked against the numerical results as shown in Fig. 10.4: numerical solutions at various h and ℓ do show good agreement with the prediction (10.10). This formula shows too that, to be observed, a cascade requires a small scale ℓ explicit in $g(y)$: if ℓ is too large, the number of generations decreases to one and there is no cascade.

Our plate model and its numerical implementation provide insights into the physical structure of the self-similar solutions. Earlier papers (S. Nechaev and R. Voituriez, 2001; E. Sharon *et al.*, 2002; M. Marder *et al.*, 2003) and a more recent experimental paper (E. Sharon, B. Roman, and H. L. Swinney, 2007) propose a mechanism for the cascade based on the idea that there is no smooth embedding associated with the

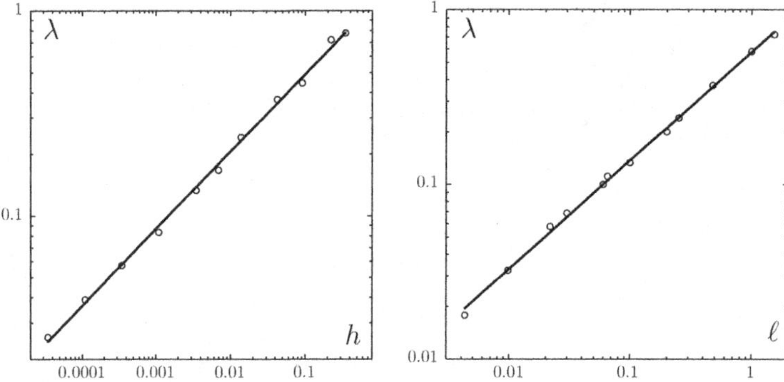

Fig. 10.4 Numerical check of scaling law (10.10) for the cutoff: smallest wavelength λ in the cascade as a function of plate thickness h (left) and of metric typical length ℓ (right). Thick lines are best power-law fits, yielding exponents 0.38 and 0.62 respectively, to be compared with the theoretical values $2/5 = 0.4$ and $3/5 = 0.6$. Reproduced from reference (B. Audoly and A. Boudaoud, 2003) with kind permission of the American Physical Society.

non-Euclidean metric (by 'smooth', we mean 'having a finite bending energy'). Our numerical results suggest a different interpretation of the cascade, as we found that there are *many* smooth embeddings available for a given non-Euclidean metric, the one observed in experiments being selected by flexural effects. To date, the debate seems to be still open.

10.1.8 Tuning the metrics

The relationship between the self-similar character of the wrinkles and the existence of a small scale ℓ in the residual strain $g(y)$ is now investigated. We study how buckling depends on this small length scale ℓ. In Fig. 10.5, a one-parameter family of functions $g_\ell(y) = 1/(1 + y/\ell)$ is considered, and minimizers of the total elastic energy are tracked when the parameter ℓ decreases, practically from 1 down to 10^{-2}. Interestingly, solutions displaying a different number of wrinkles can be observed for any given value of ℓ: the system is multistable and has many local equilibria. When the small length scale ℓ is decreased, the lowest energy configuration bifurcates from a single wavelength ('1') to a cascade of wrinkles by successive period tripling (up to '1+3+9+27' in this plot). Remarkably, all these configurations are very close to exact embeddings as both their stretching and bending energies remain very small and comparable when $h \to 0$. Therefore, for a given $g_\ell(y)$, the numerical results point to *many* geometrical embeddings, one of which is made up of oscillations at a single length scale while others are superpositions of wrinkles with different wavelengths. This numerical observation rules out a purely geometrical explanation of the self-similar pattern observed in experiments: self-similar patterns appear to be just one of the many possible geometrical embeddings. Among all embeddings, self-similar wrinkles are selected because of their small bending energy when ℓ is small. Indeed, near the boundary, the curvature along the x direction is of order w/λ^2, while that along y is of order w/ℓ^2. To minimize stretching, the edge has almost its natural length, hence $w \sim \lambda \sqrt{g(0)}$. The density of bending energy $Eh^3 g(0)(1/\lambda + \lambda/\ell^2)^2$ is the smallest when $\lambda \sim \ell$. Therefore small wavelengths are energetically favoured near the boundary, hence the cascade.

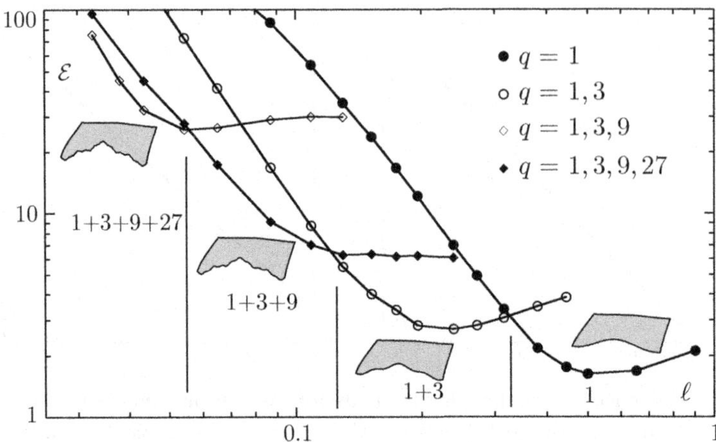

Fig. 10.5 Energy of solutions with one to four generations of wrinkles for a one-parameter family of metric $g_\ell(y) = 1/(1 + y/\ell)$. Bifurcations occur while all branches remain close to embeddings: the solution with the lowest energy is selected with the bending energy. There is a bifurcation for each value of ℓ where two curves cross. Reproduced from reference (B. Audoly and A. Boudaoud, 2003) with kind permission of the American Physical Society.

10.1.9   Discussion

This mechanism and all the main findings presented here are robust with respect to the approximations made in the stretching energy, see equation (10.7). The small slope approximation underlying the writing of the strain (10.3) is consistent, as we found self-similar solutions for arbitrarily small $g(y)$, hence for arbitrarily small slopes.

To summarize, we have explained the occurrence of self-similar wrinkles near the edge of distended plates. In a first step, configurations are restricted by the geometry to be close to embeddings so as to avoid a large stretching energy; as we found, many such embeddings are actually possible, some of which are smooth, some of which are fractal. In a second step, self-similar wrinkles are selected by elastic effects among all embeddings because their bending energy is the smallest for small ℓ. Self-similarity of the minimizers was evidenced numerically according to the theoretical fixed-point map with up to five generations of wrinkles. A prediction for the invariant magnification factor agrees well with experimental observations. The smallest and largest wavelengths of the cascade have also been found.

10.2   Case of a clamped edge

The present section deals with buckling of a plate in the limit of a large load. We already considered plate buckling problems under in-plane stress. In Chapter 7, we based our analysis on bifurcation methods, valid near the onset of buckling. At a finite distance from threshold, in Chapter 7, we minimized the elastic energy by a Ritz method using perturbations with a prescribed dependence on coordinates and an arbitrary amplitude; we compared this with experiments and found good agreement with the weakly non-linear theory and the finite amplitude Ritz approximation. In the rest of this chapter, we examine another limit of the bifurcation problem, the limit of large imposed stress. Linear stability analysis involves weak nonlinearity and is a very classical approach to buckling, mainly because of its straightforward implementation; it is applicable close the threshold. What is much less known is that the opposite limit of large loads is often amenable to analytical solutions too, as illustrated here. In this limit, the buckling parameter is assumed large and non-linear terms are not negligible. By contrast with linear stability analyses, there is no general method of solution and the approach varies from one problem to another, which makes it both more interesting and more difficult.

10.2.1   Motivation, principle of the solution, outline

The idea of a cascading structure near a boundary was first introduced by Landau (L. D. Landau, 1938) in the theory of the intermediate state of superconductors. Experiments in thin film, such as the one presented in Fig. 10.6, taken from reference (A. S. Argon, 1989), suggested that a similar phenomenon can take place in plate buckling—see also the experiments in reference (G. Gioia *et al.*, 2002). Ortiz and Gioia proposed an explanation of self-similar wrinkles in delamination blisters based on a model where any in-plane displacement is neglected (M. Ortiz and G. Gioia, 1994); however, this approximation was later shown to be inapplicable to the limit considered (W. Jin and P. Sternberg, 2002). Pomeau and Rica (Y. Pomeau and S. Rica, 1997; Y. Pomeau, 1998) gave a geometrical argument for the cascade, arguing that its function is to accommodate the presence of an edge that is effectively contracted by the clamped boundary conditions, while keeping the rest of the plate developable. They proposed a construction based on a cascade of the ridge-like singularities (ridges are studied in Chapter 9). This construction has been refined later in two close but independent papers, by Jin and Sternberg (W. Jin and P. Sternberg, 2001) and by Belgacem, Conti and collaborators (H. Ben Belgacem, 2000); these authors showed that a cascade of *smooth* folds is in fact optimal. In the rest of this chapter, we

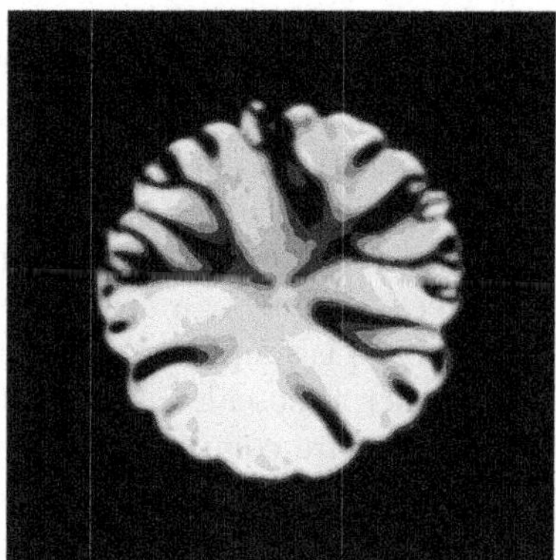

Fig. 10.6 Experimental cascade of folds near the boundary of a circular blister, redrawn from (A. S. Argon, 1989), viewed from top in a SiC film, with relative substrate compression of 1.1%. The aspect ratio of the thin film is 1/140. The overall blister shape is that of a bubble coming upwards, i.e. towards the observer. A set of connected wrinkles that have smaller and smaller wavelengths towards the boundary are superposed.

propose an informal presentation of their argument—our presentation following that given in reference (W. Jin and P. Sternberg, 2001). In these two papers, an explicit construction is proposed, which provides an upper bound on the optimal energy; this construction is shown to be optimal in the sense that any equilibrium configuration has an energy (lower bound) comparable to that of the upper bound in the limit considered. Here, we shall omit the technical aspects of this mathematical proof and present a qualitative analysis of the pattern that conveys the essential ideas.

In subsequent papers, Conti and collaborators analysed this problem in the framework of 3D elasticity (H. Ben Belgacem, 2002), motivated by the fact that the scalings of the self-similar construction are, strictly speaking, not uniformly consistent with the ones underlying the derivation of the F.–von K. equations; they also investigated the interplay of this cascade with large-scale patterns found when a plate is clamped all along its boundary (S. Conti, A. DeSimone, and S. Müller, 2005), see Fig. 10.6.

Mathematically, the problem considered here belongs to the class of penalization problems. Such problems arise when some contributions to the energy become formally large in some limit—here, the stretching energy happens to become formally large in the formal strongly post-buckled limit. The equilibrium configurations have to minimize the total energy; they are such that this type of contribution cancel altogether whenever this is possible, or at least are made small as possible. The resulting solutions often feature boundary or internal layers, as happens with the present problem. Boundary layer methods[11] are introduced in Appendix B, and illustrated in several places in this book, in Chapters 4, 9, 13 and 14. What makes the present problem interesting is that the buckling pattern that keeps this stretching energy minimum is fractal. Even more interesting is that this fractal pattern can be derived by rather simple scaling arguments. The optimal solution will be shown to feature a cascade of wrinkles of sizes that decrease in geometric progression, a feature roughly confirmed by experimental pictures of circular blisters far from threshold, see Fig. 10.6.

Before we start with a description of the problem, we shall recall the principle of dimensional analysis, as presented for instance in Barenblatt's book (G. I. Barenblatt, 1979). When a phenomenon, such as buckling, arises out of two antagonistic effects, it is possible to estimate the scale at which it will appear by formally balancing the energies associated with each one of those effects. For instance, the oscillations of a mass attached to a spring arise out of the spring elasticity and of the mass inertia. Using the same obvious notation as in Section 1.2.4, where the same example has already been used to introduce the ideas behind scaling analysis, we balance the corresponding energies, of order kx^2 and $m(x/t)^2$ respectively, to get $t \sim \sqrt{m/k}$, a correct estimate for the period of the motion. The justification of this approach is simple: at much larger timescales, the elasticity would dominate while at much smaller timescales, the inertia would dominate; there is only one timescale where the two effects are comparable. The argument for the cascade of wrinkles is an application of this principle of scaling analysis, albeit to a much more difficult problem.

---

[11] Boundary layers were first used by Prandtl for studying fast flows past wings. Different approximations have to be used for the equations of fluid mechanics: the boundary layer approximation near the wing surface, including viscous effects, and the perfect-inviscid fluid limit far from it. In this classical example, as in many similar problems, the geometry makes it obvious where one should use an approximation different from that in the bulk: near the body surface.

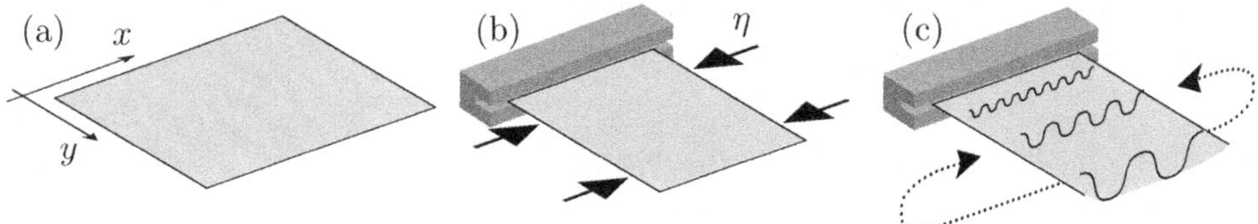

Fig. 10.7 Problem considered in Section 10.2: a square plate in its natural configuration, (a), is contracted along the x direction and then clamped along one of its edges. The resulting configuration (b) is chosen as a reference. The imposed contraction η of the plate manifests itself as a residual strain ϵ_{xx} in this reference configuration. Subsequent buckling of the plate is shown to involve a self-similar structure near the clamped end. The end opposite to the clamped end is free; periodic boundary conditions are assumed at the other edges (dotted arrow).

Paper crumpling has been studied in Chapter 9 and is a problem very similar to the one studied here: this phenomenon can be described by solutions of the equations for plates in the strongly post-buckled limit too. However, the final pattern is completely different. For crumpled paper, we end up with a polyhedral surface that is developable almost everywhere except along localized ridges, where the energy focuses. In contrast, the solution we shall derive next makes smooth, non-localized wrinkles that extend over most of the area of the plate and the energy spreads evenly along the area of the plate. The fundamental difference between the two problems comes from the boundary conditions and the details of the loading geometry: a confining 3D potential is used in crumpling, such as the hands of the operator, although in the present geometry the strain is imposed directly[12] *along* a boundary. This again illustrates the importance of the details of boundary conditions for the analysis of elasticity problems.

10.2.2   Geometry

The problem we consider is sketched in Fig. 10.7. A square plate of size $L \times L$ is contracted along the x axis and then clamped along an edge parallel to this axis. The three other edges are free. We assume periodic boundary conditions for the two other edges, parallel to the y axis, but this is not essential for the analysis. This geometry is characterized by the presence of contraction (negative residual strain) in the direction parallel to the clamped edge. This residual strain is associated with compressive stress which can make the plate buckle when this residual strain is large enough. As explained in Section 10.1.3, two components of residual strain, namely that in the perpendicular direction and that corresponding to in-plane shear, can be relaxed by a simple in-plane displacement, at least when only one of the edges is clamped:[13] we set them to zero in the analysis.

---

[12] In the problem studied here, the clamped boundary conditions impose that one of the edges has a length different from its natural length: the boundary conditions are explicitly incompatible with developability. In crumpling problems, the boundary conditions are more permissive, and are compatible with (polyhedral) developable configurations.

[13] This is not the case when the plate is clamped all along its boundary, see reference (S. Conti, A. DeSimone, and S. Müller, 2005).

Using the classical approximation underlying the F.–von K. equations, see Chapter 6, we write the in-plane strain components as

$$\epsilon_{xx} = u_{,x} + \frac{w_{,x}^{2}}{2} - \eta, \tag{10.11a}$$

$$\epsilon_{xy} = \frac{u_{,y} + v_{,x}}{2} + \frac{w_{,x}\, w_{,y}}{2} \tag{10.11b}$$

$$\epsilon_{yy} = v_{,y}^{2} + \frac{w_{,y}^{2}}{2}, \tag{10.11c}$$

where $u(x,y)$ and $v(x,y)$ are the in-plane displacement along x and y respectively, and $w(x,y)$ is the deflection. We use the compact notation $f_{,\alpha}$ for the partial derivative $\partial f/\partial\alpha$.

Here, η is for residual strain, and $\epsilon_{\alpha\beta}$ for the natural strain (i.e. measured with respect to natural configuration), as explained in Section 10.1.3. The boundary conditions on the clamped edge $y = 0$ read:

$$u(x,0) = 0, \quad v(x,0) = 0, \quad w(x,0) = 0, \quad w_{,y}(x,0) = 0. \tag{10.12}$$

Along the edges $x = 0$ and $x = L$, we impose periodic boundary conditions: we seek functions u, v and w that are periodic with respect to the variable x, i.e. $u(x + L, y) = u(x, y)$, $v(x + L, y) = v(x, y)$, $w(x + L, y) = w(x, y)$. The boundary conditions at the stress-free edge $y = L$ do not need to be imposed specifically as they will derive from the condition of energy minimum.

Since the following derivation is concerned with order of magnitudes, the value of Poisson's coefficient is irrelevant: for simplicity, we set $\nu = 0$. We seek configurations minimizing the elastic energy (6.95), which for $\nu = 0$ reads:

$$\mathcal{E}_{\mathrm{FvK}} = \iint \mathrm{d}x\,\mathrm{d}y \left(\frac{Eh}{2} \left(\epsilon_{xx}^{2} + 2\,\epsilon_{xy}^{2} + \epsilon_{yy}^{2} \right) + \frac{D}{2} \left(w_{,xx}^{2} + 2\,w_{,xy}^{2} + w_{,yy}^{2} \right) \right), \tag{10.13}$$

where $D = Eh^{3}/(12(1 - \nu^{2}))$ is the bending modulus. From now on, we drop any numerical constant of order 1, and so we write for instance $D = Eh^{3}/12 \sim Eh^{3}$.

10.2.3 Dimensionless equations

We shall first rescale the variables as explained in Section 1.2.4 so as to write the equations in dimensionless form: after suitable rescalings, there remains a single dimensionless parameter in the problem—in addition to Poisson's coefficient ν, which has arbitrarily been set to zero but is of order unity anyway. Following a notation already used many times, we introduce stars to denotes new scales: we introduce $x^{*} = y^{*} = L$ as the new scale for in-plane displacements, $u^{*} = v^{*}$ for the in-plane displacements, and w^{*} for the transverse displacements. By balancing the various terms in the definition (10.11) of in-plane strain, including the residual strain η along the clamped edge, we obtain the following suggestions for the natural units of in-plane and out-of-plane displacement: $u^{*}/L \sim \eta$ and $(w^{*}/L)^{2} \sim \eta$. Therefore we set

$$u^{*} = v^{*} = L\,\eta, \qquad w^{*} = L\,\eta^{1/2}.$$

With these scales, the ratio of the bending term to the stretching term can then be estimated as

$$\frac{Eh^3(w^*/L^2)^2}{Eh\,\eta^2} = \frac{h^2}{L^2\,\eta}.$$

We define the dimensionless number ϵ of our problem as the square root of this ratio:

$$\epsilon = \frac{h}{L\,\eta^{1/2}}. \tag{10.14}$$

Note that this scalar parameter ϵ has no indices, and cannot be confused with the in-plane strain $\epsilon_{\alpha\beta}$.

We can now rewrite the equations of our problem in terms of rescaled variables, $\overline{x} = x/x^* = x/L$, $\overline{y} = y/y^* = y/L$, $\overline{w} = w/w^*$, etc. This leads to the following definition for the rescaled strain:

$$\overline{\epsilon_{xx}} = \overline{u}_{,\overline{x}} + \frac{\overline{w}_{,\overline{x}}^{\,2}}{2} - 1, \tag{10.15a}$$

$$\overline{\epsilon_{xy}} = \frac{\overline{u}_{,\overline{y}} + \overline{v}_{,\overline{x}}}{2} + \frac{\overline{w}_{,\overline{x}}\,\overline{w}_{,\overline{y}}}{2} \tag{10.15b}$$

$$\overline{\epsilon_{yy}} = \overline{v}_{,\overline{y}}^{\,2} + \frac{\overline{w}_{,\overline{y}}^{\,2}}{2}, \tag{10.15c}$$

from which it appears that our rescalings amount to setting the residual strain to $\overline{\eta} = 1$.

The boundary conditions at the clamped edge are easily rewritten as

$$\overline{u}(\overline{x},0) = 0, \quad \overline{v}(\overline{x},0) = 0, \quad \overline{w}(\overline{x},0) = 0, \quad \overline{w}_{,\overline{y}}(\overline{x},0) = 0, \tag{10.16}$$

and periodicity for the variable $\overline{x}$ is now with a period 1, the domain being defined by $0 \leq \overline{x} \leq 1$ and $0 \leq \overline{y} \leq 1$. Here, we rescale the in-plane strains by $\epsilon_{\alpha\beta}^* = \eta$.

Rescaling the energy with the natural scale

$$\mathcal{E}^* = E\,h\,\eta^2\,L^2$$

we obtain the rescaled energy as

$$\overline{\mathcal{E}}_{\mathrm{FvK}} \sim \iint \mathrm{d}\overline{x}\,\mathrm{d}\overline{y}\ \left(\overline{\epsilon_{xx}}^{\,2} + 2\,\overline{\epsilon_{xy}}^{\,2} + \overline{\epsilon_{yy}}^{\,2}\right) + \epsilon^2\left(\overline{w}_{,\overline{xx}}^{\,2} + 2\,\overline{w}_{,\overline{xy}}^{\,2} + \overline{w}_{,\overline{yy}}^{\,2}\right), \tag{10.17}$$

where we have again dropped numerical factors of order one.

As it induces compressive stresses along the boundary, the residual strain makes the flat configuration of the plate potentially unstable. This potentially leads to a buckling instability, arising from the competition of a destabilizing stretching effect, which tends to make the energy smaller, and a stabilizing bending effect, which tends to make it larger. At buckling threshold, $\eta = \eta_c$ is such that bending and stretching effects are of comparable magnitude and ϵ, by definition, is of order 1. In the following, we consider instead the strongly post-buckled limit, when the residual strain η is much larger than η_c: this corresponds to the limit

$$\epsilon \ll 1 \tag{10.18}$$

in the previous set of equations. The rescaled form of the energy (10.17) shows that the coefficient in front of the terms that have the derivatives of highest order (second-order in space) in the energy are very small. This points to a singular limit, as mentioned several times in this book.

10.2.4   Boundary layer near the edge

For the sake of readability, we shall drop bars on top of rescaled quantities: unless specified otherwise, we will always work with rescaled quantities, and omit the bars.

As is apparent from equation (10.17), the stretching term is penalizing in the limit $\epsilon \to 0$. Now, all the components of the in-plane strain can be removed by in-plane displacement, as explained in Section 10.1.2, except for the component ϵ_{xx} in the direction parallel to the boundary: the clamp prevents the edge from sliding along the boundary. The fact that out-of-plane displacement is required to relax this strain component can be expressed mathematically by taking the average of this strain component ϵ_{xx}, along any line parallel to the clamped boundary:

$$\langle \epsilon_{xx} \rangle_x = \frac{\langle w_{,x}^{2} \rangle_x}{2} - 1,$$

where $\langle \cdot \rangle_x$ denotes averaging with respect to the variable x, the result being a function of the distance y to the edge. The term $u_{,x}$ in equation (10.15a) has disappeared since its average $\langle u_{,x} \rangle_x = \int_0^1 u_{,x}(x, y)\, \mathrm{d}x = u(1, y) - u(0, y) = 0$ is zero by periodicity.

The aim of the wrinkles is to relax the strain component ϵ_{xx}. When this relaxation is perfect, $\epsilon_{xx} = 0$ and this implies

$$\frac{\langle w_{,x}^{2} \rangle_x}{2} = 1. \tag{10.19}$$

The wrinkles are expected to have a wavevector oriented mainly along the x axis parallel to the clamped boundary and so to have crests and valleys perpendicular to the clamped boundary: this is what makes $w_{,x}^{2}$ largest. Denoting $k(y)$ as the typical value of the wavevector along x at a distance y from the edge, and $A(y)$ the amplitude of the oscillations, we shall write formally:

$$w(x, y) \sim A(y) \cos(k(y)\, x). \tag{10.20}$$

By this formal writing, we mean that a cut of the film by the plane $y = y_0$ yields a profile $w(x, y_0)$ whose typical wavelength and amplitude are comparable to $k(y)$ and $A(y)$ respectively; however, this shorthand notation is not meant to imply that the dependence on x is perfectly harmonic, see Section 10.2.6. Collecting the previous equations, we estimate the order of magnitude of the stretching term associated with ϵ_{xx} as:

$$\mathcal{E}_{xx}^{s} = \iint \epsilon_{xx}^{2}\, \mathrm{d}x\, \mathrm{d}y \sim \iint \left(\frac{\langle w_{,x}^{2} \rangle}{2} - 1 \right)^{2} \mathrm{d}x\, \mathrm{d}y \sim \int_0^1 (k^2 A^2 - 1)^2\, \mathrm{d}y. \tag{10.21}$$

In this equation like in the following, we drop all numerical factors of order 1, such as $1/2$ for instance. Recall that the domain of integration is defined by $0 \le x \le 1$ and $0 \le y \le 1$ in rescaled variables.

Unlike the bending term which comes with a very small factor ϵ^2, this stretching term is penalizing. To make the energy as small as possible, we should try to have $k(y)\, A(y) = 1$ everywhere—in this equality, the right-hand side is just residual strain η, which has been rescaled to a constant. However, $k(y)\, A(y) = 1$ is not possible at the clamped edge

$y = 0$ where the boundary conditions $w = 0$ and $w_{,y} = 0$ imposes $A(0) = 0$ and $A'(0) = 0$. Therefore, there exists a small boundary layer near the clamped edge where $k(y)\, A(y) - 1$ is of order 1. Let us call δ the width of this layer, assuming $\delta \ll 1$. The stretching energy in the layer can be estimated easily, since $(k\, A - 1) \sim 1$ and $y \sim \delta$ there by definition:

$$\mathcal{E}^{\mathrm{s}}_{xx}(\delta) \sim \int_0^\delta (k^2\, A^2 - 1)^2 \, \mathrm{d}y \sim \int_0^\delta \mathrm{d}y \sim \delta. \tag{10.22a}$$

To avoid spending a lot of stretching energy, we would like δ to be as small as possible. In order to estimate the width δ of the layer, we have to identify an energy contribution that would prefer δ not to be too small, and balance this contribution with the stretching energy. Let us call $k(\delta)$ the typical wavenumber of the oscillations in the boundary layer. The amplitude A has to start from $y = 0$ with $A(0) = 0$ and $A'(0) = 0$ by the boundary conditions, and reach $A(\delta) \sim 1/k(\delta)$ by definition of δ. Then, we can estimate the curvatures $w_{,xy} \sim k(\delta)\, A' \sim 1/\delta$ and $w_{,yy} \sim A'' \sim 1/(\delta^2\, k(\delta))$ in the boundary layer. These curvatures diverge, and so does the bending energy, if δ goes to zero. The contributions of the boundary layers in the bending energy read:

$$\mathcal{E}^{\mathrm{b}}_{xy}(\delta) = \epsilon^2 \iint w_{,xy}{}^2 \mathrm{d}x\,\mathrm{d}y \sim \epsilon^2 \int_0^1 \mathrm{d}x \int_0^\delta \mathrm{d}y \, \frac{1}{\delta^2} \sim \frac{\epsilon^2}{\delta}, \tag{10.22b}$$

$$\mathcal{E}^{\mathrm{b}}_{yy}(\delta) = \epsilon^2 \iint w_{,yy}{}^2 \mathrm{d}x\,\mathrm{d}y \sim \epsilon^2 \int_0^1 \mathrm{d}x \int_0^\delta \mathrm{d}y \, \frac{1}{k^2\delta^4} \sim \frac{\epsilon^2}{\delta^3\, k^2(\delta)}. \tag{10.22c}$$

Both these terms would like the layer to be not too small, and we have to balance one of them with the stretching energy (10.22a) in order to get δ. Since we do not know which bending contribution is dominant, we have to try both, one after the other.

Assuming that the second bending term (10.22c) wins over the first one (10.22b) in a strict sense, $\mathcal{E}^{\mathrm{b}}_{yy} \gg \mathcal{E}^{\mathrm{b}}_{xy}$, we can balance it with the stretching term, $\mathcal{E}^{\mathrm{b}}_{yy} \sim \mathcal{E}^{\mathrm{s}}_{xx}$, to determine δ. This yields $\epsilon^2/(\delta^3\, k^2(\delta)) \sim \delta$, hence $\delta \sim (\epsilon/k(\delta))^{1/2}$. Plugging back into the energy, we find that both energy terms are then of order $\mathcal{E}(\delta) \sim \delta \sim (\epsilon/k(\delta))^{1/2}$. Now, for the assumption to be consistent, we have to check $\mathcal{E}^{\mathrm{b}}_{yy} \gg \mathcal{E}^{\mathrm{b}}_{xy}$, which implies $\delta\, k(\delta) \ll 1$ and so $(\epsilon\, k(\delta))^{1/2} \ll 1$. We conclude that in this case the energy is much larger than ϵ: $\mathcal{E}(\delta) \sim (\epsilon/k(\delta))^{1/2} \sim \epsilon/(\epsilon\, k(\delta))^{1/2} \gg \epsilon$.

Consider the opposite case, when the first bending contribution $\mathcal{E}^{\mathrm{b}}_{xy}(\delta)$ is assumed to be comparable to, or larger than the second one, $\mathcal{E}^{\mathrm{b}}_{yy}(\delta)$, but not much smaller. Then we can balance it with the stretching term $\mathcal{E}^{\mathrm{b}}_{xy}(\delta) \sim \mathcal{E}^{\mathrm{s}}_{xx}(\delta)$. This selects δ such that $\epsilon^2/\delta \sim \delta$, that is $\delta \sim \epsilon$. Plugging back into the energy, we find $\mathcal{E}^{\mathrm{b}}_{xy}(\delta) \sim \mathcal{E}^{\mathrm{s}}_{xx}(\delta) \sim \epsilon$. The energy is found to be lower than under the previous assumption, where it was much larger than ϵ. We conclude that the present assumption is the correct one:

$$\delta \sim \epsilon.$$

For consistency, we have to check that $\mathcal{E}^{\mathrm{b}}_{yy}(\delta) \sim 1/(\epsilon\, k^2(\delta))$ is at most of order $\mathcal{E}^{\mathrm{b}}_{xy}(\delta) \sim \epsilon$ but not much larger; this implies that $k(\delta)$ is at least of order $1/\epsilon$ or larger. Therefore, $k(\delta)$ is a large number, meaning that there are short-scale oscillations of the film near the boundary. Now, making $k(\delta)$ much larger than $1/\epsilon$ does not bring any significant benefit (it will merely make $\mathcal{E}^{\mathrm{b}}_{yy}$ negligible in front of $\mathcal{E}^{\mathrm{b}}_{xy}$ but without changing the scalings just found), and would significantly increase the energy in the interior, as explained later in Section 10.2.5. Therefore, we shall assume

$$k(\delta) \sim \frac{1}{\epsilon}.$$

To summarize, the boundary layer has a width δ, a wavelength $\lambda(\delta)$, a wavenumber $k(\delta) = 2\pi/\lambda(\delta) \sim 1/\lambda(\delta)$ and an energy $\mathcal{E}_{\mathrm{bl}}$, all given, in rescaled units, by:

$$\delta \sim \epsilon, \qquad k(\delta) \sim \epsilon^{-1}, \qquad \lambda(\delta) \sim \epsilon, \qquad \mathcal{E}_{\mathrm{bl}} \sim \epsilon. \qquad (10.23)$$

In physical units, this translates into:

$$\delta \sim \frac{h}{\eta^{1/2}}, \qquad \lambda \sim \frac{h}{\eta^{1/2}}, \qquad \mathcal{E}_{\mathrm{bl}} \sim E\,h^2\,L\,\eta^{3/2}. \qquad (10.24)$$

Note that both the width of the boundary layer and the wavelength of the ripples is much larger than the plate's width, h, since we assume that the physical residual strain $\eta \ll 1$ is small so we can remain in the framework of Hookean elasticity. This is required for the approximations underlying the F.–von K. to be valid. Note that we assumed in addition that the residual strain is much above the (very small) residual strain at buckling threshold, namely $\eta_{\mathrm{c}} \sim (h/L)^2$. Therefore, the present analysis addresses the following range of residual strain:

$$\left(\frac{h}{L}\right)^2 \ll \eta \ll 1 \text{ in physical units, that is } \frac{h}{L} \ll \epsilon \ll 1 \text{ in rescaled form.}$$

This clarification is important as we have two small numbers ϵ and h/L; they must be such that $\epsilon \gg h/L$.

10.2.5 Cascade in the interior

In the boundary layer, the wavelength is very short, $\lambda = 2\pi/k \sim \epsilon$ and the wavenumber $k \sim \epsilon^{-1}$ is large. Further away from the boundary, such short-wave wrinkles are undesirable as they yield a large bending energy:

$$\mathcal{E}_{\mathrm{xx}}^{\mathrm{b}} = \epsilon^2 \int w_{,xx}{}^2 \, \mathrm{d}y \sim \epsilon^2 \int A^2\,k^4 \, \mathrm{d}y \sim \epsilon^2 \int (A\,k)^2\,k^2 \, \mathrm{d}y \sim \epsilon^2 \int_{\delta}^{1} k^2(y) \, \mathrm{d}y, \qquad (10.25)$$

where we have used $k\,A \sim 1$ since the in-plane strain ϵ_{xx} is assumed to be perfectly relaxed in the interior, i.e. far away from the boundary layer. In the rest of the chapter, we show that the wavelength of the wrinkles increases away from the clamped edge, by a succession of steps where the period is multiplied. This leads to the physical picture of a self-similar cascade of wrinkles in the interior of the plate.

We first have to identify what limits variations of $k(y)$ towards the interior, as $k(y)$ cannot drop abruptly from $\sim \epsilon^{-1}$ to zero at the edge of the boundary layer. When $k(y)$ is constant, the plate profile is cylindrical, i.e. invariant along the direction y, and so is developable, see Section 6.3. Therefore, we shall assume that variations of $k(y)$ with respect to y are limited by the stretching energy: we shall also assume that the structure of the solution in the interior reflects a balance between the bending energy (10.25) and the stretching energy, which we now estimate.

Let us start with a wrong but instructive estimate for this stretching energy. Assume that $k(y)$ relaxes from $\sim \epsilon^{-1}$ at the edge of the boundary layer to much smaller values in the interior, to avoid expending a lot of bending energy. Since the range of the variable y is $0 \leq y \leq 1$ in rescaled variables, this suggests that k' is also of order ϵ^{-1} in the interior. A naive estimate of $w_{,y}$ from equation (10.20) yields $w_{,y} \sim A k' \sim 1$ since $A \sim 1/k$ is at least of order ϵ. As a result, $w_{,yy}$ is at least of order 1, and so is $w_{,xx} \sim \partial w_{,x}/\partial x$, given that $w_{,x} \sim A k \sim 1$ by the inextensibility condition. We find that the Gauss curvature $K = w_{,xx}\, w_{,yy} - w_{,xy}{}^2$ is of order 1 over the whole plate. By Gauss' Theorema Egregium or by equation (6.88b) for the Airy potential, the in-plane strains, the density of stretching energy, and in consequence the stretching energy itself, are all of order 1 according to this estimate. In rescaled variables, the energy of the unbuckled configuration is of order 1 too, implying that the present construction hardly releases any stretching energy in the interior, even though we are far above threshold.

It is in fact possible to relax the stretching energy in the interior in a much more efficient manner. The key point is to make sure that the surface remains locally almost developable— almost cylindrical in fact. This is possible by changing the wavelength by successive period doubling (or tripling, etc.) rather than by a continuous change. In Section 10.2.6, we study an elementary rectangular cell of size $a \times b$ where period doubling occurs, assuming that the length b of the sides parallel to the y axis is comparable to or larger than the length a of the sides parallel to the x axis, but not much smaller; we show that the period can be doubled with a stretching energy not larger than $(a \times b) \times (a/b)^4$, see equation (10.31). We shall refer specifically to 'period doubling' as a more general formulation would be cumbersome, but it should be understood that the same estimates also hold for period tripling ($q = 3$) or for any other integer q: within the current framework of scaling analysis, it is not possible to decide which value of q is optimal.

The density of stretching energy given by this refined construction, $(a/b)^4$, is much lower than that obtained with the former naive estimate, ~ 1, when the cell has a large aspect ratio, that is for $b/a \gg 1$. This is because the surface is designed to be almost developable in this limit $b/a \gg 1$. In the following, we consider an array of such cells and consider a global construction in the interior featuring wrinkles whose wavelengths increase by successive period doubling.

Let us call $n(y)$ the generation of the fractal pattern at a distance y from the edge. In principle, n is an integer variable and every increment of n by one corresponds to a period doubling (or tripling, etc.) Since n takes on large values, the piecewise constant function $n(y)$ can be reasonably approximated by a *continuous*, smooth function. Since the solution is assumed to be periodic with respect to the variable x whose period is 1, the fundamental mode $n = 1$ corresponds to the wavevector $k = 2\pi$. The second mode $n = 2$ would correspond to $k = 2\pi q$, etc., where we recall that $q \geq 2$ is an integer that defines the type of cascade we consider. Dropping numerical constants of order 1, we obtain

$$k(y) \sim q^{n(y)}, \tag{10.26}$$

an equation that implies that the large quantities on both sides of the '$\sim$' sign are comparable.

To evaluate the stretching energy in the cascade, we anticipate the result proved in Section 10.2.6, namely that the density of stretching energy can be estimated as $(a/b)^4$, where a and b are the typical dimensions of the cell where period doubling takes place. The length a is just the wavelength of the pattern, and so $a = 2\pi/k(y) \sim q^{-n(y)}$. The parameter b is the increment of the variable y corresponding to a multiplication of the wavelength by

q; by definition, this corresponds to an increment of $n(y)$ by a unit. Therefore, b is such that $n'(y)\, b \sim 1$. We arrive at the following estimate for the stretching energy in the interior:

$$\mathcal{E}^{\mathrm{s}}_{\mathrm{int}} \sim \int \frac{a^4}{b^4}\, \mathrm{d}y \sim \int_{\delta}^{1} \frac{n'^4(y)}{(q^4)^{n(y)}}\, \mathrm{d}y. \tag{10.27}$$

We shall need to check in the end that a remains of order b or smaller, which is required for this estimate to be valid:

$$q^{-n(y)}\, n'(y) = \mathcal{O}(1). \tag{10.28}$$

Here we use the classical notation $\mathcal{O}(1)$ for a quantity that is comparable to or smaller than 1, but not much larger.

The wavevector $k(y)$ results locally from a balance of the bending energy, which tries to make it small, and from the stretching energy, which tries to prevent it from varying quickly. Therefore, the density of bending energy (10.25) and the density of stretching energy (10.27) work against each other and can be balanced. This leads to

$$\epsilon^2\, (q^2)^{n(y)} \sim \frac{n'^4(y)}{(q^4)^{n(y)}}$$

and therefore

$$n'(y) \sim \left(\epsilon^2\, (q^6)^{n(y)} \right)^{1/4} \sim \epsilon^{1/2} \left(q^{3/2} \right)^{n(y)}.$$

Instead of trying to determine $n(y)$, it is more convenient to work with the inverse function $y(n)$, whose derivative can be estimated as

$$\frac{\mathrm{d}y}{\mathrm{d}n} = \frac{1}{n'(y)} \sim \frac{\pm 1}{\epsilon^{1/2}\, (q^{3/2})^n}$$

by the previous equation.

This last equation can be viewed as a differential equation for the unknown function $n(y)$. Its integration involves an exponential function and can be carried out using the identity $\mathrm{d}n/(q^{3/2})^n = \mathrm{d}\left(-1/3/2 \ln q\, 1/(q^{3/2})^n \right)$. Dropping the factor $(3/2 \ln q)$, which is of order 1, we obtain

$$\mathrm{d}(\epsilon^{1/2}\, y) \sim \mathrm{d}\left(\frac{1}{(q^{3/2})^n} \right).$$

The constant of integration can be determined by requiring that the cascade matches with the boundary layer: for $y \sim \delta \sim \epsilon$ we expect $1/(q^{3/2})^n \sim (1/q^n)^{3/2} \sim (1/k)^{3/2} \sim \epsilon^{3/2}$. Plugging back into the above equation, this asymptotic condition is found to be consistent[14] with a vanishing constant of integration, hence the solution:

$$\epsilon^{1/2}\, y \sim \frac{1}{(q^{3/2})^n}.$$

This means that the n-th generation of the cascade can be found at a distance $y(n)$ from the edge given by

$$y(n) \sim \frac{1}{\epsilon^{1/2}}\, \frac{1}{(q^{3/2})^n}. \tag{10.29}$$

---

[14] The constant of integration is not defined uniquely since we are doing an analysis in orders of magnitude: everything is defined up to an arbitrary factor of order one.

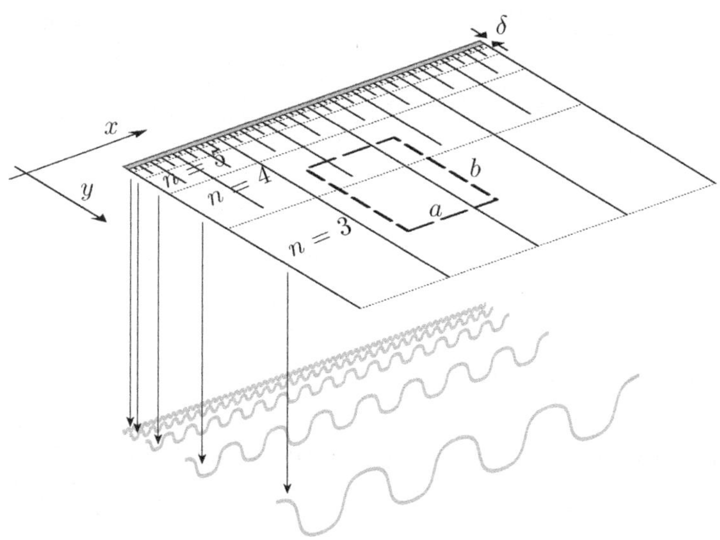

Fig. 10.8 Global picture of the solution for a strongly post-buckled plate with a clamped edge: there is a boundary layer of width δ with a large density of stretching energy and small-scale oscillations, see Section 10.2.4. Towards the interior, the wavelength of the oscillations relaxes to much larger values by a sequence of period-doubling (or tripling, etc.) steps, see Section 10.2.5. The elementary cell, of size $a \times b$, over which period doubling occurs is studied in Section 10.2.6.

We end up with the solution depicted in Fig. 10.8, featuring a cascade of wrinkles in the interior in geometric progression.

The largest wavelength in the cascade is found at the edge $y = 1$ opposite the clamped edge. Taking $y = 1$ in equation (10.8), we have $q^n \sim \epsilon^{-1/3}$ and therefore $\lambda(y = 1) \sim 1/q^n \sim \epsilon^{1/3}$. This is a short wavelength in comparison with the width of the plate,[15] which is 1 in rescaled units, but this is still a much longer wavelength that that found near the clamped end, ϵ.

We can now estimate the bending and stretching energies in the interior region, which are both of the same order of magnitude by construction:

$$\mathcal{E}_{\text{int}} \sim \int_{y=\delta}^{1} \epsilon^2 (q^n)^2 \, \mathrm{d}y \sim \int_{y=\delta}^{1} \epsilon^2 \left(\frac{1}{\epsilon^{1/3} y^{2/3}} \right)^2 \mathrm{d}y \sim \int_{y=\delta}^{1} \frac{\epsilon^{4/3}}{y^{1/3}} \, \mathrm{d}y \sim \frac{\epsilon^{4/3}}{\delta^{1/3}} \sim \epsilon.$$

Therefore, the bending energy and the stretching energy of the boundary layer, and the bending energy and the stretching energy of the interior domain are all of the same order of magnitude, namely $\mathcal{E}_{\text{bl}} \sim \mathcal{E}_{\text{int}} \sim \epsilon$.

Let us return to equation (10.28) expressing the condition $a/b = \mathcal{O}(1)$. With the solution just given, we compute

$$\frac{a}{b} \sim \frac{n'}{q^n} \sim \frac{1}{\frac{\mathrm{d}y}{\mathrm{d}n} \, q^n} \sim \frac{\epsilon^{1/2}}{\left(\frac{1}{q^{3/2}} \right)^n q^n} \sim (\epsilon \, q^n)^{1/2} \sim \left(\frac{\epsilon}{y} \right)^{1/2}.$$

---

[15] By equation (10.29), the plate would need to be much wider in the y direction, of size $1/\epsilon^{1/2}$, for the fundamental wavelength $n = 1$ to be reached.

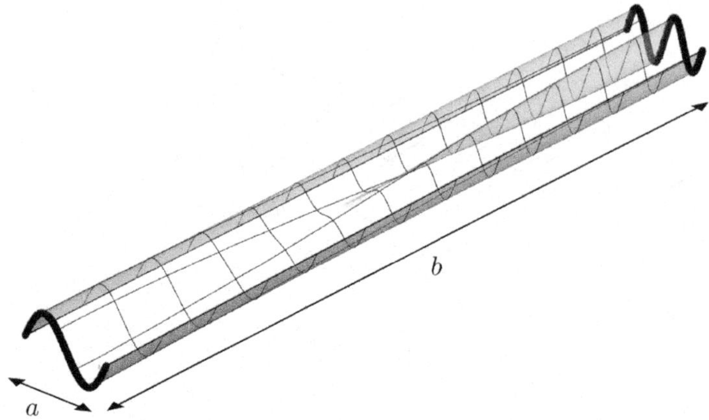

Fig. 10.9 Elementary cell, of size $a \times b$, where period doubling takes place, after reference (W. Jin and P. Sternberg, 2001).

This number is of order one in the vicinity of the boundary layer, where of course the density of stretching energy becomes of order one as in the layer. Far away from the layer, $y \gg \delta$ and a/b becomes very small. The aspect ratio a/b is nowhere much larger than 1 and so the assumption is justified.

10.2.6 Period doubling in an elementary cell

We consider a cell of size $a \times b$ over which period doubling occurs. As mentioned earlier, the same scaling argument could be applied to period tripling ($q = 3$), or to any integer value of q. We consider the case of period doubling ($q = 2$) as an illustration but do not imply that this is the optimal value of this parameter. We introduce a plate profile which is almost developable in the limit of a cell with a large aspect ratio, $b \ll a$, and so has a low stretching energy in this limit. The analysis is correct as well in the case when the sides of the cell are of comparable length, $b \sim a$, even though the construction is uninteresting as it does not make the stretching energy particularly small in this case. We exclude the case $b \gg a$, where the present construction becomes irrelevant.

The geometry is shown in Fig. 10.9: the plate profile is constructed by evolving smoothly from one oscillation on one side, to two oscillations on the other side—for details, see reference (W. Jin and P. Sternberg, 2001). In the limit $b \gg a$, the surface is close to being cylindrical, that is invariant along the direction y.

An important remark is that the derivatives along x and y follow different scalings when the cell has large aspect ratio:

$$\frac{\partial}{\partial x} \sim \frac{1}{a} \quad \gg \quad \frac{\partial}{\partial y} \sim \frac{1}{b}.$$

Calling A the typical amplitude as earlier, we have the following estimate for the Gauss curvature:

$$K = w_{,xx}\, w_{,yy} - w_{,xy}^{2} \sim \frac{A^2}{a^2\, b^2}.$$

The developability condition along the direction x imposes

$$1 \sim w_{,x} \sim \frac{A}{a}.$$

Combined with the previous equation, this yields

$$K \sim \frac{1}{b^2}.$$

Therefore, the Gauss curvature is small but non-zero. By the compatibility equation (6.12), the in-plane strain has to be non-zero too, and there is some stretching energy to pay:

$$2\frac{\partial^2 \epsilon_{xy}}{\partial x\,\partial y} - \frac{\partial^2 \epsilon_{xx}}{\partial y^2} - \frac{\partial^2 \epsilon_{yy}}{\partial x^2} = K \sim \frac{1}{b^2}. \tag{10.30}$$

Notice that balancing the various terms in the left-hand side of this equation with the right-hand side yields different scalings for the corresponding strain components: $\epsilon_{xx} \sim 1$, $\epsilon_{xy} \sim a/b$ and $\epsilon_{yy} \sim (a/b)^2$. Therefore, the term $\epsilon_{xx}{}^2$ in the density of stretching energy (10.17) is formally of order 1, the term $2\,\epsilon_{xy}{}^2$ much smaller, of order $(a/b)^2$, and the term $\epsilon_{yy}{}^2$ is even smaller, of order $(a/b)^4$. Clearly, the stretching energy subjected to the constraint (10.30) can be minimized in the limit $b \gg a$ by taking

$$\epsilon_{xx} = 0, \quad \epsilon_{xy} = 0, \quad \epsilon_{yy} \sim \frac{a^2}{b^2}$$

in the compatibility equation (10.30). Doing so, higher-order contributions to the stretching energy are avoided altogether and the density of stretching energy is as small as

$$\epsilon_{xx}{}^2 + 2\,\epsilon_{xy}{}^2 + \epsilon_{yy}{}^2 \sim \frac{a^4}{b^4}, \tag{10.31}$$

as was announced earlier.

Using the definition of strain, the equations $\epsilon_{xx} = 0$ and $\epsilon_{xy} = 0$ can be used to reconstruct the in-plane displacement u and v explicitly, see reference (W. Jin and P. Sternberg, 2001). Of course, this requires the plate profile to be chosen such that the inextensibility condition (10.19) is satisfied for all y.

The argument above is adapted to the case $b \gg a$. Now, if b and a are comparable, it is no longer optimum to cancel ϵ_{xy} and ϵ_{xx} but still that gives an estimate of the stretching energy, 1, which is correct in order of magnitude: when b/a is no longer a small parameter, equation (10.30) imposes that all components of the in-plane strain are of order 1, and so is the density of stretching energy. Our estimate then yields $(b/a)^4 \sim 1$ for the density of stretching energy.

10.2.7   Summary

We have formally constructed a solution of the F.–von K. equations describing a strongly post-buckled plate having a clamped edge, under residual strain in the direction parallel to the clamped edge. The structure of the solution was proposed based on scaling arguments: close to the boundary there is a thin layer where the stretching energy is large. In this layer, small-scale oscillations are formed so as to relax the longitudinal strain. Further away from the layer, these small-scale wrinkles relax towards larger wavelengths by a succession of period doubling. Remarkably, this rather complex construction is supported entirely by simple scaling arguments. In references (W. Jin and P. Sternberg, 2001) and (H. Ben Belgacem, 2000), this construction is complemented by a mathematical proof showing that the energy of the minimizers of the elastic energy follows the same scaling law as the construction proposed here. Although this is not a formal proof, this strongly suggests that the minimizers themselves are self-similar.

10.3 Summary and conclusion

The theory of plate is notoriously difficulty due to the non-linear character of the equations. In this chapter, we have given two illustrations where the geometrical underpinning of plate theory, related to the problem of embeddings, can be utilized to obtain solutions.

In the first part of this chapter, we studied the buckling of a plate that has a stretched edge under moderate loading. We introduced a simplified numerical model that reproduces the fractal buckling patterns observed in the experiments. In particular, we studied how the details of the metric, such as the value of the regularizing length, affect the cascade. In the second approach, the loading is concentrated along the edge via a clamping condition. In this case, the cascade can be described extensively using scaling arguments in the strongly post-buckled limit.

The work presented in Section 10.1 was done in collaboration with A. Boudaoud, see reference (B. Audoly and A. Boudaoud, 2003); the post-buckling of plates with a clamped edge has been studied in collaboration with S. Rica in an early paper (Y. Pomeau and S. Rica, 1997). Both of them are warmly thanked.

References

B. Audoly and A. Boudaoud. Self-similar structures near boundaries in strained systems. *Physical Review Letters*, 91(8):086105, Aug. 2003.

B. Audoly and A. Boudaoud. Buckling of a thin film bound to a compliant substrate (part 2). A global scenario for the formation of herringbone pattern. *Journal of the Mechanics and Physics of Solids*, 56(7):2422–2443, 2008.

A. S. Argon, V. Gupta, H. S. Landis, and J. A. Cornie. Intrinsic toughness of interfaces between SiC coatings and substrates of Si or C fibre. *Journal of Materials Science*, 24:1207–1218, 1989.

G. I. Barenblatt. *Similarity, Self-similarity and Intermediate Asymptotics*. Consultant bureau, New York, 1979.

H. Ben Belgacem, S. Conti, A. DeSimone, and S. Müller. Rigorous bounds for the Föppl–von Kármán theory of isotropically compressed plates. *Journal of Nonlinear Science*, 10(6):661–685, 2000.

H. Ben Belgacem, S. Conti, A. DeSimone, and S. Müller. Energy scaling of compressed elastic films – three-dimensional elasticity and reduced theories. *Archive for Rational Mechanics and Analysis*, 164(1):1–37, 2002.

S. Conti, A. DeSimone, and S. Müller. Self-similar folding patterns and energy scaling in compressed elastic sheets. *Computer Methods in Applied Mechanics and Engineering*, 194(21-24):2534–2549, 2005.

S. Conti. Branched microstructures: scaling and asymptotic self-similarity. *Communications on Pure and Applied Mathematics*, 53:1448–1474, 2000.

G. Gioia, A. DeSimone, M. Ortiz, and A. M. Cuitino. Folding energetics in thin-film diaphragms. *Proceedings of the Royal Society A: Mathematical, Physical and Engineering Sciences*, 458(2021):1223–1229, 2002.

W. Jin and P. Sternberg. Energy estimates for the von Kármán model of thin-film blistering. *Journal of Mathematical Physics*, 42(1):192–199, 2001.

W. Jin and P. Sternberg. In-plane displacements in thin-film blistering. *Proceedings of the Royal Society of Edinburgh: Section A Mathematics*, 132(4):911–930, 2002.

L. D. Landau. The intermediate state of supraconductors. *Nature*, 141:688, 1938.

L. D. Landau. On the theory of the intermediate state of superconductors. *Journal of Physics-USSR*, 7:99, 1943.

B. Mandelbrot. *Fractals: Form, Chance and Dimension.* W. H. Freeman and Co., 1977.

M. Marder, E. Sharon, S. Smith, and B. Roman. Theory of edges of leaves. *Europhysics Letters*, 62(4):498–504, 2003.

S. Nechaev and R. Voituriez. On the plant leaf's boundary, 'jupe à godets' and conformal embeddings. *Journal of Physics A: Mathematical and General*, 34:11069–11082, 2001.

M. Ortiz and G. Gioia. The morphology and folding patterns of buckling-driven thin-film blisters. *Journal of the Mechanics and Physics of Solids*, 42(3):531–559, 1994.

Y. Pomeau. Buckling of thin plates in the weakly and strongly nonlinear regimes. *Philosophical Magazine B*, 78(2):235–242, 1998.

Y. Pomeau and S. Rica. Plaques très comprimées. *Comptes Rendus de l'Académie des Sciences - Series IIB - Mechanics-Physics-Chemistry-Astronomy*, 325(4):181–187, 1997.

M. Spivak. *A Comprehensive Introduction to Differential Geometry*, volume 5. Publish or Perish, Inc., Houston (TX), 2nd edition, 1979.

E. Sharon, B. Roman, M. Marder, G.-S. Shin, and H. L. Swinney. Mechanics: Buckling cascades in free sheets. *Nature*, 419(6907):579–579, 2002.

E. Sharon, B. Roman, and H. L. Swinney. Geometrically driven wrinkling observed in free plastic sheets and leaves. *Physical Review E (Statistical, Nonlinear, and Soft Matter Physics)*, 75(4):046211, 2007.

H. von Koch. Sur une courbe continue sans tangente, obtenue par une construction géométrique élémentaire. *Arkiv för Matematik*, 1(681–704), 1904.

Part III

Shells

Thin elastic shells are similar to elastic plates: they are thin two-dimensional elastic objects. Unlike plates, their centre surface can be naturally curved. The elasticity of thin shells share many features in common with the elasticity of plates but the geometry plays an even greater role and is in fact even more interesting. We touched on the fundamental geometric question underlying problems in shell theories earlier in Chapter 10, namely that of isometric embeddings: we modified the plate equation in a way that makes them rather similar to the equations for shells.

In plate theory, we found that the in-plane and out-of-plane displacements contribute to in-plane stretching either directly (in-plane) or through non-linear terms (out-of-plane). As a result, we found that the in-plane and out-of-plane displacements follow different scalings, the plate being essentially weak with respect to normal displacements. These scalings do not hold in shell theory: as illustrated by the radial inflation of a spherical shell, a purely normal displacement can modify the tangential strain to first order. This means that all components of the displacements must be considered on a more equal footing.

In plate theory, the purely transverse modes of displacement involve no stretching energy to first order and so a plate is 'soft' with respect to these modes of deformation: this property explains why it is important to consider bending effects in many problems of plate elasticity, as it restores some stiffness to these soft modes. Some soft modes can be encountered in shell theory too, and they play a similar role to those found in plate theory: typically, an elastic shell displays soft modes when its centre surface is non-rigid geometrically, i.e. when it can deform without changing its metric properties. This is where geometry comes into the equations. Unlike in plate problems, the description of these inextensional deformations is not possible in closed form; the analysis of inextensional deformations of the centre surface is the topic of Chapter 11. Our presentation is restricted to the case of shells of revolution as it is not possible to find explicit solutions of the equations in the general case; the equations for non-axisymmetric elastic shells can be found in many references such as Goldenveizer's classical book (A. L. Goldenveizer, 1961). In Chapter 12 we derive some equations for the equilibrium of shells of revolution, which extend the F.–von K. equations given earlier for elastic plates. In the two remaining chapters, we study the deformations of two simple shell geometries, a torus in Chapter 13 and a spherical shell in Chapter 14. The example of an elastic torus is interesting because the centre surface is of mixed type, the outer part of the torus being elliptic although the inner part is hyperbolic; the geometrical rigidity of the centre surface is ruled by some special curves which are shown to dictate the mechanical behaviour of the elastic structure. In the last chapter, the analysis of the ball (a ping-pong ball, for instance) reveals a buckling phenomenon whereby a spherical cap pops inwards, something that can be analysed using piecewise smooth embeddings.

References

A. L. Goldenveizer. *Theory of Elastic Thin Shells*. Pergamon Press, New York, NY, USA, 1961.

11
Geometric rigidity of surfaces

An elastic shell is a piece of elastic material whose thickness is small compared with its other dimensions. In contrast to plates, a shell has a non-flat natural configuration. Plates can be seen as a very particular case of shells, if one wished to put things in a general framework. As for plates, we will consider shells of uniform thickness made of a homogeneous material. This assumption is for the sake of simplicity and the analysis is not changed in any fundamental way if the material is not homogeneous or the thickness not constant. The particular case when the shell properties, such as thickness or elastic constants, change at a small scale[1] is interesting: a general method, called homogenization, allows one to derive effective, uniform parameters characterizing the deformations of the shell at a large scale; it has been the subject of active research, in particular in problems related to elasticity of plates and shells. We refer the interested reader to a monograph on the topic (É. Sanchez Palencia and J. Sanchez-Hubert, 1992).

The elasticity of shells is *a priori* a more challenging question than the elasticity of plates, because of the underlying geometry. By definition, the natural state of a shell defines a curved surface—we shall assume that this surface is sufficiently smooth. Much fewer results are known concerning the differential geometry of curved surfaces than in the planar and developable case: for instance, one does not know in general if a given surface can be deformed in a smooth way without stretching, besides rigid-body rotations and translations.

This geometric question is central in problems for elastic shells. As was done for plates earlier in Chapter 6, it is possible to carry out a systematic expansion of the elastic energy in powers of the small parameter of the problem, the thickness h. This leads to an expansion of the equilibrium equations for shells in powers of h. The structure of their solutions for small h depends crucially on the answer to the geometrical problem mentioned above, which is not known in general.[2] Before attacking problems in shells elasticity, we consider in this chapter the geometrical problem of isometric deformations of a surface. In some sense, this corresponds to the limit of a shell with a vanishing thickness, $h = 0$. The elasticity problem, corresponding to small but non-zero thickness $h > 0$, will be considered in the following chapters.

The geometrical problem of the isometric deformation of surfaces is defined more precisely in the introductory Section 11.1 where the connection with the elasticity of shells is established. We characterize the isometric deformations of surfaces in a simple geometrical setting first: in Section 11.2, we consider the case of surfaces that remain everywhere close

---

[1] This length scale has to be much smaller than the large size of the plate and much larger than its thickness.

[2] A very similar situation was encountered in Chapter 10 when we investigated the existence of geometric embeddings associated with a particular profile of residual strain. Neither the existence of embeddings nor their explicit description was available in general.

to a plane. In this case, one can use the same simplified expression of the in-plane strain as in the analysis of plates in Chapter 6, and the analysis follows closely that of developable surfaces. The general case is studied next: the equations for the isometric deformations of a surface with arbitrary shape at rest are derived in Section 11.3. These equations are partial differential equations and have no known general solution. We consider two particular cases where the equations can be solved: minimal surfaces in Section 11.4 and surfaces of revolution in Section 11.5. The detailed analysis of surfaces of revolution will be useful in the forthcoming chapters, when we study the elasticity of toroidal and spherical shells. The isometric deformations of surfaces of revolution are amenable to complete analytical solution; this allows us to point out special curves drawn on the surface that provide additional rigidity. This important property is explained geometrically, and extended to surfaces that are not of revolution in Section 11.6.

11.1   Introduction

In Chapter 6 we derived the equations for thin elastic plates and studied in detail the developability of surfaces. This was motivated by the remark that the stretching energy, although formally much larger than the bending energy, can be made exactly zero if the centre surface of a plate remains developable. In shell problems, isometric deformations of the centre surface have exactly the same property. As a result, they show up naturally in the deformation of elastic shells, as was already known to Rayleigh. In 1877, he wrote, in his study of the vibrations of shells (J. W. S. Rayleigh, 1976, Section 235b): *[a] deformation [of a thin elastic shell of thickness h] includes, in general, both stretching and bending, and any expression for the [elastic] energy will be of the form:*

$$A\,h(\text{extension})^2 + B\,h^3(\text{bending})^2. \tag{11.1}$$

This energy is to be as small as possible. Hence, when the thickness is diminished without limit, the actual displacement will be of pure bending, if such there be... As pointed out by Rayleigh, the presence of modes of vibration with a low frequency in the spectrum is a sign of the existence of isometric modes of deformation of the centre surface. This geometric question is central not only to the analysis of the vibrations of a shell but also to its buckling behaviour, its deformation under imposed loads, etc.

The isometric rigidity problem can be stated as follows: given a surface in the Euclidean 3D space,[3] can it be deformed without being stretched? The absence of stretching means that the length of any curve drawn on the surface is unaffected by the deformation. This can be characterized by means of the metric tensor introduced in Section 10.1.2, or more accurately by its change with respect to the undeformed configuration (this defines an in-plane strain tensor).

We shall consider undeformed surfaces that are smooth, and isometric deformations of these surfaces that are also smooth. The smoothness requirements on the deformations are essential: as we shall see later, a sphere cannot be deformed in a smooth and isometric manner, but it can be deformed in a non-smooth and isometric manner as shown by the inversion of a cap in a spherical shell studied in Section 14.7. Actually any surface can

---

[3] The rigidity problem is linked to the problem of embedding of the surface in the Euclidean 3D space: the surface is rigid when the embedding problem has a unique solution.

be deformed isometrically with C^1 smoothness[4] but the deformed surfaces have in general an infinite bending energy and are therefore irrelevant to the theory of elastic shells—see reference (I. Kh. Sabitov, 1992) for a review and references therein such as (M. L. Gromov and V. A. Rokhlin, 1970).

Three types of isometric deformations can be distinguished. In the literature, they are commonly referred to as (i) warpings, (ii) bendings, and (iii) infinitesimal bendings, although the names sometimes vary. Warpings are discrete: they define a change of configuration of a surface involving a finite displacement, the final surface being disconnected from the original one in the space of configurations. An example of warping is the (smooth) mirror-reflection of a half-sphere resting on an undeformable circle. Bendings refer to a one-parameter family of surfaces that are all isometric to each other, the initial surface being in the family. In this case, the surface can be continuously deformed while remaining unstretched. An example of a bending is the bending of a plane into a cylinder. Infinitesimal bendings are infinitesimal perturbations to a surface that keep the metric tensor unchanged to first order with respect to the amplitude of the deformation. For a planar surface, any infinitesimal displacement field that is purely normal does not change the metric tensor to first order, and defines an infinitesimal bending. These three notions (warpings, bendings and infinitesimal bending) are all different. In particular, an infinitesimal bending may not be integrable into a bending with finite amplitude (I. Kh. Sabitov, 1992, §3), as shown by the example of a rectangular surface with four edges attached to an undeformable frame: no finite amplitude bending exists in this geometry, although any infinitesimal displacement field that is purely normal and vanishes on the frame yields an infinitesimal bending.

In the rest of this chapter, we shall exclusively consider infinitesimal bendings: we study infinitesimal displacements of the surface that keep its metric tensor unchanged to first order. These infinitesimal bendings are sometimes called *pure* infinitesimal bendings in the literature, the term 'pure' referring to the absence of in-plane extension. In the absence of ambiguity, we shall simply refer to *bendings*, often omitting the adjective 'infinitesimal'. The existence or lack of infinitesimal bendings will be called the (infinitesimal) *rigidity problem*.

The boundary conditions applied on the edges of a surface are essential for the rigidity problem. The examples given just above suggest that a surface may be bendable when its edges are free, but may become isometrically rigid when they are undeformable. The example of hyperbolic surfaces is interesting: they may or may not be made rigid by imposing fixed edge conditions on parts of their boundaries, depending on the relative position of these boundary conditions and of the characteristic curves (M. Spivak, 1979).

There is a vast literature on the rigidity of surfaces, and no general answer exists for an arbitrary surface. The mathematically inclined reader can refer to (M. Spivak, 1979, Ch. 12) for a presentation of some of the known results, including the case of hypersurfaces in higher-dimensional spaces; we also recommend the more accessible review M. L. Szwabowicz, 1999, Section 3). A detailed analysis of the local problem can be found in (I. Kh. Sabitov, 1992, Sections 6–7) including the case of flat points of arbitrary order. As we show in the rest of this chapter, the problem of finding infinitesimal bendings amounts to solving a set of coupled partial differential equations of the same type as the geometric type of the surface. The rigidity problem can therefore be studied using the general tools of analysis of

---

[4] The deformed surface is required to be differentiable and to have continuous unit normals but the latter may not be differentiable, and so the curvature may be ill-defined.

partial differential equations (PDE). Known results for the rigidity problem mostly concern convex surfaces, when the underlying equations are elliptic. A classical result is that closed convex surfaces are rigid, see for instance reference (M. Spivak, 1979). Various extensions of this result are available, to surfaces that are non-convex but close to convex in some sense (A. D. Alexandrov, 1938; L. Nirenberg, 1963)—however, note that when an arbitrarily small domain with positive curvature is removed from a convex surface, it is no longer rigid isometrically, see A. V. Pogorelov, 1973, Ch. 2) and (M. Spivak, 1979). The isometric rigidity of closed convex surfaces explains the impressive resistance of an egg shell to compression— the experiment can be tried with the hands, preferably after emptying the contents of the egg shell through a tiny hole.

11.2   Infinitesimal bendings of a weakly curved surface

We start by deriving the equations for the infinitesimal bendings of a surface that is weakly curved, i.e. consider the case when the surface departs only slightly from a plane (in shell theory, this geometry is often referred to as the case of a *shallow* shell). With this assumption, the equations for bendings can be derived by simple mathematical arguments, using a Cartesian parameterisation of the surface. This simplifying assumption will be relaxed in Section 11.3 when the bendings will be studied using intrinsic geometry, i.e. without making reference to a particular basis in the ambient space.

The present problem is an extension of the question of the developability of a surface— the developability problem concerns isometries between a surface and a piece of a plane. The rigidity problem is analysed here using an approach similar to our derivation of Gauss' Theorema Egregium in Section 6.2. Using the same notation, we write S as the undeformed surface and $\bar{S}$ as the infinitesimally bent surface. We use a Cartesian basis whose axes x and y are contained in the plane remaining close to S; the surface S can be described by a Cartesian equation $z = f(x, y)$. The undeformed configuration is given by the surface with parameterisation

$$(x, y) \mapsto (x, y, f(x, y)). \tag{11.2a}$$

Calling $u(x, y)$, $v(x, y)$ and $w(x, y)$ the three components of the displacement, the parameterisation of the bent surface, $\bar{S}$, reads:

$$(x, y) \mapsto (x + u(x, y), y + v(x, y), \overline{f}(x, y)), \tag{11.2b}$$

an equation similar to (6.6a). Here we have defined the deformed profile of the surface

$$\overline{f}(x, y) = f(x, y) + w(x, y). \tag{11.3}$$

We assume that the deformation is infinitesimal: the functions u, v and w will all be considered to be of order ϵ at most, where ϵ is a small number. Since we assume that S deviates slightly from a plane, we shall assume that f is of order ϵ or smaller:

$$u(x, y) = \mathcal{O}(\epsilon), \quad v(x, y) = \mathcal{O}(\epsilon), \quad w(x, y) = \mathcal{O}(\epsilon), \quad f(x, y) = \mathcal{O}(\epsilon). \tag{11.4}$$

The notation $\mathcal{O}(\epsilon)$, used in different places in this book, means that a quantity is of order ϵ or smaller.

The condition of isometry between S and $\bar{S}$ has been given in equation (6.8); it involves the strain tensor in the two states, namely $\epsilon_{\alpha\beta}(x,y)$ and $\bar{\epsilon}_{\alpha\beta}(x,y)$, which have to be equal:

$$\epsilon_{\alpha\beta}(x,y) = \bar{\epsilon}_{\alpha\beta}(x,y). \tag{11.5}$$

This equation must be satisfied for all in-plane indices $\alpha, \beta = x, y$, and for all x and y. By convention, the in-plane strain is defined here using a plane as a reference configuration. The strain has been expressed in equations (6.9) in terms of the displacements. In general it depends non-linearly on the displacement but in Section 6.2.1 we showed that it can be linearized with respect to in-plane displacement under specific assumptions. As a result, we could eliminate u and v and the equations (11.5) could be cast into an equation for $f(x,y)$ and $w(x,y)$, which we interpreted as the conservation of Gauss curvature under isometry.

A similar reasoning can be made here. By our scaling assumptions (11.4), all the non-linear terms are of order ϵ^2 or smaller in the strains (6.9). Together with the isometry condition (11.5), this implies that $u_{,x}$, $v_{,y}$ and $u_{,y} + v_{,x}/2$ are of order ϵ^2 or smaller, and not of order ϵ as could be expected by our scalings. These three quantities can be identified as the strain induced by a displacement $(u(x,y), v(x,y))$ in a 2D geometry, as if we had set $f(x,y) = 0$ and $w(x,y) = 0$ for all x and y. This implies[5] that u and v can be decomposed into a rigid-body translation and rotation, possibly of order ϵ, plus an additional displacement of order ϵ^2. We shall discard the rigid-body part of the displacement as it is irrelevant for the analysis of bendings. We have just shown that in-plane displacements are much smaller than the out-of-plane ones:

$$u(x,y) = \mathcal{O}(\epsilon^2), \quad v(x,y) = \mathcal{O}(\epsilon^2). \tag{11.6}$$

Starting from slightly different assumptions, we have recovered the same scalings for the in-plane displacement as in Section 6.2.1. As earlier, we can then neglect the non-linear terms coming from in-plane displacements in the strain; this leads again to the expressions (6.11), which are linear with respect to u and v but non-linear with respect to f and w. Elimination of the functions u and v shows that the quantity

$$\overline{K}(x,y) = \overline{f}_{,xx}\,\overline{f}_{,yy} - \overline{f}_{,xy}{}^2 \tag{11.7}$$

depends on the strains $\bar{\epsilon}_{\alpha\beta}$ only, as shown earlier in equation (6.12). The consequence is that the quantity K, called the Gauss curvature, is preserved by isometric deformations: $K(x,y) = \overline{K}(x,y)$ where $K(x,y)$ is the Gauss curvature of the undeformed surface S, obtained by substituting $\overline{f}$ by f in equation (11.7).

Since w is infinitesimal, one can linearize the equation $K = \overline{K}$ with respect to w. Using the definition of $\overline{f}$ in equation (11.3), we obtain the equation

$$f_{,yy}\,w_{,xx} - 2f_{,xy}\,w_{,xy} + f_{,xx}\,w_{,yy} = 0, \tag{11.8}$$

---

[5] To prove this, note the 2D strains $\hat{\epsilon}_{xx} = u_{,x}$, $\hat{\epsilon}_{yy} = v_{,y}$ and $\hat{\epsilon}_{xy} = (u_{,y} + v_{,x})/2$, and define $\omega = (-u_{,y} + v_{,x})/2$. Remark that $\omega_{,x} = \hat{\epsilon}_{xy,x} - \hat{\epsilon}_{xx,y}$ and $\omega_{,y} = -\hat{\epsilon}_{xy,y} + \hat{\epsilon}_{yy,x}$. Here, we assume $\hat{\epsilon}_{\alpha\beta} = 0$ at order ϵ: ω has a vanishing gradient and so is constant at this order. Integrating $u_{,x} = \hat{\epsilon}_{xx} = 0$ and $u_{,y} = -\omega + \hat{\epsilon}_{xy} = -\omega$, we find $u(x,y) = u_0 - \omega\,y + \mathcal{O}(\epsilon^2)$. By a similar argument, $v(x,y) = v_0 + \omega\,x + \mathcal{O}(\epsilon^2)$. The constants of integration u_0 and v_0 represent a rigid-body translation and ω represents a rigid-body rotation: when the strain is zero at order ϵ, the displacement is a rigid-body translation and rotation at this order.

where $f(x,y)$ is given and $w(x,y)$ is the unknown. Note that the left-hand side is just the bracket operator introduced in equation (1.13): the equation for bendings can be written $[f,w] = 0$.

Equation (11.8) is the partial differential equation (PDE) for the infinitesimal bendings $w(x,y)$ of a weakly curved surface $z = f(x,y)$. Second-order partial differential equations are classified depending on the sign of the determinant of the matrix formed by collecting the coefficients in front of the highest order derivatives. In the present case, the determinant is

$$\det \begin{pmatrix} -f_{,yy} & f_{,xy} \\ f_{,xy} & -f_{,xx} \end{pmatrix} = \det \begin{pmatrix} f_{,xx} \ f_{,xy} \\ f_{,xy} \ f_{,yy} \end{pmatrix} = \det H(x,y),$$

where $H(x,y)$ is the Hessian matrix of the undeformed surface $f(x,y)$. By equation (6.14), the determinant of this matrix $H(x,y)$ is just the Gauss curvature $K(x,y)$: the equation (11.8) for infinitesimal bendings is elliptic when the surface is itself elliptic ($K > 0$), parabolic when the surface is parabolic ($K = 0$) and hyperbolic when the surface is hyperbolic ($K < 0$). When the surface is of mixed type, the PDE is also of mixed type (see the example of the torus in Chapter 13).

11.3   Infinitesimal bendings: an intrinsic approach

We now consider the equations for infinitesimal bendings on an arbitrary, smooth surface using an intrinsic approach: in contrast to the previous section, we do not assume that the surface remains close to a plane, and do not make use of a basis in the ambient Euclidean space.

11.3.1   Parameterisation, tangent space

Let us consider a smooth surface given by a parameterisation

$$(u, v) \mapsto \mathbf{r}(u, v),$$

where (u, v) is a local system of coordinates along the surface. The surface is assumed to be smooth (the final equation that we shall obtain for bendings shows that it has to be differentiable three times). At any given point along the surface, the tangent plane is spanned by the two vectors

$$\mathbf{e}_u(u, v) = \mathbf{r}_{,u}(u, v) \quad \text{and} \quad \mathbf{e}_v(u, v) = \mathbf{r}_{,v}(u, v). \tag{11.9}$$

As in several places in this book, we use the compact notation $f_{,x}$ for the partial derivative $\partial f/\partial x$ of a function $f(x)$. The vector basis $(\mathbf{e}_u, \mathbf{e}_v)$ is naturally associated with the coordinates (u, v). The unit normal to the surface is defined by

$$\mathbf{n}(u, v) = \frac{\mathbf{e}_u \times \mathbf{e}_v}{|\mathbf{e}_u \times \mathbf{e}_v|}, \tag{11.10}$$

assuming that the coordinates (u, v) are regular, that is $\mathbf{e}_u \times \mathbf{e}_v \neq \mathbf{0}$. All these notations and the ones to come are explained in Fig. 11.1 and its captions.

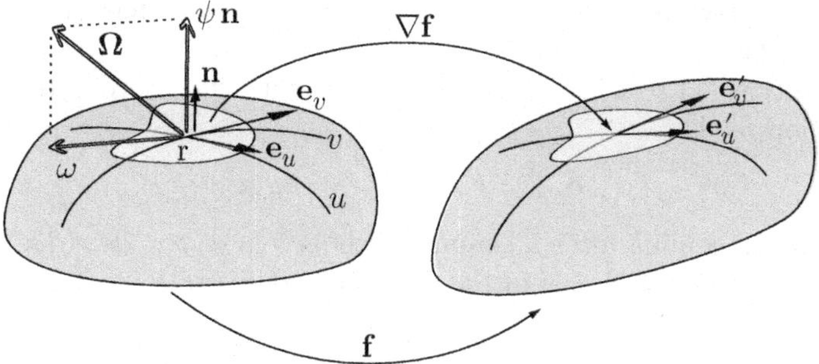

Fig. 11.1 Infinitesimal rigidity problem

Let $\mathbf{f}$ be an isometric mapping from this reference configuration to a deformed configuration: the deformed configuration is parameterised by

$$(u, v) \mapsto \mathbf{f}(\mathbf{r}(u, v)).$$

The tangent vectors are mapped to tangents vectors on the deformed surface, which can be evaluated by the chain rule:

$$\mathbf{e}'_u = \frac{\partial \mathbf{f}(\mathbf{r}(u, v))}{\partial u} = \nabla \mathbf{f} \cdot \mathbf{e}_u \quad \text{and} \quad \mathbf{e}'_v = \frac{\partial \mathbf{f}(\mathbf{r}(u, v))}{\partial v} = \nabla \mathbf{f} \cdot \mathbf{e}_v,$$

where $\nabla \mathbf{f}$ is the gradient of the transformation $\mathbf{f}$ along the surface, mapping the tangent plane in the undeformed configuration to the tangent plane in the deformed configuration.

11.3.2 Local rotation, equation of compatibility

A key remark is that when $\mathbf{f}$ is an isometric deformation, the norm of tangent vectors are preserved and so its differential $\nabla \mathbf{f}(u, v)$ defines a rigid-body transformation between the tangent plane in the reference configuration, at $\mathbf{r}(u, v)$, and the tangent plane in the deformed configuration, at $\mathbf{f}(\mathbf{r}(u, v))$: this $\nabla \mathbf{f}(u, v)$ is a rigid-body *rotation*. As explained in Section 1.3.1, if the transformation is infinitesimal, the rotation $\nabla \mathbf{f}(u, v)$ can be expressed by an infinitesimal rotation vector $\boldsymbol{\Omega}(u, v)$ as:

$$\mathbf{t}' = \nabla \mathbf{f}(u, v) \cdot \mathbf{t} = \mathbf{t} + \epsilon \, \boldsymbol{\Omega}(u, v) \times \mathbf{t} + \cdots$$

Here, $\mathbf{t}$ is an arbitrary vector tangent to the undeformed surface at $\mathbf{r}(u, v)$, $\mathbf{t}'$ its image by the isometry, ϵ is the small magnitude of the isometric deformation, and the dots stand for higher powers of ϵ that can be neglected in the context of *infinitesimal* bendings. In the following, the rotation vector $\boldsymbol{\Omega}(u, v)$ is used to describe the isometric transformation; we derive an equation satisfied by its coordinates in the local basis $(\mathbf{e}_u, \mathbf{e}_v)$.

The above relation holds in particular for any of the vector $\mathbf{t} = \mathbf{e}_\alpha$ of the coordinate basis:

$$\mathbf{e}'_\alpha(u, v) = \mathbf{e}_\alpha(u, v) + \epsilon \, \boldsymbol{\Omega}(u, v) \times \mathbf{e}_\alpha(u, v) + \cdots. \tag{11.11}$$

The letter α is a dummy index in the tangent direction, as implied by the choice of a Greek letter: it can take the values u or v.

The rotation field $\boldsymbol{\Omega}(u, v)$ cannot be arbitrary: it is directly related to the gradient of the isometric map $\mathbf{f}$, and is subjected to compatibility conditions $\mathbf{r}_{,uv}(u, v) = \mathbf{r}_{,vu}(u, v)$.

This condition is worked out below, leading us to an equation characterizing infinitesimal bendings of the surface.

From the definition (11.9) of the basis associated with the coordinates (u, v), the compatibility condition reads

$$\mathbf{e}_{u,v} = \mathbf{r}_{,uv} = \mathbf{r}_{,vu} = \mathbf{e}_{v,u}. \tag{11.12}$$

Recall that any index coming *after* a comma, such as v in $\mathbf{e}_{u,v} = \partial \mathbf{e}_u / \partial v$, implies a partial derivative. A similar equation holds in the deformed configuration:

$$\mathbf{e}'_{u,v} = \mathbf{e}'_{v,u}.$$

Using the explicit form of $\mathbf{e}'_u$ and $\mathbf{e}'_v$ given in equation (11.11), this compatibility condition can be rewritten as

$$\frac{\partial}{\partial v}\left(\mathbf{e}_u(u,v) + \epsilon\, \mathbf{\Omega}(u,v) \times \mathbf{e}_u(u,v) + \dots\right) = \frac{\partial}{\partial u}\left(\mathbf{e}_v(u,v) + \epsilon\, \mathbf{\Omega}(u,v) \times \mathbf{e}_v(u,v) + \dots\right).$$

Reading this equation to first order in ϵ and taking into account equation (11.12) we obtain

$$\mathbf{\Omega}_{,v} \times \mathbf{e}_u = \mathbf{\Omega}_{,u} \times \mathbf{e}_v, \tag{11.13}$$

after cancellation of the terms $\mathbf{\Omega} \times \mathbf{e}_{u,v}$ and $\mathbf{\Omega} \times \mathbf{e}_{v,u}$ appearing on both sides of the equal sign.

11.3.3   Equation of compatibility in tangent plane

Equation (11.13) implies that the gradient of $\mathbf{\Omega}$ lies in the tangent plane, as we show now. First, we compute

$$\mathbf{\Omega}_{,u} \cdot (\mathbf{e}_u \times \mathbf{e}_v) = -\mathbf{e}_u \cdot (\mathbf{\Omega}_{,u} \times \mathbf{e}_v) = -\mathbf{e}_u \cdot (\mathbf{\Omega}_{,v} \times \mathbf{e}_u) = -\mathbf{\Omega}_{,v} \cdot (\mathbf{e}_u \times \mathbf{e}_u) = 0,$$

where we have used the invariance of the mixed product by permutation of its arguments, see equation (1.10), as well as the projection of equation (11.13) on the tangent plane. Dividing both sides of the equation by $|\mathbf{e}_u \times \mathbf{e}_v|$ and using equation (11.10), we obtain

$$\mathbf{\Omega}_{,u} \cdot \mathbf{n} = 0.$$

A similar equation can be derived for $\mathbf{\Omega}_{,v}$, namely $\mathbf{\Omega}_{,v} \cdot \mathbf{n} = 0$. These two equations are noted collectively as

$$\mathbf{\Omega}_{,\alpha} \cdot \mathbf{n} = 0, \tag{11.14}$$

where α is again a dummy index that can take the values u or v.

We shall rewrite equation (11.14) in coordinates. To do so, we split $\mathbf{\Omega}$ into a tangential and normal part,

$$\mathbf{\Omega}(u,v) = \boldsymbol{\omega}(u,v) + \psi(u,v)\, \mathbf{n}(u,v), \tag{11.15}$$

where $\boldsymbol{\omega}$ is the projection onto the tangent plane:

$$\boldsymbol{\omega}(u,v) \cdot \mathbf{n}(u,v) = 0. \tag{11.16}$$

The normal rotation of angle $\psi(u,v)$ will be the main unknown of our rigidity equations.

We can now plug the decomposition (11.15) into equation (11.14) to get

$$\boldsymbol{\omega}_{,\alpha} \cdot \mathbf{n} + \psi_{,\alpha} = 0,$$

after dropping the term $\psi\,\mathbf{n}_{,\alpha}\cdot\mathbf{n}=\psi\,(\mathbf{n}^2)_{,\alpha}/2$ which is zero since $\mathbf{n}$ is everywhere a unit vector.

We shall further decompose the tangent rotation $\boldsymbol{\omega}$ on the basis $(\mathbf{e}_u,\mathbf{e}_v)$:

$$\boldsymbol{\omega}(u,v)=\omega^\beta(u,v)\,\mathbf{e}_\beta(u,v), \tag{11.17}$$

by introducing[6] its coordinates ω^β and using Einstein's convention (implicit summation over repeated indices). After plugging equation (11.17) into the equation above and taking into account the equation $\omega^\beta{}_{,\alpha}\,(\mathbf{e}_\beta\cdot\mathbf{n})=0$ (since $\mathbf{e}_\beta$ is perpendicular to $\mathbf{n}$), we obtain

$$(\mathbf{e}_{\beta,\alpha}\cdot\mathbf{n})\,\omega^\beta+\psi_{,\alpha}=0.$$

The factor in parenthesis has already been encountered earlier and is called the curvature 2-form: we showed in Section 6.2.2 that it defines a symmetric bilinear map from the tangent plane to real numbers. Calling $b_{\alpha\beta}$ the components of this quadratic form in the coordinate basis:

$$b_{\alpha\beta}=\mathbf{e}_{\beta,\alpha}\cdot\mathbf{n}=\mathbf{r}_{,\alpha\beta}\cdot\mathbf{n}, \tag{11.18}$$

we can rewrite the equation above as

$$\psi_{,\alpha}=-b_{\alpha\beta}\,\omega^\beta. \tag{11.19}$$

This equation expresses the gradients of the normal curvature ψ as a linear function of the in-plane components of the rotation $\boldsymbol{\omega}$. The coefficients $b_{\alpha\beta}$ of these linear relations are given by the curvature of the undeformed surface in space.

In the following, ψ will be our main unknown although the in-plane components of the rotation are treated as secondary variables. Therefore, it makes sense to invert the linear relation (11.19). To do so, we need the inverse of the 2×2 matrix with entries $b_{\alpha\beta}$. This inverse is called the adjoint curvature tensor, $d^{\alpha\beta}$ and is defined by:

$$d^{\alpha\beta}\,b_{\beta\gamma}=\delta^\alpha_\gamma, \tag{11.20}$$

where δ^α_γ stands for the Kronecker symbol, whose value is one if the two indices are equal and zero otherwise. This matrix inversion is possible only if the curvature form defined by $b_{\alpha\beta}$ is non-degenerate, i.e. has two non-zero eigenvalues. From the analysis of Section 6.2.2, this inverse $d_{\alpha\beta}$ is defined on regions of the surface *where the Gauss curvature is non-zero.*

Inversion of linear system in equation (11.19) is carried out by multiplying both sides by $d^{\gamma\alpha}$. After renaming the indices we obtain

$$\omega^\alpha=-d^{\alpha\beta}\,\psi_{,\beta}\qquad\text{(for }K\neq0\text{).} \tag{11.21}$$

Like $b_{\alpha\beta}$, the matrix $d^{\alpha\beta}$ characterizes the curvature of the undeformed surface in the embedding Euclidean space.

---

[6] We shall not explain in detail why the index β is noted in superscript in ω^β. As a rule of thumb, note that contracted indices have to go in pairs, one being contravariant (superscript) and the other covariant (superscript).

11.3.4 Equation of compatibility in normal direction

So far, we have only used the tangential projection of equation (11.13). We shall derive another equation by taking the normal component of both sides of the equation:

$$(\boldsymbol{\Omega}_{,v} \times \mathbf{e}_u) \cdot \mathbf{n} = (\boldsymbol{\Omega}_{,u} \times \mathbf{e}_v) \cdot \mathbf{n}. \tag{11.22}$$

Let us introduce the so-called *dual basis* $(\mathbf{e}^u, \mathbf{e}^v)$ of the tangent frame:

$$\mathbf{e}^u = \frac{\mathbf{e}_v \times \mathbf{n}}{|\mathbf{e}_u \times \mathbf{e}_v|} \quad \text{and} \quad \mathbf{e}^v = \frac{\mathbf{n} \times \mathbf{e}_u}{|\mathbf{e}_u \times \mathbf{e}_v|}. \tag{11.23}$$

Note the position of the letters u and v, which are superscripts for the dual basis, and subscripts for the primal basis. The name 'dual' comes from the following identity, which can be checked directly:

$$\mathbf{e}^\alpha \cdot \mathbf{e}_\beta = \delta^\alpha_\beta. \tag{11.24}$$

With the help of this dual basis, equation (11.22) can be rewritten as follows, after dividing both sides by $|\mathbf{e}_u \times \mathbf{e}_v|$ and permuting the mixed products:

$$\mathbf{e}^u \cdot \boldsymbol{\Omega}_{,u} + \mathbf{e}^v \cdot \boldsymbol{\Omega}_{,v} = 0,$$

or in compact form, with implicit summation:

$$\mathbf{e}^\alpha \cdot \boldsymbol{\Omega}_{,\alpha} = 0.$$

In terms of the in-plane and normal components of the rotation vector $\boldsymbol{\Omega}$ given by equation (11.17); this equation becomes

$$(\mathbf{e}^\alpha \cdot \boldsymbol{\omega}_{,\alpha}) + (\mathbf{e}^\alpha \cdot \mathbf{n}_{,\alpha})\,\psi = 0. \tag{11.25}$$

The two terms in parentheses have a geometrical interpretation. The first term defines what is known as the covariant divergence of the tangent vector field $\boldsymbol{\omega}$:

$$\operatorname{div}\boldsymbol{\omega} = \mathbf{e}^\alpha \cdot \boldsymbol{\omega}_{,\alpha}. \tag{11.26}$$

Although we shall not prove it, this definition of the divergence is independent of the particular coordinate system chosen, hence the term 'covariant'.

The second parenthesis in equation (11.25) is related to the mean curvature $H(u, v)$. To prove this, we first need to show the following lemma: the coordinates of the vectors of the dual basis in the primal basis are given by a matrix $g^{\alpha\beta}$ that is the inverse of the metric tensor defined by $g_{\alpha\beta} = \mathbf{e}_\alpha \cdot \mathbf{e}_\beta$. To prove the lemma, let us start by decomposing a dual vector in the primal basis, using some coordinates $g^{\alpha\beta}$ (mind the position of the indices):

$$\mathbf{e}^\alpha = g^{\alpha\beta}\,\mathbf{e}_\beta.$$

Taking the dot product of both sides with $\mathbf{e}_\gamma$, we obtain, after making use of the fundamental property (11.24) of the dual basis:

$$\delta^\alpha_\gamma = g^{\alpha\beta}\,g_{\beta\gamma}.$$

This proves the lemma: the matrices with entries $g^{\alpha\beta}$ (coordinates of dual vectors in primal basis) and $g_{\alpha\beta}$ (metric) are inverse.

We can now rewrite the second parenthesis in equation (11.25) by making use of the above decomposition of $\mathbf{e}^\alpha$ in the primal basis:

$$\mathbf{e}^\alpha \cdot \mathbf{n}_{,\alpha} = g^{\alpha\beta}\,\mathbf{e}_\beta \cdot \mathbf{n}_{,\alpha} = g^{\alpha\beta}\left((\mathbf{e}_\beta \cdot \mathbf{n})_{,\alpha} - \mathbf{e}_{\beta,\alpha} \cdot \mathbf{n}\right) = -g^{\alpha\beta}\,b_{\alpha\beta},$$

where we have used $\mathbf{e}_\beta \cdot \mathbf{n} = 0$ and identified the metric tensor $b_{\alpha\beta}$ from its definition (11.18). Now, the magic of tensorial notation[7] tells us that the value of the right-hand side, being obtained by contraction of pairs of covariant (subscript) and contravariant (superscript) indices, is actually independent of the particular basis chosen. To identify the geometric quantity appearing in the last equation, we evaluate it in an orthonomal basis aligned with the principal directions,[8] which we denote u' and v'. In this particular basis, which is orthonormal, the metric tensor $g_{\alpha\beta}$ is the identity matrix by definition, and so is its inverse $g^{\alpha\beta}$: the above equation evaluates to $-(b_{u'u'} + b_{v'v'}) = -(\kappa_{u'} + \kappa_{v'})$, which is the opposite of the sum of the principal curvatures $\kappa_{u'}$ and $\kappa_{v'}$, that is (-2) times the average of the principal curvatures. This average is just what has been defined earlier as the *mean curvature*, noted $H = (\kappa_{u'} + \kappa_{v'})/2$. We have just shown that the quantity $(\mathbf{e}^\alpha \cdot \mathbf{n}_{,\alpha})$ does not depend on the choice of coordinates and is given by

$$\mathbf{e}^\alpha \cdot \mathbf{n}_{,\alpha} = -2\,H.$$

This allows us to rewrite equation (11.25) in intrinsic form:

$$\operatorname{div}\boldsymbol{\omega} = 2\,H\,\psi,$$

where the divergence operator has been defined as $\operatorname{div} = \mathbf{e}^\alpha \cdot \partial_{,\alpha}$.

11.3.5   Summary, discussion

We have derived the equations for the compatibility of a rotation field associated with an infinitesimal isometric deformation of a surface in the form of a set of linear partial differential equations

$$\operatorname{div}\boldsymbol{\omega} = 2\,H\,\psi \tag{11.27a}$$

$$\omega^\alpha = -d^{\alpha\beta}\,\psi_{,\beta}, \tag{11.27b}$$

where the unknowns are the normal component ψ and the in-plane components ω^α of the rotation vector.

By eliminating ω^α, these two equations can be condensed into a single linear partial differential equation (PDE) for the normal rotation $\psi(u,v)$:

$$\operatorname{div}(d^{\alpha\beta}\,\psi_{,\alpha}\,\mathbf{e}_\beta) + 2\,H\,\psi = 0. \tag{11.28}$$

Note that the divergence is a first-order differential operator that depends on the undeformed configuration of the surface. The adjoint curvature $d^{\alpha\beta}$ depends only on the undeformed configuration too; it is undefined in domains where the Gauss curvature is zero (in the undeformed configuration).

The infinitesimal displacement can be found from the rotation $\boldsymbol{\Omega}$ by integrating equation (11.11), which we rewrite as $(\mathbf{r}' - \mathbf{r})_{,\alpha} = \epsilon\,\boldsymbol{\Omega} \times \mathbf{e}_\alpha$. Integration of a function of two

---

[7] This result will not be proved here but is established in many textbooks on tensor analysis, for instance in reference (R. M. Wald, 1984).

[8] Recall that the principal directions were defined in Section 6.2.2 as the eigenvectors of the curvature form, which is a symmetric quadratic form defined in any tangent space.

variables (u, v) with a prescribed gradient is possible only if the gradient satisfies some compatibility condition; this compatibility condition is just the PDE (11.28). Any solution of the PDE for $\Omega(u, v)$ gives an infinitesimal bending of the surface when the surface is simply connected. When it is not simply connected, i.e. when it has holes in it that can be enlaced by a closed path drawn on the surface, an additional condition must be satisfied to make sure that the rotation field can be integrated into a *single-valued* displacement field. Besides, some solutions of this equation may be trivial in the sense that they correspond to rigid-body infinitesimal rotations of the surface, as shown in Section 11.5.3 in the case of surfaces of revolution.

For a weakly curved surface, equations (11.27) are equivalent to equation (11.8). However, these equations use different unknowns; as a result, deriving the former equations from the latter requires some effort.

11.3.6 Qualitative analysis of the equation for bendings

Infinitesimal bendings on a surface are solutions of a second-order linear PDE given in equation (11.28). The rigidity problem can then be viewed as a particular instance of the general problem of studying the existence and of deriving solutions for linear PDEs. Second-order PDEs are classified depending on their type (elliptic, hyperbolic, parabolic or mixed); PDE's of different types have qualitatively different behaviours. Elliptic PDEs are ruled by the maximum principle and typically have a unique solution when a boundary condition is prescribed along the boundary. In hyperbolic PDEs, the information is carried along special curves called characteristics, and the well-posedness of a particular problem depends on the relative position of the boundary condition and of the characteristics. For a detailed presentation, see reference (M. Spivak, 1979, Ch. 10).

The type of the PDE can be found by looking at the higher-order derivatives in the equation. In the problem at hand, these higher-order terms are revealed by expanding the divergence operator in equation (11.28):

$$d^{\alpha\beta}\,\psi_{,\alpha\beta} + \cdots = 0,$$

where the dots stand for terms with lower-order derivatives $\psi_{,\alpha}$ and ψ. The PDE is elliptic when the determinant of the 2×2 matrix containing the coefficients of $\psi_{,\alpha\beta}$ is positive, parabolic when this determinant is zero, and hyperbolic when it is negative. If its sign is non-constant in space, the equation is said to be of mixed type. The sign of the determinant of the matrix $d^{\alpha\beta}$ is the same as the sign of the determinant of its inverse, $b_{\alpha\beta}$, and the latter is just the sign of Gauss[9] curvature $K(u, v)$. As in the case of weakly curved surfaces, we find that the equation for infinitesimal bendings is elliptic when the surface is elliptic ($K > 0$), parabolic when the surface is parabolic ($K = 0$), hyperbolic when the surface is parabolic ($K < 0$) (M. Spivak, 1979, p. V.279), and of mixed type if the surface is itself of mixed type.[10]

One of the classical results in the theory of partial differential equations is that an elliptic PDE has no non-trivial solution on a compact (ball-shaped) manifold. In the context of

---

[9] The sign of the Gauss curvature is related to the shape of the intersections of the surface with planes parallel to its tangents planes and close to it: these intersections are small ellipses for $K > 0$, and local hyperbolas for $K < 0$.

[10] The example of a torus is typical of a surface of mixed type and its geometric rigidity will be studied in detail later in Section 11.5.

infinitesimal rigidity, this translates into the following result, one of the few general results that are known: an elliptic ($K > 0$) and compact manifold, such as a sphere, is rigid. Note that this rigidity result holds for sufficiently smooth deformations only; as mentioned in the introduction, Section 11.1, all surfaces admit bending displacements having a low smoothness ($\mathcal{C}^1$). The analysis of an elastic ball pushed by a rigid plane, given in Chapter 14, will highlight an isometric deformation of the sphere with even lower smoothness: for large applied forces, a spherical cap undergoes a mirror reflection and pops into the sphere; the global elastic response of the ball largely reflects the behaviour of the circular ridge at the edge of this inverted cap.

11.4 Minimal surfaces, Weierstrass transform

The equations for infinitesimal bendings (11.27) have no general solution in closed analytical form. An interesting exception is when the mean curvature $H(u, v)$ of the surface is zero everywhere. This property defines a *minimal surface*.[11] On minimal surfaces, the equations for bendings have a simple particular solution:

$$\psi(u, v) = C, \quad \omega^u(u, v) = 0, \quad \omega^v(u, v) = 0,$$

where C is an arbitrary constant. This corresponds to a local rotation $\mathbf{\Omega}(u, v)$, which is proportional to the unit normal $\mathbf{n}(u, v)$:

$$\mathbf{\Omega}(u, v) = C\,\mathbf{n}(u, v). \tag{11.29}$$

Except in the particular case where the minimal surface is a plane, this rotation field is non-uniform and the bending deformation is not trivial (it is not a rigid-body displacement).

This particular solution provides an infinitesimal bending for any simply connected[12] minimal surface. This property is related to an elegant parameterisation of minimal surfaces due to Weierstrass (D. J. Struik, 1961), which can be used to show the existence of isometries between pairs of minimal surfaces. A classical example is the explicit, one-parameter families of isometries transforming the catenoid into a helicoid, see Fig. 11.2; note that all the intermediate shapes are also minimal surfaces.

The continuous transformation of a catenoid into a helicoid defines an infinitesimal bending at each frame that coincides with the particular solution $\mathbf{\Omega}(u, v) = C\,\mathbf{n}(u, v)$ in equation (11.29). More accurately, this transformation operates on a catenoid *torn along a meridian*. The catenoid is not a simply connected surface: the circles obtained by intersecting the surface with a plane perpendicular to the axis of revolution cannot be smoothly shrunk into a point. A cut along the meridian effectively removes the hole in the surface, makes it simply connected, and allows integration of the rotation into a single-valued displacement field.[13]

---

[11] Such surfaces are called minimal because their area is smaller that any close surface that has the same boundary. Soap bubbles held by a wire frame are instances of minimal surfaces at equilibrium since they minimize their capillary energy, proportional to their area.

[12] As mentioned earlier, integration of the rotation into a single-valued displacement field requires the surface to be simply connected, or an additional condition to be enforced.

[13] On the original (uncut) catenoid, there are infinitely many curves joining a point of reference O to a current point P. These curves have different winding numbers about the axis and cannot be deformed one into another in a smooth manner. As a result, integration of the rotation vector $\mathbf{\Omega} = C\,\mathbf{n}$ from O to P yields a result that depends on the path chosen in general, and the displacement field is multivalued.

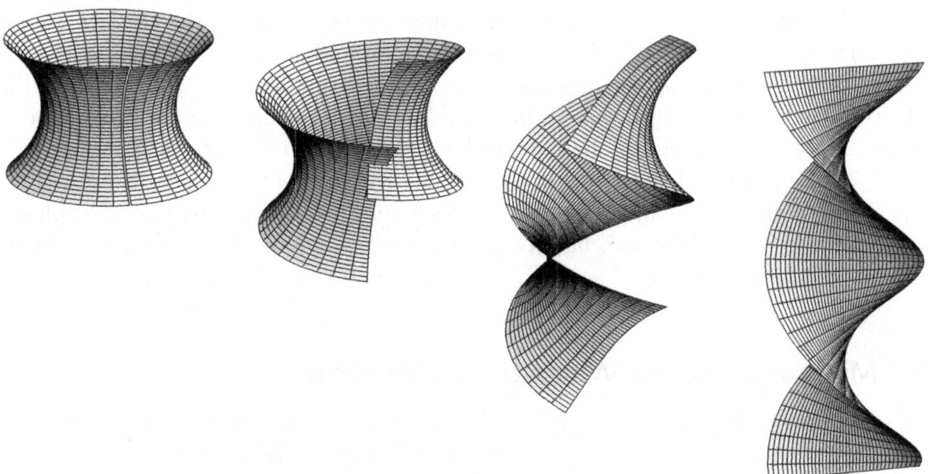

Fig. 11.2 Smooth transformation of a catenoid into a helicoid, through a one-parameter family of minimal surfaces. This transformation is based on Weierstrass parameterisation of minimal surfaces.

11.5   Surfaces of revolution

In this section, we focus on the infinitesimal rigidity of surfaces of revolution. This case is interesting as the linear partial differential equations for the rigidity derived in Section 11.3 can be rewritten as a set of *ordinary* differential equations (ODEs). The rigidity problem can then be studied using tools from the theory of ODEs, which is much simpler than the theory of PDEs. As we shall show, the ODE of the rigidity problem has singular points, and the analysis of these singular points yields the answer to the rigidity problem. This analysis was published by Cohn-Vossen in 1929 (S. Cohn-Vossen, 1929) and is presented in the rest of the current section—Cohn-Vossen's analysis can also be found in reference (I. Ivanova-Karatopraklieva and I. Kh. Sabitov, 1995).

The final result is a classification of surfaces of revolution based on their infinitesimal rigidity. Interpreting this rigidity criterion, we shall point out a particular type of curves that have the remarkable property that they make the surface more rigid, in a sense to be made accurate later. This result is derived here in the context of surfaces of revolution. In Section 11.6, we shall propose a geometrical interpretation of the rigidity brought about by these curves, thereby allowing the identification of 'rigid' curves on surfaces that are not of revolution. Recall that this chapter is concerned with the geometry of bendable but inextensible surfaces; the elasticity of some shells of revolution will be studied in the coming chapters.

11.5.1   Geometry

We apply the formalism of Section 11.3 to the case of a surface of revolution. Cylindrical coordinates (ρ, θ, z) are used. Here (ρ, θ) are the polar coordinates in the (x, y) plane, $\rho^2 = x^2 + y^2$ and $y = x \tan\theta$, and the axis z is aligned with the axis of revolution of the surface. The units vectors defined by

$$\mathbf{u}_r(\theta) = \cos\theta\, \mathbf{e}_x + \sin\theta\, \mathbf{e}_y, \quad \mathbf{u}_\theta(\theta) = -\sin\theta\, \mathbf{e}_x + \cos\theta\, \mathbf{e}_y, \tag{11.30}$$

define a moving basis, their derivatives being given by $\mathbf{u}_r'(\theta) = \mathbf{u}_\theta(\theta)$ and $\mathbf{u}_\theta'(\theta) = -\mathbf{u}_r(\theta)$.

The undeformed surface of revolution is defined by its intersection with any plane containing the z axis: this defines the so-called generating curve in the (ρ, z) plane. This curve is described by a parametric equation $(\rho = \rho(s), z = z(s))$, where the arc length s is the parameter:

$$\rho'^2(s) + z'^2(s) = 1. \tag{11.31}$$

Here, primes denote derivatives of functions of a single variable, s.

In the (ρ, z) plane, the generating curve has unit tangent $\rho'(s)\,\mathbf{u}_r + z'(z)\,\mathbf{e}_z$. The derivative of this tangent with respect to s reads $\rho''(s)\,\mathbf{u}_r + z''(z)\,\mathbf{e}_z$. Identifying with equation (1.17), we compute the curvature of the generating curve:

$$\kappa(s) = -z'(s)\,\rho''(s) + \rho'(s)\,z''(s). \tag{11.32}$$

The following identities will be useful:

$$\rho'(s)\,\rho''(s) + z'(s)\,z''(s) = 0 \tag{11.33a}$$

$$\kappa(s)\,\rho'(s) = +z''(s) \tag{11.33b}$$

$$\kappa(s)\,z'(s) = -\rho''(s). \tag{11.33c}$$

The first one comes from deriving equation (11.31) and the other two are from combining equations (11.32) and (11.33a).

On the surface of revolution, the coordinates $u = s$ and $v = \theta$ are used. The surface is obtained by revolving the generating curve about the axis z: a current point on the surface is given by

$$\mathbf{r}(s, \theta) = \rho(s)\,\mathbf{u}_r(\theta) + z(s)\,\mathbf{e}_z,. \tag{11.34}$$

as shown in Fig. 11.3.

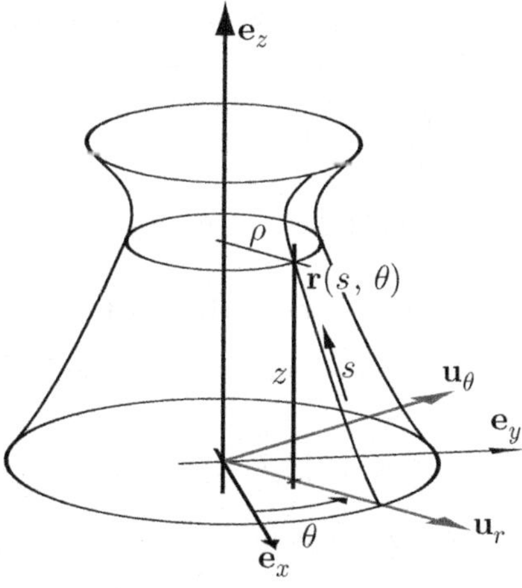

Fig. 11.3 Parameterisation of a surface of revolution.

We proceed to compute the various geometric quantities introduced in Section 11.3 when the particular coordinates $(u, v) = (\rho, \theta)$ are used. In equation (11.9), we introduced the tangent basis $(\mathbf{e}_u, \mathbf{e}_v)$, which reads:

$$\mathbf{e}_s(s, \theta) = \rho'(s)\,\mathbf{u}_r(\theta) + z'(s)\,\mathbf{e}_z, \qquad \mathbf{e}_\theta(s, \theta) = \rho(s)\,\mathbf{u}_\theta(\theta). \tag{11.35}$$

The normal to the surface is found by evaluating $\mathbf{e}_s \times \mathbf{e}_\theta$. After normalization, it reads:

$$\mathbf{n}(s, \theta) = -z'(s)\,\mathbf{u}_r(\theta) + \rho'(s)\,\mathbf{e}_z. \tag{11.36}$$

The dual basis is given by equation (11.24) as:

$$\mathbf{e}^s(s, \theta) = \mathbf{e}_s(s, \theta), \qquad \mathbf{e}^\theta(s, \theta) = \frac{\mathbf{u}_\theta(\theta)}{\rho(s)}. \tag{11.37}$$

Calculation of the metric tensor $g_{\alpha\beta} = \mathbf{e}_\alpha \cdot \mathbf{u}_\beta$,

$$g_{ss} = 1, \quad g_{\theta\theta} = \rho^2, \quad g_{s\theta} = g_{\theta s} = 0, \tag{11.38}$$

and of its inverse,

$$g^{ss} = 1, \quad g^{\theta\theta} = \frac{1}{\rho^2}, \quad g^{s\theta} = g_{\theta s} = 0, \tag{11.39}$$

is straightforward.

The curvature tensor has been defined in equation (11.18). For a surface of revolution, it reads

$$b_{ss} = \mathbf{n} \cdot \mathbf{r}_{,ss} = \kappa, \quad b_{\theta\theta} = \mathbf{n} \cdot \mathbf{r}_{,\theta\theta} = \rho\,z', \quad b_{s\theta} = b_{\theta s} = 0. \tag{11.40}$$

A second flavour of the curvature tensor can be defined by raising one of its indices, using the metric tensor: $b_\alpha{}^\beta = b_{\alpha\gamma}\,g^{\gamma\beta}$:

$$b_s{}^s = \kappa, \quad b_\theta{}^\theta = \frac{z'}{\rho}, \quad b_s{}^\theta = b_\theta{}^s = 0. \tag{11.41}$$

As shown earlier, the mean curvature H is half the trace of this curvature tensor $b_\alpha{}^\beta$ with mixed contravariant and covariant indices, while the Gauss curvature K is given by its determinant:

$$H = \frac{\mathrm{tr}(b_\alpha{}^\beta)}{2} = \frac{b_\alpha{}^\alpha}{2} = \frac{1}{2}\left(\kappa + \frac{z'}{\rho}\right), \quad K = \det(b_\alpha{}^\beta) = \kappa\,\frac{z'}{\rho} = -\frac{\rho''}{\rho}. \tag{11.42}$$

We have used some of the identities (11.33) to simplify these expressions.

11.5.2   Equations for bendings in azimuthal Fourier components

For a surface of revolution, the decomposition of the infinitesimal rotation vector introduced in Section (11.3.3) reads:

$$\boldsymbol{\Omega}(s, \theta) = \omega^s(s, \theta)\,\mathbf{e}_s(s, \theta) + \omega^\theta(s, \theta)\,\mathbf{e}_\theta(s, \theta) + \psi(s, \theta)\,\mathbf{n}(s, \theta) \tag{11.43}$$

and the equations for bendings (11.27) become

$$\omega^s(s, \theta) = -\frac{\psi_{,s}(s, \theta)}{\kappa(s)}, \quad \omega^\theta(s, \theta) = -\frac{\psi_{,\theta}(s, \theta)}{\rho(s)\,z'(s)} \tag{11.44a}$$

and

$$\omega^s{}_{,s}(s,\theta) + \omega^\theta{}_{,\theta}(s,\theta) + \frac{\rho'(s)}{\rho(s)}\,\omega^s(s,\theta) = 2\,H(s)\,\psi(s,\theta). \tag{11.44b}$$

Owing to the symmetry of revolution and the linearity of the equations, it is beneficial to decompose the unknown functions $\omega^s(s,\theta)$, $\omega^\theta(s,\theta)$ and $\psi(s,\theta)$ in Fourier modes with respect to the azimuthal variable θ. For $f = \omega^s$, $f = \omega^s$ or $f = \psi$, we introduce the amplitudes $\hat{f}(s;q)$ of the Fourier modes, the Fourier index q being a positive integer. The functions can be reconstructed on the surface using the formula

$$\begin{aligned}
f(s,\theta) &= \sum_{q\geq 0} \Re(\hat{f}_q(s)\,\exp(i\,q\,\theta)) \\
&= \sum_{q\geq 0} \left(\Re(\hat{f}_q(s))\,\cos(q\,\theta) - \Im(\hat{f}_q(s))\,\sin(q\,\theta) \right),
\end{aligned} \tag{11.45}$$

where $\Re$ and $\Im$ stand for the real and imaginary parts of a complex number, respectively.

The equations (11.44) for bendings are linear and their coefficients do not depend explicitly on θ: when the Fourier transform is applied, all Fourier modes become uncoupled. Indeed, plugging equation (11.45) into equations (11.44) for bendings, we find

$$\hat{\omega}_q^s(s) = -\frac{\hat{\psi}_q'(s)}{\kappa(s)}, \quad \hat{\omega}_q^\theta(s) = -\frac{i\,q\,\hat{\psi}_q(s)}{\rho(s)\,z'(s)} \tag{11.46a}$$

and

$$\hat{\omega}_q^{s\prime}(s) + i\,q\,\hat{\omega}_q^\theta(s) + \frac{\rho'(s)}{\rho(s)}\,\hat{\omega}_q^s(s) = 2\,H(s)\,\hat{\psi}_q(s). \tag{11.46b}$$

This set of linear ODEs has to be satisfied for all integer values of q. Using Fourier analysis, we have transformed the linear PDEs for infinitesimal bendings into a set of linear, *independent* ODEs for each Fourier mode q. Note that the undeformed surface is assumed to be of revolution, but the bendings need not be axisymmetric ($q \neq 0$).

11.5.3   Rigid-body modes of deformation ($q = 0, 1$)

We shall start by showing that the trivial bendings, corresponding to infinitesimal rigid-body displacements, correspond to the solutions of the equations (11.46) for $q = 0, 1$.

First, note that rigid-body translations yield the trivial solution $\boldsymbol{\Omega}(s,\theta) \equiv 0$ in terms of the local rotation field: these rigid-body translations are recovered as constants of integration when the displacement is integrated from the local rotation $\boldsymbol{\Omega}(u,v)$.

Infinitesimal rigid-body rotations about the axis z of the surface of revolution are represented by an axial rotation vector $\boldsymbol{\Omega}(s,\theta) = \Omega_z\,\mathbf{e}_z$, where the infinitesimal angle Ω_z is a constant. The normal component of the rotation is obtained by projection, $\psi(s,\theta) = \Omega_z\,\mathbf{e}_z \cdot \mathbf{n} = \Omega_z\,\rho'(s)$. The only non-zero Fourier mode is $\hat{\psi}_{q=0}(s) = \Omega_z\,\rho'(s)$. The in-plane components are calculated similarly: $\hat{\omega}_0^s(s) = \Omega_z\,z'(s)$ and $\hat{\omega}_0^\theta(s) = 0$. Using the identities (11.33), it can be checked that these three functions are always a solution of the equation (11.46b) for $q = 0$.

Now, consider an infinitesimal rotation about the axis perpendicular to the axis of revolution, say $\boldsymbol{\Omega}(s,\theta) = \Omega_x\,\mathbf{e}_x$. The normal rotation reads $\psi(s,\theta) = \Omega_x\,\mathbf{e}_x \cdot \mathbf{n}(s,\theta) = -\Omega_x\,z'(s)\,\cos(\theta)$. Because of the $\cos\theta$ dependence, this corresponds to the Fourier index

$q = 1$, with amplitude $\hat{\psi}_{q=1}(s) = -\Omega_x\, z'(s)$. The in-plane projections of the rotation are computed similarly: $\hat{\omega}_1^s(s) = \Omega_x\, \rho'(s)$ and $\hat{\omega}_1^\theta(s) = i\,\Omega_x/\rho(s)$. Using the identities (11.33) again, it can be checked that this defines a solution of the equation (11.46b) for $q = 1$.

A rigid-body rotation about $\mathbf{e}_y$ is described by multiplying all Fourier components of the previous solution by $(-i)$.

Conversely, any solution of equations (11.46) for $q = 0$ or $q = 1$ can be shown to yield either a rigid-body displacement, which is a linear combination of the ones above, or must be discarded as it leads to a multiply-valued displacement field by integration (a surface of revolution with one or no pole is multiply connected). We have just shown that the trivial bending modes (rigid-body translations and rotations) correspond exactly to the solutions of equations (11.46) for $q = 0$ or $q = 1$. In the following, we characterize the rigidity of a surface of revolution by studying[14] the existence of solutions for $q \geq 2$.

11.5.4   Analysis of singular points

We rewrite equations (11.46) for infinitesimal bendings in a form that is suited to the analysis of singular points. To do so, we eliminate the function $\hat{\omega}_q^\theta(s)$, multiply both sides of the second equation by $\rho(s)$, and isolate the highest-order derivatives of the two remaining unknowns $\hat{\psi}_q(s)$ and $\hat{\omega}_q^s(s)$ in the left-hand side:

$$\frac{\mathrm{d}\hat{\psi}}{\mathrm{d}s} = -\kappa\,\hat{\omega}^s \tag{11.47a}$$

$$\rho\,\frac{\mathrm{d}\hat{\omega}^s}{\mathrm{d}s} = -\hat{\omega}^s\,\rho' + \left(2\,H\,\rho - \frac{q^2}{z'}\right)\hat{\psi}. \tag{11.47b}$$

To simplify the notation, we have omitted the index q and the argument s of the functions $\hat{\psi} = \hat{\psi}_q(s)$ and $\hat{\omega}^s = \hat{\omega}_q^s(s)$. The argument s of the functions $\rho(s)$, $z(s)$, $\kappa(s)$, $H(s)$ is also dropped.

A linear ordinary differential equation is said to be regular if the coefficients in front of the higher-order term are all non-zero, and all coefficients vary smoothly. By regular, we mean that there exists n independent solutions of the equation, where n is the order of the equation, $n = 2$ here. Our equation for bendings is regular when $\rho \neq 0$ and $z' \neq 0$. The first condition comes from the coefficient in front of $(\hat{\omega}^s)'$ which must not vanish, and the second from the coefficient in the right-hand side of equation (11.47b), which has to remain bounded. When either one of these conditions is not satisfied, the solutions may diverge— but this is not always the case. The existence of smooth solutions, and their number can only be found by a detailed analysis of the equation near its singular points, which are the roots of $\rho(s) = 0$ or $z'(s) = 0$. Recall that our derivation of the equation for bendings assumed $K \neq 0$, the Gauss curvature being expressed as $K = \kappa\, z'/\rho$ in equation (11.42): in addition to the previous points, we must also study singular points such that $\kappa = 0$. This yields three possible types of singular points, as illustrated in Fig. 11.4.

---

[14] Modes with $q \geq 2$ never lead to multivalued displacements: by equation (11.11), the gradient of the infinitesimal displacement contains the Fourier modes $q - 1$, q and $q + 1$. All these Fourier modes cancel when integrated over a loop enlacing the axis z (s constant, $0 \leq \theta \leq 2\pi$). Therefore the reconstructed displacement does not depend on the winding number about the z axis of the path chosen to integrate its gradient.

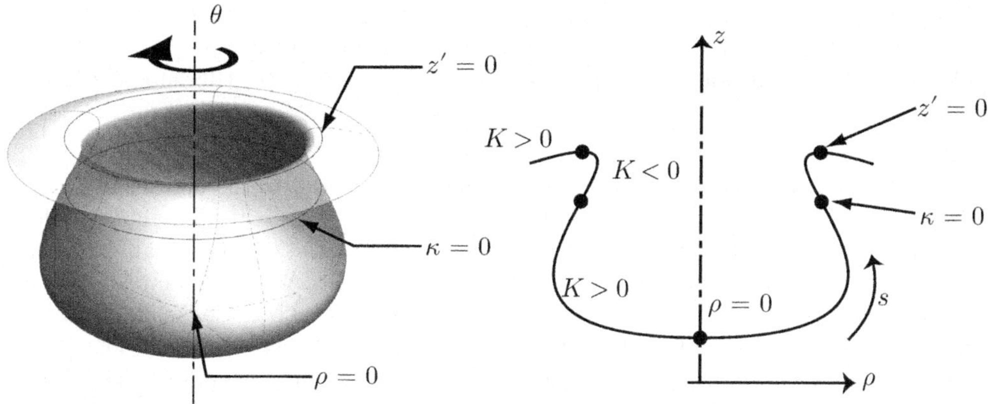

Fig. 11.4 Singular points of the differential equation for bendings on a surface of revolution are of three types: (i) poles, such that $\rho(s) = 0$; (ii) inflexion points of the generating curves, such that $\kappa(s) = 0$; (iii) or crowns, such that $z'(s) = 0$. The Gauss curvature K is given by $K = b_s^s\, b_\theta^\theta$ and vanishes at inflexion points (ii) where the surface is flat in the s direction, $b_s^s = \kappa = 0$, or at crowns (iii) where the surface is flat in the θ direction, $b_\theta^\theta \propto z' = 0$.

Let us first consider a pole, i.e. a place where $\rho = 0$. Taking the origin of the arc length parameterisation at this pole temporarily, we have $\rho \sim s$ in the vicinity of the pole. We consider the generic case when the curvature curvature $\kappa_0 = \kappa(s = 0)$ is non-zero there, $\kappa_0 \neq 0$. At the pole, the tangent plane is perpendicular to the axis and so $z'(0) = 0$. The definition of κ in equation (11.32) shows that $z''(0) = \kappa_0$ and so $z'(s) \sim \kappa_0\, s$ near the pole. In equation (11.47b), we group the two terms on the sides of the equal sign in the left-hand side, and identify the exact derivative $(\rho\,\hat{\omega}^s)'$ in the resulting expression $\rho\,(\hat{\omega}^s)' + \rho'\,\hat{\omega}^s$; we neglect the first, regular term in the parenthesis in front of the second one which is diverging, eliminate $\hat{\omega}^s$ using the equation above and use the expansions just given. This yields

$$s\,\frac{\mathrm{d}}{\mathrm{d}s}\left(s\,\frac{\mathrm{d}\hat{\psi}}{\mathrm{d}s} \right) = q^2\,\hat{\psi}$$

after simplification of the factor $-1/\kappa_0$ appearing on both sides. We seek solutions of this differential equation in the form of a Laurent expansion with unknown dominant power p near the pole:

$$\hat{\psi}(s) = a\,s^p + a'\,s^{p-1} + \cdots$$

Plugging the dominant term in this expansion into the equation above, one obtains:

$$p^2 = q^2, \quad \text{that is } p = \pm q.$$

As noted earlier, $q = 0$ and $q = 1$ correspond to rigid-body deformations of the surface, and we now assume $q \geq 2$. The Laurent expansion gives two linearly independent solutions of the equation near the pole, one corresponding to $p = -q$ and the other one to $p = q$. The solution $p = -q$ is singular as $\psi \sim 1/s^q$ and the displacement field, obtained by integration diverges like $1/s^{q-1}$: it has to be ruled out. Now, the other solution $p = q$ is regular, as can be checked easily; for this other solution, the rotation vector $\boldsymbol{\Omega}$ goes smoothly to zero as the pole is approached.

When there is no singular point, the equations for bendings is a regular linear ODE of order 2, and so the manifold of its solutions has dimension 2. We have just shown that, near a pole, regular solutions form a subspace of dimension 1. Noting D_P^q this subspace, we have

$$\dim D_P^q = 1. \tag{11.48}$$

Here, the letter P refers to the pole under consideration, and q to the Fourier index.

The second type of singular point is given by the equation $\kappa(s) = 0$. This equation defines a circle on the surface that lies in a plane perpendicular to the axis (we shall consider the generic case when κ has a root of order 1, that is $\kappa'(s) \neq 0$). Across this circle, the Gauss curvature changes sign. Near these points, the set of differential equations (11.47b) remains regular and yields a manifold of solutions of dimension 2. However, these solutions may not all be physical as equation (11.21) has been derived under the assumption $K \neq 0$, which implies $\kappa \neq 0$: when $\kappa = 0$, the linear system (11.19) is singular and some care must be taken when writing down its solutions.

Practically, we have to carefully check that a solution of the differential equations satisfies the original equation (11.19) at points where $\kappa = 0$. In the case of a surface of revolution, this equation (11.19) is equivalent to $\hat{\psi}' = -\kappa \hat{\omega}^s$ and $i q \hat{\psi} = -\rho z' \hat{\omega}^\theta$. The first equation $\hat{\psi}' = -\kappa \hat{\omega}^s$ is obviously satisfied by any solution of the set of linear equations (11.47) since this equations is part of the set—at the singular point, this equation becomes $0 = 0$. The other equation, $i q \hat{\psi} = -\rho z' \hat{\omega}^\theta$, involves quantities that vary smoothly near a singular point where κ vanishes, and so is satisfied by continuity at the point. To sum up, singular points such that $\kappa = 0$ are harmless: any solution for an infinitesimal bending remains smooth near such points.

The last type of singular points is when $z' = 0$. This corresponds to a circle on the surface of revolution lying in a plane that is everywhere tangent to the surface along the circle, see Fig. 11.4. It is convenient to redefine the origin of the arc length coordinate s to be the singular point under study: $z'(0) = 0$. Again, we consider the generic case, such that $\kappa_0 = \kappa(0) \neq 0$. Then, the differential equation is singular because the coefficient $-q^2/z'$ diverges in equation (11.47b). Seeking as earlier solutions of the differential equations (11.47) in the form of Laurent series:

$$\hat{\psi}(s) = a\, s^p + a'\, s^{p-1} + \cdots$$

we find by plugging the first term in the differential equations that the expansion can start with $p = 0$ or with $p = 1$. Although this looks like the beginning of the expansions of two smooth and independent solutions, closer examination reveals that the solution with $p = 0$ is singular. Indeed, equation (11.46a) shows that when $\hat{\psi}$ tends to a non-zero constant at the singular point $s = 0$ (as happens for $p = 0$), then $\hat{\omega}^\theta$ diverges[15] like $1/z' \sim 1/s$. As a result, the norm of the local rotation vector $\boldsymbol{\Omega}$ diverges like $1/s$ and the displacement mode, obtained by spatial integration, diverges logarithmically: solutions corresponding to $p = 0$ are not acceptable. On the other hand, the analysis of the case $p = 1$ shows that the corresponding solutions are smooth. The final result is that the infinitesimal bendings that are smooth near a singular point $z' = 0$ form a sub-manifold of dimension 1, noted D_c^q:

---

[15] This divergence is balanced in the second equation (11.46b) by the first term: $(\hat{\omega}^s)' \sim -i q \hat{\omega}^\theta \sim 1/s$, which implies in turn that $\hat{\omega}^s$ diverges logarithmically. A detailed analysis yields the (non-polynomial) expansion of $\hat{\psi}$ in the form $\hat{\psi}(s) = a\left(1 + q^2/\rho_0\, s \ln|s| + \ldots\right)$.

$$\dim D_C^q = 1, \tag{11.49}$$

where the subscript C refers to the singular point being studied, which is a root of $z'(s_C) = 0$. We refer to such points as *crowns* in the following.

11.5.5   Counting the bending modes (generic case)

Kinematical boundary conditions can be imposed on the edges $s = s_{\min}$ or $s_{\max}$ of the surface of revolution. Infinitesimal bending modes have to be consistent with these boundary conditions. For instance, if the displacement of the surface is prescribed to be purely transverse to the surface on the edge, the local rotation vector has to be tangent to the surface and one has to impose the boundary condition $\hat{\psi} = 0$. We call D_e^q the subspace of solutions compatible with all boundary conditions imposed at all edges. The equation for bendings is of order 2 for any $q \geq 2$: the dimension of D_e^q is given by

$$\dim D_e^q = 2 - n_e, \tag{11.50}$$

where n_e is the number of independent boundary conditions on the edges of the surface.

For an infinitesimal bending mode to be admissible, it has to satisfy the boundary conditions applied on the edges (if any is specified), and has to be smooth near poles (if the surface has poles) and near crowns (if it has crowns). Thus, the space of solutions D^q of the bending problem, is given by the intersection

$$D^q = D_e^q \cap D_P^q \cap \cdots \cap D_C^q \cap D_{C'}^q \cap D_{C''}^q \cap \cdots \tag{11.51}$$

As denoted by the ellipses, this intersection includes as many sets D_P^q, $D_{P'}^q$ as poles in the surface, and as many sets D_C^q, $D_{C'}^q$, etc. as crowns.

The rigidity problem is solved if we can compute the dimension of these linear subspaces D^q: if the sets D^q contain only the trivial solution ($\dim D^q = 0$) for all q, the surface is rigid; any value of q such that $\dim D^q \geq 1$ yields a non-trivial bending mode.

The dimension of D^q given by equation (11.51) can be computed easily if we assume that no pair of one-dimensional subspaces $D_P^q, \ldots, D_C^q, D_{C'}^q, \ldots$ are equal, and that none of them are contained in D_e^q. Unless the surface is specifically tailored to make these subspaces aligned, this assumption is satisfied. We refer to this as the *generic case*. Under these assumptions, the dimension of D^q reads:

$$d^q = \dim D^q = 2 - n_e - n_p - n_c \qquad \text{(generic case)},$$

where n_p denotes the number of poles and n_c the number of crowns on the surface of revolution (recall that n_e is the number of independent kinematical boundary conditions on the edges of the surface). This formula can be justified as follows: the dimension of the intersection D^q is that of D_e^q, namely $(2 - n_e)$, minus the number of sets D_P^q or D_C^q since they are all of co-dimension 1 by equations (11.48) and (11.49). By convention, a negative dimension $d^q \leq 0$ means, like $d^q = 0$, that the surface is rigid, i.e. there is no non-trivial bending modes for this values of q (the interpretation of negative value d^q will become clear when we study exceptional modes of deformation later in Section 11.5.7).

The quantity $(2 - n_p)$ can be identified as the number of (circular) boundaries of the surface—think of a sphere ($n_p = 2$) that has no boundary, or a piece of a paraboloid ($n_p = 1$) that has one boundary, or a piece of a cylinder ($n_p = 0$) that has two boundaries. We can

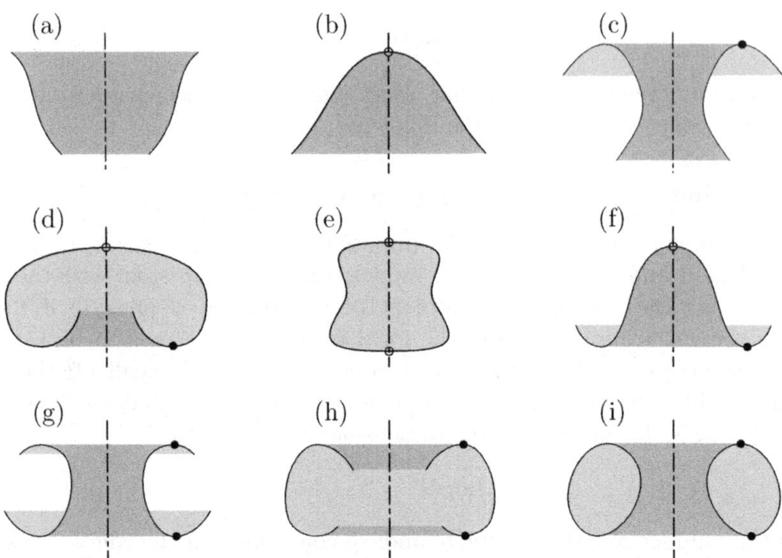

Fig. 11.5 Generic rigidity of some surfaces of revolution. Open dots denote poles and solid dots denote crowns. The dimension d^q is evaluated from equation (11.52). Surfaces having non-trivial bending modes are: (a) $n_f = 2$, $n_c = 0$, $d^q = 2$, (b) $n_f = 1$, $n_c = 0$, $d^q = 1$, (c) $n_f = 2$, $n_c = 1$, $d^q = 1$. The other surfaces are generically rigid: (d) $n_f = 1$, $n_c = 1$, $d^q = 0$, (e) $n_f = 0$, $n_c = 0$, $d^q = 0$, (f) $n_f = 1, n_c = 1, d^q = 0$, (g) $n_f = 2, n_c = 2, d^q = 0$, (h) $n_f = 2, n_c = 2, d^q = 0$, (i) $n_f = 0$, $n_c = 2$, $d^q = -2$.

further identify $n_f = (2 - n_p - n_e)$ as the number of free boundaries, i.e. the number of boundaries not subjected a kinematical constraint.[16] The generic dimension of the space of bendings with Fourier index q can be put in the form

$$d^q = n_f - n_c, \tag{11.52}$$

where n_f is the number of free boundaries just introduced.

A surface of revolution is rigid if $d^q \leq 0$. For $d^q > 0$, it has generically one ($d^q = 1$) or two ($d^q = 2$) independent modes of bending for each value of q. This criterion for the infinitesimal rigidity of surfaces of revolution is illustrated by some examples in Fig. 11.5.

Equation (11.52) points out the special role played by crowns: each crown drawn on the surface suppresses one infinitesimal bending mode for each value of q. In Section 11.6, we shall give a geometrical interpretation of the additional rigidity brought about by crowns. This geometrical interpretation will allow us to extend the notion of crowns to surfaces that are not of revolution.

---

[16] In the case of a torus, one has to impose the condition of periodicity of the solution of the differential equation after one turn in the cross-section. To this end, both $\hat{\psi}$ and $\hat{\psi}'$ have to be matched after one turn in the cross-section. This counts for two kinematical conditions, $n_e = 2$. This yields $n_f = 2 - n_p - n_e = 2 - 0 - 2 = 0$, consistent with the fact that the surface has no physical boundary.

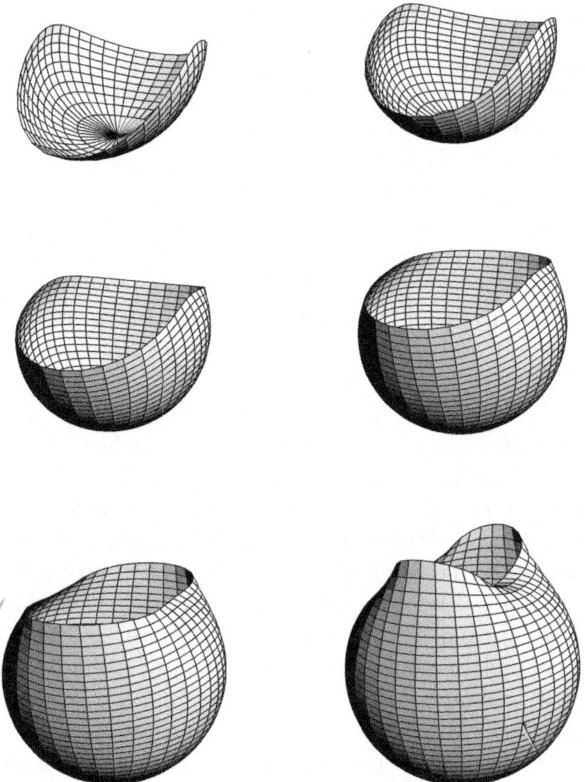

Fig. 11.6 Infinitesimal bending modes of spherical surfaces ($q = 2$). The undeformed surface is the intersection of a sphere $x^2 + y^2 + z^2 = 1$ with the lower half-space $z \leq z_0$. For these surfaces $d^q = 1$; the mode shown here is that with Fourier index $q = 2$. The plots correspond to different undeformed surfaces, i.e. to different values of the parameter z_0. When z_0 approaches $+1$ from below (bottom right), the spherical cap closes into a sphere with a tiny hole and the bending mode diverges near the hole: it disappears when the surface becomes a full sphere ($z_0 = +1$), a surface which is rigid ($d^q = 0$).

11.5.6   Illustration: closing of a spherical cap

As an illustration, we consider the bending modes of a spherical cap in Fig. 11.6. This surface has one boundary ($n_f = 1$), and no crowns ($n_c = 0$): it is not rigid, $d^q = 1$. Its bending modes are given by a closed analytical formula (F. I. Niordson, 1985). When the spherical cap closes into a sphere with a tiny hole, the bending modes become localized near the hole, and they disappear in the limit. This is consistent with the fact that the unpunctured sphere is rigid ($d^q = 0$). The related problem of the eigenfrequencies of a spherical shell with a hole has been considered in reference (P. Patrício, 1998): the frequencies of purely bending modes is very low when the hole is of finite size, but diverges when the size of the hole goes to zero; for shells of finite thickness, this divergence is regularized by stretching effects: these bending modes merge with extensional modes in this limit, and their frequencies become comparable.

11.5.7 Exceptional modes of deformation

The rigidity criterion $d^q \leq 0$ with d^q given by equation (11.52) has been derived under the generic assumption that no pair of subspaces appearing in the right-hand side of equation (11.51) are aligned. In the present section, we give two examples of surfaces that do not satisfy this assumption. In fact, those surfaces are tailored in such a way that some sub-manifolds are aligned. As a result, they admit some non-trivial bending modes, called *exceptional modes*, even though the generic dimensional count yields $d^q \leq 0$. Exceptional modes are present for specific surfaces and for specific values of q only.

The first example of an exceptional mode is from the original paper by Cohn-Vossen (S. Cohn-Vossen, 1929). The surface of revolution is defined in terms of three parameters $a > 1/4$, $b > 0$ and $c > 0$: for $|z| \leq c$, $\rho = b \cos z$; for $c \leq |z| \leq c + a - 1/4$, $\rho = 1/2 \exp(2(a + c - |z|) - 1/2)$; for $c + a - 1/4 \leq |z| \leq a + c$, $\rho = \sqrt{a + c - |z|}$. There are two conditions warranting continuity of the generating curve at $|z| = c$ and $|z| = c + a - 1/4$, and they allow elimination of b and c in favour of a. A surface in this family is shown in Fig. 11.7. Like the ovoid in Fig. 11.5(c), it has $d^q = 0$, and so is rigid for a generic value of a. However, the parameter a can be tuned in such a way that a smooth bending mode exists for a particular value of q (the equations for bendings can be solved analytically and the existence of a smooth, global solution leads ultimately to a transcendental equation for the remaining parameter a, see reference (S. Cohn-Vossen, 1929) for details). This amounts to aligning the spaces D_P^q and $D_{P'}^q$ of solutions that are smooth at either pole of the surface. Note that the undeformed surface proposed by Cohn-Vossen has non-constant Gauss curvature as the curvature of the generating curve changes sign, see Fig. 11.7; by a classical rigidity result, no bending mode, not even an exceptional one, could be found on an ovoid (i.e. on a ball-like, compact surface with positive Gauss curvature everywhere).

Our second example is a torus that has a strip cut off around its outer equator, see Fig. 11.8. We consider a family of such tori, indexed by the aspect ratio of the cross-section, denoted λ. The parametric equation of the undeformed surface reads

$$\rho(\lambda, t) = 1 + \lambda\,(1 - \cos t),$$
$$z(\lambda, t) = \sin t. \tag{11.53}$$

Here, t is the coordinate along the generating curve and varies in the range $-2 \leq t \leq +2$. It is related to the arc length s by $\mathrm{d}s = (\rho_{,t}{}^2 + z_{,t}{}^2)^{1/2}\,\mathrm{d}t$. These surfaces are of the same type as that in Fig. 11.5(g), and are all generically rigid: $d^q = 0$. The following procedure allows one to find numerical values of the parameter λ where exceptional bending modes are present.

For any fixed value of λ and q, one can integrate numerically the equation for bendings starting at the inner equator, $t = 0$. These equations are of order 2: integration requires the initial values of $\hat{\psi}_q(0)$ and $\hat{\psi}'_q(0)$. The pair $(\hat{\psi}_q(0), \hat{\psi}'_q(0))$ can be described using polar coordinates:

$$\hat{\psi}_q(0) = \mu \cos \phi, \quad \hat{\psi}'_q(0) = \mu \sin \phi.$$

The analysis at the end of Section 11.5.4 shows that the solution $\hat{\psi}_q$ always has a finite limit at the crowns $t \to \pm\pi/2$, but this limit has to be zero in order for the displacement to

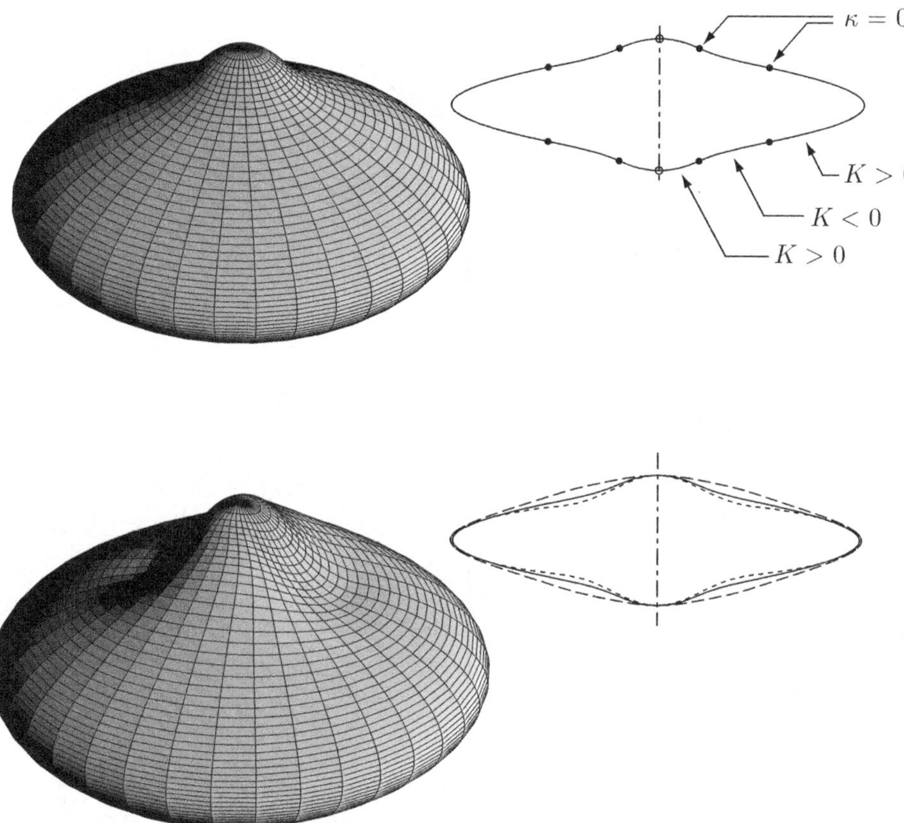

Fig. 11.7 Cohn-Vossen's example of an exceptional bending mode on a surface of revolution, $q = 2$. *Top*: undeformed surface; *Bottom*: surface deformed according to its exceptional mode, shown with arbitrary amplitude. Cuts through a plane containing the axis are shown in the insets: undeformed configuration (solid curve) and deformed configuration obtained by cutting through two planes making a right angle (dashed and doted curves); the black dots are inflexion points of the generating curve, where the Gauss curvature changes sign.

be bounded. We note $\delta^{\pm}$ this limit, $\delta^{\pm} = \hat{\psi}_q\,(\pm\pi/2)$. The equations for bendings are linear and so $\delta^{\pm}$ depends linearly on μ:

$$\delta^+ = \mu\,\Delta^+(q,\lambda,\phi) \quad\text{and}\quad \delta^- = \mu\,\Delta^-(q,\lambda,\phi).$$

The functions $\Delta^{\pm}(q,\lambda,\phi)$ can be found by numerical integration of the equations (11.46) for bendings from $t = 0$ to $t = \pm\pi/2$, using the initial condition $\hat{\psi}_q(0) = \cos\phi$ and $\hat{\psi}'_q(0) = \sin\phi$ (this amounts to setting $\mu = 1$ for the purpose of evaluating $\Delta^{\pm}$).

In the space of solutions with polar coordinates (μ,ϕ), the subspaces of solutions that are smooth near the upper or lower crown, denoted D^q_{C+} and D^q_{C-}, are defined by the implicit

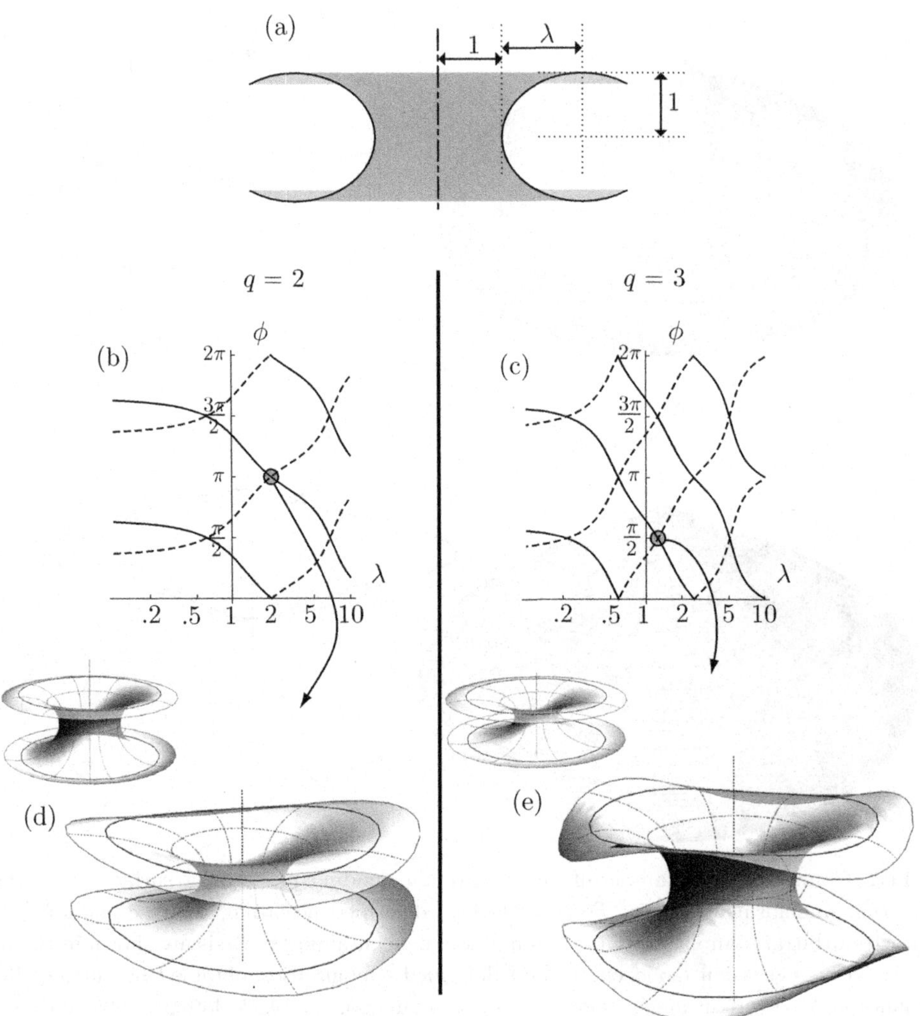

Fig. 11.8 Exceptional modes of an open torus: (a) undeformed configuration; (b–c) numerical plot of the curves defined by the implicit equation $\Delta^+(q, \lambda, \phi) = 0$ (solid curves) and $\Delta^-(q, \lambda, \phi) = 0$ (dashed curves) in the (λ, ϕ) plane. Intersection of a solid and a dashed curve corresponds to an exceptional bending mode, shown here with arbitrary amplitude: (d) $q = 2$, $\lambda = 2.155$; (e) $q = 3$, $\lambda = 1.259$.

equations $\Delta^\pm(q, \lambda, \phi) = 0$. Bending modes have to be regular at both crowns and this is expressed by the set of equations:

$$\Delta^+(q, \lambda, \phi) = 0$$

$$\Delta^-(q, \lambda, \phi) = 0.$$

For a given value of q, each equation $\Delta^+(q, \lambda, \phi) = 0$ or $\Delta^-(q, \lambda, \phi) = 0$ defines implicitly a set of curves in the plane (λ, ϕ); any crossing between curves denotes an exceptional bending mode. In Fig. 11.8, the curve $\Delta^{\pm}(q, \lambda, \phi)$ are plotted numerically for $q = 2$ and $q = 3$, and some exceptional bending modes are visualized. Note that exceptional modes can be found for discrete values of the aspect ratio λ: surfaces exhibiting exceptional bending modes are quantized.[17]

In this example, we considered tori with a strip removed around their *outer* equator. If the same procedure was repeated with tori cut along their *inner* equator, i.e. by letting t vary in the range $\pi/4 \leq t \leq 7\pi/4$ for instance, we would have found *no* exceptional bending mode, even if the parameters λ and q are tuned. When the cut is made along the inner equator, the region enclosed by the crowns is the outer part of the torus and so has positive Gauss curvature, $K > 0$. In reference (S. Cohn-Vossen, 1929), Cohn-Vossen showed that no exceptional mode is possible if a surface of revolution has an elliptic region ($K > 0$) bounded[18] by two crowns, by one crown and one pole, or by two poles with respect to the coordinate s. This result is proved by applying Sturm's comparison theorem to the solutions of the equations (11.47) for bendings: by this theorem, the solutions oscillate more and more quickly[19] for increasing values of q in regions where the Gauss curvature is positive, and this can be shown to contradict the boundary conditions to be satisfied at the crowns or poles when $q \geq 2$. This rigidity result concerns an elliptic region of the surface ($K > 0$) bounded by crowns or poles; it does not apply to the case of a torus cut along its outer equator, as in Fig. 11.8: its crowns enclose a hyperbolic region and its elliptic regions are limited by the free boundaries.

11.6 Crowns: interpretation of rigidity, extension to arbitrary surfaces

The generic rigidity of surfaces of revolution is characterized by the number d^q defined in equation (11.52); this number depends on an integer with a topological meaning, the number of free boundaries n_f, and on the number of crowns drawn on the surface, n_c. In the context of infinitesimal bendings, the crowns, defined by the equation $z'(s) = 0$, make a surface of revolution more rigid. In the present section, we propose a geometrical interpretation of this rigidity; this will allow us to extend the definition of crowns on surfaces that are not of revolution.

11.6.1 Interpretation based on the theory of PDE's

On surfaces of revolution, crowns are a place where the Gauss curvature changes its sign— recall that it can also change sign when the generating curve has an inflexion point. The equations for bendings being locally of the same type as the surface, they go from elliptic to hyperbolic across a crown.[20] The smoothness of the solutions of partial differential equations

---

[17] By 'quantized', we mean that the surfaces featuring exceptional modes form a manifold of co-dimension $(1-d^q)$ in the space of all smooth manifolds: these surfaces are specified by $(1-d^q)$ scalar constraints. In the case of the open torus, $d^q = 0$ and they form a discrete sequence within the one-parameter family of surfaces defined by equation (11.53).

[18] Recall that the Gauss curvature of a surface of revolution can change its sign either along crowns or along inflexion points of the generating curve.

[19] In the short wave limit ($q \to \infty$), WKB-like arguments (R. H. Dicke and J. P. Wittke, 1960) can be used to study the quantization condition for exceptional surfaces. Hyperbolic parts of the surface are then mapped to classically permitted regions, while elliptic ones are mapped to classically forbidden ones; circles of vanishing Gauss curvature correspond to turning points of the potential.

[20] The equations for trans-sonic flows provides another example of partial differential equations of mixed type.

near a curve where the type of the equations goes from elliptic to hyperbolic depends crucially on the way the characteristic curves merge or cross the boundary of the hyperbolic domain. As noted by Bakievich (N. I. Bakievich, 1963), a crown satisfies the following interesting property: the tangents $\mathbf{u}_\theta$ to the curve defined by $K = 0$ (the crowns) happen to define an asymptotic direction of the surface (in the non-axisymmetric case, the asymptotic direction can make an arbitrary angle with the curve $K = 0$). By asymptotic, we mean a direction $\mathbf{t}$ of the tangent plane along which the surface is locally flat, i.e. is such that $t^\alpha b_{\alpha\beta} = 0$ (Greek indices range over in-plane directions); at a crown, the curvature tensor given by equation (11.40) has only one non-zero component, b_{ss}, and so $t^\alpha b_{\alpha\beta} = 0$ for $\mathbf{t} = \mathbf{u}_\theta$, as announced. The characteristic curves of the PDE for the infinitesimal bendings, defined in the hyperbolic regions of the surface, follow the asymptotic directions. Therefore, the presence of crowns on a surface leads to the somewhat pathological situation where the PDE for infinitesimal bendings goes from hyperbolic to elliptic and *all the characteristic curves, defined in the region where $K = 0$, are tangent to the curve $K = 0$ on the edge of the parabolic region.*

This specific orientation of the characteristic (asymptotic) curves near the crowns explains their rigidity from the viewpoint of the theory of partial differential equations (N. I. Bakievich, 1963). This property can be used to propose an extension of crowns to surfaces that are not of revolution. On an arbitrary surface, any curve along which the Gauss curvature changes sign ($K = 0$) and whose tangents define an asymptotic direction of the surface, has the same rigidifying properties as a crown on a surface of revolution. We shall call *crowns* such curves. An example of a crown on a non-axisymmetric surface is shown in Fig. 11.9.

By the Beltrami-Enneper theorem (M. Spivak, 1979, p. III.291), the geometric torsion of an asymptotic curve is given by $\sqrt{-K(s)}$, and so is zero if the curve is such that $K = 0$. This implies that the normal $\mathbf{n}(s)$ to the surface along a crown is actually independent of s. A crown, defined in general as an asymptotic curve where the Gauss curvature changes sign ($K = 0$), is always contained in some plane; this plane is tangent to the surface at every point of the curve.

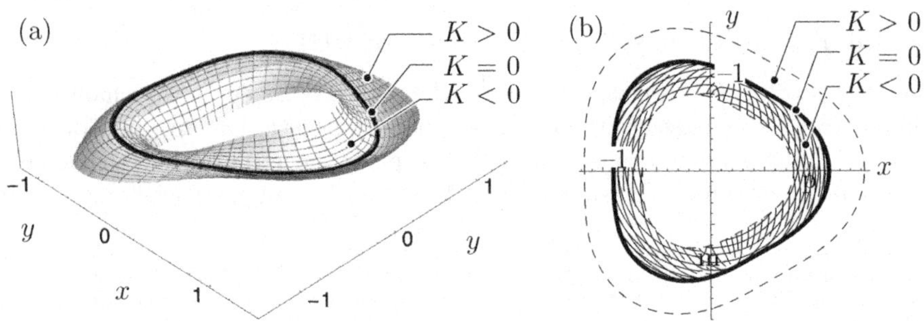

Fig. 11.9 On a general surface (not necessarily of revolution), a crown is defined as a curve across which the Gauss curvature changes sign, and whose tangents define a asymptotic direction of the surface (i.e. a direction where the surface is flat). (a) Visualization of a typical crown; (b) projection on the plane containing the crown, showing the asymptotic lines in the hyperbolic (inner) region. Note tangency of the asymptotic curves with the crown at the edge of the hyperbolic region.

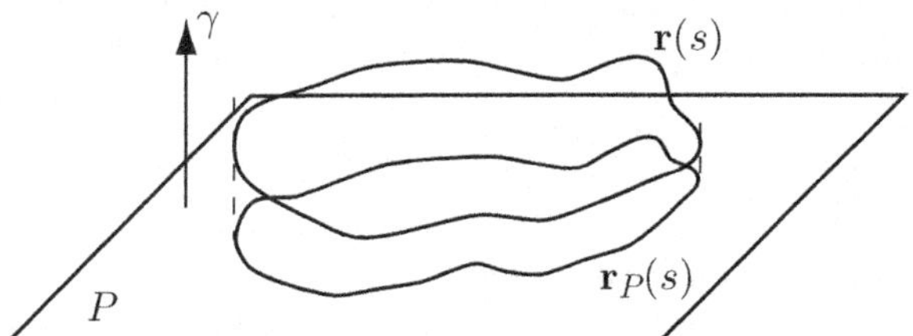

Fig. 11.10 The projection of the curve $\mathbf{r}(s)$ onto a plane P in which it was initially contained does not modify its curvature, to first order in the deformation.

11.6.2　Geometric interpretation

We propose a second interpretation of the rigidity brought about by crowns. This interpretation is geometrical and is probably more intuitive that the one given in Section 11.6.1; it was first published in reference (B. Audoly, 1999).

This interpretation is based on the decomposition of the curvature of a curve drawn on a surface S. Let $\mathbf{r}(s)$ be an arc length parameterisation of the curve. The curve being drawn on the surface, its unit tangent $\mathbf{t}(s) = \mathbf{r}'(s)$ lies in the plane tangent to the surface at $\mathbf{r}(s)$. Its curvature is the vector $\boldsymbol{\kappa}(s) = \mathbf{t}'(s)$; this quantity, which makes no reference to the surface S, is called the *extrinsic* curvature. We decompose the extrinsic curvature on the local tangent plane to surface and on the local unit normal $\mathbf{n}$ to the surface:

$$\boldsymbol{\kappa}(s) = \boldsymbol{\kappa}_i(s) + \kappa_n(s)\,\mathbf{n}(s). \tag{11.54}$$

A classical result of the geometry of curves on surfaces, which can be found in many textbooks such as in reference (M. Spivak, 1979), states that (i) the tangential projection $\boldsymbol{\kappa}_i(s)$ depends on how the curve is drawn on the surface but not on how the surface bends in the Euclidean space; this vector $\boldsymbol{\kappa}_i(s)$ is called the intrinsic curvature, and follows the surface rigidly when the latter is deformed isometrically; and (ii) the normal projection $\kappa_n(s)$ depends only on how the surfaces bends in Euclidean space along the tangent $\mathbf{t}(s)$ tangent to the curve; concretely, it is given by[21] $\kappa_n(s) = t^\alpha\,b_{\alpha\beta}\,t^\beta$, where t^α are the components of the tangent on the local coordinate frame $\mathbf{e}_\alpha$, and $b_{\alpha\beta}$ is the second fundamental form of the surface at the point (curvature form).

Now, consider the case of a curve that is asymptotic in undeformed configuration, as in Fig. 11.10: then, $\kappa_n^0(s) = 0$ for all s, where the superscript '0' refers to the undeformed configuration. Noting that the right-hand side of equation (11.54) is the sum of two orthogonal vectors, we write the squared curvature in deformed configuration as

$$|\boldsymbol{\kappa}|^2(s) = |\boldsymbol{\kappa}_i|^2(s) + {\kappa_n}^2(s).$$

Let us consider infinitesimal bendings of the surface with amplitude ϵ. Then, by the above theorem, the first term in the right-hand side remains unchanged as the surface is deformed isometrically, while the second term is a second-order quantity in ϵ (κ_n vanishes

---

[21] The quantity $t^\alpha\,b_{\alpha\beta}\,t^\beta$ is just the quadratic form C_P defined earlier in equation (6.18), evaluated on the tangent to the curve $\mathbf{t}(s)$.

in undeformed configuration and so is of order ϵ). This implies that the scalar curvature, $|\boldsymbol{\kappa}|(s)$, remains unchanged to first order by the deformation of the surface:

$$|\boldsymbol{\kappa}|(s) = |\boldsymbol{\kappa}_i^0|(s) + \mathcal{O}(\epsilon^2). \tag{11.55}$$

We shall further assume that the curve is planar in undeformed configuration. We note P the plane containing the undeformed curve, $\mathbf{n}^0$ the normal to this plane, and $\mathbf{r}_P(s)$ the curve obtained by projecting the deformed curve $\mathbf{r}(s)$ back into the plane P, see Fig. 11.10. Let $\gamma(s)$ be the normal displacement of a point on the curve $\mathbf{r}(s)$ with respect to the plane P, normalized by the small amplitude ϵ of the displacement:

$$\mathbf{r}(s) = \mathbf{r}_P(s) + \epsilon\,\gamma(s)\,\mathbf{n}^0.$$

Differentiating this equation twice with respect to s, we find $\boldsymbol{\kappa}(s) = \mathbf{r}_P''(s) + \epsilon\,\gamma''(s)\,\mathbf{n}^0$. Since $\mathbf{r}_P''(s)$ lies in the plane P, it is perpendicular to $\mathbf{n}^0$ and we have

$$|\boldsymbol{\kappa}|(s) = |\mathbf{r}_P''|(s) + \mathcal{O}(\epsilon^2). \tag{11.56}$$

Comparison of equations (11.55) and (11.56) reveals that the curvature of the planar curve $\mathbf{r}_P$ is unaffected by the deformation of the surface, to first order. This implies that the curve obtained by projection in the plane P deforms in a rigid manner, i.e. by a combined infinitesimal rotation and translation.

We have just shown that the infinitesimal bending displacements of a surface are purely transverse in the neighbourhood of a curve that is asymptotic and planar. In-plane displacements are geometrically forbidden near such curves:[22] this property severely constraints the motion of the surface in the vicinity of the curve, and tends to make the surface more rigid. This rigidity is ultimately a consequence of equation (11.54): when the curve is asymptotic in undeformed configuration, $\kappa_n^0(s) = 0$, the scalar curvature $|\boldsymbol{\kappa}|(s)$ is at minimum in undeformed configuration.

Using again Beltrami-Enneper theorem (M. Spivak, 1979, p. III.291), we find that an asymptotic and planar curve is also a curve where the Gauss curvature of the surface vanishes: the geometrical rigidity result of the present section is based on assumptions that are exactly equivalent to those of the previous section.

11.7   Conclusion

We have derived the equation for the infinitesimal rigidity of inextensible shells. Studying surfaces of revolution first, we have solved the problem of the existence of infinitesimal bending modes and shown afterwards that these modes are affected by the presence of special curves, called crowns, where the Gauss curvature changes sign; we have also discussed the presence of exceptional modes for exceptional surfaces.

We extended this rigidity result to arbitrary surfaces of mixed type, relying on an extension of the notion of stiffening curves, called crowns, to non-axisymmetric surfaces, as in Fig. 11.10. These curves are expected to have a strong influence on the vibration spectra of the surface, as well as on the crumpling and buckling properties of thin shells.

---

[22] Obviously, this does not apply to rigid-body in-plane displacements.

In the following chapters, we present explicit calculations of the deformations of elastic shells under various loads. The case of spherical shells, studied in Chapter 14, shows that considerations based on smooth deformation of surfaces may give only a partial answer to the elasticity problem: beyond a well-defined critical load, the deformation becomes singular in the zero-thickness limit. Deformations of an elastic toroidal shell are studied in Chapter 13. A torus makes an interesting topic of study because its centre surface contains two crowns (in the sense of the present chapter) which make it isometrically rigid. The equations for the inextensible deformations of the torus change from elliptic to hyperbolic along these curves. As a consequence, it is shown that the deformation becomes very large near these curves; bending effects are required locally to describe accurately the deformations of the toroidal shell. Because of geometry, a vibrating elastic torus has two natural frequencies that are much smaller than all the other ones.

We thank A. Hamdouni for pointing out useful references.

References

A. D. Alexandrov. On a class of closed surfaces. *Recueil de la société mathématique de Moscou*, 4:67, 1938.

B. Audoly. Rigidifying curves on surfaces. *Comptes Rendus de l'Académie des Sciences - Series I - Mathematics*, 328(4):313–316, 1999.

N. I. Bakievich. On a certain equation of mixed type in the theory of infinitely small bendings of surfaces. *Volzhsk. Mat. Sb., Teor. Ser.*, 1:32–42, 1963.

S. Cohn-Vossen. Unstarre geschlossene Flächen. *Mathematische Annalen*, 102:10, 1929.

R. H. Dicke and J. P. Wittke. *Introduction to Quantum Mechanics*. Addison Wesley, 1960.

M. L. Gromov and V. A. Rokhlin. Immersions and embeddings of Riemannian manifolds. *Russian Mathematical Surveys*, 25(5):1–57, 1970.

I. Ivanova-Karatopraklieva and I. Kh. Sabitov. Bending of surface. Part II. *Journal of Mathematical Sciences*, 74(3):997–1043, 1995.

F. I. Niordson. *Shell Theory*, volume 29 of North-Holland series in applied mathematics and mechanics. North-Holland, Amsterdam, Holland, 1985.

L. Nirenberg. Rigidity of a class of closed surfaces. In *Nonlinear Problems (Proc. Sympos., Madison, Wis., 1962)*, page 177, Madison, WI, 1963. University of Wisconsin Press.

P. Patrício. *Instabilités géométriques en élasticité*. PhD thesis, Université de Paris VII, Paris, France, 1998.

A. V. Pogorelov. *Extrinsic Geometry of Convex Surfaces*. American Mathematical Society, Providence, RI, 1973.

J. W. S. Rayleigh. *The Theory of Sound*, volume 1. Dover Publications, 2nd edition, 1976.

I. Kh. Sabitov. Local theory of bendings of surfaces. In *Geometry III. Theory of surfaces*, volume 48 of *Encyclopaedia of Mathematical Sciences*, chapter 3, pages 179–250. Springer, 1992.

M. Spivak. *A Comprehensive Introduction to Differential Geometry*, volume 5. Publish or Perish, Inc., Houston (TX), 2nd edition, 1979.

É. Sanchez Palencia and J. Sanchez-Hubert. *Introduction aux méthodes asymptotiques et à l'homogénéisation; application à la mécanique des milieux continus.* Masson, 1992.

D. J. Struik. *Lectures on Classical Differential Geometry.* Addison-Wesley, Reading, MA, 2nd edition, 1961.

M. L. Szwabowicz. *Deformable surfaces and almost inextensional deflections of thin shells (Habilitation dissertation).* PhD thesis, Zeszyty Naukowe IMP PAN, Gdansk, 1999.

R. M. Wald. *General Relativity.* University of Chicago Press, 1984.

12

Shells of revolution

In this chapter, we derive the equations for static equilibrium of thin elastic shells, when the shell is of revolution both in its reference and deformed configurations. More specifically we consider *axisymmetric, twistless deformations* of shells of revolution: by assumption[1] the displacement expressed in cylindrical coordinates has a zero azimuthal component, and both the radial and axial components are independent of the azimuthal coordinate θ. The equations for axisymmetric twistless shells take the form of ordinary differential equations with respect to the meridional coordinate s, and do not depend on the azimuthal (circumferential) coordinate θ. As a result they are much simpler than the equations for general shells which are partial differential of two variables. Still, all the salient mechanical features of shells are preserved: the axisymmetric, twistless case provides a nice introduction to the theory of shells. The equations of axisymmetric shells are classical and are presented in many books, such as (A. L. Goldenveizer, 1961; V. V. Novozhilov, 1970; W. Flügge, 1973; F. I. Niordson, 1985). Like Libai and Simmonds (A. Libai and J. G. Simmonds, 1998), we obtain the equations for axisymmetric shells by following three distinct steps, namely (i) perform a kinematical analysis, (ii) define an energy together with constitutive laws, and (iii) derive the equations of equilibrium using the calculus of variation.

Following the general theme of this book we shall emphasize the connection between geometry and the mechanics of shells. To make this connection clearer, we shall not try to derive the equations in their most general setting—this would be hopeless anyway since there are many variants of the equations for shells, based on various kinematical approximations (small or finite strain, small or finite rotations, presence of non-classical effects such as normal shear or normal curvature into account, etc.), the final equations being in some cases extremely cumbersome to write down. Instead, we focus on a simple setting where the important mechanical effects can be illustrated without unnecessary complications: we consider the case of small strains and moderate rotations. By 'moderate', we mean that the rotations angles are given by small numbers but may still be much larger than the typical measure of strain, which is another very small quantity (the Föppl–von Kármán equations for plates were derived in Chapter 6 by exactly the same kind of approximation). The corresponding set of equations has been derived by Koiter (W. T. Koiter, 1966) under the heading of 'small finite deflections' and Sanders (J. L. Sanders, 1963) as the 'approximation of small strain and moderately small rotations'.

The equations derived in this chapter are applied to two particular shell problems later, the deformation of an elastic torus in Chapter 13 and the contact of a thin spherical shell with a perfectly rigid plane in Chapter 14.

---

[1] Using the notations used throughout this chapter, the cylindrical components of the displacement read $v_r(s)$, $v_\theta = 0$ and $v_z(s)$.

Technically, we follow an approach very similar to that used in Chapter 6 to derive the Föppl–von Kármán equations for elastic plates. One writes the elastic energy as the sum of a stretching energy and a bending energy. The equations of equilibrium are obtained by variation of this energy functional—for a complete presentation of variational methods applied to shell theory, see reference (R. Valid, 1995). The difference with the derivation of plate equations has its origin in geometry: the connection between strain and displacement is sensitive to the initial curvature, which is non-zero for shells but zero for plates.

12.1   Geometry

Our notation will be similar to that used in Section 11.5 for the analysis of isometric deformations of surfaces of revolution. This notation is summarized in Fig. 12.1. Cylindrical coordinates (ρ, θ, z) are used, ρ and θ being the polar coordinates in the plane (x, y) perpendicular to the axis of revolution z. From the unit vectors $(\mathbf{e}_x, \mathbf{e}_y, \mathbf{e}_z)$ associated with Cartesian coordinates, we define the unit polar vectors $\mathbf{u}_r(\theta) = \cos\theta\,\mathbf{e}_x + \sin\theta\,\mathbf{e}_y$ and $\mathbf{u}_\theta(\theta) = -\sin\theta\,\mathbf{e}_x + \cos\theta\,\mathbf{e}_y$ as in equation (11.30). Their derivatives are given by $\mathbf{u}_r'(\theta) = \mathbf{u}_\theta(\theta)$ and $\mathbf{u}_\theta'(\theta) = -\mathbf{u}_r(\theta)$.

12.1.1   Undeformed configuration

The undeformed configuration of the shell is defined by a so-called *generating plane curve*, given by its parametric equation $\rho = \rho(s)$ and $z = z(s)$, $\rho(s)$ and $z(s)$ being given functions of s, the arc length, such that:

$$\rho'^2(s) + z'^2(s) = 1. \tag{12.1}$$

Here and in the rest of this chapter, primes refer to derivation with respect to the parameter of a function, s here—note that the shell can stretch, and as a result s will not necessarily remain the arc length after deformation. In view of equation (12.1), we can introduce the angle $\gamma(s)$ of the undeformed tangent with the x axis, defined by

$$\rho'(s) = \cos\gamma(s), \qquad z'(s) = \sin\gamma(s). \tag{12.2}$$

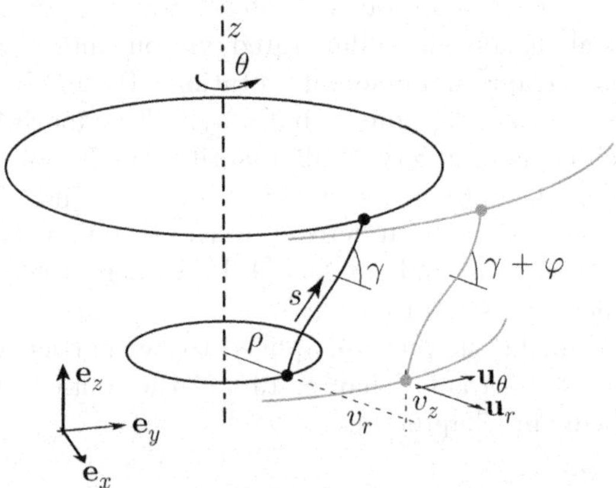

Fig. 12.1 Notation used for the analysis of an twistless axisymmetric shell.

The undeformed surface is spanned by rotating the generating plane curve about the axis z of revolution, and a current point in reference configuration is given by

$$\mathbf{r}(s, \theta) = \rho(s)\, \mathbf{u}_r(\theta) + z(s)\, \mathbf{e}_z. \tag{12.3}$$

The local tangent plane is spanned by the unit tangent vectors defined as

$$\mathbf{e}_s(s, \theta) = \frac{\partial \mathbf{r}(s, \theta)}{\partial s} \quad \text{and} \quad \mathbf{e}_\theta(s, \theta) = \frac{1}{\rho(s)} \frac{\partial \mathbf{r}(s, \theta)}{\partial \theta}.$$

These tangent vectors can be computed explicitly from equation (12.3) as

$$\mathbf{e}_s(s, \theta) = \rho'(s)\, \mathbf{u}_r(\theta) + z'(s)\, \mathbf{e}_z = \cos\gamma(s)\, \mathbf{u}_r(\theta) + \sin\gamma(s)\, \mathbf{e}_z,$$

$$\mathbf{e}_\theta(s, \theta) = \mathbf{u}_\theta(s).$$

12.1.2   Displacement

By assumption, the displacement of any material point in an axisymmetric, twistless shell lies within the plane passing through the axis of revolution, which is spanned by the vectors $\mathbf{u}_r(\theta)$ and $\mathbf{e}_z$. In addition, the components of displacement in the cylindrical basis are independent of θ by assumption. Let $v_r(s)$ and $v_z(s)$ denote the two non-zero components of the displacement. With this notation, the equation of the centre surface in the actual configuration[2] reads

$$\overline{\rho}(s) = \rho(s) + v_r(s), \quad \overline{\theta}(\theta) = \theta, \quad \overline{z}(s) = z(s) + v_z(s), \tag{12.4}$$

and a current point on the deformed surface is given by

$$\overline{\mathbf{r}}(s, \theta) = (\rho(s) + v_r(s))\, \mathbf{u}_r(\theta) + (z(s) + v_z(s))\, \mathbf{e}_z. \tag{12.5}$$

12.1.3   Membrane strain

Consider two neighbouring material points, with Lagrangian coordinates (s, θ) and $(s + \mathrm{d}s, \theta + \mathrm{d}\theta)$. The squared length of the infinitesimal (material) line connecting them reads

$$(\mathrm{d}\mathbf{r})^2 = \left(\frac{\partial \mathbf{r}}{\partial s} \mathrm{d}s + \frac{\partial \mathbf{r}}{\partial \theta} \mathrm{d}\theta \right)^2 = (\mathbf{e}_s\, \mathrm{d}s + \rho\, \mathbf{e}_\theta\, \mathrm{d}\theta)^2 = \mathrm{d}s^2 + \rho^2(s)\, \mathrm{d}\theta^2 \tag{12.6a}$$

in the reference configuration. In the deformed configuration, a similar calculation yields:

$$(\overline{\mathrm{d}\mathbf{r}})^2 = \left((\rho'(s) + v'_r(s))^2 + (z'(s) + v'_z(s))^2 \right)\, \mathrm{d}s^2 + (\rho(s) + v_r(s))^2\, \mathrm{d}\theta^2. \tag{12.6b}$$

Following the general equation (2.8), we define the strain by the relative increment of length of a material line upon deformation:

$$(\overline{\mathrm{d}\mathbf{r}})^2 = (1 + 2\,\epsilon_s)\, \mathrm{d}s^2 + (1 + 2\,\epsilon_\theta)\, \rho^2(s)\, \mathrm{d}\theta^2. \tag{12.6c}$$

The quantities ϵ_s and ϵ_θ introduced here define the so-called *membrane* or *in-plane* strain. Note that setting $\epsilon_s = 0$ and $\epsilon_\theta = 0$ in the right-hand side yields back the metric tensor in the reference configuration, as expected.

---

[2] To avoid any ambiguity with the derivation with respect to s, we shall not use the prime for deformed quantities. Instead, we use a bar.

By identification of equations (12.6b) and (12.6c), we find the membrane strains as

$$\epsilon_s(s) = \rho'(s)\, v_r'(s) + z'(s)\, v_z'(s) + \frac{1}{2}\left(v_r'^{\,2}(s) + v_z'^{\,2}(s)\right) \tag{12.7a}$$

$$\epsilon_\theta(s) = \frac{v_r(s)}{\rho(s)} + \frac{1}{2}\left(\frac{v_r(s)}{\rho(s)}\right)^2. \tag{12.7b}$$

Here, ϵ_s and ϵ_θ denote the membrane strains denoted ϵ_{ss} and $\epsilon_{\theta\theta}$ earlier. For axisymmetric, twistless shells, the in-plane shear strain is zero, $\epsilon_{s\theta} = 0$.

Equation (12.7) defines the exact, geometrically non-linear strain. Let us define the *linear* strain ϵ_s^{lin} and $\epsilon_\theta^{\text{lin}}$ by linearizing with respect to the displacement v_r and v_z:

$$\epsilon_s^{\text{lin}} = \rho'(s)\, v_r'(s) + z'(s)\, v_z'(s), \qquad \epsilon_\theta^{\text{lin}} = \frac{v_r(s)}{\rho(s)}. \tag{12.8}$$

We could build a shell theory on this linearized strain ϵ_s^{lin} and $\epsilon_\theta^{\text{lin}}$. However, this book is concerned with geometric nonlinearity associated with rotations, and we would like to do better: we shall retain some non-linear terms in the definition of strain, as we did earlier in the theory of plates. For this purpose we shall decompose the full non-linear strain (12.7a) as the sum of (i) a linear contribution, denoted as ϵ_s^{lin} below, (ii) a non-linear contribution proportional to the square of ϵ_s^{lin}, and (iii) a remainder made of a non-linear term that cannot be expressed as a function of ϵ_s^{lin} or its powers:

$$\epsilon_s = \epsilon_s^{\text{lin}} + \frac{1}{2}\left(v_r'^{\,2} + v_z'^{\,2}\right) = \epsilon_s^{\text{lin}} + \frac{1}{2}\left(\left(\rho'^2\, v_r'^{\,2} + z'^2\, v_z'^{\,2}\right) + \left(z'^2\, v_r'^{\,2} + \rho'^2\, v_z'^{\,2}\right)\right)$$

$$= \epsilon_s^{\text{lin}} + \frac{1}{2}\left(\left((\epsilon_s^{\text{lin}})^2 - 2\rho'\, z'\, v_r'\, v_z'\right) + z'^2\, v_r'^{\,2} + \rho'^2\, v_z'^{\,2}\right) = \epsilon_s^{\text{lin}} + \frac{1}{2}(\epsilon_s^{\text{lin}})^2 + \frac{1}{2}\phi^2.$$

To write this series of equations, we have made use of the identity (12.1), and have defined the new quantity

$$\phi(s) = -z'(s)\, v_r'(s) + \rho'(s)\, v_z'(s). \tag{12.9}$$

This quantity ϕ can be interpreted as the first-order rotation of the tangent to the generating curve about the azimuthal vector $\mathbf{u}_\theta$, see Fig. 12.1.

To sum up, we have established the following relation between the exact strain ϵ_s and ϵ_θ, the linearized strain ϵ_s^{lin} and $\epsilon_\theta^{\text{lin}}$ defined in equation (12.8), and the linearized rotation ϕ defined in equation (12.9):

$$\epsilon_s(s) = \epsilon_s^{\text{lin}}(s) + \frac{\left(\epsilon_s^{\text{lin}}\right)^2(s)}{2} + \frac{\phi^2(s)}{2} \tag{12.10a}$$

$$\epsilon_\theta(s) = \epsilon_\theta^{\text{lin}}(s) + \frac{\left(\epsilon_\theta^{\text{lin}}\right)^2(s)}{2}. \tag{12.10b}$$

Note that there is no approximation involved so far.

12.1.4   Small strain, moderate rotation

We shall now introduce an approximation of the strain that is valid for small strain and moderate rotations. By moderate rotations, we mean that rotation angles are much smaller than 1 when measured in radians, but may still be much larger[3] than the membrane strain $(\epsilon_s, \epsilon_\theta)$, another small quantity.

We shall not only assume that the strain ϵ_s and ϵ_θ is small, but also that the *linearized* strain $\epsilon_s^{\mathrm{lin}}$, $\epsilon_\theta^{\mathrm{lin}}$ is small.[4] The non-linear terms $(\epsilon_s^{\mathrm{lin}})^2$ and $(\epsilon_\theta^{\mathrm{lin}})^2$ are then negligible in front of the linear ones $\epsilon_s^{\mathrm{lin}}$ and $\epsilon_\theta^{\mathrm{lin}}$ in equations (12.10). For the rotation $\phi(s)$, the picture is different: in the absence of a linear term $\phi(s)$ in the equations, we have to keep the non-linear term ϕ^2, which is not negligible *a priori*, even if $\epsilon_s^{\mathrm{lin}}$ is small. In view of this, our presentation of the theory of shells will be based on the following approximate definition of the membrane strain

$$e_s(s) = \epsilon_s^{\mathrm{lin}}(s) + \frac{\phi^2(s)}{2},$$

$$e_\theta(s) = \epsilon_\theta^{\mathrm{lin}}(s),$$

in the place of the geometrically exact one given earlier in equation (12.10).

Plugging the definition (12.8) of the linearized strain into these equations, we have

$$e_s(s) = \rho'(s)\, v_r'(s) + z'(s)\, v_z'(s) + \frac{\phi^2(s)}{2}, \tag{12.11a}$$

$$e_\theta(s) = \frac{v_r(s)}{\rho(s)}. \tag{12.11b}$$

Note that equation (12.11a) is similar to the definition (6.63a) of the strain used in plate theory, when the linearized in-plane strain $\partial u_x / \partial x$ is identified as the quantity $\rho'\, v_r' + z'\, v_z'$ and the rotation $\partial w / \partial x$ as ϕ defined by (12.9). The interpretation of equation (12.11b) is simply that any radial displacement v_r implies an expansion (when $v_r > 0$) or contraction (when $v_r < 0$) of the circumference of the shell; this e_θ is called hoop strain.

A non-linear term, ϕ^2, has been retained in equation (12.11a). This is the only nonlinearity that we shall include in our derivation of the equations for shells. The problems studied in the following chapters provide two examples of the interesting mechanical behaviour of shells that can arise out of this nonlinearity, namely localization of deformation and buckling.

12.1.5 Curvature strains

The membrane strains introduced in the previous section will be used to define the stretching energy. This stretching energy is all we need to build up a membrane theory, and this is the goal of the forthcoming Section 12.3. The shell theory includes bending terms in addition (see Section 12.4), and we still have to define some relevant measure of curvature strain. A natural definition is based on the change in curvature of the surface from reference to actual configuration.

For axisymmetric, twistless shells, the principal directions of curvature are always aligned with the local s and θ directions. The calculation of the curvature tensor given

---

[3] If we simply linearized the equations, we would not capture the situation where rotations are larger than strain.

[4] Note that the strain ϵ_s, ϵ_θ and the linearized strain $\epsilon_s^{\mathrm{lin}}$, $\epsilon_\theta^{\mathrm{lin}}$ are not always of the same order of magnitude. What we have in mind is the following typical situation, relevant to buckling: when a plate or a shell buckles and makes wrinkles but does not stretch longitudinally, the exact strain ϵ is zero and by equations (12.10) the linearized strain $\epsilon^{\mathrm{lin}} \sim -\phi^2/2$ is small but non-zero.

in equation (11.41) for a surface of revolution is applicable: the principal curvatures read $b_{ss} = \rho'(s)\, z''(s) - \rho''(s)\, z'(s)$ and $b_{\theta\theta} = z'(s)/\rho(s)$ in the reference configuration, and $\bar{b}_{ss} = \bar{\rho}'\bar{z}'' - \bar{\rho}''\bar{z}'/(\bar{\rho}'^2 + \bar{z}'^2)^{3/2}$ and $\bar{b}_{\theta\theta} = \bar{z}'\bar{\rho}/(\bar{\rho}'^2 + \bar{z}'^2)^{1/2}$ in the deformed configuration;[5] because of the axial symmetry the off-diagonal entry is always zero, $b_{s\theta} = 0$.

Circumferential bending strain

The changes in principal curvatures define the so-called curvature strains, which are denoted K_s and K_θ. Starting with the circumferential curvature, we introduce

$$K_\theta = \bar{b}_{\theta\theta} - b_{\theta\theta} = \frac{z' + v_z'}{(\bar{\rho}'^2 + \bar{z}'^2)^{3/2}\,(\rho + v_r)} - \frac{z'}{\rho}.$$

Using the definition of the strain ϵ_s, which implies $\bar{\rho}'^2 + \bar{z}'^2 = 1 + 2\epsilon_s$, and the expression (12.11b) for the hoop strain ϵ_θ, we can rewrite this expression as

$$K_\theta = \frac{1}{(1 + 2\,\epsilon_s)^{1/2}\,(1 + \epsilon_\theta)}\,\frac{z' + v_z'}{\rho} - \frac{z'}{\rho}.$$

It is useful to split this expression in two terms, one, denoted k_θ, that does not depend on membrane strain, and the other, denoted $k_\theta^\dagger$, that represents the change in curvature strain that is due to membrane strain. We define

$$K_\theta(s) = k_\theta(s) + k_\theta^\dagger(s), \tag{12.12a}$$

where

$$k_\theta = \frac{v_z'}{\rho} \tag{12.12b}$$

$$k_\theta^\dagger = \left(\frac{1}{(1 + 2\,\epsilon_s)^{1/2}\,(1 + \epsilon_\theta)} - 1\right)\frac{z' + v_z'}{\rho}. \tag{12.12c}$$

The second contribution, $k_\theta^\dagger$, vanishes for inextensible deformations of the centre surface, when $\epsilon_s = 0$ and $e_\theta = 0$. This implies that it can be neglected, as explained in full details in Section 12.4.

Meridional bending strain

A meridional bending strain can be defined in a similar fashion. Let us introduce the exact increment of geometric curvature K_s as:

$$K_s = \bar{b}_{ss} - b_{ss} = \frac{\bar{\rho}'\,\bar{z}'' - \bar{\rho}''\,\bar{z}'}{(1 + 2\epsilon_s)^{3/2}} - (\rho'\,z'' - \rho''\,z'). \tag{12.13}$$

In the expression above, $\bar{b}_{ss}$ is for the actual curvature, and b_{ss} for the curvature at rest, a notation used throughout the coming development. We can work out this expression so as to make a contribution appear, denoted $k_s^\dagger$, that cancels when $\epsilon_s = 0$:

$$K_s = k_s^\dagger + \hat{K}_s,$$

---

[5] As noted earlier, s is not an arc length parameter in the actual configuration. This is taken care of by the new factors proportional to powers of $(\bar{\rho}'^2 + \bar{z}'^2)^{-1/2}$ in the expressions of the principal curvature.

where

$$k_s^\dagger = \left[\frac{1}{(1+2\epsilon_s)^{3/2}} - 1\right] (\overline{\rho}'\,\overline{z}'' - \overline{\rho}''\,\overline{z}'). \tag{12.14a}$$

The remainder $\hat{K}_s$ is defined by $\hat{K}_s = (\overline{\rho}'\,\overline{z}'' - \overline{\rho}''\,\overline{z}') - (\rho'\,z'' - \rho''\,z')$. This remainder can be further simplified by solving equations (12.8) and (12.9) for v_r' and v_z'; after identifying the rotation $\gamma(s)$ defined by equation (12.2) in the resulting expressions, this yields

$$v_r'(s) = \epsilon_s^{\mathrm{lin}}(s)\cos\gamma(s) - \phi(s)\sin\gamma(s), \qquad v_z'(s) = \epsilon_s^{\mathrm{lin}}(s)\sin\gamma(s) + \phi(s)\cos\gamma(s).$$

Plugging then $\overline{\rho} = \rho + v_r$ and $\overline{z} = z + v_z$ into the expression for $\hat{K}_s$ above we obtain

$$\hat{K}_s(s) = k_s(s) + k_s^\ddagger(s),$$

where

$$k_s = \phi' \tag{12.14b}$$

$$k_s^\ddagger = \left(2\,\epsilon_s^{\mathrm{lin}} + \left(\epsilon_s^{\mathrm{lin}}\right)^2\right)\gamma' + (\epsilon_s^{\mathrm{lin}}\,\phi' - \epsilon_s^{\mathrm{lin}'}\,\phi) + \phi^2\,\gamma'. \tag{12.14c}$$

We have therefore decomposed the curvature strain, K_s, in three contributions:

$$K_s(s) = k_s^\dagger(s) + \hat{K}_s(s)$$
$$= k_s(s) + k_s^\dagger(s) + k_s^\ddagger(s). \tag{12.14d}$$

Simplified curvature strains

In Section 12.4, we shall justify that the contributions $k_\theta^\dagger$, $k_s^\dagger$ and $k_s^\ddagger$ can be neglected—this approximation is fully justified in the present context of thin shell theory. Therefore, our measure of the curvature strains will be k_s and k_θ

Therefore, we will measure the curvature strain using k_s and k_θ in place of the original strain K_s and K_θ:

$$k_s(s) = \phi'(s). \tag{12.15a}$$

and

$$k_\theta(s) = \frac{v_z'(s)}{\rho(s)}. \tag{12.15b}$$

The strain k_s associated with meridional curvature, defined in equation (12.15a), has a simple interpretation: the curvature b_{ss} along meridians is the rate of change of the tangent orientation per arc length, $b_{ss} = \rho'\,z'' - \rho''\,z' = \gamma'$, in reference configuration. Upon deformation, the angle γ is changed to $\gamma + \phi$ and so the curvature changes by $(\gamma + \phi)' - (\gamma)' = \phi'$, up to corrections proportional to membrane strain, which are neglected here.

The other curvature strain, k_θ, given in equation (12.15b), can be interpreted as follows. First, recall that the curvature $b_{\theta\theta} = z'/\rho$ given in equation (11.41) measures the normal projection of the curvature vector measured along the local s-line. By s-line, we mean a curve with equation s equals constant, which in the present case is a circle with radius ρ. Its curvature vector has norm $1/\rho$ and is pointing towards the axis; the additional factor $z' = \sin\gamma$ in $b_{\theta\theta}$ accounts for the projection of this vector onto the local normal to the surface. Now, in equation (12.15b) for the curvature *strain*, the factor $v'_z(s)$ accounts for the change of orientation of the normal upon deformation, which affects the value of curvature.[6]

12.2   Constitutive relations

To derive a complete set of equations for membranes or shells, we need to complement the geometrical definition of strain given in Section 12.1 with constitutive laws and equations of equilibrium. The present section is devoted to constitutive laws, both for stretching and bending deformations. Note that constitutive laws for bending, derived in Section 12.2.2, are irrelevant for membranes that by definition have a zero bending stiffness. Later on, the equations of equilibrium will be derived, for membranes in Section 12.3 and for shells in Section 12.4.

12.2.1   Stretching

The constitutive laws for a shell can be derived from 3D elasticity by asymptotic analysis, see for instance the book by Ciarlet (P. G. Ciarlet, 2000). Here, we shall justify them qualitatively. As earlier, in the theory of plates, we shall assume that the elastic energy of the shell is the sum of a stretching contribution and a bending contribution. The stretching part $\mathcal{E}_s$ is first considered here. By definition, it depends on membrane strain but not on curvature strain:[7]

$$\mathcal{E}_s = \iint \Phi(e_s, e_\theta, s)\, \mathrm{d}a, \tag{12.16}$$

where Φ denotes the density of stretching energy per unit area. The explicit dependence of Φ on s allows one to account for inhomogeneous thickness, or for material properties depending on s (but not on θ by assumption). By convention, the density of energy Φ is given per unit area of the centre surface *measured in reference configuration*:[8] the quantity $\mathrm{d}a$ denotes the element of area in reference configuration,

$$\mathrm{d}a = \rho(s)\, \mathrm{d}s\, \mathrm{d}\theta. \tag{12.17}$$

Some constitutive relations can be derived from the stretching energy $\Phi(e_s, e_\theta, s)$: according to the fundamental relation (2.55) for an elastic continuum, the stress is the first variation of the local density of elastic energy with respect to the strain. A stress N_r and N_θ can then be defined by the equations

---

[6] There is another way that deformation can change the value of $b_{\theta\theta}$, which is by changing the value of the curvature $1/\rho$ of the local s-curve. However, the curvature strain can be evaluated under the assumption that the membrane strain is zero, as we shall show later. By equation (12.11b), $e_\theta = 0$ implies that $\bar{\rho} = \rho + v_r = (1 + e_\theta)\rho$ remains equal to ρ, and so the change of curvature $1/\rho$ upon deformation can be neglected. For almost isometric deformations, the circumferential curvature $b_{\theta\theta}$ can only change as a result of the change of the orientation of the local normal, not as the result of a change of the algebraic curvature $1/\rho$ of the local s-curve.

[7] Note that in non-classical shell theories, the stretching energy of the shell may depend on additional kinematical degrees of freedom such as normal shear. Non-classical theories are not considered here. They can be derived by the same variational approach as used here, see for instance the book by Libai and Simmonds (A. Libai and J. G. Simmonds, 1998).

[8] The other convention, whereby this area is measured in actual configuration, is possible although this is slightly less convenient.

$$\delta \mathcal{E}_\mathrm{s} = \iint (N_s \, \delta e_s + N_\theta \, \delta e_\theta) \, \mathrm{d}a \tag{12.18}$$

where

$$N_s = \frac{\partial \Phi(e_s, e_\theta, s)}{\partial e_s}, \qquad N_\theta = \frac{\partial \Phi(e_s, e_\theta, s)}{\partial e_\theta}. \tag{12.19}$$

The quantities N_s and N_θ are by definition the membrane stress. As we shall show in the following section, N_s and N_θ can be interpreted as the integrals over the thickness of the 3D stress σ_{ss} and $\sigma_{\theta\theta}$, respectively; as a result, N_α has the physical dimension of a 3D stress ($\sigma_{\alpha\beta}$, i.e. a pressure) *times a length*.

Equation (12.19) is the general form of a constitutive law for a membrane. In particular, the dependence of N_s and N_θ on e_s and e_θ may be non-linear. In the particular case of a Hookean, isotropic material with Young's modulus E, Poisson's ratio ν and thickness h, the elastic stretching (or membrane) energy is quadratic with respect to the strain and is given by

$$\Phi_{\mathrm{Hooke}}(e_s, e_\theta, s) = \frac{E\,h}{2\,(1 - \nu^2)}\left({e_s}^2 + {e_\theta}^2 + 2\,\nu\,e_s\,e_\theta\right). \tag{12.20}$$

This can be shown by computing the thickness-integrated density of the 3D energy density, given earlier in equation (6.91a), after identifying $\epsilon_{xx} = e_s$, $\epsilon_{yy} = e_\theta$ and setting $\epsilon_{xy} = 0$. By equation (12.19), the corresponding constitutive relations read

$$N_s^{\mathrm{Hooke}} = \frac{E\,h}{1 - \nu^2}\,(e_s + \nu\,e_\theta), \tag{12.21a}$$

$$N_\theta^{\mathrm{Hooke}} = \frac{E\,h}{1 - \nu^2}\,(e_\theta + \nu\,e_s). \tag{12.21b}$$

Because of the symmetry about the axis, there is no in-plane shear stress, $\sigma_{s\theta} = 0$.

12.2.2   Bending

We carry out a similar procedure and derive the constitutive relations for bending starting from a density of bending energy $\mathcal{E}_\mathrm{b}$. As for elastic plates, the elastic energy splits into two contributions in the limit of small thickness, namely a stretching energy and a bending energy. The limit of small thickness implies that the principal curvatures of the shell remain everywhere much smaller than the inverse thickness $1/h$.

As explained later in Section 12.4, bending effects can be computed under the assumption that the membrane strain is zero[9]; if this is not the case, the curvature energy is negligible in front of the stretching energy anyway. Therefore, the bending energy can be written as a function of the curvature strain that does not depend on membrane strain:

$$\mathcal{E}_\mathrm{b} = \iint \Psi(k_s, k_\theta, s) \, \mathrm{d}a.$$

---

[9] This does not imply that stretching is neglected as we minimize the sum of this bending energy and a stretching energy, see Chapters 13 and 14. It just means that bending energy will be negligible in front of stretching energy whenever the latter is non-zero.

The stress components M_s and M_θ are conjugate to the curvature strain k_s and k_θ and are defined by

$$\delta\mathcal{E}_{\rm b} = \iint (M_s\,\delta k_s + M_\theta\,\delta k_\theta)\,{\rm d}a \tag{12.22}$$

where

$$M_s = \frac{\partial\Psi(k_s,k_\theta,s)}{\partial k_s}, \qquad M_\theta = \frac{\partial\Psi(k_s,k_\theta,s)}{\partial k_\theta}.$$

They will be interpreted later as bending moments, similar to the internal moment $\mathbf{M}$ of an elastic rod.

Under the key assumption that the shell's radii of curvature are much larger than its thickness, the analysis of the bending of plates done in Chapter 6 continues to hold, as the natural curvature of the shell is not visible at the relevant length scale, which is a few times the thickness of the shell. In particular, in the case of an elastic shell made out of an isotropic material, equation (6.95b) yields, after identifying $w_{,xx}$ and $w_{,yy}$ with the principal curvatures k_s and k_θ and setting $w_{,xy} = 0$:

$$\Psi_{\rm Hooke}(k_s,k_\theta,s) = \frac{D}{2}\left((k_s + k_\theta)^2 - 2\,(1-\nu)\,k_s\,k_\theta\right) = \frac{D}{2}\,({k_s}^2 + {k_\theta}^2 + 2\,\nu\,k_s\,k_\theta). \tag{12.23}$$

The bending modulus reads $D = E\,h^3/12\,(1-\nu^2)$, as derived in equation (6.83). By equation (12.22) this leads to the following constitutive relations:

$$M_s^{\rm Hooke} = D\,(k_s + \nu\,k_\theta), \qquad M_\theta^{\rm Hooke} = D\,(\nu\,k_s + k_\theta). \tag{12.24}$$

12.3 Equilibrium of membranes

Having derived the geometric expression of strain and found the constitutive laws, we now proceed to the equations of equilibrium. This is done by computing the first-order variation of total energy (principle of virtual work), like in Section 2.3.4. The resulting equations will be interpreted as a balance of forces and moments.

12.3.1 Integration by parts in axisymmetric geometry

To make the forthcoming calculation easier, we shall establish a formula for integrating by parts in axisymmetric geometry. In an annular domain $s_1 \le s \le s_2$, $0 \le \theta \le 2\pi$ bounded by the circles $s = s_1$ and $s = s_2$, the integral of a function $f(s)\,g'(s)$ can be integrated by parts as follows:

$$\iint f(s)\,g'(s)\,{\rm d}a = \int_{s_1}^{s_2} f(s)\,g'(s)\,2\pi\,\rho(s)\,{\rm d}s$$

$$= \left[f(s)\,g(s)\,2\pi\,\rho(s)\right]_{s_1}^{s_2} - \int_{s_1}^{s_2} \frac{{\rm d}(\rho(s)\,f(s))}{{\rm d}s}\,g(s)\,2\,\pi\,{\rm d}s$$

$$= \left[f(s)\,g(s)\,2\pi\,\rho(s)\right]_{s_1}^{s_2} - \int_{s_1}^{s_2} \left\{\frac{1}{\rho(s)}\,\frac{{\rm d}(\rho(s)\,f(s))}{{\rm d}s}\right\}\,g(s)\,2\,\pi\,\rho(s)\,{\rm d}s, \tag{12.25}$$

where square brackets denote boundary terms, $[q(s)]_{s_1}^{s_2} = q(s_2) - q(s_1)$. The last integrand in the right-hand side has been rewritten so as to make the area element $da = 2\pi\,\rho(s)\,ds$ appear, see equation (12.17).

12.3.2 External load

We consider external forces applied on the shell. Their density per unit area has components $f_r(s)$ in the radial direction and $f_z(s)$ in the axial direction, the element of area being da. We also consider a lineic density of force $F_r^i(s)$ and $F_z^i(s)$ applied on the boundaries $s = s_i$ with $i = 1, 2$. By convention, the force density f_r and f_z are measured per *unit area in reference configuration*; similarly, the force density F_r^i and F_z^i are measured per unit length of the boundaries *in reference configuration*. If these forces are conservative, they are associated with a potential energy $\mathcal{E}_\mathrm{p}$ whose first variation reads

$$\delta\mathcal{E}_\mathrm{p} = -\sum_{i=1}^{2} 2\pi\,\rho(s_i)\,\left(F_r^i(s_i)\,\delta v_r(s_i) + F_z^i(s_i)\,\delta v_z(s_i)\right)$$

$$- \int_{s_1}^{s_2} \left(f_r(s)\,\delta v_r(s) + f_z(s)\,\delta v_z(s)\right)\,2\pi\,\rho(s)\,ds \tag{12.26}$$

upon an infinitesimal change of configuration of the shell. If the forces are not conservative, as happens when they arise from dissipative processes for instance, there is no potential energy $\mathcal{E}_\mathrm{p}$ and the present minimization framework must be replaced with the Principle of virtual work, which is incremental; in practice this is a minor change as the linear form $\delta\mathcal{E}_\mathrm{p}$ is replaced by some other linear form, which is no longer the first variation of a potential energy. Note that the boundary forces F_r^i and F_z^i amount to a Dirac contribution $\delta_D(s - s_1)$ or $\delta_D(s - s_2)$ to the functions f_r and f_z.

12.3.3 Virtual work of membrane forces

The variation of membrane strain e_s due to an infinitesimal change of configuration of the shell follows from its definition (12.11a), and from that of $\phi(s)$ in equation (12.9):

$$\delta e_s = (\rho'(s) - \phi(s)\,z'(s))\,\delta v_r'(s) + (z'(s) + \phi(s)\,\rho'(s))\,\delta v_z'(\theta).$$

Using equation (12.2), this can be written as

$$\delta e_s(s) = T_r(s)\,\delta v_r'(s) + T_z(s)\,\delta v_z'(\theta),$$

where the vector $\mathbf{T}(s)$ is defined by

$$\mathbf{T}(s, \theta) = T_r(s)\,\mathbf{u}_r(\theta) + T_z(s)\,\mathbf{e}_z \tag{12.27a}$$

and

$$T_r(s) = \cos\gamma(s) - \phi(s)\,\sin\gamma(s), \tag{12.27b}$$

$$T_z(s) = \sin\gamma(s) + \phi(s)\,\cos\gamma(s). \tag{12.27c}$$

Recall that $\gamma(s)$ is the orientation of the tangent to the s-curves in reference configuration, and $\phi(s)$ is the change of orientation of these tangents upon deformation: the vector $\mathbf{T}(s)$ is an approximation to the unit tangent to the *deformed* s-curves, to first order in the deformation—see Fig. 12.2. This vector comes naturally into the equations of equilibrium

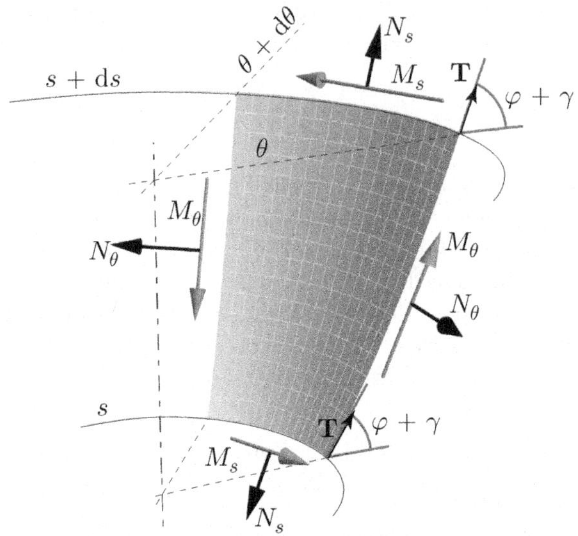

Fig. 12.2 Membrane forces acting on a small piece of an axisymmetric, twistless shell, cut between azimuthal angles θ and $\theta + d\theta$, and arc lengths s and $s + ds$. Note that the top and bottom black arrows are not exactly aligned, nor are those on the left and right-hand sides. The net membrane force is captured by the left-hand sides of the equations of equilibrium (12.31).

as the membrane stress N_s gives rise to a force that is aligned with the *deformed* tangent $\mathbf{T}(s)$ (this comes from the interpretation of the equations of equilibrium given later in Section 12.3.4).

The variation of the hoop strain e_θ can be computed from equation (12.11b):

$$\delta e_\theta = \frac{1}{\rho(s)}\, \delta v_r(s).$$

To sum up, the variation of membrane strain reads

$$\delta e_s(s) = T_r(s)\, \delta v_r'(s) + T_z(s)\, \delta v_z'(s) \tag{12.28a}$$

$$\delta e_\theta(s) = \frac{1}{\rho(s)}\, \delta v_r(s). \tag{12.28b}$$

These equations reflect the specific geometry of axisymmetric, *curved* surfaces, and are distinctive of shells elasticity—in fact, the equations for elastic plates are derived from the same set of equations as for shells, except precisely for the two equations above.

Combining with equation (12.18), we can express the increment of stretching energy in terms of the infinitesimal displacement:

$$\delta \mathcal{E}_{\mathrm{s}} = \int_{s_1}^{s_2} \left(N_s\, T_r\, \delta v_r' + N_s\, T_z\, \delta v_z' + \frac{N_\theta}{\rho(s)}\, \delta v_r(s) \right)\, 2\pi\, \rho(s)\, \mathrm{d}s. \tag{12.29}$$

Integration by parts using equation (12.25) allows one to remove any derivative of δv_r and δv_z in the integrand:

$$\delta \mathcal{E}_{\mathrm{s}} = \left[2\pi\,\rho\,\left(N_s\,T_r\,\delta v_r + N_s\,T_z\,\delta v_z \right) \right]_{s_1}^{s_2} \cdots$$

$$- \int_{s_1}^{s_2} \left[\left(\left\{ \frac{1}{\rho}\frac{\mathrm{d}(\rho\,N_s\,T_r)}{\mathrm{d}s} \right\} - \frac{N_\theta}{\rho} \right) \delta v_r + \left\{ \frac{1}{\rho}\frac{\mathrm{d}(\rho\,N_s\,T_z)}{\mathrm{d}s} \right\} \delta v_z \right] (2\pi\,\rho\,\mathrm{d}s). \quad (12.30)$$

The coefficients in parentheses in factor of δv_r and δv_z can be interpreted as the radial and axial components of the net force on a shell element arising out of membrane stress, per unit area. This net force has a simple geometric interpretation: it is the divergence of the membrane stress[10]; see the book by Niordson (F. I. Niordson, 1985) for a derivation of the shell equations emphasizing these geometrical aspects, in a coordinate-free formulation.

12.3.4   Equilibrium of an axisymmetric membrane

In the case of a membrane, which by definition has zero bending rigidity, the equations of equilibrium are found by cancelling the variation of the sum of the stretching energy and of the potential energy associated with the applied forces: the quantity $(\mathcal{E}_{\mathrm{s}} + \mathcal{E}_{\mathrm{p}})$ has to be stationary for arbitrary, small variations of the displacements $\delta v_r(s)$ and $\delta v_z(s)$. Combining equations (12.26) and (12.30) above and reading off the coefficients of δv_r and δv_z in the expression for $(\delta\mathcal{E}_{\mathrm{s}} + \delta\mathcal{E}_{\mathrm{p}})$, one obtains two equations:

$$\left(\frac{1}{\rho}\frac{\mathrm{d}(\rho\,N_s\,T_r)}{\mathrm{d}s} - \frac{N_\theta}{\rho} \right) + f_r = 0 \qquad\qquad (12.31\mathrm{a})$$

$$\left(\frac{1}{\rho}\frac{\mathrm{d}(\rho\,N_s\,T_z)}{\mathrm{d}s} \right) + f_z = 0 \qquad\qquad (12.31\mathrm{b})$$

in the interior of the annular domain, $s_1 < s < s_2$. The first equation is for radial equilibrium and the second one for axial equilibrium. As noted earlier, the terms in parentheses are the divergence of the internal, biaxial stress (N_s, N_θ), expressed here in the non-Cartesian coordinates (s, θ); they are similar to the Cartesian divergence in the first term of equation (2.51) for the equilibrium of a 3D elastic body.

In addition to the above condition in the interior, cancellation of the variation $(\delta\mathcal{E}_{\mathrm{s}} + \delta\mathcal{E}_{\mathrm{p}})$ implies some conditions coming from boundary terms. Setting the boundary terms to zero in the sum of equations (12.26) and (12.30), one is led to the conditions of equilibrium of the boundaries:

$$-N_s(s_2)\,\mathbf{T}(s_2, \theta) + \mathbf{F}^2(\theta) = 0 \qquad\qquad (12.32\mathrm{a})$$

$$+N_s(s_1)\,\mathbf{T}(s_1, \theta) + \mathbf{F}^1(\theta) = 0, \qquad\qquad (12.32\mathrm{b})$$

where the deformed tangent $\mathbf{T}$ has been defined in equation (12.27a), and the applied force on the boundaries $\mathbf{F}^i$ is defined as $\mathbf{F}^i(\theta) = F_r^i\,\mathbf{u}_r(\theta) + F_z^i\,\mathbf{e}_z$, where $i = 1$ for the first boundary $s = s_1$ and $i = 2$ for the second boundary $s = s_2$.

---

[10] The first term in equation (2.51) for the equilibrium of a 3D elastic body is similar: the net force is given by the divergence of the Cauchy stress tensor.

All these equations can be interpreted by considering a 3D biaxial stress field with $\sigma_{ss} = N_s/h$, $\sigma_{\theta\theta} = N_\theta/h$, the other component $\sigma_{s\theta}$ being zero by the axisymmetry. As shown in Fig. 12.2, the membrane force transmitted across an infinitesimal length $(\rho(s_0)\mathrm{d}\theta)$, measured in reference configuration, of some circle $s = s_0$, is $[N_s(s_0)\,\mathbf{T}(s_0)\,\rho(s_o)\,\mathrm{d}\theta]$. More accurately, this is the force applied by the side $s > s_0$ of the circle on the side $s < s_0$; by the action–reaction principle, the force applied by the other side $s < s_0$ on the side $s > s_0$ is simply the opposite.[11] The magnitude of this internal membrane force is given by the function N_s and its direction by the deformed tangent $\mathbf{T}(s_0)$, and both quantities are unknowns of the elasticity problem. The interpretation of equations (12.32) as a balance of forces on a infinitesimal length of the circles $s = s_1, s_2$ bounding the domain is now straightforward: on the boundary $s = s_1$, the membrane force is applied by the rest of the domain which is 'downstream', $s > s_1$, hence the $(+)$ sign by our convention in the expression for the force $[+N_s(s_1)\,\mathbf{T}(s_1, \theta)\,\rho(s_1)\,\mathrm{d}\theta]$; on the other boundary $s = s_2$ the force is applied by the 'upstream' part $s < s_2$ and so has a minus sign, $[-N_s(s_2)\,\mathbf{T}(s_2, \theta)\,\rho(s_2)\,\mathrm{d}\theta]$.

Similarly, the force transmitted along an infinitesimal cut of length $\mathrm{d}s$ in reference configuration on a meridian line $(\theta = \theta_0)$ reads $[N_\theta(s)\,\mathbf{u}_\theta(\theta_0)\,\mathrm{d}s]$; again, this is by convention the force applied by the 'downstream' part, $\theta > \theta_0$ on the 'upstream' part $\theta < \theta_0$. These forces are sketched in Fig. 12.2. With the help of the figure, the left-hand sides of equations (12.3.4) can be interpreted as the radial and axial components of the net force per unit area (measured in reference configuration) of a shell element. Note in particular the term $(-N_\theta/\rho)$, which describes a normal net force arising out of a combination of tangential (hoop) stress N_θ and curvature $1/\rho$; a similar term has been encountered in equation (2.77) already, and was discussed graphically in Fig. 2.11 in the case of an inflating balloon.

12.4 Equilibrium of shells

As we did for plate theory in Chapter 6, we complement this membrane model by adding flexural effects. This is done by considering the bending energy of the shell.

The following important remark will considerably simplify the description of bending effects. First note that shell theories, like plate theories, are only asymptotically valid for small thicknesses. In the formal expansion of the equations of 3D elasticity with respect to this small thickness, the formally dominant term is given by stretching; bending is described by higher-order terms. The reason for retaining bending terms is that the stretching energy may cancel exactly, when the centre surface deforms isometrically. The role of bending is to map a non-zero energy to isometric—or almost isometric—deformations. As a result, *the calculation of bending energy needs to be accurate for deformations close to inextensible ones only.* This implies that it is legal to compute the bending energy in a manner that is inaccurate when there is a significant amount of stretching (as the bending energy will be dominated by stretching in this case), provided it remains accurate where there is little or no stretching. Concretely, one can get rid of any term in the bending energy that vanishes when $\epsilon_s = 0$ and/or $\epsilon_\theta = 0$.

---

[11] We used similar sign conventions when defining the internal force in an elastic rod in Chapter 3.

This justifies neglecting the contributions $k_\theta^\dagger$ and $k_s^\dagger$ in our definition (12.15) of the curvature strains, as both $k_\theta^\dagger$ and $k_s^\dagger$ cancel when $\epsilon_s = 0$ and $\epsilon_\theta = 0$ by equations (12.12c) and (12.14a). The other contribution $k_s^\ddagger$ introduced in equation (12.14c) can also be neglected: by the same argument, all terms depending on ϵ_s^{lin} can be neglected; the remaining term $\phi^2\,\gamma'$, which is second order with respect to the rotation ϕ, is negligible as well[12] by our approximation of moderate rotations. This justifies our description of curvature strain based on the quantities k_s and k_θ defined in equations (12.15). By the same argument, any dependence of the bending energy on the membrane strain can been neglected, hence the form of equation (12.16).

12.4.1 Virtual work of bending forces

With the aim of rewriting the first variation of the bending energy given in equation (12.22) in terms of the infinitesimal displacement δv_r and δv_z, we first use equation (12.15a) to write $\delta k_s = \delta \phi'$, and integrate by parts:

$$\delta\mathcal{E}_\text{b} = \left[M_s\,\delta\phi\,2\pi\,\rho \right]_{s_1}^{s_2} - \int_{s_1}^{s_2} \left(\frac{1}{\rho}\frac{\mathrm{d}\,(\rho\,M_s)}{\mathrm{d}\rho}\,\delta\phi - M_\theta\,\delta k_\theta \right)\,2\pi\,\rho\,\mathrm{d}s. \tag{12.33}$$

Second, we use the definitions (12.9) of ϕ and (12.15b) of k_θ, to make the variations of the unknowns δv_r and δv_z appear in the expression of $\delta\mathcal{E}_\text{b}$ above: $\delta\phi = -(\sin\gamma)\,\delta v_r' + (\cos\gamma)\,\delta v_z'$ and $\delta k_\theta = \delta v_z'/\rho$. The resulting expression for $\delta\mathcal{E}_\text{b}$ has a structure very similar to the intermediate expression (12.33) obtained earlier in the derivation of membrane forces: the integrand becomes a linear function of $\delta v_r'$ and $\delta v_z'$ although with different coefficients. We can then carry out an integration by parts as earlier. This calculation is not difficult and yields the expression of the net bending forces that must be included in the equations of equilibrium to complement the membrane terms already found in equations (12.31). These bending forces have quite long expressions, which are not given here.

It turns out that in the specific shell problems studied in the remaining chapters of this book,[13] bending effects are restricted to parts of the shell where it is almost perpendicular to its axis of revolution; therefore, we shall content ourselves with an approximate derivation of the bending forces valid in those regions. This derivation is technically much simpler than in the general case. Conceptually, it is not very different from the general case: the approximation is just a means of avoiding lengthy calculations which are unnecessary for the problems at hand.

---

[12] In the analysis of stretching, we argued that it is important to retain non-linear terms in ϕ; the reason was that there is no linear term in ϕ in the membrane strain. The picture is different for the analysis of bending, as a linear term $k_s = \phi'$ is present in the curvature strain; this justifies dropping the non-linear terms in ϕ here.

[13] The elastic torus, studied in Chapter 13, has two crowns where the tangent plane is perpendicular to the axis. There, the membrane stress is unable to sustain axial forces; axial forces are transmitted across the crowns by a boundary layer near the crowns where bending forces must be considered. In Chapter 14, we study the axisymmetric dimpling of a spherical shell pushed on to a plate. For moderate pushing forces, a circular ridge forms near the region of contact; again, in the ridge region, the tangent to the shell is almost perpendicular to the axis of the dimple; this is the only place where bending needs to be considered because it is not dominated by stretching.

12.4.2 Bending effects near flat regions

We shall work out the general derivation of bending forces in the vicinity of a circle or pole $s = s_0$ where the tangent plane to the shell is perpendicular to the axis z of revolution: s_0 is such that $z'(s_0) = 0$, that is $\sin \phi(s_0) = 0$. The angle $\phi(s_0)$ is then an integer multiple of π, and by equation (12.2) we have $\rho'(s_0) = \cos \phi(s_0) = \pm 1$. We call $\eta = \pm 1$ this sign. To sum up,

$$z'(s_0) = \sin \phi(s_0) = 0, \qquad \rho'(s_0) = \cos \phi(s_0) = \eta \text{ where } \eta = \pm 1. \tag{12.34}$$

Then, the angle ϕ, defined in equation (12.9), can be approximated by

$$\phi(s) = -z'(s)\, v'_r(s) + \rho'(s)\, v'_z(s) \approx \eta\, v'_z(s) \tag{12.35}$$

in the vicinity of the point s_0.

The variations $\delta\phi$ and δk_θ can be found from equations (12.35) and (12.15b). Near a flat region, they depend on the increment of the axial displacement, δv_z, but no longer on δv_r:

$$\delta\phi(s) = \eta\, \delta v'_z(s), \qquad \delta k_\theta(s) = \frac{\delta v'_z(s)}{\rho(s)}.$$

Plugging this into the variation of the bending energy (12.33), we find

$$\delta\mathcal{E}_{\mathrm{b}} = \left[M_s\, \eta\, \delta v'_z\, (2\pi\, \rho) \right]_{s_1}^{s_2} + \int_{s_1}^{s_2} q(s)\, \delta v'_z(s)\, 2\pi\, \rho(s)\, \mathrm{d}s, \tag{12.36}$$

where we have introduced a new quantity $q(s)$,

$$q(s) = -\left(\frac{\eta}{\rho} \frac{\mathrm{d}\,(\rho\, M_s)}{\mathrm{d}s} - \frac{M_\theta}{\rho} \right).$$

The mechanical interpretation of the quantity $q(s)$ will be given soon.

An integration by parts in equation (12.36) yields

$$\delta\mathcal{E}_{\mathrm{b}} = \left[(M_s\, \eta\, \delta v'_z + q\, \delta v_z)\, 2\pi\, \rho \right]_{s_1}^{s_2} - \int_{s_1}^{s_2} \frac{1}{\rho} \frac{\mathrm{d}(\rho\, q)}{\mathrm{d}s}\, \delta v_z\, 2\pi\, \rho\, \mathrm{d}s. \tag{12.37}$$

An explicit form of the quantity $q(s)$ just defined can be found by plugging the definition (12.15a) for k_s together with that for k_θ, namely $k_\theta = \eta\, \phi/\rho$, into the constitutive relations (12.24). This yields

$$q(s) = -\frac{\eta\, D}{\rho(s)} \left(\frac{\mathrm{d}\,(\rho(s)\, \phi'(s))}{\mathrm{d}s} - \frac{\phi(s)}{\rho(s)} \right), \tag{12.38}$$

an expression that happens to be independent of Poisson's ratio ν. By expanding the derivatives, we find, among other terms, $q = -\eta\, D\, \phi'' + \cdots = -D\, v'''_z + \cdots$. This is very similar to the definition (6.76) of the shear force $q_\alpha = -D\, w_{,\beta\beta\alpha}$ found in the context of plates, when we identify v_z with the deflection w of a plate. It is also very similar to the normal shear force $F_z = -EI\, w'''$ introduced in equation (6.84) in the context of an elastic rod undergoing small deformations. These quantities q, q_α and F_z all represent an internal shear force whose role is to balance bending forces, be it in an axisymmetric shell, in a plate or in a rod.

Near a flat region of a shell, bending can be taken into account by adding the energy variation $\delta\mathcal{E}_\mathrm{b}$ in equation (12.37) to the variation $(\delta\mathcal{E}_\mathrm{s} + \delta\mathcal{E}_\mathrm{p})$ considered in the membrane approximation earlier. This leads to a single new term f_b in the equations of equilibrium:

$$\left(\frac{1}{\rho}\frac{\mathrm{d}(\rho\,N_s\,T_r)}{\mathrm{d}s} - \frac{N_\theta}{\rho}\right) + f_r = 0 \tag{12.39a}$$

$$\left(\frac{1}{\rho}\frac{\mathrm{d}(\rho\,N_s\,T_z)}{\mathrm{d}s}\right) + f_\mathrm{b}(s) + f_z = 0, \tag{12.39b}$$

where

$$f_\mathrm{b}(s) = \frac{1}{\rho}\frac{\mathrm{d}(\rho\,q)}{\mathrm{d}s}. \tag{12.40}$$

Comparison with the membrane equations (12.31) reveals that the effect of bending on the transverse equilibrium is to add the new term f_b representing a net bending force per unit area. This force is normal to the local tangent plane to the shell.

Using equation (12.38), we find an explicit expression for the bending force

$$f_\mathrm{b}(s) = -\frac{\eta\,D}{\rho}\frac{\mathrm{d}}{\mathrm{d}s}\left((\rho\,\phi')' - \frac{\phi}{\rho}\right). \tag{12.41}$$

The bending force is proportional to D, and vanishes in the membrane theory ($D = 0$). By expanding the derivatives in the equation above we find, among other terms, $f_\mathrm{b} = -\eta\,D\,\phi''' + \cdots = -D\,v_z'''' + \cdots$; in this form, the bending force for an axisymmetric shell appears to be very similar to that for a plate or for a rod, given in equations (6.82) and (6.85) respectively; in particular it depends on the fourth derivatives of the displacement.[14] We refer the reader to the interpretation of the origin of the bending force given earlier in Section 6.5.3 for a plate and in Section 6.5.5 for a rod, the argument being similar for the bending of a shell.

For later reference, we list the boundary terms arising in the variation of the elastic energy of the shell, including those arising from bending:

$$\delta\mathcal{E}_\mathrm{s} + \delta\mathcal{E}_\mathrm{b} = \left[2\pi\,\rho\left(N_s\,T_r\,\delta v_r + (N_\theta\,T_z + q)\,\delta v_z + \eta\,M_s\,\delta v_z'\right)\right]_{s_1}^{s_2} + \int_{s_1}^{s_2}\cdots\,\mathrm{d}z, \tag{12.42}$$

as given by equations (12.30) and (12.37). In this equation, the integrand denoted by an ellipsis is made up of the left-hand sides of the equations of equilibrium (12.39), and vanishes when these equations are satisfied.

The following exercise, left to the reader, provides a more direct (but also less rigorous) derivation of the bending force f_b: one can show that variation of the alternative bending energy

---

[14] This is consistent with the fact that the curvature energy is proportional in all cases to the square of the curvature, which is a second derivative of the displacement. The bending force is obtained by two integration by parts, hence the fourth derivative.

$$\hat{\mathcal{E}}_{b} = \int \frac{D}{2} \left(\Delta v_z\right)^2 2\pi\, \rho(s)\, ds$$

leads to the same bending force in the interior as defined earlier in equation (12.41). Here the Laplacian is defined in polar coordinates as

$$\Delta v_z = v_z'' - \frac{v_z'}{\rho}.$$

This bending energy $\hat{\mathcal{E}}_{b}$ is inspired by that of a plate, given in equation (6.95b), after dropping the second term in the integrand, proportional to Gauss curvature, whose variation vanishes by Gauss–Bonnet theorem. This shows that the energy $\hat{\mathcal{E}}_{b}$ must yield the correct bending energy in flat regions of the shell.

12.5   Conclusion, extensions

In this chapter, we have derived the equations for the equilibrium of axisymmetric, twistless shells, under the simplifying assumption of small displacement but moderate rotations: we have identified and retained the same non-linear terms as those associated with buckling in plate theory, and dropped the others.

We have first obtained a definition of the strain tensor based on approximations supported by geometrical insights. Given the expressions (12.11) and (12.15), the rest of the derivation involved no other approximation;[15] we assumed that the energy of the shell is given as the sum of a stretching and a bending contribution, and worked out the equations of equilibrium by variation. Therefore, all approximations are done at the kinematical level, and are wrapped in our initial definitions of strain, and of stretching and bending energies; the equations of equilibrium follow from a systematic procedure. Henceforth the variational structure is preserved by the approximations, as it should; this may not be the case if the approximation is carried out differently.

The derivation of the shell equations, presented above in one of its simplest forms, can be extended to twisted configurations, to non-axisymmetric rest and/or actual configurations (the resulting equations then become partial differential equations), to non-isotropic materials, to finite rotations and to finite strain (a non-linear constitutive law must then be specified through the functions Φ and Ψ). All these extensions can be carried out by the same general procedure as followed here: derivation of the relevant definition of strain, definition of the constitutive laws, definition of a suitable shell energy and derivation of the equilibrium equations by variation.

We have restricted our presentation to the so-called *classical* shell models. Non-classical shell models can be introduced by considering additional kinematical modes, such as normal shear, and mapping an energy to these modes; the rest of the procedure is identical to what we have just proposed. Non-classical models have been proposed similarly in the context of rods—some of which question the Euler–Bernoulli hypothesis, for instance. The major problem with non-classical models is that it is unclear whether they can be justified from 3D elasticity. The motivation behind these models is to circumvent some limitations of

---

[15] We outlined how the general expression of the bending forces can be obtained, but worked out the specific case of flat regions only. This additional approximation was introduced simply to save us from lengthy calculations; it can be relaxed without difficulty.

the classical theory (to describe 'thick' shells for instance, a concept that lacks a clear mathematical definition); however, they do so by modifying the kinematics in an *ad hoc* manner. In addition, some subtle points can arise.[16]

References

P. G. Ciarlet. *Mathematical Elasticity. Vol. 3: Theory of Shells*, volume 3 of *Studies in Mathematics and its Applications*. North Holland, 2000.

W. Flügge. *Stresses in Shells*. Springer-Verlag, Berlin, 2nd edition, 1973.

A. L. Goldenveizer. *Theory of Elastic Thin Shells*. Pergamon Press, New York, NY, USA, 1961.

W. T. Koiter. On the nonlinear theory of thin elastic shells, I–III. *Proceedings of the Koninklijke Nederlandse Akademie van Wetenschappen, Series B*, 69:1–54, 1966.

A. Libai and J. G. Simmonds. *The Nonlinear Theory of Elastic Shells*. Cambridge University Press, 2nd edition, 1998.

F. I. Niordson. *Shell Theory*, volume 29 of *North-Holland Series in Applied Mathematics and Mechanics*. North-Holland, Amsterdam, Holland, 1985.

V. V. Novozhilov. *Thin Shell Theory*. Kluwer Academic Publishers, 2nd edition, 1970.

J. L. Sanders. Nonlinear theories for thin shells. *Quarterly of Applied Mathematics*, 21:177–187, 1963.

R. Valid. *The Nonlinear Theory of Shells Through Variational Principles: From Elementary Algebra to Differential Geometry*. John Wiley & Sons Ltd, Chichester, 1995.

[16] In their remarkable book on shells (A. Libai and J. G. Simmonds, 1998), Libai and Simmonds present a derivation of shell theory where they try to avoid any *a priori* assumption on how the displacement depends on the transverse coordinate. Instead, they propose to carry out weighted averages throughout the thickness to reduce the 3D equations to 2D directly. While this is fine from a formal viewpoint, a closer look reveals what appears to us as a major problem. The authors are forced to define the orientation of the tangent plane by dividing the time integral of the angular momentum of a cross-section, by its moment of inertia. However, knowing nothing about the displacement through the thickness, they are forced to use the moment of inertia in *reference configuration*; the latter may be significantly different from the actual moment of inertia when the shell undergoes finite strain. As a result, their definition does not always correctly track the orientation of the tangent plane. This affects the decomposition of the physical forces into membrane and bending contributions, and ultimately the writing of the constitutive laws. The result may be a shell theory that is self-consistent but not fully consistent with the underlying 3D elasticity (in this particular example, the problem may arise for finite strains, or even for small strains combined with large rotations).

13

The elastic torus

13.1 Introduction

As a first illustration of the theory of axisymmetric twistless shells derived in Chapter 12, we study the deformation of a hollow[1] elastic torus subjected to shear: an axial, distributed force is applied along its inner equator and the opposite, axial force is applied along its outer equator. The mechanical problem is defined in Section 13.2 and in Fig. 13.1. Toroidal shells find many applications, ranging from fuel tanks on rockets to turbine shrouds (A. Libai and J. G. Simmonds, 1998). They are interesting from an academic perspective as the torus is a surface of mixed type, and little is known in general about elastic shells of mixed type. By 'mixed type', we mean that the torus is an elliptic surface in its outer half, and hyperbolic in its inner half where it looks locally like a saddle (the type of a surface has been defined in Section 6.2.2). As we shall show, a toroidal shell deforms in a very specific manner, as the strain and stress concentrate in two very narrow regions. We carry out a detailed analysis of these internal layers. The concentration of stress and strain has a geometric origin: it takes place at the crowns of torus, a class of curves that has been introduced (and defined) in our analysis of the inextensible deformations of surfaces in Chapter 11. These crowns are where the Gauss curvature of the surface changes sign.

In Section 13.3, we start by addressing this problem of shell elasticity using the simplest framework, namely the linearized membrane theory. In membrane theory, bending is neglected; only stretching effects are considered as they formally yield the dominant contribution to the elastic energy when the thickness is much smaller than the radius of curvature of the surface. As we shall see, the linearized membrane theory yields solutions that are divergent near the crowns (these crowns are located at the top-most and bottom-most circles drawn on the torus, when it is looked at with its axis in the vertical direction). This divergence points to a failure of the approximations underlying the linearized membrane theory, which is derived under the assumption that some parameters are small. One of those parameters is the magnitude of strain: for small strain, one can use linear constitutive laws (Hookean elasticity) but more general constitutive laws may be required for finite strain. Two other parameters have to be small for the linearized membrane theory to hold, namely the ratio of the thickness to the radius of curvature of the surface, and the angle of rotation of the material. A problem arises when these parameters fail to be *uniformly* small over the shell surface, and this is what happens in the case of an elastic torus. Some terms that had been neglected in the linearized membrane approximation are actually of the same order of magnitude as those that were retained. This is a typical situation where boundary layers

---

[1] The deformation of a *full* toroidal elastic body has been considered by Kutsenko (G. V. Kutsenko, 1979), based on the equations of 3D elasticity.

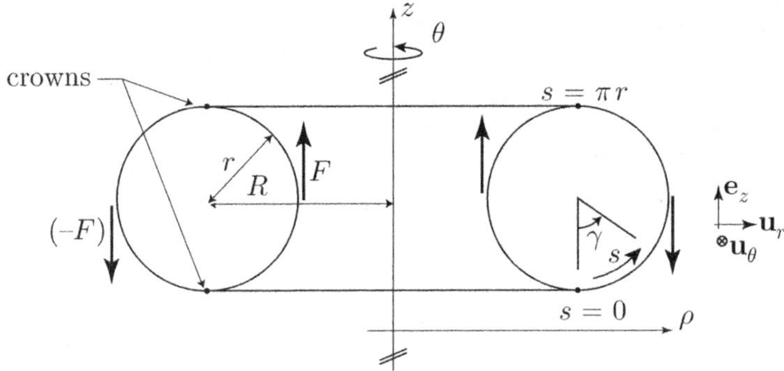

Fig. 13.1 The geometry studied in this chapter is known as the Goldenveizer problem: a toroidal shell is loaded under axial forces. The outer and inner equators are loaded with opposite forces, so that the net force applied on the shell is zero. Membrane theory is unable to predict deformation in this geometry, as it yields diverging stress near the crowns $s = 0$ and $s = \pi r$.

occur:[2] the membrane theory is valid almost everywhere except in a neighbourhood of the crowns, where a different scheme of approximation must be used. The goal of Sections 13.4 and following will be to derive and solve the equations that hold in the inner layer near the crowns. We consider two possible cases of regularization, namely by bending or by geometrical nonlinearity.

For an excellent review on the analysis of axisymmetric toroidal shells, including a historical perspective, we refer the reader to Libai and Simmonds' book (A. Libai and J. G. Simmonds, 1998), where most of the references given here are discussed.[3]

Table 13.1 summarizes the most important results obtained in previous works in terms of the two dimensionless combinations of parameters that are relevant for the problem. The rows correspond to the value of the shear force transmitted by the crowns, Ξ, which will be defined later in equation (13.26). The shearless case $\Xi = 0$ has been extensively studied in previous work, and applies to the special case of circular toroidal shells under pressure loading; our interest will be mostly in the general case $\Xi \neq 0$, which is obtained for arbitrary (non-circular) sections and/or arbitrary (non-pressure) loading. The columns in the table are associated with the different possible types of regularization, which can be purely flexural, purely non-linear, or mixed. In the shearless case $\Xi = 0$, Reissner's parameter ρ_R determines which kind of regularization applies (E. Reissner, 1963); in the case $\Xi \neq 0$, the parameter Ψ introduced later plays a similar role: the purely flexural, purely non-linear and mixed regimes correspond to $\Psi \to 0$, $\Psi \to \infty$ and $\Psi = \mathcal{O}(1)$ respectively.

The need for a careful analysis of toroidal shells was seemingly first raised in 1939 by Dean (D. R. Dean, 1939), who noticed that the stress previously computed by Föppl (L. Föppl, 1907) using a linearized membrane theory yields a non-smooth displacement field near the crowns (Föppl's analysis applied to the particular case of circular toroidal shells

---

[2] See Appendix B on the boundary layer theory in general, and the work by É. Sanchez and collaborators (P. Karamian-Surville, J. Sanchez-Hubert, and É. Sanchez Palencia, 2002; P. Karamian, J. Sanchez-Hubert, and É. Sanchez Palencia, 2000; J. Sanchez-Hubert and É. Sanchez Palencia, 1997) on applications to the theory of elastic shells.

[3] We thank J. G. Simmonds for pointing out these references to us at an early stage of this work.

Table 13.1 Classification of the boundary layers in toroidal shells. Reference are given to the work of Clark (R. A. Clark, 1950), Rossettos and Sanders (J. N. Rossettos and J. L. Sanders, 1965), Jordan (P. F. Jordan, 1962), and Sanders and Liepins (J. L. Jr. Sanders and A. A. Liepins, 1963). The sign § refers to a section of the present book

	purely flexural	mixed	purely non-linear
no shear force ($\Xi = 0$)	R. A. Clark (1950) and §13.7.1	J. N. Rossettos and J. L. Sanders (1965)	P. F. Jordan (1962), J. L. Jr. Sanders and A. A. Liepins (1963)
with shear force ($\Xi \neq 0$)	R. A. Clark (1950) and 13.7.1	§13.6	§13.8

under pressure loading). Dean pointed out that these inconsistencies can be eliminated by taking *bending* into account. This suggestion was put in practice about ten years later, in a paper by Clark (R. A. Clark, 1950); a treatment of the purely flexural layers, equivalent to Clark's original solution, is presented in Section 13.7.1. As apparent from Table 13.1, Clark's solution fully solves the problem in the linear case (i.e. when regularization is purely flexural); his solution applies to arbitrary section shapes (both circular and non-circular) and to arbitrary loading (pressure-type or not).

Later on, Jordan (P. F. Jordan, 1962) argued that the divergences found in the linear membrane theory can be suppressed by a different ingredient, namely the *geometric nonlinearity* associated with rotations of the tangent plane. Sanders and Liepins (J. L. Jr. Sanders and A. A. Liepins, 1963) proposed a detailed solution for the non-linear case, which applies to the case of toroidal shells under pressure loading ($\Xi = 0$) only—see Table 13.1. Their solution is remarkable as non-linear membrane effects are captured by *linear* boundary layer equations, thereby allowing an analytical solution.

This raises the question of which kind of regularization is applicable for a given shell and a given loading. In the absence of shear forces ($\Xi = 0$), Reissner (E. Reissner, 1963) showed that the regularizations by bending or by geometric nonlinearity can be viewed as the limiting cases of a more general problem; these limits being obtained when a parameter related to the intensity of loading, which he calls[4] ρ_R, becomes very small or very large. Shortly thereafter, Rossettos and Sanders (J. N. Rossettos and J. L. Sanders, 1965) derived a solution to Reissner's equations, and solved the so-called intermediate regime where both bending effects and geometric nonlinearity are significant. Their theory applies to the shearless case only ($\Xi = 0$).

As apparent from Table 13.1, previous work has focused on the shearless case ($\Xi = 0$); the only analysis of the general case $\Xi \neq 0$ is Clark's work (R. A. Clark, 1950) on the purely flexural case. Little is known on the deformations of toroidal shells subjected to shear ($\Xi \neq 0$) in the regimes of moderate or strong geometric nonlinearity (Ψ of order 1, or $\Psi \gg 1$); the analysis of these regimes is a new contribution of the present chapter.

As mentioned earlier, when a toroidal shell has a circular cross-section *and* is loaded by internal *pressure* force, its crowns support no shear force and so $\Xi = 0$. This will be shown

---

[4] Reissner's parameter ρ_R applies to the shearless case $\Xi = 0$, and plays a role similar to the parameter Ψ, which we introduce in the general case $\Xi \neq 0$ below.

in Section 13.5. Pressurized toroidal shells of circular cross-sections are interesting for some applications such as fuel tanks on rockets, but many other applications involve shells that either do not have a circular cross-section, or are subjected to a more complex loading, as in the case of Omega-joints[5] for instance. Apart from the special case of a torus with a circular cross section subjected to pressure loading, crowns in a toroidal shell are subjected to a non-zero axial shear, $\Xi \neq 0$. In their book (A. Libai and J. G. Simmonds, 1998 p. 252), Libai and Simmonds note that 'remarkably, no one to [their] knowledge has tackled this obvious question [of what happens in the case $\Xi \neq 0$]'. This motivates the analysis of the simple shear-type loading introduced in Fig. 13.1, which is studied in detail in the rest of this chapter.

This chapter is organized as follows: in Section 13.2 we define the mechanical problem (shell geometry and loading). In Section 13.3, we specify to a toroidal geometry the membrane equations obtained in Chapter 12 for general shells of revolution; we analyse the elastic deformations using the linearized membrane theory, and point out some divergences near crowns, that is where the Gauss curvature K changes sign. The rest of the chapter is devoted to the study of the boundary layers[6] near these crowns: the layer equations are derived in Section 13.4. Then, we derive the solutions in various limits: when both nonlinearity and bending effects are important, we show in Section 13.6 how a numerical solution can be obtained; the limit of a purely flexural boundary layer is studied in Section 13.7 (where we recover Clark's solution), and that of a purely non-linear one in Section 13.8.

13.2   Mechanical problem

We consider a thin elastic torus with circular section subjected to surface loads parallel to the axis, and distributed along the inner-most and outer-most equators. Our results can be extended in a number of ways (to arbitrary sections, to arbitrary applied forces, and even to shells that have the topology of the torus but are not of revolution) but, again, our aim is to get rid of any unnecessary difficulty and to point out generic mechanisms by considering the simplest possible geometries.

13.2.1   Geometry of undeformed configuration

Let us start by defining the geometry of the torus in its rest state, used as the reference configuration. Let r be the radius of the circular cross-section, and R the distance of the centre of this circle to the axis of revolution; the inequality

$$r < R$$

warrants that the torus does not self-intersect. We use the same notation as for the analysis of general surfaces of revolution in Chapter 11 and shells of revolution in Chapter 12. The

---

[5] Such joints are obtained by drawing a horizontal axis below the letter Ω and by rotating the letter Ω around the axis. They are used to connect two cylindrical pipes, the purpose being to take advantage of the elasticity of the joint to absorb axial stretching or contraction of the pipes. An analysis of such joints can be found in (R. A. Clark, 1950), but only in the pure bending case ($\Psi = 0$).

[6] Even though the layers are *internal*—the torus has no boundaries—we shall call them *boundary layers*, following a classical word abuse. Historically, the theory of boundary layers was developed in fluid mechanics after the work of Prandtl in 1904. In fluids, such layers are indeed most often seen near fluid boundaries, e.g. near an obstacle. See Appendix B for an introduction to the topic.

cylindrical coordinates are denoted (ρ, θ, z), and the local cylindrical basis $(\mathbf{u}_r(\theta), \mathbf{u}_\theta(\theta), \mathbf{e}_z)$ is defined in equation (11.30).

Let us denote s the arc length along the circular cross-section, the origin $s = 0$ being by convention at the bottom of the circle. Then, parametric equation of the undeformed shell reads

$$\rho(s) = R + r \, \sin\left(\frac{s}{r}\right) \tag{13.1a}$$

$$z(s) = -r \, \cos\left(\frac{s}{r}\right). \tag{13.1b}$$

The undeformed tangent to this circle, defined by setting $\phi = 0$ in equation (12.27), is given by

$$\begin{aligned}
\mathbf{T}^0(s, \theta) &= \rho'(s) \, \mathbf{u}_r(\theta) + z'(s) \, \mathbf{u}_\theta(\theta) \\
&= \cos\gamma(s) \, \mathbf{u}_r(\theta) + \sin\gamma(s) \, \mathbf{u}_\theta(\theta),
\end{aligned} \tag{13.2}$$

where the angle γ was defined in equation (12.2) and gives the orientation of the undeformed tangent:

$$\gamma(s) = \frac{s}{r}. \tag{13.3}$$

The principal curvatures of a general surface of revolution were computed in equation (11.41). Noting R_s^0 and R_θ^0 the principal radii of curvature along the s-curves and the θ-curves in undeformed configuration (as implied by the superscript 0), we have

$$\frac{1}{R_s^0} = b_s{}^s = -z' \rho'' + \rho' z'' = \frac{1}{r} \tag{13.4a}$$

$$\frac{1}{R_\theta^0} = b_\theta{}^\theta = \frac{z'}{\rho} = \frac{\sin\gamma}{R + r \sin\gamma}. \tag{13.4b}$$

Note the cancellation of the principal azimuthal curvature $1/R_\theta^0$ when $\sin\gamma(s) = 0$. This cancellation takes place along two curves, with equations $s = 0$ and $s = \pi$. These curves are the crowns discussed in the context of the isometric deformations of surfaces in Chapter 11. Along these crowns, the surface is flat in the azimuthal direction.[7]

The Gauss curvature of the undeformed surface reads

$$K(s) = \frac{1}{R_s^0} \frac{1}{R_\theta^0} = \frac{\sin\gamma(s)}{r \, R \left(1 + \frac{r}{R} \sin\gamma(s)\right)}.$$

It vanishes and changes sign at the crowns, that is for $s = 0$ ($\gamma = 0$) and $s = \pi r$ ($\gamma = \pi$). It is positive along the outer part of the torus $0 < s < \pi r$, where the surface is elliptic; it is negative along the inner half $\pi r < s < 2\pi r$, where the surface is hyperbolic.

Recall that the partial differential equations governing the elasticity of elliptic shells ($K > 0$) are themselves elliptic while those for hyperbolic shells ($K < 0$) are hyperbolic.

---

[7] At the crowns, the surface is flat in the direction tangent to the crowns. However, the crowns have nonzero curvature when viewed as 3D curves, since they are both circles of radius $R < \infty$. This is consistent: the curvature vector of these crowns, viewed as 3D curves, lies within the local tangent plane to the surface; it gives the null vector by projection onto the local normal to the surface, and this is what defines the curvature of the surface along a local direction.

This means that the equations for a toroidal shell are of a mixed type, namely elliptic in the outer half of the torus, and hyperbolic in the inner one. This is reminiscent of the mechanics of compressible fluids in the situation of the so-called *transsonic* flows, where part of the fluid moves below the velocity of sound and is described by elliptic equations, while the rest of the flow moves faster than the velocity of sound and obeys hyperbolic equations. Owing to the difficulty of solving partial differential equations with both hyperbolic and elliptic regions, few general results are known for shells with mixed type. The toroidal geometry provides the simplest example of the more general problem of solving shell equations for shells of mixed type. As we shall show, the mixed type of the torus has dramatic and directly observable consequences, as the deformation is mostly concentrated in boundary layers where $K = 0$, and the overall stiffness of the shell is entirely determined by what happens near these layers. These layers arise in connection with the change of sign of Gauss curvature, and are essentially *geometric*. This contrasts with other boundary layers classically found in elastic shells, which are caused by boundary conditions at the edges that are incompatible with the membrane approximation, or by a loading that is not smooth enough: those layers are caused directly by the loading or the boundary conditions.

One may wonder how this classification of *partial* differential equations (hyperbolic versus elliptic) can be relevant when we deal with axisymmetric deformations, since the latter are ultimately governed by *ordinary* differential equations—the azimuthal coordinate θ disappears from the equations. A word of explanation may help here. The following example illustrates how the change of type of a partial differential equation (PDE) influences the behaviour of a related ordinary differential equation (ODE). Let us consider the PDE

$$
A(s)\,\frac{\partial^2 u}{\partial s^2} + B(s)\,\frac{\partial^2 u}{\partial \theta^2} = f(s),
$$

where the right-hand side f represents the loading, the unknown $u(s,\theta)$ represents the displacement, and (s,θ) are the meridional and azimuthal coordinates as earlier. This equation is elliptic whenever the product $A(s)\,B(s)$ is positive, and hyperbolic when it is negative. The change of type of the equation occurs when this product changes sign. If we consider θ-independent solutions, this PDE becomes an ODE, much like in the analysis of axisymmetric perturbations of a torus. The change of type of the equation of equilibrium occurs at values of s, say s_0, such that $A(s_0) = 0$, and where the product $A(s)\,B(s)$ changes sign as well. It appears that the transition from hyperbolic to elliptic in the PDE manifests itself as a *singular point*[8] in the ODE. Note that higher degeneracies may be found if the functions A and B happen to vanish for the same value of s.

13.2.2   Loading

The loading is shown by thick arrows in Fig. 13.1. It consists of a force with total intensity $(F\,\mathbf{e}_z)$ uniformly distributed along the inner equator $s = 3\pi\,r/2$, and an opposite force with total intensity $(-F\,\mathbf{e}_z)$ uniformly distributed along the outer equator $s = \pi\,r/2$. This is described by the loading density f_r and f_z introduced in Section 12.3.2:

---

[8] In the theory of ordinary differential equations, singular points are defined as points where the coefficient of the highest derivative of an ODE cancels. At singular points, the smoothness of solutions of the ODE is not warranted.

$$f_r(s) = 0, \qquad f_z(s) = F\left(-\frac{\delta_D\left(s - \dfrac{\pi r}{2}\right)}{2\pi\,(R+r)} + \frac{\delta_D\left(s - \dfrac{3\pi r}{2}\right)}{2\pi\,(R-r)}\right), \tag{13.5}$$

where δ_D denotes the Dirac function. These functions f_r and f_z are the components of the applied force per unit area, and have the physical dimension of a pressure.

We first carry out a local analysis of the crowns in the following section; this local analysis is independent of the details of the loading. As a result, the above form (13.5) of the loading will not be used until Section 13.7.2, and the analysis of the crown is quite general.

13.3   Linearized membrane theory

The inconsistencies arising from the application of the linearized membrane approximation to toroidal shells have been pointed out by Dean (D. R. Dean, 1939). They are also mentioned by Goldenveizer in his monumental treatise (A. L. Goldenveizer, 1961) on the elasticity of shells and by others scientists as well—see for instance (W. Flügge, 1973).

The first reason for the absence of solution to the linearized membrane equations for thin shells is that bending forces are neglected although they are the only ones capable of transmitting shear forces across the two crowns. There is even a second, perhaps more subtle, cause of failure of membrane theory in this case, related to the absence of infinitesimal isometric deformations (W. Flügge, 1973, p. 30,94) near the crowns, where the Gauss curvature changes sign. This will be discussed below.

13.3.1   Linearized equations for membranes

The equations for a smooth axisymmetric membrane undergoing axisymmetric deformations have been derived in Chapter 12. The equations of equilibrium (12.31) are non-linear as they were derived assuming moderate rotations. The present section is concerned with a more severe assumption, namely infinitesimally small rotations (linearized membrane equations).

Note that the membrane stress N_s is a first order quantity (like the strain, it cancels in undeformed configuration). As a result, the term $(\rho\,N_s\,T_r)$ in equation (12.31a) can be approximated as $(\rho\,N_s\,\cos\gamma)$ at the linear order, since $T_r = \cos\gamma$ by equation (12.27), up a small correction.[9] Similarly, it is legal to set $(\rho\,N_s\,T_z) = (\rho\,N_s\,\sin\gamma)$ in the linearized theory. Therefore, the linearized form of the equations of equilibrium (12.31) is

$$\frac{\mathrm{d}(\rho\,N_s\,\cos\gamma)}{\mathrm{d}s} - N_\theta + \rho\,f_r = 0 \tag{13.6a}$$

$$\frac{\mathrm{d}(\rho\,N_s\,\sin\gamma)}{\mathrm{d}s} + \rho\,f_z = 0. \tag{13.6b}$$

Later, we shall restore two effects that have been omitted here, namely geometric nonlinearity and bending forces.

---

[9] Here, we recover the classical rule that there is no need to distinguish between the actual and reference configurations when writing the equations of elasticity linearized near a stress-free configuration: T_r is defined as the radial component of the deformed tangent in general, but here we can use the tangent in reference configuration, $T_r^0 = \cos\gamma$.

In these equations, N_s and N_θ denote the the biaxial membrane stress, $N_s = h\,\sigma_{ss}$ and $N_\theta = h\,\sigma_{\theta\theta}$; the tangential shear stress is zero for axisymmetric, twistless deformations of shells of revolution. The angle γ was defined in equation (13.3), and the loading density f_r and f_z in equation (13.5).

These equations are complemented by a constitutive law. We shall assume that the torus is made of a homogeneous, isotropic material of constant thickness. By equations (12.21), the constitutive law for such a membrane reads

$$N_s = \frac{E\,h}{1-\nu^2}(e_s + \nu\,e_\theta), \qquad N_\theta = \frac{E\,h}{1-\nu^2}(e_\theta + \nu\,e_s), \tag{13.7}$$

or, the other way around:

$$e_s = \frac{N_s - \nu\,N_\theta}{E\,h}, \qquad e_\theta = \frac{N_\theta - \nu\,N_s}{E\,h}. \tag{13.8}$$

Finally, the strain is defined in terms of the displacement $(v_r(s), v_z(s))$ by the relations (12.7), which can be linearized as follows:

$$e_s = \rho'\,v_r' + z'\,v_z' \tag{13.9a}$$

$$e_\theta = \frac{v_r}{\rho}. \tag{13.9b}$$

The quantities $v_r(s)$ and $v_z(s)$ were introduced in Chapter 11; they denote the radial and axial components of the displacement, and primes denote derivatives with respect to s.

The elastic deformation of the torus due to an external load can in principle be computed as follows. From the stress balance (13.6), one computes the stress field (N_s, N_θ) using the prescribed external loading (f_z, f_r). Once the stress is known, the strain can be computed by the constitutive law (13.8). Finally, by solving the equations (13.9), one computes the displacement. As we shall see shortly, this approach yields solutions that are divergent near the crowns, $s = 0$ and $s = \pi r$ (that is $\gamma = 0$ and $\gamma = \pi$).

13.3.2   Divergence of solutions near crowns

We shall now show that the solutions of the set of equations established in the previous section is singular near the crowns. Unless the loading satisfies a non-generic condition (essentially, that the coefficient of the divergence is zero), the linearized membrane approximation is inadequate near the crowns, and it should be supplemented by a boundary layer analysis there. This will be the aim of the following Sections. For the moment, we study the divergence in the framework of the linearized membrane theory; this will be useful later for setting up a theory describing the internal layers near the crowns.

We proceed to show that the solution $N_s(s)$ of the equilibrium equation (13.6b) diverges in general[10] whenever $\sin\gamma = 0$, that is at the crowns. Let us consider the neighbourhood of the lower crown $s = 0$, where $\gamma = s/R$ is a small quantity—the neighbourhood of the upper crown, $s \approx \pi r$, is treated similarly. Integrating equation (13.6b) with respect to s we find

---

[10] Again, the solution may not be singular if the coefficient of this divergence, noted Ξ later, happens to be zero. This case, which is not the generic case, will be studied later in Section 13.5.

$$\rho(s)\,N_s(s)\,\sin\gamma(s) = -\int^{s} \rho(s')\,f_z(s')\,\mathrm{d}s', \tag{13.10}$$

where the constant of integration is implicit. We shall assume that the applied loading is smooth near the crown; this is indeed the case for the particular loading considered here, see equation (13.5). Then, $\rho\,f_z$ is bounded near the lower crown, and the right-hand side of equation above is smooth at $s = 0$. Let Ξ be its value at $s = 0$, divided by 2π; we can then rewrite equation (13.10) including the constant of integration as follows:

$$\rho(s)\,N_s(s)\,\sin\gamma(s) = \frac{\Xi}{2\pi} - \int_0^{s} \rho(s')\,f_z(s')\,\mathrm{d}s'. \tag{13.11}$$

In the neighbourhood of the crown, that is for $|\gamma| \ll 1$ or $s \ll r$, this yields

$$N_s(s) \approx \frac{\Xi/(2\pi)}{\rho\,\sin\gamma} \approx \frac{\Xi}{2\pi\,R\,\gamma} \approx \frac{\Xi\,r}{2\pi\,R\,s}, \tag{13.12}$$

where we have used $\sin\gamma \approx \gamma = s/r$ and $\rho \approx R$. In the generic case $\Xi \neq 0$, the above equation shows that the stress N_s predicted by the linearized membrane theory is divergent. This divergence is not physical: the above equation cannot be expected to be valid along the crown, and the theory has to be refined there. However, at a moderate distance from the crown, the stress has to follow the above law. We shall now reconstruct the other mechanical quantities (hoop stress, strain and displacement) based on equation (13.12) above, as this will be required to build up the boundary layer equations.

The constant Ξ introduced in equation (13.11) depends on the distribution of the axial applied forces f_z; it will be interpreted below as the shear force supported by the crown. For the moment, we shall only assume that Ξ is not zero. This is a restriction on the class of loading considered, which excludes in particular the case of a torus under internal pressure—this particular case is discussed in Section 13.5.

The other component of the stress, N_θ, diverges near $s = 0$ as well. Indeed, plugging the value of N_s given by equation (13.12) into the balance equation for the radial forces (13.6a), one finds:

$$N_\theta \approx \frac{\mathrm{d}(\rho\,N_s\,\cos\gamma)}{\mathrm{d}s} + \rho\,f_r \approx R\,\frac{\mathrm{d}N_s}{\mathrm{d}s} \approx -\frac{\Xi\,r}{2\pi\,s^2}. \tag{13.13}$$

Here, we have assumed that the applied loading is smooth, which implies that f_r remains bounded near $s = 0$; the term proportional to f_r has been neglected in front the divergent terms. We have also set $\cos\gamma \approx 1$. We have just found that N_θ diverges more severely than N_s:

$$|N_\theta| \sim \frac{1}{s^2} \quad \gg \quad |N_s| \sim \frac{1}{s}. \tag{13.14}$$

Even though the present theory does not hold in the small layer near the crowns, the membrane stress N_s will remain negligible there in front of N_θ: we shall assume that the relation just derived, $|N_\theta| \gg |N_s|$, remains applicable in the boundary layer theory. It will be extremely useful to derive boundary layer equations, and in fact it justifies carrying out a detailed analysis of the linearized membrane theory first even though the latter has some obvious limitations.

Thanks to the analysis of the stress just made, the constitutive law (13.8) can be simplified in the vicinity of the crowns by neglecting N_s in front of N_θ:

$$e_s \approx -\frac{\nu}{E\,h}\,N_\theta \quad \text{and} \quad e_\theta \approx \frac{1}{E\,h}\,N_\theta. \tag{13.15}$$

Plugging the asymptotic form of N_θ found in equation (13.13), we find:

$$e_s \approx \frac{\nu\,r\,\Xi}{2\pi\,E\,h}\,\frac{1}{s^2} \quad \text{and} \quad e_\theta \approx -\frac{r\,\Xi}{2\pi\,E\,h}\,\frac{1}{s^2}. \tag{13.16}$$

Equation (13.9b) with $\rho \approx R$ yields the following divergence of the radial displacement:

$$v_r \approx R\,e_\theta \approx -\frac{r\,R\,\Xi}{2\pi\,E\,h}\,\frac{1}{s^2}. \tag{13.17}$$

The vertical component of the displacement can then be found from equation (13.9a) which yields

$$v_z' = \frac{e_s - \rho'\,v_r'}{z'} \approx -\frac{v_r'\,\cos\gamma}{\sin\gamma} \approx -\frac{r\,v_r'}{s} \approx -2\,\frac{r^2\,R\,\Xi}{2\pi\,E\,h}\,\frac{1}{s^4} \tag{13.18}$$

after neglecting $e_s \sim 1/s^2$ in front of $\rho'\,v_r' \sim 1/s^2$. A integration with respect to s yields

$$v_z \approx \frac{2}{3}\,\frac{r^2\,R\,\Xi}{2\pi\,E\,h}\,\frac{1}{s^3}. \tag{13.19}$$

We have just derived the divergent behaviour of the solution to the linearized membrane theory near the lower crown $s = 0$. The upper one can be worked out similarly and leads to the same divergences, as can be expected by the invariance of the torus under the mirror symmetry $z \mapsto (-z)$.

13.3.3   Discussion

The linearized membrane theory predicts divergent strain, stress and displacement near the crowns when $\Xi \neq 0$. This framework is unsuited for the toroidal geometry unless the axial loading f_z satisfies a non-generic condition, $\Xi = 0$, to be interpreted later; this was pointed out by Goldenveizer in his book (A. L. Goldenveizer, 1961).

The divergence obtained in the linearized membrane theory has a simple interpretation. By definition, the internal stress of a membrane is along the local tangent plane; in addition, the linearized theory neglects rotation of the tangent plane, forcing the internal stress to remain in the plane that is tangent to the membrane in *reference configuration*. Near crowns this plane is perpendicular to the axis. As a result, no axial internal force can be transmitted across the crowns, hence the divergences when the crowns are subjected to shear.

This simple interpretation of the divergences is in fact incomplete; there is a second, independent source of divergence. Recall that our analysis of isometric deformations in Chapter 11 revealed the special role played by crowns: near such curves, the geometrical problem of integrating strains into an infinitesimal displacement field is singular.[11] As

---

[11] The analysis of infinitesimal isometric deformations is ruled by solution of equations (13.9) in the particular case where the imposed strain is zero, $e_s = 0$ and $e_\theta = 0$. The geometric rigidity brought about by crowns is a consequence of the fact that this equation for $v_r(s)$ and $v_z(s)$ is singular near crowns

a result, there does not exist in general a solution for the displacement v_r, v_θ to the equations (13.9) when the strain e_s, e_θ is prescribed, even when this strain is smooth. As noted by Flügge (W. Flügge, 1973, p. 30, 94), it is impossible in general to reconstruct the displacement from the strain field obtained by solving the equations of equilibrium and the constitutive equations.

There are therefore two independent sources of difficulties with the membrane theory: on the one hand, membrane stress diverges near crowns when they are loaded with shear ($\Xi \neq 0$); on the other, reconstruction of the displacement from a given linearized strain is usually not possible, even if this strain is smooth near crowns (both for $\Xi = 0$ and $\Xi \neq 0$). The two causes of divergence have not always been clearly recognized in the literature. That the failure of the membrane approximation is twofold has practical consequences: there are two different classes of boundary layers depending on whether some axial force is transmitted across the layer or not. Here, our interest is mostly in the generic case, $\Xi \neq 0$, in which the divergence of the stress points to the *mechanical* inability of the crowns to sustain shear. Now, when the loading is such that $\Xi = 0$, the stress divergence disappears, but the displacement remains singular by the geometric argument, and this leads to a different type of boundary layer. The case $\Xi = 0$, which has already received much attention in the literature, will be discussed briefly in Section 13.5.

We shall now proceed to remove the divergences found in the preceding section. Two mechanisms allowing axial stress to be transmitted across the crowns are investigated: one is bending effects, the other one is geometrical nonlinearity. The diverging asymptotic laws for the stress and displacement, derived earlier in equations (13.12), (13.13), (13.17) and (13.19), will be revisited in the context of a boundary layer theory: these laws are valid at a distance from the crowns that is small compared with the size r of the shell, but much larger than the width of the boundary layer, yet to be found. Within the layer, the divergences are regularized by including new terms in the equations, which are corrections to the linearized membrane theory—these corrections are small everywhere except in the interior the layer. The coming Section 13.4 is devoted to the analysis of these layers.

13.4   Boundary layer equations

This section aims at deriving the equations for the boundary layers near the crowns.[12] These equations will extend those obtained by Reissner in the special case $\Xi = 0$, to the generic case $\Xi \neq 0$. Near the crowns we make use of the power of the boundary layer analysis, which saves us from resorting to the full theory of shells. The diverging behaviours predicted by the linearized membrane approximation in Section 13.3 are extremely important as they can be used to infer the relative magnitude of the various mechanical quantities within the layer. As a result, most of the terms coming from the full set of equations for elastic shells can be dropped. In fact, only two new terms are needed to regularize the neighbourhood of the crowns: one is a bending term, the other is a non-linear term coming from rotations (the diverging strain obtained in the linearized theory is a sure indication that a neglected nonlinearity becomes relevant, at least locally).

---

[12] We recall that these layers are called *boundary layers* following a standard abuse of the words, even though they should be called *internal layers* as the crowns are located in the interior of the shell.

The term coming from rotations is related to the angle $\phi(s)$ the definition of which we recall here. In reference configuration, the direction of the tangent to a meridian curve is given by the angle $\gamma(s)$ defined in equation (13.3). Deformation involves a small rotation $\phi(s)$ of the tangent plane, and in actual configuration this angle becomes $\gamma(s) + \phi(s)$—see equation (12.9) and Fig. 12.1. The quantity $\phi(s)$ was dropped when we linearized the equations of equilibrium to obtain equation (13.6); as a result the unknown $\phi(s)$ disappeared from the linearized membrane theory. It has to be retained in the boundary layer as it becomes comparable to the other small angle $\gamma(s) = s/r$ there.

In the following sections, we review the general equations for axisymmetric twistless shells derived in Chapter 12, as we did earlier for the analysis based on the linearized membrane equations. This time, we take care to include any contribution that may be relevant in the boundary layer.

13.4.1   Kinematics of tangent rotation

According to equation (12.9) valid for axisymmetric twistless shells, the rotation ϕ is defined in terms of the displacement (v_r, v_z) by:

$$\phi(s) = -v_r'(s)\,\sin\gamma(s) + v_z'(s)\,\cos\gamma(s). \tag{13.20}$$

Let us first estimate the relative magnitude of the two terms in the right-hand side using the linearized membrane theory: by equations (13.18) and (13.17), we find

$$\frac{|v_z'(s)\,\cos\gamma(s)|}{|-v_r'(s)\,\sin\gamma(s)|} \approx \frac{\left|2\,\dfrac{r^2\,R\,\Xi}{2\pi\,E\,h}\,\dfrac{1}{s^4}\times 1\right|}{\left|\dfrac{2\,r\,R\,\Xi}{2\pi\,E\,h}\,\dfrac{1}{s^3}\times\dfrac{s}{r}\right|} = \frac{r^2}{s^2},$$

where we have used $\sin\gamma \approx \gamma = s/r$ and $\cos\gamma \approx 1$ for $|s| \ll r$. Therefore, in the framework of the linearized membrane theory, the term $(v_z'\,\cos\gamma)$ is much larger than the term $(-v_r'\,\sin\gamma)$ in the vicinity of the crowns, that is for $|s| \ll r$.

Following the general approach presented in Appendix B for solving boundary layer problems, we shall assume that *terms that are negligible at the edge of the outer domain remain negligible within the boundary layer.*[13] Therefore, we shall assume that $|v_r'\,\sin\gamma| \ll |v_z'\,\cos\gamma|$ holds in the boundary layer as well: then, equation (13.20) implies

$$\phi(s) = v_z'(s) \tag{13.21}$$

at dominant order in the layer—here, we have again used $\cos\gamma \approx 1$. This relation will allow us to eliminate v_z in favour of ϕ.

The tangent $\mathbf{T} = T_r\,\mathbf{u}_r + T_z\,\mathbf{e}_z$ is defined by equations (12.27). With the approximations $\cos\gamma \approx 1$ and $\sin\gamma \approx \gamma$, this yields, at dominant order:

$$T_r(s) = 1 \tag{13.22a}$$

$$T_z(s) = \gamma(s) + \phi(s). \tag{13.22b}$$

---

[13] Note that the boundary layer equations can be written without any small or large parameter, and that their solutions have to match with those coming from the outer region. In consequence, any scaling relation established at the edge of the outer domain remains valid in the layer when the coordinate s is replaced with the typical width of the layer. In particular, terms that are asymptotically negligible for $s \to 0$ in the outer domain will remain negligible in the entire layer.

13.4.2   Mechanical equilibrium

Let us consider the equations of equilibrium for general axisymmetric shells. Expanding the derivative of $(\rho\, N_s\, T_r)$ in equation (12.39a), one has

$$\left(\rho'\, N_s\, T_r + \rho\, N_s'\, T_r + \rho\, N_s\, T_r'\right) - N_\theta - \rho\, f_r = 0.$$

If we plug into this equation the divergent behaviour obtained in the linearized membrane theory, we find that the second term is much larger than the first and third ones when s is small. As earlier, we shall assume that this hierarchy is preserved in the boundary layer; then the first and third terms can be neglected in the boundary layer as well, and we have

$$\rho\, N_s'\, T_r - N_\theta - \rho\, f_r = 0.$$

Assuming that the loading f_r is smooth, we can neglect the last term, which is bounded, in front of the two other ones, which are very large when the thickness of the shell is small. In addition, $\rho \approx R$ and $T_r \approx 1$ in the boundary layer ($|s| \ll r$) and the above equation yields

$$R\, N_s'(s) - N_\theta(s) = 0. \tag{13.23}$$

This is our first boundary layer equation. Note that this equation is satisfied by the outer solution given by equations (13.12) and (13.13).

Proceeding to the axial equilibrium, we rewrite equation (12.39b) as

$$2\pi\, \frac{\mathrm{d}(\rho\, N_s\, T_z)}{\mathrm{d}s} + 2\pi\, \rho\, f_{\mathrm{b}} + 2\pi\, \rho\, f_z = 0.$$

We have multiplied both sides of the original equation by 2π in order to ease the interpretation of the equation later. In this equation, f_{b} is the bending force defined in equation (12.41). Expanding the derivatives in the right-hand side of this equation, noting that $\rho\phi'' \gg \rho'\phi' \gg \phi/\rho$ by equations (13.21) and (13.18), and again using the fact that these scaling relations carry over to the interior of the layer, we find that the dominant contribution to the bending force reads

$$f_{\mathrm{b}}(s) = -\frac{D}{\rho}\, \frac{\mathrm{d}(\rho\,\phi''(s))}{\mathrm{d}s}.$$

Plugging into the above equation for axial equilibrium, we have

$$\frac{\mathrm{d}}{\mathrm{d}s}\left(2\pi\, \rho\, (N_s\, T_z - D\,\phi'')\right) + 2\pi\, \rho\, f_z = 0. \tag{13.24}$$

The first term is (strongly) divergent in the linearized membrane theory, while the loading f_z is assumed to be smooth and therefore bounded: we can therefore neglect the inhomogeneous term proportional to f_z in the neighbourhood of the layer. Then, the argument of the derivative in equation above, namely $2\pi\, \rho\, (N_s\, T_z - D\,\phi'')$, is approximately constant throughout the layer:

$$2\pi\, \rho\, (N_s\, T_z - D\,\phi'') = \text{Constant}. \tag{13.25}$$

This quantity can be interpreted by returning to equation (13.24), and noticing that it is a balance of axial forces: the derivative yields the net axial force applied on the region located between coordinates s and $s + \mathrm{d}s$ by the rest of the shell (interior stress), while $2\pi\, \rho\, f_z$ is by the definition of f_z the external force applied on this region, divided by $\mathrm{d}s$.

This shows that the left-hand side of equation (13.25) is the axial component of the internal force transmitted through any meridian circle (s constant).

As announced earlier, we call Ξ the axial force transmitted through the crown $s = 0$. Therefore, we identify the constant in the right-hand side of equation (13.25) above with Ξ. Then, the equation above writes

$$N_s(s)\, T_z(s) - D\, \phi''(s) = \frac{\Xi}{2\,\pi\,R} \tag{13.26}$$

after using $\rho(s) \approx R$ one more time. Note that this equation is compatible with the linearized membrane theory, whereby N_s is given by equation (13.12), $T_z \approx \sin\gamma = s/r$ and $D = 0$. Still, note that the above equation is more general than the linearized membrane theory since we have retained bending and have not dropped the geometrically non-linear terms in T_z.

13.4.3 Constitutive laws for membrane stress

The general form of constitutive law for membrane stress has been given in equation (13.8). Close to the crowns, the stress becomes uniaxial, $|N_\theta| \gg |N_s|$ by equation (13.14); this prediction of the linearized membrane theory carries over to the interior of the boundary layer, by the same argument as earlier. Therefore, the special form (13.15) of the constitutive law applicable to uniaxial stress N_θ holds at dominant order in the layer. In particular,

$$e_\theta = \frac{N_\theta}{E\,h}. \tag{13.27}$$

13.4.4 Strain

The general definition of strain in terms of the displacement is given in equations (12.11). With $\rho' = \cos\gamma \approx 1$, $z' = \sin\gamma \approx \gamma$ and $v_z' = \phi$ by equation (13.21), equation (12.11a) yields

$$e_s = v_r' + \gamma\,\phi + \frac{\phi^2}{2}.$$

In the linearized membrane theory, we noted that the left-hand side e_s can be neglected, see equation (13.18). This is confirmed by computing the ratio of the left-hand side to the first term in the right-hand side, $e_s/v_r' = v_s/(2R) \ll 1$, as implied by the asymptotic behaviour of the outer solution, equations (13.15) and (13.17). In consequence, the left-hand side e_s of the equation above can be neglected in the boundary layer too; at the dominant order, the definition of the strain e_s yields a condition of compatibility between v_r and ϕ:

$$0 = v_r'(s) + \gamma(s)\,\phi(s) + \frac{\phi^2(s)}{2}. \tag{13.28a}$$

The hoop strain e_θ is given by equation (12.11b) with $\rho \approx R$:

$$e_\theta(s) = \frac{v_r(s)}{R}. \tag{13.28b}$$

13.4.5 Boundary layer equations

We have written a complete set of equations for the boundary layer. We shall now put these equations in compact form by carrying out simple eliminations.

Elimination of v_r in equations (13.28a) and (13.28b) yields

$$R\, e_\theta' + \gamma\, \phi + \frac{\phi^2}{2} = 0.$$

The quantity e_θ can then be eliminated in favour of N_θ using equation (13.27); in the resulting equation, N_θ can be eliminated using the radial equilibrium (13.23). This yields, after using the definition of $\gamma(s)$ in equation (13.3):

$$\frac{R^2}{E\,h}\, N_s''(s) + \frac{s\,\phi(s)}{r} + \frac{\phi^2(s)}{2} = 0. \tag{13.29a}$$

A second equations is found by plugging the definition (13.22b) of T_z into the other equation of equilibrium (13.26):

$$N_s(s)\left(\frac{s}{r} + \phi(s)\right) - D\,\phi''(s) = \frac{\Xi}{2\pi R}. \tag{13.29b}$$

The two equations (13.29a) and (13.29b) form a set of differential equations for the two main unknowns of the boundary layer, the stress $N_s(s)$ and the rotation $\phi(s)$. Note that the quantity $\gamma(s)$ is known, and is given by equation (13.3); the parameter Ξ is fixed by the external loading.

These equations extend the linearized membrane theory by taking into account non-linear effects due to rotations (term ϕ^2 in the first equation, and cross-term $N_s\,\phi$ in the second equation) and bending (term $[-D\,\phi'']$ in the second equation). These new terms become negligible far away from the layer, and the behaviour predicted by the linearized membrane theory is recovered; this property will make it possible to match the layer solutions to those in the outer region where the linearized membrane theory will be used.

Our layer equations (13.29) are *non-linear*. This is in contrast with the layer equations obtained earlier in the special case $\Xi = 0$, which happened to be linear—see Reissner's or Rossettos and Sanders' papers (E. Reissner, 1963; J. N. Rossettos and J. L. Sanders, 1965). As a result, our equations are more difficult to solve.

Recall that the cancellation of $\sin\gamma \approx s/r$ caused a divergence of N_s in equation (13.12). This divergence is cured in the boundary layer theory, as the new bending term in equation (13.29b) can now balance the right-hand side when the first term vanishes.

13.4.6 Scaling analysis of the boundary layer

It is possible to remove of the parameters that enter into equations (13.29a) and (13.29b) by using the natural scales associated with the layer. Let s^* be the typical width of the layer, whose final expression is given later in equation (13.32). The length s^* is expected to be very small in the thin shell limit $h \to 0$, something that will have to be confirmed when the expression for s^* is available. Balancing the terms s/r and $\phi(s)$ in equation (13.29b) we find

$$\gamma \sim \frac{s^*}{r} \sim \phi. \tag{13.30a}$$

We use the notation $\sim$ between quantities that are of the same order of magnitude. Balancing the various terms in equations (13.29a) and (13.29b), we have:

$$\frac{R^2}{E\,h}\frac{N_s}{(s^*)^2}\sim\frac{(s^*)^2}{r^2}\qquad\text{and}\qquad N_s\,\frac{s^*}{r}\sim D\,\frac{s^*/r}{(s^*)^2}. \tag{13.30b}$$

After elimination of N_s between these scaling relations one obtains:

$$(s^*)^6\sim\frac{D}{E\,h}\,R^2\,r^2. \tag{13.31}$$

Using the definition (6.83) of the bending modulus D, we obtain the typical width of the boundary layer as:

$$s^*=\frac{(h\,R\,r)^{1/3}}{\left(12\,(1-\nu^2)\right)^{1/6}}. \tag{13.32}$$

This length scale s^* is intermediate between the large scale $(R\,r)^{1/2}$ and the thickness h of the shell as we assume that h is much smaller than both r and R:

$$h\ll s^*\ll R. \tag{13.33}$$

As the layer is large compared with the thickness $(s^*\gg h)$, the equations for elastic shells are applicable. The validity of the boundary layer analysis also requires this width s^* to be much smaller than $2\pi\,r$. Therefore the aspect ratio of the torus must satisfy the inequality:

$$\frac{r}{R}\gg\frac{h}{r}. \tag{13.34}$$

For a small but fixed value of the thickness h, this fails if the great radius R is too large compared with r, that is if the torus is too close to a cylinder. Obviously, a cylinder, obtained here by taking the limit $R\to\infty$, being a developable surface has an elastic response completely different from that of a torus. Stated differently, if R is too large, s^* can become of order r and the boundary layers invade the whole surface of the torus: it is not surprising that the present theory breaks down in this limit.

Returning to equations (13.30), we get the order of magnitude of the various quantities in the boundary layer as:

$$N^*=\frac{E\,h}{\left(12\,(1-\nu^2)\right)^{2/3}}\left(\frac{h^2}{r\,R}\right)^{2/3} \tag{13.35a}$$

$$\phi^*=\gamma^*=\frac{s^*}{r}=\frac{1}{\left(12\,(1-\nu^2)\right)^{1/6}}\left(\frac{h\,R}{r^2}\right)^{1/3} \tag{13.35b}$$

$$\Xi^*=2\pi\,R\,\phi^*\,N^*=\frac{2\pi\,E\,h^2}{\left(12\,(1-\nu^2)\right)^{5/6}}\left(\frac{h\,R}{r^2}\right)^{2/3} \tag{13.35c}$$

$$v_r^*=R\,\frac{N^*}{E\,h}\qquad\text{by equation (13.28b)} \tag{13.35d}$$

$$v_z^*=\phi^*\,s^*\qquad\text{by equation (13.21).} \tag{13.35e}$$

The stiffness of the layer under shear force is defined as the ratio of the typical axial force to the resulting displacement; it will be measured in units of

$$k^* = \frac{\Xi^*}{v_z^*} = \frac{2\pi}{\left(12\left(1 - \nu^2\right)\right)^{1/2}} \frac{E\,h^2}{r}.$$

$$(13.36)$$

Note that the typical stiffness k^* of the shell is proportional to h^2, and so is intermediate between a deformation dominated by bending (stiffness proportional to h^3) or by stretching (stiffness proportional to h), a result on which we shall comment later.

13.4.7   Dimensionless boundary layer equations

Using the scales defined in equation (13.35) and the general notation introduced in Section 1.2.4, we introduce the rescaled variable

$$\overline{s} = \frac{s}{s^*}$$

and rescaled unknown functions

$$\overline{N}_s(\overline{s}) = \frac{N_s\left(s = (s^*\,\overline{s})\right)}{N^*}, \qquad \overline{\phi}(\overline{s}) = \frac{\phi\left(s = (s^*\,\overline{s})\right)}{\phi^*}.$$

The rescaled shear force applied on the crown is denoted:

$$\Psi = \frac{\Xi}{\Xi^*},$$

$$(13.37)$$

instead of $\overline{\Xi}$, which is more difficult to read. This parameter measures the shear loading on the crown in units pertinent to the boundary layer. It is similar to the parameter ρ_R that Reissner introduces (E. Reissner, 1963) in the shearless case $\Xi = 0$.

In terms of rescaled variables, the boundary layer equations (13.29) write:

$$\overline{N}_s''(x) + x\,\overline{\phi}(x) + \frac{\overline{\phi}^2(x)}{2} = 0$$

$$(13.38a)$$

$$\left(x + \overline{\phi}(x)\right)\overline{N}_s(x) - \overline{\phi}''(x) = \Psi,$$

$$(13.38b)$$

where we have introduced a shorthand notation for the rescaled arc length, $x = \overline{s}$. Note that in front of rescaled functions, primes imply derivatives with respect to the rescaled arc length $x = \overline{s}$, and not the physical one s.

We introduce another rescaling that will be useful when we study the limit of small Ψ later: we rescale the membrane stress $\overline{N}_s$ and the angle $\overline{\phi}$ by the intensity of the applied loading Ψ, introducing new functions $f(x)$ and $g(x)$ such that

$$\overline{N}_s(x) = \Psi\,g(x) \quad \text{and} \quad \overline{\phi}(x) = \Psi\,f(x).$$

$$(13.39)$$

With these new unknowns, the boundary layer equations become:

$$g''(x) + x\,f(x) + \frac{1}{2}\,\Psi\,f^2(x) = 0$$

$$(13.40a)$$

$$-f''(x) + x\,g(x) + \Psi\,f(x)\,g(x) = 1,$$

$$(13.40b)$$

the parameter Ψ now appearing in front of non-linear terms.

The equations (13.40b) have the following symmetry: any solution remains a solution if the signs of x, Ψ and g are flipped together, the sign of f remaining unchanged. This symmetry holds in the boundary layer but does not apply to the original set of equations.

13.4.8   Asymptotic conditions far away from the layer

As usual in the boundary layer theory, some asymptotic conditions must be satisfied by the solutions of equations (13.40) for large values of the rescaled variable $x = \bar{s}$, to allow matching with the outer region, where the linearized membrane theory is applicable. This matching of the boundary layer solution with the outer solution takes place in the intermediate domain, $s^* \ll s \ll r$.

Rewriting the asymptotic behaviour of the outer solution given by equations (13.12) and (13.18) in terms of rescaled variables using equation (13.21), we find:

$$g(x) \approx \frac{1}{x} \quad \text{and} \quad f(x) \approx -\frac{2}{x^4} \qquad \text{for } x \to \pm\infty. \tag{13.41}$$

The same linearized membrane solution can be recovered by setting $\Psi = 0$ and dropping the bending term $-f''$ in the layer equations (13.40). Then, the second equation yields $g(x) = 1/x$, which is indeed the rescaled version of equation (13.12). Plugging into the first one we have $f(x) = -2/x^4$, which is the rescaled version of equation (13.18).

This warrants that the expansion of boundary layer solution for $x \to \infty$ matches the expansion of the outer solution for $s \to 0$, as explained in Appendix B. The condition (13.41) is equivalent to requiring that both the non-linear terms and the bending term $(-f'')$ become negligible far from the layer, i.e. that the layer is limited to a neighbourhood of the crowns.

Before trying to solve these equations, we note that the above set of equations can be derived by variation of the following functional:

$$\mathcal{E} = \int_{-\infty}^{+\infty} \mathrm{d}x \left(\frac{1}{2}(f')^2 - \frac{1}{2}(g')^2 + x\,f\,g + \frac{\Psi}{2}\,f^2\,g - f \right). \tag{13.42}$$

The Euler–Lagrange method is explained in Appendix A. Note that the two squared derivatives in the formal energy have opposite sign, so that the sought solution is a saddle point of the energy, and not an absolute minimum or maximum (the cubic term $\Psi/2\,f^2\,g$ also prevents the existence of an absolute minimum or maximum).

13.5   The curious case of pressurized circular toroidal shells

In this section, we show that the case of a toroidal shell with circular cross-section under internal pressure is very specific, as the parameter Ξ vanishes. This case has been extensively studied in the literature.

Let p be the uniform internal pressure in the toroidal shell. The axial projection P_z^{outer} of the total pressure force acting on the outer part of the torus $0 \le s \le \pi r$ reads

$$P_z^{\mathrm{outer}} = \iint p\,\mathrm{d}a\,(\mathbf{n} \cdot \mathbf{e}_z),$$

where $\mathbf{n}$ is the outer unit normal to the shell, and $\mathrm{d}a = 2\pi\,\rho\,\mathrm{d}s$ is the element of area in reference configuration.

Using the equality

$$\mathbf{n} \cdot \mathbf{e}_z = -\cos\gamma(s) = -\rho'(s),$$

one can rewrite the axial force P_z^{outer} as

$$P_z^{\text{outer}} = \int_{\rho(0)}^{\rho(\pi r)} 2\pi\,\rho\,p(-\mathrm{d}\rho) = -\pi\,p\left(\rho^2(\pi r) - \rho^2(0)\right).$$

This formula is valid for a toroidal shell with arbitrary cross-section. In the case of a *circular* toroidal shell, the crowns $s = 0$ and $s = \pi r$ have the same radius,

$$\rho(0) = \rho(\pi r),$$

and we have

$$P_z^{\text{outer}} = 0.$$

Since by symmetry the other components of the total pressure force applied on the outer half of the torus is zero too, we find that the net pressure forces applied on the outer half of the torus is zero (note that by the global equilibrium of the torus, the total pressure force applied on the inner half of the torus has to be zero too).

Consider now the equilibrium of the outer half of the torus, $0 \le s \le \pi r$. This subsystem is subjected to internal elastic stress transmitted across the lower crown by the rest (inner half) of the shell, to internal elastic stress transmitted across the upper crown, and to pressure forces. We have earlier defined Ξ as the axial component of the force applied by the outer torus on the inner torus across the lower crown, $s = 0$; by the principle of action–reaction, the resultant of the internal elastic stress applied by the inner part of the torus on the outer part across this lower crown is $(-\Xi)$. By the up–down symmetry, the force applied across the upper crown $s = \pi r$ is $(-\Xi)$ too. Therefore, the equilibrium of the outer half of the torus implies

$$-2\,\Xi + P_z^{\text{outer}} = 0.$$

Now, we have just shown that $P_z^{\text{outer}} = 0$ in the geometry considered, and so

$$\Xi = 0 \qquad \text{(in pressurized, circular-toroidal shells).} \tag{13.43}$$

In such shells, the boundary layers are free of shear loading. This makes this class of shells rather specific, and leads to a special type of divergence and to a special type of boundary layers. These boundary layers have been studied in detail in the literature (see first line in Table 13.1) and will not be further discussed here. In the rest of this chapter, we return to the generic case $\Xi \ne 0$, which is obtained generically for non-pressure loads (even in circular-toroidal shells), or for toroidal shells with non-circular cross sections (even for pressure loads).

We refer to Table 13.1 for classical references on circular toroidal shells under pressure loading ($\Xi = 0$).

13.6   Boundary layer solution for moderate nonlinearity

This section is devoted to the analysis of the boundary layer subjected to a shear force[14] ($\Xi \neq 0$), when the applied loading is such that the non-linear effects and the flexural terms are all of the same order of magnitude: this happens when the rescaled intensity of the applied loading, Ψ, is of order one. The other case $\Psi = 0$ is classical and corresponds to a purely flexural regularization, as will be studied in Section 13.7; the more difficult case $\Psi \gg 1$ of strong nonlinearity will be discussed in Section 13.8.

13.6.1   Uniqueness of the solution: a simple counting argument

Before calculating the solutions of the boundary layer equations, we show by a simple counting argument that the boundary layer solutions are discrete (and probably unique, as we argue later).

Let us assume that the functions (f, g) are solutions of the boundary layer equations having the correct asymptotic behaviour. We show below that it is impossible for a perturbed solution $(f + \delta f, g + \delta g)$ to satisfy all the equations—unless of course $\delta f(x) = 0$ and $\delta g(x) = 0$ for all values of x. As a result, the solutions (f, x) of the boundary layer problems are isolated.

Since the perturbations δf and δg are small, the boundary layer equations (13.40) can be linearized around the solution (f, g):

$$\delta g''(x) + x\,\delta f(x) + \Psi\, f(x)\,\delta f(x) = 0,$$

$$-\delta f''(x) + +x\,\delta g(x) + \Psi\, g(x)\,\delta f(x) + \Psi\, f(x)\,\delta g(x) = 0.$$

In the matching region, that is for large $|x|$, f and g are given by equation (13.41). Plugging this into the equations above and retaining the dominant terms only, we have

$$\delta g''(x) + x\,\delta f(x) = 0, \tag{13.44a}$$

$$-\delta f''(x) + x\,\delta g(x) = 0. \tag{13.44b}$$

This happens to be the same set of equation as obtained by Clark (R. A. Clark, 1950) in the purely flexural case ($\Psi = 0$). He noticed that these equations can be simplified by introducing the complex function

$$\tau(x) = -\delta f(x) + i\,\delta g(x).$$

This transforms the equation into an Airy-like differential equation:

$$\tau''(x) - \imath\,x\,\tau(x) = 0, \tag{13.45}$$

where $\imath$ is the purely imaginary number $\sqrt{-1}$.

A WKB analysis of this Airy equation gives the asymptotic behaviour of a generic solution $\tau(x)$ for large $|x|$: plugging into the equation a solution of the form $\tau(x) = \exp(\omega(x))$ yields at dominant order $\omega'^2(x) = i\,x$. Including the next order correction, one obtains the following WKB modes:

$$\tau(x) = -\delta f + i\,\delta g = \sum_{\pm} c_{\pm}\left(|x|^{-1/4} + \cdots\right)\exp\left(\pm\frac{2}{3}\frac{1 + i\,\mathrm{sgn}(x)}{\sqrt{2}}\,|x|^{3/2}\right), \tag{13.46}$$

---

[14] In the shearless case, $\Xi = 0$, the boundary layer equations happen to be linear and have been solved analytically by Rossettos and Sanders (J. N. Rossettos and J. L. Sanders, 1965).

where the ellipsis stands for smaller terms (when $|x|$ is large), c_+ and c_- are two arbitrary *complex* constants. The symbol 'sgn' stands for the sign function, and its value is -1 if the argument is negative, and $+1$ if it is positive. As this WKB expansion can be performed at either plus or minus infinity, there are two pairs of constants (c_+, c_-), one for the expansion at $-\infty$, and another one for the expansion $x \to +\infty$.

Being of total order 4, the differential system (13.44) has a space of solutions of dimension 4; this corresponds to the choice of the real and imaginary parts of the two constants c_+ and c_-. Now, if the complex constant c_+ is non-zero, the function $\tau(x)$ diverges exponentially for $x \to +\infty$. On the other hand, if $c_+ = 0$, the modes corresponding to the real and imaginary parts of c_- are exponentially small. This shows that there is a subspace of dimension 2 where the solutions to the linearized equations have the correct asymptotic behaviour for $x \to +\infty$. The same argument can be repeated on the other side, and there exists another subspace of dimension 2 where the linearized equations have the correct asymptotic behaviour for $x \to -\infty$. The solutions of the linearized boundary layer equations satisfying the asymptotic conditions on both sides $x \to \pm\infty$ are therefore given by the intersection of two subspaces of dimension 2 in a space of dimension 4. Generically, this intersection is of dimension $4 - (4 - 2) - (4 - 2) = 0$, and so it is trivial: it contains only the solution $\delta f = 0$ and $\delta g = 0$. This shows that the solutions of the equations (13.40) with boundary conditions (13.41) are discrete.

The numerical investigation below suggests that this solution is in fact unique.

13.6.2   Numerical solution

The linearized membrane equations are valid far away from the crowns. By solving these equations, one can compute the deformation of the torus everywhere except in the neighbourhood of the crowns (this is done for instance in Section 13.7.2 in a particular geometry). The deformation near the crowns can be found by solving equations (13.40), and this is the aim of the present section. Then, it remains to match this boundary layer solution with the so-called outer solution; this will be done later in Section 13.7.2 in a particular geometry too.

In the absence of an analytical solution for the boundary layer equations (13.40) and (13.41), one has to solve the equations numerically. The system of differential equations (13.40) define a non-linear boundary value problem (BVP) as the boundary conditions are distributed on both sides of the domain.

One possible approach is to use numerical shooting, a technique introduced in Section 8.7.3. The idea is to integrate the equations numerically, starting from a large negative value of x, say $x_{\min} = -10$; the initial conditions for $f(x_{\min})$, $f'(x_{\min})$, $g(x_{\min})$ and $g'(x_{\min})$ are obtained by combining the asymptotic form of the solution given by equation (13.41) for $x \to -\infty$, with the two exponentially small modes given by $(c_+ = 0, c_- = 1)$ and $(c_+ = 0, c_- = i)$ in equation (13.46) using two small arbitrary amplitudes α and β. This yields a solution indexed by (α, β) living in a space of dimension 4. For generic values of the parameters α and β, this solution diverges exponentially for $x \to +\infty$. The counting argument given in Section 13.6.1 shows that the subspace of solutions that have the correct asymptotic behaviour for $x \to +\infty$ is of co-dimension 2. Therefore it is possible to tune α and β to satisfy the asymptotic conditions (13.41) at $+\infty$. The coefficients of the exponentially large modes are estimated at a large positive x, say $x_{\max} = 10$, and α and β are adjusted to cancel these two coefficients. Practically, this is done by setting up a non-linear root finding procedure on top of the numerical integration from x_{min} to x_{max}.

In practice, it is very difficult to make this scheme work. The reason is that any numerical error is amplified exponentially when x increases towards x_{max} as this is the generic behaviour of a solution of the differential system. As a result, the root-finding problem is very stiff, i.e. is very sensitive to initial data. To circumvent this problem, one classical approach is shoot from both ends x_{min} and x_{max} towards the centre $x = 0$ independently, and then match the solutions at $x = 0$. As numerical integration is always carried out towards the centre $x = 0$ of the interval, one avoids the exponential amplification of numerical round-off errors—instead, they are damped exponentially. There is one last difficulty: most of the time, the non-linear root finding algorithm does not converge because the initial guess is not close enough to the solution. To work around this difficulty, one can start from $\Psi = 0$, for which the boundary layer equation has an analytic solution, given in Section 13.7 below, and set up a simple path-following technique: the value of Ψ is changed by small increments, and the previous numerical root is used as an initial guess for the next root-finding procedure.

We implemented a different approach for solving this non-linear boundary value problem, which is based on relaxation and not on shooting. As a rule, relaxation is more robust than shooting when the generic behaviour of the solution introduces some spurious exponential divergences. In contrast with the shooting method discussed above, the differential equation is not integrated. Instead, one starts from an initial guess for the functions $(f(x), g(x))$, discretized over values x_i of the variable x. This initial guess does not have to satisfy the differential equation or even the boundary conditions. An iterative procedure is applied to this set of values $\{\ldots, f_i, g_i, \ldots\}$, with the aim of making it converge towards a solution of the equation with the correct boundary conditions. The main idea of the algorithm is to rewrite the differential equations using finite differences, and to consider the resulting equations together with the boundary conditions as a set of non-linear equations to be satisfied by the unknowns f_i and g_i. The trick is that this non-linear problem is solved iteratively using a linear solver, and the linear problem *involves a sparse matrix* which provides good efficiency. We refer to the excellent presentation in reference (W. H. Press *et al.*, 2002, Ch. 17) for details. Another advantage of this method is that it is very efficient when the initial guess is close to the solution; this makes it well suited to problems such as this one where one seeks a *family* of solutions depending on a parameter, Ψ in the present case. Increasing Ψ step by step and recycling the previous solution as the initial guess for the following step leads to a simple but efficient path-following method.

Using both the shooting and the relaxation methods, we could compute the solution of the non-linear boundary layer for various values of Ψ. We could not track solutions for values of Ψ larger than about 3, as none of the algorithms would converge; this point is discussed in Section 13.8 below. As an illustration, the solutions f and g are shown in Fig. 13.2 for $\Psi = 2.5$. Using this solution, we determined the axial displacement v_z in the boundary layer by integration of equation (13.21). This yields the profile of a deformed torus plotted in Fig. 13.3.

13.7   Boundary layer solution for weak nonlinearity

This section is devoted to the analysis of the limit of small Ψ, that is when the applied force is much smaller than the typical force Ξ^* introduced in equation (13.35c). Then, bending

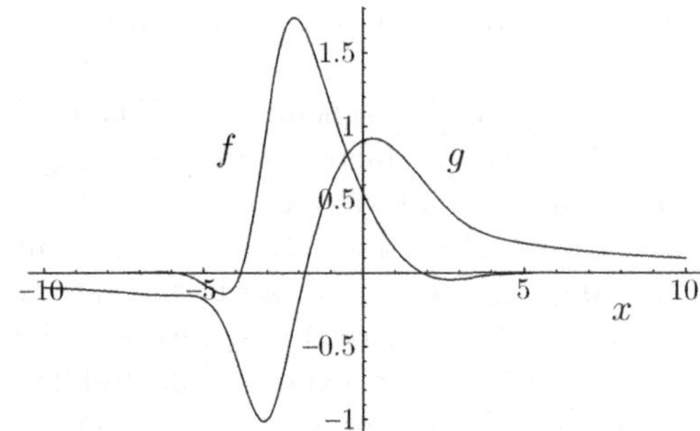

Fig. 13.2 Numerical solution of the boundary layer equations (13.40) with asymptotic conditions (13.41) for $\Psi = 2.5$, obtained by a numerical relaxation.

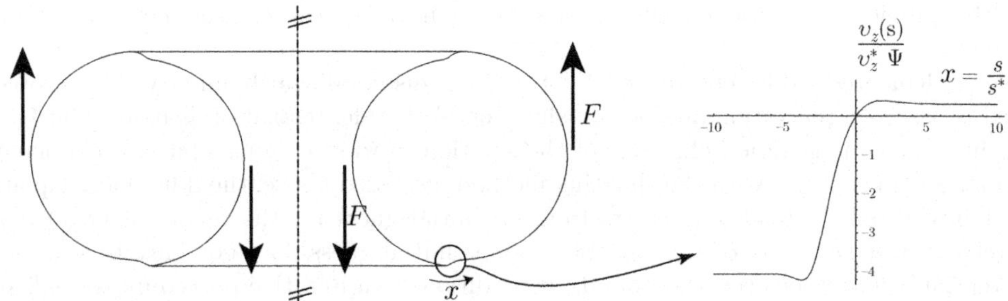

Fig. 13.3 Deformation of an elastic torus in the case of mixed regularization (both bending and nonlinearity are important along the crowns): $\Psi = 2.5$. Amplitude of displacement is magnified to aid visualization. The shear force Ξ across the crowns is such that $\Xi = 2.5\,\Xi^*$, where Ξ^* is the typical force, of order $E\,h^2/R$, defined in (13.35c); the actual loading parameter F is related to Ξ by equation (13.70), with a change of sign due to different sign conventions used in the present figure: $\Xi = +F/2$. The inset shows the rescaled vertical component of the displacement $v_z(s = s^*\,x)/v_z^*\,\Psi = \int_0^x f$ in the boundary layer. This figure is plotted using the numerical solution shown in Fig. 13.2.

effects are dominant in the boundary layer while non-linear effects remain negligible. This limit is interesting because there exists an explicit solution of the boundary layer equations, derived first by Clark (R. A. Clark, 1950). The present derivation of the solution of the boundary layer equation for $\Psi = 0$ differs from Clark's original presentation.[15]

---

[15] Clark's original method is based on Langer's method of asymptotic integration (R. E. Langer, 1931). This involves a change of the variable s for zooming into the boundary layer (the boundary layer becomes of size of order one with the new variable); the membrane equations (with bending terms added) are transformed accordingly and solved and there is no matching. In contrast, here we solve the boundary layer equations (inner region) and the membrane equations (outer region) separately and match them later. The approach based on asymptotic integration yields an analytical expression of the solution over the entire domain, although at the price of difficult calculations; we prefer the approach based on matched expansions since it is quite simple and can be done by hand, while the more technical and less illuminating ones can be solved on the computer using a symbolic calculation language (to obtain a solution of the outer equations for a complicated section shape for instance).

13.7.1   Explicit solution of the flexural boundary layer

We start from the dimensionless equations (13.40) for the boundary layer:

$$g''(x) + x\, f(x) + \frac{1}{2}\, \Psi\, f(x)^2 = 0 \tag{13.47a}$$

$$-f''(x) + x\, g(x) + \Psi\, f(x)\, g(x) = 1, \tag{13.47b}$$

together with the asymptotic conditions (13.41) for $x \to \pm\infty$:

$$g(x) \sim \frac{1}{x}, \qquad f(x) \sim -\frac{2}{x^4}. \tag{13.48}$$

Recall that, in these equations, x is the dimensionless coordinate along the cross-section, with origin $x = 0$ at the crown. The physical distance is obtained from x by multiplication by the small length s^* defined in equation (13.32). All quantities investigated below are expected to change over a typical scale $x \sim 1$. The rescaled membrane stress $g(x)$ and tangent rotation $f(x)$ were defined in equation (13.39) using the units introduced in Section 13.4.6.

In equation (13.40) the parameter Ψ appears in factors of the non-linear terms. It is therefore an obvious yardstick for the non-linear effects: for $|\Psi| \ll 1$, non-linear effects will be negligible. Below, we set $\Psi = 0$ and derive Clark's solution. Equation (13.47) becomes:

$$g''(x) + x\, f(x) = 0 \tag{13.49a}$$

$$-f''(x) + x\, g(x) = 1. \tag{13.49b}$$

Note that this boundary layer equation in the limit $\Psi \to 0$ is universal, i.e. it is free of parameter. In particular it does not depend on the aspect ratio r/R of the torus (as long as $r \gg \sqrt{h\,R}$), this aspect ratio being *a priori* of order 1. We argued in Section 13.6.1 that the solutions of (13.50) with the boundary conditions (13.48) are discrete, and in fact probably unique—the analysis of the case $\Psi = 0$ presented below confirms that it is unique in this case. In addition note that, by equation (13.39), f and g are proportional to N_s and ϕ divided by the loading parameter Ψ; as a result, the linear response of the torus for a small loading (when N_s and ϕ are both proportional to $\Psi \sim \Xi \sim F$, which should happen for $|\Psi| \ll 1$) is attained when the two functions $f(x)$ and $g(x)$ become independent of Ψ.

Equation (13.49) happens to be the same as equations (13.44) for small perturbations near a solution of the layer equation for *finite* Ψ. As already noted, this equation can be transformed into an Airy-like equation by introducing $\tau(x) = -f(x) + \imath\, g(x)$. This allows one to rewrite the boundary layer equations as a single ODE with complex coefficients:

$$\tau''(x) - \imath\, x\, \tau(x) = 1. \tag{13.50}$$

This form of the boundary layer equation was the starting point of Clark's solution, its solutions being known in terms of special functions. With the aim of making the derivation self-contained, we shall make as little use of the theory of special functions as possible. We start by seeking f and g in the form of power series of their argument x:

$$f(x) = \sum_{k=0}^{\infty} f_{2k}\, x^{2k}, \quad g(x) = \sum_{k=0}^{\infty} g_{2k+1}\, x^{2k+1}, \tag{13.51}$$

where we have anticipated that f is an even function of x and g an odd function due to the symmetries of both the equation (13.49) and the asymptotic conditions (13.48).

By inserting the series into equation (13.49), one gets the following recursion relations:

$$(2k+3)(2k+2)\,g_{2k+3} + f_{2k} = 0 \qquad\qquad \text{for } k \geq 0 \qquad\qquad (13.52\text{a})$$

$$-(2k+2)(2k+1)\,f_{2k+2} + g_{2k-1} = 0 \qquad\qquad \text{for } k \geq 1 \qquad\qquad (13.52\text{b})$$

$$-2 \cdot 1 \cdot f_2 = 1, \qquad\qquad (13.52\text{c})$$

where the last equation (13.52c) is a degenerate form of (13.52b) for $k = 0$, the number 1 in the right-hand side coming from the inhomogeneous term in (13.49b).

By eliminating g_{2k+1} and solving these recursion relations, one finds an explicit formula for the coefficients f_{2k}:

$$f_{6k+2p} = \frac{(-1)^k\, f_{2p}}{\prod_{k'=1}^{k}[(6k'+2p)(6k'+2p-1)(6k'+2p-3)(6k'+2p-4)]}. \qquad (13.53)$$

This relation holds for $k \geq 1$ and for any positive integer p. For $p = 0$, it gives f_6, f_{12}, f_{18}, $\ldots$ in terms of f_0. For $p = 1$, it gives the f_{6k+2}'s in terms of f_2. Finally for $p = 3$, it gives f_{6k+4}'s in terms of f_4. Higher values of p do not yield any new relation. Therefore, by collecting terms proportional to f_0, f_2 and f_4 respectively, we find that f writes:

$$f(x) = f_0\, F^0(x) + f_2\, F^2(x) + f_4\, F^4(x), \qquad (13.54)$$

where the coefficient f_2 is given by (13.52c), but f_0 and f_4 are constants to be determined. The three functions $F^{[2p]}$ for $p \in \{0, 1, 2\}$ are defined by collecting the relevant terms in the series expansion (13.51) with the help of the explicit formula (13.53) for the coefficients f_{6k+2p}:

$$F^{[2p]}(x) = \sum_{k=0}^{\infty} \frac{(-1)^k\, x^{6k+2p}}{\prod_{k'=1}^{k}[(6k'+2p)(6k'+2p-1)(6k'+2p-3)(6k'+2p-4)]} \qquad (13.55\text{a})$$

$$= x^{2p} \sum_{k=0}^{\infty} \frac{-\left(\frac{x^6}{6^4}\right)^k}{\prod_{k'=1}^{k}\left[\left(k'+\frac{2p}{6}\right)\left(k'+\frac{2p-1}{6}\right)\left(k'+\frac{2p-3}{6}\right)\left(k'+\frac{2p-4}{6}\right)\right]} \qquad (13.55\text{b})$$

$$= x^{2p} \sum_{k=0}^{\infty} \frac{-\left(\frac{x^6}{6^4}\right)^k}{\left(1+\frac{2p}{6}\right)_k \left(1+\frac{2p-1}{6}\right)_k \left(1+\frac{2p-3}{6}\right)_k \left(1+\frac{2p-4}{6}\right)_k} \qquad (13.55\text{c})$$

$$= x^{2p}\, {}_1F_4\left(1; \frac{2p+2}{6}, \frac{2p+3}{6}, \frac{2p+5}{6}, \frac{2p+6}{6}; -\frac{x^6}{6^4}\right). \qquad (13.55\text{d})$$

In equation (13.55c), obtained by reordering the denominator in (13.55b), we used the notation r_k, where r is a rational number, $r = 1 + 2p/6$, $r = 1 + (2p-1)/6$, etc., and k an integer: r_k denotes the product of the k terms, $[r\,(r+1)\ldots(r+k-1)]$. This number r_k can eventually be expressed in terms of Euler gamma function Γ. In equation (13.55d), we have introduced the *generalized hypergeometric function* ${}_1F_4$, which is defined as the series on the line above, i.e. in equation (13.55c).

Equation (13.54) gives the general solution $f(x)$ of the boundary layer equations (13.49) that is even with respect to x. It has two free parameters, f_0 and f_4, since f_2 is imposed by (13.52c). As explained in Section 13.6.1, these two remaining constants are imposed by the asymptotic condition $f \sim -2/x^4$ for $x \to \pm\infty$. Indeed, the three functions $F^{[2p]}$ defined in (13.55) diverge exponentially with an oscillatory factor for $x \to \pm\infty$. Cancellation of the exponentially large terms in the right-hand side of equation (13.54) fixes the value of the coefficients f_0 and f_4, as we show below.

Derivation of the asymptotic behaviour of the functions $F^{[2p]}(x)$ for $x \to \pm\infty$ is a bit technical and we shall simply outline it. It follows a method that is standard when investigating the asymptotic behaviour of power series, as explained for instance in the book by Ince (E. L. Ince, 1956). In a first step, the series for $f(x)$ is rewritten as a contour integral in the complex plane by interpreting each term in (13.55c) as a residue in Cauchy formula for an analytic function $\phi(z)$. The function $\phi_x(z)$ that has singularities at every $z = -k$ and yields the correct values of the residues reads:

$$\phi_x(z) = \Gamma(b_1)\ldots\Gamma(b_4)\, \frac{\pi}{\sin(\pi z)}\, \frac{\left(\frac{x^6}{64}\right)^{-z} x^{2p}}{\Gamma(b_1 - z)\ldots\Gamma(b_4 - z)}, \qquad (13.56)$$

where Γ stands for the Euler gamma function, and $b_1 \ldots b_4$ denote the four rational numbers $(2p+2)/6, \ldots, (2p+6)/6$ in the arguments of the hypergeometric function in equation (13.55d). That this ϕ_x yields the correct formula by complex integration can be checked directly by computing the residues, or simply by using a general formula found on-line (Wolfram, undated). The integration contour for ϕ_x is drawn around the half-line $\mathbb{R}^-$ in such a way that all residues $z = -k$ are enclosed. Again, all the terms in (13.55b) are found by application of the Cauchy formula.

Now, by deforming this contour and making use of the method of stationary phase, one can derive the asymptotic behaviour of $F^{[2p]}(x)$ for $x \to \pm\infty$. This yields:

$$F^{[2p]}(x) \sim x^{2p}\, \frac{\Gamma(b_1)\ldots\Gamma(b_4)}{(2\pi)^{3/2}}\, \frac{\exp(2\sqrt{2}\,t^{1/4})}{t^{\frac{B_p}{4} - \frac{5}{8}}}\, \cos\left(2\sqrt{2}\,t^{1/4} - \pi\left(\frac{11}{8} + \frac{B_p}{4}\right)\right), \qquad (13.57)$$

where

$$B = b_1 + \cdots + b_4 = \frac{8 + 4p}{3} \quad \text{and} \quad t = \frac{x^6}{64}. \qquad (13.58)$$

By expanding the cosine in the right-hand side of equation (13.57), it can be checked that the above asymptotic behaviour of the three functions $F^{[2p]}$ is always a linear combination of the same two WKB modes derived earlier in equation (13.46):

$$\exp(2\sqrt{2}\,t^{1/4})\,t^{-1/24}\,\cos(2\sqrt{2}\,t^{1/4}) \quad \text{and} \quad \exp(2\sqrt{2}\,t^{1/4})\,t^{-1/24}\,\sin(2\sqrt{2}\,t^{1/4}). \qquad (13.59)$$

Here we have more, as equation (13.57) provides an explicit formula for the coefficients of the asymptotic expansions of $F^{[2p]}$ on each WKB mode.

Plugging the expansion of $F^{[2p]}$ given in equation (13.59) into equation (13.54) and expanding the cosine, we find two explicit conditions for cancelling the two exponentially large components of $f(x)$ at infinity. These two conditions are linear with respect to f_0 and f_4. Solution of this linear system yields:

$$f_0 = \frac{\Gamma\left(\frac{1}{3}\right)}{\sqrt[3]{9}} \quad \text{and} \quad f_4 = \frac{\Gamma\left(\frac{5}{3}\right)}{8\sqrt[3]{3}}. \tag{13.60a}$$

In addition, recall that f_2 is given by equation (13.52c) as

$$f_2 = -\frac{1}{2}. \tag{13.60b}$$

This fixes the constants in equation (13.54), and the boundary layer solution $f(x)$ is now known explicitly for $\Psi = 0$. This function f is plotted in Fig. 13.4, together with $g(x) = (f''(x) + 1)/x$.

This does not end the story of the boundary layer as we still have to compute the vertical displacement v_z in the boundary layer. Using the relation (13.21), $v_z'(s) = \phi$, one derives, after restoring the physical units using equation (13.35e):

$$v_z(s) = v_z^* \, \Psi \int_0^x f(x') \, \mathrm{d}x', \qquad \text{where } x = \frac{s}{s^*}. \tag{13.61}$$

This function is plotted in Fig. 13.5.

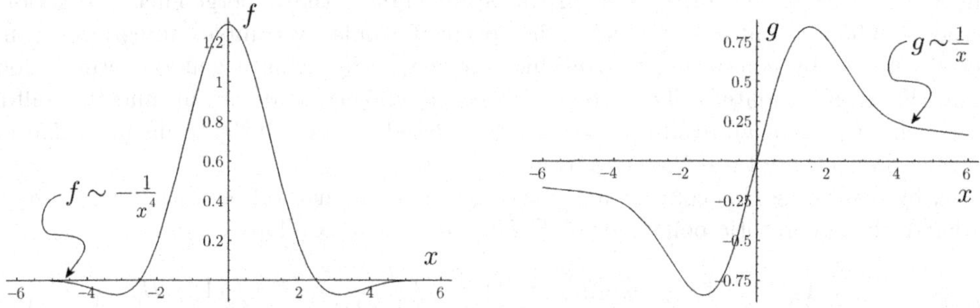

Fig. 13.4 Plot of the analytical solution (f, g) of the flexural boundary layer equations (13.49) for $\Psi = 0$ (linearized response of the crowns). By equation (13.39), f is the rescaled rotation ϕ, and g is the rescaled membrane stress N_s.

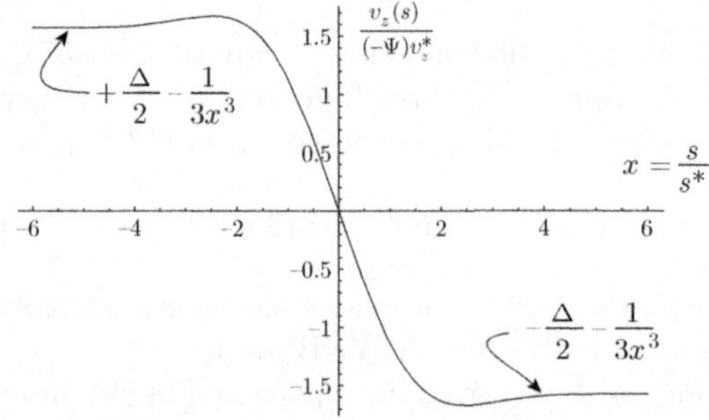

Fig. 13.5 Rescaled axial displacement $v_z(s)$ in the flexural boundary layer for $\Psi = 0$. By equation (13.61), $v_z(s)/(-\Psi)\,v_z^* = -\int_0^x f(x')\,\mathrm{d}x'$.

From equation (13.61), we find that there is a step of vertical displacement Δ across the boundary layer, defined as:

$$\Delta = v_z(x = +\infty) - v_z(x = -\infty)$$
$$= v_z^* \, \Psi \int_{-\infty}^{+\infty} f(x') \, \mathrm{d}x = 2 \, v_z^* \, \Psi \int_0^{+\infty} f(x') \, \mathrm{d}x, \tag{13.62}$$

where we have used the fact that f is an even function of x. This quantity Δ is the discontinuity in the axial displacement $v_z(s)$ across the layer as seen from scales much larger than the width s^* of the layer. Indeed, $|x| \to \infty$ corresponds to $|s| \ll s^*$. At this scale, the shell looks like a staircase near the crown, and Δ is its height. The integral in the right-hand side can be evaluated as

$$\int_0^{+\infty} f(x') \, \mathrm{d}x' = \frac{\pi}{2},$$

and therefore

$$\Delta = \pi \, v_z^*. \tag{13.63}$$

This yields, after restoring the physical units using (13.35):

$$\Delta = \frac{(12(1 - \nu^2))^{1/2} \, r}{2 \, E \, h^2} \, \Xi \tag{13.64}$$

The first factor in the right-hand side is the compliance of the flexural boundary layer: it is the proportionality constant in the linear relation between the shear displacement Δ to the shear force Ξ. This compliance is the inverse of the typical rigidity k^* introduced in equation (13.36) based on dimensional analysis, up to a numerical factor of order 1.

Finally, by plugging an expansion of $f(x)$ into equation (13.61) and restoring physical variables, we have

$$v_z(s) = v_z^* \, \Psi \left(\mathrm{sgn}(s) \frac{\Delta}{2} + \frac{s^{*3}}{3 \, s^3} + \cdots \right), \tag{13.65a}$$

an expansion which is valid in the intermediate region, $s^* \ll |s| \ll R$. Similarly, for the membrane stress N_s we have

$$N_s = \Psi \, N^* \left(\frac{s^*}{s} + \cdots \right). \tag{13.65b}$$

In the intermediate regions, these expansions must match the outer solution derived using the linearized membrane theory, as explained in Appendix B. This is possible thanks to the consistent behaviours $v_z \sim 1/s^3$ and $N_s \sim 1/s$, which have also been obtained[16] in the analysis of the outer solution in Section 13.3.2. The constant term proportional to Δ in equation (13.65a) can be interpreted as a rigid-body, axial translation of the inner half of the torus with respect to the outer half.

---

[16] From the point of view of the outer solution, the $1/s^3$ behaviour is a divergence, although from the point of view of the inner solution, it is a small correction. This is the usual situation in boundary layer theory. This reflects the fact that the inner solution has a prefactor diverging as h tends to zero, and is thus seen as infinite with the scaling of the outer solution.

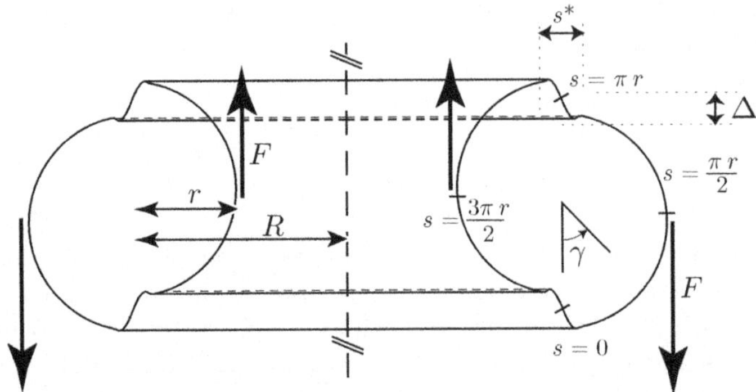

Fig. 13.6 Deformation of an elastic torus under axial forces applied along the equators. Shear forces cannot be balanced by stretching the mean surface near the top and bottom parallels but only by flexural effects, which are weaker. This leads to elastic boundary layers of width $s^* \propto h^{1/3}\, r^{2/3}$ where the elastic energy concentrates. These boundary layers coincide with the 'rigidifying' lines in the problem of isometric deformations. Displacement is enhanced for the sake of clarity, although the linear boundary layer theory underlying the analysis applies only when the slopes in the boundary layers remain small.

In the coming section, we show how this local description of the layer can be matched with a global solution defined over the rest of the toroidal shell, and to do so we consider a particular loading geometry.

13.7.2   Example: deformation under axial forces applied at the equators

In order to show a 'practical' application of the general analysis of the layer, we explicitly solve the problem of a torus submitted to vertical forces applied along its two equators,[17] as shown in Fig. 13.6. This loading is described by the exterior force field given earlier in equation (13.5):

$$
f_r(s) = 0, \qquad f_z(s) = F\left(-\frac{\delta_D\left(s - \frac{\pi r}{2}\right)}{2\pi\,(R+r)} + \frac{\delta_D\left(s - \frac{3\pi r}{2}\right)}{2\pi\,(R-r)} \right), \tag{13.66}
$$

where F is the force applied on both halves of the torus, $s = \pi r/2$ is the exterior equator and $s = 3\pi r/2$ is the interior equator—recall that the crowns are located at $s = 0$ and $s = \pi r$.

The first step is to solve the equations for membrane stress. Equation (13.6b) yields:

$$
\frac{\mathrm{d}(2\pi\,\rho\,N_s\,\sin\gamma)}{\mathrm{d}s} = -2\pi\,\rho\,f_z.
$$

Using equation (13.66), we can write the integral of the right-hand side as

$$
\int \left(-2\pi\,\rho(s)\,f_z\left(\frac{s}{s^*}\right) \right)\,\mathrm{d}s = \Sigma + \frac{\epsilon(s)}{2}\,F,
$$

<hr>

[17] Similar geometries are solved in two papers by Steele (C. R. Steele, 1965, 1964).

where Σ is a constant of integration, which will be determined later, and

$$\epsilon(s) = -\operatorname{sgn}(\cos\gamma(s))$$

is the piecewise constant function with value $(+1)$ on the upper half of the torus ($0 < s < \pi r$) and (-1) on the lower half of the torus ($\pi r < s < 2\pi r$). This function $\epsilon(s)$ arises by integration of the Dirac weight located at the inner and outer equators. Combining these equations, we have

$$N_s(s) = \frac{\Sigma + \frac{\epsilon(s)}{2}\, F}{2\pi\, \rho(s)\, \sin\gamma(s)}. \tag{13.67a}$$

The other equation (13.6a) yields:

$$N_\theta(s) = -\frac{\Sigma + \frac{\epsilon(s)}{2}\, F}{2\pi\, r\, \sin^2\gamma(s)}. \tag{13.67b}$$

Because of the factor $\sin\gamma$ in the denominators, these solutions diverge near the crowns, as expected. There, they should be replaced by the boundary layer solutions.

The membrane stress and the displacement in the outer regions still have to be computed. This can be done using the equations derived in Section 13.3. Plugging the expressions of the membrane stress obtained above into equations (13.8), one can find the strain e_s and e_θ. By equation (13.9b) one can then determine the radial displacement v_r directly. The next step is to solve equation (13.9a) for $v_z'(s)$, and to integrate the result to find $v_z(s)$. In general, the function v_z determined in this way will be multivalued; in order for v_z to take on the same values after one turn (when s is increased by $2\pi r$), one has to adjust the constant Σ which has not yet been found. These calculations raise no particular difficulty but are quite cumbersome. They are not given here, but it can be checked that the condition that v_z is single-valued indeed fixes the constant Σ unambiguously; the result is $\Sigma = 0$.

This result can be established without calculations, as it is due to the symmetries of the undeformed torus with respect to a mirror symmetry about the plane $z = 0$. To show this, let us define an even function to be any function $q(s)$ such that $q(s) = q(\pi r - \gamma)$, and an odd function to be any function $q(s)$ such that $q(s) = -q(\pi r - \gamma)$ (note that the transformation $s \mapsto (\pi r - s)$ is precisely the mirror symmetry about the plane $z = 0$). Next, let us introduce the following property: a function $q(s)$ will be said to be *quasi-even* if it can be written as $(\Sigma + \epsilon(s)/2\, F)$ times an even function of s. Note that we do not know yet whether $(\Sigma + \epsilon(s)/2\, F)$ itself is odd, even, or neither odd nor even. It is clear from equations (13.67) that both $N_s(s)$ and $N_\theta(s)$ are quasi-even—in fact, the definition is tailored to them. By the constitutive law (13.8), it follows that the principal strain components $e_s(s)$ and $e_\theta(s)$ are quasi-even as well. In equation (13.9b), $\rho(s)$ is even, and so $v_r = \rho\, e_\theta$ is quasi-even. Since ρ' is odd (the derivative of an even function is odd and *vice-versa*), this implies that $(\rho'\, v_r')$ is quasi-even too. Then, equation (13.9a) shows that $v_z'\, z' = e_s - \rho'\, v_r'$ is another quasi-even function. The function $z'(s) = \sin\frac{s}{r}$ is even in our sense, and so we have shown that $v_z'(s)$ is quasi-even. Now, given the symmetry of the loading and of the undeformed shape of the torus, one can seek $v_z(s)$ in the form of an even function: according to Fig. 13.6, we have $v_z(\pi r - \gamma) = v_z(\gamma)$. This implies that v_z', which we know is quasi-even, is also odd. Returning to the definition of a quasi-even function, this shows that $(\Sigma + \epsilon(s)/2\, F)$ is an odd function in our sense. Since the value of $\epsilon(s)$ flips from $(+1)$ to (-1) by the mirror symmetry, this yields

$$\left(\Sigma + \frac{(-1)}{2}\,F\right) = -\left(\Sigma + \frac{(+1)}{2}\,F\right).$$

As announced earlier, this implies

$$\Sigma = 0. \tag{13.68}$$

Having found Σ, the solution is now complete in the outer region. It remains to match with the inner solution in the layer. To do so, we just have to compute the expansion of N_s given in equation (13.67a) near the lower crown $s = 0$. This yields:

$$N_s \approx \frac{-\dfrac{F}{2}}{2\pi\,R\,\dfrac{s}{r}} \qquad \text{for } |s| \ll r. \tag{13.69}$$

A similar expansion can be found near the upper crown $s = \pi r$. By identification with equation (13.12), we find

$$\Xi = -\frac{F}{2}. \tag{13.70}$$

This is consistent with our interpretation of Ξ as the shear force supported by each crown: there are two crowns, and each of them supports half the applied force F. The minus sign is due to our convention for the orientation of the applied force.

Equation (13.70) allows the diverging membrane stress (13.69) to match smoothly with the asymptotic behaviour of the boundary layer stress (13.65b) in the intermediate region $s^* \ll s \ll r$, as is explained graphically in Fig. 13.7. The displacement (v_r, v_z) can be matched similarly. As noted earlier, the constant Δ in equation (13.65a) can be interpreted as a rigid-body displacement with amplitude Δ of the outer half-torus with respect to the inner one, as sketched in Fig. 13.6. This rigid-body displacement is the main source of deformation far away from the boundary layer. This justifies defining the stiffness k of the shell by

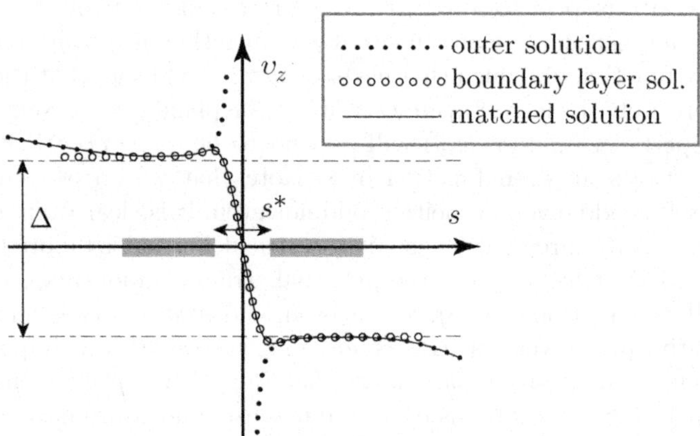

Fig. 13.7 Matching of the outer membrane solution and the inner boundary layer solution in the intermediate region (depicted by the grey boxes along the axes). The matched solution for the axial displacement, v_z, is shown by the solid curve. The membrane stress and the rotation ϕ of the tangent match similarly.

$$k = \frac{(-F)}{\Delta},$$

where the minus sign accounts for the fact that a positive force F yields a negative displacement Δ with our conventions. Combining with equations (13.64) and (13.70), we find

$$k = \frac{(-F)}{\Delta} = \frac{4\,E\,h^2}{(12(1-\nu^2))^{1/2}\,r}. \tag{13.71}$$

Again, this stiffness is of order k^*, the typical stiffness proposed in equation (13.36) based on dimensional analysis; this scaling law remains true for any loading geometry as long as the boundary layer remains in the linear regime ($\Psi \ll 1$), or moderately non-linear (Ψ of order 1). However, the range of applicability of the detailed solution given above is limited by non-linear effects, to loadings such that $F \ll \Xi^*$, as defined in (13.35c).

Classically, the stiffness k of an elastic shell can follow two types of behaviour. For inhibited shells, the stretching energy dominates bending everywhere and dimensional analysis shows that the stiffness is of order $k^* \sim E\,h$, while for shells that have isometric modes of deformation, the modes with lowest elastic energy only involve bending and their typical stiffness reads $k^* \sim E\,h^3/r^2$, r being the typical size of the shell. Now, equation (13.71) shows that toroidal shells follow a different regime: their rigidity $k^* \sim E\,h^2/r$ is halfway between these two classical cases. This has to do with the particular structure of the deformation that concentrates in the vicinity of the two layers, while away from these layers, it is essentially a rigid-body deformation. This 'intermediate' stiffness concerns all shells whose mean surface is made rigid by the presence of rigidifying curves. Such surfaces are not necessarily surfaces of revolution, as discussed in Chapter 11. The stiffness of a torus is a measurable quantity whose order of magnitude ultimately reflects a property of differential geometry. This is another example of the deep connection between elasticity and geometry.

For a similar reason, the mode of vibration of the elastic torus with lowest frequency is very different from that of classical shells. Indeed it is related to the mode of deformation with the lowest stiffness, which is the one studied just above: this mode is made of a vertical rigid-body displacement of the outer half-torus with respect to the inner one, without appreciable deformation except near the crowns. Therefore the kinetic energy in the vibrating motion will be $M(d\Delta/dt)^2/2$, where M is the mass of the torus, so that the frequency of the lowest mode will be of order $2\pi\,(k/M)^{1/2}$, the actual formula being $2\pi\,(k(m_1 + m_2)/m_1 m_2)^{1/2}$, where $m_{1,2}$ are the masses of the inner and outer halves of the torus. Because k^* scales like h^2 while it scales like h for regular inhibited shells, this mode of vibration has a particularly low frequency, as a consequence of the presence of the boundary layers near the crowns. This situation is reminiscent of a classical Helmholtz resonator, where the kinetic and potential energy are stored in different places. Here the potential energy is stored in the boundary layer while the kinetic energy is in the bulk of the resonator, although the Helmholtz resonator has its kinetic energy near the neck (a boundary layer in some sense) and the potential energy in the bulk of the container.

13.8 Boundary layer solution for strong nonlinearity

We now consider the opposite limit, $\Psi \to \infty$, where the non-linear effects dominate bending effects. This limit has been considered by Jordans (P. F. Jordan, 1962) and by Sanders and Liepins (J. L. Jr. Sanders and A. A. Liepins, 1963) in the particular case of pressure loading ($\Xi = 0$). They found that nonlinearity can regularize the membrane divergences. Here, we consider the generic case, $\Xi \neq 0$, and reach a different conclusion: geometric nonlinearity is not sufficient to regularize the membrane divergences, and one still has to resort to flexural effects to get a smooth solution. Geometric nonlinearity can only change the type of singularity near crowns but not regularize them; in this sense, they are less severe than bending effects.

It might seem paradoxical to study the limit of strong nonlinearity, $\Psi \gg 1$, using a theory based on Hookean elasticity and on the assumption of small strain. This is in fact perfectly consistent. The typical value of the maximum strain can be estimated by setting $s = s^*$ in the expansion (13.16) valid in the intermediate region:

$$
e_s \sim e_\theta \sim \frac{r\, R\, \Xi}{E\, h\, (s^*)^2} \sim \frac{\Xi}{\dfrac{E\, h^{5/3}\, R^{2/3}}{r^{1/3}}}.
$$

The small strain approximation therefore requires:

$$
\Xi \ll \Xi^{**}, \qquad \text{where } \Xi^{**} = \frac{E\, h^{5/3}\, R^{2/3}}{r^{1/3}}.
$$

Since the ratio Ξ^{**}/Ξ^* is large, of order R/h, this condition is compatible with $\Psi \gg 1$ (which is equivalent to $\Xi \gg \Xi^*$). A similar situation has been encountered in the buckling of flat plates, whereby geometrical nonlinearity might become significant while the approximation of small strain continues to be valid.

We start again from the rescaled boundary layer equations (13.40). Since nonlinearity dominates bending effects in the limit that we are studying, $\Psi \gg 1$, it is natural to try to drop the bending term $(-f'')$ in (13.40b). Then, f can be eliminated from equation (13.40b):

$$
f(x) = \frac{1}{\Psi}\left(-x + \frac{1}{g(x)}\right), \tag{13.72}
$$

and the other equation (13.40a) can be written in terms of g (B. Audoly, 2004):

$$
2\,\Psi\, g''(x) + \frac{1}{g^2} = x^2. \tag{13.73a}
$$

The solution of this non-linear differential equation has to satisfy the asymptotic condition given earlier in equation (13.48):

$$
g(x) \approx \frac{1}{x} \qquad \text{for } x \to \pm\infty. \tag{13.73b}
$$

This condition implies that the term g'' becomes negligible asymptotically in equation (13.73a).

Note that the parameter Ψ could be eliminated from the non-linear boundary layer equation (13.73a), by taking as a new unknown function $\Psi^{1/5}\, g(x\, \Psi^{1/5})$; this yields a parameterless equation.

However, one can easily show that the set of equations (13.73) has no solution: nonlinearity by itself cannot be used to build a boundary layer in this limit $\Psi \to \infty$. The reason is fairly simple: from the asymptotic condition (13.73b), the solution $g(x)$ has different signs at $-\infty$ and $+\infty$ and must therefore cross zero somewhere, say at $x = x_0$. Its behaviour near x_0 is found at once by balancing the terms in the left-hand sides of equation (13.73a), since the right-hand side becomes negligible; this yields

$$g(x) \approx \pm \left(\frac{9}{4\Psi} \right)^{1/3} |x - x_0|^{2/3}. \tag{13.74}$$

Now, if $g(x)$ vanishes, then equation (13.72) shows that f becomes infinite, and so does f''. Near x_0, our assumption that the bending term $-f''$ is negligible breaks down.

Several track can be explored to build a consistent solution in the limit $\Psi \to \infty$. The first is to take bending into account: the argument above suggests that in the vicinity of x_0, there is a secondary, much smaller boundary layer, enclosed in the purely non-linear boundary layer governed by (13.73a), where bending becomes important. Another possibility is that some non-linear terms that have been dropped by our assumption of moderate rotations, become important only in the limit $\Psi \to \infty$ and should be restored in the boundary layer equation, but this seems unlikely. A third possibility is that a 'catastrophe' takes place in the boundary layer equations for some *finite* value of Ψ, and that the boundary layer equations (13.40) are not valid beyond a critical value of Ψ— making pointless the analysis of the limit $\Psi \to \infty$ of these equations. The description of the elastic torus in this limit of strong nonlinearity is still an open question. Tracking the numerical solution derived in Section 13.6.2 for larger and larger values of Ψ should give a first hint of what is actually happening in the limit $\Psi \to \infty$. So far, we could only track the numerical solution up to values of Ψ close to $\Psi = 3$. At larger numerical values numerical convergence is lost, which could either be due to a catastrophe for finite Ψ close to three, or to the fact that the solution continues to exist, but changes too rapidly to be tracked numerically.

13.9 Conclusion

The case of shearless boundary layers ($\Xi = 0$) has been extensively discussed in the literature, as it is relevant for pressurized thin toroidal shells with circular cross-section, which is probably the simplest geometry for a toroidal shell. As we showed in Section 13.5, this geometry is in fact quite specific. For this reason, we focused on the generic case $\Xi \neq 0$, and derived original solutions to the equations for toroidal shells. We shall now briefly compare our results with the classical results obtained earlier in the shearless case ($\Xi = 0$).

In the purely flexural case ($\Psi \to 0$), the deformation of the crowns does not depend essentially on the presence of a shear force: Clark's solution (R. A. Clark, 1950), rederived in Section 13.7, holds both for $\Xi = 0$ and for $\Xi \neq 0$.

In the mixed regime, that is when both flexural and non-linear effects are important ($\Psi = \mathcal{O}(1)$), both the analysis of Rossettos and Sanders (J. N. Rossettos and J. L. Sanders, 1965) for $\Xi = 0$ and the one presented in Section 13.6 for $\Xi \neq 0$ yield a boundary layer of size of order $R^{2/3}h^{1/3}$. The boundary layer equations, however, are different: Rossettos and Sanders obtain a set of coupled, *linear*[18] differential equations, while it does not seem possible to rewrite the boundary layer equations as linear ODEs when $\Xi \neq 0$. The essential difference between the layers comes from the fact that the orders of magnitude of the various quantities in the vicinity of the boundary layer that impose which terms are to be kept in the boundary layer equations, are given by different outer membrane solutions depending on whether Ξ vanishes or not. Recall, for instance, that the stress of this outer solution is smooth when $\Xi = 0$, but not when $\Xi \neq 0$.

Regarding the limit of strong nonlinearity, $\Psi \to \infty$, the difference between the two types of boundary layers is even more striking: Sanders and Liepins (J. L. Jr. Sanders and A. A. Liepins, 1963) have shown that the boundary layer equations for $\Xi = 0$ are still linear, and that geometric nonlinearity is sufficient to eliminate the divergences of the linear membrane theory. In contrast, the argument of Section 13.8 suggests that nonlinearity is no longer sufficient to regularize divergences when $\Xi \neq 0$: either additional sources of nonlinearity have to be taken into account or, more probably, one still has to resort to bending effects within a secondary boundary layer enclosed into the first one.

In this chapter, we have shown how to handle divergences found when using the membrane approximation. The approach was illustrated on a particular geometry, namely toroidal shells under axisymmetric loading. The main result is that the deformation concentrates in boundary layers that are located along the crowns, i.e. along the circles whose tangent plane is perpendicular to the axis of the torus and that, away from these boundary layers, the inner and outer halves of the torus essentially undergo rigid-body motion. We have applied the general ideas of the boundary layer theory, introduced in Appendix B, which applies to many problems in both fluid and solid mechanics and in many other fields. Divergent solutions indicate that a different sort of approximation must be used in some regions of the shell. In these so-called boundary layers, there is no need to resort to the full theory, as the orders of magnitude derived from the outer solution allows one to discard most of the terms in the layers. This yields ultimately boundary layer equations that are often solvable analytically or at least depend on fewer parameters than the general problem (a single parameter Ψ in the present case).

We have also discussed the geometrical origin of these boundary layers, located along the lines where the Gauss curvature changes, and their connection with the rigidifying curves studied in Chapter 11. The analysis of what is probably the simplest example of a shell with mixed type has therefore revealed the dramatic influence of some *geometrical* properties of surfaces on the *mechanical* behaviour of thin shells.

References

B. Audoly. Geometric boundary layers in shells with mixed type. In *Theories of Plates and Shells, Critical Review and New Application*, volume 16 of *Lecture Notes in Applied and Computational Mechanics*, chapter 2, pages 13–20. Springer, 2004.

---

[18] This surprising feature (recall that $\Psi \neq 0$) has to do with the fact that the stress is smooth in the membrane approximation when $\Xi = 0$, with the consequence that the unknown in the boundary layer equations is a *correction* to the membrane stress rather than the physical stress itself. This correction is small in front of the dominant, smooth membrane stress, and it makes sense to linearize the former with respect to the latter. In contrast, when $\Xi \neq 0$, it is the dominant contribution to the stress that enters the boundary layer equations, making it essentially non-linear.

R. A. Clark. On the theory of thin elastic toroidal shells. *Journal of Mathematics and Physics*, 29:146–178, 1950.

D. R. Dean. The distortion of a curved tube due to internal pressure. *Philosophical Magazine*, 28:452–464, 1939.

W. Flügge. *Stresses in Shells*. Springer-Verlag, Berlin, 2nd edition, 1973.

L. Föppl. *Vorlesungen über technische Mechanik*, volume 5. B. G. Teubner, Leipzig, Germany, 1907.

A. L. Goldenveizer. *Theory of Elastic Thin Shells*. Pergamon Press, New York, NY, USA, 1961.

E. L. Ince. *Ordinary Differential Equations*. Dover Publications, 1956.

P. F. Jordan. Stresses and deformations of the thin-walled pressurized torus. *Journal of Aerospace Engineering*, 29:231–225, 1962.

P. Karamian, J. Sanchez-Hubert, and É. Sanchez Palencia. A model problem for boundary layers of thin elastic shells. *Mathematical Modelling and Numerical Analysis*, 34(1):1–30, 2000.

P. Karamian-Surville, J. Sanchez-Hubert, and É. Sanchez Palencia. Propagation of singularities and structure of layers in shells: Hyperbolic case. *Computers and Structures*, 80(9-10):747–768, 2002.

G. V. Kutsenko. Axisymmetric deformation of a circular torus. *International Applied Mechanics*, 15(11):1049–1053, 1979.

R. E. Langer. On the asymptotic solution of ordinary differential equations. *Transactions of the American Mathematical Society*, 33:23–64, 1931.

A. Libai and J. G. Simmonds. *The Nonlinear Theory of Elastic Shells*. Cambridge University Press, 2nd edition, 1998.

W. H. Press, S. A. Teukolsky, W. T. Vetterling, and B. P. Flannery. *Numerical Recipes in C*. Cambridge University Press, 2002.

E. Reissner. On stresses and deformations in toroidal shells of circular cross sections which are acted upon by uniform normal pressure. *Quarterly of Applied Mathematics*, 21:177–188, 1963.

J. N. Rossettos and J. L. Sanders. Toroidal shells under internal pressure in the transition range. *AIAA Journal*, 3:1901–1909, 1965.

J. Sanchez-Hubert and É. Sanchez Palencia. *Coques élastiques minces: Propriétés asymptotiques*. Masson, Paris, 1997.

J. L. Jr. Sanders and A. A. Liepins. Toroidal membrane under internal pressure. *AIAA Journal*, 1(2105–2110), 1963.

C. R. Steele. Orthotropic pressure vessels with axial constraint. *AIAA Journal*, 2(4):703–709, 1964.

C. R. Steele. Toroidal pressure vessels. *Journal of Spacecraft and Rockets*, 2(6):937–943, 1965.

Wolfram. Generalized hypergeometric function: integral representations (formula 07.31.07.0003.01). http://functions.wolfram.com/07.31.07.0003.01.

14
Spherical shell pushed by a wall

14.1 Introduction

We discuss the contact of a thin elastic shell, such as a ping-pong or a tennis ball, with a perfectly rigid plane. Our aim is to extend to *thin* shells Hertz' famous theory for the contact of a *full* elastic ball. When thin shells are considered, geometry plays an important role: one of the configurations of the shell that we shall encounter involves an inverted spherical cap, shown in Fig. 14.1, right. This cap merges with the rest of the shell along a circular ridge, which we investigate in full details below. The spherical cap, obtained by reflection of the original shell, is ultimately related to the absence of smooth isometric deformations of a sphere. This discussion emphasizes the geometric basis of many concepts in the theory of elastic shells.

Our presentation is supported by simple experiments, carried out originally on tennis and ping-pong balls in reference (L. Pauchard, Y. Pomeau, and S. Rica, 1997). Similar experiments with the transparent, spherical packaging used for wrapping some ping-pong balls are shown in Fig. 14.1. The ball was cut into two halves and then one piece was slowly pushed against a flat plane with a force that was smoothly increased. This force was applied using another flat plate. A half-ball was used in order to get rid of the pressure force of the fluid trapped in the ball. The displacement of the top of the half-ball was measured. This showed a first-order (i.e. discontinuous) transition between two different configurations: the first involves a flat contact between the shell and the rigid plane, whereas, in the second case, contact is limited to a circle, the shell being partially inverted. For a review of earlier experimental and numerical work on the compression of spherical shells by rigid planes, see (D. P. Updike and A. Kalnins, 1970, 1972; J. Nowinka and S. Lukasiewicz, 1994) and references therein.

Fig. 14.1 Experimental compression of a transparent spherical shell showing a disc of contact at small forces and an inverted cap at larger forces. Images courtesy of L. Pauchard, FAST Laboratory, Université Paris-Sud, Orsay (France).

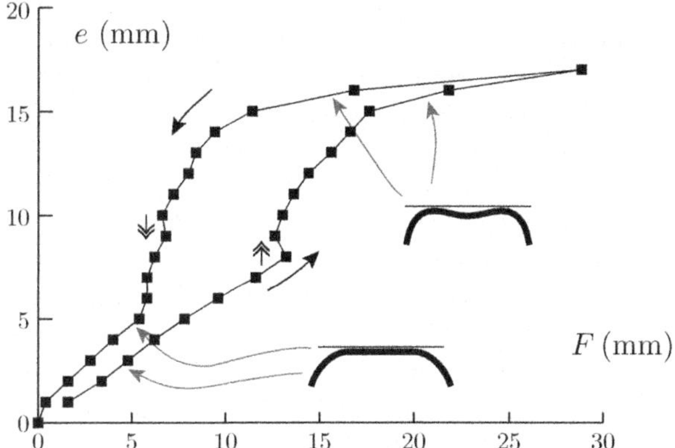

Fig. 14.2 Experimental force–displacement curve of a half-tennis ball compressed by a rigid plane under cyclic loading, from references (L. Pauchard, Y. Pomeau, and S. Rica, 1997; L. Pauchard and S. Rica, 1998). Characteristics of this shell are: Young's modulus $E = 10^7$ MPa, Poisson's ratio $\nu \approx 0.3$, radius $R = 35$ mm, thickness $h = 3.7$ mm. At increasing force, the lower branch is followed. At some critical value of the force, a dynamic buckling process (double arrow) transforms the flat region of contact into an inverted cap. When the force is decreased, the inverted cap configuration is followed over a wider range of forces, leading to an hysteresis.

In the experiments, this first-order transition was observed together with a hysteresis: at increasing load a sharp drop of the displacement occurs at a critical value of the force. The inverse phenomenon happens for decreasing forces, but with a lower critical force. This yields two jumps when the displacement is plotted vs. the applied force, as shown in Fig. 14.2. The area drawn by the hysteresis loop gives the energy lost during one cycle of loading. Some energy is indeed dissipated by the friction between the ball and the squeezing planes, but most of the dissipation was seemingly due to plastic deformations of the material of the shell, which is not accounted for in the Hookean approach that follows. Nevertheless, we shall be able to recover some striking experimental results such as the hysteresis and the first-order transition by an analysis of the equations for shells. This is related to the existence of two types of solutions in the equations: with disc-like contact on one hand, and with circular contact on the other hand. We shall first explain this phenomenon in qualitative terms, using dimensional analysis and geometrical arguments, and then present a fully quantitative approach.

Before proceeding to the analysis of a thin spherical shell in contact with a perfectly rigid plane, we present both classical Hertz' contact theory (contact of a *full* elastic sphere with a rigid plane) and Pogorelov's analysis of a thin elastic shell pushed upon by a *point force* at the pole. Although Hertz' results for the contact of a full ball is rather different from those we shall derive for a thin shell (no transition, different scaling laws for the force, etc.), it is interesting both from the historical viewpoint and because it shows how one can balance the local deformation near the contact region with the imposed force, an approach that will be extended to the case of the thin spherical shell. Pogorelov's results are also interesting as they show how an order of magnitude analysis can give physical insight into the problem with relatively light work—an approach that we shall complement in a second step by explicitly solving the shell equations.

Deformation of spherical shells or spherical caps subjected to an inward point force, or to external pressure has been studied by many authors. Biezeno proposed the first solution to the problem based on non-linear equations for shells, and calculated load–deflection curves exhibiting some unstable equilibria (C. B. Biezeno, 1935). Experiments were carried out by Evan-Iwanovski and Berke (R. M. Evan-Iwanowski, H. S. Cheng, and T. C. Loo, 1962; L. Berke and R. L. Carlson, 1968), on both shallow and deep shell segments, for clamped and simply supported caps, with particular focus on stability. Ashwell noticed the formation of a region with inverted curvature (dimple) whose size increases with the load, and proposed a solution based on two approximate solutions of the non-linear equations, one for the inverted region and one for the outer one (D. G. Ashwell, 1959). A more accurate description of the dimple can be obtained based on singular perturbations methods. This approach was first proposed by Koiter (W. T. Koiter, 1969), and then refined and adapted to different loading configurations (G. V. Ranjan and C. R. Steele, 1977; G. A. Kriegsmann and C. G. Lange, 1980; F. Y. M. Wan, 1980), F. Y. M. Wan (1980a) (D. F. Parker and F. Y. M Wan, 1984; M. Gräff *et al.*, 1985); the analysis of the circular fold in Section 14.7 is based on these papers. An excellent review can be found in the book by Libai and Simmonds (A. Libai and J. G. Simmonds, 1998). In the forthcoming analysis, we will be mostly concerned with axisymmetric configurations; their stability and the possible transitions towards asymmetric configurations are discussed in Section 14.8.6.

14.2 A short account of Hertz' contact theory

In this section, we reconsider, however briefly, Hertz' well-known contact problem. When he was less than thirty years old, Hertz completely solved the problem of contact of a full elastic sphere (actually even more generally of a full ellipsoid) with another (full) sphere, of unequal diameter. We can only urge the interested reader to look at this amazing piece of mathematical and physical ingenuity—Landau and Lifshitz (L. D. Landau and E. M. Lifshitz, 1981) reproduce rather closely the original derivation by Hertz. We did not feel the need to repeat it, since it cannot be improved upon, only uselessly rephrased. We shall content ourselves with a mere derivation of the order of magnitude that is pertinent for Hertz' contact problem, and limited to the case of a (full) elastic sphere squeezed on a plane.

The main assumptions by Hertz is that the contact is along a flat disc, and that the ball is only slightly deformed ($e \ll r \ll R$). Here, R is the radius of the elastic ball, e is the lowering of the pushing plane (and of the apex of the ball), and r is the radius of the disc of contact, while $\gamma \approx r/R$ is half the angle of the cone generated by the region of contact (see Fig. 14.3). A geometric relation between r, e and R, the radius of the sphere, can be obtained by writing $\cos\gamma$ in two different ways. Indeed, for small e,

$$\cos\gamma \approx 1 - \frac{1}{2}\left(\frac{r}{R}\right)^2 \approx \frac{R-e}{R}, \tag{14.1}$$

hence, after dropping the numerical constant $1/2$ (we are only interested in orders of magnitude so far):

$$r \sim (e\,R)^{1/2}. \tag{14.2}$$

The elastic energy stored in the ball is of the order of the work done to flatten the sphere, i.e. the product $e\,F$. The estimation of the elastic energy rests on two rather subtle

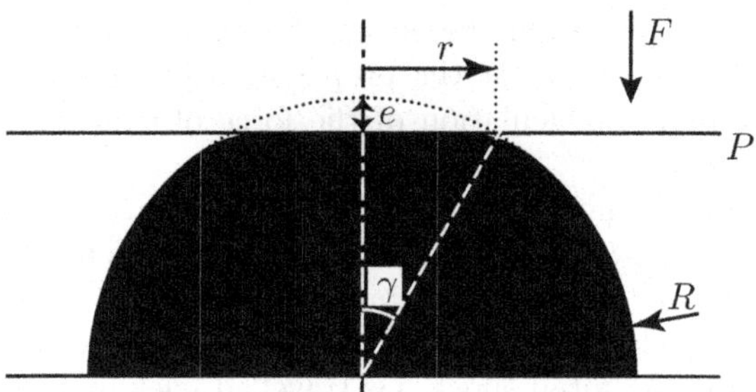

Fig. 14.3 Hertz' contact: a downward vertical force F is applied through a plane P on a full elastic ball.

assumptions: first, the perturbed volume near the disc of contact has a size of order r, because there is no intrinsic length in Hookean elasticity. Furthermore, the displacement inside the sphere is of order e. Therefore the strain i.e. the gradient of this displacement is of order e/r. Now the elastic energy is Young's modulus E times the volume of the perturbed area, r^3, times the square of the strain, $(e/r)^2$. This yields:

$$\mathcal{E}_{\text{Hertz}} \sim E\, r^3 \left(\frac{e}{r}\right)^2 \sim F\, e. \tag{14.3}$$

Eliminating r by equation (14.2) and solving for the displacement e of the apex, we obtain the following scaling between force and displacement, which is valid up to numerical constants of order unity:

$$e \sim \left(\frac{F^2}{E^2 R}\right)^{1/3}. \tag{14.4}$$

This characterizes the elastic response of the contact between the ball and the rigid plane.

Another remarkable observation by Hertz is that the bouncing time in a collision of a sphere with a plane can be predicted using this formula, as far as the velocity of the incoming sphere is less than the speed of sound in the elastic material it is made of. Let U be this incident velocity, and τ be the duration of the collision. Since the velocity of the ball changes from $+U$ to $-U$ during the collision, it always remains of order U. Therefore, $e \sim U\,\tau$. From Newton's law of dynamics, M, the mass of the ball, times its acceleration is equal to the force F. The order of magnitude of the acceleration is U/τ. This yields.

$$\tau^{5/2} \sim \frac{M}{E\, R^{1/2}\, U^{1/2}}.$$

Supposing now that the sphere is made of an homogeneous material of mass density ρ, one gets $M \sim \rho\, R^3$ and

$$\tau^{5/2} \sim \frac{R^{5/2}}{c^2\, U^{1/2}},$$

where $c = \sqrt{E/\rho}$ is the speed of sound in the elastic material. This can be written as

$$\tau \sim \frac{R}{c} \left(\frac{c}{U}\right)^{1/5}.$$

For a velocity U much smaller than the speed of sound, c, the duration of the collision is much longer than the time R/c for the propagation of sound across the solid sphere. This validates our quasi-static calculation of the force of contact between the plane and the ball.

It is of interest to notice that the same assumption of an adiabatic collision does not hold in 2D and even less in 1D. The one-dimensional problem of a rod hitting a plane and bouncing out of it was treated by Saint-Venant a long time ago, and is described in (A. E. H Love, 1927). The rod remains in contact with the reflecting plane until the sound wave generated by the initial shock has travelled back and forth from the hit point. The 2D case is made more complicated by the presence of logarithms.

14.3   Point-like force on a spherical shell (Pogorelov)

In their book of elasticity, Landau and Lifshitz (L. D. Landau and E. M. Lifshitz, 1981) consider a thin spherical shell deformed by a localized radial force pointing inwards. This problem was solved by Pogorelov (A. V. Pogorelov, 1988). When the intensity of the force is small, the deformation is localized near the point of application and grows linearly with the force (see Fig. 14.4). For large forces, a circular fold appears around the point of application and the displacement becomes proportional to the square of the applied force. The transition between the two regimes is continuous, contrary to the case of a pushing plane investigated later on. For large forces, Pogorelov derived the scaling laws for the radius of the fold. As presented by Landau and Lifshitz, this analysis is based upon the remark that the work done by the localized force is equal to the elastic energy stored in the deformed shell. In addition, this energy concentrates in the ridge. We shall recover Pogorelov's results below.

Let e be the displacement of the point of application of the force (see Fig. 14.4). The two regimes found by Pogorelov are described by the scaling laws:

$$F \sim \frac{E\,h^2}{R}\,e \qquad\qquad\qquad \text{for}\quad e < (e_{\mathrm{c}} \sim h) \qquad\qquad (14.5\mathrm{a})$$

$$F \sim \frac{E\,h^{5/2}}{R}\,e^{1/2} \qquad\qquad\qquad \text{for}\quad e > (e_{\mathrm{c}} \sim h), \qquad\qquad (14.5\mathrm{b})$$

respectively, where E is Young's modulus, R the radius of the shell at rest, h its thickness and F the applied force. The scaling laws are based on Pogorelov's estimates of the elastic energy of the shell, which are rederived below. The (continuous) transition between these behaviours

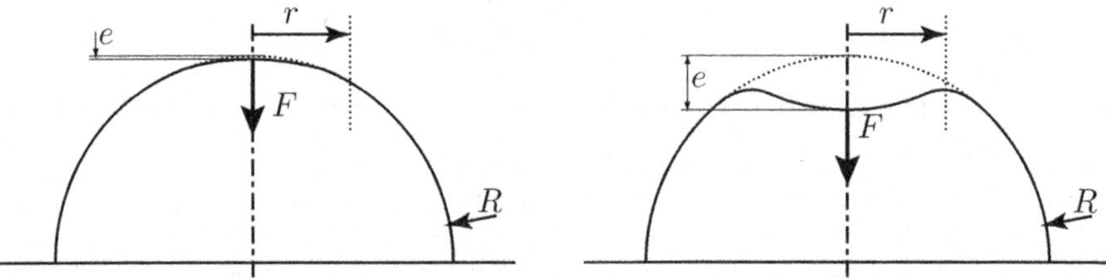

Fig. 14.4 Pogorelov's geometry: a (hollow) spherical shell with a point force F. Two regimes are possible: displacement e proportional to the force F at small forces (left), and displacement quadratic in F with an inverted cap at larger forces (right).

can be found by balancing the two expressions for the force, yielding $F_c \sim Eh^3/R$, that is for $e_c \sim h$.

14.3.1 Small forces: smooth deformation

The linear behaviour of e as a function of F, as given by equation (14.5a), is found as follows. The work done by the applied force is of order[1] (Fe). This work is equal to the elastic energy stored in the shell, which is concentrated in a disc of radius of order r around the point of application of the force. Rather than using the full shell equations to estimate this energy, we present a simplified analysis based on the F.–von K. equations for plates, with suitable modifications to account for the rest curvature $1/R$. This is justified by the fact that the perturbed region is small, $r \ll R$, in such a way that the F.–von K. can be used to find the correct orders of magnitude.

Let us use the first F.–von K. equation to derive a relationship between the order of magnitude of e and χ, the Airy potential. Recall that the curvature is the second derivative of the deflection w. In the presence of a natural curvature $1/R$, the term $[\chi, w]$ in the first F.–von K. equation (6.86a) can be estimated as $(1/R)(\chi/r^2)$, where r is the typical radius of the perturbed region along the shell surface, which is unknown for the moment. In this expression, we replaced second derivatives of χ simply by χ/r^2, something that gives the correct order of magnitude, since $\partial/\partial x \sim 1/r$, where x is the radial coordinate. Similarly the biharmonic operator Δ^2 appearing in the first F.–von K. equation is of order $1/r^4$ so that $\Delta^2 w \sim e/r^4$. Finally, the first F.–von K. equation reads

$$h^2 \frac{e}{r^4} \sim \frac{\chi}{r^2} \frac{1}{R},$$

and the second yields similarly:

$$\frac{\chi}{r^4} \sim \frac{e}{r^2} \frac{1}{R}.$$

Combining these two relations, one has

$$r^2 \sim hR,$$

which gives the size r of the perturbed region near the point of application of the force. For small forces, a linear regime is observed whereby the size r of the perturbed region is independent of the force and of the depression e: of course, the amplitude of the perturbation e depends linearly on the force F. This linear regime takes place as long as the perturbation to curvature remains much smaller than the natural curvature $1/R$. For larger forces, the picture is different as the size of the perturbed region grows with the applied force.

The bending contribution $(\Delta^2 w)$ in the first F.–von K. equation and the stretching one $[\chi, w]$ have to be balanced. If one assumed that the elastic energy is dominated by the bending contribution, one would obtain no deformation at all because of the geometric rigidity of the closed sphere (see Chapter 11); assuming instead that stretching dominates bending would lead one to a singular ridge whose bending energy is infinite, hence a contradiction. Now, since the bending and stretching energies are of the same order of

---

[1] (Fe) is just an order of magnitude estimate of the work done by the applied force, since e depends on F. In the case of a linear dependence for instance, the stored energy is $1/2Fe$.

magnitude, we can estimate the total energy by computing just the bending energy, as done below.

Up to multiplicative constants of order 1, the bending energy is E times h^3 times r^2 (the perturbed area) times the biharmonic operator applied on the deflection, of order e, that is $(e/r^2)^2$. This yields the bending energy $Eh^3r^2e^2/r^4$. It is indeed equal to the work Fe done by the force F, provided F is given by (14.5a): $F \sim Eh^2e/R$. Again, this derivation assumes that the curvature of the shell is close to its natural value $1/R$, including in the perturbed region. This is necessary to approximate $[\chi, w]$ by $1/R\chi/r^2$.

A detailed analytical solution of this problem based on the linearized equations for shells, including prefactors, can be found in references (W. T. Koiter, 1963; A. Libai and J. G. Simmonds, 1998).

14.3.2   Larger forces: inverted cap

We can now study the second regime, corresponding to a force F that is too large for the assumption of a linear response to be valid. The perturbation to the curvature, which was e/r^2 in the first regime, becomes of order of (or larger than) the rest curvature $1/R$ when e becomes of order or larger than $r^2/R = h$. This defines the transition from the domain of validity of equation (14.5a) (small values of F) to equation (14.5b) (large values of F). In the case of a point force, the transition from linear to square root behaviour of F as a function of e is smooth, and happens roughly when e is of order h. Incidentally, we can remark that this theory of a point force is valid as long as F is applied on a domain of size smaller than the radius of the perturbed area, namely $r \sim (hR)^{1/2}$. In the case of large forces, for which the square root behaviour of equation (14.5b) applies, the structure of the solution is similar, be it obtained through point forces (Pogorelov) or extended forces (pushing plane, as investigated below). This is the matter of the forthcoming discussion.

The assumption, to be checked at the end, is that, for large forces (i.e. for forces much larger than $F_{\mathrm{c}} = Eh^3/R$), the sphere deforms into an inverted cap. This configuration, shown in Fig. 14.5, is obtained by horizontal reflection of a small cap centred at the point of application of the force. Reflections conserve lengths. This is why this configuration, isometric to the spherical one, is favourable: it yields almost no stretching energy. A ridge forms in an annular domain between the unperturbed surface of the sphere, outside, and the inverted cap, inside (see experimental picture '1' in Fig. 14.24). Without this ridge, the deformation would be isometric but singular and the bending energy would become infinite. The elastic energy is concentrated near this ridge. The inner structure of the ridge makes a compromise between antagonistic stretching and bending effects (stretching energy is minimized by making the width of the ridge zero since the shell then becomes exactly developable, although bending energy is minimized by making the ridge thicker as this decreases the transverse curvature of the shell along the ridge); as a result, the bending and stretching energies stored in the ridge are of the same order of magnitude.

Let α be the kink angle, i.e. the change of orientation of the tangent plane across the ridge. The energy of the ridge is computed in two steps. First one shows that the Gauss curvature along the ridge is much larger than the curvature $1/R^2$ of the sphere at rest. In a first approximation this allows one to neglect the rest curvature in the calculation of the stretching energy, by again using a plate theory (F.–von K. equations).

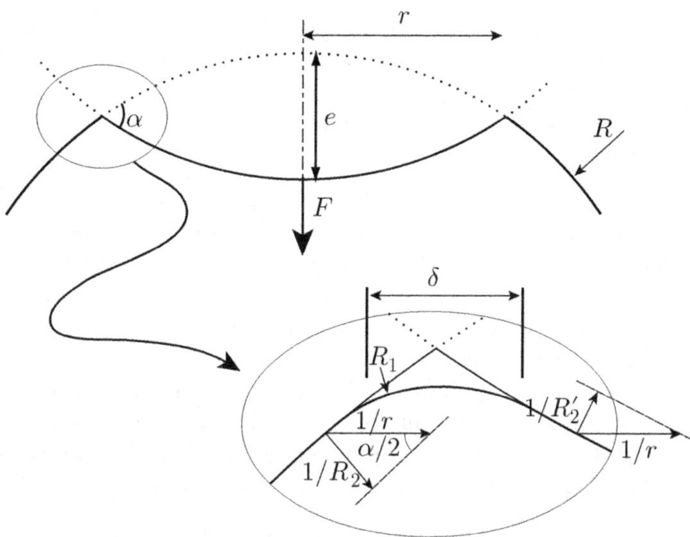

Fig. 14.5 The spherical ball with a point force (Pogorelov's set-up) in the large force limit. A small cap of radius $r \ll R$ gets inverted near the point of application of the force, and a very thin ridge of width $\delta \ll r$ connects the unperturbed region outside with the reflected cap. The kink angle over the ridge is denoted α.

The next step is to estimate the order of magnitude of the Gauss curvature in the ridge area.

Let us estimate the two principal radii of curvature near the ridge. By symmetry of revolution, the principal directions are the radial and azimuthal ones. One of the principal radius of curvature, R_1, is the radius of curvature of the deformed meridian. It is smaller than the second radius R_2 and its order of magnitude is δ/α where δ is the width of the ridge, a length scale that should tend to zero as h tends to zero at fixed R. The other principal radius of curvature, R_2, is found as follows. Differential geometry of surfaces defines the principal curvature (the inverse of the sought *radius* of curvature R_2) as the length of the projection on to the surface of the curvature vector of any curve drawn on the surface that is tangent to the principal direction (see Section 11.5.1 for a detailed justification). The tangent to a parallel circle points to the azimuthal direction everywhere, and these circles can be used to compute R_2. Their 3D curvature vector points toward the axis and has length $1/r$, where r is the distance to axis (see Fig. 14.5). This vector has to be projected onto the surface (see vector labelled $1/R_2$ and $1/R_2'$ on the same figure). After projection, $1/R_2 = 1/(r \sin \alpha/2) \sim \alpha/r \sim e/r^2 \sim 1/R$. We recover Pogorelov's idea that, across the ridge, R_2 goes from $+R$ on the surface of the undeformed sphere to $-R$ inside the inverted cap. Therefore it remains of order R everywhere near the ridge.

Using these estimates for the principal radii of curvature, the Gauss curvature, defined as the product $(1/R_1)(1/R_2)$, is of order $\alpha/R\,\delta$. To estimate δ, the width of the ridge, one has to balance the stretching and bending energies: as noted earlier, the shape of this ridge results precisely from the optimum balance between the two effects. The bending energy per unit area is of order

$$Eh^3(\frac{1}{R_1})^2 \sim Eh^3\frac{\alpha^2}{\delta^2},$$

while the density of stretching energy is:

$$Eh\left(\Delta^{-1}(\frac{1}{R_1 R_2})\right)^2 \sim Eh\frac{\delta^4}{(\frac{\delta}{\alpha}R)^2}.$$

This expression is borrowed from our analysis of the stretching energy in the F.–von K. theory, see Chapter 6, equation (6.2). Using a plate theory to estimate the ridge energy of a spherical shell, as we do, is justified by the fact that Gauss curvature along the ridge is much larger than the curvature of the undistorted shell, i.e. $1/R^2$. Recall that we are only seeking order of magnitude estimates for now.

The two energies above (stretching and bending) depend on δ with different powers (bending like δ^{-2}, stretching like δ^2), therefore the optimum of the sum of the two energies is reached when δ is such that the two energies are equal (up to numerical constants irrelevant for this analysis). This happens for:

$$Eh^3\frac{\alpha^2}{\delta^2} \sim Eh\left(\frac{\alpha\delta}{R}\right)^2,$$

that is for $\delta \sim (hR)^{1/2}$. Putting this last result into the expression of the F.–von K. energy, one obtains the total elastic energy of the circular ridge:

$$\mathcal{E}_{\text{ridge}} \sim \frac{E\,h^{5/2}\,e^{3/2}}{R}. \tag{14.6}$$

Setting this energy to be equal to Fe, which is the work done by the pushing force, one obtains equation (14.5b), as announced.

14.4   Spherical shell pushed by a plane: overview

14.4.1   Outline

The case of a force applied through a flat plane has some similarities with that of a point force (Pogorelov's geometry) but with some important differences: there are now three regimes depending on the intensity of the pushing force, as discussed in the next section, Section 14.4.2. Unlike the case of Pogorelov, the transition from the flat configuration to the inverted one happens to be sub-critical, that is discontinuous: experimentally, there is an abrupt jump from one configuration to the other. In the rest of this chapter, we analyse rather thoroughly the contact of a spherical shell with a perfectly rigid plane. We shall arrive at a prediction for the dimple depth e and the size r of the contact in function of the applied force F (see Fig. 14.6). This gives us the opportunity to show how order of magnitude analyses such as those presented in the previous sections can be used as a starting point toward a fully quantitative analysis of an elasticity problem.

The main results that we shall obtain are outlined below (see Fig. 14.7).

At small forces, the contact between the sphere surface and the pushing plane takes place along a (flat) disc. By the same geometrical argument as earlier (Pogorelov's case of a point force) its radius r will be shown to be of order $r \sim (hR)^{1/2}$. The orders of magnitude pertinent to this weak-force limit remain roughly the same as in Pogorelov's geometry.

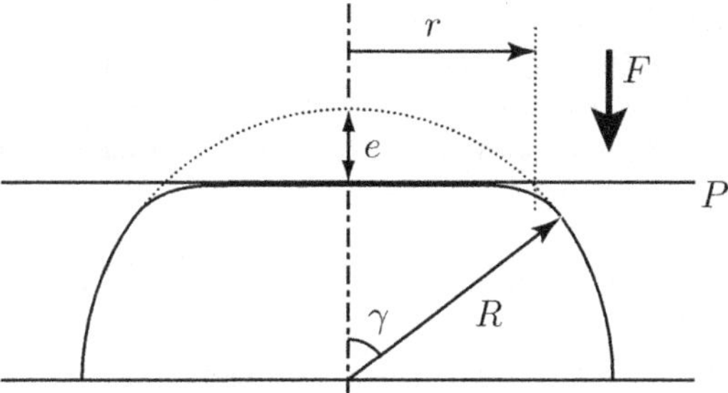

Fig. 14.6 Cross-section of the half-ball. A downward vertical force F is applied through a plane P. For small forces, the ball flattens, with a downward shift e at the top. e, α, R and r are the geometrical quantities entering into the estimation of the flattening. Our theory deals with thin shells ($h \ll R$), for which deformation is concentrated near the pole ($|\gamma| \ll 1$), although for the sake of clarity rather large values of γ are shown here.

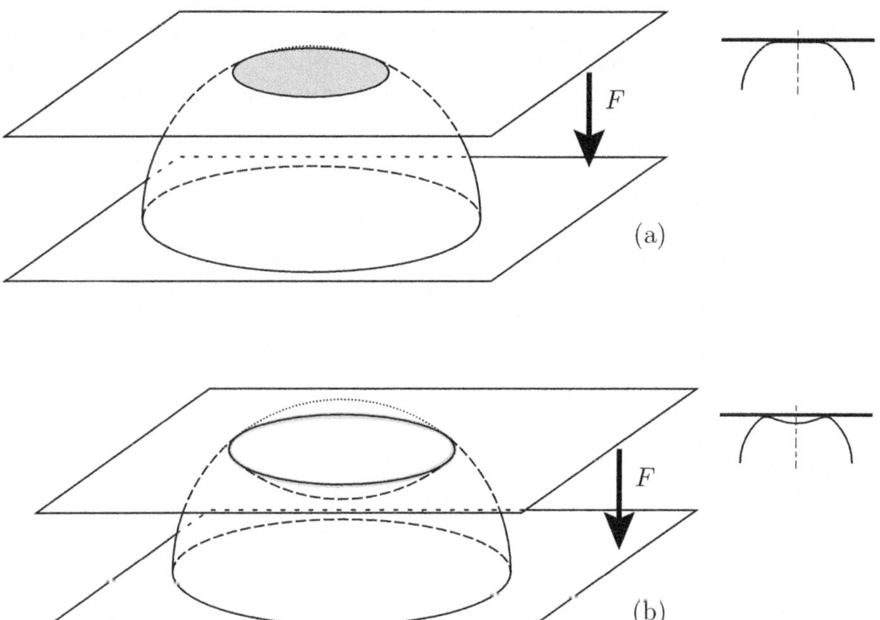

Fig. 14.7 Two configurations of the spherical shell, as observed in the experiments (see Fig. 14.1) are analysed in the rest of this chapter: at small squeezing forces F, the top of the thin sphere contacts the pushing plane along a flat disc (shaded area in a); beyond a critical squeezing force F, the top of the sphere bumps downward and contacts the plane along a circular ridge (b).

However, there is a difference: in Pogorelov's geometry, the value of the force can be scaled out, which ultimately yields a single parameterless problem for the determination of stress. On the other hand, a dimensionless parameter remains in the case of a flat pushing plane, which is the ratio of the applied force to the characteristic force $F_{\rm c} = Eh^3/R$. Therefore, instead of having one single mathematical problem to solve at the end, the dimensional reduction yields a one-parameter family of numerical equations. When this dimensionless force F/F_c becomes large, the stretching energy stored in the disc grows faster than the bending energy, until a critical value of the force is reached where the inverted cap configuration is energetically more beneficial than the flat disc of contact. This leads to the transition shown in the experiments (see Fig. 14.2, p. 491).

At larger forces, it becomes more beneficial to form an inverted cap near the pole separated from the outer part of the sphere by a ridge, as in Pogorelov's geometry. The elastic energy is again stored almost exclusively in the ridge. Contact with the pushing plane takes place along the circular ridge. From the viewpoint of membrane or shell theory, contact takes place along a singular circle of zero width, although a refined 3D analysis could be used to study the locus of contact, whose thickness is of order the thickness of the shell.

14.4.2 Order of magnitude estimates of the different regimes

In this section, we present in a unified way the three different regimes of flattening of a spherical shell by a pushing plate. The aim is to obtain, in a single sweep, all three regimes with the pertinent scaling laws. The net result is formulated in terms of the order of magnitude of the pushing force where the different regimes are relevant. The rest of this chapter is concerned with a detailed discussion of the equations and their solutions in those different limits.

We shall discuss three different possibilities: at very small loading the deformation of the shell is small enough to be insensitive to its finite thickness: a Hertz-like contact deforms the shell in a region too small to cross the thickness of the shell. At larger pushing forces, the shell still flattens, as in Hertz' contact, but along a disc of contact that is far larger than the thickness of the shell. This intermediate regime already poses a shell-like problem and the equilibrium situation involves a balance between bending and stretching. As the forces increases, the stretching energy due to the flattening of the sphere becomes too large and the optimal geometry becomes an inverted shell, with similarities and differences with respect to Pogorelov's problem.

Hertzian contact with a shell

When discussing Hertz' contact, we noticed that the extent of the region perturbed by the pushing is of the order of magnitude of the diameter of the disc of contact. According to the equation (14.4), the vertical displacement e is of order $\left(F^2/E^2R\right)^{1/3}$, where F is the pushing force. The radius r of the disc of contact is $(eR)^{1/2}$. This radius also defines the order of magnitude of the region perturbed inside the solid (three-dimensional) ball. Therefore the contact between the pushing plane and the shell is Hertz-like if this radius r is less than h, the thickness of the shell. A little algebra shows that, for this to happen, the pushing force has to be far less than a force $F_c \sim Eh^3/R$. When this is true, the shell is not different from a full elastic sphere from the viewpoint of elasticity theory.

Let us estimate the maximum incident velocity of a ping-pong ball, such that it bounces in this Hertz regime. As we did for Hertz' contact, we estimate this speed by writing Newton's law of dynamics during the bouncing. The elastic force is of the order of M, the mass of the ball, times the acceleration of the ball when it reverses the direction of its motion. In the present case, putting $r \sim h$, one obtains as an order of magnitude of the maximum possible speed for a Hertz-like contact

$$U_1 \sim \left(\frac{Eh^5}{MR^2} \right)^{1/2}.$$

Practically, taking $M = 3$ g, $E = 1$ GPa, $h = 0.5$ mm, $R = 1$ cm, one obtains the rather low speed $U_1 = 0.3$ m/s.

In the next two subsections we shall push harder and see what happens at higher impact velocities.

Disc-like contact between shell and plate

This situation is actually not so different from the one we just looked at. It is characterized by values of the force of order F_c, the upper bound for the range of applicability of Hertz-like contact. In this range of forces the spherical shell touches the plane along a flat disc and the stretching and bending energies have the same order of magnitude. The bending energy is found by analogy with the case of a bent plate, which is flat at equilibrium. Then the bending energy per unit area of the plate is $Eh^3 \left(1/R_1 + 1/R_2\right)^2$, where $(1/R_1 + 1/R_2)$ is the mean curvature of the surface. This mean-curvature can be seen as the difference between the equilibrium mean-curvature (zero for a plane) and the actual curvature in the bent state. Therefore, the flattening of a spherical shell that has curvature $2/R$ at equilibrium brings a density of bending energy that is proportional to the square of the difference between the mean-curvature in the rest state (namely $2/R$) and in the perturbed state (that is zero) with the same coefficient Eh^3 as for the F.–von K. bending energy of a plate. This bending energy per unit area is (up to numerical constants, irrelevant for the present developments) $Eh^3 \left(1/R\right)^2$.

The stretching energy per unit area of a shell of thickness h is $Eh \left(\partial u/\partial x\right)^2$, where $(\partial u/\partial x)$ is a general expression for the strain. As in Hertz' contact problem, the radius of the disc of contact, r, depends on radial displacement e by the geometrical relation: $r = (eR)^{1/2}$. The work done by the pushing force is of order of the product Fe and it will ultimately be assumed to be of the same order of magnitude as the elastic energy stored in the deformed shell. The material length along the flat disc has been changed—actually shortened—from $R \sin^{-1}(r/R)$ in the rest state with curvature $1/R$, to r. Therefore the typical displacement along the surface is the difference between these two length, which is of order r^3/R^2 (the result is proportional to the third power of r because one has to expand the $\sin^{-1}$ function for small values of its arguments, assuming $r \ll R$, to the first non-zero term, and because the contribution of the leading order linear contribution vanishes). This gives the order of magnitude of the numerator in the derivative $(\partial u/\partial x)$ for the strain. The denominator is itself of order r, the size of the perturbed domain on the shell surface. Therefore $(\partial u/\partial x) \sim r^2/R^2$. Balancing now the bending and stretching energy one obtains

$$r^2 \sim hR. \tag{14.7}$$

This shows that this regime is found when the flattening is of the same order of magnitude as the thickness h, since this flattening e is related to r and R in the same way as is h from equation (14.7). Therefore in this range of parameters, this force is of order of magnitude F_c. This domain is considered in details in Section 14.6 below, where the full equations of shell theory are solved. The stretching energy is a rapidly growing function of the radius r of the disc of contact. As the pushing force becomes larger than a value

F_1, another type of equilibrium is found that lowers the stretching energy, as discussed in the next subsection.

Contact along a circular ridge with inverted cap

As in the case studied by Pogorelov, the shell inverts itself and forms a dimple (hollow inverted cap), as a way of accommodating the increasing force applied by the pushing plane. From the point of view of the order of magnitude estimates, there are many analogies between this case and the one considered by Pogorelov. Up to a factor of two, the work done by the pushing force is the same in the two cases (this factor 2 comes from the fact that the point of application of the force is at the bottom of the inverted cap in Pogorelov's geometry, which is lower than the ridge).

The order of magnitude of the energy in the ridge is the same as in the case of Pogorelov, it is $\mathcal{E} \sim E\,e^{3/2}\,h^{5/2}/R$. Assuming that it is equal to the work done by the pushing force, one obtains the same relation between the force and the displacement e as in equation (14.5b): $F \sim E\,h^{5/2}e^{1/2}/R$. It is relevant at this step to compare the energy in the ridge and the bending energy coming from the inversion of the cap. This bending energy comes from the change of mean curvature from $+2/R$ at rest to $-2/R$ when inverted. The general expression for the bending energy is area times Eh^3 times the square of the change of mean curvature. This yields in the present case

$$\mathcal{E}_{\mathrm{b(cap)}} \sim Eh^3 r^2 \left(\frac{1}{R}\right)^2. \tag{14.8}$$

From this equation the bending energy of the inverted cap is of order Eh^3r^2/R^2. It is dominated, as expected, by the energy of the ridge, as soon as $hR \ll r$. It is important to notice that the cross-over between the regime of the flat disc of contact and of the inverted cap happens in the range where this inequality is marginally satisfied, namely when $r \sim (hR)^{1/2}$. Let us notice too that the pushing force becomes comparable to the one computed for the flat disc when $e \sim h$, as expected. A detailed analysis of this regime, based on the theory of shells is given in Section 14.7. It fully agrees with the order of magnitude estimation we have just given.

The transition from the flat disc to the inverted cap can occur on bouncing. The typical speed of a ping-pong ball for which this occurs is such that the kinetic energy of the ball is of the order of magnitude of the elastic energy in the borderline case. With the estimation above, this yields $1/2Mu^2 \sim Eh^4/R$. For a a ball with $R \approx 2$ cm, $h \approx 0.5$ mm, $E = 1$ GPa and $M = 3$ g, this gives a critical speed $u \approx 1$ m s^{-1}, very much in the range of the speed reached in the ping-pong game.

This completes the estimation of the order of magnitude relationship between the pushing force and the displacement e. This will be refined in the coming sections by solving explicitly the relevant equation of shell elasticity in the various ranges of parameters that have just been found.

14.5   Equation for spherical shells

In this section, we specialize the general equations for shells of revolution derived in Chapter 12 to the problem of a spherical shell that is compressed by a rigid plane. Leaving aside the non-axisymmetric buckling that can occur at higher stresses (see Section 14.8.6

and Fig. 14.24 at the end of this chapter), we carry out all the simplifications that can be done under the assumption of deformations localized near the apex, when the dimple size is much smaller than the radius of the shell (shallow shell equations).

14.5.1   Kinematics

In these general equations for shells of revolution, the spherical reference shape is specified by:

$$\rho(s) = R \sin \frac{s}{R}, \quad z(s) = R \cos \frac{s}{R}. \tag{14.9}$$

The derivatives of these functions read $\rho'(s) = \cos s/R$ and $z'(s) = -\sin s/R$. Identification with equation (12.2) defining the orientation of the undeformed tangent $\gamma(s)$ yields

$$\gamma(s) = -\frac{s}{R}, \tag{14.10}$$

the minus sign being a consequence of our geometrical conventions.

We will be concerned with the deformations near the apex where the force is applied. In this region, the slope is small and the trigonometric functions appearing in the equations for shells can be linearized; still, some important non-linear terms associated with buckling will be retained. These approximations lead to the so-called *shallow* shell equations of Reissner (E. Reissner, 1947a,b). In the neighbourhood of the apex,

$$s \ll R, \tag{14.11}$$

where $|\gamma| \ll 1$, one can use the approximations

$$\rho(s) \approx s, \quad \rho'(s) \approx 1, \quad z'(s) \approx -\frac{s}{R}. \tag{14.12}$$

Similar approximations were introduced in Chapter 13 for the analysis of the elastic torus near its crowns.

The unknowns of the elasticity problem are the radial and axial components of the displacements, $v_r(s)$ and $v_z(s)$. As everywhere in the present chapter, we will use the notation of Chapter 12, which is recalled in Fig. 14.8. The rotation ϕ of the tangent is

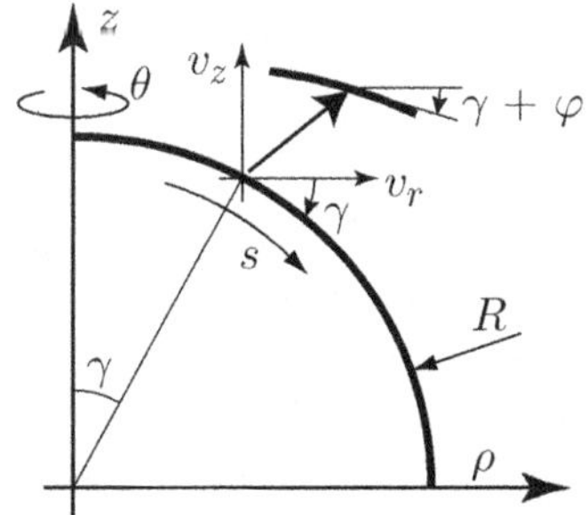

Fig. 14.8 Conventions used for the analysis of a thin spherical shell of revolution. Note that the angles γ and $(\gamma + \phi)$ are negative on the figure as the arrows point in the clockwise (anti-trigonometric) direction.

given by equation (12.9) in terms of the displacement; near the pole, this definition takes the simplified form:

$$\phi(s) = v'_z(s). \tag{14.13}$$

This equation will be used to eliminate v_z in favour of ϕ later on.

For axisymmetric deformations, the shear component $e_{s\theta}$ of the strain vanishes ($e_{s\theta} = 0$). The two remaining components e_{ss} and $e_{\theta\theta}$, which are simply denoted e_s and e_θ, were given in equation (12.11b). With the approximations (14.12) holding near the pole, the definition of membrane strain reads

$$e_s(s) = v'_r(s) - \frac{s}{R}\,\phi(s) + \frac{\phi^2(s)}{2}, \tag{14.14a}$$

$$e_\theta(s) = \frac{v_r(s)}{s}. \tag{14.14b}$$

These expressions were obtained under the assumption of small strain and moderate rotations, as explained in Chapter 12. Primes are used to denote derivatives of a function with respect to its argument, the arc length s in the present case.

14.5.2 Constitutive relations

The constitutive relations connecting membrane stress and strain were given in equation (12.21). Here it will be convenient to invert them, i.e. to express the strain as a function of the stress:

$$e_s = \frac{1}{E\,h}\,(N_s - \nu\,N_\theta) \tag{14.15a}$$

$$e_\theta = \frac{1}{E\,h}\,(N_\theta - \nu\,N_s). \tag{14.15b}$$

14.5.3 Mechanical equilibrium

Near the pole, the approximations (14.12) allow one to simplify the expressions (12.27) for the deformed tangent as follows:

$$T_r(s) = 1 \tag{14.16a}$$

$$T_z(s) = -\frac{s}{R} + \phi(s). \tag{14.16b}$$

These equations are similar to equations (13.22) used in the analysis of the elastic torus near its crowns in Chapter 13. They were derived assuming small strain and moderate rotations. In particular note that we retain some dependence of the tangent on the deformation through the unknown contribution ϕ in equation (14.16b); in contrast, a linear analysis would drop this term and identify the reference and deformed configurations of the tangent for the purpose of writing down the equation of equilibrium.

In the vicinity of the pole, $|\gamma \ll 1|$, our simplified account of bending is applicable. We shall use this simplified framework to study the buckling of a spherical shell since this buckling is localized near the pole for moderate forces. Plugging the expressions (14.16) above for the deformed tangent into the equations of equilibrium for shells derived in equations (12.39), we have

$$\left(\frac{1}{s} \frac{\mathrm{d}(s\, N_s(s))}{\mathrm{d}s} - \frac{N_\theta(s)}{s} \right) + f_r(s) = 0 \tag{14.17a}$$

$$\frac{1}{s} \frac{\mathrm{d}}{\mathrm{d}s} \left(s\, N_s(s) \left(-\frac{s}{R} + \phi(s) \right) \right) - \frac{D}{s} \frac{\mathrm{d}}{\mathrm{d}s} \left(\frac{\mathrm{d}(s\, \phi'(s))}{\mathrm{d}s} - \frac{\phi(s)}{s} \right) + f_z(s) = 0. \tag{14.17b}$$

Here, we have expressed the bending force f_b using equation (12.41), which yields the term proportional to the bending modulus D in the above equation. This bending modulus was defined earlier as

$$D = \frac{E\, h^3}{12\,(1 - \nu^2)}.$$

We shall assume a frictionless contact: the reaction force is exactly normal to the pushing plane,

$$f_r(s) = 0. \tag{14.18}$$

As usual in contact problems, the expression of the normal reaction force f_z is not known in advance, and will be determined as part of the solution of the problem—in particular, it has to be found in a manner that is self-consistent with the yet-unknown locus of contact.

14.5.4 Equations for membrane stress and rotation

In this section, we rewrite the equations by eliminating some of the unknowns in favour of two main unknowns, the membrane stress $N_s(s)$ and the rotation $\phi(s)$. These equations will later be cast in dimensionless form and the resulting equations will serve as a basis for the analysis of the contact of a thin elastic shell with a rigid plane.

The strain e_θ can be eliminated by combining equations (14.14b) and (14.15b):

$$v_r(s) = \frac{s}{E\, h}\, (N_\theta - \nu\, N_s). \tag{14.19}$$

Plugging this into the right-hand side of equation (14.14a), and eliminating e_s in the left-hand side using the constitutive relation (14.15a) we obtain

$$\frac{N_s - \nu\, N_\theta}{E\, h} = \frac{\mathrm{d}}{\mathrm{d}s} \left(\frac{s}{E\, h}\, (N_\theta - \nu\, N_s) \right) - \frac{s\, \phi(s)}{R} + \frac{\phi^2(s)}{2}.$$

In this equation, we can eliminate N_θ using equation (14.17a) for radial equilibrium, recalling that $f_r(s) = 0$:

$$N_\theta(s) = N_s(s) + s\, N_s'(s). \tag{14.20}$$

After a little algebra, this yields a first equation relating the unknown quantities N_s and ϕ:

$$\frac{3\, s\, N_s'(s) + s^2\, N_s''(s)}{E\, h} - \frac{s\, \phi(s)}{R} + \frac{\phi^2(s)}{2} = 0. \tag{14.21}$$

A second equation for the same set of unknowns N_s and ϕ is provided by equation (14.17b) for axial equilibrium. This equation can be integrated by introducing the total reaction force $F(s)$ integrated over a disc of radius s, defined as

$$F(s) = -\int_0^s f_z(s')\, 2\pi\, s'\, \mathrm{d}s', \tag{14.22}$$

where $\mathrm{d}a = 2\pi\, s'\, \mathrm{d}s'$ is the area element in cylindrical coordinates. A minus sign has been included by convention, making $F(s)$ positive when the ball is pushed downwards, as happens in the current geometry. From equation (14.22), we have

$$f_z(s) = -\frac{1}{s}\frac{\mathrm{d}}{\mathrm{d}s}\left(\frac{F(s)}{2\pi}\right). \tag{14.23}$$

Plugging this into equation (14.17b) and integrating with respect to s, we obtain

$$s\, N_s(s)\left(-\frac{s}{R} + \phi(s)\right) - D\left(\frac{\mathrm{d}(s\,\phi'(s))}{\mathrm{d}s} - \frac{\phi(s)}{s}\right) = \frac{F(s)}{2\pi}. \tag{14.24}$$

There is no need to add a constant of integration in this expression as it can be checked that both[2] sides of the equation go to 0 for $s \to 0$.

This set of equations (14.21) and (14.24) describe the equilibrium of a spherical shell undergoing a localized axisymmetric perturbation near its pole—recall that by 'pole' we mean its intersection with the z axis of the polar coordinates. Before solving these equations, we shall work out the associated boundary conditions in Section 14.5.5, and then write them in dimensionless form in Section 14.5.6. This makes a small parameter, ϵ, appear, which provides a way of solving the equations by asymptotic methods.

14.5.5 Edge of a region with contact

Later on in the analysis, we shall encounter forces localized along a circle—as we will show, such a concentration of contact force takes place both in the case of disc-like and of circular contact, at the edge of the region of contact. Here, we work out the jump conditions holding near such a circle. To do this, we return to the variational formulation of the equations for shells in Chapter 12: as earlier in the analysis of the Elastica pushed on to a plane in Section 9.2.2, we derive the jump conditions near a singular point directly from the condition of stationarity of the energy.

Let us assume that a vertical downward force is applied along the circle $s = r$ with radius r, and let F_r be the total magnitude of the singular force (integrated over the circle):

$$f_r(s) = -\frac{F_r}{2\pi\, r}\, \delta_D(s - r), \tag{14.25}$$

where δ_D denotes the Dirac distribution. The minus sign is such that a positive force F_r pushes downwards (namely towards the centre of the sphere), and the normalization factor such that the integrated force $\iint f_r(s)\,\mathrm{d}a$ is equal to $(-F_r)$ as announced. The potential energy associated with this force reads $\mathcal{E}_{\mathrm{p}} = F_r\, v_z(r)$, and its first variation is

$$\delta\mathcal{E}_{\mathrm{p}} = F_r\, \delta v_z(r). \tag{14.26}$$

The variation of the elastic energy of the shell on a domain $s_1 \le s \le s_2$ has been given in equation (12.42) as

$$\delta\mathcal{E}_{\mathrm{s}} + \delta\mathcal{E}_{\mathrm{b}} = \left[\delta B\right]_{s_1}^{s_2} + \cdots \tag{14.27a}$$

---

[2] The right-hand side vanishes at $s = 0$ by our particular choice of bounds in the integral defining $F(s)$ in equation (14.22). The first term in the left-hand side clearly goes to zero for $s \to 0$. The second term goes to zero too, as it can be identified with $(\rho(s)\, q(s))$ using equation (12.38), and this quantity appears to be the coefficient of $\delta v_z(0)$ in the boundary term coming from equation (12.37): in the absence of axial point force applied at the pole, this coefficient vanishes by the condition of equilibrium.

where the ellipsis stands for a contribution coming from an integral contribution from the interior of the domain, which will not be needed here. The detailed expression for the boundary term has been given earlier:

$$\delta B = 2\pi \rho \left(N_s\, T_r\, \delta v_r + (N_s\, T_z + q)\, \delta v_z + M_s\, \delta v_z' \right). \tag{14.27b}$$

The brackets with subscript and superscript in the right-hand side of equation (14.27a) stand for the variation coming from integration by parts:

$$\left[g \right]_{s_1}^{s_2} = g(s_2) - g(s_1). \tag{14.27c}$$

The variation in equation (14.27a) was calculated under the assumption that all functions are smooth in the domain $s_1 \le s \le s_2$. Here, some functions might be discontinuous at $s = r$, where the singularity takes place. In consequence, we have to split the domain into smooth sub-domains prior to computing the variation of elastic energy: the energy is the sum of a contribution (A) accounting for the strain of the inner part of the shell $s < r$ plus another contribution (B) for the outer part of the shell $s > r$, plus the potential energy associated with the force localized at $s = r$. The total variation to be considered reads

$$\delta \mathcal{E}_t = (\delta \mathcal{E}_s + \delta \mathcal{E}_b)^A + (\delta \mathcal{E}_s + \delta \mathcal{E}_b)^B + \delta \mathcal{E}_p. \tag{14.28}$$

Applying now equation (14.27a) to the sub-domain A, whose upper bound is the point r approached from the left-hand side, $s_2^A = r^-$, we have

$$(\delta \mathcal{E}_s + \delta \mathcal{E}_b)^A = \delta B(r^-) + \cdots,$$

where the ellipsis stands for boundary terms coming from points other than r, and for integral terms that we do not consider here. Similarly, the right-hand side of the point r defines the lower bound of domain B, $s_1^B = r^+$, and so

$$(\delta \mathcal{E}_s + \delta \mathcal{E}_b)^B = -\delta B(r^+) + \cdots$$

Plugging this into equation (14.28) and using the expression (14.26) for $\delta \mathcal{E}_p$, we find the following terms relevant to the point r in the variation:

$$\delta \mathcal{E}_t = \delta B(r^-) - \delta B(r^+) + F_r\, \delta v_z(r) + \cdots,$$

where again the ellipsis stands for terms coming from other boundaries and from integral contributions, which are not needed here.

Using now the explicit form of δB provided in equation (14.27b), we have

$$\delta \mathcal{E}_t = -2\pi\, r\, [\![N_s\, T_r]\!]_r\, \delta v_r(r) + \left(-2\pi\, r\, [\![N_s\, T_z + q]\!]_r + F_r \right) \delta v_z(r)$$

$$-2\pi\, r\, [\![M_s]\!]_r\, \delta v_z'(r) + \cdots \tag{14.29}$$

Here, we have used the classical double-bracket with subscript r for the jump of a function across the singularity at $s = r$:

$$[\![g]\!]_r = g(r^+) - g(r^-). \tag{14.30}$$

In equation (14.29) above, we have implicitly assumed that the functions v_r, v_z and $\phi = v_z'$ are smooth at r:

$$[\![v_r]\!]_r = 0, \quad [\![v_z]\!]_r = 0, \quad [\![\phi]\!]_r = 0, \tag{14.31}$$

which makes it licit to write expressions such as $v_r(r)$ in equation (14.29) above. Those continuity assumptions are natural: continuity of the displacement means that a cut is not allowed; continuity of $\phi = v_z'$ warrants that the deformed tangent varies smoothly across the circle $s = r$ and prevents the formation of a sharp fold with infinite bending energy.

The continuity conditions that we are seeking can now be obtained by cancelling the variation $\delta\mathcal{E}_t = 0$ in equation (14.29) for arbitrary localized perturbations $\delta v_r(r)$, $\delta v_z(r)$ and $\delta v_z'(r)$, as required at equilibrium. This corresponding coefficients must therefore vanish:

$$[\![N_s\,T_r]\!]_r = 0, \quad [\![N_s\,T_z + q]\!]_r = \frac{F_r}{2\pi\,r}, \quad [\![M_s]\!]_r = 0. \tag{14.32}$$

Since $T_r = 1$ is continuous everywhere near the pole, the first equality implies continuity of the membrane stress N_s across the circle of application of the localized force:

$$[\![N_s]\!]_r = 0.$$

In equation (14.19), v_r and N_s are known to be continuous already and so N_θ must be continuous too:

$$[\![N_\theta]\!]_r = 0.$$

By a similar argument, note that in equation (14.20) both N_s and N_θ have been shown to be continuous; this implies that N_s' is continuous too across the singularity:

$$[\![N_s']\!]_r = 0.$$

Plugging the definition of M_s derived in Chapter 12—equations (12.24) and (12.15)—and using its continuity condition (14.32), we find another continuity condition:

$$[\![\phi']\!]_r = 0.$$

It remains to use the second equality in equation (14.32), which contains two terms. The term $N_s\,T_z$ is continuous. The other term q was defined in equation (12.38); expanding the derivatives in this definition and using the continuity conditions available so far, we find

$$[\![q]\!]_r = -D\,[\![\phi'']\!]_r.$$

Combining with equation (14.32), this yields

$$[\![\phi'']\!]_r = -\frac{F_r}{2\pi\,D\,r}.$$

To summarize, assuming the continuity of displacement and slope near a circle where a localized force is imposed, see equation (14.31), we have derived the following conditions:

$$[\![N_s(s)]\!]_r = 0, \tag{14.33a}$$

$$[\![N_s'(s)]\!]_r = 0, \tag{14.33b}$$

$$[\![\phi'(s)]\!]_r = 0, \tag{14.33c}$$

$$[\![\phi''(s)]\!]_r = -\frac{F_r}{2\pi\,D\,r}. \tag{14.33d}$$

It should be recalled that F_r is the total intensity of a localized force applied along the parallel circle of radius r.

Alternatively, the jump condition for ϕ'' can be derived by integration of equation (14.17b) for axial equilibrium across the singularity, assuming the continuity of $\phi(s)$ and $\phi'(s)$ at $s = r$.

14.5.6 Dimensionless equations

To write the equations obtained so far in dimensionless form, we use the typical curvilinear length s^*, the typical angle ϕ^*, the typical membrane stress N_s^* and the typical force F^* built using the constants $(E\,h)$ and R:

$$s^* = R \tag{14.34a}$$

$$\phi^* = 1 \tag{14.34b}$$

$$N_s^* = E\,h \tag{14.34c}$$

$$F^* = E\,h\,R. \tag{14.34d}$$

Here, we follow the notation of Section 1.2.4 whereby a star denotes new units used to define dimensionless quantities.

Based on this set of units, we introduce the rescaled curvilinear coordinate x, the rescaled radius of the region of contact X, the rescaled rotation $\Psi(x)$, the rescaled membrane stress $\Sigma(x)$ and the rescaled integrated force of contact $N(x)$ as:

$$x = \frac{s}{R} \tag{14.34e}$$

$$X = \frac{r}{R} \tag{14.34f}$$

$$\Psi(x) = -\phi(s) \tag{14.34g}$$

$$\Sigma(x) = \frac{N_s(s)}{E\,h} \tag{14.34h}$$

$$N(x) = \frac{F(s)}{E\,h\,R}. \tag{14.34i}$$

Note the convention to include a minus sign in the rescaled rotation $\Psi(x)$, which will introduce some changes in the signs when the equations written in terms of ϕ will be rewritten in terms of Ψ. We emphasize that these rescaled quantities will not all be of order one as a small parameter ϵ remains in the rescaled equations, see below—this is consistent as the membrane stress, rescaled using $E\,h$ here, must be very small for the assumption of small strain (Hookean elasticity) to remain valid. Later on, we shall introduce improved scaling laws where all quantities will actually be of order 1.

In terms of these rescaled variables, equations (14.21) and (14.24) derived in Section 14.5.4 become

$$3\,x\,\Sigma'(x) + x^2\,\Sigma''(x) + x\,\Psi(x) + \frac{\Psi^2(x)}{2} = 0 \tag{14.35a}$$

and

$$x\,\Sigma(x)\,(x + \Psi(x)) - \epsilon^2 \left(\frac{\mathrm{d}(x\,\Psi'(x))}{\mathrm{d}x} - \frac{\Psi(x)}{x} \right) = -\frac{N(x)}{2\pi}, \tag{14.35b}$$

where the quantity $N(x)$, written without subscript, is the rescaled integrated force of contact $F(s)$, applied by pushing a plane over a circular cap of radius $s = R\,x$. In the equation above, a small parameter ϵ has appeared, which is defined by

$$\epsilon^2 = \frac{D}{E\,h\,R^2} = \frac{h^2}{12\,(1 - \nu^2)\,R^2}. \tag{14.36}$$

This parameter ϵ is of order of the aspect ratio of the shell h/R. What follows is a study of solutions of equations (14.35) in the limit of small ϵ, relevant for thin shells.

The jump and continuity conditions across a circle where a concentrated force is applied have been worked out in Section 14.5.5. In terms of the rescaled quantities they are:

$$[\![\Sigma(x)]\!]_X = 0, \tag{14.37a}$$

$$[\![\Sigma'(x)]\!]_X = 0, \tag{14.37b}$$

$$[\![\Psi(x)]\!]_X = 0, \tag{14.37c}$$

$$[\![\Psi'(x)]\!]_X = 0, \tag{14.37d}$$

$$[\![\Psi''(x)]\!]_X = \frac{N_X}{2\pi\,\epsilon^2\,X}, \tag{14.37e}$$

as shown by applying our rescalings to equations (14.33). The radius of the circle where the concentrated force is applied and the total intensity of the force are denoted X and N_X in rescaled variables, respectively, and their physical value is $r = R\,X$ and $F_r = E\,h\,R\,N_X$, see equations (14.34).

In the coming sections, we go on to solve the set of equations (14.35).

14.6   Spherical shell pushed by a plane: disc-like contact

Deformations of spherical shells subjected to point forces, or to forces applied over a disc-like or annular regions have been studied by many authors (see references at the end of Section 14.1). The particular case of a spherical shell pushed upon by rigid walls, which we consider here, was investigated by Updike and Kalnins Kalnins D. P. Updike and A.Kalnins (1970, 1972a) in part with the purpose of rationalizing tonometric measurements.[3] The analysis presented in the present section is based on these papers.

As discussed earlier, two topologies of contact can be expected during the compression of a spherical shell by a rigid plane. We first consider the case of a disc of contact (see Fig. 14.1, left), which occurs for small forces. More accurately, the force F is small in the sense that, when measured in units EhR, it must be of the order of the small dimensionless number ϵ raised to some positive power—the power law is derived below.

The solution with a disc of contact involves two regions, which are dealt with separately: first the region of contact, corresponding to the range of values of s such that $0 < s < r$,

---

[3] The 'applanation tonometry' is a method of investigation of the inner pressure inside an human eye (intra-ocular pressure) by pushing a flat surface against it. It requires one to discriminate between the response of the shell surrounding the eye and the one of the liquid inside. This can be done, at least partially, by looking at the distribution of pressure along the disc of contact.

second the outer free part of the shell, $s > r$. The radius[4] of the disc of contact, r, will be determined using the boundary conditions (14.37) at the edge of the region of contact.

14.6.1   Solution in the interior of the disc of contact

For $0 < s < r$, the angle $(\gamma + \phi)$ measuring the direction of the tangent along a deformed meridian has to vanish since this tangent is parallel to the pushing plane. By equation (14.10), $\gamma = -s/R$ and so $\phi = s/R$ for $0 \leq s \leq r$. In terms of the rescaled angle Ψ defined in equation (14.34g), contact with the plane becomes

$$\Psi(x) = -x \quad \text{for } 0 \leq x \leq X, \tag{14.38}$$

where $X = r/R$ is the rescaled radius of the disc of contact.

The other unknown $\Sigma(x)$ can be found by inserting this expression for $\Psi(x)$ into equation (14.35a) for the radial forces. This yields the linear ordinary differential equation

$$3x\,\Sigma'(x) + x^2\,\Sigma''(x) = \frac{x^2}{2} \tag{14.39}$$

whose general solution is:

$$\Sigma(x) = \frac{x^2}{16} + C - \frac{C_1}{2\,x^2}, \tag{14.40}$$

where C and C_1 are two constants of integration. The stress singularity at the pole $x = 0$ is due to our choice of cylindrical parameters. This singularity of the stress is removed by setting $C_1 = 0$:

$$\Sigma(x) = \frac{x^2}{16} + C. \tag{14.41}$$

The remaining constant C will be determined later by matching this solution with the solution in the free region.

The other equation (14.35b) has not yet been used. Plugging the solution (14.38) for the rotation angle $\Psi(x)$ into it, we have

$$N(x) = 0 \quad \text{for } 0 \leq x \leq X.$$

The integrated force vanishes in physical units too, $F(s) = 0$ for $0 \leq s \leq r$, and equation (14.23) shows that the contact pressure is zero along the disc of contact:

$$f_z(s) = 0 \quad \text{for } 0 \leq s < r.$$

Therefore, the contact pressure is zero in the interior of the disc of contact (F. Essenburg, 1960).

However, the total reaction force is non-zero as the contact pressure concentrates along the *edge* of the disc of contact. This situation is very similar to that discussed in Chapter 9 for the d-cone pressed against a cylinder, or the Elastica pressed against a plane, two examples wherein singular contact forces build up at the edge of regions of contact. Such a concentration of the pressure in contact problems has been pointed out by several authors

---

[4] Our notation is consistent as we will show that there is a concentrated force of contact applied by the plane on the shell along the circular edge of the region of contact, which is of radius r.

(F. Essenburg, 1960; D. P. Updike and A. Kalnins, 1970). Because of this concentration, the pressure of contact predicted by the thin shell theory is infinite along the edge of the contact region. This infinite pressure can be removed by using a three-dimensional description of the shell in a neighbourhood of the edge of the disc of contact, or by considering the effect of transverse shear stress in a refined shell theory (F. Essenburg, 1960; D. P. Updike and A. Kalnins, 1972). Divergences are then regularized over a typical length of order of the shell thickness, h. Another regularizing mechanism, which has seemingly not been discussed so far, is the action of molecular forces, such as van der Waals forces, between the plate and substrate at the edge of the region of contact. Simple estimates show that this regularizes the pressure of contact over a typical length given by $(a\,R^2)^{1/3}$, where a is the range of the molecular forces, typically 10^{-9} m (one nanometre). Depending on the relative magnitude of h and $(a\,R^2)^{1/3}$—which may vary since $a \ll h \ll R$—any effect can be dominant. Note, however, that this kind or regularization has only a weak influence on the macroscopic quantities like r, F, etc. Therefore, we shall simply accept Dirac distributions of the pressure of contact in the following. In this framework, the contact pressure is concentrated along a circle of radius r:

$$f_z(s) = -\frac{F}{2\pi\,r}\,\delta_D(s-r), \tag{14.42}$$

where δ_D represents the Dirac distribution, and $F = \iint(-f_z)\,da$ is the resultant of the force of contact, a free parameter of the problem. By equation (14.22), the integrated force $F(s)$ is

$$F(s) = F\,H(s-r), \tag{14.43}$$

where $H(s)$ stands for the Heaviside step function, defined by

$$H(u) = \begin{cases} 0 & \text{if } u < 0 \\ 1 & \text{if } u > 0 \end{cases}. \tag{14.44}$$

Note that the function $F(s)$ and the constant F (without argument) refer to different entities.

In terms of rescaled variables, equation (14.43) becomes

$$N(x) = N\,H(x-X), \tag{14.45}$$

where again the constant N (total reaction force) and the function $N(x)$ (internal force) are different entities. The piecewise constant function $N(s)$ is 0 inside the disc of contact (for $x < X$), and N outside (for $x > X$).

The edge of the disc of contact will be considered after the analysis of the free part of the shell: the corresponding boundary conditions will be used to match the solutions obtained in the region of contact and in the free part of the shell.

14.6.2   Equations for the free part

The free part of the shell is located outside the area of contact, and corresponds to $s > r$. As we shall see, the solution of the shell equations in this free part describes a narrow ridge connecting the flat disc of contact to the unperturbed part of the shell, far from the region of contact.

In the free region, the tangent rotation $\Psi(x)$ is no longer known in advance, and one has to determine the two unknown functions $\Sigma(x)$ and $\Psi(x)$. However, we know that there is no contact and the contact pressure has to vanish, see equation (14.42):

$$f_z(s) = 0 \quad \text{for } s > r.$$

This implies that the integrated force of contact $F(s)$ and its rescaled equivalent $N(X)$ are independent of their argument in the free region, see equation (14.45):

$$N(X) = N \quad \text{for } x > X,$$

where the constant N in the right-hand side denotes the rescaled, total force of contact and will be determined as an outcome of the analysis.

Before solving the set of equations (14.35), we shall introduce a secondary rescaling, which allows us to remove the small parameter ϵ from the equations. Since the slope of the deformed tangent $(x + \Psi(x))$ is zero on the left boundary $x = X$ of the domain, $\Psi(X) = -X$, we shall assume the scaling $\Psi \sim x \sim X$. In addition, balancing the first two and last two terms in equation (14.35a), we have $\Sigma \sim x^2$; balancing the terms in the other equation yields $x\,\Sigma \sim \epsilon^2 \sim N$. Combining these relations, we have $x \sim \epsilon^{1/2}$, $X \sim \epsilon^{1/2}$, $\Sigma \sim \epsilon$, $\Psi \sim \epsilon^{1/2}$, $N \sim \epsilon^2$. This suggests the following secondary rescaling:

$$\hat{x} = \frac{x}{\epsilon^{1/2}} \tag{14.46a}$$

$$\hat{X} = \frac{X}{\epsilon^{1/2}} \tag{14.46b}$$

$$\hat{\Sigma}(\hat{x}) = \frac{\Sigma(x)}{\epsilon} \tag{14.46c}$$

$$\hat{\Psi}(\hat{x}) = \frac{\Psi(x)}{\epsilon^{1/2}} \tag{14.46d}$$

$$\hat{N} = \frac{N}{\epsilon^2}, \tag{14.46e}$$

where it should be noted that functions that have a hat depend on $\hat{x}$ and not on x.

In terms of these new variables, equations (14.35b) and (14.35a) take the form

$$3\,\hat{x}\,\hat{\Sigma}'(x) + \hat{x}^2\,\hat{\Sigma}''(\hat{x}) + \hat{x}\,\hat{\Psi}(\hat{x}) + \frac{\hat{\Psi}^2(\hat{x})}{2} - 0, \tag{14.47a}$$

$$\hat{x}\,\hat{\Sigma}(\hat{x})\,(\hat{x} + \hat{\Psi}(\hat{x})) - \left[\frac{\mathrm{d}(\hat{x}\,\hat{\Psi}'(\hat{x}))}{\mathrm{d}\hat{x}} - \frac{\hat{\Psi}(\hat{x})}{\hat{x}} \right] = -\frac{\hat{N}}{2\pi}, \tag{14.47b}$$

where the small parameter ϵ is gone. These equations hold in the free part, $\hat{x} > \hat{X}$.

This set of coupled, non-linear ordinary differential equations is subjected to boundary conditions. At the merging with the disc of contact, $\hat{x} = \hat{X}$, some boundary conditions follow from the continuity conditions (14.37): since $\Sigma(x)$ and $\Psi(x)$ are continuous with first derivatives across the singular circle $x = X$, the values of these functions and their derivatives have to match those imposed by the solution in the interior of the disc of contact, given in equations (14.38) and (14.41):

$$\hat{\Sigma}(\hat{X}) = \hat{C} + \frac{\hat{X}^2}{16} \tag{14.48a}$$

$$\hat{\Sigma}'(\hat{X}) = \frac{\hat{X}}{8} \tag{14.48b}$$

$$\hat{\Psi}(\hat{X}) = -\hat{X} \tag{14.48c}$$

$$\hat{\Psi}'(\hat{X}) = -1. \tag{14.48d}$$

Here $\hat{C}$ is still an arbitrary constant, related to the original constant C in equation (14.41) by the same rescaling as Σ:

$$\hat{C} = \frac{C}{\epsilon}.$$

Boundary conditions (14.48) will be complemented by asymptotic conditions for $\hat{X} \to +\infty$ describing the merging of the bump with the rest of the shell, which is unperturbed far away from the region of contact. These asymptotic conditions are worked out in the next section.

In Section 14.6.4, we shall proceed to the numerical solution of the differential equations (14.47) together with the boundary conditions (14.48) and the relevant asymptotic conditions. To do so, we will have to find the values of the two auxiliary unknowns (the scaled radius $\hat{X}$ of the disc of contact and the constant $\hat{C}$), as a function of the applied force $\hat{N}$.

Before proceeding to this numerical solution, we point out that the new rescaling introduced in equation (14.46) is equivalent to Pogorelov's dimensional analysis. To show this, we combine this rescaling with the first rescaling (14.34), so as to revert to the physical quantities. For instance, the physical value of the pushing force is

$$F = E\,h\,R\,N = E\,h\,R\,\epsilon^2\,\hat{N} = \frac{E\,h^3}{R}\,\frac{\hat{N}}{12\,(1-\nu^2)}. \tag{14.49}$$

Since the equations involve no large or small parameter when expressed in terms of hat variables, we anticipate that their solution will be of order 1. The second factor in the right-hand side of equation above is then of order 1, and $F \sim E\,h^3/R$ which is Pogorelov's scaling (14.5a). Similarly, one can show $e \sim v_z \sim h$, something that is left as an exercise to the reader.

14.6.3 Asymptotic conditions far away from the ridge

The asymptotic conditions holding for $\hat{x} \to +\infty$ are investigated. This limit corresponds to distances that are much larger than 1 in rescaled variables, i.e. much larger than the radius of the disc of contact, $R\,\epsilon^{1/2} \sim (h\,R)^{1/2}$ in physical variables. Since $(h\,R)^{1/2} \ll R$, this limit is still consistent with our assumption of a perturbation localized near the apex of the shell, where the geometrical approximations (14.12) are valid: the asymptotic conditions derived below hold in the intermediate region $(h\,R)^{1/2} \ll s \ll R$.

Far away from the region of contact, the shape of the shell must become insensitive to the details of the geometry of contact, and it can feel the resultant[5] of the contact force only,

---

[5] This is known as Saint-Venant's principle.

which is F in physical variables and $\hat{N}$ in rescaled variables. The linearized membrane theory, which neglects both bending effects and nonlinearity associated with moderate rotations, must be applicable for $\hat{x} \gg 1$. As a result, the square brackets in equation (14.47b) must be negligible for large $\hat{x}$ (bending terms), and the term $\hat{\Psi}(\hat{x})$ must be negligible in front of $\hat{x}$ (non-linear membrane effects). Noticing that the following solution cancels both these bending and non-linear terms exactly:

$$\hat{\Sigma}_0(\hat{x}) = -\frac{\hat{N}}{2\pi\,\hat{x}^2}, \qquad \hat{\Psi}_0(\hat{x}) = 0, \tag{14.50}$$

we shall assume that this is the asymptotic behaviour of the actual solution $(\hat{\Sigma}, \hat{\Phi})$. In other terms, one can write

$$\hat{\Sigma}(\hat{x}) = \hat{\Sigma}_0(\hat{x}) + \hat{\Sigma}_1(\hat{x}), \qquad \hat{\Psi}(\hat{x}) = \hat{\Psi}_0(\hat{x}) + \hat{\Psi}_1(\hat{x}),$$

where the correction $(\hat{\Sigma}_1, \hat{\Psi}_1)$ is very small for large values of $\hat{x}$.

Plugging this expansion into the differential equations (14.47) and linearizing with respect to $\hat{\Sigma}_1$ and $\hat{\Psi}_1$, we obtain the following equations:

$$3\,\hat{x}\,\hat{\Sigma}_1'(\hat{x}) + \hat{x}^2\,\hat{\Sigma}_1''(\hat{x}) + \hat{x}\,\hat{\Psi}_1(\hat{x}) = 0, \tag{14.51a}$$

$$\hat{x}^2\,\hat{\Sigma}_1(\hat{x}) - \frac{\hat{N}}{2\pi\,\hat{x}}\,\hat{\Psi}_1(\hat{x}) - \left[\frac{\mathrm{d}(\hat{x}\,\hat{\Psi}_1'(\hat{x}))}{\mathrm{d}\hat{x}} - \frac{\hat{\Psi}_1(\hat{x})}{\hat{x}}\right] = 0. \tag{14.51b}$$

A key problem is to determine how many independent solution of the linearized system (14.51) are physically acceptable. To answer this question, we shall list the possible asymptotic behaviours of these solutions $(\hat{\Sigma}_1, \hat{\Psi}_1)$, as we have several times already in a similar situation. This is done by applying the WKB method, an acronym for 'Wentzel–Kramers–Brillouin'. Counting how many solutions of the linearized equations (14.51) diverge and how many go to zero for $\hat{x} \to \infty$ will allow us to make sure that the problem is well-posed. In addition, the WKB expansion will be used to set up the numerical solution of the boundary layer equations in Section 14.6.4.

The WKB theory assumes first that the unknown function can be written as the exponential of a rapidly varying argument $\Phi(\hat{x})$: $\hat{\Sigma}_1(\hat{x}) = A \exp(\Phi(\hat{x}))$ and $\hat{\Psi}_1(\hat{x}) = B \exp(\Phi(\hat{x}))$, where A and B are slowly varying amplitudes. When these expressions are plugged into the linearized equations (14.51), the following terms dominate the others (and balance each other) in the limit of a rapidly varying phase, $|\Phi'(x)| \gg 1$:

$$\hat{x}^2\,\hat{\Sigma}_1''(\hat{x}) + \hat{x}\,\hat{\Psi}_1(\hat{x}) = 0$$

$$\hat{x}^2\,\hat{\Sigma}_1(\hat{x}) - \hat{x}\,\hat{\Psi}_1''(\hat{x}) = 0.$$

In terms of the phase $\Phi(\hat{x})$, this yields

$$A\,\hat{x}^2\,\Phi'^2 + B\,\hat{x} = 0$$

$$A\,\hat{x}^2 - B\,\hat{x}\,\Phi'^2 = 0.$$

This linear system for the unknown amplitudes A and B has a non-zero solution provided that its determinant is zero. This yields the following condition on the phase function:

$$\Phi'^4 + 1 = 0,$$

the solution of which is

$$\Phi(\hat{x}) = \frac{\pm 1 \pm i}{\sqrt{2}}\,\hat{x} + \cdots$$

Here the ellipsis stands for a constant of integration and other higher-order terms.

The function $\hat{\Psi}_1$ can be reconstructed by applying the exponential to $\Phi(\hat{x})$, multiplying by B, and then taking the real part. Depending on the choice of signs above, this yields four linearly independent functions, which are the real and imaginary parts of:

$$\exp\left(+\frac{1 \pm i}{\sqrt{2}}\,\hat{x} + \cdots\right) = \left(\cos\frac{\hat{x}}{\sqrt{2}} \pm i \sin\frac{\hat{x}}{\sqrt{2}}\right)\exp\left(+\frac{\hat{x}}{\sqrt{2}} + \cdots\right) \tag{14.52a}$$

and

$$\exp\left(-\frac{1 \pm i}{\sqrt{2}}\,\hat{x} + \cdots\right) = \left(\cos\frac{\hat{x}}{\sqrt{2}} \mp i \sin\frac{\hat{x}}{\sqrt{2}}\right)\exp\left(-\frac{\hat{x}}{\sqrt{2}} + \cdots\right). \tag{14.52b}$$

Because of the plus sign in the exponential, the first line (14.52a) corresponds to two exponentially large modes (one with the cosine and the other one with the sine function); the second line (14.52b) has a minus in the exponential, and yields two exponentially decaying modes.

As the WKB analysis just showed, two modes are exponentially large at infinity. This behaviour is clearly incompatible with the physical requirement that non-linear effects should become negligible far away from the disc of contact: $|\hat{\Psi}(\hat{x})| \ll \hat{x}$ for $\hat{x} \to \infty$. Cancellation of these two modes counts as *two asymptotic conditions* for $\hat{x} \to \infty$. These two asymptotic conditions provide two relations between the three free parameters of the equations, which are the force $\hat{N}$, the size of the region of contact $\hat{X}$ and the stress constant $\hat{C}$: for instance, they allow one to express $\hat{X}$ and $\hat{C}$ (and all other unknowns of the problem, including the displacement of the pushing plane) in terms of the main parameter, the force $\hat{N}$ applied by the pushing plane. In this sense, the problem appears to be well-posed.

The numerical implementation of this idea will be explained in the coming section. We shall then need a slightly refined form of the asymptotic form of $\hat{\Psi} = \hat{\Psi}_1$ given in equation (14.52b):

$$\hat{\Psi}(\hat{x}) = e^{-\hat{x}/\sqrt{2}}\,\frac{1}{\sqrt{\hat{x}}}\left(-a\sin\frac{\hat{x}}{\sqrt{2}} + b\cos\frac{\hat{x}}{\sqrt{2}}\right) \quad \text{for } \hat{x} \to \infty. \tag{14.53}$$

The two real coefficients a and b are the amplitude of the two exponentially small modes, and will be determined later on (they are essentially the real and imaginary parts of the complex constant B introduced earlier). The correcting factor $1/\sqrt{\hat{x}}$ in the right-hand side has been obtained by pushing the WKB expansion one step beyond the dominant order (the details of this calculation are omitted here).

14.6.4 Numerical solution

We are almost done with the analysis of a disc-like contact. All equations pertaining to the interior of the disc of contact have been solved in Section 14.6.1; all equations for the outer region (free part of the shell) have been written down—specifically, the set of differential equations (14.47) must be solved with the boundary conditions (14.48) on the left-hand side, and the asymptotic condition (14.53) on the right-hand side. In this section, we solve the equations for this outer region numerically.

Equations (14.47) are integrated numerically backwards, starting from $\hat{x} = +\infty$ as this avoids amplifying numerical errors exponentially (this was already noted in Section 13.6.2). More accurately, we start from a large value of $\hat{x}$, say $\hat{x} = 10$, and use the asymptotic expansion (14.53) to set up initial data for our numerical integration. In this expansion there are two unknown coefficients, a and b: numerical integration of equations (14.47) is carried out inside a procedure, whose arguments are $\hat{N}$, a and b.

Let us start by choosing an arbitrary numerical value of the rescaled force $\hat{N}$, such as $\hat{N} = 5\,\pi$ for instance. For this value of $\hat{N}$, we scan a range of values of a and b and integrate numerically the equations (14.47). This yields numerical solutions, denoted $\hat{\Sigma}(\hat{x}; a, b, \hat{N})$ and $\hat{\Psi}(\hat{x}; a, b, \hat{N})$, for discrete but finely spaced values of a and b. For every value of a and b, we collect the roots $\hat{X} = \hat{X}_{\hat{N},a,b,i}$ of equation (14.48d) defining the radius $\hat{X}$ of the disc of contact:

$$\left. \frac{\mathrm{d}\hat{\Psi}(\hat{x}; a, b, \hat{N})}{\mathrm{d}\hat{x}} \right|_{\hat{x}=\hat{X}} = -1.$$

These roots are simple to find numerically as the function $\hat{\Psi}(\cdot; a, b, \hat{N})$ is available by numerical integration in the form of a piecewise polynomial interpolation. The index i in the root $\hat{X}_{\hat{N},a,b,i}$ refers to the fact that there may be more than one root $\hat{X}$ for a given value of the parameters $\hat{N}$, a and b, as show for instance in Fig. 14.9 for $\hat{N} = 5\pi$, $a = -6$ and $b = 2$.

For a prescribed value of $\hat{N}$, iteration of this process over many values of a and b leads to the surface plot shown in Fig. 14.9 in the space $(a, b, \hat{X})$.

It remains to solve equations (14.48b) and (14.48c). These equations are satisfied when the quantities e and f defined below vanish:

$$e(\hat{N}, a, b, \hat{X}) = \hat{\Sigma}'(\hat{X}; \hat{N}, a, b) - \frac{\hat{X}}{8}, \qquad f(\hat{N}, a, b, \hat{X}) = \hat{\Psi}(\hat{X}; \hat{N}, a, b) + \hat{X}.$$

For a fixed value of $\hat{N}$, the equations $e(\hat{N}, a, b, \hat{X}) = 0$ and $f(\hat{N}, a, b, \hat{X}) = 0$ have discrete roots $(a, b, \hat{X})$ over the manifold with implicit equation $\hat{\Psi}'(\hat{X}) = -1$ (this manifold is shown in Fig. 14.9), as we have two equations on a manifold of dimension two—for the range of parameters scanned numerically, we even found a *unique* solution. These equations $e = 0$ and $f = 0$ can be solved graphically by colouring the manifold with grey levels according to the values of the function

$$\frac{1}{2} + \frac{\mathrm{sgn}(e\,f)}{\frac{1}{2} + e^2 + f^2} \in [0, 1],$$

where 0 stands for black and 1 for white. The result is shown in Fig. 14.9. Because of the discontinuity of the sign function (denoted sgn in the last equation) when its argument is

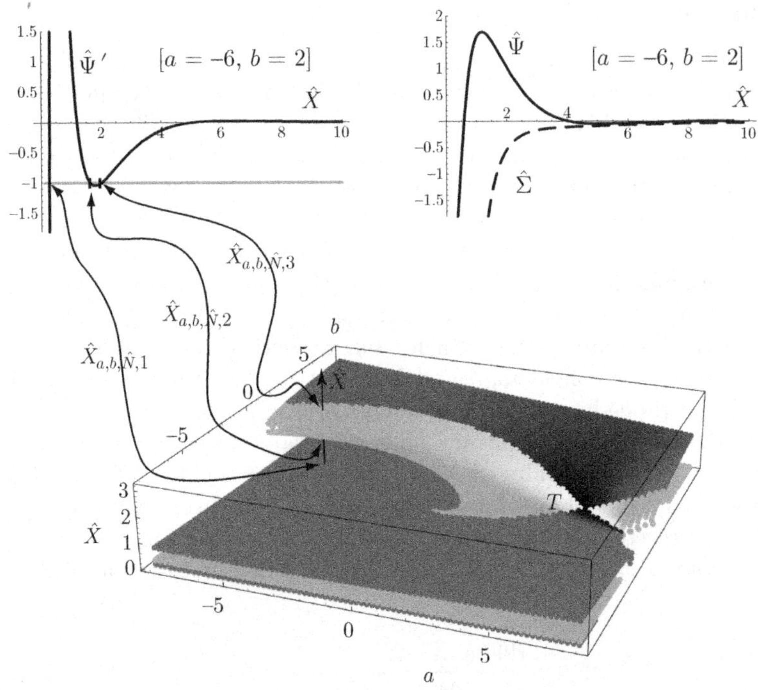

Fig. 14.9 Numerical solution of the geometry with disc-like contact, with rescaled pushing force set to $\hat{N} = 5\pi$. Equation (14.47) is integrated backwards, using particular values of a and b in equation (14.53) as initial data: numerical solutions for $\hat{\Psi}$ and $\hat{\Psi}'$ are shown at the top for particular values of a and b. Points $\hat{X}_{\hat{N},a,b,i}$ satisfying the implicit equation (14.48d) are recorded. By scanning over different values of a and b, we obtain a manifold in the space $(a, b, \hat{X})$, shown bottom. The two remaining boundary condition in (14.48) are visualized by colouring the manifold. This leaves out one physical solution, labelled T in the figure, at the intersection of the coloured regions; the exact numerical values for point T are listed in equation (14.54).

zero, the roots of $e = 0$ and $f = 0$ show up at the crossing of four domains with different levels of gray. For $\hat{N} = 5\pi$, this root is unique and is denoted by the symbol T in Fig. 14.9; the numerical values are

$$\hat{N} = 5\pi, \quad a = 5.594, \quad b = 0.812, \quad \hat{X}_{\hat{N},a,b,1} = 0.770. \tag{14.54}$$

This numerical procedure, illustrated in Fig. 14.9 for $\hat{N} = 5\pi$, yields a solution to the equations for the spherical shell with disc-like contact for any particular value of the rescaled contact force $\hat{N}$. As an outcome of this procedure, we find the values a and b of the shooting parameters, the functions $\hat{\Sigma}(\hat{x})$ and $\hat{\Psi}(\hat{x})$—shown in Fig. 14.10 for $\hat{N} = 5\pi$, and the radius of the disc of contact $\hat{X}$. The stress constant can then be found using the last equation (14.48a):

$$\hat{C} = \hat{\Sigma}(\hat{X}) - \frac{\hat{X}^2}{16}.$$

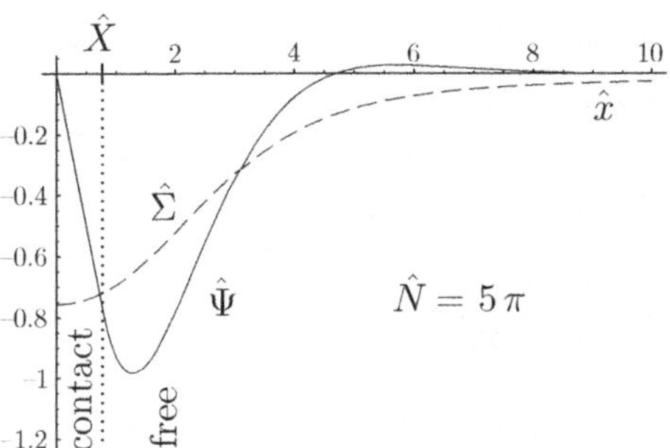

Fig. 14.10 Solutions $\hat{\Sigma}(\hat{x})$ and $\hat{\Psi}(\hat{x})$ of the equations for a spherical shell with disc-like contact, for $\hat{N} = 5\pi$. This solution corresponds to the point labelled T in Fig. 14.9 and to the set of parameters in equation (14.54). The functions $\hat{\Psi}$ and $\hat{\Sigma}$ were plotted by matching the solution in the interior of the disc of contact (Section 14.6.1), with the solution in the free region (Section 14.6.2).

Repeating this procedure for different values of the applied force $\hat{N}$, one can follow[6] numerically the solution with disc-like contact. This yields a family of configurations indexed by the rescaled value of the force of contact $\hat{N}$. The values of the parameters $a(\hat{N})$ and $b(\hat{N})$, the rescaled size of the region of contact $\hat{X}(\hat{N})$, and the stress constant $\hat{C}(\hat{N})$ found in this way, are plotted in Figs 14.11, 14.12 and 14.13.

In order to reconstruct the profile of the deformed shell, one can use equation (14.13); in terms of the rescaled quantities defined in equations (14.34) and (14.46), we have:

$$v_z(s) = -\int_s^{+\infty} \phi(s')\,\mathrm{d}s' = R \int_{x=\frac{s}{R}}^{+\infty} \Psi(x')\,\mathrm{d}x' = R\,\epsilon \int_{\hat{x}=\frac{s}{R\,\epsilon^{1/2}}}^{+\infty} \hat{\Psi}(\hat{x}')\,\mathrm{d}\hat{x}'.$$

The sum of this deflection and the unperturbed profile in equation (14.9), $z(s) \approx R - s^2/2\,R$, yields the profile of the shell in actual configuration:

$$z(s) + v_z(s) = R + \epsilon\,R\,Z\left(\hat{N}, \frac{s}{R\,\epsilon^{1/2}}\right), \tag{14.55a}$$

where the rescaled profile $Z(\hat{N}, \hat{x})$ is defined in terms of the numerical solution $\hat{\Psi}_{\hat{N}}$ found earlier by

$$Z(\hat{N}, \hat{x}) = -\frac{\hat{x}^2}{2} + \int_{\hat{x}}^{+\infty} \hat{\Psi}_{\hat{N}}(\hat{x}')\,\mathrm{d}\hat{x}'. \tag{14.55b}$$

Meridional cuts of the disc of contact and of the annular ridge connecting it to the outer part of the shell are shown in Fig. 14.14. Cuts of the deformed shell were computed by equation (14.55) for various values of the force $\hat{N}$, and were superimposed.

Focusing on the apex $s = 0$ ($\hat{x} = 0$) in equations (14.55), one obtains the load–displacement relation of the shell with disc-like contact; this relation connects the displacement e of the plane and the rescaled contact force $\hat{N}$:

---

[6] The parameter $\hat{N}$ is changed by small increments or decrements, and the solution obtained at the previous iteration is used as a guess for the non-linear root-finding procedure at the next iteration. This numerical continuation starts from the solution T for $\hat{N} = 5\pi$.

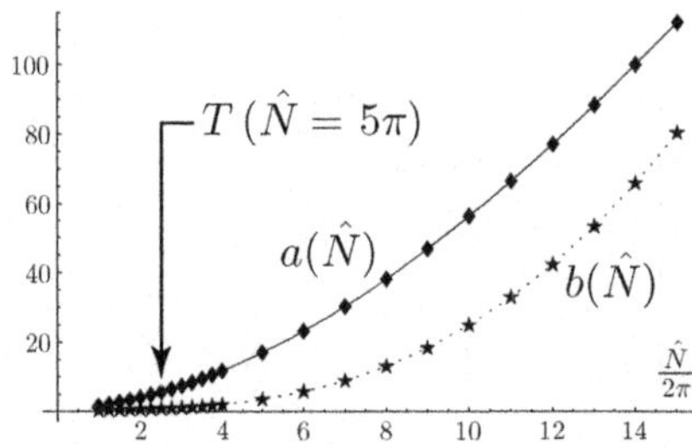

Fig. 14.11 Numerical continuation of the shooting parameters $a(\hat{N})$ and $b(\hat{N})$. Initial values, $a(\hat{N} = 5\pi)$ and $b(\hat{N} = 5\pi)$, are provided by equation (14.54).

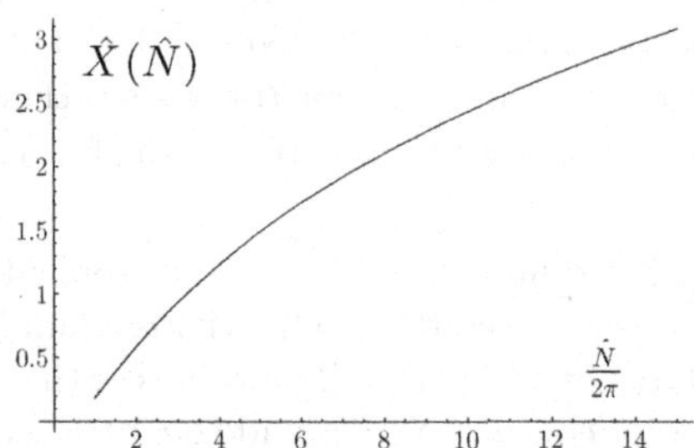

Fig. 14.12 Radius of the region of contact $\hat{X}(\hat{N})$ as a function of the force $\hat{N}$.

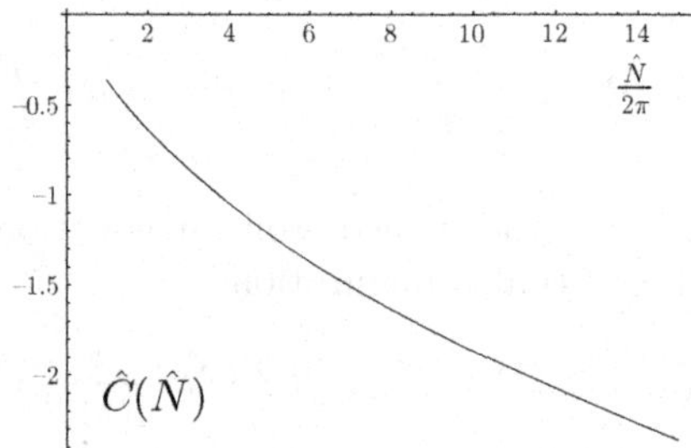

Fig. 14.13 Constant $\hat{C}(\hat{N})$ entering in the expression (14.41) for the stress in the interior of the region of contact, as a function of the rescaled force $\hat{N}$.

$$e = -v_z(0) = -R\,\epsilon\, Z(\hat{N}, 0) = R\,\epsilon\, \hat{e}(\hat{N}), \quad \text{where } \hat{e}(\hat{N}) = -\int_0^{+\infty} \hat{\Psi}_{\hat{N}}(\hat{x}')\, \mathrm{d}\hat{x}'. \tag{14.56}$$

This load–displacement curve is plotted in Fig. 14.15.

In the present section, deformations of a spherical shell pushed by a rigid wall have been found by means of asymptotic analysis, in the case of a disc-like contact and under the assumption of cylindrical symmetry. In the next section, we consider the configuration with a region of inverted curvature surrounded by a circle of contact—this is the quantitative version of Pogorelov's order of magnitude analysis of the ridge. Later, in Section 14.8, we study the competition between these two types of configurations: an analysis of stability of the configuration with flat contact reveals that this configuration becomes unstable above a critical value of the force, and this instability leads to the other configuration.

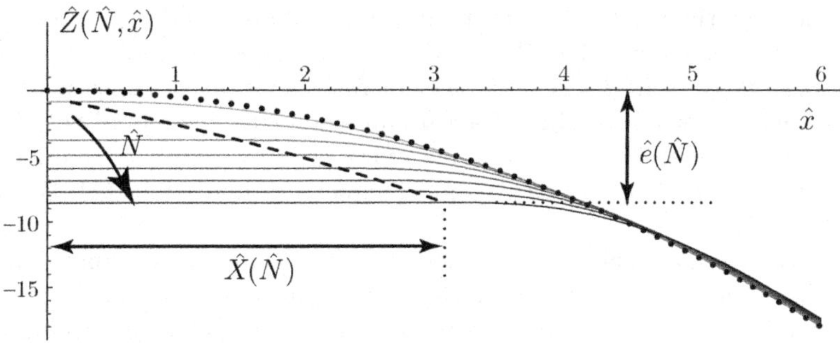

Fig. 14.14 Progressive flattening of a spherical shell with disc-like contact: meridional cuts of the deformed shell for various values of the force of contact were computed using equation (14.55): from lighter to darker greys, $\hat{N} = 6.3,\ 18.8,\ 31.4,\ 44.0,\ 56.6,\ 69.1,\ 81.7,\ 94.2$. In dimensionless variables, the undeformed configuration (dotted curve) is the parabola with unit curvature at the origin, $Z = -\hat{x}^2/2$. The dashed curve is the parametric curve $(\hat{X}(\hat{N}), -\hat{e}(\hat{N}))$ traced out by the edge of the disc of contact.

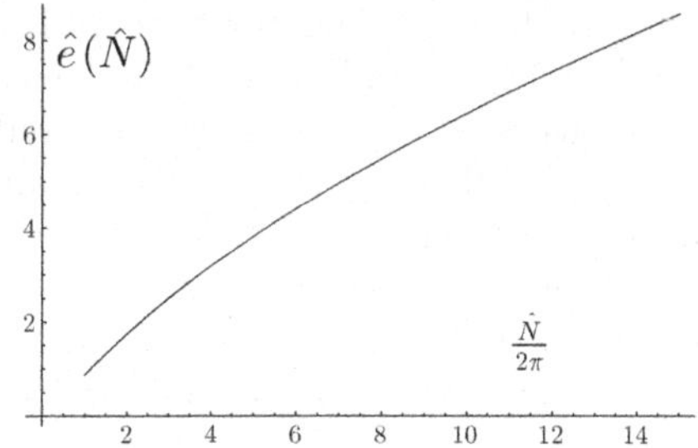

Fig. 14.15 Force-displacement curve of the spherical shell with a disc of contact, as predicted by equation (14.56). Note that this curve, when expressed in rescaled parameters (displacement of the plane $\hat{e} = e/R\epsilon$, and contact force $\hat{N}/2\pi = F/2\pi\,E\,R\,h\,\epsilon^2$) depends on no parameter. Stability of this solution will be studied later.

14.7   Spherical shell pushed by a plane: circular contact

In this section, we examine what happens for values of the applied force F that are much larger than in the previous section, that is $F \gg F_{\mathrm{c}} \sim (EhR\epsilon^2)$ in physical units. As noted in Section 14.4, an extended contact with the pushing plane is no longer optimal from the viewpoint of energy minimization. A spherical cap has Gauss curvature $1/R^2$ and is not isometric to a plane. As a result, the flattening of the spherical shell near its apex, as in the case of a disc-like contact, involves an increasing energy of stretching as the radius of the disc becomes larger and larger. A configuration with far less stretching energy is obtained by replacing the flat disc of contact by an inverted cap: a dimple forms, obtained by reflection from the original cap. Having the same Gauss curvature as the shell in its natural configuration, this dimple gets rid of the stretching energy.

Dimples in spherical shells subjected to a point force, to load concentrated over a small area, or to pressure loading have been reported experimentally (R. M. Evan-Iwanowski, H. S. Cheng, and T. C. Loo, 1962; L. Berke and R. L. Carlson, 1968), and studied both numerically and analytically (W. T. Koiter, 1969; G. V. Ranjan and C. R. Steele, 1977; G. A. Kriegsmann and C. G. Lange, 1980; F. Y. M. Wan, 1980a, b; M. Gräff *et al.*, 1985). More recently, dimples have been observed at the microscopic scale, for instance in composite membrane obtained by self-assembly of actin filaments with giant fluid vesicles (E. Helfer *et al.*, 2008) and in nano-indentation tests on a viral shell by atomic force microscopy, both experimentally (W. S. Klug *et al.*, 2006) and in numerical simulations (M. Buenemann and P. Lenz, 2007).

When a dimple forms, the shell is in contact with the pushing plane along a circle of radius r. We shall consider a range of forces such that this radius r remains much smaller than the radius of the shell, $r \ll R$. There is a price to be paid for cancelling the stretching energy in the interior of the dimple, namely the elastic energy of the ridge connecting the inverted cap with the outer, non-inverted part of the shell. Let δ be the width of this ridge. A fundamental assumption, to be checked at the end, is that the radius of the circle of contact, r, is much larger than the width δ of the ridge,[7] $\delta \ll r$. Technically, the radius r and the displacement e of the plane will be related to the applied force via a solubility condition; the ridge can be described as an inner layer solution of the shell equations—boundary layer methods are introduced in Appendix B.

A key point of our analysis is that, at leading order, there is a solution of the equations for the ridge with a *vanishing* pushing force ($F = 0$). This does not mean of course that an inverted cap could be obtained without applied force, but only that this force appears at next order in the expansion. In physical terms, the small parameter on which the expansion is based is the ratio δ/r (width of the ridge divided by the radius of the circle of contact). There is an important difference with the case of a flat disc of contact, as the force F can be scaled out in the equations, which ultimately do not contain any parameters. In stretched variables, the ridge has a universal profile, which we determine numerically.

14.7.1   Equations for the ridge

In the case of a circular contact, the applied pressure $f_z(s)$ is localized along a circle of radius r. This pressure happens to be given by exactly the same function as in the case of disc-like contact, given earlier in equations (14.42) for the contact pressure, (14.43) for the internal force, and (14.45) for the rescaled internal force. In terms of the set of dimensionless variables bearing a hat, introduced in equation (14.46), the internal force becomes

$$\hat{N}(\hat{x}) = \hat{N}\, H(\hat{x} - \hat{X}), \tag{14.57}$$

where $\hat{N}$ is the rescaled resultant of the contact force, a parameter of the problem, and H is the usual Heaviside step function, defined in equation (14.44).

We start again from the set of equations (14.47), with two minor modifications: we use the internal force given by equation (14.57) in the right-hand side of the second equation, and change the unknown $\hat{\Psi}$ in favour of $\hat{\Gamma}$ defined by:

$$\hat{\Gamma}(\hat{x}) = \hat{x} + \hat{\Psi}(\hat{x}). \tag{14.58}$$

---

[7] This contrasts with the case of the disc-like contact, studied in the previous section for much lower values of the force F. Then the radius r of the disc and the width δ of the ridge were found to be of the same order of magnitude.

Applying the rescalings (14.34) and (14.46) to the definition (14.10) of the natural shape of the shell shows that the direction of the undeformed tangent is $\hat{x}$ in rescaled variables. The variable $\hat{\Gamma}(\hat{x})$, defined as the sum of this $\hat{x}$ and the rotation $\hat{\Psi}$ of the tangent, is therefore the rescaled angle measuring the direction of the tangent in actual configuration. The benefit of replacing the unknown $\hat{\Psi}$ by $\hat{\Gamma}$ is that the equations of equilibrium (14.47) become invariant under a mirror-reflection that changes the signs of both $\hat{\Gamma}$ and $\hat{N}$:

$$3\,\hat{x}\,\hat{\Sigma}'(\hat{x}) + \hat{x}^2\,\hat{\Sigma}''(\hat{x}) + \frac{\hat{\Gamma}^2(\hat{x})}{2} = \frac{\hat{x}^2}{2}, \tag{14.59a}$$

$$\hat{x}\,\hat{\Sigma}(\hat{x})\,\hat{\Gamma}(\hat{x}) - \left[\frac{\mathrm{d}(\hat{x}\,\hat{\Gamma}'(\hat{x}))}{\mathrm{d}\hat{x}} - \frac{\hat{\Gamma}(\hat{x})}{\hat{x}}\right] = -\frac{\hat{N}\,H(\hat{x} - \hat{X})}{2\pi}. \tag{14.59b}$$

These are two coupled non-linear ordinary differential equations, with unknown functions $\hat{\Sigma}(\hat{x})$ and $\hat{\Gamma}(\hat{x})$, and a parameter $\hat{N}$.

Equations (14.59) are written in terms of the hat variables that were introduced earlier in the analysis of the disc-like contact, when we considered the range of forces $F \sim F_{\rm c}$, i.e. $\hat{N}$ of order 1. Here, we study much larger forces, $F \gg F_{\rm c}$, i.e. $\hat{N} \gg 1$, and consequently much larger values of the radius, $\hat{X} \gg 1$ (recall that $\hat{X}$ was of order 1 for disc-like contact). Therefore, we have to consider some limit in the previous set of equations, whereby both $\hat{N}$ and $\hat{X}$ are large numbers. These parameters being related, we choose $\hat{X}$ as the primary unknown and seek an expansion of the solution $(\hat{\Sigma}, \hat{\Gamma})$ in powers of the small parameter[8]

$$\frac{1}{\hat{X}} \to 0. \tag{14.60}$$

Let us first analyse the orders of magnitude in equation (14.59). More specifically, we derive those orders of magnitude that are consistent with a solution describing a ridge of radius $\hat{X}$ and width $\hat{\delta}$, with $\hat{X} \gg 1$ and $\hat{\delta} \ll \hat{X}$. Recall that this ridge connects an inverted cap, for which $\hat{\Psi} \approx -2\,\hat{x}$ and so $\hat{\Gamma} \approx -\hat{x}$, to the non-inverted part of the shell on the exterior, for which $\hat{\Psi}$ goes to zero, and so $\hat{\Gamma} \approx +\hat{x}$. This shows that $\hat{\Gamma}$ is necessarily of order $\hat{x} \approx \hat{X}$ in the ridge:

$$\hat{\Gamma} \sim \hat{X},$$

where the sign $\sim$ implies that its two sides are of comparable orders of magnitude, even though they can differ by a numerical factor of order 1.

In the first equation (14.59a), the first term $3\,\hat{x}\,\hat{\Sigma}' \sim \hat{X}\,\hat{\Sigma}/\hat{\delta}$ appears to be negligible in front of the second one, $\hat{x}^2\,\hat{\Sigma}'' \sim \hat{X}^2\,\hat{\Sigma}/\hat{\delta}^2$, in the limit considered, $\hat{\delta} \ll \hat{X}$. Balancing the second and third terms in the left-hand side (the right-hand side being negligible), we have

$$\hat{X}^2\,\frac{\hat{\Sigma}^2}{\hat{\delta}^2} \sim \hat{\Gamma}^2.$$

----

[8] We are considering three different limits at the same time: first, the limit of a thin shell, $h \ll R$; second, the limit of a shallow shell (perturbation is localized near the apex of the shell where the non-linear rotation terms are moderate), $e \ll R$; and, third, the limit $\hat{X} \to \infty$ of equation (14.60), that is $e \gg h$. Doing so, we shall solve a problem involving sub-dominant terms (second-order problem studied in Section 14.7.5), even though the shell theory we start from is valid at the dominant order only. There is no paradox as the shell theory retains the dominant order *with respect to* h/R, although the solubility condition for $n_{(2)}$ studied in Section 14.7.5 involves sub-dominant terms in the expansion with respect to a *different* small parameter, namely $1/\hat{X} \sim (e/h)^{1/2}$.

By a similar argument, we retain the dominant contribution $-\hat{x}\,\hat{\Gamma}''(\hat{x})$ to the bending terms only, and balance the remaining terms in the second equation (14.59b):

$$\hat{X}\,\hat{\Sigma}\,\hat{\Gamma} \sim \hat{X}\,\frac{\hat{\Gamma}}{\hat{\delta}^2} \sim \hat{N}.$$

Combining these relations, we obtain the orders of magnitude relevant to the limit $\hat{X} \gg 1$:

$$\hat{\delta} \sim 1, \quad \hat{\Sigma} \sim 1, \quad \hat{\Gamma} \sim \hat{X}, \quad \hat{N} \sim \hat{X}^2. \tag{14.61}$$

So far what we did is consistent, as the ridge width, $\hat{\delta}$, is of order 1, and is therefore much smaller than the radius $\hat{X}$. In addition, note that the width of the ridge in physical variables, $\delta = R\,\epsilon^{1/2}\,\hat{\delta} \sim R\,\epsilon^{1/2}$, remains small compared with R.

The scaling laws (14.61) suggest eliminating the variable $\hat{x}$ in favour of a new variable t defined by:

$$\hat{x} = \hat{X} + t, \tag{14.62a}$$

as this new variable t takes on values of order 1 in the ridge region (recall that $\hat{\delta} \sim 1$). Note that the circle of contact, $\hat{x} = \hat{X}$ is at the origin $t = 0$ of the new variable.

Similarly, we expand the unknowns $\hat{\Sigma}$ and $\hat{\Gamma}$ in powers of the large parameter $\hat{X}$, guessing the form of the first term in each expansion from the scaling laws in equation (14.61):

$$\hat{\Sigma}(\hat{x}) = f_{(1)}(t) + \frac{1}{\hat{X}}\,f_{(2)}(t) + \cdots \tag{14.62b}$$

$$\hat{\Gamma}(\hat{x}) = \hat{X}\,g_{(1)}(t) + g_{(2)}(t) + \cdots \tag{14.62c}$$

These expansions are relevant to the limit of large $\hat{X}$. We shall solve the problem order by order, and compute the functions $f_{(1)}$ and $g_{(1)}$ first, and then $f_{(2)}$ and $g_{(2)}$, etc.

The scalar $\hat{N}$, which is the rescaled resultant of the contact force is expanded in powers of $\hat{X}$ as well, the leading order being $\hat{X}^2$ by equation (14.61):

$$\hat{N} = \hat{X}^2\,n_{(1)} + \hat{X}\,n_{(2)} + \cdots \tag{14.62d}$$

Our main goal in the present section is to compute the numerical constants $n_{(1)}$ and $n_{(2)}$; then the equation above will yield the loading–displacement relation for the spherical shell with a dimple.

Plugging these expansions (14.62) into the equations of equilibrium (14.59), we obtain at leading order, which is order $\hat{X}^2$ in both equations:

$$f''_{(1)}(t) + \frac{g_{(1)}^{\;2}(t)}{2} = \frac{1}{2} \tag{14.63a}$$

$$f_{(1)}(t)\,g_{(1)}(t) - g''_{(1)}(t) = -\frac{n_{(1)}\,H(t)}{2\pi}. \tag{14.63b}$$

In this set of coupled non-linear differential equations, the quantity $\hat{X}$ has disappeared. This illustrates the power of the asymptotic analysis: by focusing on the small scale, here the ridge width, it is sufficient to solve the equations order by order, a much simpler task than the direct solution of the original equations. This first-order problem (14.63) will be solved in Section 14.7.4, and we shall show that the constant $n_{(1)}$ is actually zero.

To derive the force–displacement relation for the dimple, we shall need to push the expansion to the next order. Plugging now the expansions (14.62) into the equations of equilibrium (14.59) and reading off the sub-dominant terms of order $\hat{X}$, we have:

$$f''_{(2)}(t) + g_{(1)}(t)\, g_{(2)}(t) = I_{(1)}(t) \tag{14.64a}$$

$$g_{(1)}(t)\, f_{(2)}(t) + f_{(1)}(t)\, g_{(2)}(t) - g''_{(2)}(t) = -\frac{n_{(2)}\, H(t)}{2\pi} + J_{(1)}(t), \tag{14.64b}$$

where the quantities $I_{(1)}(t)$ and $J_{(1)}(t)$ in the right-hand sides depend only on first-order solution:

$$I_{(1)}(t) = t - 3\, f'_{(1)}(t) - 2\, t\, f''_{(1)}(t) \tag{14.64c}$$

$$J_{(1)}(t) = -t\, f_{(1)}(t)\, g_{(1)}(t) + g'_{(1)}(t) + t\, g''_{(1)}(t). \tag{14.64d}$$

Equations (14.64a) and (14.64b) for a set of *linear* differential equations for the unknowns $f_{(2)}$ and $g_{(2)}$. The problem of finding the value of $n_{(2)}$ such that these equations (14.64), together with suitable asymptotic conditions for $t \to \pm\infty$, have a solution (eigenvalue problem) will be studied in Section 14.7.5 by using a solubility condition.

14.7.2   Asymptotic and boundary conditions

Far away from the core of ridge, the shell should be close to its natural shape on the outside, and close to an inverted cap inside. This gives $|\hat{\Psi}| \to 0$ and so $\hat{\Gamma}(\hat{x}) \approx \hat{x}$ on the outside; on the inside, we have $\hat{\Psi} \approx -2\,\hat{x}$ and so $\hat{\Gamma}(\hat{x}) \approx -\hat{x}$ for $\hat{X} - \hat{x} \gg \hat{\delta}$. In terms of the unknown $g_{(1)}(t)$, these asymptotic conditions become:

$$g_{(1)}(t) \to \pm 1 \quad \text{for } t \to \pm\infty. \tag{14.65a}$$

In equation (14.63b), the term $(-g''_{(1)})$ comes from bending and must therefore become negligible far from the ridge. In this equation, the first term proportional to $f_{(1)}$ must therefore balance the right-hand side, and we have:

$$f_{(1)}(t) \to \begin{cases} -\frac{n_{(1)}}{2\pi} & \text{for } t \to +\infty \\ 0 & \text{for } t \to -\infty. \end{cases} \tag{14.65b}$$

In addition, the shell has to be tangent to the plane at the circle of contact, $\hat{x} = \hat{X}$ (which is $t = 0$ in the new set of variables). There, the tangent must be orthogonal to the axis of revolution in actual configuration and so $\hat{\Gamma}(\hat{X}) = 0$. This implies:

$$g_{(1)}(0) = 0. \tag{14.66}$$

By symmetry, the equations need to be solved over one half of the ridge only, say $0 \leq t < +\infty$. Equation (14.66) then yields a boundary condition on the left endpoint of the domain, $t = 0$.

14.7.3   Dominant order: cancellation of the force

We proceed to solve the first-order problem, which consists of the non-linear differential equations (14.63) for the unknowns $f_{(1)}$ and $g_{(1)}$, together with the asymptotic conditions (14.65). It turns out that these equations have a solution when $n_{(1)} = 0$ only.

The cancellation of $n_{(1)}$ can be explained as follows. Assume that the rescaled force $\hat{N}$ scales like a power k of $\hat{X}$, that is $\hat{N} \sim \hat{X}^k$: by equation (14.62d), we have $k = 2$ if $n_{(1)} \neq 0$, but $k = 1$ if $n_{(1)} = 0$ and $n_{(2)} \neq 0$. Equation (14.49) shows that the pushing force is then, in physical variables

$$F \sim \frac{E\,h^3}{R}\,\hat{X}^k. \tag{14.67}$$

Now, $\hat{X}$ is the rescaled radius and is connected to the physical one, r, by $\hat{X} = r/R\,\epsilon^{1/2} \sim r/R^{1/2}\,h^{1/2}$. Using the geometric relation (14.2), $r \sim R^{1/2}\,e^{1/2}$, to eliminate r in favour of the vertical displacement, we find $\hat{X} \sim e^{1/2}/h^{1/2}$. Combining with equation (14.67), we obtain the following scaling law for the force:

$$F \sim \frac{E\,h^{3-k/2}}{R}\,e^{k/2}. \tag{14.68}$$

Comparison with Pogorelov's scaling law in equation (14.5b) reveals that the regime studied here corresponds to $k = 1$, that is $\hat{N} \sim \hat{X}$; this is not what we anticipated based on a formal balance of the terms in the equations of equilibrium, see equation (14.61). This suggests that the coefficient $n_{(1)}$ in equation (14.62d) vanishes.

We shall now give a more rigorous proof that the leading-order problem has a solution if $n_{(1)} = 0$ only. This proof starts by multiplying by $f'_{(1)}$ the first equation (14.63a), by $g'_{(1)}$ the second equation (14.63b), and adding the results:

$$f'_{(1)}(t)\left(f''_{(1)}(t) + \frac{g_{(1)}{}^2(t)}{2} - \frac{1}{2}\right)\cdots$$

$$+\, g'_{(1)}(t)\left(f_{(1)}(t)\,g_{(1)}(t) - g''_{(1)}(t) + \frac{n_{(1)}\,H(t)}{2\pi}\right) = 0. \tag{14.69}$$

This expression happens to be an exact derivative. For instance, the sum of the term $f'_{(1)}\,g_{(1)}{}^2/2$ from the first scalar product and the term $g'_{(1)}\,f_{(1)}\,g_{(1)}$ from the second one is the derivative of $f_{(1)}g_{(1)}{}^2/2$. Working out the other terms similarly, one can rewrite equation (14.69) as

$$\frac{\mathrm{d}\ell_{(1)}(t)}{\mathrm{d}t} = 0, \tag{14.70}$$

where

$$\ell_{(1)}(t) = \frac{f'_{(1)}{}^2(t)}{2} - \frac{g'_{(1)}{}^2(t)}{2} + f_{(1)}(t)\,\frac{g_{(1)}{}^2(t) - 1}{2} + \frac{n_{(1)}}{2\pi}\,H(t)\,g_{(1)}(t). \tag{14.71}$$

Equation (14.70) shows that $\ell_{(1)}(t)$ is an invariant for any solution $f_{(1)}$, $g_{(1)}$. The existence of this invariant comes from the variational structure of equations (14.63): they can be obtained as the condition of stationarity of some function by the Euler–Lagrange principle— for details, see Appendix A on the calculus of variations for functions. The above method for deriving the invariant is the standard proof of energy conservation for a Lagrangian system with no explicit dependence on the time variable.

Since $f_{(1)}(t)$ and $g_{(1)}(t)$ are expected to converge smoothly to their asymptotic behaviour for $t \to \pm\infty$ given in equation (14.65), we shall assume $f'_{(1)} \to 0$ and $g'_{(1)} \to 0$. Together with equations (14.65), this yields

$$\ell_{(1)}(t) \rightarrow \begin{cases} \frac{n_{(1)}}{2\pi} & \text{for } t \rightarrow +\infty \\ 0 & \text{for } t \rightarrow -\infty \end{cases}. \tag{14.72}$$

Now, equation (14.70) shows that $\ell_{(1)}(t)$ is a conserved quantity for any solution $f_{(1)}(t)$ and $g_{(1)}(t)$ of the differential equations.[9] Therefore, both its limits for $t \rightarrow \pm\infty$ have to be equal: a necessary condition for the existence of solutions to the leading-order problem is

$$n_{(1)} = 0. \tag{14.73}$$

As shown in the beginning of this section, this is consistent with Pogorelov's scaling laws (14.5b) as this implies $k = 1$ in equation (14.68). Therefore, the cancellation of $n_{(1)}$ is the result of the existence of an invariant as noted by several authors (G. A. Kriegsmann and C. G. Lange, 1980; A. Libai and J. G. Simmonds, 1998).

Having set to zero the parameter $n_{(1)} = 0$, we can proceed to solving the first-order problem; this is the aim of the following section. Because $n_{(1)} = 0$, the force–displacement relation (14.62d) is dominated by the second term, proportional to $n_{(2)}$. This constant $n_{(2)}$, will be computed in Section 14.7.5 by looking at the second-order problem.

14.7.4 Dominant order: numerical solution

With $n_{(1)} = 0$, equations (14.63) become

$$f''_{(1)}(t) + \frac{g_{(1)}{}^2(t)}{2} = \frac{1}{2} \tag{14.74a}$$

$$f_{(1)}(t)\, g_{(1)}(t) - g''_{(1)}(t) = 0. \tag{14.74b}$$

These equations are invariant under the mapping:

$$f_{(1)} \rightarrow f_{(1)}, \quad g_{(1)} \rightarrow (-g_{(1)}), \quad t \rightarrow (-t). \tag{14.75}$$

The asymptotic conditions (14.65) and the boundary condition (14.66) are invariant under the same mapping as well. This symmetry[10] helps simplifying the numerical solution of the problem, as we now show.

Because the application of the symmetry (14.75) to a solution of the first-order problem yields another solution of the same problem, there are two possibilities: either this solution is invariant under the symmetry or it is degenerate and comes with a twin (different from itself), the two twins being exchanged by the symmetry. We shall not consider the second possibility, which is less common and can in fact be ruled out by a detailed analysis: we seek a solution of the first-order problem which is invariant under the symmetry (14.75). In words, the function $f_{(1)}$ is assumed to be an even function of t, and the function $g_{(1)}$ an odd function of t. This implies

$$f'_{(1)}(0) = 0, \qquad g_{(1)}(0) = 0. \tag{14.76a}$$

---

[9] Some care has to be taken at $t = 0$, where these solutions are not perfectly smooth. Solutions $(f_{(1)}, g_{(1)})$ of equations (14.63) are continuous with continuous first-order derivatives ($\mathcal{C}^1$), as can be shown by integrating these equations across the singularity. As a result, all terms in $\ell_{(1)}$ are continuous across the singularity at $t = 0$ (for the last term, use equation (14.66)) and $\ell_{(1)}$ has indeed no discontinuity at $t = 0$.

[10] In addition, equations (14.74) are autonomous, i.e. are invariant by translation of the variable t. However, this other symmetry is suppressed by the condition (14.66) and so does not make the numerical solution easier.

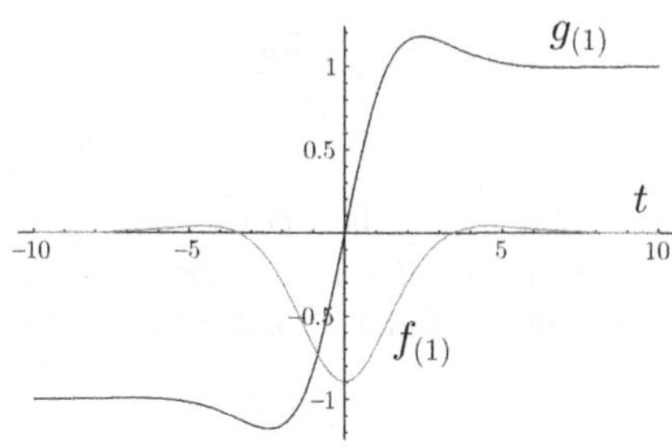

Fig. 14.16 Solution of the first-order problem for the ridge with circular contact topology. The pushing force is zero at this order, $n_{(1)} = 0$.

In addition, equation (14.72) shows that the value of the invariant is actually $\ell_{(1)}(t) = 0$. Inserting its definition (14.71) and setting $t = 0$, we obtain a new relation, $f_{(1)}(0) = -g_{(1)}'^2(0)$. Setting $\tau = g_{(1)}'(0)$, we can rewrite this as:

$$f_{(1)}(0) = -\tau^2, \quad g_{(1)}'(0) = \tau. \tag{14.76b}$$

Here is how the first-order problem can be solved numerically: for any value of τ, equations (14.76) provide a set of four initial values, for $f_{(1)}(0)$, $f_{(1)}'(0)$, $g_{(1)}(0)$ and $g_{(1)}'(0)$, which are those required to start the numerical integration of equations (14.74) on the interval $t > 0$. Recall that these differential equations are second-order with respect to both $f_{(1)}$ and $g_{(1)}$. As we did earlier in Section 14.6.4, one can adjust τ by a root-finding algorithm so as to cancel the exponentially growing modes for $t \to \infty$. This yields[11]

$$\tau = 0.947.$$

To sum up, the solution is obtained by integration of equations (14.74) over the interval $t > 0$ with the initial values

$$f_{(1)}(0) = -0.896, \quad f_{(1)}'(0) = 0, \quad g_{(1)}(0) = 0, \quad g_{(1)}'(0) = 0.947. \tag{14.76c}$$

Values on the negative part of the axis, $t < 0$, are reconstructed using the parity of $f_{(1)}$, which is even, and $g_{(1)}$, which is odd. The solution obtained in this way is plotted in Fig. 14.16. From now on, the symbols $f_{(1)}$ and $g_{(1)}$ will refer to this particular solution.

14.7.5   Calculation of reaction force by a solubility condition

So far, we have solved the leading-order problem for the ridge, and have found the expressions for the profile ($g_{(1)}$ is the tangent rotation $\phi(s)$ after rescaling and shifting by a constant to make the equations symmetric) and the membrane stress ($f_{(1)}$ is $N_s(s)$ rescaled). However, we have not yet obtained the curve–displacement relation connecting $\hat{N}$ to $\hat{X}$ (that is F to e in physical variables); since $n_{(1)} = 0$, this requires knowledge of the constant $n_{(2)}$ in equation (14.62d). Calculation of $n_{(2)}$ is the goal of the present section.

Our task will be greatly simplified by noting that we do not actually need to compute $f_{(2)}$ and $g_{(2)}$ but simply to determine the values of $n_{(2)}$ such that the equations for $f_{(2)}$

---

[11] Other discrete values of τ are possible but they involve more oscillations of the solutions $f_{(1)}$ and $g_{(1)}$, and so are very likely unstable. The value of τ cited in text has the lowest absolute value.

and $g_{(2)}$ have a solution: we seek a *solubility condition*[12] for these equations, which is much simpler than computing a solution.

Derivation of the solubility conditions starts by multiplying equation (14.64a) by the known function $f'_{(1)}$, and equation (14.64b) by $g'_{(1)}$, as we did earlier. This yields

$$f'_{(1)}(t)\left(f''_{(2)}(t)+g_{(1)}(t)\,g_{(2)}(t)\right)+g'_{(1)}(t)\left(g_{(1)}(t)\,f_{(2)}(t)+f_{(1)}(t)\,g_{(2)}(t)-g''_{(2)}(t)\right)\cdots$$
$$=f'_{(1)}(t)\,I_{(1)}(t)+g'_{(1)}(t)\left(-\frac{n_{(2)}\,H(t)}{2\pi}+J_{(1)}(t)\right). \tag{14.77}$$

The left-hand side of this equation depends on the functions $f_{(2)}$ and $g_{(2)}$, which we would like to avoid computing—this is possible as the left-hand side of the equation is zero, as we shall now show.

We expand the left-hand side of equation (14.77), and integrate each term so as to eliminate the derivatives of $f_{(2)}$ and $g_{(2)}$ in favour of the functions $f_{(2)}$ and $g_{(2)}$ themselves, without derivative. For instance, the first term is worked out as follows:

$$f'_{(1)}\,f''_{(2)}=\frac{d(f'_{(1)}\,f'_{(2)})}{dt}-f''_{(1)}\,f'_{(2)}=\frac{d(f'_{(1)}\,f'_{(2)}-f''_{(1)}\,f_{(2)})}{dt}+f'''_{(1)}\,f_{(2)}.$$

Treating the other terms similarly, one can rewrite the left-hand side of equation (14.77) as

$$\frac{d\ell_{(2)}(t)}{dt}+f_{(2)}(t)\,\frac{d\left(f''_{(1)}+\frac{g_{(1)}{}^{2}}{2}\right)}{dt}+g_{(2)}(t)\,\frac{d\left(f_{(1)}\,g_{(1)}-g''_{(1)}\right)}{dt}, \tag{14.78}$$

where

$$\ell_{(2)}(t)=f'_{(1)}(t)\,f'_{(2)}(t)-f''_{(1)}(t)\,f_{(2)}(t)-g'_{(1)}(t)\,g'_{(2)}(t)+g''_{(1)}(t)\,g_{(2)}(t). \tag{14.79}$$

In the expression (14.78), the second and third terms are zero and will be dropped: this follows from the remark that the arguments of the derivatives in factors of $f_{(2)}$ and $g_{(2)}$ are the left-hand sides of equations (14.74a) and (14.74b) for the first-order problem.

Now, replacing the left-hand side of equation (14.77) with the only remaining term in equation (14.78), the first one, and integrating from $-\infty$ to $+\infty$, we have

$$\ell_{(2)}(+\infty)-\ell_{(2)}(-\infty)=\int_{-\infty}^{+\infty}\left(f'_{(1)}(t)\,I_{(1)}(t)+g'_{(1)}(t)\left(-\frac{n_{(2)}\,H(t)}{2\pi}+J_{(1)}(t)\right)\right)dt.$$
$$\tag{14.80}$$

We shall assume that $f'_{(2)}$, $f''_{(2)}$, $g'_{(2)}$, and $g''_{(2)}$ all remain bounded for $t\to\pm\infty$; this is indeed a mild requirement that we make on the solutions of the second-order problem. Since $f'_{(1)}$, $f''_{(1)}$, $g'_{(1)}$, and $g''_{(1)}$ all go zero for $t\to\pm\infty$, this implies by equation (14.79):

$$\ell_{(2)}(t)\to 0\quad\text{for }t\to\pm\infty. \tag{14.81}$$

[12] Solubility conditions have been encountered earlier in Section 7.5.3 in the post-buckling analysis of a long elastic plate clamped on its edges. In the present chapter, another solubility condition was worked out in Section 14.7.3, when we showed that $n_{(1)}=0$ is a necessary condition for the first-order problem to have a solution.

Then, the left-hand side of equation (14.80) is zero and the latter takes the form of a linear equation for $n_{(2)}$:

$$\frac{n_{(2)}}{2\pi} \int_{-\infty}^{+\infty} H \, g'_{(1)} \, dt = \int_{-\infty}^{+\infty} \left(f'_{(1)}(t) \, I_{(1)}(t) + g'_{(1)}(t) \, J_{(1)}(t) \right) dt. \tag{14.82}$$

This equation is the solubility condition for $n_{(2)}$ that we were seeking: we have managed to get rid of the functions $f_{(2)}$ and $g_{(2)}$ of the second-order problem.

The coefficient in the left-hand side can be evaluated as follows:

$$\int_{-\infty}^{+\infty} H \, g'_{(1)} dt = \int_{0}^{+\infty} g'_{(1)} dt = g_{(1)}(+\infty) - g_{(1)}(0) = 1,$$

and so the value of $n_{(2)}$ is known directly in terms of the solution of the first-order problem calculated in Section 14.7.4:

$$n_{(2)} = 2\pi \int_{-\infty}^{+\infty} \left(f'_{(1)}(t) \, I_{(1)}(t) + g'_{(1)}(t) \, J_{(1)}(t) \right) dt. \tag{14.83}$$

The integrand in the right-hand side can be written explicitly using the definitions (14.64c) and (14.64d) of $I_{(1)}$ and $J_{(1)}$; the integral can be evaluated numerically using our numerical solution $(f_{(1)}, g_{(1)})$. This yields

$$n_{(2)} = 20.83. \tag{14.84}$$

We emphasize that all parameters of the problem could be absorbed by rescalings in the regime considered. As a result the value of $n_{(2)}$ is a constant.

Although this may not look obvious, this derivation of the solubility condition for $n_{(2)}$ is quite natural. In Section 14.7.3 we mentioned the existence of the invariant ℓ_1, in connection with the variational structure of the first-order problem. Now, it should be noticed that the left-hand side of the second-order problem is obtained by linearization of the equations for the first-order problem; to check this, replace $f_{(1)}$ by $\tilde{f}_{(1)} = f_{(1)} + \eta f_{(2)}$ and $\tilde{g}_{(1)} = g_{(1)} + \eta g_{(2)}$, expand equations (14.74) to first order with respect to η, and compare with equations (14.64). In view of this, the second invariant $\ell_{(2)}(t)$ appears to derive from the first invariant $\tilde{\ell}_{(1)}$ by linearization to first order in η, when $\tilde{f}_{(1)}$ and $\tilde{g}_{(1)}$ are inserted in the first invariant: the solubility condition for $n_{(2)}$ follows from the existence of the invariant $n_{(2)}$, which is essentially the same as the invariant $n_{(1)}$.

As an epilogue to the analysis of the circular topology of contact, we shall express the load–displacement relation in terms of the physical parameters of the problem, the force F and the displacement e of the apex with respect to the fixed equator of the shell. For that purpose, we use the expansion (14.62d) with $n_{(1)} = 0$:

$$\hat{N} = \hat{X} \, n_{(2)}, \tag{14.85}$$

and restore the physical units in this equation using (14.46). This yields:

$$\frac{F}{E \, h \, R \, \epsilon^2} = \frac{r}{R \, \epsilon^{1/2}} \, n_{(2)}. \tag{14.86}$$

The geometrical relation between the radius r of the circle of contact and the amount of squeezing of the ball e can be rewritten, including the numerical factor that had been discarded in (14.2):

$$r = \sqrt{2\,e\,R}. \tag{14.87}$$

Finally, we obtain the reaction force in physical units as:

$$N = \sqrt{2}\,\epsilon^{\frac{3}{2}}\,n_{(2)}\,E\,h\,R^{\frac{1}{2}}\,e^{\frac{1}{2}}$$

$$= \frac{\sqrt{2}\,n_{(2)}}{[12\,(1-\nu^2)]^{3/4}}\,\frac{E\,h^{5/2}}{R}\,e^{1/2}, \tag{14.88}$$

where the value of the constant $n_{(2)}$ is available from equation (14.84) above. This formula captures exactly the dependence of the force N on the various parameters of the problem, such as Poisson's ratio ν, Young's modulus E and thickness h, in the regime considered here, $h \ll e \ll R$. It complements Pogorelov's scaling analysis in equation (14.5b) by providing the exact value of the prefactor. The value of this prefactor is consistent with that found in Libai and Simmonds' book (A. Libai and J. G. Simmonds, 1998) in the regime that they call $p \to \infty$.

14.7.6 Other regimes

In the present analysis of the dimple, we have focused on the regime where the radius of the circle of contact is large compared with the ridge width, $r \ll \delta$, but small compared with the shell radius, $r \ll R$. This regime is the simplest to analyse as all physical parameters can be removed by rescaling.

Other regimes can be studied. For a review, see the book by Libai and Simmonds (A. Libai and J. G. Simmonds, 1998). For smaller values of the force, a nascent dimple is formed; this corresponds to the scaling laws of Section 14.6, when the radius r of the region of contact becomes comparable with the width δ of the ridge. Then, both the bending and stretching energies become evenly distributed over a neighbourhood of the apex, and there is no longer an internal layer in the sense of Appendix B. For even smaller forces, i.e. before a dimple appears, the deformation can be described by a linearized membrane theory, which has an analytical solution (W. T. Koiter, 1963; A. Libai and J. G. Simmonds, 1998) for the case of a point force.

On the contrary, if the force (or the displacement of the wall) is increased, the radius of the dimple may not remain small compared with the radius of the shell, and may become such that $r \sim R$. If this happens, shallow shell equations are not longer applicable, and one has to return to the general equations for (deep) shells. If the shell is thin enough to make the width of the ridge such that $\delta \ll r \sim R$, the ridge can be still described by an internal layer, and the approach of the previous section can be adapted: treatment of the deep and the shallow cases are similar, the only difference being that the trigonometric functions that have been removed by linearization in the shallow case have to be retained in the deep case. The layer equations depend on an additional geometric parameter in the deep case, which is the angle $\gamma = \sin^{-1}(r/R)$ of the ridge. For details, see Fulton's PhD thesis (J. P. Fulton, 1988) or Libai and Simmonds' book (A. Libai and J. G. Simmonds, 1998); in this book, an interesting calculation is presented: the first corrections to the load–displacement curve predicted by the shallow theory are found by perturbation with respect to the small parameter γ.

14.8   Stability of disc-like contact, transition to circular contact

In the present section, the competition between configurations with flat contact and those with an inverted cap is studied; we show that the configuration with flat contact becomes linearly unstable for large enough forces. The stability analysis presented here largely follows the work of Updike and Kalnins (D. P. Updike and A. Kalnins, 1970) but refines their approximate treatment of the edge of the region of contact.

The stability of spherical shells subjected to a point force directed inwards has been addressed by several authors. Using an approximate solution of the non-linear equations for shells, Biezeno calculated load–displacement curves for axisymmetric deformations of the shell and identified some regions with unstable equilibria (C. B. Biezeno, 1935). Evan-Iwanovski experimentally measured the critical loads of spherical shell caps subjected to concentrated loads, depending on the shallowness of the shell (R. M. Evan-Iwanowski, H. S. Cheng, and T. C. Loo, 1962). Based on a numerical solution of the non-linear equations for deep, axisymmetric shells, Mescall (J. F. Mescall, 1965) computed load–deflection curves for spherical caps of various depths;[13] typically, these curves are non-monotonic: the load increases, and then decreases, and then increases again as the displacement of the apex increases. At the first maximum of the load, there is a discontinuous (snap-through) bifurcation in a load-controlled experiment, leading to a configuration with an inverted cap—see also (J. R. Fitch, 1968; J. R. Fitch and B. Budiansky, 1970). As noted by Budiansky (B. Budiansky, 1959) the nature of the bifurcation (continuous versus discontinuous) may be different when the load is increased (dimple is formed) or decreased (dimple disappears). This concerns stability with respect to axisymmetric configurations of the shell; instabilities breaking the symmetry of revolution are also possible, and are discussed later in Section 14.8.6.

The section is organized as follows. In Section 14.8.1, we return to the solution with flat, disk-like contact, and consider the case of large pushing forces. Based on a simple dimensional analysis, we show that this solution involves more elastic energy than that with an inverted cap when the force is large enough. This suggests a bifurcation from a disc-like contact to a circular one at increasing force. This bifurcation is studied quantitatively from Section 14.8.2 on: we compute the critical force at which the disc-like contact becomes linearly unstable, and show that a dynamic instability (snap-through) occurs, leading to an inverted cap.

14.8.1   A simple order of magnitude argument

In this subsection, we show that the flat disc of contact cannot remain the configuration with the lowest energy when the displacement e of the pushing plane is large enough: the elastic energy stored in the flat disk is estimated by dimensional analysis, and is found to be larger than that for the inverted cap configuration. This suggests an instability from the flat disc of contact to the inverted cap for some critical value of the force (this instability will be studied in detail next).

Let us then compare the elastic energy of the inverted cap configuration to that of the flat disc of contact when the compression e is in the range $h \ll e \ll R$. Pogorelov's order

---

[13] By depth, we refer to the case of a deep shell, versus a shallow one. The depth of a spherical shell cap is measured by a parameter λ that depends on the opening angle γ of the cap.

of magnitude estimate (14.6) of the inverted cap configuration, recovered in section 14.7, applies:

$$\mathcal{E}_{\text{circ}} \sim F\,e \sim \frac{E\,h^{5/2}\,e^{3/2}}{R}. \tag{14.89}$$

Now, a lower bound for the energy of the configuration with a flat contact can be obtained by considering the contribution from the region of contact only. There, the stress is given by equation (14.41): $N_s = Eh\left(C + (s/R)^2/16\right)$ in physical units. The value of the constant C is of the same order, $(r/R)^2$ as the other term, and so the membrane stress N_s is of order $E h\, r^2/R^2$ on average in the disc of contact. As a result, the energy contribution of this region is of order Young's modulus, E, times the volume, of order $h\,r^2$, times the square of the typical strain, $(r^2/R^2)^2$. Using the geometrical relation $r \sim R^{1/2}e^{1/2}$, we have:

$$\mathcal{E}_{\text{disc}} \gtrsim \frac{E\,h\,e^3}{R}. \tag{14.90}$$

This lower bound is valid under the condition $e \ll R$, which we used as a starting point of this chapter. Note that the lower bound (14.90) is compatible with the scaling laws of Section 14.6 as both sides of the equations are then of order Eh^4/R, e being then of order h.

Comparison of (14.89) with the lower bound in equation (14.90) shows that the energy of the configuration with a disc of contact grows much faster (like e^3 versus $e^{3/2}$) for large e than for an inverted cap. This suggests an instability of the former towards the latter when the pushing force is increased. This instability is studied next.

14.8.2 Linear stability of the flat disc of contact

This argument suggests an instability of the configuration with disc-like contact, leading to the formation of an inverted cap. This instability has been studied numerically by Updike and Kalnins (D. P. Updike and A. Kalnins, 1970) using approximate boundary conditions (they consider a disc of contact with a prescribed radius while the boundary is actually expanding during the instability). In addition, they solve the problem numerically for small but finite values of the thickness, and do not properly study the limit of a thin shell. Here, we investigate this instability with the exact boundary conditions at the edge of the region of contact. Our asymptotic analysis, relevant to the limit $h \to 0$, enables us to characterize the instability for a wider range of parameters, by means of analytical formulae.

We study the linear stability of the family of solutions with a flat contact, obtained in Section 14.6. We consider stability with respect to perturbations that allow the disc of contact to lift off from the rigid plane, thereby leading to a circular contact. We find that this bifurcation is trans-critical and should result in a dynamical instability (snap-through), as has been observed in experiments—see Fig. 14.2.

Let us consider a small perturbation of the shell near an equilibrium configuration with a flat contact:

$$\Sigma(x) = \Sigma^{(0)}(x) + \Sigma^{(1)}(x) \qquad \Psi(x) = \Psi^{(0)}(x) + \Psi^{(1)}(x), \tag{14.91}$$

where $\Sigma^{(0)}$ and $\Psi^{(0)}$ denote the solution with a flat (disc-like) contact derived earlier, and $(\Sigma^{(1)}, \Psi^{(1)})$ is a small perturbation. We denote the solutions in the inner ($x < X$) and outer part $(x > X)$ by using subscripts 'in' and 'out'. Moreover, we use the same rescalings as in equation (14.46). All symbols in the present section should bear a hat, even though these hats are omitted for the sake of readability: we write x, $\Sigma^{(0)}$, $\Psi^{(0)}$, $\Psi^{(0)}$, etc., in place of $\hat{x}$, $\hat{\Sigma}^{(0)}$, $\hat{\Psi}^{(0)}$, $\hat{\Psi}^{(0)}$, etc. The use of hat variables defined by this scaling (14.46) is relevant to the regime

$$e \sim h,$$

on which we focus here. Note that the scaling argument of Section 14.8.1 indeed predicts that the instability takes place in this regime—this is where the right-hand sides of equations (14.89) and (14.90) become of comparable magnitude.

The linear stability analysis aims at finding perturbations that satisfy the equations of equilibrium to first order, as this allows one to track continuous bifurcations. Let us first consider the inner region, $x < X$. There, $\Sigma_{\mathrm{in}}^{(0)}$ is given by equation (14.41) and $\Psi_{\mathrm{in}}^{(0)}$ by (14.38). The perturbation lifts off the shell and changes the topology of contact, turning the interior of the disc of contact into a free cap. The equations for the perturbation are found by setting[14] $\hat{N} = 0$ in the equations (14.47) for a free cap, and linearizing near the base solution $(\Sigma_{\mathrm{in}}^{(0)}, \Psi_{\mathrm{in}}^{(0)})$:

$$3\, x\, \Sigma_{\mathrm{in}}^{(1)'}(x) + x^2\, \Sigma_{\mathrm{in}}^{(1)''}(x) = 0, \tag{14.92a}$$

$$x\left(C^{(0)} + \frac{x^2}{16}\right)\Psi_{\mathrm{in}}^{(1)}(x) - \left(x\,\Psi_{\mathrm{in}}^{(1)''}(x) + \Psi_{\mathrm{in}}^{(1)'}(x) - \frac{\Psi_{\mathrm{in}}^{(1)}(x)}{x}\right) = 0, \tag{14.92b}$$

where the first term in the second equation comes from the explicit expression for $\Sigma_{\mathrm{in}}^{(0)}$ in equation (14.92b). Remarkably,[15] these equation for $\Sigma_{\mathrm{in}}^{(1)}$ and $\Psi_{\mathrm{in}}^{(1)}$ decouple.

The first equation (14.92a) is the linearized version of equation (14.39). Its solutions that are smooth at the origin $x = 0$ can be found either directly, or by linearizing the general solution (14.41); this yields

$$\Sigma_{\mathrm{in}}^{(1)}(x) = C^{(1)}, \tag{14.93a}$$

for some constant $C^{(1)}$ to be determined.

The second equation (14.92b) is singular at the origin, $\Psi_{\mathrm{in}}^{(1)}(x)$ diverging generically like $1/x$ for $x \to 0$. Being a differential equation of order 2, its space of solutions that are regular at $x = 0$ is one dimensional. This means that the solution $\Psi_{\mathrm{in}}^{(1)}(x)$ is a fixed function, denoted Ψ_{p}, times some multiplicative constant $G^{(1)}$:

$$\Psi_{\mathrm{in}}^{(1)}(x) = G^{(1)}\, \Psi_{\mathrm{p}}(x). \tag{14.93b}$$

The function $\Psi_{\mathrm{p}}(x)$ is a particular solution of equation (14.92b) that is regular at the origin. It can be computed numerically by numerical integration of the equation, starting off with initial data corresponding to the smooth expansion of $\Psi_{\mathrm{in}}^{(1)}$ near the origin, for any value

---

[14] In the absence of an external, vertical, point force applied at the pole, the equilibrium of this point imposes $\hat{N} = 0$ in the interior region, $x < X$.

[15] The uncoupling of equations (14.92) is a consequence of the up–down symmetry of the base solution in the (former) region of contact. A similar remark has been made in Section 7.4 in the analysis of buckling of a plate near a planar configuration.

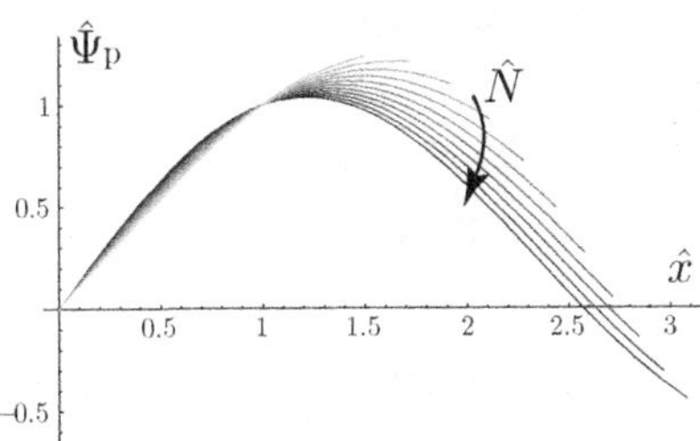

Fig. 14.17 Particular solution $\hat{\Psi}_{\mathrm{p}}(\hat{x})$ of equation (14.92b). The function $\hat{\Psi}_{\mathrm{p}}$ is plotted in the range $0 \le \hat{x} \le \hat{X}^{(0)}(\hat{N})$ for 15 different values of rescaled force $\hat{N}$, evenly spaced in the range 6.28–94.25. Here $\hat{X}^{(0)}(\hat{N})$ denotes the radius of the disc of contact in the base solution, a quantity which was plotted earlier in figure 14.12.

of the parameter $C^{(0)}$ (recall that $C^{(0)}$ is known as a function of the applied force by the results of Section 14.6.4, see Fig. 14.13). This function $\Psi_{\mathrm{p}}(x)$ is plotted in Fig. 14.17) for various values of the applied force N.

We now proceed to the analysis of the outer region $(x > X)$. Let $\Sigma_{\mathrm{out}}^{(0)}$ and $\Psi_{\mathrm{out}}^{(0)}$ denote the solution of equations (14.47), which were computed in Section 14.6.4. The perturbations $\Sigma_{\mathrm{out}}^{(1)}$ and $\Psi_{\mathrm{out}}^{(1)}$ are solutions of a linearized version of the same equations. Now, recall that the unperturbed solutions $\Sigma_{\mathrm{out}}^{(0)}$ and $\Psi_{\mathrm{out}}^{(0)}$ were determined by backward numerical integration of (14.47), starting from $x = +\infty$: for large x the generic, regular asymptotic behaviour of $\Sigma_{\mathrm{out}}^{(0)}$ and $\Psi_{\mathrm{out}}^{(0)}$ was expressed in terms of two parameters, now denoted $a^{(0)}$ and $b^{(0)}$; this generic asymptotic expansion was used to provide initial values for the backward integration. Next order solutions $\Sigma_{\mathrm{out}}^{(1)}$ and $\Psi_{\mathrm{out}}^{(1)}$ of the linearized problem can be determined by exactly the same method, using two expansion parameters $a^{(1)}$ and $b^{(1)}$. The equations now being linear, their solutions are obtained by linear superposition of particular solutions $(\Sigma_{\mathrm{a}}, \Psi_{\mathrm{a}})$ and $(\Sigma_{\mathrm{b}}, \Psi_{\mathrm{b}})$, corresponding to $(a^{(1)} = 1, b^{(1)} = 0)$, and to $(a^{(1)} = 0, b^{(1)} = 1)$ respectively:

$$\Sigma_{\mathrm{out}}^{(1)}(x) = a^{(1)}\,\Sigma_{\mathrm{a}}(x) + b^{(1)}\,\Sigma_{\mathrm{b}}(x) \tag{14.94a}$$

$$\Psi_{\mathrm{out}}^{(1)}(x) = a^{(1)}\,\Psi_{\mathrm{a}}(x) + b^{(1)}\,\Psi_{\mathrm{b}}(x). \tag{14.94b}$$

The functions Σ_{a}, Ψ_{a}, Σ_{b} and Ψ_{b} can be computed numerically for any value of the applied force N.

This linear stability analysis has to be complemented by matching the solutions constructed above in the interior $x < X$, see equation (14.93), and from the exterior region $x > X$, see equation (14.94). We shall now work out these matching conditions, which hold at the edge, $x = X$. These conditions will allow us to fix the remaining parameters of the problem, $C^{(1)}$, $G^{(1)}$, $a^{(1)}$, $b^{(1)}$ and, more importantly, to compute the threshold of instability.

Let us introduce $X^{(1)}$, which is the perturbation to the radius of the circle of contact—taking into account the fact that the disc of contact may expand (or shrink) during the instability, we release Updike and Kalnins' approximation (D. P. Updike and A. Kalnins, 1970) of a *clamped* disc of contact. The perturbed radius of the circle reads

$$X = X^{(0)} + X^{(1)}. \tag{14.95}$$

The first matching condition (14.37a) writes $\Sigma(X^{+}) - \Sigma(X^{-}) = 0$, that is

$$\left(\Sigma_{\text{out}}^{(0)} + \Sigma_{\text{out}}^{(1)}\right)_{X^{(0)}+X^{(1)}} - \left(\Sigma_{\text{in}}^{(0)} + \Sigma_{\text{in}}^{(1)}\right)_{X^{(0)}+X^{(1)}} = 0, \tag{14.96}$$

and there are three other similar conditions for the continuity of Σ', Ψ and Ψ'. By linearizing the equation above for small $\Sigma^{(1)}$, $\Psi^{(1)}$ and $X^{(1)}$, one obtains:

$$\left(\Sigma_{\text{out}}^{(0)} - \Sigma_{\text{in}}^{(0)}\right)_{X^{(0)}} + X^{(1)} \left.\frac{\mathrm{d}\left(\Sigma_{\text{out}}^{(0)} - \Sigma_{\text{in}}^{(0)}\right)}{\mathrm{d}x}\right|_{X^{(0)}} + \left(\Sigma_{\text{out}}^{(1)} - \Sigma_{\text{in}}^{(1)}\right)_{X^{(0)}} = 0. \tag{14.97}$$

Note that the last term is already a first-order correction, and so can be evaluated indifferently at $X^{(0)}$ or at $X = X^{(0)} + X^{(1)}$ at this order. The first term vanishes since the base solution $(\Sigma^{(0)}, \Psi^{(0)})$ satisfies the continuity condition at $x = X^{(0)}$ by construction. The last term can be expressed in terms of Σ_{a}, Σ_{b} and $C^{(1)}$ using equations (14.93) and (14.94). Finally, the second term is $X^{(1)}$ times the discontinuity of $\Sigma^{(0)'}$ at the unperturbed edge of the disc of contact $x = X$. This discontinuity is known from equations (14.37) and is zero in the present case, $[\![\Sigma^{(0)'}]\!] = 0$. However, since $\Psi^{(0)''}$ makes a jump $N/(2\pi X^{(0)})$ at $x = X$, this yields a non-zero contribution for the condition of continuity of Ψ'—see the first term in the right-hand side of equation (14.98) below.

The four matching conditions for the linear stability problem are worked out similarly:

$$\begin{pmatrix} [\![\Sigma]\!] \\ [\![\Sigma']\!] \\ [\![\Psi]\!] \\ [\![\Psi']\!] \end{pmatrix}_X = \begin{pmatrix} 0 \\ 0 \\ 0 \\ X^{(1)} \dfrac{N}{2\pi X^{(0)}} \end{pmatrix} + \begin{pmatrix} a^{(1)} \Sigma_{\text{a}} + b^{(1)} \Sigma_{\text{b}} \\ a^{(1)} \Sigma_{\text{a}}' + b^{(1)} \Sigma_{\text{b}}' \\ a^{(1)} \Psi_{\text{a}} + b^{(1)} \Psi_{\text{b}} \\ a^{(1)} \Psi_{\text{a}}' + b^{(1)} \Psi_{\text{b}}' \end{pmatrix}_{X^{(0)}} - \begin{pmatrix} C^{(1)} \\ 0 \\ G^{(1)} \Psi_{\text{p}} \\ G^{(1)} \Psi_{\text{p}}' \end{pmatrix}_{X^{(0)}}. \tag{14.98}$$

The five unknowns $r^{(1)}$, $a^{(1)}$, $b^{(1)}$, $C^{(1)}$, $G^{(1)}$ have to be adjusted in such a way that both sides of the equation are zero. Equation (14.98) provides a set of four linear equations for these five unknowns.

The missing equation comes from the condition that the tangent to the shell has to be horizontal at the circle of contact, something we expressed earlier in equation (14.66). In terms of the current variables, this condition reads $\Psi(X) + X = 0$. Recall that the orientation of the deformed tangent is given by the rescaled angle $x + \Psi(x)$. By expanding this equation with respect to the perturbation, one obtains:

$$\Psi^{(0)}(X^{(0)}) + X^{(0)} + \Psi^{(1)}(X^{(0)}) + X^{(1)} \Psi^{(0)'}(X^{(0)}) + X^{(1)} = 0.$$

Now, $\Psi^{(0)'}(X^{(0)}) = -1$ by equation (14.48), and so the last two terms cancel. In addition the first two terms cancel since the base solution satisfies the condition of tangent contact at $X = X^{(0)}$ by construction. We are left with the term $\Psi^{(1)}(X^{(0)})$, that is, from equation (14.93b),

$$G^{(1)} \Psi_{\text{p}}(X^{(0)}) = 0. \tag{14.99}$$

Together, equations (14.98) and (14.99) govern the stability of flat contact. There are five linear equations with five adjustable parameters. These equations have generically only one trivial solution: $X^{(1)} = a^{(1)} = b^{(1)} = C^{(1)} = G^{(1)} = 0$. However, for some discrete values of the force N, called *critical* values, the linear system is singular, as we show. Then,

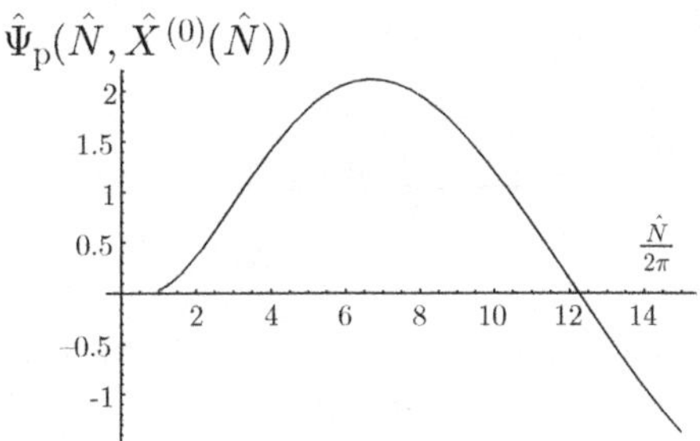

Fig. 14.18 Graphical solution of the implicit equation (14.101) defining the critical force $\hat{N}_{\mathrm{c}}$ associated with the loss of linear stability for the configuration with a disc-like topology of contact. This curve is drawn by the endpoints of the family of curves plotted in Fig. 14.17. The root is near $\hat{N}/2\pi = 12$; the exact value found by a root-finding algorithm appears in equation (14.102).

there exists non-vanishing solutions $(\Sigma^{(1)}, \Psi^{(1)})$ that satisfy the linearized equations of equilibrium, and the shell is linearly unstable. This critical value of the force is investigated below.

14.8.3 Numerical calculation of the critical force

By putting equations (14.98) and (14.99) into matrix form, the marginal stability condition can be expressed by the vanishing of the following five by five determinant:

$$\det \begin{pmatrix} 0 & \Sigma_{\mathrm{a}} & \Sigma_{\mathrm{b}} & -1 & 0 \\ 0 & \Sigma_{\mathrm{a}}' & \Sigma_{\mathrm{b}}' & 0 & 0 \\ 0 & \Psi_{\mathrm{a}} & \Psi_{\mathrm{b}} & 0 & -\Psi_{\mathrm{p}} \\ \dfrac{N}{2\pi X^{(0)}} & \Psi_{\mathrm{a}}' & \Psi_{\mathrm{b}}' & 0 & -\Psi_{\mathrm{p}}' \\ 0 & 0 & 0 & 0 & \Psi_{\mathrm{p}} \end{pmatrix}_{X^{(0)}} = 0,$$

where all functions are now evaluated at $X^{(0)}$. Once expanded with respect to its first and fourth columns and finally with respect to its last row, this determinant simplifies into:

$$\frac{N}{2\pi X^{(0)}} \, \Psi_{\mathrm{p}}(X^{(0)}) \, \det \begin{pmatrix} \Sigma_{\mathrm{a}}'(X^{(0)}) & \Sigma_{\mathrm{b}}'(X^{(0)}) \\ \Psi_{\mathrm{a}}(X^{(0)}) & \Psi_{\mathrm{b}}(X^{(0)}) \end{pmatrix} = 0. \tag{14.100}$$

This equation must be seen as an equation for the critical force N at which the configuration with a flat contact becomes linearly unstable. Even though this dependence is not made explicit in this equation, all the functions Ψ_{p}, Σ_{a} and Σ_{b}, as well as the number $X^{(0)}$, depend implicitly on the applied force N: they were obtained by solving the linearized equations for a particular value of N—see Figs 14.12 and 14.17 for instance.

We explained earlier how the functions Ψ_{p}, Σ_{a}, Σ_{b}, Ψ_{a} and Ψ_{b} appearing in equation (14.100) can be computed numerically for an arbitrary value of the force N. We can therefore evaluate these functions at $x = X^{(0)}$, something that yields the numerical values of the left-hand side of equation (14.100) for different N. We did so, and found that the determinant does not change sign over a broad range of values of N. However, the term $\Psi_{\mathrm{p}}(X^{(0)})$ does change sign, as shown by Figs 14.17 and 14.18. The root N_{c} of the implicit equation

$$\Psi_{\mathrm{p}}(N_{\mathrm{c}}, X^{(0)}(N_{\mathrm{c}})) = 0 \tag{14.101}$$

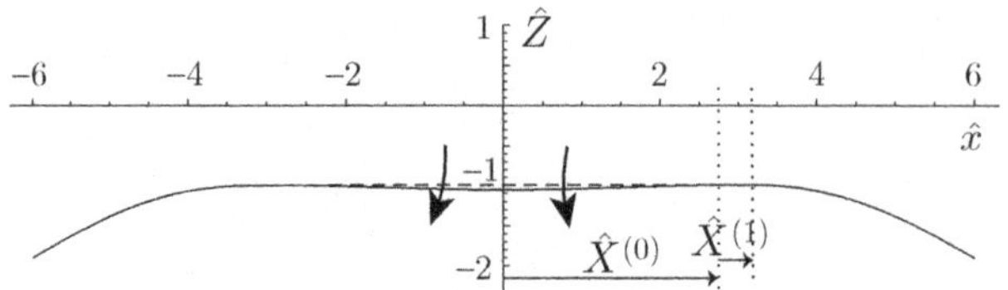

Fig. 14.19 Visualization of the marginally unstable mode near a solution with disc-like contact by a cut through the shell containing the axis of revolution, for $\hat{N} = \hat{N}_{\mathrm{c}}$. This mode is plotted with an arbitrary, small amplitude for the purpose of visualization.

defines the critical force at which the flat contact looses stability. This equation is solved graphically in Fig. 14.18. The most accurate value obtained by numerical root-finding is

$$N_{\mathrm{c}} = 76.94. \tag{14.102a}$$

The critical force F_{c} in physical units can be found from equation (14.49), recalling that what is called N_{c} in equation above is in fact $\hat{N}_{\mathrm{c}}$, as hats are omitted in the present section:

$$F_{\mathrm{c}} = \frac{N_{\mathrm{c}}}{1 - \nu^2} \frac{E\,h^3}{R}. \tag{14.102b}$$

The value of N_{c} is given just above in equation (14.102a). Since all parameters of the problem could be scaled out, this value is universal; this calculation of the buckling threshold exactly captures the dependence on the thickness h, Poisson's ratio ν, Young's modulus of the shell. The corresponding mode of instability is shown in Fig. 14.19.

14.8.4 Nature of the bifurcation

This analysis of stability shows that the configuration with a flat contact becomes linearly unstable for a critical value of the force F_{c}. In this subsection, we investigate whether the transition is continuous or discontinuous—recall that hysteresis was observed in the experiments.

In Fig. 14.15, we plotted only the branch corresponding to the configuration with flat contact in the force–displacement curve. We shall now try to complement this graph: at threshold F_{c}, this configuration becomes linearly unstable by crossing another branch corresponding to the inverted cap configuration. Depending on the direction of crossing, the bifurcation of the shell could be smooth *without* hysteresis (left), or dynamical *with* hysteresis (right), as sketched in Fig. 14.20.

This figure should be understood as follows. First, consider the part of the branch corresponding to an inverted cap is located *below* the linearly unstable point I. Configurations on this part of the curve are such that the lowering e of the pushing plane, shown along the vertical axis, is less in the buckled configuration than in the unbuckled one: this corresponds to a disc of contact buckling *upwards*, thereby penetrating the rigid plane. Such configurations must be discarded (they are shown with dotted curves).

Now, let $\beta = \partial e_{\mathrm{inv}}/\partial F$ be the slope of the branch with an inverted cap at the crossing point I, for which $F = F_{\mathrm{c}}$. In the absence of particular symmetry, β has no reason to vanish. In Fig. 14.20, the case $\beta > 0$ is shown at left, while $\beta < 0$ is at right. If $\beta > 0$,

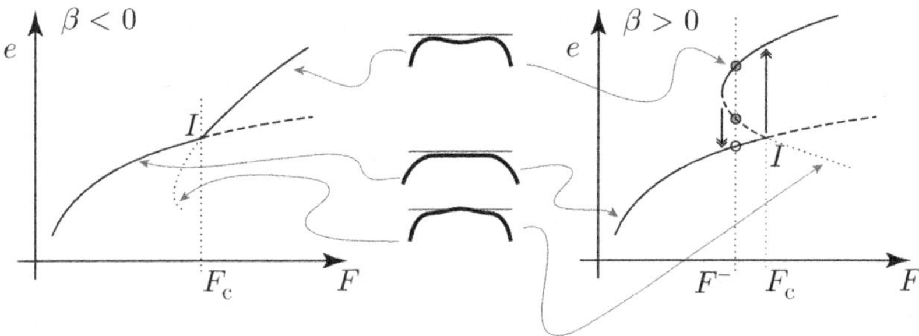

Fig. 14.20 Sketch of the possible configurations of the force–displacement curve of the spherical shell. One branch corresponds to the unbuckled solution with flat contact, and the other one to the solution with an inverted cap. When $\beta > 0$ (left), instability of the flat disc is smooth, while when $\beta < 0$ (right), it is discontinuous (snap-through is denoted by double arrows) and leads to a hysteresis in the force–displacement curve. Dashed curves denote the unstable parts of the branches, while dotted ones are for unphysical configurations that penetrate into the rigid plane. The case $\beta < 0$ (right) can be identified by the fact that two equilibrium configurations with an inverted cap (grey discs) can be found for $N = N^{-}$ just below the threshold for linear stability F_{c}, in addition to the solution with a disc-like contact (circle).

behaviour of the shell above F_{c} is smooth: the buckling amplitude of the disc of contact increases gradually from zero above $F = F_{\mathrm{c}}$ and the same path is followed when F is decreased down to $F = F_{\mathrm{c}}$. Now, when $\beta < 0$, there exists no steady configuration in the vicinity of the flat contact as soon as F becomes slightly larger than F_{c}. As a result, the deflection e changes abruptly while an inverted cap is formed in a dynamic buckling process. When F is decreased, the upper branch can be followed even when the loading remains well below the linear threshold F_{c} and load–displacement curve displays some hysteresis.

The actual value of β could be computed by going beyond the linear stability analysis, that is by studying the weakly non-linear regime as done in Chapter 7. The following trick will allow us to find the sign of β more easily. Indeed, if we set the force to $F = F^{-}$, a value chosen slightly below F_{c}, then Fig. 14.20 reveals that no inverted cap configurations can be found if $\beta > 0$, although there are two such inverted configuration if $\beta < 0$ (depicted by small circles on the figure) By solving the full inverted cap problem, that is by using the non-linear equations in both the outer and inner free parts, by matching them at the circle of contact, and by using a shooting method to enforce boundary conditions at the pole and for $s \to \infty$, we were able to find two inverted cap configurations. We used $F^{-} = 24\,\pi \approx 75.4$, which is indeed just below N_{c}—compare with (14.102a). The corresponding solutions Ψ are shown in Fig. 14.21, while the corresponding configurations of the shell are shown in Fig. 14.22.

This proves that for $\beta < 0$: the actual force–displacement curve therefore undergoes a discontinuous transition when the flat disc of contact buckles into an inverted cap, as in Fig. 14.20, right. Moreover, there is some hysteresis because two configurations are stable for a range of values of the force below the threshold F_{c} corresponding to linear stability. The second critical force, relevant to *decreasing* applied forces, could be computed by a full numerical solution of the non-linear equations.

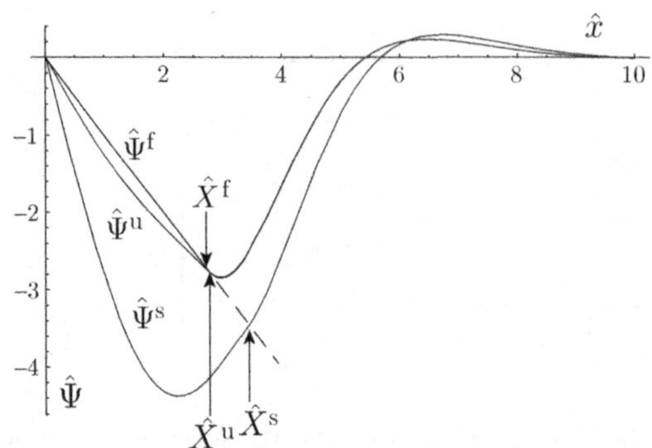

Fig. 14.21 Just below F_c (here $\hat{N} = \hat{N}^- = 75.4$), two equilibrium configurations with an inverted cap can be found: one, labelled u, that is very close to the flat configuration, labelled f, and another one, labelled s, which is farther away. These two solutions correspond to the grey disks in Fig. 14.20 right. This show that the sign of β is $\beta < 0$. The solution 'u' is unstable while the solution 's' is stable. The dashed line corresponds to $\hat{\Psi} = -\hat{x}$, which holds in the presence of contact.

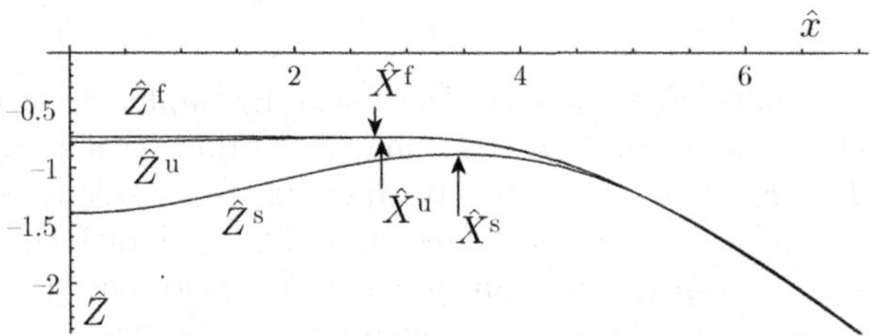

Fig. 14.22 Equilibrium configurations just below the critical force F_c, shown by a cut through the shell in rescaled units. As in Fig. 14.21, $\hat{N} = \hat{N}^- = 75.4$. These solutions are obtained by a non-linear shooting method: there are two stable solutions: 'f' with a flat contact and 's' with an inverted cap, plus one unstable inverted cap configuration, 'u'. Since $\beta < 0$, the two stable configurations are those that can be followed in an hysteresis cycle. When F is increased beyond F_c, for instance, there is an abrupt jump for a flat configuration to the stable inverted cap, 's'.

14.8.5 Comparison with experiments

A detailed comparison with the experiments presented at the beginning of this chapter is shown in Fig. 14.23. The main sources of discrepancy between theory and experiment are the friction and the non-Hookean behaviour of the shell in the experiments. These sources of dissipation delay the transitions and widen the observed hysteresis loop—for instance, one can expect that the two experimental paths in the far-left region of this diagram would collapse in the absence of friction. Nevertheless, the agreement between theory and experiment is quite good and concerns both the location of the branch with a flat disc of contact, and the loss of stability of this branch ($F = F_\text{c}$). Note that there is no adjustable parameter in Fig. 14.23.

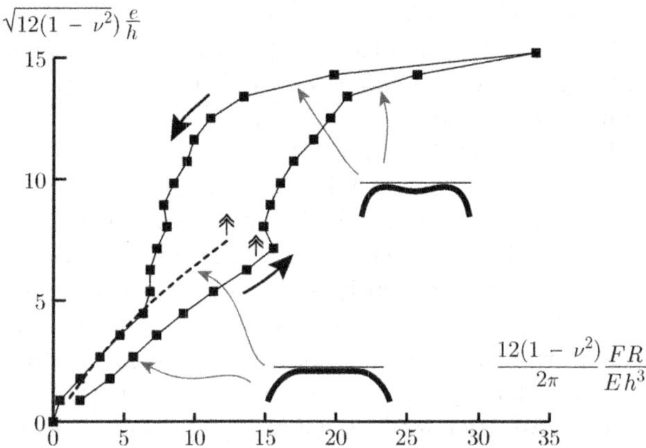

Fig. 14.23 Comparison between theory and experiments in dimensionless variables. Experimental results plotted earlier in Fig. 14.2 are shown as solid curves after proper rescaling, from the data in reference (L. Pauchard, Y. Pomeau, and S. Rica, 1997). They are superimposed with the theoretical results of section 14.6 for the flat contact (dashed curve). The latter branch is plotted up to the theoretical value of the critical force F_c where dynamic buckling was predicted. There is no adjustable parameter.

14.8.6 Other stability considerations

The analysis of stability of the configuration with flat contact has been restricted to axisymmetric perturbations. Updike and Kalnins extended this analysis of linear stability to non-symmetric perturbations (D. P. Updike and A. Kalnins, 1972), and considered as a base state both the configuration with flat contact and the configuration with disc-like contact. They found that a dimple appears first by an instability affecting the configuration with flat contact; at a larger value of the deflection, this dimpled solution becomes linearly unstable with respect to configurations without axial symmetry, displaying three or four lobes.

In the case of a point force applied on a spherical shell, bifurcations towards non-symmetric shapes are also possible, and have in fact been studied earlier. The early experimental work of Evan-Iwanovski and Penning (R. M. Evan-Iwanowski, H. S. Cheng, and T. C. Loo, 1962; F. A. Penning, 1966) revealed that the number of lobes varies with the load. For instance Penning (F. A. Penning, 1966) reports three lobes at the onset of asymmetric bifurcation, a number increasing first with the load up to five, and then decreasing down to three lobes at still larger loads. Budiansky and Hutchinson, and then Bushnell explained these experimental results by numerically studying the linear stability problem of an axisymmetric configuration with respect to non-symmetric perturbations (B. Budiansky and J. W. Hutchinson, 1966; D. Bushnell, 1967). Fitch and Budiansky studied more specifically the competition between (smooth) non-symmetric bifurcations, and the discontinuous (snap-through) axisymmetric bifurcations, as a function of the shell thickness, shell depth, and size of the region where the force is applied in the case of a spatially extended loading (J. R. Fitch, 1968; J. R. Fitch and B. Budiansky, 1970). Instabilities involving symmetric patterns (dimpling) and those involving non-symmetric patterns (formation of lobes) are of a different nature: axisymmetric buckling take place at a maximum of the loading curve and leads to snap-through as the load-carrying capacity of the shell starts to decrease; in contrast, asymmetric buckling is continuous as the load-carrying capacity of the shell typically continues to increase above threshold. For a more detailed discussion, see references (J. R. Fitch, 1968), (J. R. Fitch and B. Budiansky, 1970).

To complement of these considerations on stability, we shall briefly mention sensitivity to imperfections, a detailed discussion of which is unfortunately beyond the scope of this book. In some (but not so rare) cases, the critical load of a real shell is substantially decreased by small imperfections, of the order of the thickness of the shell. As a result the buckling load calculated for an ideal shell is far too optimistic which can lead to potentially disastrous design errors. Sensitivity to imperfections occurs when a bifurcation is such that the load that can be sustained by the shell starts to decrease past the bifurcation threshold, and Koiter (W. T. Koiter, 1965) has shown that sensitivity to imperfections is largely dictated by the weakly non-linear post-buckling behaviour of a perfect shell: it can be assessed by pushing the analysis of linear stability to second order. Spherical shells under external pressure are imperfection sensitive, as shown by Hutchinson (J. W. Hutchinson, 1967)—for a review see (A. Libai and J. G. Simmonds, 1998) and references therein.

Far above the initial buckling threshold, these smooth lobes evolve towards a network of polygonal creases. Such networks are shown in Fig. 14.24. They have been observed in qualitative indentation experiments with elliptic shells (L. Pauchard and S. Rica, 1998), in experiments with colloidal particles (C. Quilliet *et al.*, 2008) and in numerical simulations (S. Komura, K. Tamura, and T. Kato, 2005; C. Quilliet, 2006; A. Vaziri and L. Mahadevan, 2008; A. Vaziri, 2009). Even though some scaling arguments support the formation of a network of sharp creases for large loads (A. E. Lobkovsky, 1996; L. Pauchard and S. Rica, 1998), this intriguing and robust pattern is far from being fully understood.

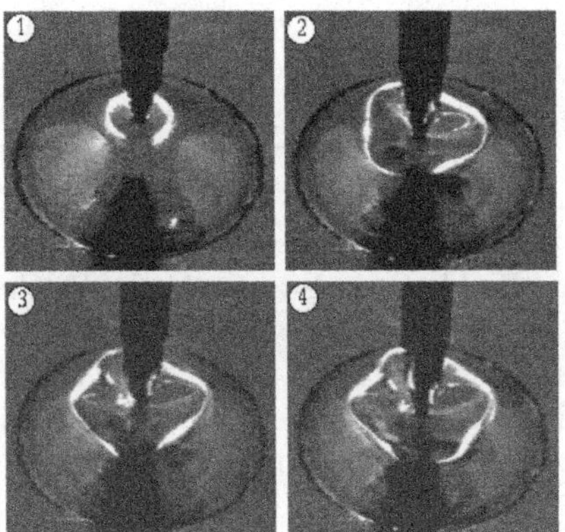 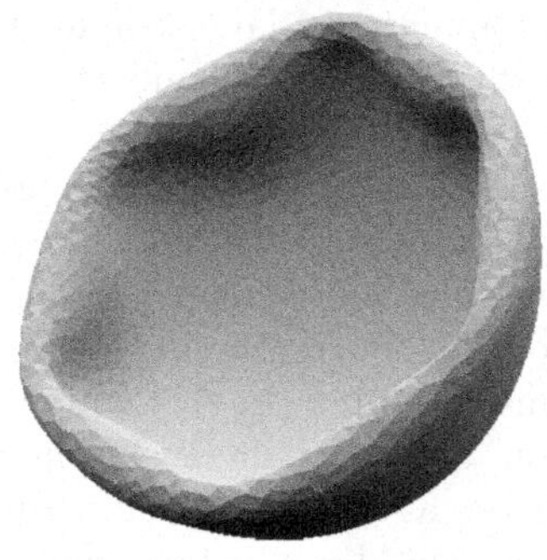

Fig. 14.24 *Left:* indentation of a spherical shell by a pen reveals that, at higher forces, the circular fold looses its symmetry of revolution by making polygons with more and more edges. Image courtesy of L. Pauchard (FAST laboratory, Univ. Paris-Sud, Orsay), reproduced from reference (L. Pauchard, Y. Pomeau, and S. Rica, 1999) with kind permission of Société Française de Physique. *Right:* numerical simulation of a thin spherical shell under large external pressure, showing the same type of polygonal ridge (C. Quilliet *et al.*, 2008). Image courtesy of L. Quilliet (lab. de Spectrométrie Physique, Univ. Joseph Fourier, Grenoble).

The problem of the elastic ball pushed by a rigid plane is an interesting example of application the of many ideas, from dimensional analysis to boundary layers methods, and to numerical methods for solving boundary value problems (BVP). The formation of singular structures (ridges) is a consequence of the lack of smooth isometric deformations of a sphere.

References

D. G. Ashwell. On the large deflection of a spherical shell with an inward point load. In W. T. Koiter, editor, *The Theory of Thin Elastic Shells, Proc. IUTAM Symposium, Delft, 1959*, pages 43–63. North-Holland, 1960.

L. Berke and R. L. Carlson. Experimental studies of the postbuckling behavior of complete spherical shells. *Experimental Mechanics*, 8:548–553, 1968.

B. Budiansky and J. W. Hutchinson. A survey of some buckling problems. *AIAA Journal*, 4:1505–1510, 1966.

C. B. Biezeno. Über die Bestimmung der 'Durchschlagkraft' einer schwachgekrümmten, kreisförmigen Platte. *Zeitschrift für angewandte Mathematik und Physik (ZAMP)*, 15:10–22, 1935.

M. Buenemann and P. Lenz. Mechanical limits of viral capsids. *PNAS*, 104(24):9925–9930, 2007.

B. Budiansky. Buckling of clamped shallow spherical shells. In *Proceedings, IUTAM Symposium on the Theory of Thin Elastic Shells, Delft, The Netherlands*, page 64, 1959.

D. Bushnell. Bifurcation phenomena in spherical shells under concentrated and ring loads. *AIAA Journal*, 5:2034–2040, 1967.

R. M. Evan-Iwanowski, H. S. Cheng, and T. C. Loo. Experimental investigation of deformations and stability of spherical shells subjected to concentrated load at the apex. In Americal Society Society of Mechanical Engineers, editor, *Proceedings of the Fourth U. S. National Congress of Applied Mechanics*, volume 1, pages 563–575, New York, 1962.

F. Essenburg. On a class of nonlinear axisymmetric plate problems. *Journal of Applied Mechanics*, 27:677–680, 1960.

J. R. Fitch and B. Budiansky. Buckling and postbuckling of spherical caps under axisymmetric loads. *AIAA Journal*, 8:686–693, 1970.

J. R. Fitch. The buckling and post-buckling behavior of spherical caps under concentrated load. *International Journal of Solids and Structures*, 4:421–446, 1968.

J. P. Fulton. *Rubber-like spherical shells and circular plates under vertical point-loads*. PhD thesis, University of Virginia, 1988.

M. Gräff, R. Scheidl, H. Troger, and E. Weinmüller. An investigation of the complete post-buckling behavior of axisymmetric spherical shells. *Zeitschrift für angewandte Mathematik und Physik (ZAMP)*, 36(6):803–821, 1985.

E. Helfer, S. Harlepp, L. Bourdieu, J. Robert, F. C. MacKintosh, and D. Chatenay. Buckling of actin-coated membranes under application of a local force. *Physical Review Letters*, 87(8):088103, Aug 2001.

J. W. Hutchinson. Imperfection sensitivity of externally pressurized shells. *Journal of Applied Mechanics*, 34(49–55), 1967.

W. S. Klug, R. F. Bruinsma, J.-P. Michel, C. M. Knobler, R. L. Ivanovska, C. F. Schmidt, and G. J. L. Wuite. Failure of viral shells. *Physical Review Letters*, 97(22):228101, 2006.

G. A. Kriegsmann and C. G. Lange. On large axisymmetrical deflection states of spherical shells. *Journal of Elasticity*, 10(2):179–192, 1980.

W. T. Koiter. A spherical shell under point loads at its poles. In D. C. Drucker, editor, *Progress in Applied Mechanics (The Prager Anniversary Volume)*, pages 155–169. MacMillan, New York, 1963.

W. T. Koiter. *On the stability of elastic equilibrium*. PhD thesis, Delft, Holland, 1965.

W. T. Koiter. The nonlinear buckling problem of a complete spherical shell under uniform external pressure. *Proceedings of the Koninklijke Nederlandse Akademie van Wetenschappen, Series B*, 72:40–123, 1969.

S. Komura, K. Tamura, and T. Kato. Buckling of spherical shells adhering onto a rigid substrate. *The European Physical Journal E - Soft Matter*, 18(3):343–358, 2005.

L. D. Landau and E. M. Lifshitz. *Theory of Elasticity (Course of Theoretical Physics)*. Pergamon Press, 2*nd* edition, 1981.

A. E. Lobkovsky. *Structure of crumpled thin elastic membranes*. PhD thesis, University of Chicago, Dept of Physics, aug. 1996.

A. E. H Love. *A Treatise on the Mathematical Theory of Elasticity*. (Reprinted by Dover, New York, 1944). Cambridge University Press, 1927.

A. Libai and J. G. Simmonds. *The Nonlinear Theory of Elastic Shells*. Cambridge University Press, 2*nd* edition, 1998.

J. F. Mescall. Large deflections of spherical shells under concentrated loads. *Journal of Applied Mechanics*, 32:936–938, 1965.

J. Nowinka and S. Lukasiewicz. Hemisphere compressed by rigid plates. *International Journal of Non-Linear Mechanics*, 29(1):23–29, 1994.

F. A. Penning. Nonaxisymmetric behavior of shallow shells loaded at the apex. *Journal of Applied Mechanics*, 33:699–700, 1966.

A. V. Pogorelov. *Bendings of Surfaces and Stability of Shells*. Number 72 in Translation of mathematical monographs. American Mathematical Society, Providence, RI, 1988.

L. Pauchard, Y. Pomeau, and S. Rica. Déformations des coques élastiques. *Comptes Rendus de l'Académie des Sciences - Series II - Mécanique*, 324:411–418, 1997.

L. Pauchard, Y. Pomeau, and S. Rica. Contact et compression de coques sphériques : la physique non linéaire des balles de ping-pong. *Bulletin de la Société Française de Physique*, 119:27, 1999.

L. Pauchard and S. Rica. Contact and compression of elastic spherical shells: the physics of a 'ping-pong' ball. *Philosophical Magazine B*, 78(2):225–233, 1998.

D. F. Parker and F. Y. M Wan. Finite polar dimpling of shallow caps under sub-buckling axisymmetric pressure distributions. *SIAM Journal on Applied Mathematicsa*, 44(2):301–326, 1984.

C. Quilliet. Depressions at the surface of an elastic spherical shell submitted to external pressure. *Physical Review E (Statistical, Nonlinear and Soft Matter Physics)*, 74(4):046608, 2006.

C. Quilliet, C. Zoldesi, C. S. Riera, A. van Blaaderen, and A. Imhof. Anisotropic colloids through non-trivial buckling. *European Physical Journal E*, 27:13–20, 2008.

E. Reissner. Stresses and small displacements of shallow spherical shells, I. *Journal of Mathematics and Physics*, 25:80–85, 1947.

E. Reissner. Stresses and small displacements of shallow spherical shells, II. *Journal of Mathematics and Physics*, 25:279–300, 1947.

G. V. Ranjan and C. R. Steele. Large deflection of deep spherical shells under concentrated load. In *Proceedings, AIAA/ASME 18th Structures, Structural Dynamics and Materials Conference, San Diego*, pages 269–278, March 1977.

D. P. Updike and A. Kalnins. Axisymmetric behavior of an elastic spherical shell compressed between rigid planes. *Journal of Applied Mechanics*, 37:635–640, 1970.

D. P. Updike and A. Kalnins. Axisymmetric postbuckling and nonsymmetric buckling of a spherical shell compressed between rigid plates. *Journal of Applied Mechanics*, 39:172–178, 1972.

D. P. Updike and A. Kalnins. Contact pressure between an elastic spherical shell and a rigid plate. *Journal of Applied Mechanics*, 39:1110–1114, 1972.

A. Vaziri. Mechanics of highly deformed elastic shells. *Thin-Walled Structures*, 47:692–700, 2009.

A. Vaziri and L. Mahadevan. Localized and extended deformations of elastic shells. *Proceedings of the National Academy of Sciences*, page 7913–7918, 2008.

F. Y. M. Wan. The dimpling of spherical caps. *Mechanics Today*, 5:495–508, 1980.

F. Y. M. Wan. Polar dimpling of complete spherical shells. In W. T. Koiter and G. K. Mikhailov, editors, *Theory of shells, Proc. 3rd IUTAM Shell Symposium Tbilissi, 1978*, pages 589–605. North-Holland, 1980.

Appendix A
Calculus of variations: a worked example

This appendix is devoted to a short outline of the calculus of variations, a mathematical technique used in many places in this book.

The physical background is the following. Consider a ball, pulled down by the gravitational field, which is rolling on a hard surface with valleys and crests. If friction is small, this ball will stop after a while at the bottom of a valley, which is technically a local minimum of the potential energy of the ball in the gravity field. The potential energy is directly proportional to the elevation z. Given a Cartesian equation $z = f(x, y)$ of the surface, where z is the vertical coordinate and (x, y) the horizontal coordinates, this local minimum is such[1] that $\partial z/\partial x = \partial z/\partial y = 0$. The search for the equilibrium positions of the ball has been reduced to the minimization of a function of two variables, $z(x, y)$. This idea of minimizing an energy was extended by Euler to far more general situations, namely to cases where a state of the system is defined by a *continuum* of variables. In the Elastica, which we are going to discuss now, the list of coordinates (x, y) is replaced by a *function* $\theta(s)$, which describes the shape of a planar inextensible rod.

A.1   Model problem: the Elastica

The variational approach to the search for equilibrium will follow the same strategy as for the ball on a surface: we write first an energy, i.e. a function of the actual state of the system, and then characterize the minimum or minima of this function (often called a 'functional' in this field). This abstract way of characterizing mechanical equilibrium is often simpler than the usual method of writing the balance of forces and torques acting on every piece of material—the two methods are nevertheless equivalent, as demonstrated in various places in this book. To illustrate the power of the variational approach, we shall describe the traditional Elastica problem. Later we shall introduce another important theoretical item, the method of the Lagrange multiplier. For an excellent historical discussion of the Elastica problem, and its connections with mechanics, calculus of variation and numerics, see reference (R. Levien, 2008), which is freely available on-line.

Let us consider an elastic rod clamped vertically in the ground, see Fig. A.1 left. The rod is assumed to be inextensible, and is constrained to deform in a plane. Its shape is then given by the function $\theta(s)$ that defines the local orientation of the rod, where s is the

---

[1] This is not enough to guarantee that this point is a *minimum* of the elevation: any small change of position should not only make the energy stationary to first order, but should also *increase* it to the next order. This requires that the quadratic form $(\partial^2 z/\partial x^2)u^2 + (\partial^2 z/\partial y^2)v^2 + (2\partial^2 z/\partial x\partial y)\,u\,v$ is positive and definite.

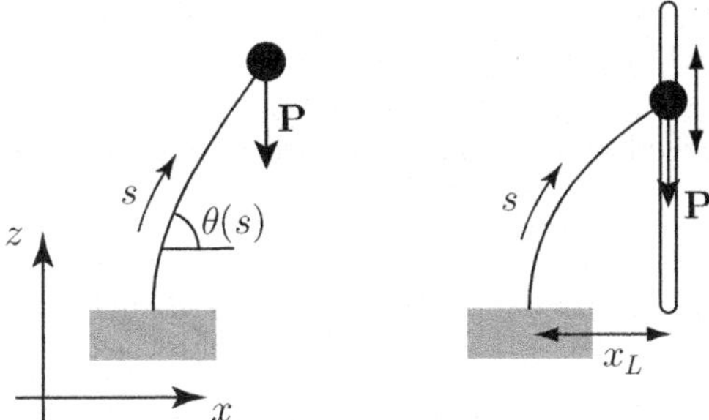

Fig. A.1 Geometry of the unconstrained (left) and constrained (right) Elastica. In the constrained case, the upper endpoint slides along a vertical line without friction.

curvilinear distance along the rod. By convention, $\theta = 0$ represents a horizontal tangent and $\theta = \pi/2$ a vertical one pointing upwards. The boundary condition for the orientation of the rod at the clamped end $s = 0$ is $\theta(0) = \pi/2$. On top of the rod of length L, that is for $s = L$, a mass M is attached.

Let $(\mathbf{e}_x, \mathbf{e}_y, \mathbf{e}_z)$ be a Cartesian frame with $\mathbf{e}_z$ pointing upwards and $\mathbf{e}_y$ perpendicular to the plane of the deformation. A parametric equation $\mathbf{r}(s) = (x(s)\,\mathbf{e}_x + z(s)\,\mathbf{e}_z)$ for the centre line of the Elastica can be derived from the function $\theta(s)$: the Cartesian coordinates x and z are given by the geometrical relation

$$\frac{\mathrm{d}\mathbf{r}(s)}{\mathrm{d}s} = \frac{\mathrm{d}(x(s)\,\mathbf{e}_x + z(s)\,\mathbf{e}_z)}{\mathrm{d}s} = \mathbf{t}(\theta(s)) = \cos(\theta(s))\,\mathbf{e}_x + \sin(\theta(s))\,\mathbf{e}_z,$$

where $\mathbf{t}(\theta)$ is the local tangent. By using the clamped end of the rod $s = 0$ as the origin for those coordinates, we get:

$$x(s) = \int_0^s \mathrm{d}s'\,\cos\theta(s'), \qquad z(s) = \int_0^s \mathrm{d}s'\,\sin\theta(s'). \tag{A.1}$$

The bending energy of the rod, derived in Chapter 3, is proportional to the square of its curvature. For planar curves, the curvature is expressed as $\mathrm{d}\theta/\mathrm{d}s$, as explained in Section 1.3.2. This yields

$$\mathcal{E}_{\mathrm{el}} = \frac{EI}{2} \int_0^L \mathrm{d}s \left(\frac{\mathrm{d}\theta}{\mathrm{d}s}\right)^2. \tag{A.2}$$

In equation (A.2), E is Young's modulus of the rod material, and I the momentum of inertia of its cross-section, see Chapter 3. We consider the case of a uniform Elastica, for which EI is independent of s.

Of course the equilibrium state of the rod that makes the elastic energy the lowest is just the straight-up configuration, $\theta(s) = \pi/2$, which is compatible with the boundary condition at $s = 0$. Things become more interesting by attaching a mass M with weight $\mathbf{P} = -M\,g\,\mathbf{e}_z$ to the top of the rod. It adds the potential energy:

$$M\,g\,z(L) = M\,g \int_0^L \mathrm{d}s\,\sin\theta(s). \tag{A.3}$$

To lower this particular contribution to the energy one should minimize $z(L)$, the elevation of the top-end of the rod. This can be achieved making $\theta(s)$ depart from its uniform value

$\pi/2$. Owing to the boundary condition $\theta(0) = \pi/2$, $\theta(s)$ cannot remain uniform, and this will involves some bending energy. Therefore the minimum of total energy derives from a balance between the elastic and gravitational energies.

What has to be minimized at the end is the full energy including the gravitational and bending terms:

$$\mathcal{E}_{\text{tot}} = \int_0^L \mathrm{d}s \left[\frac{EI}{2} \left(\frac{\mathrm{d}\theta}{\mathrm{d}s} \right)^2 + Mg \sin\theta(s) \right]. \tag{A.4}$$

At the extremum (minimum or maximum) of a functional $\mathcal{E}_{\text{tot}}$, much like at the extremum of a regular function, any variation of the 'variable', here the function $\theta(s)$, brings no change of $\mathcal{E}_{\text{tot}}$ at first order to the amplitude of the variation.

A.2   Discretization of energy using a Riemann sum

To make things more familiar to the reader who is not used to the calculus of variations, we start by writing the integral in (A.4) as a Riemann sum:

$$\mathcal{E}_{\text{tot}} = \frac{L}{N} \left[\sum_{i=0}^{N-1} \frac{EI}{2} \left(\frac{\theta_{i+1} - \theta_i}{L/N} \right)^2 + \sum_{i=0}^{N} Mg \sin\theta_i \right]. \tag{A.5}$$

In this sum, θ_i is for $\theta(s = i\,L/N)$, i is an integer between 0 and N, where N is the number of segments after discretization. In the limit of large N, the Riemann sum in equation (A.5) tends to the integral in (A.4) for a regular $\theta(s)$. The finite sum brings us back to the usual case of an energy that is expressed in terms of discrete variables θ_i. The condition that this sum is at an extremum (hopefully a minimum) is just the usual condition for the extremum of a function of N discrete variables $\theta_1, \ldots, \theta_N$:

$$\frac{\partial \mathcal{E}_{\text{tot}}(\theta_1, \ldots, \theta_N)}{\partial \theta_i} = 0 \quad \text{for } i = 1, 2, \ldots, N. \tag{A.6}$$

Note that the condition of stationarity of the energy with respect to θ_0 is not written: this variable is fixed to $\pi/2$ by the boundary condition at $s = 0$,

$$\theta_0 = \frac{\pi}{2}. \tag{A.7a}$$

When plugged into the actual form of the energy (A.5), the conditions of stationarity (A.6) yield:

$$-EI \frac{N}{L} (\theta_{i+1} + \theta_{i-1} - 2\theta_i) + Mg \frac{L}{N} \cos\theta_i = 0 \quad \text{for } 1 \leq i \leq (N-1) \tag{A.7b}$$

$$EI \frac{N}{L} (\theta_N - \theta_{N-1}) + Mg \frac{L}{N} \cos\theta_N = 0 \quad \text{for } i = N. \tag{A.7c}$$

In the bulk $(1 \leq i \leq N-1)$, the condition of stationary energy yields equation (A.7b). Note the special condition (A.7c) for $i = N$, which requires a particular treatment. We shall come back soon to this tricky question of the boundary conditions in the variational calculation. Note that, altogether, equations (A.7) yield N equations for the N unknowns $(\theta_i)_{1 \leq i \leq N}$.

To make the Riemann sum close to the integral, the discrete mesh along the rod must be fine enough to make the difference $(\theta_{i+1} - \theta_i)$ small. Then, this finite difference becomes almost equal to $L/N \, d\theta/ds$. Similarly, $(\theta_{i+1} + \theta_{i-1} - 2\theta_i)$ converges to the second derivative $d^2\theta/ds^2$ times L^2/N^2. In the limit N very large, the condition of stationarity of the total energy (A.7b) becomes a differential equation for the smooth function $\theta(s)$:

$$EI\,\frac{d^2\theta}{ds^2} - Mg\,\cos\theta(s) = 0. \tag{A.8a}$$

Note that the number of discretization points, N, has completely disappeared in the limit $N \to \infty$. The boundary conditions (A.7a) and (A.7c) can also be rewritten as

$$\theta(0) = \frac{\pi}{2}, \qquad EI\,\theta'(L) = 0, \tag{A.8b}$$

where we have used the fact that the second term in equation (A.7c) becomes negligible in the large N limit. The set of equations (A.8) is made of an ordinary differential equation of order two, with two boundary conditions. This set is complete, and fully characterizes the equilibrium positions of the rod, as we shall show next.

Two things remain to be done. In Section A.3, we shall explain how to obtain the equations (A.8) without going the long way round, that is by directly deriving the energy with respect to the function $\theta(s)$. In Section A.4, we show how additional constraints can be handled in the minimization and introduce the method of Lagrange multipliers.

A.3   Calculus of variations: the Euler–Lagrange method

It is not necessary to transform a functional to be minimized into a Riemann sum, as we did just above, and then to minimize this sum and make the mesh size tend to zero. As shown in the present section, a formal derivation of a functional with respect to *values* of the function yields a direct derivation of the equations of equilibrium.

The condition that the energy $\mathcal{E}_{\text{tot}}$ given by the integral expression (A.4) is stationary under changes of the function $\theta(s)$ is derived by adding a small arbitrary perturbation $\delta\theta(s)$ to $\theta(s)$. This brings a change of $\mathcal{E}_{\text{tot}}$ that is at the dominant order linear with respect to $\delta\theta(s)$. Imposing that this linear variation is zero for *any* $\delta\theta(s)$ yields the condition of stationarity of the energy functional. This extends the familiar result that, to first order in the shift of a variable, a function is at an extremum if its linear variation, proportional to its first derivative, is zero. Carrying out this first variation in the expression (A.4) one obtains:

$$\delta\mathcal{E}_{\text{tot}} = EI \int_0^L ds\,\frac{d\theta}{ds}\frac{d\delta\theta}{ds} + Mg \int_0^L ds\,\cos\theta(s)\,\delta\theta(s). \tag{A.9}$$

The problem now is to reduce this integral to an integral with an integrand proportional to $\delta\theta(s)$ by getting rid—in one way or another—of the derivative $d\delta\theta/ds$. In the Riemann sum there was no need to do this explicitly, because this sum was proportional to the value of θ_i, without any derivative (in fact, this was done implicitly by regrouping all terms proportional to θ_i in the condition on the derivative of the sum). In the present case, the integral depending on $d\delta\theta/ds$ can be transformed into an integral depending on $\delta\theta$ only by integration by parts (an exercise performed very often in the calculation of variations). This uses the relation:

$$\int_0^L \mathrm{d}s \, \frac{\mathrm{d}\theta}{\mathrm{d}s} \frac{\mathrm{d}\delta\theta}{\mathrm{d}s} = \left[\frac{\mathrm{d}\theta}{\mathrm{d}s} \delta\theta\right]_0^L - \int_0^L \mathrm{d}s \, \frac{\mathrm{d}^2\theta}{\mathrm{d}s^2} \delta\theta(s), \tag{A.10}$$

where the bracket denotes the variation of $\mathrm{d}\theta/\mathrm{d}s \, \delta\theta$ between $s = 0$ and $s = L$. Now the first variation of the full energy reads:

$$\delta\mathcal{E}_{\mathrm{tot}} = \left[EI \frac{\mathrm{d}\theta}{\mathrm{d}s} \delta\theta\right]_0^L + \int_0^L \mathrm{d}s \, \delta\theta(s) \left[-EI \frac{\mathrm{d}^2\theta}{\mathrm{d}s^2} + Mg \cos\theta(s)\right]. \tag{A.11}$$

The condition that this variation is zero for an arbitrary perturbation $\delta\theta(s)$ splits in two: first the coefficient of $\delta\theta(s)$ in the integral must vanish. This yields back the second-order differential equation (A.8). There is however another condition, namely that the boundary term of the integration by parts

$$\left[\frac{\mathrm{d}\theta}{\mathrm{d}s} \delta\theta\right]_0^L = \left.\frac{\mathrm{d}\theta}{\mathrm{d}s}\right|_{s=L} \delta\theta(L) - \left.\frac{\mathrm{d}\theta}{\mathrm{d}s}\right|_{s=0} \delta\theta(0)$$

is zero, for any $\delta\theta(s)$ compatible with the boundary conditions. Owing to the boundary condition $\theta(0) = \pi/2$ at the end $s = 0$, $\theta(s)$ cannot be changed there, that is $\delta\theta(0) = 0$. The second boundary term (at $s = 0$) is therefore zero in the equation above. In contrast, the value of $\theta(L)$ at the other end is free, as the mass can freely rotate. As a result, $\delta\theta(L)$ is arbitrary and the vanishing of the remaining boundary term in equation above yields:

$$\left.\frac{\mathrm{d}\theta}{\mathrm{d}s}\right|_{s=L} = 0, \tag{A.12}$$

a condition that can be interpreted as the vanishing of the internal moment near the end $s = L$, whose orientation is free.[2] We have therefore recovered the full set of equations (A.8) directly by what is known as the Euler–Lagrange method.

A.4   Handling additional constraints

Now we shall consider a slightly more difficult problem for the same physical system. We shall assume now that the top end of the Elastica is held in such a way that it has a prescribed horizontal displacement:

$$x(L) = \int_0^L \mathrm{d}s \, \cos\theta(s) = x_L \tag{A.13}$$

with respect to the straight configuration of the Elastica, see Fig. A.1 (right). This will provide an opportunity to apply the method of the Lagrange multiplier. This method has been introduced in Section 1.3.5 in the case of a function of two variables. The principle is similar in the case of a functional. We shall first put the constraint in the form $C\{\theta(s)\} = 0$ by introducing

$$C\{\theta(s)\} = -x_L + \int_0^L \mathrm{d}s \, \cos\theta(s),$$

---

[2] Minimization of the energy imposes that the *curvature* is zero at this free end, $\mathrm{d}\theta/\mathrm{d}s = 0$ at $s = L$, whilst the equations for an Elastica were derived by assuming that the internal moment in the bar is proportional to its local curvature.

where the braces indicate that C depends on the function $\theta(s)$ globally. As explained in Section 1.3.5, a constrained minimization problem can be solved by requiring that the first-order variation of the functional $\delta\mathcal{E}_{tot}$, *combined linearly with the variation of the constraint,* δC, vanishes:

$$\delta\mathcal{E}_{\text{tot}} - \mu\,\delta C = 0,$$

where

$$\delta C = -\int_0^L \mathrm{d}s\,\delta\theta(s)\,\sin\theta(s).$$

This new term in the variation modifies equation (A.11) as follows:

$$\delta\mathcal{E}_{\text{Lag}} = \left[EI\,\frac{\mathrm{d}\theta}{\mathrm{d}s}\delta\theta\right]_0^L$$
$$+ \int_0^L \mathrm{d}s\,\delta\theta(s)\left[-EI\,\frac{\mathrm{d}^2\theta}{\mathrm{d}s^2} + Mg\,\cos\theta(s) + \mu\sin\theta(s)\right]. \tag{A.14}$$

The boundary terms vanish again by the boundary conditions $\theta(0) = \pi/2$ and $\theta'(L) = 0$. The function $\theta(s)$ that is a solution of the constrained problem satisfies the same boundary conditions as earlier but the differential equation (A.8a) is replaced by:

$$EI\frac{\mathrm{d}^2\theta}{\mathrm{d}s^2} - Mg\,\cos\theta(s) - \mu\,\sin\theta(s) = 0. \tag{A.15}$$

In the present case, the Lagrange multiplier has a direct physical meaning. From a dimensional viewpoint, $\mu\,\delta C$ is an energy, while δC measures a displacement along x: the quantity μ is therefore a force along x. This force is just the force applied by the guiding mechanism that constrains the x position of the endpoint of the Elastica. Equation (A.15) is just the equation of equilibrium for an unconstrained Elastica loaded by an internal force $\mathbf{F} = \mu\,\mathbf{e}_x - M\,g\,\mathbf{e}_z$ (see Chapter 3 for the definition of the internal force, and for the equations for rods in the presence of a non-zero internal force). For more complex constraints, however, Lagrange multipliers may lack a simple physical interpretation.

A.5   Linear stability analysis

This section and the next are not directly concerned with the calculus of variations. They complement the analysis of the Elastica by deriving explicit solutions of the equations of equilibrium derived above. We study the linearized solution at the onset of bifurcation in the present section, and the exact solution in the next one.

Returning to the unconstrained case, the condition of stationarity of the energy functional $\mathcal{E}_{\text{tot}}$ yields the ordinary differential equation (A.15) and the boundary conditions (A.8b). As is well known, this set of equations has the unique, unbuckled solution $\theta(s) = \pi/2$ for all s when M is small. There is a bifurcation toward a non-trivial solution at a critical value of the mass M, which is found by looking at a solution of the linearized equation. To this end, we set

$$\theta(s) = \frac{\pi}{2} + \theta_{\text{lin}}(s),$$

where $|\theta_{\text{lin}}| \ll 1$ measures the small deflection[3] from the vertical configuration, close to the buckling threshold. By linearizing equation (A.8a), we obtain

$$EI\,\frac{\mathrm{d}^2\theta_{\text{lin}}}{\mathrm{d}s^2} + Mg\,\theta_{\text{lin}}(s) = 0.$$

The solution compatible with the boundary condition $\theta_{\text{lin}}(0) = 0$ reads

$$\theta_{\text{lin}}(s) = A\,\sin\left(\sqrt{\frac{Mg}{EI}}\,s\right),$$

where the amplitude A is a real constant. The other boundary condition, $\theta'(L) = 0$ yields

$$\sqrt{\frac{Mg}{EI}}\,L = \frac{\pi}{2} + k\pi,$$

where k is any non-negative integer. The first bifurcation encountered when increasing L (or increasing M at fixed L) occurs for $k = 0$, with a threshold L_{c} defined by:

$$L_{\text{c}}{}^2 = \frac{\pi^2}{4}\frac{EI}{Mg}. \tag{A.16}$$

Beyond the bifurcation, that is for $L > L_{\text{c}}$, the solution can be computed explicitly because the differential equation (A.8) is integrable. This solution, due to Euler, is well-known and reproduced in many textbooks (see also Section A.6 below). Comparison of the energies of the unbuckled and buckled configurations reveals that the energy is decreased upon buckling, as expected, so that the unbuckled configuration becomes unstable beyond the threshold.

A.6 Exact solution

The straight-up configuration $\theta(s) = \pi/2$ is an equilibrium solution of the unconstrained problem. For the rest of this section, we shall change our conventions for measuring the angles and introduce:

$$\psi(s) = \theta(s) - \frac{\pi}{2},$$

which is such that $\psi = 0$ when the rod is pointing upwards. The equilibrium equation (A.8a) becomes

$$EI\,\psi''(s) + Mg\,\sin\psi(s) - \mu\,\cos\psi(s) = 0. \tag{A.17}$$

By multiplying the equation by the first derivative $\psi'(s)$ and by integrating with respect to s, we find a conserved quantity:

$$\frac{EI}{2}\left(\frac{\mathrm{d}\psi}{\mathrm{d}s}\right)^2 - Mg\,\cos\psi(s) - \mu\,\sin\psi(s) = C'. \tag{A.18}$$

---

[3] By assuming that the deflection is small close to the buckling threshold, we are implicitly assuming that the bifurcation is super-critical. If the bifurcation were sub-critical, the equilibrium configuration could change by a finite amount at the threshold. The bifurcation is in fact super-critical, as shown by the exact post-buckling calculation outlined in Section A.6.

In this solution, C' is a constant of integration. It is found by imposing $\theta'(L) = 0$. This yields:

$$C' = -M\,g\,\cos\psi(L) - \mu\,\sin\psi(L).$$

Plugging back into equation (A.18), we obtain the curvature as

$$\frac{\mathrm{d}\psi}{\mathrm{d}s} = \pm\sqrt{\frac{2}{EI}}\,\left[M\,g\,(\cos\psi(s) - \cos\psi(L)) + \mu\,(\sin\psi(s) - \sin\psi(L))\right]^{1/2}.$$

From now on, we focus on buckled solutions, for which the right-hand side of the equation above does not vanish identically. For simplicity, we shall consider only configurations with positive curvature, and choose the plus sign above (configurations with negative curvature are obtained by symmetry, after changing the sign of μ).

This formula yields a relationship between the length L and the angle $\psi(L)$:

$$L = \int_0^L \mathrm{d}s = \int_0^{\psi(L)} \frac{\mathrm{d}\psi}{\sqrt{\frac{2}{EI}\left[Mg(\cos\psi - \cos\psi(L)) + \mu(\sin\psi - \sin\psi(L))\right]}}. \tag{A.19}$$

A.6.1 Unconstrained Elastica

Let us first study the case without transverse loading ($\mu = 0$). Following the method given in Section 1.2.4, we write the equations in dimensionless form. All the physical parameters can be effectively set to one by introducing the dimensionless length $\overline{L} = L/L^*$, where the chosen unit length, $L^* = \sqrt{EI/2Mg}$, coincides with the critical length given in equation (A.16) up to a numerical factor. From the equation above, this dimensionless length is related to the orientation of the terminal cross-section $\psi(\overline{L})$ by

$$\overline{L} = \int_0^{\psi(\overline{L})} \frac{\mathrm{d}\psi}{\sqrt{(\cos\psi - \cos\psi(\overline{L}))}}. \tag{A.20}$$

The right-hand side is a numerical function of $\psi(\overline{L})$, plotted in Fig. A.2. This equation yields a relationship between $\overline{L}$ and $\psi(\overline{L})$. The integral can be carried out in terms of

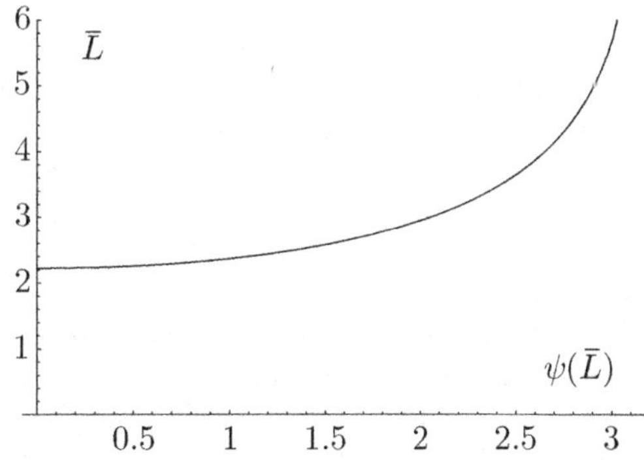

Fig. A.2 Graph of the function given by the right-hand side of equation (A.20), which maps $\psi(\overline{L})$ to $\overline{L}$.

elliptic integrals, leading to an explicit solution of the buckled configuration. We shall not follow this approach here: we simply use the exact formulation above to gain more insight into what happens at the onset of buckling.

The right-hand side of equation (A.20) appears to be a monotonic increasing function of its argument, with a minimum at $\psi(\overline{L}) = 0$: this corresponds to the straight-up configuration. The minimum value of the function is found by replacing $\cos\psi$ by the first two terms of its Taylor expansion near $\psi = 0$:

$$\min_{\psi(\overline{L})} \int_0^{\psi(\overline{L})} \frac{\mathrm{d}\psi}{\sqrt{(\cos\psi - \cos\psi(\overline{L}))}} = \lim_{\psi(\overline{L}) \to 0} \int_0^{\psi(\overline{L})} \frac{\mathrm{d}\psi}{\sqrt{\frac{\psi^2(\overline{L}) - \psi^2}{2}}} = \frac{\pi\sqrt{2}}{2}.$$

This value $\overline{L}_c = \pi\sqrt{2}/2$ corresponds exactly to the threshold $L = L_c$ already computed in equation (A.16) by the analysis of linear stability. From the present analysis, we recover that there is no solution other than $\psi(s) = 0$ of the minimization problem for $\overline{L} \leq \pi\sqrt{2}/2$. For larger values of $\overline{L}$, there are two other solutions, symmetric under the change of sign $\psi(s) \to -\psi(s)$. For $\overline{L}$ slightly larger than the critical value $\pi\sqrt{2}/2 = \overline{L}_c$, the non-zero solutions can be found by expansion in powers of the small parameter $(\overline{L} - \overline{L}_c)$, by following the principles outlined in Chapter 7 for the buckling of a plate under transverse loading. In the present case, a more direct approach is possible: it is enough to expand the integral on the right-hand side of equation (A.20) for ψ small to second order in $\psi(L)$. This relies on the expansion

$$\frac{1}{\sqrt{(\cos\psi - \cos\psi(\overline{L}))}} \approx \frac{\sqrt{2}}{\sqrt{[\psi^2(\overline{L}) - \psi^2]}} \left[1 + \frac{\psi^2 + \psi^2(\overline{L})}{24} + \cdots\right].$$

Putting this into equation (A.20), one gets the beginning of the expansion of $\overline{L}$ as a function of $\psi(\overline{L})$ for $\psi(\overline{L})$ small:

$$\overline{L} \approx \frac{\pi\sqrt{2}}{2} + \frac{\pi\sqrt{2}}{32}\, \psi^2(\overline{L}) + \cdots$$

Inverting this to get $\psi(\overline{L})$ as a function of $\overline{L}$, one has at the lowest non-trivial order:

$$\psi^2(\overline{L}) \approx \frac{32\,(\overline{L} - \overline{L}_c)}{\sqrt{2}\,\pi}. \tag{A.21}$$

This is the familiar square-root behaviour[4] of the amplitude near a super-critical bifurcation. For small amplitudes the solution $\psi(s)$ is very close to be

$$\psi(s) \approx \psi(\overline{L}) \sin\left(\frac{\pi s}{2L}\right).$$

This shows that $\psi(\overline{L})$ is the amplitude of the solution in the usual sense. For $\overline{L}$ larger than $\overline{L}_c$, the solution with the lowest energy is this non-trivial solution, not the unbuckled one. This can be shown in a number of ways: for instance, one can compute near the bifurcation threshold the energy of the various solutions, the trivial one $\psi(s) = 0$ and the bifurcated ones $\psi(s) \approx \psi(\overline{L}) \sin \pi s/2L$.

---

[4] Here the square-root is $\sqrt{(\overline{L} - \overline{L}_c)}$, $\overline{L}$ being the control parameter, and $\overline{L}_c$ its critical value.

The equation (A.17) with $\mu = 0$ is the equation for the non-linear oscillations of a pendulum when the variable s is identified with the time. All the above results can then be reinterpreted. For instance, the graph in Fig. A.2 yields the period of the pendulum as a function of the amplitude of oscillation. This period increases when one goes from small, linear oscillations to large, non-linear ones. By this analogy, the limit of small oscillations is connected with the linear stability analysis of the Elastica.

A.6.2   Constrained Elastica

Let us return to the study of the minima of the energy with a transverse loading at the top of the rod. The method of the Lagrange multiplier has given the integral relation:

$$L = \int_0^{\psi(L)} \frac{\mathrm{d}\psi}{\sqrt{\frac{2}{EI}\left[Mg(\cos\psi - \cos\psi(L)) + \mu(\sin\psi - \sin\psi(L))\right]}}. \tag{A.22}$$

The Lagrange multiplier μ itself is related to the imposed horizontal displacement $x(L)$ by the condition:

$$x(L) = \int_0^L \cos\theta(s)\,\mathrm{d}s = -\int_0^L \sin\psi(s)\,\mathrm{d}s$$

$$= -\int_0^{\psi(L)} \frac{\mathrm{d}\psi\,\sin\psi}{\sqrt{\frac{2}{EI}\left[Mg(\cos\psi - \cos\psi(L)) + \mu(\sin\psi - \sin\psi(L))\right]}}. \tag{A.23}$$

Recall that for $\mu = 0$, the angle $\psi(L)$ has three possible values, $\psi(L) = 0$ and two non-zero values $\pm\psi(L)$ for $L > L_\mathrm{c}$; there is a special degeneracy in the problem because all three solutions merge continuously to a single one, $\psi(L) = 0$, when $L < L_\mathrm{c}$. For $\mu \neq 0$, this degeneracy is lifted and no more than two branches of solutions can merge as L varies. According to Catastrophe theory (E. C. Zeeman, 1997; R. Thom, 1976), this general phenomenon is explained by the following analogy. Let $U(X')$ be a quartic polynomial:

$$U(X') = X'^4 + a_1\,X'^3 + a_2\,X'^2 + a_3\,X',$$

all the numbers being real, including the control parameters a_1, a_2 and a_3. Catastrophe theory is concerned with the behaviour of the extrema of this potential as a function of the parameters $(a_i)_{i=1,2,3}$. The number of parameters can be reduced to two by changing X' into $X = X' + a_1/4$. The new polynomial reads, after an irrelevant constant term has been dropped:

$$U(X) = X^4 + b_1\,X^2 + b_2\,X.$$

The extrema of $U(X)$ are the root(s) of

$$\frac{\mathrm{d}U}{\mathrm{d}X} = 4\,X^3 + 2b_1\,X + b_2 = 0.$$

This model polynomial describes our constrained Elastica problem if $\psi(\overline{L})$, μ and $(L - L_\mathrm{c})$ are identified with X, b_2 and $(-b_1)$ respectively.

The polynomial is even if $b_2 = 0$; then, one recovers the symmetry of the Elastica energy $\mathcal{E}_{\mathrm{tot}}$ under change of sign of ψ when $\mu = 0$. If $b_2 = 0$ there are three extrema of $U(X)$ at

$X = 0$ and $X = \pm\sqrt{-b_2/2}$ for b_2 negative and $X = 0$ only for b_2 positive. The three extrema merge continuously into a single one at $b_2 = 0$.

For $b_1 \neq 0$, there are three extrema (two minima and one maximum) for $27\,b_2{}^2 + 8\,b_1{}^3 < 0$, two if $27\,b_2{}^2 + 8\,b_1{}^3 = 0$, and only one if $27\,b_2{}^2 + 8\,b_1{}^3 > 0$. In the marginal case, $27\,b_2{}^2 + 8\,b_1{}^3 = 0$ there is one single root and one double root, unless $b_1 = 0$: it is only when $b_1 = b_2 = 0$ that there is a triple root $X = 0$, which corresponds to the merging of three solutions, as in the bifurcation problem of the Elastica: the solution $\psi(L) = 0$ is triply degenerated when $L = L_c$ and $\mu = 0$. The behaviour of the solutions in the parameter space (L, μ) is sketched in Fig. A.3. On the same graph is sketched the shape of a potential $U(X)$, which has the same properties as the energy of an Elastica. Figure A.4 shows the corresponding bifurcation diagrams for positive, zero and negative values of the asymmetry μ.

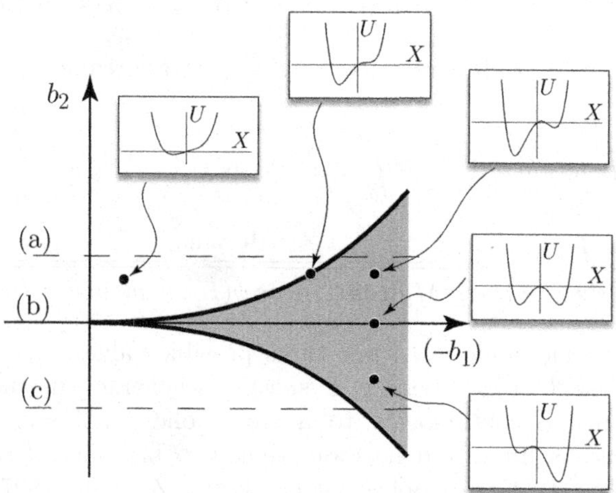

Fig. A.3 Number of equilibrium solutions of the rod in the parameter space $(-b_1, b_2) \sim (L - L_c, \mu)$. There is one solution in the white domain, two solutions including a double one along the black curve, a triple solution at the tip of the black curve, and three solutions in the grey domain. The shape of the potential $U(X)$ is represented in insets, where $X \sim \psi(L)$ is the amplitude of bifurcation. The horizontal cuts labelled (a), (b), (c) correspond to the three plots in Fig. A.4.

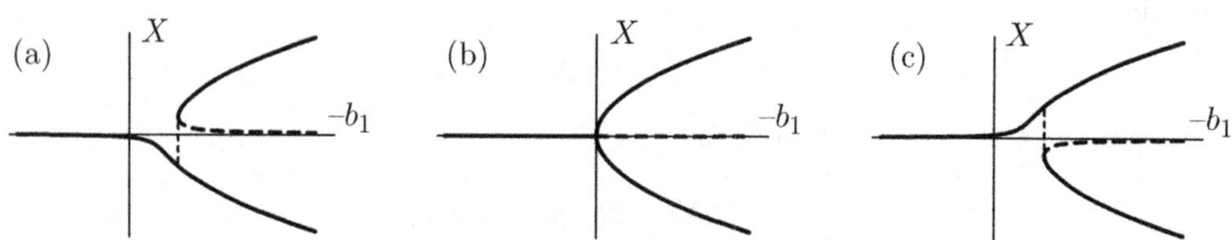

Fig. A.4 Bifurcation diagrams at constant μ showing the bifurcation amplitude X as a function of the bifurcation parameter $(-b_1) \sim (L - L_c)$: (a) $\mu > 0$, (b) $\mu = 0$, (c) $\mu < 0$. The dashed lines are for linearly unstable solutions. Each diagram corresponds to a horizontal cut through Fig. A.3.

References

R. Levien. The Elastica: a mathematical theory. Technical Report UCB/EECS-2008-103, Electrical Engineering and Computer Sciences, University of California at Berkeley, 2008.

R. Thom. *Structural Stability and Morphogenesis*. Benjamin, 1976.

E. C. Zeeman. *Catastrophe Theory (Selected Papers, 1972–1977)*. Addison-Wesley, 1997.

Appendix B
Boundary and interior layers

This appendix is devoted to a short outline of the ideas and methods of boundary (or interior) layer theory. Boundary layer theory is a set of methods used for solving ODEs or PDEs that have a small parameter, in cases where this parameter cannot be assumed to be negligible over the whole range of the variable. This requires one to use different methods of approximations of the solution in different ranges of the variable. Instead of presenting a general theory, we explicitly derive a boundary layer solution for a particular equation, equation (B.1) below. There are several reasons for basing the presentation on a particular example. It is quite difficult to explain how boundary layers work in general and a specific example provides a better insight. Examples can always be found that contradict almost any general statement about the boundary layer method, usually (but not always) because of the occurrence of logarithms. Many textbooks are devoted in part or in totality to this topic, see for instance (E. J. Hinch, 1991; W. Eckhaus, 1979; J. Kevorkian and J. D. Cole, 1981). It is often forgotten that the very idea of asymptotic expansion, variation of the coefficients, etc. takes its root in a few paragraphs of Newton's *Principia* (Book 3, prop. 22–35). There he obtained the secular variation of the parameters of the Moon's orbit by an outstandingly clever method... which is quite difficult to follow.

B.1 Layer at an interior point

B.1.1 Model equation

We consider the following bifurcation problem, which can be seen as a model equation for the buckling of a slightly inhomogeneous, long strip—the simpler case of a *homogeneous*, long strip has been examined in Chapter 7. Let $A(x)$ be a smooth function, solution of the differential equation for real variables and real valued functions:[1]

$$\frac{\mathrm{d}^2 A(x)}{\mathrm{d}x^2} + Q(x)\,A(x) - A^3(x) = 0, \tag{B.1}$$

where $Q(x)$ is a given function. Up to the coefficient in front of the non-linear term, which can be set to one by proper rescalings, this equation is similar to the complex amplitude equation (7.47) for a long elastic strip under compression with $Q(x) = \epsilon^{-2}$. In the analysis of the strip, the function $A(x)$ has been defined as the envelope of the buckled solution. Here, we consider a real $A(x)$, although the general theory introduces complex valued amplitudes. This does not matter as long as one is interested only in the 'ground state' amplitude: the

---

[1] The extension to the complex domain of this type of problem may bring very crucial information including in the real domain. This was discovered by Poincaré in his study of the refraction of electromagnetic waves by a conducting sphere (H. Poincaré, 1910) and has been extended since by Kruskal and Segur to the 'Asymptotics beyond all orders' (M. D. Kruskal and H. Segur, 1991).

solution that minimizes the energy has constant arbitrary phase and so can be assumed to be real.

The asymptotic conditions for large $|x|$ that we impose on the solutions of equation (B.1) will be detailed later. Loosely speaking, we require that the function $A(x)$ goes to zero at infinities—more accurately, we shall in fact require that the energy of the system is finite.

The bifurcation parameter in equation (B.1) is the function $Q(x)$. In Chapter 7, this function was a constant number $Q = \epsilon^{-2}$, independent of x. This constant was interpreted as the bifurcation parameter as it is positive above the buckling threshold and negative below it. Here, variations of Q with x allow for inhomogeneities of the strip properties. For instance, if the strip is thicker close to its ends and thinner in the middle, it can happen that, for a given longitudinal compression, the strip is above the buckling threshold near the ends but below in the middle. This is represented by a function $Q(x)$ that is negative near the ends and positive in the middle. The corresponding solution $A(x)$ will eventually display localized buckling, as discussed below.

When $Q(x)$ is a constant and the solution is independent of x, the bifurcation is found easily:[2] for negative Q, $A = 0$ is the only solution, and for positive Q there are three real solutions, $A = 0$, $A = \sqrt{Q}$ and $A = -\sqrt{Q}$. When they exist, the solutions $\pm\sqrt{Q}$ minimize the density of energy, the energy itself being in general:

$$\mathcal{E} = \int \mathrm{d}x \left[\frac{1}{2}\left(\frac{\mathrm{d}A}{\mathrm{d}x}\right)^2 - \frac{Q(x)}{2}A^2 + \frac{1}{4}A^4 \right].$$

(B.2)

By writing the condition of stationarity of the above energy (see Appendix A), one indeed recovers equation (B.1) above.

Boundary layer methods are needed to solve equation (B.1) when the solution $A(x)$ varies on length scales that are much shorter than the typical scale of variation of $Q(x)$. By balancing the first two terms in our model equation

$$\frac{A}{x} \sim Q\,A,$$

the typical scale of variation of A is found to be $x \sim 1/\sqrt{Q}$: we shall consider the limit where *the function Q has large values.*

To be more specific, let us take the particular example[3] of the function $Q(x)$, namely $Q(x) = \omega - x^2$, where ω is a constant. When plugged back into the model equation, this yields:

$$\frac{\mathrm{d}^2 A}{\mathrm{d}x^2} + (\omega - x^2)\,A - A^3 = 0.$$

(B.3)

We are going to look at the solution(s) of this equation that go to zero as x tends to $\pm\infty$, indexed by the parameter ω. This equation is not integrable by quadrature.

The equation (B.3) always has the solution $A(x) = 0$ for all x but this is not the only one as soon as ω gets larger than a critical ω_c. This critical value can be found by the property

---

[2] Taking into account the boundary conditions can be difficult, however. This is the topic of Chapter 7.

[3] This example is less specific than it might seem, as any function $Q(x)$ that is quadratic can be put in the form $Q(x) = \omega_0 - \omega_1 x^2$ by a change of origin, and further reduced to the form $Q(x) = \omega - x^2$ by the rescaling $\overline{x} = x\,\omega_1^{1/4}$, $\overline{A} = A\,Q_1^{-1/4}$ and $\omega = \omega_0/\sqrt{\omega_1}$.

that, at the threshold $\omega = \omega_c$, there exists a non-trivial solution of the linearized system, a so-called marginal mode A_{lin}. This linearized equation near $A(x) = 0$ is the Hermite equation:

$$\frac{\mathrm{d}^2 A_{\text{lin}}}{\mathrm{d}x^2} + (\omega - x^2)\, A_{\text{lin}} = 0. \tag{B.4}$$

It has solutions decaying at infinity for discrete values of ω. The smallest value for which this happens is $\omega_c = 1$, the solution then being the Gaussian $A_{\text{lin}}(x) = \exp(-x^2/2)$. For ω slightly larger than this critical value, it is possible to find by expansion a solution of the full non-linear equation (B.3). We do not undertake this calculation here.

B.1.2 Outer solution

The boundary layer theory concerns the regime where ω is very large positive, $\omega \gg \omega_c = 1$, which we study below. In this limit a rescaling is useful, which is obtained by dropping the second derivative in (B.3) and balancing the two remaining terms. This suggests a new choice of 'units', $x^* = \omega^{1/2}$ and $A^* = \omega^{1/2}$. The rescaled variables are defined by $\overline{x} = x/x^*$ and $\overline{A} = A/A^*$. As usual, the rescaled function $\overline{A}$ is considered to be a function of $\overline{x}$, not of x, something that is important for the calculation of derivatives. When inserted into (B.3), these rescalings reveal that the second derivative term is formally negligible in the large ω limit:

$$\frac{1}{\omega^2} \frac{\mathrm{d}^2 \overline{A}}{\mathrm{d}\overline{x}^2} + (1 - \overline{x}^2)\, \overline{A} - \overline{A}^3 = 0. \tag{B.5}$$

Dropping this second derivative yields the algebraic equation:

$$(1 - \overline{x}^2)\, \overline{A} - \overline{A}^3 = 0. \tag{B.6}$$

However, with the aim of keeping the notation simple, we shall not use these rescaled quantities $\overline{x}$ and $\overline{A}$: we start again from the writing (B.3) of the model equation and keep in mind that, in the large ω limit, the second derivative is formally small. Without this second derivative, equation (B.3) becomes:

$$(\omega - x^2)\, A - A^3 = 0, \tag{B.7}$$

which is the unscaled version of (B.6). The approximation underlying equation (B.7) is valid if A varies over a typical length $x \sim \omega^{1/2}$ or greater.

The equation (B.7) has three continuous solutions that are consistent with the boundary condition at infinity, A_{out}^0, A_{out}^+ and A_{out}^-, called the outer[4] solutions:

1. $A_{\text{out}}^0(x) = 0$,
2. $A_{\text{out}}^+(x) = 0$ for $x^2 \geq \omega$ and $A_{\text{out}}^+(x) = +\sqrt{\omega - x^2}$ for $x^2 \leq \omega$,
3. $A_{\text{out}}^-(x) = 0$ for $x^2 \geq \omega$ and $A_{\text{out}}^-(x) = -\sqrt{\omega - x^2}$ for $x^2 \leq \omega$.

---

[4] For historical reasons, the 'boundary', 'inner' and 'outer' terminology refers to the geometry of boundary layers in fluids, which is often different from the geometry of layers in elasticity: the outer solution (far away from walls in low viscosity flows) is obtained by setting the small parameter (the fluid viscosity) to zero while the inner solution is where it is effectively set to a number of order one by rescalings (this happens close to the walls in fluids). Unfortunately, when the singularity is located at an interior point of the domain, like $x = \pm\sqrt{\omega}$ here, the resulting 'boundary' layer is actually *inside* the domain—such boundary layers are sometimes called *internal boundary layers*.

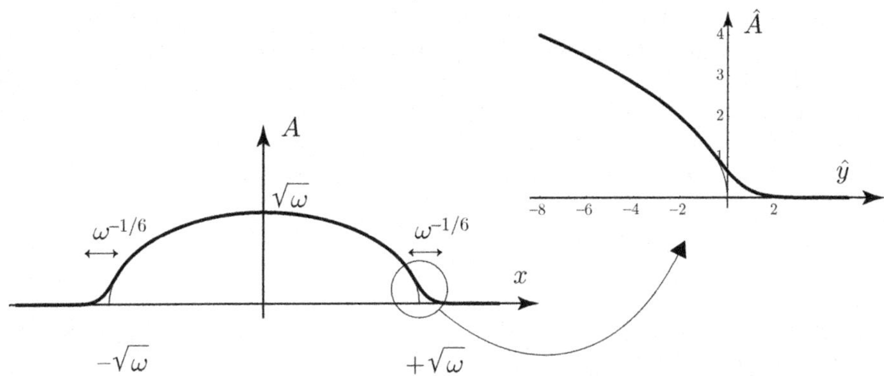

Fig. B.1 Solution to equation (B.1) in the limit of large ω (left). The outer solution (thin curve), has a slope discontinuity at $x = \pm\sqrt{\omega}$, where two boundary layers develop. The solution of the inner equation (B.10) with the asymptotic boundary conditions (B.11a) and (B.11b), formulated in stretched variables $\hat{y}$ and $\hat{A}$, is shown right.

The first solution, A_{out}^{0}, represents the unbuckled solution that is unstable (it is a local maximum of energy since $\omega \gg \omega_{\text{c}} = 1$) and is not considered here. The two other solutions, $A_{\text{out}}^{\pm}$, are symmetric with respect to the x-axis. They both play a similar role. The positive one, for instance, is shown in Fig. B.1 by a thin curve. It is called the 'outer solution' of our boundary layer construction, for a reason that will be explained below. This solution is continuous but has a discontinuous first derivative at $x = \pm\sqrt{\omega}$. If one plugs this function A_{out}^{+}, which is a solution of the approximated model equation (B.7), back into the full model equation (B.3), the second derivative, which had been neglected, becomes infinite near $x = \pm\sqrt{\omega}$ as the first derivative is unbounded. Therefore, $A_{\text{out}}^{\pm}$ cannot be a good solution of the full original equation.

B.1.3 Boundary layer equation

Actually the function $A_{\text{out}}^{+}(x)$ is almost everywhere a good solution to (B.3), in the sense that the neglected term with a second derivative is much smaller than the two other ones, *except in the vicinity* of the points $x = \pm\sqrt{\omega}$. Near these points, the approximation to drop the second derivative term breaks down and has to be replaced by a different approximation. This brings us the idea of a boundary layer: near $x = \pm\sqrt{\omega}$, there is a small 'inner' region where the solution by itself does not change much but where its first derivative does change a lot. On both sides of this boundary layer, the inner solution merges smoothly with the outer solution $A_{\text{out}}^{+}(x)$. This is the so-called matching problem that will be considered after the boundary layer equation has been written.

The boundary layer is *a priori* a narrow region of transition between the zero solution outside the interval $x^2 \le \omega$ and the solution $A(x) = \sqrt{\omega - x^2}$ inside. There are actually two boundary layers, one at $x = \sqrt{\omega}$, another one at $x = -\sqrt{\omega}$. We shall look at the layer near $x = +\sqrt{\omega}$, the other being obtained by symmetry. We define a local coordinate $y = x - \sqrt{\omega}$ and make the approximation:

$$\omega - x^2 = -2\sqrt{\omega}\,y - y^2 \approx -2\sqrt{\omega}\,y. \tag{B.8}$$

Since we expect the boundary layer to be narrow, we have neglected y^2 compared with the product $\sqrt{\omega}\, y$ in the expansion above—this approximation is valid as long as $y \ll \sqrt{\omega}$, that is as long as the boundary layer has a width much smaller than $\sqrt{\omega}$, something that we indeed check at the end.

Using this new coordinate, y, together with the approximation (B.8) that is valid for small y, the original ODE (B.3) yields the so-called inner equation:

$$\frac{\mathrm{d}^2 A}{\mathrm{d}y^2} - 2\sqrt{\omega}\, y\, A - A^3 = 0. \tag{B.9}$$

Although this is not totally obvious with this writing, this inner equation is simpler than the original ODE. This is because one can eliminate the large parameter ω from the inner equation (B.9) by rescaling. Balancing the various terms as explained in Section 1.2.4, one is led to suggest the stretched variables $\hat{y} = y\,\omega^{1/6}$ and $\hat{A}(\hat{y}) = \omega^{-1/6}\, A(y)$. This yields a numerical equation without any small or large parameter:

$$\frac{\mathrm{d}^2 \hat{A}}{\mathrm{d}\hat{y}^2} - 2\hat{y}\, \hat{A} - \hat{A}^3 = 0. \tag{B.10}$$

This boundary layer equation is an approximation of the original equation (B.3), which is designed to be accurate in the boundary layer, where one expects $|y| = |x - \sqrt{\omega}| \ll \sqrt{\omega}$. It complements equation (B.6), which is another approximation of the same equation, to be used far away from the boundary layers.

Incidentally, the last rescaling provides an estimation for the width of the boundary layer as $\omega^{-1/6}$; this is much smaller than $\sqrt{\omega}$, which makes the boundary layer analysis consistent. Indeed, the solution $\hat{A}$ of the inner equation (B.10) will vary over a typical length scale $\hat{y} \sim 1$, because this equation has no large or small parameter, and the original variable is related to the rescaled one by $y = \hat{y}\,\omega^{-1/6}$. Therefore, the function A will vary over a typical length scale $\Delta x = \Delta y \sim \omega^{-1/6}$ in the layer.

B.1.4   Matching problem

The general solution of the inner equation (B.10) is known as the second Painlevé transcendent (E. L. Ince, 1956). Being of order two, this equation has in fact a two-parameter family of such solutions. The relevant solution is selected by imposing boundary conditions for $\hat{y} \to \pm\infty$. They are dictated by the matching problem: their role is simply to allow the inner solution to match with the outer solution A_{out}^{+} given above.

We shall therefore explain how the solution of the inner equation (B.10) can be matched with the 'outer' solution, that is the solution of the algebraic equation (B.7) valid outside of this inner region. This is linked to the meaning of the last rescaling: we have focused on the neighbourhood of the transition, by taking a typical length scale for y of order $\omega^{-1/6}$, although the outer algebraic solution changes over scales of order $\omega^{1/2}$, much longer than $\omega^{-1/6}$ when ω is large. Therefore, with respect to the typical length scale of the outer algebraic solution, a 'large' $\hat{y}$ is still a small quantity. As a result, the large $\hat{y}$ behaviour of the inner solution should coincide with the behaviour of the outer solution near $x = \sqrt{\omega}$.

The behaviour of the outer solution near $x = \sqrt{\omega}$ is given, on the left-hand side of singularity $(x < \sqrt{\omega})$, by

$$A(y) = \sqrt{\omega - x^2} \approx \sqrt{2}\,\omega^{1/4}\,(-y)^{1/2} = \sqrt{2}\,\omega^{1/6}\,(-\hat{y})^{1/2}.$$

This approximation is valid for $|y| \ll \sqrt{\omega}$, as explained earlier, that is for $|\hat{y}| \ll \omega^{2/3}$ (in fact, it is only valid on the left-hand side of a singularity—the other side will be studied next). For $|\hat{y}| \gg 1$, we expect that the inner solution has reached its asymptotic behaviour for large $\hat{y}$—recall that the inner equation is expressed without any small or large parameter. The key remark is that the two conditions $|\hat{y}| \ll \omega^{2/3}$ and $|\hat{y}| \gg 1$ are consistent for large ω and they define a range of values of $\hat{y}$, called the *intermediate range*, where *both the inner and outer solutions accurately represent the solution*. We should emphasize that, by this construction, these two approximations match over a domain and not simply at a point.[5]

For large negative $\hat{y}$, that is for $|\hat{y}| \gg 1$ and $\hat{y} < 0$, the behaviour of the inner solution should therefore match that of the outer solution near the singularity. Using inner variables, this writes:

$$\hat{A}(\hat{y}) = A/\omega^{1/6} \approx \sqrt{2}\,(-\hat{y})^{1/2} \text{ for } \hat{y} \to -\infty. \tag{B.11a}$$

This condition allows the inner and the outer solutions to overlap[6] in the so-called intermediate range $\omega^{-1/6} \ll (-y) \ll \omega^{1/2}$, that is $1 \ll (-\hat{y}) \ll \omega^{2/3}$ in rescaled variables.

Equation (B.11a) fixes the asymptotic behaviour of the solution of the inner equation (B.10) and plays the role of a boundary condition for $\hat{y} \to -\infty$. The other side of the singularity, for $\hat{y} \to +\infty$, is derived similarly: there, the inner solution has to merge with a vanishing function. Therefore, the matching condition reads:

$$\hat{A}(\hat{y}) \to 0 \text{ for } \hat{y} \to +\infty. \tag{B.11b}$$

Like the differential equation (B.10) itself, the two asymptotic conditions (B.11a) and (B.11b) are expressed in terms of numbers of order one in rescaled variables. This validates our assumption that the inner solution varies over a typical scale $\Delta\hat{y} \sim 1$.

B.1.5   Uniqueness of the boundary layer solution

There remains a last question, which is whether the inner equation (B.10) together with the boundary conditions (B.11a) and (B.11b) forms a well-posed mathematical problem.[7] We shall not attempt to prove that there exists a unique solution to this inner equation. Instead, we present a simple counting argument to show that its solutions are discrete. We use a reasoning by contradiction.

---

[5] Matching the inner and outer at a single point is usually a bad idea. Indeed, the point of matching has to be chosen somewhat arbitrarily, and one must then carefully check for the independence of the resulting solution with respect to it. It is therefore safer to require that the matching happens over an entire range of values.

[6] That the inner and outer equations share exactly the same asymptotic behaviour is a consequence of the fact that this behaviour is dictated by the terms of the original equation that have been retained both in the outer and in the inner equation.

[7] The existence and uniqueness of solutions of an ordinary differential equation with asymptotic conditions cannot be determined simply by comparing the order of the ODE with the number of asymptotic conditions as the infinity is a singular point. Instead, one has to plug the generic expansion of the solution of the ODE into the asymptotic conditions and see whether they are compatible or not.

Let us first consider the boundary conditions at $\hat{y} \to +\infty$ only. There, we seek the asymptotic behaviour of $\hat{A}(\hat{y})$. From the condition (B.11b), the cubic term in (B.10) will become negligible in front of the linear one, and can be discarded as long as one is interested in the dominant order. Dropping this cubic term yields the Airy equation:

$$\frac{\mathrm{d}^2 \hat{A}_{\mathrm{lin}}^{+\infty}(\hat{y})}{\mathrm{d}\hat{y}^2} - 2\hat{y}\,\hat{A}_{\mathrm{lin}}^{+\infty}(\hat{y}) = 0. \tag{B.12}$$

It has one exponentially decreasing and one exponentially increasing solution at large $\hat{y}$, which go like $\exp((\pm 2\sqrt{2}/3)\,\hat{y}^{3/2})$. The exponentially large mode should be discarded,[8] as it contradicts the boundary condition (B.11b). Out of the two-parameter family of solutions of the inner equation, this condition leaves out a *one-parameter* family of solutions consistent with the asymptotic condition.

Another condition is derived similarly by imposing the other asymptotic condition (B.11b), for $\hat{y} \to -\infty$. In this limit, the solution must be such that $\hat{A} \approx \sqrt{-2\hat{y}}$. Let us linearize the equation (B.10) near this solution and let $\hat{A}_{\mathrm{lin}}^{-\infty}$ be the small perturbation that is the solution of the homogeneous equation:

$$\frac{\mathrm{d}^2 A_{\mathrm{lin}}^{-\infty}(\hat{y})}{\mathrm{d}\hat{y}^2} + 4\,\hat{y}\,A_{\mathrm{lin}}^{-\infty}(\hat{y}) = 0. \tag{B.13}$$

Somehow this solution of the linear homogeneous equation is to be added 'freely' to the regular Laurent expansion (B.11a). One of its possible behaviours is exponentially growing and so must be discarded. This yields the other boundary condition for the solution of the inner equation. It is of interest too to notice that this shows as well that the Laurent expansion of the outer solution beginning like

$$a \approx \sqrt{-2y}\left(1 - \frac{1}{8x^3} + \cdots\right) \tag{B.14}$$

has zero radius of convergence.[9]

Again, we find that there is a one-parameter family of solutions consistent with the asymptotic condition for large negative $\hat{y}$. The solutions of the full problem are therefore given by the intersection of two curves (two one-parameter families of solution satisfying either asymptotic condition) on a surface (the two-dimensional space of solutions of the equation without boundary conditions). The intersection is made generically by discrete points—a numerical study shows that this intersection is in fact reduced to a single point, the solution shown in Fig. B.1, right.

B.1.6 Summary

We have shown how to handle the matching of the inner and outer solution in the particular case of a model buckling equation. The general methodology that should be derived from this example is as follows.

---

[8] The appearance of such spurious, exponentially growing modes can be understood as a consequence of the fact that the inner equation is of order two, while the outer equation is algebraic (of order zero). This provides the inner equation with two additional degrees of freedom corresponding to solutions that do not satisfy all the imposed boundary conditions.

[9] This is an indirect consequence of the fact that, when computing the terms of this expansion one order after the other, no condition of cancellation of the exponentially growing solution of the Airy equation is ever written. This very important property is rather general and has recently been the object of detailed studies (H. Segur, S. Tanveer, and H. Levine, 1991).

1. Write the full original equation so as to make obvious what term is small in the appropriate limit (here, this is the second derivative in equation (B.5) in the limit of large ω).
2. Solve the resulting equation by setting the small parameter to zero.
3. This solution at the dominant order may, once plugged back into the original equation, make the neglected term diverge for some value(s) of the variable(s) located inside the relevant range. In the example above, the second derivative of a function diverges near $x = \pm\sqrt{\omega}$. When this happens, a boundary layer exists near these values of the variable.
4. There, one should resort to the full equation and devise a different approximation that applies to a narrow domain near the point of divergence. The width of the boundary layer is found by appropriate rescaling of this boundary layer equation (when rescaled, the latter should have no large or small parameter). In any case, the boundary layer equation should include the term, the second derivative here, that diverges in the outer approximation. The solution of this inner problem tends asymptotically to a function merging continuously with the outer solution: the asymptotic behaviour of the outer solution for small values of the physical variable yields an asymptotic condition for the inner solution, for large values of the rescaled variable.

Although we tried to give rules that are as general as possible, concrete problems quite often require us to use a few other tricks that are difficult to put into a general formal schema.[10]

B.2   Layers near a boundary

We are now going to use this general approach to solve another problem, built on the same equation as before but with different boundary conditions: we now restrict the variable x to the range $0 < x < +\infty$, and impose a new boundary condition at $x = 0$. As we show, this condition generates a boundary layer. In the previous example, the layers were caused by a discontinuity of the derivative of the outer solution at interior points of the domain, namely $x = \pm\sqrt{\omega}$. In contrast the boundary layer below arises at the boundary and is a consequence of the condition imposed there to the solution of the equation (B.3). For convenience this equation is rewritten below

$$\frac{\mathrm{d}^2 A}{\mathrm{d}x^2} + \left(\omega - x^2\right) A - A^3 = 0, \tag{B.15}$$

[10] One of these tricks, called the 'composite approximation', is to write the full solution (inner and outer piece) in a single compact form. To explain how it works, let $A_{\mathrm{out}}(x) = \sqrt{\omega - x^2}\,\Theta(\sqrt{\omega} - x)$ be the outer solution, $A_{\mathrm{inn}}(x)$ be the inner solution, valid near $x = \sqrt{\omega}$ and $A_{\mathrm{match}}(x) = \sqrt{2}\,\omega^{1/4}(\sqrt{\omega} - x)^{1/2}\Theta(\sqrt{\omega} - x)$ be the solution in the matching domain, $\Theta(.)$ being the Heaviside function equal to one for positive arguments and to zero otherwise. A_{match} represents the solution in the matching domain on both sides of $\sqrt{\omega} = x$, which it zero for $x > \sqrt{\omega}$ and equal to $\sqrt{2}\,\omega^{1/4}(\sqrt{\omega} - x)^{1/2}$ on the side $x < \sqrt{\omega}$. The composite solution (valid near $x \sim \sqrt{\omega}$) is just the sum $A_{\mathrm{comp}}(x) = A_{\mathrm{inn}}(x) + A_{\mathrm{out}}(x) - A_{\mathrm{match}}(x)$. It is a uniform approximation to the exact solution, valid everywhere at the dominant order. In the outer domain, $A_{\mathrm{inn}}(x)$ and $A_{\mathrm{match}}(x)$ are almost equal and cancel so that $A_{\mathrm{comp}}(x)$ is equal to A_{out}, a good approximation of the true solution. In the inner domain a very similar argument shows that $A_{\mathrm{comp}}(x)$ is equal to A_{inn}. In the matching domains on both side of the boundary layer, all three functions A_{out}, A_{inn} and A_{match} are almost equal. Therefore they are also equal to A_{comp}, which approximate $A(x)$ well in this domain. This composite solution is not very useful in general. It may help sometimes when one has to compute integrals depending on the solution in the full domain.

where ω is a large parameter. We now take as boundary conditions for $A(x)$:

$$A(0) = 0, \qquad A(x) \to 0 \text{ for } x \to +\infty. \tag{B.16}$$

The condition $A(0) = 0$ is incompatible with the outer solutions found previously, for which $A_{\mathrm{out}}(0) = \pm\sqrt{\omega}$. Again, we discard the solution that is zero everywhere, $A(x) = 0$ for all x, which describes an unstable equilibrium and does not have the lowest energy.

By introducing a new layer near the boundary of the domain, $x = 0$, one can construct a solution that, far from $x = 0$ (in a sense to be made precise below), is close to the solution with lowest energy, $A_{\mathrm{out}}^{+}(x) = \sqrt{\omega - x^2}$ for $0 < x < \sqrt{\omega}$ and $A_{\mathrm{out}}^{+}(x) = 0$ for $x > \sqrt{\omega}$, and satisfies the boundary condition $A(0) = 0$. Near $x = \sqrt{\omega}$, this solution has another boundary layer of thickness of order $\omega^{-1/6}$, which is constructed as before.

Below, we show how to build the new boundary layer near $x = 0$. In this layer, the solution should go in one way or another from the boundary value $A(0) = 0$ to a value close to $A_{\mathrm{out}}^{+}(0) = \sqrt{\omega}$. We expect this transition to take place in a layer much narrower than $\sqrt{\omega}$, where one can neglect the variation of $(\omega - x^2)$ in the original equation, and simply replace it by its limit value at $x = 0$, namely ω. Therefore the equation for the boundary layer is derived from the full original equation by replacing $(\omega - x^2)$ by ω:

$$\frac{\mathrm{d}^2 A}{\mathrm{d}x^2} + \omega A - A^3 = 0. \tag{B.17}$$

As earlier, this inner equation can be rescaled so as to get rid of any small or large parameter. This is achieved by introducing $\overline{x} = x/\omega^{-1/2}$ and $\overline{A} = A/\sqrt{\omega}$. This means that the width of this boundary layer will be of order $\omega^{-1/2}$, which is indeed much smaller than $\sqrt{\omega}$, the typical scale of the outer solution. Using these variables, the inner equation can be put in a dimensionless form:

$$\frac{\mathrm{d}^2 \overline{A}}{\mathrm{d}\overline{x}^2} + \overline{A} - \overline{A}^3 = 0. \tag{B.18}$$

The boundary conditions for this inner equation are simply:

$$\overline{A}(0) = 0, \qquad \overline{A}(\overline{x}) \to 1 \text{ for } \overline{x} \to +\infty. \tag{B.19}$$

The first one derives directly from the boundary condition $A(x) = 0$, while the second one is the rescaled version of the matching condition $A(x) = \sqrt{\omega}$ when x lies in the intermediate range $\omega^{-1/2} \ll x \ll \omega^{+1/2}$. This intermediate range is such that $\overline{x} \gg 1$ while $x \ll \omega^{1/2}$: there, the outer solution reaches its small x limit, while the inner solution reaches its large $\overline{x}$ limit.

The inner equation (B.18) with boundary conditions (B.19) has a simple analytical solution:

$$\overline{A} = \tanh\left(\frac{\overline{x}}{\sqrt{2}}\right). \tag{B.20}$$

This is the inner solution at $x = 0$. Using the original variables, the value of A in this boundary layer reads:

$$A(x) = \omega^{1/2} \tanh\left(\frac{x\,\omega^{1/2}}{\sqrt{2}}\right).$$

The full boundary layer solution based on this explicit form is shown in Fig. B.2.

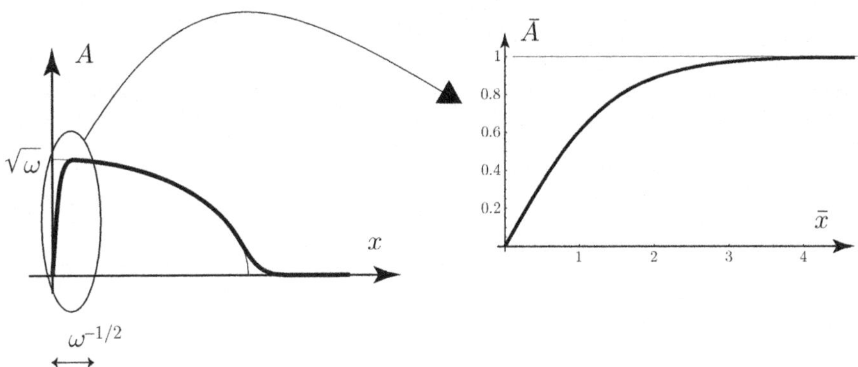

Fig. B.2 Layer near a boundary. A layer develops near $x = 0$ to accommodate the boundary condition $A(0) = 0$, which is incompatible with the outer solution $A_{\mathrm{out}}^{+}(x)$ shown by a thin curve. The solution (B.20) of the inner problem in stretched variables is shown right.

To sum up, we have presented a method to solve an ODE containing a large parameter, called ω. A case of particular interest is when the highest derivative of the equation comes with a prefactor that goes to zero for large ω: simply neglecting this term leads to inconsistencies, which are physically connected to the existence of boundary layers. We have presented an example of a general method for handling this class of problem. It builds on the idea that this highest derivative may become large inside the boundary layer, which compensates its small prefactor.

This simple presentation provides a first insight into boundary layer methods and only gives a flavour of their power: the differential equation that we studied was rather simple, and its boundary layers affected the solution only in the vicinity of boundaries or singular points. In many problems, however, the effect of boundary layers is much more severe—in our analysis of Hertz' contact for a thin shell in Chapter 14, for instance, we find that the reaction force can only be determined by the analysis of boundary layers. Other examples of boundary layers found in this book include the hair problem in various limits, as studied in Chapter 4, the buckling of plates in the large load limit in Chapter 7, etc. The small parameter in these cases, which plays a role similar to $1/\omega$ above, is the ratio of the bending modulus to the stretching modulus, a ratio that is proportional to the square of the small thickness.

References

W. Eckhaus. *Asymptotic Analysis of Singular Perturbations*, volume 9 of *Studies in Mathematics and its Applications*. North-Holland, 1979.

E. J. Hinch. *Perturbation Methods*. Cambridge Texts in Applied Mathematics. Cambridge University Press, 1991.

E. L. Ince. *Ordinary Differential Equations*. Dover Publications, 1956.

J. Kevorkian and J. D. Cole. *Perturbation Methods in Applied Mathematics*, volume 34 of *Applied Mathematical Sciences*. Springer-Verlag, 1981.

M. D. Kruskal and H. Segur. Asymptotics beyond all orders in a model of crystal growth. *Studies in Applied Mathematics*, 85:129–181, 1991.

H. Poincaré. Sur la diffraction des ondes hertziennes. *Rendiconti del Circolo Matematico di Palermo*, 29:169, 1910.

H. Segur, S. Tanveer, and H. Levine, editors. *Asymptotics Beyond all Orders*. Proceedings of the NATO Advanced Research Workshop. Plenum Press, New York, 1991.

Appendix C
The geometry of helices

This appendix shows that a Cosserat curve, such as the centre line of a Kirchhoff rod, is a helix when the prescribed curvatures and torsion are constant.

Therefore, we consider the set of equations (3.2) and (3.3). The centre line is reconstructed from the tangent $\mathbf{d}_3(s)$ by integration of

$$\frac{\mathrm{d}\mathbf{r}(s)}{\mathrm{d}s} = \mathbf{d}_3(s). \tag{C.1}$$

Here, we consider the case of constant curvatures and twist:

$$\frac{\mathrm{d}\kappa^{(1)}}{\mathrm{d}s} = 0, \qquad \frac{\mathrm{d}\kappa^{(2)}}{\mathrm{d}s} = 0, \qquad \frac{\mathrm{d}\tau}{\mathrm{d}s} = 0.$$

We shall also assume clamped boundary conditions at the scalp, $s = 0$, where both the position and the material frames are fixed:

$$\mathbf{r}(0) = \mathbf{r}^{\mathrm{s}}, \qquad \mathbf{d}_i(0) = \mathbf{d}_i^{\mathrm{s}} \tag{C.2}$$

for $i = 1, 2, 3$. Here, the superscript s refer to the known quantities at the end clamped in the scalp, $s = 0$.

As we now show, this set of kinematical equations have an analytical solution which describes an helix.

The first step is to show that the Darboux vector $\boldsymbol{\Omega}$ is independent of s. Its derivative reads

$$\frac{\mathrm{d}\boldsymbol{\Omega}(s)}{\mathrm{d}s} = \frac{\mathrm{d}(\kappa^{(1)}(s)\,\mathbf{d}_1(s) + \kappa^{(2)}(s)\,\mathbf{d}_2(s) + \tau\,\mathbf{d}_3(s))}{\mathrm{d}s}$$

$$= \kappa^{(1)'}\,\mathbf{d}_1 + \kappa^{(2)'}\,\mathbf{d}_2 + \tau'\,\mathbf{d}_3 + \kappa^{(1)}\,\mathbf{d}_1' + \kappa^{(2)}\,\mathbf{d}_2' + \tau\,\mathbf{d}_3'.$$

In the right-hand side, $\kappa^{(1)'}(s) = 0$, $\kappa^{(2)'}(s) = 0$ and $\tau'(s) = 0$ by assumption. The remaining terms in the equation above are expressed using the definition (3.2) of the Darboux vector:

$$\frac{\mathrm{d}\boldsymbol{\Omega}(s)}{\mathrm{d}s} = \boldsymbol{\Omega} \times (\kappa^{(1)}\,\mathbf{d}_1 + \kappa^{(2)}\,\mathbf{d}_2 + \tau\,\mathbf{d}_3) = \boldsymbol{\Omega} \times \boldsymbol{\Omega} = \mathbf{0}. \tag{C.3}$$

In the case of uniform curvatures and twist, the Darboux vector is thus a constant vector.

This remark brings an important simplification to the kinematical equations (3.2) as the material frame $(\mathbf{d}_1(s), \mathbf{d}_2(s), \mathbf{d}_3(s))$ rotates with a *constant* angular 'velocity' $|\boldsymbol{\Omega}|$ around the *fixed* axis when s varies. Consider first the easy case $\boldsymbol{\Omega} = \mathbf{0}$. Then, $\mathbf{d}_1(s)$, $\mathbf{d}_2(s)$ and $\mathbf{d}_3(s)$ are independent of s. Since $\mathbf{d}_3$ is the tangent to the centre line, integration of $\mathbf{r}'(s) = \mathbf{d}_3(s)$ yields a straight line for the centre line $\mathbf{r}(s)$.

When the vector $\boldsymbol{\Omega}$ is non-zero, we write $\Omega = |\boldsymbol{\Omega}|$ as its norm and introduce the parallel and perpendicular projections of $\mathbf{d}_3(s)$ onto the axis spanned by $\boldsymbol{\Omega}$:

$$\mathbf{d}_3^{\parallel}(s) = \frac{(\mathbf{d}_3(s) \cdot \boldsymbol{\Omega})\,\boldsymbol{\Omega}}{\Omega^2}, \qquad \mathbf{d}_3^{\perp}(s) = \mathbf{d}_3(s) - \mathbf{d}_3^{\parallel}(s).$$

By definition, $\boldsymbol{\Omega}$ has the coordinate τ onto $\mathbf{d}_3(s)$ and so $(\mathbf{d}_3(s) \cdot \boldsymbol{\Omega}) = \tau$ is a constant. As a result,

$$\frac{\mathrm{d}(\mathbf{d}_3^{\parallel}(s))}{\mathrm{d}s} = \mathbf{0}. \tag{C.4a}$$

This shows that the projection of $\mathbf{d}_3(s)$ along the axis spanned by $\boldsymbol{\Omega}$ is a constant vector. With $\mathbf{r}'(s) = \mathbf{d}_3(s)$, this means that the motion of a point $\mathbf{r}(s)$ on the centre line has a constant 'velocity' along this axis with respect to the 'time' s.

A similar calculation for the transverse component of the motion yields

$$\frac{\mathrm{d}(\mathbf{d}_3^{\perp}(s))}{\mathrm{d}s} = \frac{\mathrm{d}(\mathbf{d}_3 - \mathbf{d}_3^{\parallel})}{\mathrm{d}s} = \boldsymbol{\Omega} \times \mathbf{d}_3 - \mathbf{0} = \boldsymbol{\Omega} \times (\mathbf{d}_3 - \mathbf{d}_3^{\parallel}) = \boldsymbol{\Omega} \times \mathbf{d}_3^{\perp}(s). \tag{C.4b}$$

This equation shows that the vector $\mathbf{d}_3^{\perp}(s)$ has a uniform rotating motion in the plane perpendicular to the axis spanned by $\boldsymbol{\Omega}$. Upon projection of the equation $\mathbf{r}'(s) = \mathbf{d}_3(s)$ onto this plane and integration with respect to the 'time' s, it appears that the projection of the motion of a point $\mathbf{r}(s)$ of the centre line yields a circle in the plane perpendicular to $\boldsymbol{\Omega}$. This circle is followed at constant angular 'velocity'.

Therefore the equations (C.4) characterize a helix of axis parallel to the vector $\boldsymbol{\Omega}$. Any Cosserat curve with constant curvatures and twist is therefore a helix. Some degenerate cases are possible: for $\kappa^{(1)} = \kappa^{(2)}$, the centre line is a straight line, which is twisted if $\tau \neq 0$ and untwisted otherwise ($\tau = 0$). For $\tau = 0$, the centre line is a circle since $\boldsymbol{\Omega} \cdot \mathbf{d}_3 = 0$. We shall everywhere consider straight lines and circles as particular cases of a helix.

Cosserat curves with constant curvatures and twist appear for instance when the Kirchhoff equations are discretized by taking piecewise constant curvatures and twist, such as in the Super-Helix model, see Chapter 4.

Appendix D
Derivation of the plate equations by formal expansion from 3D elasticity

---

D.1   Introduction

This appendix presents a systematic derivation of the equation for elastic plates, for normal deviations that can be many times the plate thickness. Bending is considered, with the restriction that the radius of curvature is much larger than the thickness h of the plate. This derivation is not really necessary for establishing the F.–von K. equations: they have been derived directly in Section 6.5 from a set of heuristic arguments, such as the Kirchhoff hypothesis and assumptions on the orders of magnitude of the various quantities involved. This appendix presents a more rigorous derivation in the sense that both the Kirchhoff hypothesis and the orders of magnitude of the various terms are shown to derive from a limited set of assumptions, and from the general equations for elasticity with finite displacements.

The original reference on these equations is by von Kármán (T. von Kármán, 1910). The assumptions underlying the equations are discussed in the classical textbook of Landau (L. D. Landau and E. M. Lifshitz, 1981). Using these scalings, Ciarlet showed (Ph. G. Ciarlet, 1980) that the F.–von K. equations can indeed be obtained as the dominant order of a formal expansion in the equations of 3D elasticity, with the small plate thickness as the expansion parameter. The coming appendix is a reformulation of Ciarlet's argument using elementary mathematical tools. The same tools have be applied to derive the equations for viscous sheets (P. D. Howell, 1996; N. M. Ribe, 2002),which have many similarities to those for elastic plates. As in Ciarlet's paper, this presentation is based on a formal expansion and makes no attempt to establish rigorously the convergence of the solution when the thickness goes to zero. Important results concerning this convergence have been obtained recently based on the ideas of Γ convergence (G. Friesecke, R. D. James, and S. Müller, 2006).

In the following, we take advantage of the fact that the thickness is much smaller than the longitudinal size of the plate, to solve the full 3D equations of elasticity by a formal expansion. The equations we solve include appropriate boundary conditions on the two surfaces limiting the plate. Unlike the presentation in Chapter 6, we separate as much as possible the geometrical assumptions from the mechanical ones. Whenever possible, we avoid repeating the calculations presented in the main body of the text and provide references to the relevant sections. In addition to improved rigour, the interest of the derivation is to show that the Kirchhoff assumption[1] is not a starting point but, instead, follows from a consistent schema of expansion in powers of a small parameter.

---

[1] Another somewhat arbitrary starting point in the derivation of F.–von K. equations given in Chapter 6 was the custom 'definition' of strain in which some non-linear terms were dropped out. This point too is clarified in the present appendix.

We shall first introduce the notations and conventions. Everything will be based upon the existence of a small parameter ϵ, the ratio of the thickness, h, to the horizontal length scale, L:

$$\epsilon = \frac{h}{L}.$$

By assumption, every quantity depending on the in-plane coordinates (x,y) changes over distances of order L. Many quantities are going to be expanded with respect to the small parameter whose presence will become explicit after convenient rescaling. This rescaling is such that the new transverse coordinate, $\overline{z}$, is related to the physical one, z, by

$$\overline{z} = \frac{z}{\epsilon}.$$

Quantities varying across the plate vary with z over a typical length h. In terms of the rescaled transverse coordinate $\overline{z}$, they vary *formally* over the typical length $L = h/\epsilon$, like the in-plane coordinates. In particular, the rescaled variable $\overline{z}$ lies between $-L/2$ and $+L/2$. There remains only one length scale, L, after rescaling, and it can be chosen as the unit length—this amounts to taking $L = 1$ in what follows.

As explained above, the derivation of the equations for plates is based on the existence of a small parameter ϵ. Using standard notation, an arbitrary quantity is expanded as:

$$f(x,y,z) = \overline{f}(x,y,\overline{z}) = \overline{f}^{[0]}(x,y,\overline{z}) + \epsilon\,\overline{f}^{[1]}(x,y,\overline{z}) + \cdots + \epsilon^n\,\overline{f}^{[n]}(x,y,\overline{z}) + \mathcal{O}(\epsilon^{n+1}),$$

where bars denote rescaled quantities, square brackets are for the order in the expansion, and the symbol $\mathcal{O}$ means that the remainder is at most of order its argument, here ϵ^{n+1}. In what follows, we will tend to omit the bars when the context makes it obvious that one is dealing with a rescaled quantity, but write them explicitly for the variable z whenever needed.

As noticed in Section 1.2.4, some care must be exercised when deriving with respect to variables that have a non-trivial scaling with ϵ. Indeed, taking the derivative with respect to z in the equation above:

$$\frac{\partial f}{\partial z} = \frac{1}{\epsilon}\frac{\partial \overline{f}^{[0]}}{\partial \overline{z}} + \cdots + \epsilon^{n-1}\frac{\partial \overline{f}^{[n-1]}}{\partial \overline{z}} + \mathcal{O}(\epsilon^n),$$

we find that the order $(i - 1)$ of $\partial f/\partial z$ is given by the order i of f:

$$\overline{\left(\frac{\partial f}{\partial z}\right)}^{[i-1]} = \frac{\partial(\overline{f}^{[i]})}{\partial \overline{z}}.$$

In rescaled variables, any derivation with respect to z shifts the order of the expansion by one. This is not true with the variables x and y, which do not scale with ϵ. Omitting the bars for terms in the expansion (that is writing $f^{[i]}$ for $\overline{f}^{[i]}$) and using the notation $\partial_{x_j} f = \partial_j f$ for $\partial f/\partial x_j$, this can be summarized as:

$$(\partial_z f(x,y,z))^{[i-1]} = \partial_{\overline{z}} f^{[i]}(x,y,\overline{z}) \qquad (\partial_\alpha f(x,y,z))^{[i]} = \partial_\alpha f^{[i]}(x,y,\overline{z}). \tag{D.1}$$

In what follows, we shall first recall the general equations for 3D elasticity to be solved (Section D.3), list the basic scaling assumptions underlying the F.–von K. equations (in Section D.2), write out the expansion for all the unknown quantities (Section D.4), and finally plug the expansions into the general equations to derive the F.–von K. equations at the dominant order (Section D.5).

D.2 Scaling assumptions

The derivation of the F.–von K. equations makes use of two ingredients: some basic assumptions regarding the orders of magnitude of stress and displacement, detailed in the present section, and the general equations of elasticity, recalled in the following section. It can seem awkward to start the derivation with *a priori* assumptions concerning the orders of magnitude of the unknowns: after all, these unknowns should be determined by solving the problem. In fact, these assumptions about the unknowns derive from implicit assumptions on the magnitude of the physical parameters of the problem, such as the applied loading and the imposed displacement. The situation is very similar to that discussed in Section 3.7 for extensible rods: the same geometrical object, a thin rod or a thin plate, obeying the general equations of 3D elasticity, can exhibit very different behaviours depending on the magnitude[2] of the applied forces. As a matter of fact, different scaling assumptions would lead to a different limit problem, describing a different mechanical behaviour of the plate (G. Friesecke, R. D. James, and S. Müller, 2006). Therefore, it should not be too surprising that one has to start with some scaling assumptions—the goal of the game is to keep their number to a bare minimum.

The orders of magnitude for the stress and for the deflection associated with the F.–von K. equations have been found by qualitative arguments in equation (6.111):

$$\sigma_{ij} = \mathcal{O}(\epsilon^2), \qquad u_z \sim w = \mathcal{O}(\epsilon). \tag{D.2}$$

They will be our starting point for the derivation of the F.–von K. equations below.

The first scaling assumption that we make above amounts to assuming that in-plane stress and bending have comparable effects.[3] The second scaling assumption is very different in nature as it is geometrical and not mechanical. It involves the displacement, not the strain or the stress, and assumes that rotations are small—the subtle point being that these rotations are not neglected altogether, something that explains that the final equations will be non-linear. Because of this geometric assumption that limits the magnitude of rotations, the F.–von K. equations are not geometrically exact, as already mentioned in many places in this book. This geometric assumption could be relaxed at the price of a greater notational burden, leading to a generalization of the F.–von K. equations for finite rotations.[4]

D.3 Basic equations

It has already been stressed in Chapter 6 that the nonlinearity present in the F.–von K. equations has a geometrical origin. Not surprisingly, the forthcoming derivation is based on

---

[2] Since we are solving the equations of elasticity by an expansion with respect to a parameter ϵ, the absolute magnitude of the loading is not relevant. What is important is, in mathematical terms, how the applied loading or the imposed displacement scale with the small thickness when the latter goes to zero.

[3] If applied forces are stronger, bending becomes negligible, and this amounts to setting $D = 0$ in the F.–von K. equations. As explained in Chapter 9, this membrane approximation is very singular and often lead to mathematically ill-posed problems.

[4] This involves using the covariant harmonic operator, and the exact expression for the Gauss curvature in place of the Monge–Ampère operator.

the equations of elasticity for finite displacements[5] and not on the linearized equations of elasticity. In this framework, the stress tensor is actually the Piola–Kirchhoff stress tensor, introduced in Section 2.5. As usual, (x, y, z) refers to the position in reference configuration, and we use the Lagrangian description of displacement: this triple labels a given material point, and is not its actual position in physical space.

The general definition of the strain in terms of the displacement in equation (2.9) can be rewritten in a way that distinguishes the in-plane and transverse non-linear contributions:

$$\epsilon_{\alpha\beta} = \frac{1}{2} \left(\partial_\beta u_\alpha + \partial_\alpha u_\beta \right) + \frac{1}{2} \partial_\alpha u_\gamma \, \partial_\beta u_\gamma + \frac{1}{2} \partial_\alpha u_z \, \partial_\beta u_z \tag{D.3a}$$

$$\epsilon_{\alpha z} = \frac{1}{2} \left(\partial_z u_\alpha + \partial_\alpha u_z \right) + \frac{1}{2} \partial_\alpha u_\gamma \, \partial_z u_\gamma + \frac{1}{2} \partial_\alpha u_z \, \partial_z u_z \tag{D.3b}$$

$$\epsilon_{zz} = \partial_z u_z + \frac{1}{2} \left(\partial_z u_\gamma \right)^2 + \frac{1}{2} \left(\partial_z u_z \right)^2 . \tag{D.3c}$$

The equations of equilibrium[6] for finite displacements, given earlier in equation (2.77), can be written as:

$$\partial_j(\sigma_{ij}) + \partial_j(\partial_k u_i \, \sigma_{kj}) = 0.$$

Again, in order to make apparent the separation between in-plane and out-of-plane (transverse) indices this can be rearranged, along a tangent direction $(i = \alpha)$, as:

$$\partial_\beta(\sigma_{\alpha\beta}) + \partial_z(\sigma_{\alpha z}) + \partial_\beta(\partial_\gamma u_\alpha \, \sigma_{\gamma\beta}) + \partial_\beta(\partial_z u_\alpha \, \sigma_{z\beta}) \cdots$$
$$+ \partial_z(\partial_\gamma u_\alpha \, \sigma_{\gamma z}) + \partial_z(\partial_z u_\alpha \, \sigma_{zz}) = 0 \tag{D.4a}$$

and along the transverse direction as $(i = z)$:

$$\partial_\beta(\sigma_{z\beta}) + \partial_z(\sigma_{zz}) + \partial_\beta(\partial_\gamma u_z \, \sigma_{\gamma\beta}) + \partial_\beta(\partial_z u_z \, \sigma_{z\beta}) \cdots$$
$$+ \partial_z(\partial_\gamma u_z \, \sigma_{\gamma z}) + \partial_z(\partial_z u_z \, \sigma_{zz}) = 0. \tag{D.4b}$$

In the absence of applied forces on the lower and upper faces of the plate, the following free boundary conditions hold:

$$\sigma_{\alpha z}\left(x, y, \overline{z} = \pm \frac{L}{2} \right) = 0, \qquad \sigma_{zz}\left(x, y, \overline{z} = \pm \frac{L}{2} \right) = 0. \tag{D.5}$$

To make this system of equations complete, one has to specify constitutive relations for the material. We shall only consider the case of small strain, and so linear constitutive relations are used. If we further restrict ourselves to the case of an isotropic material, an assumption that can be relaxed without major difficulty, one is led to Hookean elasticity, defined by the

---

[5] The consequence of the approximation of small displacement would be to get rid of all effects associated with rotations and geometric nonlinearity. We shall see later that the F.–von K. equations lie somewhere in-between the general framework of finite displacements and that of small displacements, as some of the non-linear terms are retained, but only those that are important at the dominant order.

[6] For the sake of simplicity, we do not consider the case of applied forces or moments on the plate. This case can be handled by adding appropriate source terms in the F.–von K. equations, as shown by a simple extension of the present derivation.

constitutive relations (2.61)—these equations are used in the following derivation but are not repeated here.

D.4   Expansion of the basic quantities

Since we have assumed σ_{ij} to be of order 2 with respect to ϵ, its expansion can be written, using the notations introduced above:

$$\sigma_{\alpha\beta}(x,y,z) = \epsilon^2\,\sigma^{[2]}_{\alpha\beta}(x,y,\overline{z}) + \epsilon^3\,\sigma^{[3]}_{\alpha\beta}(x,y,\overline{z}) + \cdots \tag{D.6a}$$

$$\sigma_{\alpha z}(x,y,z) = \epsilon^2\,\sigma^{[2]}_{\alpha z}(x,y,\overline{z}) + \epsilon^3\,\sigma^{[3]}_{\alpha z}(x,y,\overline{z}) + \cdots \tag{D.6b}$$

$$\sigma_{zz}(x,y,z) = \epsilon^2\,\sigma^{[2]}_{zz}(x,y,\overline{z}) + \epsilon^3\,\sigma^{[3]}_{zz}(x,y,\overline{z}) + \cdots \tag{D.6c}$$

In this equation, we have omitted the bar in $\overline{\sigma}^{[i]}_{ij}$: since there is no ambiguity, $\sigma^{[i]}_{ij}$ has to be understood as a rescaled quantity, function of the rescaled transverse coordinate $\overline{z}$.

Although this is not obvious, the scaling assumptions given in equation (D.2) imply that the in-plane displacement u_α and the deflection u_z are of different orders of magnitude. Indeed, by the constitutive relations (Hooke's relation), strain is directly proportional to stress and so $\epsilon_{\alpha\beta} = \mathcal{O}(\epsilon^2)$. We can then rewrite the definition (D.3a) for $\epsilon_{\alpha\beta}$ so as to make a purely 2D strain $\epsilon^{2D}_{\alpha\beta}$ appear:

$$\epsilon_{\alpha\beta} - \frac{1}{2}\,\partial_\alpha u_z\,\partial_\beta u_z = \epsilon^{2D}_{\alpha\beta}, \tag{D.7a}$$

where

$$\epsilon^{2D}_{\alpha\beta} = \frac{1}{2}\left(\partial_\beta u_\alpha + \partial_\beta u_\alpha\right) + \frac{1}{2}\,\partial_\alpha u_\gamma\,\partial_\beta u_\gamma. \tag{D.7b}$$

Now, the left-hand side in equation (D.7a) is of order ϵ^2 by assumption, while the right-hand side, defined in the second equation (D.7b), appears to be the strain associated with the following 'virtual' two-dimensional problem: fix z and consider the *planar* displacement of a planar domain defined by $(x,y) \mapsto u_\alpha(x,y,z)$. For this 2D virtual problem, the strain is everywhere of order ϵ^2. As a result, the displacement u_α, which can be computed by integration of the strain, must be of order ϵ^2 times the typical size of the domain, up to a rigid-body motion. By assumption, the typical size of the domain is L and does not vary with ϵ. A rigid body motion for this 2D problem corresponds to a translation along the (x,y) plane of the plate in real space, combined with a rotation along the axis z. Such rigid-body motions are not considered, as one of our assumptions is that the displacement remains small. Finally, we conclude that the in-plane displacement u_α has to be of second order, unlike the deflection u_z which is first order:

$$u_\alpha = \mathcal{O}(\epsilon^2).$$

This leads to the following expansion for the displacement:

$$u_\alpha(x,y,z) = \epsilon^2\,u^{[2]}_\alpha(x,y,\overline{z}) + \cdots \tag{D.8a}$$

$$u_z(x,y,z) = \epsilon\,u^{[1]}_z(x,y,\overline{z}) + \epsilon^2\,u^{[2]}_z(x,y,\overline{z}) + \cdots \tag{D.8b}$$

D.5   Solution at leading order

So far, we have written down the equations to be solved, as well as the expansions of the various quantities. We can now proceed to plug these expansions into the equations of 3D elasticity, and derive the equations for plates order by order.

D.5.1   Vanishing of normal shear stress at order [2]

Let us examine the equation for in-plane equilibrium (D.4a). Recalling the rule (D.1) that derivations with respect to z shift the order with respect to ϵ, we find that the dominant order in this equation is [1], with a single term:

$$\partial_{\bar{z}}\left(\sigma^{[2]}_{\alpha z}\right) = 0.$$

This can be integrated with respect to $\bar{z}$ as:

$$\sigma^{[2]}_{\alpha z}(x, y, \bar{z}) = A(x, y),$$

where $A(x, y)$ is the unknown 'constant' of integration, independent of $\bar{z}$.

Using the free boundary conditions (D.5), we find that A must cancel, that is $A(x, y) = 0$ for all (x, y):

$$\sigma^{[2]}_{\alpha z}(x, y, \bar{z}) = 0, \quad \text{hence} \quad \sigma_{\alpha z} = \mathcal{O}(\epsilon^3). \tag{D.9}$$

The normal shear stress $\sigma_{\alpha z}$ turns out to be smaller than expected. This does not make our scaling assumptions inconsistent, as we have in fact assumed that the stress was $\mathcal{O}(\epsilon^2)$, that is of order ϵ^2 *or smaller*. Moreover, the in-plane stress $\sigma_{\alpha\beta}$ will appear to be of order ϵ^2 as expected.

D.5.2   Derivation of Kirchhoff 'hypothesis'

Since σ_{ij} is a quantity of order ϵ^2 or smaller by assumption, Hooke's law shows that the strains ϵ_{kl} are $\mathcal{O}(\epsilon^2)$ as well. Consider the definition (D.3c) for ϵ_{zz}, and analyse the orders of magnitude of the various terms. The left-hand side is $\mathcal{O}(\epsilon^2)$. The second term in the right-hand side, $(\partial_z u_\gamma)^2/2$ is also a second order quantity since $u_\gamma = \mathcal{O}(\epsilon^2)$ and so $\partial_z u_\gamma = \mathcal{O}(\epsilon)$. Then, equation (D.3c) yields

$$\partial_z u_z + \frac{1}{2}\left(\partial_z u_z\right)^2 = \epsilon_{zz} - \frac{1}{2}(\partial_{\bar{z}} u_\gamma)^2 = \mathcal{O}(\epsilon^2).$$

The polynomial with variable $\partial_z u_z$ in the left-hand side has two roots, namely $\partial_z u_z = 0$ and $\partial_z u_z = -2$. The second root is unphysical: it describes a mirror symmetry with respect to the (x, y) plane bringing the upper side of the plate below and the lower side on top,[7] which is clearly inconsistent with the assumption that u_z is a small quantity. As a result, the solution $\partial_z u_z$ of the above polynomial with a small right-hand side must be close to the root $\partial_z u_z = 0$. One can write the root of this second order polynomial explicitly, and this shows that

$$\partial_z u_z = \mathcal{O}(\epsilon^2). \tag{D.10}$$

This scaling may seem surprising. From equation (D.8b), one would expect instead $\partial_z u_z = \mathcal{O}(1)$. The solution to this apparent paradox is that the higher order terms in u_z are

---

[7] For this non-physical root, the elevation in actual configuration would then be $z + u_z \sim z - 2z \sim -z$.

independent of the variable z, and disappear upon derivation with respect to z. Therefore, the scaling in equation (D.10) above imposes that the expansion for u_z is of a more particular form than that proposed in equation (D.8b):

$$u_z(x, y, z) = \epsilon\, w^{[1]}(x, y) + \epsilon^2\, w^{[2]}(x, y) + \epsilon^3\, u_z^{[3]}(x, y, \overline{z}) + \cdots \tag{D.11}$$

Here, we have introduced the notation $w(x, y)$ for the deflection of the centre surface. By definition,

$$w(x, y) = u_z(x, y, 0). \tag{D.12a}$$

Using a specific letter, w, allows us to stress the fact that this function does not depend on the transverse coordinate, and is a good candidate for appearing in the final equations. We shall later use a similar notation for the in-plane displacement of the centre surface:

$$v_\alpha(x, y) = u_\alpha(x, y, 0). \tag{D.12b}$$

We shall now analyse equation (D.3b), defining the normal shear strain components. We have shown that $\sigma_{\alpha z}$ is of order [3] or higher, and so the left-hand side $\epsilon_{\alpha z}$, which is directly proportional to $\sigma_{\alpha z}$ by Hooke's relations, is also of order ϵ^3, or smaller. The first non-linear term in equation (D.3b) is of order $2 + (2 - 1) = 3$; the other non-linear term is of order $1 + 2 = 3$, using the scaling for $\partial_z u_z = \mathcal{O}(\epsilon^2)$ just established. As a result, the sum of the remaining terms, which are linear, has to be of order ϵ^3 or smaller, $\partial_z u_\alpha + \partial_\alpha u_z = \mathcal{O}(\epsilon^3)$. By writing this equality at order [1] and [2], one obtains:

$$\partial_{\overline{z}}\left(u_\alpha^{[2]}\right) = -\partial_\alpha\left(u_z^{[1]}\right). \quad \text{and} \quad \partial_{\overline{z}}\left(u_\alpha^{[3]}\right) = -\partial_\alpha\left(u_z^{[2]}\right)$$

Now, by using the specific form of u_z found in equation (D.11), we obtain

$$\frac{\partial u_\alpha^{[2]}(x, y, \overline{z})}{\partial \overline{z}} = -\frac{\partial w^{[1]}(x, y)}{\partial x_\alpha} \quad \text{and} \quad \frac{\partial u_\alpha^{[3]}(x, y, \overline{z})}{\partial \overline{z}} = -\frac{\partial w^{[2]}(x, y)}{\partial x_\alpha}.$$

Since the right-hand sides do not depend on $\overline{z}$, integration with respect to $\overline{z}$ is straightforward, and leads to

$$u_\alpha^{[2]}(x, y, \overline{z}) = v_\alpha^{[2]}(x, y) - \overline{z}\,\frac{\partial w^{[1]}(x, y)}{\partial x_\alpha}$$

$$u_\alpha^{[3]}(x, y, \overline{z}) = v_\alpha^{[3]}(x, y) - \overline{z}\,\frac{\partial w^{[2]}(x, y)}{\partial x_\alpha}, \tag{D.13}$$

where the constant of integration is the displacement along the centre surface ($\overline{z} = 0$), which was defined $v_\alpha^{[2]}(x, y)$ in equation (D.12b). In the equation above, we have just established what is known as the Kirchhoff 'hypothesis',[8] see equation (6.68) in the main text.

D.5.3   Vanishing of transverse stress at orders [2] and [3]

We can now consider the equation (D.4b) for the equilibrium in the transverse direction. Collecting all the previous results, and recalling in particular that $\sigma_{\alpha z} = \mathcal{O}(\epsilon^3)$ and $\partial_z u_z = \mathcal{O}(\epsilon^2)$, one finds that all terms in this equation of equilibrium are of order [3] or higher but the second term, $\partial_z \sigma_{zz}$, which is formally of order [1] since σ_{zz} is of order [2]. We

---

[8] In what follows, we shall only make use of the specific form of $u_\alpha^{[2]}$: the quantity $u_\alpha^{[3]}$ in (D.13) does not appear in the F.–von K. equations.

can then repeat word to word the argument of Section D.5.1 as follows. The equation $\partial_{\bar{z}}(\sigma_{zz}^{[2]}) = 0$ implies that $\sigma_{zz}^{[2]}$ depends only on the in-plane coordinates x and y, while the stress-free boundary conditions (D.5) impose that $\sigma_{zz}^{[2]}$ vanishes along the edges. As a result, $\sigma_{zz}^{[2]}(x, y, \bar{z})$ vanishes everywhere. Reading the same equation (D.4b) at order [2], one can similarly show that $\sigma_{zz}^{[3]}(x, y, \bar{z})$ vanishes everywhere. As a result, σ_{zz} is in fact a quantity of order [4] at least:

$$\sigma_{zz}^{[2]}(x, y, \bar{z}) = 0 \text{ and } \sigma_{zz}^{[3]}(x, y, \bar{z}) = 0 \quad \text{hence} \quad \sigma_{zz} = \mathcal{O}(\epsilon^4). \tag{D.14}$$

D.5.4   Affine dependence of in-plane stress on transverse coordinate

We can now compute the in-plane stress at dominant order $\epsilon_{\alpha\beta}^{[2]}$ using equation (D.3a). Noticing that the first non-linear term, $(\partial_\alpha u_\gamma \, \partial_\beta u_\gamma)$ is of order [4] or higher, we conclude that we must retain the nonlinearity involving the transverse displacement u_z, although we can drop that involving the in-plane displacement u_γ at this order, as assumed in Chapter 6:

$$\epsilon_{\alpha\beta}^{[2]}(x, y, \bar{z}) = \frac{1}{2}\left(\partial_\beta u_\alpha^{[2]}(x, y, \bar{z}) + \partial_\alpha u_\beta^{[2]}(x, y, \bar{z})\right) \cdots$$

$$+ \frac{1}{2}\,\partial_\alpha w^{[1]}(x, y)\,\partial_\beta w^{[1]}(x, y).$$

Note that the non-linear term does not depend on $\bar{z}$. By plugging in the Kirchhoff 'hypothesis' (now proved, see equation (D.13) above), one obtains the following expression for the in-plane strain:

$$\epsilon_{\alpha\beta}^{[2]}(x, y, \bar{z}) = \left[\frac{1}{2}\left(\partial_\beta v_\alpha^{[2]} + \partial_\alpha v_\beta^{[2]}\right) + \frac{1}{2}\,\partial_\alpha w^{[1]}\,\partial_\beta w^{[1]}\right]_{(x,y)} \cdots$$

$$- \bar{z}\,\partial_{\alpha\beta} w^{[1]}(x, y). \tag{D.15}$$

The term in square brackets depends only on x and y, and can be identified with the in-plane strain $\epsilon_{\alpha\beta}^{[2]}(x, y, \bar{z} = 0)$, *measured along the centre surface*. Therefore, we have just established equation (6.69) as a result of our formal expansion method:

$$\epsilon_{\alpha\beta}^{[2]}(x, y, \bar{z}) = \epsilon_{\alpha\beta}^{[2]}(x, y, \bar{z} = 0) - \bar{z}\,\partial_{\alpha\beta} w^{[1]}(x, y).$$

The argument given at the end of Section 6.5.2 can then be repeated to establish the affine dependence of $\sigma_{\alpha\beta}^{[2]}$ proposed earlier in equation (6.70). Indeed, this argument makes use of the effective two-dimensional constitutive relations for a plate, which hold at the order dominant [2] since $\sigma_{\alpha z}$ and σ_{zz} indeed vanish at this order, see equations (D.9) and (D.14).

D.5.5   Dependence of $\sigma_{\alpha z}$ on transverse coordinate at dominant order

Examination of the equation for in-plane equilibrium reveals that only the linear terms are present in $\sigma_{\alpha z}$ at order [2]. Equation (6.71) is recovered. By the argument given in Chapter 6, one is then led to the expression (6.74) for the dominant contribution to the normal shear stress $\sigma_{\alpha z}^{[3]}$:

$$\sigma_{\alpha z}^{[3]} = \frac{E}{2\,(1 - \nu^2)} \left(\overline{z}^2 - \frac{1}{2^2}\right) \frac{\partial(\Delta w^{[1]})}{\partial x_\alpha}.$$ (D.16)

D.5.6 First F.–von K. equation

The equation for out-of-plane equilibrium yields, at order [3]:

$$\partial_{\overline{z}}\left(\sigma_{zz}^{[4]}\right) = -\partial_\alpha\left(\sigma_{\alpha z}^{[3]}\right) - \partial_\beta\left(\partial_\alpha w^{[1]}\,\sigma_{\alpha\beta}^{[2]}\right) - \partial_{\overline{z}}\left(\partial_\alpha w^{[1]}\,\sigma_{\alpha\beta}^{[3]}\right),$$ (D.17)

where we have pulled one term on the other side to stress the fact that this is an equation for $\sigma_{zz}^{[4]}$.

This equation is solved by integration with respect to $\overline{z}$ with the free boundary conditions $\sigma_{zz}^{[4]}(x, y, \overline{z} = \pm 1/2) = 0$. This makes two boundary conditions for a first-order ordinary differential equation, an overdetermined problem. For a solution to exist, a solubility condition must be verified. This solubility condition can be worked out very easily: the variation of $\sigma_{zz}^{[4]}$ across the plate thickness should be zero, meaning that the integral of the right-hand side of equation (D.17) is zero,

$$\int_{-\frac{1}{2}}^{\frac{1}{2}} \left[\partial_\alpha\left(\sigma_{\alpha z}^{[3]}\right) + \partial_\beta\left(\partial_\alpha w^{[1]}\,\sigma_{\alpha\beta}^{[2]}\right) + \partial_{\overline{z}}\left(\partial_\alpha w^{[1]}\,\sigma_{\alpha\beta}^{[3]}\right)\right]\,\mathrm{d}\overline{z} = 0.$$ (D.18)

Now, the last term in the integrand is an exact derivative that, upon integration, yields the variation of $(\partial_\alpha w^{[1]}\,\sigma_{\alpha\beta}^{[3]})$ across the plate thickness. This variation cancels since $\sigma_{\alpha\beta}^{[3]}$ has to vanish on the free edges. The integral of the first term is what has been defined as $-(f_{\mathrm{bend}})$ in the main chapter. Its value is given in equation (6.82).

To compute the second term, one may notice that $w^{[1]}$ does not depend on $\overline{z}$ and swap the in-plane derivatives and the transverse integration:

$$\int_{-\frac{1}{2}}^{\frac{1}{2}} \partial_\beta\left(\partial_\alpha w^{[1]}\,\sigma_{\alpha\beta}^{[2]}\right)\,\mathrm{d}\overline{z} = \partial_\beta\left(\partial_\alpha w^{[1]} \int_{-\frac{1}{2}}^{\frac{1}{2}} \sigma_{\alpha\beta}^{[2]}\,\mathrm{d}\overline{z}\right).$$

Now, recall that $\sigma_{\alpha\beta}^{[2]}$ is given by equation (6.70) as an affine function of $\overline{z}$. Upon integration over a symmetric interval $\overline{z} \in [-1/2, +1/2]$, the term that is linear with respect to $\overline{z}$ cancels. Therefore, the second term in equation (D.18) reads

$$\int_{-\frac{1}{2}}^{\frac{1}{2}} \partial_\beta\left(\partial_\alpha w^{[1]}\,\sigma_{\alpha\beta}^{[2]}\right)\,\mathrm{d}\overline{z} = \partial_\beta\left(\partial_\alpha w^{[1]}\,[h\,s_{\alpha\beta}^{[2]}(x, y)]\right),$$

where we have introduced $s_{\alpha\beta}(x, y)$ as the in-plane stress along the centre surface:

$$s_{\alpha\beta}(x, y) = \sigma_{\alpha\beta}(x, y, \overline{z} = 0).$$ (D.19)

Because of the affine dependence of $\sigma_{\alpha\beta}^{[2]}$ with respect to $\overline{z}$, $s_{\alpha\beta}^{[2]}$ happens to be equal to the average of $\sigma_{\alpha\beta}^{[2]}$ over the plate thickness, noted $\langle\cdot\rangle_{\overline{z}}$

$$s_{\alpha\beta}^{[2]}(x, y) = \left\langle\sigma_{\alpha\beta}^{[2]}\right\rangle_{\overline{z}}.$$

Plugging back into equation (D.18), this finally leads to

$$-D\,\Delta^2 w^{[1]} + h\,\partial_\beta\left(\partial_\alpha w^{[1]}\,s_{\alpha\beta}^{[2]}\right) = 0,$$ (D.20)

where the bending modulus D was defined in equation (6.83). Upon expansion of the derivative of the product in the last term, one of the terms vanishes since

$$\partial_\beta \left(s_{\alpha\beta}^{[2]} \right) = \partial_\beta \left(\left\langle \sigma_{\alpha\beta}^{[2]} \right\rangle_z \right) = \left\langle \partial_\beta \left(\sigma_{\alpha\beta}^{[2]} \right) \right\rangle_z = - \left\langle \partial_{\bar z} \left(\sigma_{\alpha z}^{[3]} \right) \right\rangle_z = 0. \qquad (\text{D.21})$$

The right-hand side above is zero as the average of the exact derivative is proportional to the variation of $\sigma_{\alpha z}^{[3]}$ across the thickness, which vanishes by the free boundary conditions. This allows one to rewrite equation (D.20) as

$$D\,\Delta^2 w^{[1]} - h\,s_{\alpha\beta}^{[2]}\,\partial_{\alpha\beta} w^{[1]} = 0. \qquad (\text{D.22})$$

This partial differential equation involves the variables (x, y) only. It is the F.–von K. equation for transverse equilibrium given, earlier, in equation (6.88a).

D.5.7   Second F.–von K. equation

To make the system complete, we have to specify how the in-plane stress $s_{\alpha\beta}(x, y)$ can be computed in the above equation. To do so, we start from the expression for the average in-plane strain in equation (D.15). The in-plane stress being affine with respect to $\bar z$, its average is just the constant term in square brackets:

$$\left\langle \epsilon_{\alpha\beta}^{[2]} \right\rangle_{\bar z} = \frac{1}{2} \left(\partial_\beta v_\alpha^{[2]} + \partial_\alpha v_\beta^{[2]} \right) + \frac{1}{2} \partial_\alpha w^{[1]} \, \partial_\beta w^{[1]}, \qquad (\text{D.23})$$

this equality being between two functions of two variables, x and y.

The equation (D.23) above allows one to compute the in-plane strain $\langle \epsilon_{\alpha\beta}^{[2]} \rangle (x, y)$ along the centre surface of the plate, given the displacement $(v_x^{[2]}, v_y^{[2]}, w^{[1]})$ of this centre surface. The in-plane stress $s_{\alpha\beta}^{[2]}(x, y)$ can then be computed[9] using the effective constitutive relations (6.59). The conditions of equilibrium in the three directions is then enforced by equations (D.21) ($\beta = x, y$) and (D.22). This makes the set of equations (D.23), (6.59), (D.21) and (D.22) a complete system of equations for the unknown displacement $(v_\alpha^{[2]}(x, y), w^{[1]}(x, y))$ of the centre surface (appropriate boundary conditions need of course to be specified).

An alternative presentation of the same equations can be obtained by taking the Airy potential $\chi^{[2]}(x, y)$ as unknown, instead of the in-plane displacement $v_\alpha^{[2]}$. Indeed, as explained in Section 6.4.5, the equations (D.21) for in-plane equilibrium are equivalent to the existence of a potential, called the Airy potential, defined by equation (6.65). By eliminating the in-plane displacement $v_\alpha^{[2]}$ from the mean in-plane strain in equation (D.23), and by making use of the effective constitutive relations (6.59), one can derive as earlier the second F.–von K. equation (6.88b), which expresses the compatibility of in-plane displacement:

$$\Delta^2 \chi^{[2]} + E \left(\partial_{xx} w^{[1]} \, \partial_{yy} w^{[1]} - \left(\partial_{xy} w^{[1]} \right)^2 \right) = 0. \qquad (\text{D.24})$$

In this second approach, the set of equations (D.22) and (D.24), together with appropriate boundary conditions, constitutes a complete set of equations, known as the F.–von K. equations for elastic plates.

---

[9] Recall that the Kirchhoff hypothesis has been validated at the dominant order. As a result, the effective constitutive relations hold.

D.6   Conclusion

Based solely on the scaling assumptions given in equations (D.2) and on the general equations for 3D Hookean elasticity, one can justify the F.–von K. equations for elastic plates. In particular, the Kirchhoff hypothesis (filaments normal to the centre surface remain normal upon deformation) has been established at the dominant orders in the expansion with respect to the small parameter ϵ. We have also been able to give a precise meaning to the unknowns in the F.–von K. equations, which are $w^{[1]}(x,y)$, the transverse displacement of the centre surface, and $\chi^{[2]}(x,y)$, the Airy potential for the in-plane stress $s^{[2]}_{\alpha\beta}$ averaged over the plate thickness,[10] at the relevant orders.

In order to keep the presentation simple, we have not considered the possibility of applied forces. It is a relatively simple task, left to the interested reader, to extend the equations to this case—one just has to keep the remote force term in the equations for 3D equilibrium. Obviously, this will only work provided the applied forces scale consistently with the parameter ϵ.

References

Ph. G. Ciarlet. A justification of the von Kármán equations. *Archive for Rational Mechanics and Analysis*, 73(4):349–389, 1980.

G. Friesecke, R. D. James, and S. Müller. A hierarchy of plate models derived from nonlinear elasticity by Gamma-convergence. *Archive for Rational Mechanics and Analysis*, 180:183–236, 2006.

P. D. Howell. Models for thin viscous sheets. *European Journal of Applied Mathematics*, 7:321–343, 1996.

L. D. Landau and E. M. Lifshitz. *Theory of Elasticity (Course of Theoretical Physics)*. Pergamon Press, 2nd edition, 1981.

N. M. Ribe. A general theory for the dynamics of thin viscous sheets. *Journal of Fluid Mechanics*, 457:255–283, 2002.

T. von Kármán. Festigkeitsprobleme in Maschinenbau. In F. Klein and C. Müller, editors, *Encyklopädie der Mathematischen Wissenschaften*, volume 4, pages 311–385. Teubner, 1910.

---

[10] As noted earlier, $\sigma^{[2]}_{\alpha\beta}$ is an affine function of $\bar{z}$ at this order and so $s^{[2]}_{\alpha\beta}$ is also the value of the stress measured along the centre surface $\bar{z} = 0$.

Index

Made in the USA
Monee, IL
07 July 2026

56545369R00332